Adaptive Filtering

Algorithms and Practical Implementation

Third Edition

Paulo S.R. Diniz

Adaptive Filtering
Algorithms and Practical Implementation

Third Edition

Springer

Paulo S. R. Diniz
Federal University of Rio de Janerio
Rio de Janerio
Brazil

ISBN 978-0-387-31274-3 e-ISBN 978-0-387-68606-6
DOI: 10.1007/978-0-387-68606-6

Library of Congress Control Number: 2008923554

springer.com

Diniz: Adaptive Filtering: Algorithms & Practical Implementation, 3rd ed.

First Indian Reprint 2012

ISBN 978-81-322-0672-9

This edition is published by Springer (India) Private Limited, (a part of Springer Science+Business Media), Registered Office: 7th Floor, Vijaya Building, 17 Barakhamba Road, New Delhi – 110 001, India

Printed in India by Rakmo Press Pvt. Ltd., New Delhi.

Note to Instructors

For the instructors this book has a solution manual for the problems written by Dr. L. W. P. Biscainho available from the publisher. Also available, upon request to the author, is a set of master transparencies as well as the MATLAB®[1] codes for all the algorithms described in the text.

[1]MATLAB is a registered trademark of The MathWorks, Inc.

To: My Parents,

Mariza,

Paula,

and Luiza.

PREFACE

The field of *Digital Signal Processing* has developed so fast in the last three decades that it can be found in the graduate and undergraduate programs of most universities. This development is related to the increasingly available technologies for implementing digital signal processing algorithms. The tremendous growth of development in the digital signal processing area has turned some of its specialized areas into fields themselves. If accurate information of the signals to be processed is available, the designer call easily choose the most appropriate algorithm to process the signal. When dealing with signals whose statistical properties are unknown, fixed algorithms do not process these signals efficiently. The solution is to use an adaptive filter that automatically changes its characteristics by optimizing the internal parameters. The adaptive filtering algorithms are essential in many statistical signal processing applications.

Although the field of adaptive signal processing has been subject of research for over four decades, it was in the eighties that a major growth occurred in research and applications. Two main reasons can be credited to this growth, the availability of implementation tools and the appearance of early textbooks exposing the subject in an organized manner. Still today it is possible to observe many research developments in the area of adaptive filtering, particularly addressing specific applications. In fact, the theory of linear adaptive filtering has reached a maturity that justifies a text treating the various methods in a unified way, emphasizing the algorithms suitable for practical implementation. This text concentrates on studying on-line algorithms, those whose adaptation occurs whenever a new sample of each environment signal is available. The so-called block algorithms, those whose adaptation occurs when a new block of data is available, are also included using the subband filtering framework. Usually, block algorithms require different implementation resources than the on-line algorithms. This edition also includes basic introductions to nonlinear adaptive filtering and blind signal processing as natural extensions of the algorithms treated in the earlier chapters. The understanding of the introductory material presented is fundamental for further studies in these fields which are described in more detail in some specialized texts.

The idea of writing this book started while teaching the adaptive signal processing course at the graduate school of the Federal University of Rio de Janeiro (UFRJ). The request of the students to cover as many algorithms as possible made me think how to organize this subject such that not much time is lost in adapting notations and derivations related to different algorithms. Another common question was which algorithms really work in a finite-precision implementation. These issues led me to conclude that a new text on this subject could be written with these objectives in mind. Also, considering that most graduate and undergraduate programs include a single adaptive filtering course, this book should not be lengthy. Another objective to seek is to provide an easy access to the working algorithms for the practitioner.

It was not until I spent a sabbatical year and a half at University of Victoria, Canada, that this project actually started. In the leisure hours, I slowly started this project. Parts of the early chapters of this book were used in short courses on adaptive signal processing taught at different institutions, namely: Helsinki University of Technology, Espoo, Finland; University Menendez Pelayo in Seville, Spain; and at the Victoria Micronet Center, University of Victoria, Canada. The remaining parts of the book were written based on notes of the graduate course in adaptive signal processing taught at COPPE (the graduate engineering school of UFRJ).

The philosophy of the presentation is to expose the material with a solid theoretical foundation, while avoiding straightforward derivations and repetition. The idea is to keep the text with a manageable size, without sacrificing clarity and without omitting important subjects. Another objective is to bring the reader up to the point where implementation can be tried and research can begin. A number of references are included at the end of the chapters in order to aid the reader to proceed on learning the subject.

It is assumed the reader has previous background on the basic principles of digital signal processing and stochastic processes, including: discrete-time Fourier- and $\mathcal{Z}$-transforms, finite impulse response (FIR) and infinite impulse response (IIR) digital filter realizations, multirate systems, random variables and processes, first- and second-order statistics, moments, and filtering of random signals. Assuming that the reader has this background, I believe the book is self contained.

Chapter 1 introduces the basic concepts of adaptive filtering and sets a general framework that all the methods presented in the following chapters fall under. A brief introduction to the typical applications of adaptive filtering are also presented.

In Chapter 2, the basic concepts of discrete-time stochastic processes are reviewed with special emphasis to the results that are useful to analyze the behavior of adaptive filtering algorithms. In addition, the Wiener filter is presented, establishing the optimum linear filter that can be sought in stationary environments. Appendix A briefly describes the concepts of complex differentiation mainly applied to the Wiener solution. The case of linearly constrained Wiener filter is also discussed, motivated by its wide use in antenna array processing. The transformation of the constrained minimization problem into an unconstrained one is also presented. The concept of mean-square error surface is then introduced, another useful tool to analyze adaptive filters. The classical Newton and steepest-descent algorithms are briefly introduced. Since the use of these algorithms would require a complete knowledge of the stochastic environment, the adaptive filtering algorithms introduced in the following chapters come into play. Practical applications of the adaptive filtering algorithms are revisited in more detail at the end of Chapter 2 where some examples with closed form solutions are included in order to allow the correct interpretation of what is expected from each application.

Chapter 3 presents and analyses of the least-mean-square (LMS) algorithm in some depth. Several aspects are discussed, such as convergence behavior in stationary and nonstationary environments. This chapter also includes a number of theoretical as well as simulation examples to illustrate how the LMS algorithm performs in different setups. Appendix B addresses the quantization effects on the LMS algorithm when implemented in fixed- and floating-point arithmetics.

Chapter 4 deals with some algorithms that are in a sense related to the LMS algorithm. In particular, the algorithms introduced are the quantized-error algorithms, the LMS-Newton algorithm, the normalized LMS algorithm, the transform-domain LMS algorithm, and the affine projection algorithm. Some properties of these algorithms are also discussed in Chapter 4, with special emphasis to the analysis of the fine projection algorithm.

Chapter 5 introduces the conventional recursive least-squares (RLS) algorithm. This algorithm minimizes a deterministic objective function, differing in this sense from most LMS-based algorithms. Following the same pattern of presentation of Chapter 3, several aspects of the conventional RLS algorithm are discussed, such as convergence behavior in stationary and nonstationary environments, along with a number of simulation results. Appendix C, deals with stability issues and quantization effects related to the RLS algorithm when implemented in fixed- and floating-point arithmetics. The results presented, except for the quantization effects, are also valid for the RLS algorithms presented in Chapters 7, 8, and 9. As as complement to Chapter 5, Appendix D presents the discrete-time Kalman filter formulation which despite being considered an extension of the Wiener filter has some relation with the RLS algorithm.

Chapter 6 discusses some techniques to reduce the overall computational complexity of adaptive filtering algorithms. The chapter first introduces the so called set-membership algorithms that update only when the output estimation error is higher than the prescribed upper bound. However, since set-membership algorithms require frequent updates during the early iterations in stationary environments, we introduce the concept of partial update to reduce the computational complexity in order to deal with situations where the available computational resources are not sufficient. This chapter presents several forms of set-membership algorithms related to the affine projection algorithms and their special cases. Chapter 6 also includes some simulation examples addressing standard as well as application oriented problems, where the algorithms of this and previous chapters are compared in some detail.

In Chapter 7, a family of fast RLS algorithms based on the FIR lattice realization is introduced. These algorithms represent interesting alternatives to the computationally complex conventional RLS algorithm. In particular, the unnormalized, the normalized and the error-feedback algorithms are presented.

Chapter 8 deals with the fast transversal RLS algorithms, which are very attractive due to their low computational complexity. However, these algorithms are known to face stability problems in practical implementation. As a consequence, special attention is given to the stabilized fast transversal RLS algorithm.

Chapter 9 is devoted to a family of RLS algorithms based on the QR decomposition. The conventional and a fast version of the QR-based algorithms are presented in this chapter.

Chapter 10 addresses the subject of adaptive filters using IIR digital filter realizations. The chapter includes a discussion on how to compute the gradient and how to derive the adaptive algorithms. The cascade, the parallel, and the lattice realizations are presented as interesting alternatives to the direct-form realization for the IIR adaptive filter. The characteristics of the mean-square error surface are also discussed in this chapter, for the IIR adaptive filtering case. Algorithms based on alternative error formulations, such as the equation error and Steiglitz-McBride methods are also introduced.

Chapter 11 deals with nonlinear adaptive filtering which consists of utilizing a nonlinear structure for the adaptive filter. The motivation is to use nonlinear adaptive filtering structures to better model some nonlinear phenomena commonly found in communications applications, such as nonlinear characteristics of power amplifier at transmitters. In particular, we introduce the Volterra series LMS and RLS algorithms, and the adaptive algorithms based on bilinear filters. Also, a brief introduction is given to some nonlinear adaptive filtering algorithms based on the concepts of neural networks, namely, the multilayer perceptron and the radial basis function algorithms. Some examples of DFE equalization are included in this chapter.

Chapter 12 deals with adaptive filtering in subbands mainly to address the applications where the required adaptive filter order is high, as for example in acoustic echo cancellation where the unknown system (echo) model has long impulse response. In subband adaptive filtering, some signals are split in frequency subbands via an analysis filter bank. Chapter 12 provides a brief review of multirate systems, and presents the basic structures for adaptive filtering in subbands. The concept of delayless subband adaptive filtering is also addressed, where the adaptive filter coefficients are updated in subbands and mapped to an equivalent fullband filter. The chapter also includes a discussion on the relation between subband and block adaptive filtering (also known as frequency-domain adaptive filters) algorithms.

Chapter 13 describes some adaptive filtering algorithms suitable for situations where no reference signal is available which are known as blind adaptive filtering algorithms. In particular, this chapter introduces some blind algorithms utilizing high-order statistics implicitly for the single-input single-output (SISO) equalization applications. In order to address some drawbacks of the SISO equalization systems, we discuss some algorithms using second-order statistics for the single-input multi-output (SIMO) equalization. The SIMO algorithms are naturally applicable in cases of oversampled received signal and multiple receive antennas. This chapter also discusses some issues related to blind signal processing not directly detailed here.

I decided to use some standard examples to present a number of simulation results, in order to test and compare different algorithms. This way, frequent repetition was avoided while allowing the reader to easily compare the performance of the algorithms. Most of the end of chapters problems are simulation oriented, however, some theoretical ones are included to complement the text.

The second edition differed from the first one mainly by the inclusion of chapters on nonlinear and subband adaptive filtering. Many other smaller changes were performed throughout the remaining chapters. In this edition, we introduced a number of derivations and explanations requested by students and suggested by colleagues. In addition, two new chapters on data-selective algorithms and blind adaptive filtering are included along with a large number of new examples and problems. Major changes take place in the first five chapters in order to make the technical details more accessible and to improve the ability of the reader in deciding where and how to use the concepts. The analysis of the fine projection algorithm is now presented in detail due to its growing practical importance. Several practical and theoretical examples are included aiming at comparing the families of algorithms introduced in the book.

In a trimester course, I usually cover Chapters 1 to 6 sometimes skipping parts of Chapter 2 and the analyses of quantization effects in Appendices B and C. In the remaining time, I try to cover as much as possible of the remaining chapters, usually consulting the audience to what they would prefer to study. This book can also be used for self-study where the reader can examine Chapters 1 to 6, and those not involved with specialized implementations can skip the Appendices B and C, without loss of continuity. The remaining chapters can be followed separately, except for Chapter 8 that requires reading Chapter 7. Chapters 7, 8, and 9 deal with alternative and fast implementations of RLS algorithms and the following chapters do not use their results.

ACKNOWLEDGMENTS

The support of the Department of Electronics and Computer Engineering of the Polytechnic School (undergraduate school of engineering) of UFRJ and of the Program of Electrical Engineering of COPPE have been fundamental to complete this work.

I was lucky enough to have contact with a number of creative professors and researchers that by taking their time to discuss technical matters with me, raised many interesting questions and provided me with enthusiasm to write the first, second, and third editions of this book. In that sense, I would like to thank Prof. Pan Agathoklis, University of Victoria; Prof. R. C. de Lamare, University of York; Prof. M. Gerken, University of São Paulo; Prof. A. Hjørungnes, UniK-University of Oslo; Prof. T. I. Laakso, Helsinki Uuiversity of Technology; Prof. J. P. Leblanc, Luleå University of Technology; Prof. W. S. Lu, University of Victoria; Dr. H. S. Malvar, Microsoft Research; Prof. V. H. Nascimento, University of São Paulo; Prof. J. M. T. Romano, State University of Campinas; Prof. E. Sanchez Sinencio, Texas A&M University; Prof. Trac D. Tran, John Hopkins University.

My M.Sc. supervisor, my friend and colleague, Prof. L. P. Calôba has been a source of inspiration and encouragement not only for this work but for my entire career. Prof. A. Antoniou, my Ph.D. supervisor, has also been an invaluable friend and advisor, I learned a lot by writing papers with him. I was very fortunate to have these guys as Professors.

The good students that attend engineering at UFRJ are for sure another source of inspiration. In particular, I have been lucky to attract good and dedicated graduate students, who have participated in the research related to adaptive filtering. Some of them are: Dr. R. G. Alves, Prof. J. A. Apolinário, Jr., Prof. L. W. P. Biscainho, Prof. M. L. R. Campos, Prof. J. E. Cousseau, T. N. Ferreira, M. V. S. Lima, T. C. Macedo, Jr., W. A. Martins, Prof. R. Merched, Prof. S. L. Netto, G. O. Pinto, C. B. Ribeiro, A. D. Santana, Jr., Dr. M. G. Siqueira, Dr. S. Subramanian (Anna University), Prof. F. G. V. Resende Jr., M. R. Vassali, Dr. S. Werner (Helsinki University of Technology). Most of them took time from their M.Sc. and Ph.D. work to read parts of the manuscript and provided me with invaluable suggestions. Some parts of this book have been influenced by my interactions with these and other former students.

I am particularly grateful to Profs. L. W. P. Biscainho, M. L. R. Campos and J. E. Cousseau, for their support in producing some of the examples of the book. Profs. L. W. P. Biscainho, M. L. R. Campos, and S. L. Netto also read every inch of the manuscript and provided numerous suggestions for improvements.

I am most grateful to Profs. E. A. B. da Silva, UFRJ, and R. Merched, UFRJ, for their critical inputs on parts of the manuscript. Prof. E. A. B. da Silva seems to be always around in difficult times to lay a helping hand.

Indeed the friendly and harmonious work environment of the LPS, the Signal Processing Laboratory of UFRJ, has been an enormous source of inspiration and challenge. From its manager Michelle to the Professors, undergraduate and graduate students, and staff, I always find support that goes beyond the professional obligation. Jane made many of the drawings with care, I really appreciate it.

I am also thankful to Prof. I. Hartimo, Helsinki University of Technology; Prof. J. L. Huertas, University of Seville; Prof. A. Antoniou, University of Victoria; Prof. J. E. Cousseau, Universidad Nacional del Sur; Prof. Y.-F. Huang, University of Notre Dame; Prof. A. Hjørungnes, UniK-University of Oslo, for giving me the opportunity to teach at the institutions they work for.

In recent years, I have been working as consultant to INdT (NOKIA Institute of Technology) where its President G. Feitoza and their researchers have team up with me in challenging endeavors. They are always posing me with problems, not necessarily technical, which widen my way of thinking.

The earlier support of Catherine Chang, Prof. J. E. Cousseau, Prof. F. G. V. Resende Jr., and Dr. S. Sunder for solving my problems with the text editor is also deeply appreciated.

The financial supports of the Brazilian research councils CNJPq, CAPES, and FAPERJ were fundamental for the completion of this book.

The friendship and trust of my editor Alex Greene, from Springer, have been crucial to make this third edition a reality.

My parents provided me with the moral and educational support needed to pursue any project, including this one. My mother's patience, love and understanding seem to be endless.

My brother Fernando always says yes, what else do I want? He also awarded me with my nephews Fernandinho and Daniel.

My family deserves special thanks. My daughters Paula and Luiza have been extremely understanding, and always forgive daddy for being busy. They are wonderful young ladies. My wife Mariza deserves my deepest gratitude for her endless love, support, and friendship. She always does her best to provide me with the conditions to develop this and other projects.

Prof. Paulo S. R. Diniz

Niterói, Brazil

CONTENTS

1

INTRODUCTION TO ADAPTIVE FILTERING

1.1 INTRODUCTION

In this section, we define the kind of signal processing systems that will be treated in this text.

In the last thirty years significant contributions have been made in the signal processing field. The advances in digital circuit design have been the key technological development that sparked a growing interest in the field of digital signal processing. The resulting digital signal processing systems are attractive due to their low cost, reliability, accuracy, small physical sizes, and flexibility.

One example of a digital signal processing system is called *filter*. Filtering is a signal processing operation whose objective is to process a signal in order to manipulate the information contained in the signal. In other words, a filter is a device that maps its input signal to another output signal facilitating the extraction of the desired information contained in the input signal. A digital filter is the one that processes discrete-time signals represented in digital format. For time-invariant filters the internal parameters and the structure of the filter are fixed, and if the filter is linear the output signal is a linear function of the input signal. Once prescribed specifications are given, the design of time-invariant linear filters entails three basic steps, namely: the approximation of the specifications by a rational transfer function, the choice of an appropriate structure defining the algorithm, and the choice of the form of implementation for the algorithm.

An adaptive filter is required when either the fixed specifications are unknown or the specifications cannot be satisfied by time-invariant filters. Strictly speaking an adaptive filter is a nonlinear filter since its characteristics are dependent on the input signal and consequently the homogeneity and additivity conditions are not satisfied. However, if we freeze the filter parameters at a given instant of time, most adaptive filters considered in this text are linear in the sense that their output signals are linear functions of their input signals. The exceptions are the adaptive filters discussed in Chapter 11.

The adaptive filters are time-varying since their parameters are continually changing in order to meet a performance requirement. In this sense, we can interpret an adaptive filter as a filter that performs the approximation step on-line. Usually, the definition of the performance criterion requires the existence of a reference signal that is usually hidden in the approximation step of fixed-filter

P.S.R. Diniz, *Adaptive Filtering*, DOI: 10.1007/978-0-387-68606-6_1,

design. This discussion brings the feeling that in the design of fixed (nonadaptive) filters a complete characterization of the input and reference signals is required in order to design the most appropriate filter that meets a prescribed performance. Unfortunately, this is not the usual situation encountered in practice, where the environment is not well defined. The signals that compose the environment are the input and the reference signals, and in cases where any of them is not well defined, the design procedure is to model the signals and subsequently design the filter. This procedure could be costly and difficult to implement on-line. The solution to this problem is to employ an adaptive filter that performs on-line updating of its parameters through a rather simple algorithm, using only the information available in the environment. In other words, the adaptive filter performs a data-driven approximation step.

The subject of this book is adaptive filtering, which concerns the choice of structures and algorithms for a filter that has its parameters (or coefficients) adapted, in order to improve a prescribed performance criterion. The coefficient updating is performed using the information available at a given time.

The development of digital very large scale integration (VLSI) technology allowed the widespread use of adaptive signal processing techniques in a large number of applications. This is the reason why in this book only discrete-time implementations of adaptive filters are considered. Obviously, we assume that continuous-time signals taken from the real world are properly sampled, i.e., they are represented by discrete-time signals with sampling rate higher than twice their highest frequency. Basically, it is assumed that when generating a discrete-time signal by sampling a continuous-time signal, the Nyquist or sampling theorem is satisfied [1]-[9].

1.2 ADAPTIVE SIGNAL PROCESSING

As previously discussed, the design of digital filters with fixed coefficients requires well defined prescribed specifications. However, there are situations where the specifications are not available, or are time varying. The solution in these cases is to employ a digital filter with adaptive coefficients, known as adaptive filters [10]-[17].

Since no specifications are available, the adaptive algorithm that determines the updating of the filter coefficients, requires extra information that is usually given in the form of a signal. This signal is in general called a desired or reference signal, whose choice is normally a tricky task that depends on the application.

Adaptive filters are considered nonlinear systems, therefore their behavior analysis is more complicated than for fixed filters. On the other hand, because the adaptive filters are self designing filters, from the practitioner's point of view their design can be considered less involved than in the case of digital filters with fixed coefficients.

The general set up of an adaptive-filtering environment is illustrated in Fig. 1.1, where k is the iteration number, $x(k)$ denotes the input signal, $y(k)$ is the adaptive-filter output signal, and $d(k)$ defines the desired signal. The error signal $e(k)$ is calculated as $d(k) - y(k)$. The error signal is then used to form a performance (or objective) function that is required by the adaptation algorithm in order to determine the appropriate updating of the filter coefficients. The minimization of the objective function implies that the adaptive-filter output signal is matching the desired signal in some sense.

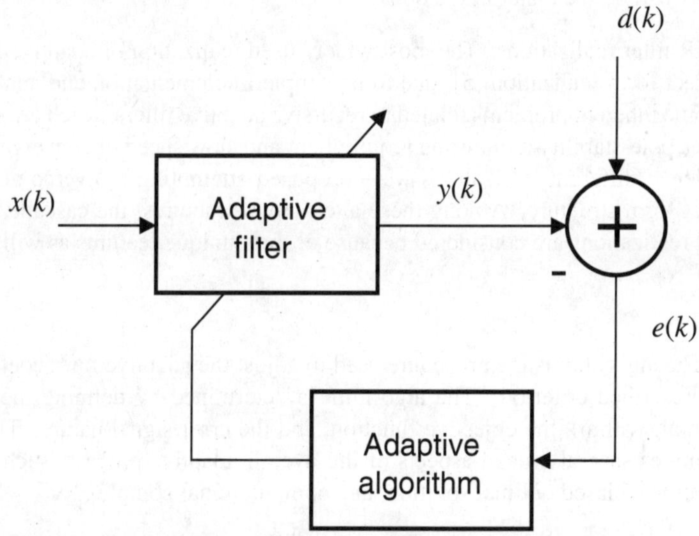

Figure 1.1 General adaptive-filter configuration.

The complete specification of an adaptive system, as shown in Fig. 1.1, consists of three items:

1) **Application**: The type of application is defined by the choice of the signals acquired from the environment to be the input and desired-output signals. The number of different applications in which adaptive techniques are being successfully used has increased enormously during the last two decades. Some examples are echo cancellation, equalization of dispersive channels, system identification, signal enhancement, adaptive beamforming, noise cancelling, and control [14]-[20]. The study of different applications is not the main scope of this book. However, some applications are considered in some detail.

2) **Adaptive-Filter Structure**: The adaptive filter can be implemented in a number of different structures or realizations. The choice of the structure can influence the computational complexity (amount of arithmetic operations per iteration) of the process and also the necessary number of iterations to achieve a desired performance level. Basically, there are two major classes of adaptive digital filter realizations, distinguished by the form of the impulse response, namely the finite-duration impulse response (FIR) filter and the infinite-duration impulse response (IIR) filters. FIR filters are usually implemented with nonrecursive structures, whereas IIR filters utilize recursive realizations.

- Adaptive FIR filter realizations: The most widely used adaptive FIR filter structure is the transversal filter, also called tapped delay line, that implements an all-zero transfer function with a canonic direct form realization without feedback. For this realization, the output signal

$y(k)$ is a linear combination of the filter coefficients, that yields a quadratic mean-square error (MSE $= E[|e(k)|^2]$) function with a unique optimal solution. Other alternative adaptive FIR realizations are also used in order to obtain improvements as compared to the transversal filter structure, in terms of computational complexity, speed of convergence, and finite wordlength properties as will be seen later in the book.

- Adaptive IIR filter realizations: The most widely used realization of adaptive IIR filters is the canonic direct form realization [5], due to its simple implementation and analysis. However, there are some inherent problems related to recursive adaptive filters which are structure dependent, such as pole-stability monitoring requirement and slow speed of convergence. To address these problems, different realizations were proposed attempting to overcome the limitations of the direct form structure. Among these alternative structures, the cascade, the lattice, and the parallel realizations are considered because of their unique features as will be discussed in Chapter 10.

3) **Algorithm**: The algorithm is the procedure used to adjust the adaptive filter coefficients in order to minimize a prescribed criterion. The algorithm is determined by defining the search method (or minimization algorithm), the objective function, and the error signal nature. The choice of the algorithm determines several crucial aspects of the overall adaptive process, such as existence of sub-optimal solutions, biased optimal solution, and computational complexity.

1.3 INTRODUCTION TO ADAPTIVE ALGORITHMS

The basic objective of the adaptive filter is to set its parameters, $\boldsymbol{\theta}(k)$, in such a way that its output tries to minimize a meaningful objective function involving the reference signal. Usually, the objective function F is a function of the input, the reference, and adaptive-filter output signals, i.e., $F = F[x(k), d(k), y(k)]$. A consistent definition of the objective function must satisfy the following properties:

- Non-negativity: $F[x(k), d(k), y(k)] \geq 0, \forall y(k), x(k)$, and $d(k)$;
- Optimality: $F[x(k), d(k), d(k)] = 0$.

One should understand that in an adaptive process, the adaptive algorithm attempts to minimize the function F, in such a way that $y(k)$ approximates $d(k)$, and as a consequence, $\boldsymbol{\theta}(k)$ converges to $\boldsymbol{\theta}_o$, where $\boldsymbol{\theta}_o$ is the optimum set of coefficients that leads to the minimization of the objective function.

Another way to interpret the objective function is to consider it a direct function of a generic error signal $e(k)$, which in turn is a function of the signals $x(k)$, $y(k)$, and $d(k)$, i.e., $F = F[e(k)] = F[e(x(k), y(k), d(k))]$. Using this framework, we can consider that an adaptive algorithm is composed of three basic items: definition of the minimization algorithm, definition of the objective function form, and definition of the error signal.

1) **Definition of the minimization algorithm for the function** F: This item is the main subject of Optimization Theory [22]-[23], and it essentially affects the speed of convergence and computational complexity of the adaptive process.

In practice any continuous function having high-order model of the parameters can be approximated around a given point $\boldsymbol{\theta}(k)$ by a truncated Taylor series as follows

$$F[\boldsymbol{\theta}(k)+\Delta\boldsymbol{\theta}(k)] \approx F[\boldsymbol{\theta}(k)] + \mathbf{g}_{\boldsymbol{\theta}}^T\{F[\boldsymbol{\theta}(k)]\}\Delta\boldsymbol{\theta}(k) + \frac{1}{2}\Delta\boldsymbol{\theta}^T(k)\mathbf{H}_{\boldsymbol{\theta}}\{F[\boldsymbol{\theta}(k)]\}\Delta\boldsymbol{\theta}(k) \tag{1.1}$$

where $\mathbf{H}_{\boldsymbol{\theta}}\{F[\boldsymbol{\theta}(k)]\}$ is the Hessian matrix of the objective function, and $\mathbf{g}_{\boldsymbol{\theta}}\{F[\boldsymbol{\theta}(k)]\}$ is the gradient vector, further details about the Hessian matrix and gradient vector are presented along the text. The aim is to minimize the objective function with respect to the set of parameters by iterating

$$\boldsymbol{\theta}(k+1) = \boldsymbol{\theta}(k) + \Delta\boldsymbol{\theta}(k) \tag{1.2}$$

where the step or correction term $\Delta\boldsymbol{\theta}(k)$ is meant to minimize the quadratic approximation of the objective function $F[\boldsymbol{\theta}(k)]$. The so-called Newton method requires the first and second-order derivatives of $F[\boldsymbol{\theta}(k)]$ to be available at any point, as well as the function value. These informations are required in order to evaluate equation (1.1). If $\mathbf{H}_{\boldsymbol{\theta}}(\boldsymbol{\theta}(k))$ is a positive definite matrix, then the quadratic approximation has a unique and well defined minimum point. Such a solution can be found by setting the gradient of the quadratic function with respect to the parameters correction terms, at instant $k+1$, to zero which leads to

$$\mathbf{g}_{\boldsymbol{\theta}}\{F[\boldsymbol{\theta}(k)]\} = -\mathbf{H}_{\boldsymbol{\theta}}\{F[\boldsymbol{\theta}(k)]\}\Delta\boldsymbol{\theta}(k) \tag{1.3}$$

The most commonly used optimization methods in the adaptive signal processing field are:

- Newton's method: This method seeks the minimum of a second-order approximation of the objective function using an iterative updating formula for the parameter vector given by

  $$\boldsymbol{\theta}(k+1) = \boldsymbol{\theta}(k) - \mu\mathbf{H}_{\boldsymbol{\theta}}^{-1}\{F[e(k)]\}\mathbf{g}_{\boldsymbol{\theta}}\{F[e(k)]\} \tag{1.4}$$

  where μ is a factor that controls the step size of the algorithm, i.e., it determines how fast the parameter vector will be changed. The reader should note that the direction of the correction term $\Delta\boldsymbol{\theta}(k)$ is chosen according to equation (1.3). The matrix of second derivatives of $F[e(k)]$, $\mathbf{H}_{\boldsymbol{\theta}}\{F[e(k)]\}$ is the Hessian matrix of the objective function, and $\mathbf{g}_{\boldsymbol{\theta}}\{F[e(k)]\}$ is the gradient of the objective function with respect to the adaptive filter coefficients. It should be noted that the error $e(k)$ depends on the parameters $\boldsymbol{\theta}(k)$. If the function $F[e(k)]$ is originally quadratic, there is no approximation in the model of equation (1.1) and the global minimum of the objective function would be reached in one step if $\mu = 1$. For nonquadratic functions the value of μ should be reduced.

- *Quasi*-Newton methods: This class of algorithms is a simplified version of the method above described, as it attempts to minimize the objective function using a recursively calculated estimate of the inverse of the Hessian matrix, i.e.,

  $$\boldsymbol{\theta}(k+1) = \boldsymbol{\theta}(k) - \mu\mathbf{S}(k)\mathbf{g}_{\boldsymbol{\theta}}\{F[e(k)]\} \tag{1.5}$$

where $\mathbf{S}(k)$ is an estimate of $\mathbf{H}_{\boldsymbol{\theta}}^{-1}\{F[e(k)]\}$, such that

$$\lim_{k\to\infty} \mathbf{S}(k) = \mathbf{H}_{\boldsymbol{\theta}}^{-1}\{F[e(k)]\}$$

A usual way to calculate the inverse of the Hessian estimate is through the matrix inversion lemma (see, for example [21] and some chapters to come). Also, the gradient vector is usually replaced by a computationally efficient estimate.

- Steepest-descent method: This type of algorithm searches the objective function minimum point following the opposite direction of the gradient vector of this function. Consequently, the updating equation assumes the form

$$\boldsymbol{\theta}(k+1) = \boldsymbol{\theta}(k) - \mu \mathbf{g}_{\boldsymbol{\theta}}\{F[e(k)]\} \tag{1.6}$$

Here and in the open literature, the steepest-descent method is often also referred to as gradient method.

In general, gradient methods are easier to implement, but on the other hand, the Newton method usually requires a smaller number of iterations to reach a neighborhood of the minimum point. In many cases, *Quasi*-Newton methods can be considered a good compromise between the computational efficiency of the gradient methods and the fast convergence of the Newton method. However, the *Quasi*-Newton algorithms are susceptible to instability problems due to the recursive form used to generate the estimate of the inverse Hessian matrix. A detailed study of the most widely used minimization algorithms can be found in [22]-[23].

It should be pointed out that with any minimization method, the convergence factor μ controls the stability, speed of convergence, and some characteristics of residual error of the overall adaptive process. Usually, an appropriate choice of this parameter requires a reasonable amount of knowledge of the specific adaptive problem of interest. Consequently, there is no general solution to accomplish this task. In practice, computational simulations play an important role and are, in fact, the most used tool to address the problem.

2) **Definition of the objective function** $F[e(k)]$: There are many ways to define an objective function that satisfies the optimality and non-negativity properties formerly described. This definition affects the complexity of the gradient vector and the Hessian matrix calculation. Using the algorithm's computational complexity as a criterion, we can list the following forms for the objective function as the most commonly used in the derivation of an adaptive algorithm:

- Mean-Square Error (MSE): $F[e(k)] = E[|e(k)|^2]$;
- Least Squares (LS): $F[e(k)] = \frac{1}{k+1}\sum_{i=0}^{k} |e(k-i)|^2$;
- Weighted Least Squares (WLS): $F[e(k)] = \sum_{i=0}^{k} \lambda^i |e(k-i)|^2$, λ is a constant smaller than 1;
- Instantaneous Squared Value (ISV): $F[e(k)] = |e(k)|^2$.

The MSE, in a strict sense, is only of theoretical value, since it requires an infinite amount of information to be measured. In practice, this ideal objective function can be approximated by the other three listed. The LS, WLS, and ISV functions differ in the implementation complexity and in the convergence behavior characteristics; in general, the ISV is easier to implement but presents noisy convergence properties, since it represents a greatly simplified objective function. The LS is convenient to be used in stationary environment, whereas the WLS is useful in applications where the environment is slowly varying.

3) **Definition of the error signal** $e(k)$: The choice of the error signal is crucial for the algorithm definition, since it can affect several characteristics of the overall algorithm including computational complexity, speed of convergence, robustness, and most importantly for the IIR adaptive filtering case, the occurrence of biased and multiple solutions.

The minimization algorithm, the objective function, and the error signal as presented give us a structured and simple way to interpret, analyze, and study an adaptive algorithm. In fact, almost all known adaptive algorithms can be visualized in this form, or in a slight variation of this organization. In the remaining parts of this book, using this framework, we present the principles of adaptive algorithms. It may be observed that the minimization algorithm and the objective function affect the convergence speed of the adaptive process. An important step in the definition of an adaptive algorithm is the choice of the error signal, since this task exercises direct influence in many aspects of the overall convergence process.

1.4 APPLICATIONS

In this section, we discuss some possible choices for the input and desired signals and how these choices are related to the applications. Some of the classical applications of adaptive filtering are system identification, channel equalization, signal enhancement, and prediction.

In the system identification application, the desired signal is the output of the unknown system when excited by a broadband signal, in most cases a white-noise signal. The broadband signal is also used as input for the adaptive filter as illustrated in Fig. 1.2. When the output MSE is minimized, the adaptive filter represents a model for the unknown system.

The channel equalization scheme consists of applying the originally transmitted signal distorted by the channel plus environment noise as the input signal to an adaptive filter, whereas the desired signal is a delayed version of the original signal as depicted in Fig. 1.3. This delayed version of the input signal is in general available at the receiver in a form of standard training signal. In a noiseless case, the minimization of the MSE indicates that the adaptive filter represents an inverse model (equalizer) of the channel.

In the signal enhancement case, a signal $x(k)$ is corrupted by noise $n_1(k)$, and a signal $n_2(k)$ correlated to the noise is available (measurable). If $n_2(k)$ is used as an input to the adaptive filter with the signal corrupted by noise playing the role of the desired signal, after convergence the output

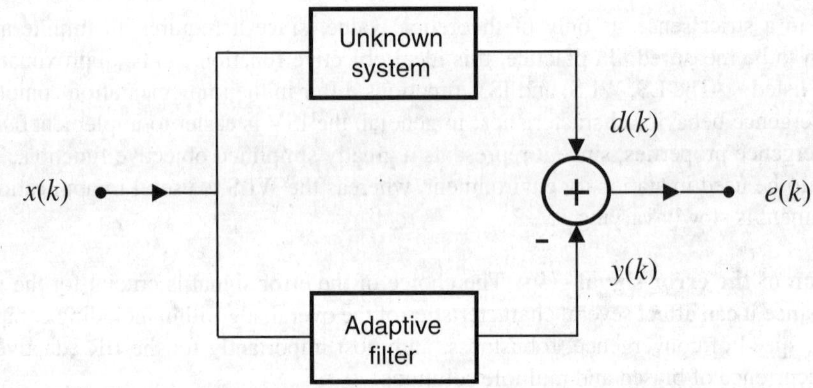

Figure 1.2 System identification.

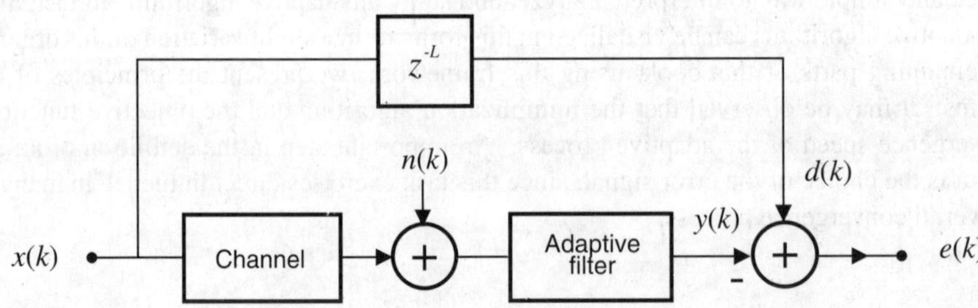

Figure 1.3 Channel equalization.

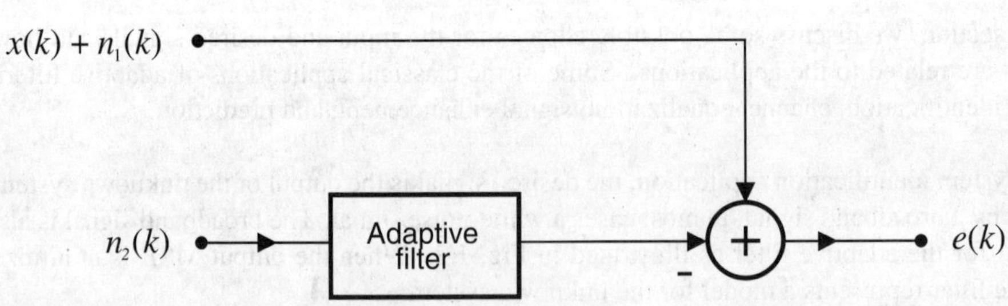

Figure 1.4 Signal enhancement ($n_1(k)$ and $n_2(k)$ are noise signals correlated to each other).

error will be an enhanced version of the signal. Fig. 1.4 illustrates a typical signal enhancement setup.

Finally, in the prediction case the desired signal is a forward (or eventually a backward) version of the adaptive-filter input signal as shown in Fig. 1.5. After convergence, the adaptive filter represents a model for the input signal, and can be used as a predictor model for the input signal.

Further details regarding the applications discussed here will be given in the following chapters.

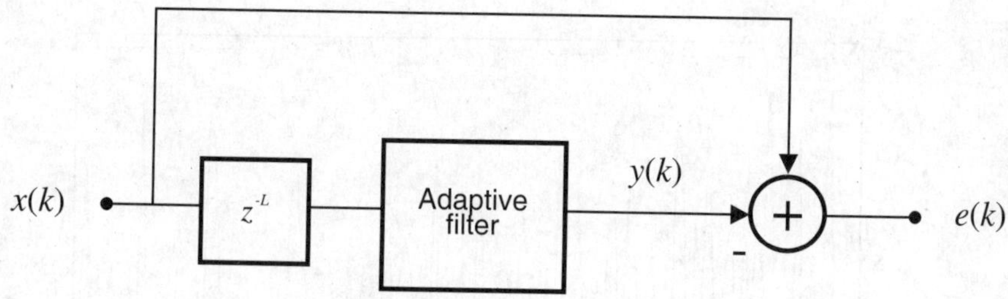

Figure 1.5 Signal prediction.

Example 1.1

Before concluding this chapter, we present a simple example in order to illustrate how an adaptive filter can be useful in solving problems that lie in the general framework represented by Fig. 1.1. We chose the signal enhancement application illustrated in Fig. 1.4.

In this example, the reference (or desired) signal consists of a discrete-time triangular waveform corrupted by a colored noise. Fig. 1.6 shows the desired signal. The adaptive-filter input signal is a white noise correlated with the noise signal that corrupted the triangular waveform, as shown in Fig. 1.7.

The coefficients of the adaptive filter are adjusted in order to keep the squared value of the output error as small as possible. As can be noticed in Fig. 1.8, as the number of iterations increase the error signal resembles the discrete-time triangular waveform shown in the same figure (dashed curve).

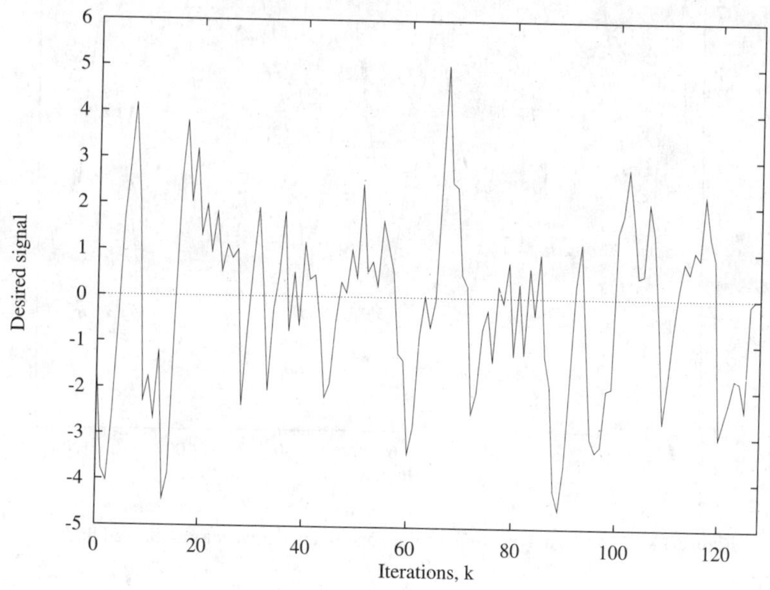

Figure 1.6 Desired signal.

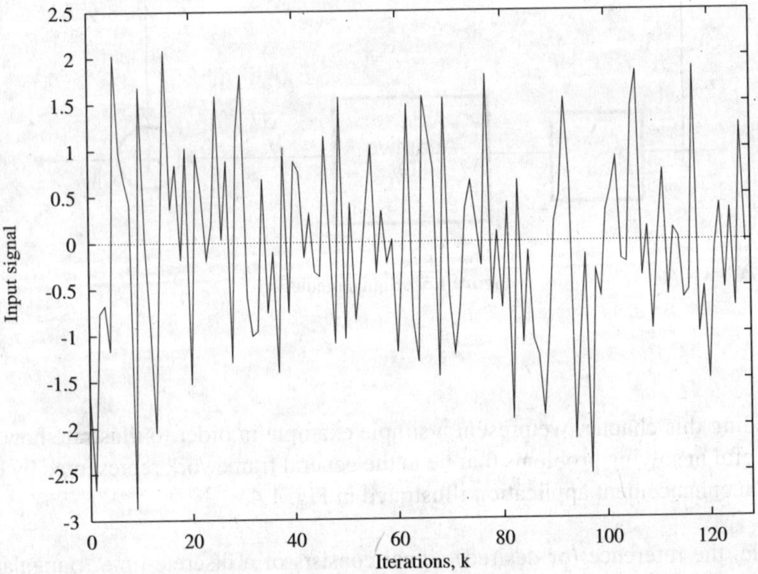

Figure 1.7 Input signal.

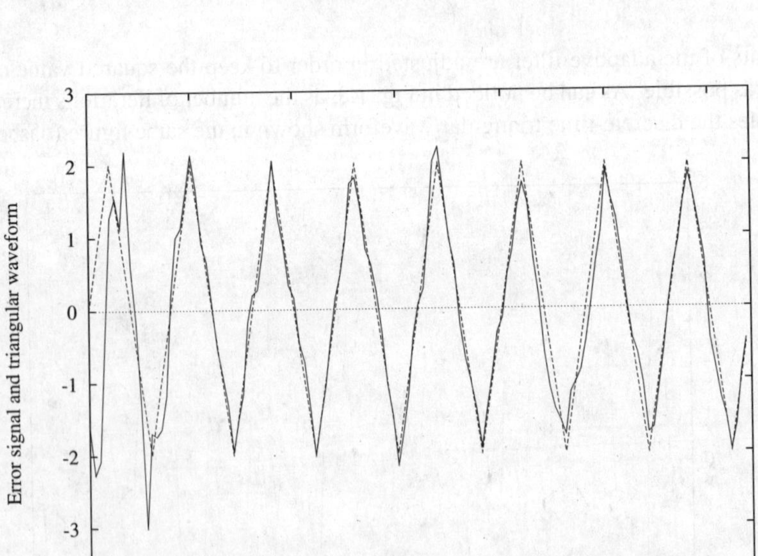

Figure 1.8 Error signal (continuous line) and triangular waveform (dashed line).

□

1.5 REFERENCES

1. P. S. R. Diniz, E. A. B. da Silva, and S. L. Netto, *Digital Signal Processing: System Analysis and Design*, Cambridge University Press, Cambridge, UK, 2002.
2. A. Papoulis, *Signal Analysis*, McGraw Hill, New York, NY, 1977.
3. A. V. Oppenheim, A. S. Willsky, and S. H. Nawab, *Signals and Systems*, Prentice Hall, Englewood Cliffs, NJ, 2nd edition, 1997.
4. A. V. Oppenheim and R. W. Schaffer, *Discrete-Time Signal Processing*, Prentice Hall, Englewood Cliffs, NJ, 1989.
5. A. Antoniou, *Digital Signal Processing: Signals, Systems, and Filters*, McGraw Hill, New York, NY, 2005.
6. L. B. Jackson, *Digital Filters and Signal Processing*, Kluwer Academic Publishers, Norwell, MA, 3rd edition, 1996.
7. R. A. Roberts and C. T. Mullis, *Digital Signal Processing*, Addison-Wesley, Reading, MA, 1987.
8. J. G. Proakis and D. G. Manolakis, *Digital Signal Processing*, Prentice Hall, Englewood Cliffs, NJ, 4th edition, 2007.
9. T. Bose, *Digital Signal and Image Processing*, John Wiley & Sons, New York, NY, 2004.
10. M. L. Honig and D. G. Messerschmitt, *Adaptive Filters: Structures, algorithms, and Applications*, Kluwer Academic Publishers, Boston, MA, 1984.
11. S. T. Alexander, *Adaptive Signal Processing*, Springer Verlag, New York, NY, 1986.
12. M. Bellanger, *Adaptive Digital Filters*, Marcel Dekker, Inc., New York, NY, 2nd edition, 2001.
13. P. Strobach, *Linear Prediction Theory*, Springer Verlag, New York, NY, 1990.
14. B. Widrow and S. D. Stearns, *Adaptive Signal Processing*, Prentice Hall, Englewood Cliffs, NJ, 1985.
15. J. R. Treichler, C. R. Johnson, Jr., and M. G. Larimore, *Theory and Design of Adaptive Filters*, John Wiley & Sons, New York, NY, 1987.
16. B. Farhang-Boroujeny, *Adaptive Filters: Theory and Applications*, John Wiley & Sons, New York, NY, 1998.
17. S. Haykin, *Adaptive Filter Theory*, Prentice Hall, Englewood Cliffs, NJ, 4th edition, 2002.
18. A. H. Sayed, *Fundamentals of Adaptive Filtering*, John Wiley & Sons, Hoboken, NJ, 2003.
19. L. R. Rabiner and R. W. Schaffer, *Digital Processing of Speech Signals*, Prentice Hall, Englewood Cliffs, NJ, 1978.
20. D. H. Johnson and D. E. Dudgeon, *Array Signal Processing*, Prentice Hall, Englewood Cliffs, NJ, 1993.

21. T. Kailath, *Linear Systems*, Prentice Hall, Englewood Cliffs, NJ, 1980.

22. D. G. Luenberger, *Introduction to Linear and Nonlinear Programming*, Addison Wesley, Reading, MA, 2nd edition, 1984.

23. A. Antoniou and W.-S. Lu, *Practical Optimization: Algorithms and Engineering Applications*, Springer, New York, NY, 2007.

2

FUNDAMENTALS OF ADAPTIVE FILTERING

2.1 INTRODUCTION

This chapter includes a brief review of deterministic and random signal representations. Due to the extent of those subjects, our review is limited to the concepts that are directly relevant to adaptive filtering. The properties of the correlation matrix of the input signal vector are investigated in some detail, since they play a key role in the statistical analysis of the adaptive-filtering algorithms.

The Wiener solution that represents the minimum mean-square error (MSE) solution of discrete-time filters realized through a linear combiner is also introduced. This solution depends on the input signal correlation matrix as well as on the the cross-correlation between the elements of the input signal vector and the reference signal. The values of these correlations form the parameters of the MSE surface, which is a quadratic function of the adaptive-filter coefficients. The linearly constrained Wiener filter is also presented, a technique commonly used in antenna array processing applications. The transformation of the constrained minimization problem into an unconstrained one is also discussed. Motivated by the importance of the properties of the MSE surface, we analyze them using some results related to the input signal correlation matrix.

In practice the parameters that determine the MSE surface shape are not available. What is left is to directly or indirectly estimate these parameters using the available data and to develop adaptive algorithms that use these estimates to search the MSE surface, such that the adaptive-filter coefficients converge to the Wiener solution in some sense. The starting point to obtain an estimation procedure is to investigate the convenience of using the classical searching methods of optimization theory [1]-[3] to adaptive filtering. The Newton and steepest-descent algorithms are investigated as possible searching methods for adaptive filtering. Although both methods are not directly applicable to practical adaptive filtering, smart reflections inspired on them led to practical algorithms such as the least-mean-square (LMS) [4]-[5] and Newton-based algorithms. The Newton and steepest-descent algorithms are introduced in this chapter, whereas the LMS algorithm is treated in the next chapter.

Also, in the present chapter, the main applications of adaptive filters are revisited and discussed in greater detail.

P.S.R. Diniz, *Adaptive Filtering*, DOI: 10.1007/978-0-387-68606-6_2,

2.2 SIGNAL REPRESENTATION

In this section, we briefly review some concepts related to deterministic and random discrete-time signals. Only specific results essential to the understanding of adaptive filtering are reviewed. For further details on signals and digital signal processing we refer to [6]-[13].

2.2.1 Deterministic Signals

A deterministic discrete-time signal is characterized by a defined mathematical function of the time index k[1], with $k = 0, \pm1, \pm2, \pm3, \ldots$. An example of a deterministic signal (or sequence) is

$$x(k) = \mathrm{e}^{-\alpha\, k} \cos(\omega k) + u(k) \tag{2.1}$$

where $u(k)$ is the unit step sequence.

The response of a linear time-invariant filter to an input $x(k)$ is given by the convolution summation, as follows [7]:

$$\begin{aligned} y(k) = x(k) * h(k) &= \sum_{n=-\infty}^{\infty} x(n)h(k-n) \\ &= \sum_{n=-\infty}^{\infty} h(n)x(k-n) = h(k) * x(k) \end{aligned} \tag{2.2}$$

where $h(k)$ is the impulse response of the filter[2].

The $\mathcal{Z}$-transform of a given sequence $x(k)$ is defined as

$$\mathcal{Z}\{x(k)\} = X(z) = \sum_{k=-\infty}^{\infty} x(k)z^{-k} \tag{2.3}$$

for regions in the $\mathcal{Z}$-plane such that this summation converges. If the $\mathcal{Z}$-transform is defined for a given region of the $\mathcal{Z}$-plane, in other words the above summation converges in that region, the convolution operation can be replaced by a product of the $\mathcal{Z}$-transforms as follows [7]:

$$Y(z) = H(z)\, X(z) \tag{2.4}$$

where $Y(z)$, $X(z)$, and $H(z)$ are the $\mathcal{Z}$-transforms of $y(k)$, $x(k)$, and $h(k)$, respectively. Considering only waveforms that start at an instant $k \geq 0$ and have finite power, their $\mathcal{Z}$-transforms will always be defined outside the unit circle.

---

[1] The index k can also denote space in some applications.

[2] An alternative and more accurate notation for the convolution summation would be $(x * h)(k)$ instead of $x(k) * h(k)$, since in the latter the index k appears twice whereas the resulting convolution is simply a function of k. We will keep the former notation since it is more widely used.

For finite-energy waveforms, it is convenient to use the discrete-time Fourier transform defined as

$$\mathcal{F}\{x(k)\} = X(e^{\jmath\omega}) = \sum_{k=-\infty}^{\infty} x(k)e^{-\jmath\omega k} \tag{2.5}$$

Although the discrete-time Fourier transform does not exist for a signal with infinite energy, if the signal has finite-power, a generalized discrete-time Fourier transform exists and is largely used for deterministic signals [16].

2.2.2 Random Signals

A random variable X is a function that assigns a number to every outcome, denoted by ϱ, of a given experiment. A stochastic process is a rule to describe the time evolution of the random variable depending on ϱ, therefore it is a function of two variables $\mathrm{X}(k, \varrho)$. The set of all experimental outcomes, i.e., the ensemble, is the domain of ϱ. We denote $x(k)$ as a sample of the given process with ϱ fixed, where in this case if k is also fixed, $x(k)$ is a number. When any statistical operator is applied to $x(k)$ it is implied that $x(k)$ is a random variable, k is fixed, and ϱ is variable. In this book, $x(k)$ represents a random signal.

Random signals do not have a precise description of their waveforms. What is possible is to characterize them via measured statistics or through a probabilistic model. For random signals, the first- and second-order statistics are most of the time sufficient for characterization of the stochastic process. The first- and second-order statistics are also convenient for measurements. In addition, the effect on these statistics caused by linear filtering can be easily accounted for as shown below.

Let's consider for the time being that the random signals are real. We start to introduce some tools to deal with random signals by defining the distribution function of a random variable as

$$P_{x(k)}(y) \stackrel{\triangle}{=} probability\, of\; x(k)\; being\; smaller\; or\; equal\; to\; y$$

or

$$P_{x(k)}(y) = \int_{-\infty}^{y} p_{x(k)}(z)dz \tag{2.6}$$

The derivative of the distribution function is the probability density function (pdf)

$$p_{x(k)}(y) = \frac{dP_{x(k)}(y)}{dy} \tag{2.7}$$

The expected value, or mean value, of the process is defined by

$$m_x(k) = E[x(k)] \tag{2.8}$$

The definition of the expected value is expressed as

$$E[x(k)] = \int_{-\infty}^{\infty} y\, p_{x(k)}(y)dy \tag{2.9}$$

where $p_{x(k)}(y)$ is the pdf of $x(k)$ at the point y.

The autocorrelation function of the process $x(k)$ is defined by

$$r_x(k,l) = E[x(k)x(l)] = \int_{-\infty}^{\infty}\int_{-\infty}^{\infty} yz p_{x(k),x(l)}(y,z)dydz \tag{2.10}$$

where $p_{x(k),x(l)}(y,z)$ is the joint probability density of the random variables $x(k)$ and $x(l)$ defined as

$$p_{x(k),x(l)}(y,z) = \frac{\partial^2 P_{x(k),x(l)}(y,z)}{\partial y \partial z} \tag{2.11}$$

where

$$P_{x(k),x(l)}(y,z) \triangleq probability\,of\,\{x(k) \leq y\,and\,x(l) \leq z\}$$

The autocovariance function is defined as

$$\sigma_x^2(k,l) = E\{[x(k)-m_x(k)][x(l)-m_x(l)]\} = r_x(k,l) - m_x(k)m_x(l) \tag{2.12}$$

where the second equality follows from the definitions of mean value and autocorrelation. For $k = l$, $\sigma_x^2(k,l) = \sigma_x^2(k)$ which is the variance of $x(k)$.

The most important specific example of probability density function is the Gaussian density function, also known as normal density function [14]-[15]. The Gaussian pdf is defined by

$$p_{x(k)}(y) = \frac{1}{\sqrt{2\pi\sigma_x^2(k)}} e^{-\frac{(y-m_x(k))^2}{2\sigma_x^2(k)}} \tag{2.13}$$

where $m_x(k)$ and $\sigma_x^2(k)$ are the mean and variance of $x(k)$, respectively.

One justification for the importance of the Gaussian distribution is the central limit theorem. Given a random variable x composed by the sum of n independent random variables x_i as follows:

$$x = \sum_{i=1}^{n} x_i \tag{2.14}$$

the central limit theorem states that under certain general conditions, the probability density function of x approaches a Gaussian density function for large n. The mean and variance of x are given, respectively, by

$$m_x = \sum_{i=1}^{n} m_{x_i} \tag{2.15}$$

$$\sigma_x^2 = \sum_{i=1}^{n} \sigma_{x_i}^2 \tag{2.16}$$

Considering that the values of the mean and variance of x can grow, define

$$x' = \frac{x - m_x}{\sigma_x} \tag{2.17}$$

In this case, for $n \to \infty$ it follows that

$$p_{x'}(y) = \frac{1}{\sqrt{2\pi}} \mathrm{e}^{-\frac{y^2}{2}} \tag{2.18}$$

In a number of situations we require the calculation of conditional distributions, where the probability of a certain event to occur is calculated assuming that another event B has occurred. In this case, we define

$$\begin{aligned} P_{x(k)}(y|B) &= \frac{P(\{x(k) \leq y\} \cap B)}{P(B)} \\ &\triangleq probability\ of\ x(k) \leq y\ assuming\ B\ has\ occurred \end{aligned} \tag{2.19}$$

This joint event consists of all outcomes $\varrho \in B$ such that $x(k) = x(k, \varrho) \leq y$[3]. The definition of the conditional mean is given by

$$m_{x|B}(k) = E[x(k)|B] = \int_{-\infty}^{\infty} y p_{x(k)}(y|B) dy \tag{2.20}$$

where $p_{x(k)}(y|B)$ is the pdf of $x(k)$ conditioned on B.

The conditional variance is defined as

$$\sigma_{x|B}^2(k) = E\{[x(k) - m_{x|B}(k)]^2|B\} = \int_{-\infty}^{\infty} [y - m_{x|B}(k)]^2 p_{x(k)}(y|B) dy \tag{2.21}$$

There are processes for which the mean and autocorrelation functions are shift (or time) invariant, i.e.,

$$m_x(k-i) = m_x(k) = E[x(k)] = m_x \tag{2.22}$$

$$r_x(k,i) = E[x(k-j)x(i-j)] = r_x(k-i) = r_x(l) \tag{2.23}$$

and as a consequence

$$\sigma_x^2(l) = r_x(l) - m_x^2 \tag{2.24}$$

These processes are said to be wide-sense stationary (WSS). If the nth-order statistics of a process is shift invariant, the process is said to be nth-order stationary. Also if the process is nth-order stationary for any value of n the process is stationary in strict sense.

Two processes are considered jointly WSS if and only if any linear combination of them is also WSS. This is equivalent to state that

$$y(k) = k_1\, x_1(k) + k_2\, x_2(k) \tag{2.25}$$

must be WSS, for any constants k_1 and k_2, if $x_1(k)$ and $x_2(k)$ are jointly WSS. This property implies that both $x_1(k)$ and $x_2(k)$ have shift-invariant means and autocorrelations, and that their cross-correlation is also shift invariant.

[3] Or equivalently, such that $\mathrm{X}(\varrho) \leq y$.

For complex signals where $x(k) = x_r(k) + \jmath x_i(k)$, $y = y_r + \jmath y_i$, and $z = z_r + \jmath z_i$, we have the following definition of the expected value

$$E[x(k)] = \int_{-\infty}^{\infty}\int_{-\infty}^{\infty} y p_{x_r(k),x_i(k)}(y_r, y_i) dy_r dy_i \tag{2.26}$$

where $p_{x_r(k),x_i(k)}(y_r, y_i)$ is the joint probability density function (pdf) of $x_r(k)$ and $x_i(k)$.

The autocorrelation function of the complex random signal $x(k)$ is defined by

$$\begin{aligned} r_x(k,l) &= E[x(k)x^*(l)] \\ &= \int_{-\infty}^{\infty}\int_{-\infty}^{\infty}\int_{-\infty}^{\infty}\int_{-\infty}^{\infty} yz^* p_{x_r(k),x_i(k),x_r(l),x_i(l)}(y_r, y_i, z_r, z_i) dy_r dy_i dz_r dz_i \end{aligned} \tag{2.27}$$

where $*$ denotes complex conjugate, since we assume for now that we are dealing with complex signals, and $p_{x_r(k),x_i(k),x_r(l),x_i(l)}(y_r, y_i, z_r, z_i)$ is the joint probability density function of the random variables $x(k)$ and $x(l)$.

For complex signals the autocovariance function is defined as

$$\sigma_x^2(k,l) = E\{[x(k) - m_x(k)][x(l) - m_x(l)]^*\} = r_x(k,l) - m_x(k)m_x^*(l) \tag{2.28}$$

2.2.2.1 *Autoregressive Moving Average Process*

The process resulting from the output of a system described by a general linear difference equation given by

$$y(k) = \sum_{j=0}^{M} b_j x(k-j) + \sum_{i=1}^{N} a_i y(k-i) \tag{2.29}$$

where $x(k)$ is a white noise, is called autoregressive moving average (ARMA) process. The coefficients a_i and b_j are the parameters of the ARMA process. The output signal $y(k)$ is also said to be a colored noise since the autocorrelation function of $y(k)$ is nonzero for a lag different from zero, i.e., $r(l) \neq 0$ for some $l \neq 0$.

For the special case where $b_j = 0$ for $j = 1, 2, \ldots, M$, the resulting process is called autoregressive (AR) process. The terminology means that the process depends on the present value of the input signal and on a linear combination of past samples of the process. This indicates the presence of a feedback of the output signal.

For the special case where $a_i = 0$ for $i = 1, 2, \ldots, N$, the process is identified as a moving average (MA) process. This terminology indicates that the process depends on a linear combination of the present and past samples of the input signal. In summary, an ARMA process can be generated by applying a white noise to the input of a digital filter with poles and zeros, whereas for the AR and MA cases the digital filters are all-pole and all-zero filters, respectively.

2.2.2.2 Markov Process

A stochastic process is called a Markov process if its past has no influence in the future when the present is specified [14], [16]. In other words, the present behavior of the process depends only on the most recent past, i.e., all behavior previous to the most recent past is not required. A first-order AR process is a first-order Markov process, whereas an Nth-order AR process is considered an Nth-order Markov process. Take as an example the sequence

$$y(k) = ay(k-1) + n(k) \tag{2.30}$$

where $n(k)$ is a white noise process. The process represented by $y(k)$ is determined by $y(k-1)$ and $n(k)$, and no information before the instant $k-1$ is required. We conclude that $y(k)$ represents a Markov process. In the previous example, if $a = 1$ and $y(-1) = 0$ the signal $y(k)$, for $k \geq 0$, is a sum of white noise samples, usually called random walk sequence.

Formally, an mth-order Markov process satisfies the following condition: for all $k \geq 0$, and for a fixed m, it follows that

$$P_{x(k)}\left(y|x(k-1), x(k-2), \ldots, x(0)\right) = P_{x(k)}\left(y|x(k-1), x(k-2), \ldots, x(k-m)\right) \tag{2.31}$$

2.2.2.3 Wold Decomposition

Another important result related to any wide-sense stationary process $x(k)$ is the Wold decomposition, which states that $x(k)$ can be decomposed as

$$x(k) = x_r(k) + x_p(k) \tag{2.32}$$

where $x_r(k)$ is a regular process that is equivalent to the response of a stable, linear, time-invariant, and causal filter to a white noise [16], and $x_p(k)$ is a perfectly predictable (deterministic or singular) process. Also, $x_p(k)$ and $x_r(k)$ are orthogonal processes, i.e., $E[x_r(k)x_p(k)] = 0$. The key factor here is that the regular process can be modeled through a stable autoregressive model [24] with a stable and causal inverse. The importance of Wold decomposition lies on the observation that a WSS process can in part be represented by an AR process of adequate order, with the remaining part consisting of a perfectly predictable process. Obviously, the perfectly predictable process part of $x(k)$ also admits an AR model with zero excitation.

2.2.2.4 Power Spectral Density

Stochastic signals that are wide-sense stationary are persistent and therefore are not finite-energy signals. On the other hand, they have finite-power such that the generalized discrete-time Fourier transform can be applied to them. When the generalized discrete-time Fourier transform is applied to a WSS process it leads to a random function of the frequency [16]. On the other hand, the autocorrelation functions of most practical stationary processes have discrete-time Fourier transform. Therefore, the discrete-time Fourier transform of the autocorrelation function of a stationary random process can be very useful in many situations. This transform, called power spectral density, is defined as

$$R_x(e^{\jmath\omega}) = \sum_{l=-\infty}^{\infty} r_x(l)e^{-\jmath\omega l} = \mathcal{F}[r_x(l)] \tag{2.33}$$

where $r_x(l)$ is the autocorrelation of the process represented by $x(k)$. The inverse discrete-time Fourier transform allows us to recover $r_x(l)$ from $R_x(e^{\jmath\omega})$, through the relation

$$r_x(l) = \frac{1}{2\pi}\int_{-\pi}^{\pi} R_x(e^{\jmath\omega})e^{\jmath\omega l}\, d\omega = \mathcal{F}^{-1}[R_x(e^{\jmath\omega})] \tag{2.34}$$

It should be mentioned that $R_x(e^{\jmath\omega})$ is a deterministic function of ω, and can be interpreted as the power density of the random process at a given frequency in the ensemble[4], i.e., considering the average outcome of all possible realizations of the process. In particular, the mean squared value of the process represented by $x(k)$ is given by

$$r_x(0) = \frac{1}{2\pi}\int_{-\pi}^{\pi} R_x(e^{\jmath\omega})d\omega \tag{2.35}$$

If the random signal representing any single realization of a stationary process is applied as input to a linear and time-invariant filter, with impulse response $h(k)$, the following equalities are valid and can be easily verified:

$$y(k) = \sum_{n=-\infty}^{\infty} x(n)h(k-n) = x(k) * h(k) \tag{2.36}$$

$$r_y(l) = r_x(l) * r_h(l) \tag{2.37}$$

$$R_y(e^{\jmath\omega}) = R_x(e^{\jmath\omega})|H(e^{\jmath\omega})|^2 \tag{2.38}$$

$$r_{yx}(l) = r_x(l) * h(l) = E[x^*(l)y(l)] \tag{2.39}$$

$$R_{yx}(e^{\jmath\omega}) = R_x(e^{\jmath\omega})H(e^{\jmath\omega}) \tag{2.40}$$

where $r_h(l) = h(l) * h(-l)$, $R_y(e^{\jmath\omega})$ is the power spectral density of the output signal, $r_{yx}(k)$ is the cross-correlation of $x(k)$ and $y(k)$, and $R_{yx}(e^{\jmath\omega})$ is the corresponding cross-power spectral density.

The main feature of the spectral density function is to allow a simple analysis of the correlation behavior of WSS random signals processed with linear time-invariant systems. As an illustration, suppose a white noise is applied as input to a lowpass filter with impulse response $h(k)$ and sharp cutoff at a given frequency ω_l. The autocorrelation function of the output signal $y(k)$ will not be a single impulse, it will be $h(k) * h(-k)$. Therefore, the signal $y(k)$ will look like a band-limited random signal, in this case, a slow-varying noise. Some properties of the function $R_x(e^{\jmath\omega})$ of a discrete-time and stationary stochastic process are worth mentioning. The power spectrum density is a periodic function of ω, with period 2π, as can be verified from its definition. Also, since for a stationary and complex random process we have $r_x(-l) = r_x^*(l)$, $R_x(e^{\jmath\omega})$ is real. Despite the usefulness of the power spectrum density function in dealing with WSS processes, it will not be widely used in this book since usually the filters considered here are time varying. However, it should be noted its important role in areas such as spectrum estimation [25]-[26].

[4]The average signal power at a given sufficiently small frequency range, $\Delta\omega$, around a center frequency ω_0 is approximately given by $\frac{\Delta\omega}{2\pi}R_x(e^{\jmath\omega_0})$.

If the $\mathcal{Z}$-transforms of the autocorrelation and cross-correlation functions exist, we can generalize the definition of power spectral density. In particular, the definition of equation (2.33) corresponds to the following relation

$$\mathcal{Z}[r_x(k)] = R_x(z) = \sum_{k=-\infty}^{\infty} r_x(k) z^{-k} \tag{2.41}$$

If the random signal representing any single realization of a stationary process is applied as input to a linear and time-invariant filter with impulse response $h(k)$, the following equalities are valid:

$$R_y(z) = R_x(z)H(z)H(z^{-1}) \tag{2.42}$$

and

$$R_{yx}(z) = R_x(z)H(z) \tag{2.43}$$

where $H(z) = \mathcal{Z}[h(l)]$. If we wish to calculate the cross-correlation of $y(k)$ and $x(k)$, namely $r_{yx}(0)$, we can use the inverse $\mathcal{Z}$-transform formula as follows:

$$\begin{aligned} E[y(k)x^*(k)] &= \frac{1}{2\pi \jmath} \oint R_{yx}(z) \frac{dz}{z} \\ &= \frac{1}{2\pi \jmath} \oint H(z) R_x(z) \frac{dz}{z} \end{aligned} \tag{2.44}$$

where the integration path is a counterclockwise closed contour in the region of convergence of $R_{yx}(z)$. The contour integral above equation is usually solved through the Cauchy's residue theorem [8].

2.2.3 Ergodicity

In the probabilistic approach, the statistical parameters of the real data are obtained through ensemble averages (or expected values). The estimation of any parameter of the stochastic process can be obtained by averaging a large number of realizations of the given process, at each instant of time. However, in many applications only a few or even a single sample of the process is available. In these situations, we need to find out in which cases the statistical parameters of the process can be estimated by using time average of a single sample (or ensemble member) of the process. This is obviously not possible if the desired parameter is time varying. The equivalence between the ensemble average and time average is called ergodicity [14], [16].

The time average of a given stationary process represented by $x(k)$ is calculated by

$$\hat{m}_{x_N} = \frac{1}{2N+1} \sum_{k=-N}^{N} x(k) \tag{2.45}$$

If

$$\sigma^2_{\hat{m}_{x_N}} = \lim_{N\to\infty} E\{|\hat{m}_{x_N} - m_x|^2\} = 0$$

the process is said to be mean-ergodic in the mean-square sense. Therefore, the mean-ergodic process has time average that approximates the ensemble average as $N \to \infty$. Obviously, $\hat{m}_{x_N}$ is an unbiased estimate of m_x since

$$E[\hat{m}_{x_N}] = \frac{1}{2N+1} \sum_{k=-N}^{N} E[x(k)] = m_x \tag{2.46}$$

Therefore, the process will be considered *ergodic* if the variance of $\hat{m}_{x_N}$ tends to zero ($\sigma^2_{\hat{m}_{x_N}} \to 0$) when $N \to \infty$. The variance $\sigma^2_{\hat{m}_{x_N}}$ can be expressed after some manipulations as

$$\sigma^2_{\hat{m}_{x_N}} = \frac{1}{2N+1} \sum_{l=-2N}^{2N} \sigma_x^2(k+l,k) \left(1 - \frac{|l|}{2N+1}\right) \tag{2.47}$$

where $\sigma_x^2(k+l,k)$ is the autocovariance of the stochastic process $x(k)$. The variance of $\hat{m}_{x_N}$ tends to zero if and only if

$$\lim_{N \to \infty} \frac{1}{N} \sum_{l=0}^{N} \sigma_x^2(k+l,k) \to 0$$

The above condition is necessary and sufficient to guarantee that the process is mean-ergodic.

The ergodicity concept can be extended to higher order statistics. In particular, for second-order statistics we can define the process

$$x_l(k) = x(k+l)x^*(k) \tag{2.48}$$

where the mean of this process corresponds to the autocorrelation of $x(k)$, i.e., $r_x(l)$. Mean-ergodicity of $x_l(k)$ implies mean-square ergodicity of the autocorrelation of $x(k)$.

The time average of $x_l(k)$ is given by

$$\hat{m}_{x_{l,N}} = \frac{1}{2N+1} \sum_{k=-N}^{N} x_l(k) \tag{2.49}$$

that is an unbiased estimate of $r_x(l)$. If the variance of $\hat{m}_{x_{l,N}}$ tends to zero as N tends to infinity, the process $x(k)$ is said to be mean-square ergodic of the autocorrelation, i.e.,

$$\lim_{N \to \infty} E\{|\hat{m}_{x_{l,N}} - r_x(l)|^2\} = 0 \tag{2.50}$$

The above condition is satisfied if and only if

$$\lim_{N \to \infty} \frac{1}{N} \sum_{i=0}^{N} E\{x(k+l)x^*(k)x(k+l+i)x^*(k+i)\} - r_x^2(l) = 0 \tag{2.51}$$

where it is assumed that $x(n)$ has stationary fourth-order moments. The concept of ergodicity can be extended to nonstationary processes [16], however, that is beyond the scope of this book.

2.3 THE CORRELATION MATRIX

Usually, adaptive filters utilize the available input signals at instant k in their updating equations. These inputs are the elements of the input signal vector denoted by

$$\mathbf{x}(k) = [x_0(k)\, x_1(k) \ldots x_N(k)]^T$$

The correlation matrix is defined as $\mathbf{R} = E[\mathbf{x}(k)\mathbf{x}^H(k)]$, where $\mathbf{x}^H(k)$ is the Hermitian transposition of $\mathbf{x}(k)$, that means transposition followed by complex conjugation or vice versa. As will be noted, the characteristics of the correlation matrix play a key role in the understanding of properties of most adaptive-filtering algorithms. As a consequence, it is important to examine the main properties of the matrix $\mathbf{R}$. Some properties of the correlation matrix come from the statistical nature of the adaptive-filtering problem, whereas other properties derive from the linear algebra theory.

For a given input vector, the correlation matrix is given by

$$\begin{aligned} \mathbf{R} &= \begin{bmatrix} E[|x_0(k)|^2] & E[x_0(k)x_1^*(k)] & \cdots & E[x_0(k)x_N^*(k)] \\ E[x_1(k)x_0^*(k)] & E[|x_1(k)|^2] & \cdots & E[x_1(k)x_N^*(k)] \\ \vdots & \vdots & \ddots & \vdots \\ E[x_N(k)x_0^*(k)] & E[x_N(k)x_1^*(k)] & \cdots & E[|x_N(k)|^2] \end{bmatrix} \\ &= E[\mathbf{x}(k)\mathbf{x}^H(k)] \end{aligned} \tag{2.52}$$

The main properties of the $\mathbf{R}$ matrix are listed below:

1. The matrix $\mathbf{R}$ is positive semidefinite.

   Proof:

   Given an arbitrary complex weight vector $\mathbf{w}$, we can form a signal given by

   $$y(k) = \mathbf{w}^H\mathbf{x}(k)$$

   The magnitude squared of $y(k)$ is

   $$y(k)y^*(k) = |y(k)|^2 = \mathbf{w}^H\mathbf{x}(k)\mathbf{x}^H(k)\mathbf{w} \geq 0$$

   The mean-square (MS) value of $y(k)$ is then given by

   $$\text{MS}[y(k)] = E[|y(k)|^2] = \mathbf{w}^H E[\mathbf{x}(k)\mathbf{x}^H(k)]\mathbf{w} = \mathbf{w}^H\mathbf{R}\mathbf{w} \geq 0$$

   Therefore, the matrix $\mathbf{R}$ is positive semidefinite.

   □

   Usually, the matrix $\mathbf{R}$ is positive definite, unless the signals that compose the input vector are linearly dependent. Linear dependent signals are rarely found in practice.

2. The matrix **R** is Hermitian, i.e.,

$$\mathbf{R} = \mathbf{R}^H \tag{2.53}$$

Proof:

$$\mathbf{R}^H = E\{[\mathbf{x}(k)\mathbf{x}^H(k)]^H\} = E[\mathbf{x}(k)\mathbf{x}^H(k)] = \mathbf{R}$$

□

3. A matrix is Toeplitz if the elements of the main diagonal and of any secondary diagonal are equal. When the input signal vector is composed of delayed versions of the same signal (that is, $x_i(k) = x_0(k-i)$, for $i = 1, 2, \ldots, N$) taken from a WSS process, matrix **R** is Toeplitz.

Proof:

For the delayed signal input vector, with $x(k)$ WSS, matrix **R** has the following form

$$\mathbf{R} = \begin{bmatrix} r_x(0) & r_x(1) & \cdots & r_x(N) \\ r_x(-1) & r_x(0) & \cdots & r_x(N-1) \\ \vdots & \vdots & \ddots & \vdots \\ r_x(-N) & r_x(-N+1) & \cdots & r_x(0) \end{bmatrix} \tag{2.54}$$

By examining the right-hand side of the above equation, we can easily conclude that **R** is Toeplitz.

□

Note that $r_x^*(i) = r_x(-i)$, what also follows from the fact that the matrix **R** is Hermitian.

If matrix **R** given by equation (2.54) is nonsingular for a given N, the input signal is said to be *persistently exciting* of order $N+1$. This means that the power spectral density $R_x(e^{j\omega})$ is different from zero at least at $N+1$ points in the interval $0 < \omega \leq 2\pi$. It also means that a nontrivial Nth-order FIR filter (with at least one nonzero coefficient) cannot filter $x(k)$ to zero. Note that a nontrivial filter, with $x(k)$ as input, would require at least $N+1$ zeros in order to generate an output with all samples equal to zero. The absence of persistence of excitation implies the misbehavior of some adaptive algorithms [17], [18]. The definition of persistence of excitation is not unique, and it is algorithm dependent (see the book by Johnson [17] for further details).

From now on in this section, we discuss some properties of the correlation matrix related to its eigenvalues and eigenvectors. A number λ is an eigenvalue of the matrix **R**, with a corresponding eigenvector **q**, if and only if

$$\mathbf{R}\mathbf{q} = \lambda\mathbf{q} \tag{2.55}$$

or equivalently

$$\det(\mathbf{R} - \lambda\mathbf{I}) = 0 \tag{2.56}$$

where **I** is the $(N+1)$ by $(N+1)$ identity matrix. Equation (2.56) is called characteristic equation of **R**, and has $(N+1)$ solutions for λ. We denote the $(N+1)$ eigenvalues of **R** by $\lambda_0, \lambda_1, \ldots, \lambda_N$.

Note also that for every value of λ, the vector $\mathbf{q} = \mathbf{0}$ satisfies equation (2.55), however we consider only those particular values of λ that are linked to a nonzero eigenvector $\mathbf{q}$.

Some important properties related to the eigenvalues and eigenvectors of $\mathbf{R}$, that will be useful in the following chapters, are listed below.

1. The eigenvalues of $\mathbf{R}^m$ are λ_i^m, for $i = 0, 1, 2, \ldots, N$.

   Proof:

   By premultiplying equation (2.55) by $\mathbf{R}^{m-1}$, we obtain

$$\begin{aligned}\mathbf{R}^{m-1}\mathbf{R}\mathbf{q}_i &= \mathbf{R}^{m-1}\lambda_i\mathbf{q}_i = \lambda_i\mathbf{R}^{m-2}\mathbf{R}\mathbf{q}_i \\ &= \lambda_i\mathbf{R}^{m-2}\lambda_i\mathbf{q}_i = \lambda_i^2\mathbf{R}^{m-3}\mathbf{R}\mathbf{q}_i \\ &= \cdots = \lambda_i^m\mathbf{q}_i \end{aligned} \tag{2.57}$$

□

2. Suppose $\mathbf{R}$ has $N+1$ linearly independent eigenvectors $\mathbf{q}_i$; then if we form a matrix $\mathbf{Q}$ with columns consisting of the $\mathbf{q}_i$'s, it follows that

$$\mathbf{Q}^{-1}\mathbf{R}\mathbf{Q} = \begin{bmatrix} \lambda_0 & 0 & \cdots & 0 \\ 0 & \lambda_1 & & \vdots \\ \vdots & 0 & \cdots & \vdots \\ \vdots & \vdots & & 0 \\ 0 & 0 & \cdots & \lambda_N \end{bmatrix} = \mathbf{\Lambda} \tag{2.58}$$

Proof:

$$\begin{aligned}\mathbf{R}\mathbf{Q} &= \mathbf{R}[\mathbf{q}_0\ \mathbf{q}_1 \cdots \mathbf{q}_N] = [\lambda_0\mathbf{q}_0\ \lambda_1\mathbf{q}_1 \cdots \lambda_N\mathbf{q}_N] \\ &= \mathbf{Q}\begin{bmatrix} \lambda_0 & 0 & \cdots & 0 \\ 0 & \lambda_1 & & \vdots \\ \vdots & 0 & \cdots & \vdots \\ \vdots & \vdots & & 0 \\ 0 & 0 & \cdots & \lambda_N \end{bmatrix} = \mathbf{Q}\mathbf{\Lambda}\end{aligned}$$

Therefore, since $\mathbf{Q}$ is invertible because the $\mathbf{q}_i$'s are linearly independent, we can show that

$$\mathbf{Q}^{-1}\mathbf{R}\mathbf{Q} = \mathbf{\Lambda}$$

3. The nonzero eigenvectors $\mathbf{q}_0$, $\mathbf{q}_1$, $\ldots \mathbf{q}_N$ that correspond to different eigenvalues are linearly independent.

Proof:

If we form a linear combination of the eigenvectors such that

$$a_0\mathbf{q}_0 + a_1\mathbf{q}_1 + \cdots + a_N\mathbf{q}_N = \mathbf{0} \tag{2.59}$$

By multiplying the above equation by $\mathbf{R}$ we have

$$a_0\mathbf{R}\mathbf{q}_0 + a_1\mathbf{R}\mathbf{q}_1 + \cdots + a_N\mathbf{R}\mathbf{q}_N = a_0\lambda_0\mathbf{q}_0 + a_1\lambda_1\mathbf{q}_1 + \cdots + a_N\lambda_N\mathbf{q}_N = \mathbf{0} \tag{2.60}$$

Now by multiplying equation (2.59) by λ_N and subtracting the result from equation (2.60), we obtain

$$a_0(\lambda_0 - \lambda_N)\mathbf{q}_0 + a_1(\lambda_1 - \lambda_N)\mathbf{q}_1 + \cdots + a_{N-1}(\lambda_{N-1} - \lambda_N)\mathbf{q}_{N-1} = \mathbf{0}$$

By repeating the above steps, i.e., multiplying the above equation by $\mathbf{R}$ in one instance and by λ_{N-1} on the other instance, and subtracting the results, it yields

$$a_0(\lambda_0 - \lambda_N)(\lambda_0 - \lambda_{N-1})\mathbf{q}_0 + a_1(\lambda_1 - \lambda_N)(\lambda_1 - \lambda_{N-1})\mathbf{q}_1$$
$$+ \cdots + a_{N-2}(\lambda_{N-2} - \lambda_{N-1})\mathbf{q}_{N-2} = \mathbf{0}$$

By repeating the same above steps several times, we end up with

$$a_0(\lambda_0 - \lambda_N)(\lambda_0 - \lambda_{N-1}) \cdots (\lambda_0 - \lambda_1)\mathbf{q}_0 = \mathbf{0}$$

Since we assumed $\lambda_0 \neq \lambda_1, \lambda_0 \neq \lambda_2, \ldots \lambda_0 \neq \lambda_N$, and $\mathbf{q}_0$ was assumed nonzero, then $a_0 = 0$. The same line of thought can be used to show that $a_0 = a_1 = a_2 = \cdots = a_N = 0$ is the only solution for equation (2.59). Therefore, the eigenvectors corresponding to different eigenvalues are linearly independent.

□

Not all matrices are diagonalizable. A matrix of order $(N+1)$ is diagonalizable if it possesses $(N+1)$ linearly independent eigenvectors. A matrix with repeated eigenvalues can be diagonalized or not, depending on the linear dependency of the eigenvectors. A nondiagonalizable matrix is called defective [19].

4. Since the correlation matrix $\mathbf{R}$ is Hermitian, i.e., $\mathbf{R}^H = \mathbf{R}$, its eigenvalues are real. These eigenvalues are equal to or greater than zero given that $\mathbf{R}$ is positive semidefinite.

Proof:

First note that given an arbitrary complex vector $\mathbf{w}$,

$$(\mathbf{w}^H\mathbf{R}\mathbf{w})^H = \mathbf{w}^H\mathbf{R}^H(\mathbf{w}^H)^H = \mathbf{w}^H\mathbf{R}\mathbf{w}$$

Therefore, $\mathbf{w}^H\mathbf{R}\mathbf{w}$ is a real number. Assume now that λ_i is an eigenvalue of $\mathbf{R}$ corresponding to the eigenvector $\mathbf{q}_i$, i.e., $\mathbf{R}\mathbf{q}_i = \lambda_i\mathbf{q}_i$. By premultiplying this equation by $\mathbf{q}_i^H$, it follows that

$$\mathbf{q}_i^H\mathbf{R}\mathbf{q}_i = \lambda_i\mathbf{q}_i^H\mathbf{q}_i = \lambda_i\|\mathbf{q}_i\|^2$$

where the operation $\|\mathbf{a}\|^2 = |a_0|^2 + |a_1|^2 + \cdots + |a_N|^2$ is the Euclidean norm squared of the vector $\mathbf{a}$, that is always real. Since the term on the left hand is also real, $\|\mathbf{q}_i\|^2 \neq 0$, and $\mathbf{R}$ is positive semidefinite, we can conclude that λ_i is real and nonnegative.

Note that $\mathbf{Q}$ is not unique since each $\mathbf{q}_i$ can be multiplied by an arbitrary nonzero constant, and the resulting vector continues to be an eigenvector[5]. For practical reasons, we consider only normalized eigenvectors having length one, that is

$$\mathbf{q}_i^H \mathbf{q}_i = 1 \quad \text{for } i = 0, 1, \ldots, N \tag{2.61}$$

5. If $\mathbf{R}$ is a Hermitian matrix with different eigenvalues, the eigenvectors are orthogonal to each other. As a consequence, there is a diagonalizing matrix $\mathbf{Q}$ that is unitary, i.e., $\mathbf{Q}^H\mathbf{Q} = \mathbf{I}$.

   Proof:

   Given two eigenvalues λ_i and λ_j, it follows that

$$\mathbf{R}\mathbf{q}_i = \lambda_i \mathbf{q}_i$$

   and

$$\mathbf{R}\mathbf{q}_j = \lambda_j \mathbf{q}_j \tag{2.62}$$

   Using the fact that $\mathbf{R}$ is Hermitian and that λ_i and λ_j are real, then

$$\mathbf{q}_i^H \mathbf{R} = \lambda_i \mathbf{q}_i^H$$

   and by multiplying this equation on the right by $\mathbf{q}_j$, we get

$$\mathbf{q}_i^H \mathbf{R}\mathbf{q}_j = \lambda_i \mathbf{q}_i^H \mathbf{q}_j$$

   Now by premultiplying equation (2.62) by $\mathbf{q}_i^H$, it follows that

$$\mathbf{q}_i^H \mathbf{R}\mathbf{q}_j = \lambda_j \mathbf{q}_i^H \mathbf{q}_j$$

   Therefore,

$$\lambda_i \mathbf{q}_i^H \mathbf{q}_j = \lambda_j \mathbf{q}_i^H \mathbf{q}_j$$

   Since $\lambda_i \neq \lambda_j$, it can be concluded that

$$\mathbf{q}_i^H \mathbf{q}_j = 0 \quad \text{for } i \neq j$$

   If we form matrix $\mathbf{Q}$ with normalized eigenvectors, matrix $\mathbf{Q}$ is a unitary matrix.

   □

[5] We can also change the order in which the $\mathbf{q}_i$'s compose matrix $\mathbf{Q}$, but this fact is not relevant for the present discussion.

An important result is that any Hermitian matrix **R** can be diagonalized by a suitable unitary matrix **Q**, even if the eigenvalues of **R** are not distinct. The proof is omitted here and can be found in [19]. Therefore, for Hermitian matrices with repeated eigenvalues it is always possible to find a complete set of orthonormal eigenvectors.

A useful form to decompose a Hermitian matrix that results from the last property is

$$\mathbf{R} = \mathbf{Q}\boldsymbol{\Lambda}\mathbf{Q}^H = \sum_{i=0}^{N} \lambda_i \mathbf{q}_i \mathbf{q}_i^H \tag{2.63}$$

that is known as *spectral decomposition*. From this decomposition, one can easily derive the following relation

$$\mathbf{w}^H\mathbf{R}\mathbf{w} = \sum_{i=0}^{N} \lambda_i \mathbf{w}^H \mathbf{q}_i \mathbf{q}_i^H \mathbf{w} = \sum_{i=0}^{N} \lambda_i |\mathbf{w}^H \mathbf{q}_i|^2 \tag{2.64}$$

In addition, since $\mathbf{q}_i = \lambda_i \mathbf{R}^{-1}\mathbf{q}_i$, the eigenvectors of a matrix and of its inverse coincide, whereas the eigenvalues are reciprocals of each other. As a consequence,

$$\mathbf{R}^{-1} = \sum_{i=0}^{N} \frac{1}{\lambda_i} \mathbf{q}_i \mathbf{q}_i^H \tag{2.65}$$

Another consequence of the unitary property of **Q** for Hermitian matrices is that any Hermitian matrix can be written in the form

$$\mathbf{R} = \left[\sqrt{\lambda_0}\mathbf{q}_0 \ \sqrt{\lambda_1}\mathbf{q}_1 \ldots \sqrt{\lambda_N}\mathbf{q}_N\right] \begin{bmatrix} \sqrt{\lambda_0}\mathbf{q}_0^H \\ \sqrt{\lambda_1}\mathbf{q}_1^H \\ \vdots \\ \sqrt{\lambda_N}\mathbf{q}_N^H \end{bmatrix}$$

$$= \mathbf{L}\mathbf{L}^H \tag{2.66}$$

6. The sum of the eigenvalues of **R** is equal to the trace of **R**, and the product of the eigenvalues of **R** is equal to the determinant of **R**[6].

Proof:

$$\text{tr}[\mathbf{Q}^{-1}\mathbf{R}\mathbf{Q}] = \text{tr}[\boldsymbol{\Lambda}]$$

where, $\text{tr}[\mathbf{A}] = \sum_{i=0}^{N} a_{ii}$. Since $\text{tr}[\mathbf{A}'\mathbf{A}] = \text{tr}[\mathbf{A}\mathbf{A}']$, we have

$$\text{tr}[\mathbf{Q}^{-1}\mathbf{R}\mathbf{Q}] = \text{tr}[\mathbf{R}\mathbf{Q}\mathbf{Q}^{-1}] = \text{tr}[\mathbf{R}\mathbf{I}] = \text{tr}[\mathbf{R}] = \sum_{i=0}^{N} \lambda_i$$

Also $\det[\mathbf{Q}^{-1}\ \mathbf{R}\ \mathbf{Q}] = \det[\mathbf{R}]\det[\mathbf{Q}]\ \det[\mathbf{Q}^{-1}] = \det[\mathbf{R}] = \det[\boldsymbol{\Lambda}] = \prod_{i=0}^{N} \lambda_i$

□

[6]This property is valid for any square matrix, but for more general matrices the proof differs from the one presented here.

7. The Rayleigh's quotient defined as

$$\mathcal{R} = \frac{\mathbf{w}^H \mathbf{R} \mathbf{w}}{\mathbf{w}^H \mathbf{w}} \tag{2.67}$$

of a Hermitian matrix is bounded by the minimum and maximum eigenvalues, i.e.,

$$\lambda_{\min} \leq \mathcal{R} \leq \lambda_{\max} \tag{2.68}$$

where the minimum and maximum values are reached when the vector **w** is chosen to be the eigenvector corresponding to the minimum and maximum eigenvalues, respectively.

Proof:

Suppose $\mathbf{w} = \mathbf{Q}\mathbf{w}'$, where **Q** is the matrix that diagonalizes **R**, then

$$\begin{aligned} \mathcal{R} &= \frac{\mathbf{w}'^H \mathbf{Q}^H \mathbf{R} \mathbf{Q} \mathbf{w}'}{\mathbf{w}'^H \mathbf{Q}^H \mathbf{Q} \mathbf{w}'} \\ &= \frac{\mathbf{w}'^H \mathbf{\Lambda} \mathbf{w}'}{\mathbf{w}'^H \mathbf{w}'} \\ &= \frac{\sum_{i=0}^{N} \lambda_i w_i'^2}{\sum_{i=0}^{N} w_i'^2} \end{aligned} \tag{2.69}$$

It is then possible to show, see problem 14, that the minimum value for the above equation occurs when $w_i = 0$ for $i \neq j$ and λ_j is the smallest eigenvalue. Identically, the maximum value for $\mathcal{R}$ occurs when $w_i = 0$ for $i \neq l$, where λ_l is the largest eigenvalue.

□

There are several ways to define the norm of a matrix. In this book the norm of a matrix **R**, denoted by $\|\mathbf{R}\|$, is defined by

$$\begin{aligned} \|\mathbf{R}\|^2 &= \max_{\mathbf{w} \neq 0} \frac{\|\mathbf{R}\mathbf{w}\|^2}{\|\mathbf{w}\|^2} \\ &= \max_{\mathbf{w} \neq 0} \frac{\mathbf{w}^H \mathbf{R}^H \mathbf{R} \mathbf{w}}{\mathbf{w}^H \mathbf{w}} \end{aligned} \tag{2.70}$$

Note that the norm of **R** is a measure of how a vector **w** grows in magnitude, when it is multiplied by **R**.

When the matrix **R** is Hermitian, the norm of **R** is easily obtained by using the results of equations (2.57) and (2.68). The result is

$$\|\mathbf{R}\| = \lambda_{\max} \tag{2.71}$$

where $\lambda_{\max}$ is the maximum eigenvalue of **R**.

A common problem that we encounter in adaptive filtering is the solution of a system of linear equations such as

$$\mathbf{R}\mathbf{w} = \mathbf{p} \tag{2.72}$$

In case there is an error in the vector $\mathbf{p}$, originated by quantization or estimation, how does it affect the solution of the system of linear equations? For a positive definite Hermitian matrix $\mathbf{R}$, it can be shown [19] that the relative error in the solution of the above linear system of equations is bounded by

$$\frac{\|\Delta\mathbf{w}\|}{\|\mathbf{w}\|} \leq \frac{\lambda_{\max}}{\lambda_{\min}} \frac{\|\Delta\mathbf{p}\|}{\|\mathbf{p}\|} \tag{2.73}$$

where $\lambda_{\max}$ and $\lambda_{\min}$ are the maximum and minimum values of the eigenvalues of $\mathbf{R}$, respectively. The ratio $\lambda_{\max}/\lambda_{\min}$ is called condition number of a matrix, that is

$$C = \frac{\lambda_{\max}}{\lambda_{\min}} = \|\mathbf{R}\|\|\mathbf{R}^{-1}\| \tag{2.74}$$

The value of C influences the convergence behavior of a number of adaptive-filtering algorithms, as will be seen in the following chapters. Large value of C indicates that the matrix $\mathbf{R}$ is ill-conditioned, and that errors introduced by the manipulation of $\mathbf{R}$ may be largely amplified. When $C = 1$, the matrix is perfectly conditioned. In case $\mathbf{R}$ represents the correlation matrix of the input signal of an adaptive filter, with the input vector composed by uncorrelated elements of a delay line (see Fig. 2.1.b, and the discussions around it), then $C = 1$.

Example 2.1

Suppose the input signal vector is composed by a delay line with a single input signal, i.e.,

$$\mathbf{x}(k) = [x(k)\, x(k-1) \ldots x(k-N)]^T$$

Given the following input signals:

(a)

$$x(k) = n(k)$$

(b)

$$x(k) = a \cos \omega_0 k + n(k)$$

(c)

$$x(k) = \sum_{i=0}^{M} b_i n(k-i)$$

(d)

$$x(k) = -a_1 x(k-1) + n(k)$$

(e)

$$x(k) = a e^{\jmath(\omega_0 k + n(k))}$$

where $n(k)$ is a white noise with zero mean and variance σ_n^2; in case (e) $n(k)$ is uniformly distributed in the range $-\pi$ to π.

Calculate the autocorrelation matrix $\mathbf{R}$ for $N = 3$.

Solution:

(a) In this case, we have that $E[x(k)x(k-l)] = \sigma_n^2\delta(l)$, where $\delta(l)$ denotes an impulse sequence. Therefore,

$$\mathbf{R} = E[\mathbf{x}(k)\mathbf{x}^T(k)] = \sigma_n^2 \begin{bmatrix} 1 & 0 & \cdots & 0 \\ 0 & 1 & \cdots & 0 \\ \vdots & \vdots & \ddots & \vdots \\ 0 & 0 & \cdots & 1 \end{bmatrix} \tag{2.75}$$

(b) In this example, $n(k)$ is zero mean and uncorrelated with the deterministic cosine. The autocorrelation function can then be expressed as

$$\begin{aligned} r(k, k-l) &= E[a^2 \cos(\omega_0 k)\cos(\omega_0 k - \omega_0 l) + n(k)n(k-l)] \\ &= a^2 E[\cos(\omega_0 k)\cos(\omega_0 k - \omega_0 l)] + \sigma_n^2\delta(l) \\ &= \frac{a^2}{2}[\cos(\omega_0 l) + \cos(2\omega_0 k - \omega_0 l)] + \sigma_n^2\delta(l) \end{aligned} \tag{2.76}$$

where $\delta(l)$ again denotes an impulse sequence. Since part of the input signal is deterministic and nonstationary, the autocorrelation is time dependent.

For the 3×3 case the input signal correlation matrix $\mathbf{R}(k)$ becomes

$$\frac{a^2}{2}\begin{bmatrix} 1 + \cos 2\omega_0 k + \frac{2}{a^2}.\sigma_n^2 & \cos\omega_0 + \cos\omega_0(2k-1) & \cos 2\omega_0 + \cos 2\omega_0(k-1) \\ \cos\omega_0 + \cos\omega_0(2k-1) & 1 + \cos 2\omega_0(k-1) + \frac{2}{a^2}\sigma_n^2 & \cos\omega_0 + \cos\omega_0(2(k-1)-1) \\ \cos 2\omega_0 + \cos 2\omega_0(k-1) & \cos\omega_0 + \cos\omega_0(2(k-1)-1) & 1 + \cos 2\omega_0(k-2) + \frac{2}{a^2}\sigma_n^2 \end{bmatrix}$$

(c) By exploring the fact that $n(k)$ is a white noise, we can perform the following simplifications:

$$\begin{aligned} r(l) &= E[x(k)x(k-l)] = E\left[\sum_{j=0}^{M-l}\sum_{i=0}^{M} b_i b_j n(k-i)n(k-l-j)\right] \\ &= \sum_{j=0}^{M-l} b_j b_{l+j} E[n^2(k-l-j)] = \sigma_n^2 \sum_{j=0}^{M} b_j b_{l+j} \\ & 0 \leq l+j \leq M \end{aligned} \tag{2.77}$$

where from the third to the fourth relation we used the fact that $E[n(k-i)n(k-l-j)] = 0$ for $i \neq l+j$. For $M = 3$, the correlation matrix has the following form

$$\mathbf{R} = \sigma_n^2 \begin{bmatrix} \sum_{i=0}^{3} b_i^2 & \sum_{i=0}^{2} b_i b_{i+1} & \sum_{i=0}^{1} b_i b_{i+2} & b_0 b_3 \\ \sum_{i=0}^{2} b_i b_{i+1} & \sum_{i=0}^{3} b_i^2 & \sum_{i=0}^{2} b_i b_{i+1} & \sum_{i=0}^{1} b_i b_{i+2} \\ \sum_{i=0}^{1} b_i b_{i+2} & \sum_{i=0}^{2} b_i b_{i+1} & \sum_{i=0}^{3} b_i^2 & \sum_{i=0}^{2} b_i b_{i+1} \\ b_0 b_3 & \sum_{i=0}^{1} b_i b_{i+2} & \sum_{i=0}^{2} b_i b_{i+1} & \sum_{i=0}^{3} b_i^2 \end{bmatrix} \tag{2.78}$$

(d) By solving the difference equation, we can obtain the correlation between $x(k)$ and $x(k-l)$, that is

$$x(k) = (-a_1)^l x(k-l) + \sum_{j=0}^{l-1} (-a_1)^j n(k-j) \tag{2.79}$$

Multiplying $x(k-l)$ on both sides of the above equation and taking the expected value of the result, we obtain

$$E[x(k)x(k-l)] = (-a_1)^l E[x^2(k-l)] \tag{2.80}$$

since $x(k-l)$ is independent of $n(k-j)$ for $j \leq l-1$.

For $l = 0$, just calculate $x^2(k)$ and apply the expectation operation to the result. The partial result is

$$E[x^2(k)] = a_1^2 E[x^2(k-1)] + E[n^2(k)] \tag{2.81}$$

therefore,

$$E[x^2(k)] = \frac{\sigma_n^2}{1-a_1^2} \tag{2.82}$$

assuming $x(k)$ is WSS.

The elements of **R** are then given by

$$r(l) = \frac{(-a_1)^{|l|}}{1-a_1^2} \sigma_n^2 \tag{2.83}$$

and the 3×3 autocorrelation matrix becomes

$$\mathbf{R} = \frac{\sigma_n^2}{1-a_1^2} \begin{bmatrix} 1 & -a_1 & a_1^2 \\ -a_1 & 1 & -a_1 \\ a_1^2 & -a_1 & 1 \end{bmatrix}$$

(e) In this case, we are interested in calculating the autocorrelation of a complex sequence, that is

$$\begin{aligned} r(l) &= E[x(k)x^*(k-l)] \\ &= a^2 E[\mathrm{e}^{-\jmath(-\omega_0 l - n(k) + n(k-l))}] \end{aligned} \tag{2.84}$$

By recalling the definition of expected value in equation (2.9), for $l \neq 0$,

$$\begin{aligned}
r(l) &= a^2 e^{\jmath\omega_0 l}\int_{-\infty}^{\infty}\int_{-\infty}^{\infty} e^{-\jmath(-n_0+n_1)} p_{n(k),n(k-l)}(n_0, n_1) dn_0 dn_1 \\
&= a^2 e^{\jmath\omega_0 l}\int_{-\pi}^{\pi}\int_{-\pi}^{\pi} e^{-\jmath(-n_0+n_1)} p_{n(k)}(n_0) p_{n(k-l)}(n_1) dn_0 dn_1 \\
&= a^2 e^{\jmath\omega_0 l}\int_{-\pi}^{\pi}\int_{-\pi}^{\pi} e^{-\jmath(-n_0+n_1)} \frac{1}{2\pi}\frac{1}{2\pi} dn_0 dn_1 \\
&= a^2 e^{\jmath\omega_0 l}\frac{1}{4\pi^2}\int_{-\pi}^{\pi}\int_{-\pi}^{\pi} e^{-\jmath(-n_0+n_1)} dn_0 dn_1 \\
&= a^2 e^{\jmath\omega_0 l}\frac{1}{4\pi^2}\left[\int_{-\pi}^{\pi} e^{\jmath n_0}\, dn_0\right]\left[\int_{-\pi}^{\pi} e^{-\jmath n_1}\, dn_1\right] \\
&= a^2 e^{\jmath\omega_0 l}\frac{1}{4\pi^2}\left[\frac{e^{\jmath\pi} - e^{-\jmath\pi}}{\jmath}\right]\left[\frac{-e^{-\jmath\pi} + e^{\jmath\pi}}{\jmath}\right] \\
&= -a^2 e^{\jmath\omega_0 l}\frac{1}{\pi^2}(\sin\pi)(\sin\pi) = 0
\end{aligned} \tag{2.85}$$

where in the fifth equality it is used the fact that $n(k)$ and $n(k-l)$, for $l \neq 0$, are independent.

For $l = 0$

$$r(0) = E[x(k)x^*(k)] = a^2 e^{\jmath(\omega_0 0)} = a^2$$

Therefore,

$$r(l) = E[x(k)x^*(k-l)] = a^2 e^{\jmath(\omega_0 l)}\delta(l)$$

where in the 3×3 case

$$\mathbf{R} = \begin{bmatrix} a^2 & 0 & 0 \\ 0 & a^2 & 0 \\ 0 & 0 & a^2 \end{bmatrix}$$

At the end it was verified the fact that when we have two exponential functions ($l \neq 0$) with uniformly distributed white noise in the range of $-k\pi$ to $k\pi$ as exponents, these exponentials are nonorthogonal only if $l = 0$, where k is a positive integer.

□

In the remaining part of this chapter and in the following chapters, we will treat the algorithms for real and complex signals separately. The derivations of the adaptive-filtering algorithms for complex signals are usually straightforward extensions of the real signal cases, and some of them are left as exercises.

2.4 WIENER FILTER

One of the most widely used objective function in adaptive filtering is the mean-square error (MSE) defined as

$$F[e(k)] = \xi(k) = E[e^2(k)] = E[d^2(k) - 2d(k)y(k) + y^2(k)] \tag{2.86}$$

where $d(k)$ is the reference signal as illustrated in Fig. 1.1.

Suppose the adaptive filter consists of a linear combiner, i.e., the output signal is composed by a linear combination of signals coming from an array as depicted in Fig. 2.1.a. In this case,

$$y(k) = \sum_{i=0}^{N} w_i(k)x_i(k) = \mathbf{w}^T(k)\mathbf{x}(k) \tag{2.87}$$

where $\mathbf{x}(k) = [x_0(k)\, x_1(k) \ldots x_N(k)]^T$ and $\mathbf{w}(k) = [w_0(k)\, w_1(k) \ldots w_N(k)]^T$ are the input signal and the adaptive-filter coefficient vectors, respectively.

In many applications, each element of the input signal vector consists of a delayed version of the same signal, that is: $x_0(k) = x(k), x_1(k) = x(k-1), \ldots, x_N(k) = x(k-N)$. Note that in this case the signal $y(k)$ is the result of applying an FIR filter to the input signal $x(k)$.

Since most of the analyses and algorithms presented in this book apply equally to the linear combiner and the FIR filter cases, we will mostly consider the latter case throughout the rest of the book. The main reason for this decision is that the fast algorithms for the recursive least-squares solution, to be discussed in the forthcoming chapters, explore the fact that the input signal vector consists of the output of a delay line with a single input signal, and, as a consequence, are not applicable to the linear combiner case.

The most straightforward realization for the adaptive filter is through the direct-form FIR structure as illustrated in Fig. 2.1.b, with the output given by

$$y(k) = \sum_{i=0}^{N} w_i(k)x(k-i) = \mathbf{w}^T(k)\mathbf{x}(k) \tag{2.88}$$

where $\mathbf{x}(k) = [x(k)\, x(k-1) \ldots x(k-N)]^T$ is the input vector representing a tapped-delay line, and $\mathbf{w}(k) = [w_0(k)\, w_1(k) \ldots w_N(k)]^T$ is the tap-weight vector.

In both the linear combiner and FIR filter cases, the objective function can be rewritten as

$$\begin{aligned} E[e^2(k)] &= \xi(k) \\ &= E\left[d^2(k) - 2d(k)\mathbf{w}^T(k)\mathbf{x}(k) + \mathbf{w}^T(k)\mathbf{x}(k)\mathbf{x}^T(k)\mathbf{w}(k)\right] \\ &= E[d^2(k)] - 2E[d(k)\mathbf{w}^T(k)\mathbf{x}(k)] + E[\mathbf{w}^T(k)\mathbf{x}(k)\mathbf{x}^T(k)\mathbf{w}(k)] \end{aligned} \tag{2.89}$$

For a filter with fixed coefficients, the MSE function in a stationary environment is given by

$$\begin{aligned} \xi &= E[d^2(k)] - 2\mathbf{w}^T E[d(k)\mathbf{x}(k)] + \mathbf{w}^T E[\mathbf{x}(k)\mathbf{x}^T(k)]\mathbf{w} \\ &= E[d^2(k)] - 2\mathbf{w}^T\mathbf{p} + \mathbf{w}^T\mathbf{R}\mathbf{w} \end{aligned} \tag{2.90}$$

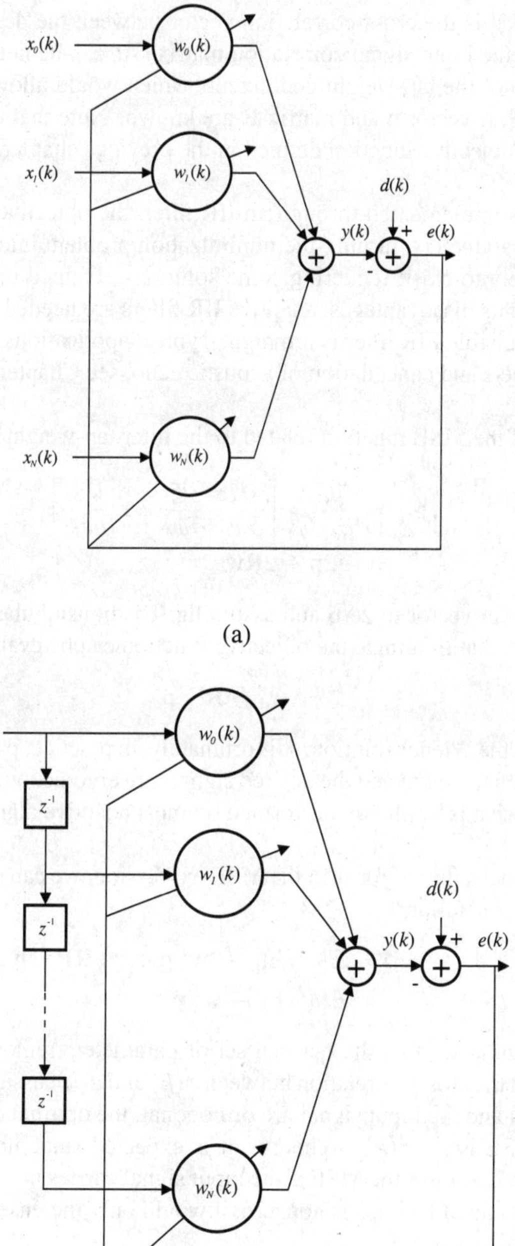

Figure 2.1 (a) Linear combiner; (b) Adaptive FIR filter.

where $\mathbf{p} = E[d(k)\mathbf{x}(k)]$ is the cross-correlation vector between the desired and input signals, and $\mathbf{R} = E[\mathbf{x}(k)\mathbf{x}^T(k)]$ is the input signal correlation matrix. As can be noted, the objective function ξ is a quadratic function of the tap-weight coefficients which would allow a straightforward solution for $\mathbf{w}$ that minimizes ξ, if vector $\mathbf{p}$ and matrix $\mathbf{R}$ are known. Note that matrix $\mathbf{R}$ corresponds to the Hessian matrix of the objective function defined in the previous chapter.

If the adaptive filter is implemented through an IIR filter, the objective function is a nonquadratic function of the filter parameters, turning the minimization problem into a much more difficult one. Local minima are likely to exist, rendering some solutions obtained by gradient-based algorithms unacceptable. Despite its disadvantages, adaptive IIR filters are needed in a number of applications where the order of a suitable FIR filter is too high. Typical applications include data equalization in communication channels and cancellation of acoustic echo, see Chapter 10.

The gradient vector of the MSE function related to the filter tap-weight coefficients is given by[7]

$$\begin{aligned}\mathbf{g_w} &= \frac{\partial \xi}{\partial \mathbf{w}} = \left[\frac{\partial \xi}{\partial w_0}\ \frac{\partial \xi}{\partial w_1} \cdots \frac{\partial \xi}{\partial w_N}\right]^T \\ &= -2\mathbf{p} + 2\mathbf{R}\mathbf{w}\end{aligned} \tag{2.91}$$

By equating the gradient vector to zero and assuming $\mathbf{R}$ is nonsingular, the optimal values for the tap-weight coefficients that minimize the objective function can be evaluated as follows:

$$\mathbf{w}_o = \mathbf{R}^{-1}\mathbf{p} \tag{2.92}$$

This solution is called the Wiener solution. Unfortunately, in practice, precise estimations of $\mathbf{R}$ and $\mathbf{p}$ are not available. When the input and the desired signals are ergodic, one is able to use time averages to estimate $\mathbf{R}$ and $\mathbf{p}$, what is implicitly performed by most adaptive algorithms.

If we replace the optimal solution for $\mathbf{w}$ in the MSE expression, we can calculate the minimum MSE provided by the Wiener solution:

$$\begin{aligned}\xi_{\min} &= E[d^2(k)] - 2\mathbf{w}_o^T\mathbf{p} + \mathbf{w}_o^T\mathbf{R}\mathbf{R}^{-1}\mathbf{p} \\ &= E[d^2(k)] - \mathbf{w}_o^T\mathbf{p}\end{aligned} \tag{2.93}$$

The above equation indicates that the optimal set of parameters removes part of the power of the desired signal through the cross-correlation between $x(k)$ and $d(k)$, assuming both signals stationary. If the reference signal and the input signal are orthogonal, the optimal coefficients are equal to zero and the minimum MSE is $E[d^2(k)]$. This result is expected since nothing can be done with the parameters in order to minimize the MSE if the input signal carries no information about the desired signal. In this case, if any of the taps is nonzero, it would only increase the MSE.

An important property of the Wiener filter can be deduced if we analyze the gradient of the error surface at the optimal solution. The gradient vector can be expressed as follows:

$$\mathbf{g_w} = \frac{\partial E[e^2(k)]}{\partial \mathbf{w}} = E[2e(k)\frac{\partial e(k)}{\partial \mathbf{w}}] = -E[2e(k)\mathbf{x}(k)] \tag{2.94}$$

[7] Some books define $\mathbf{g_w}$ as $\left[\frac{\partial \xi}{\partial \mathbf{w}}\right]^T$, here we follow the notation more widely used in the subject matter.

With the coefficients set at their optimal values, i.e., at the Wiener solution, the gradient vector is equal to zero, implying that

$$E[e(k)\mathbf{x}(k)] = \mathbf{0} \tag{2.95}$$

or

$$E[e(k)x(k-i)] = 0 \tag{2.96}$$

for $i = 0, 1, \ldots, N$. This means that the error signal is orthogonal to the elements of the input signal vector. In case either the error or the input signal has zero mean, the orthogonality property implies that $e(k)$ and $x(k)$ are uncorrelated.

The orthogonality principle also applies to the correlation between the output signal $y(k)$ and the error $e(k)$, when the tap weights are given by $\mathbf{w} = \mathbf{w}_o$. By premultiplying the equation (2.95) by $\mathbf{w}_o^T$, the desired result follows:

$$E[e(k)\mathbf{w}_o^T\mathbf{x}(k)] = E[e(k)y(k)] = 0 \tag{2.97}$$

The gradient with respect to a complex parameter has not been defined. For our purposes the complex gradient vector can be defined as [25]

$$\mathbf{g}_{\mathbf{w}(k)}\{F(e(k))\} = \frac{1}{2}\left\{\frac{\partial F[e(k)]}{\partial \text{re}[\mathbf{w}(k)]} - \jmath\frac{\partial F[e(k)]}{\partial \text{im}[\mathbf{w}(k)]}\right\}$$

where re[·] and im[·] indicate real and imaginary parts of [·], respectively. Note that the partial derivatives are calculated for each element of $\mathbf{w}(k)$.

For the complex case the error signal and the MSE are, respectively, described by, see Appendix A for details,

$$e(k) = d(k) - \mathbf{w}^H(k)\mathbf{x}(k) \tag{2.98}$$

and

$$\begin{aligned}\xi &= E[|e(k)|^2]\\ &= E[|d(k)|^2] - 2\text{re}\{\mathbf{w}^H E[d^*(k)\mathbf{x}(k)]\} + \mathbf{w}^H E[\mathbf{x}(k)\mathbf{x}^H(k)]\mathbf{w}\\ &= E[|d(k)|^2] - 2\text{re}[\mathbf{w}^H\mathbf{p}] + \mathbf{w}^H\mathbf{R}\mathbf{w}\end{aligned} \tag{2.99}$$

where $\mathbf{p} = E[d^*(k)\mathbf{x}(k)]$ is the cross-correlation vector between the desired and input signals, and $\mathbf{R} = E[\mathbf{x}(k)\mathbf{x}^H(k)]$ is the input signal correlation matrix. The Wiener solution in this case is also given by equation (2.92).

Example 2.2

The input signal of a first-order adaptive filter is described by

$$x(k) = \alpha_1 x_1(k) + \alpha_2 x_2(k)$$

where $x_1(k)$ and $x_2(k)$ are first-order AR processes and mutually uncorrelated having both unit variance. These signals are generated by applying distinct white noises to first-order filters whose poles are placed at $-s_1$ and $-s_2$, respectively.

(a) Calculate the autocorrelation matrix of the input signal.

(b) If the desired signal consists of $x_2(k)$, calculate the Wiener solution.

Solution:

(a)

The models for the signals involved are described by

$$x_i(k) = -s_i x_i(k-1) + \kappa_i n_i(k)$$

for $i = 1, 2$. According to equation (2.83) the autocorrelation of either $x_i(k)$ is given by

$$E[x_i(k)x_i(k-l)] = \kappa_i^2 \frac{(-s_i)^{|l|}}{1-s_i^2}\sigma_{n,i}^2 \tag{2.100}$$

where $\sigma_{n,i}^2$ is the variance of $n_i(k)$. Since each signal $x_i(k)$ has unit variance, then by applying $l = 0$ to the above equation

$$\kappa_i^2 = \frac{1-s_i^2}{\sigma_{n,i}^2} \tag{2.101}$$

Now by utilizing the fact that $x_1(k)$ and $x_2(k)$ are uncorrelated, the autocorrelation of the input signal is

$$\mathbf{R} = \begin{bmatrix} \alpha_1^2 + \alpha_2^2 & -\alpha_1^2 s_1 - \alpha_2^2 s_2 \\ -\alpha_1^2 s_1 - \alpha_2^2 s_2 & \alpha_1^2 + \alpha_2^2 \end{bmatrix}$$

$$\mathbf{p} = \begin{bmatrix} \alpha_2 \\ -\alpha_2 s_2 \end{bmatrix}$$

(b)

The Wiener solution can then be expressed as

$$\begin{aligned}
\mathbf{w}_o &= \mathbf{R}^{-1}\mathbf{p} \\
&= \frac{1}{(\alpha_1^2+\alpha_2^2)^2 - (\alpha_1^2 s_1 + \alpha_2^2 s_2)^2} \begin{bmatrix} \alpha_1^2+\alpha_2^2 & \alpha_1^2 s_1 + \alpha_2^2 s_2 \\ \alpha_1^2 s_1 + \alpha_2^2 s_2 & \alpha_1^2+\alpha_2^2 \end{bmatrix} \begin{bmatrix} \alpha_2 \\ -\alpha_2 s_2 \end{bmatrix} \\
&= \frac{1}{(1+\frac{\alpha_2^2}{\alpha_1^2})^2 - (s_1 + \frac{\alpha_2^2}{\alpha_1^2} s_2)^2} \begin{bmatrix} 1+\frac{\alpha_2^2}{\alpha_1^2} & s_1+\frac{\alpha_2^2}{\alpha_1^2} s_2 \\ s_1+\frac{\alpha_2^2}{\alpha_1^2} s_2 & 1+\frac{\alpha_2^2}{\alpha_1^2} \end{bmatrix} \begin{bmatrix} \frac{\alpha_2}{\alpha_1^2} \\ -\frac{\alpha_2}{\alpha_1^2} s_2 \end{bmatrix} \\
&= \alpha_2 \begin{bmatrix} \frac{1}{\alpha_1^2+\alpha_2^2-s_1\alpha_1^2-s_2\alpha_2^2} & 0 \\ 0 & \frac{1}{\alpha_1^2+\alpha_2^2+s_1\alpha_1^2+s_2\alpha_2^2} \end{bmatrix} \begin{bmatrix} \frac{1-s_2}{2} \\ -\frac{1+s_2}{2} \end{bmatrix}
\end{aligned}$$

Let's assume that in this example our task was to detect the presence of $x_2(k)$ in the input signal. For a fixed input-signal power, from this solution it is possible to observe that lower signal to interference at the input, that is lower $\frac{\alpha_2^2}{\alpha_1^2}$, leads to a Wiener solution vector with lower norm. This result reflects the fact that the Wiener solution tries to detect the desired signal at the same time it avoids enhancing the undesired signal, i.e., the interference $x_1(k)$.

□

2.5 LINEARLY CONSTRAINED WIENER FILTER

In a number of applications, it is required to impose some linear constraints on the filter coefficients such that the optimal solution is the one that achieves the minimum MSE, provided the constraints are met. Typical constraints are: unity norm of the parameter vector; linear phase of the adaptive filter; prescribed gains at given frequencies.

In the particular case of an array of antennas the measured signals can be linearly combined to form a directional beam, where the signal impinging on the array in the desired direction will have higher gain. This application is called beamforming, where we specify gains at certain directions of arrival. It is clear that the array is introducing another dimension to the received data, namely spatial information. The weights in the antennas can be made adaptive leading to the so-called adaptive antenna arrays. This is the principle behind the concept of smart antennas, where a set of adaptive array processors filter the signals coming from the array, and direct the beam to several different directions where a potential communication is required. For example, in a wireless communication system we are able to form a beam for each subscriber according to its position, ultimately leading to minimization of noise from the environment and interference from other subscribers.

In order to develop the theory of linearly constrained optimal filters, let us consider the particular application of a narrowband beamformer required to pass without distortion all signals arriving at 90° with respect to the array of antennas. All other sources of signals shall be treated as interferers and must be attenuated as much as possible. Fig. 2.2 illustrates the application. Note that in case the signal of interest does not impinge the array at 90° with respect to the array, a steering operation in the constraint vector $\mathbf{c}$ (to be defined) has to be performed [22].

The optimal filter that satisfies the linear constraints is called the *linearly-constrained minimum-variance* (LCMV) filter.

If the desired signal source is sufficiently far from the array of antennas, then we may assume that the wavefronts are planar at the array. Therefore, the wavefront from the desired source will reach all antennas at the same instant, whereas the wavefront from the interferer will reach each antenna at different time instants. Taking the antenna with input signal x_0 as a time reference t_0, the wavefront will reach the ith antenna at [22]

$$t_i = t_0 + i\frac{d\cos\theta}{c}$$

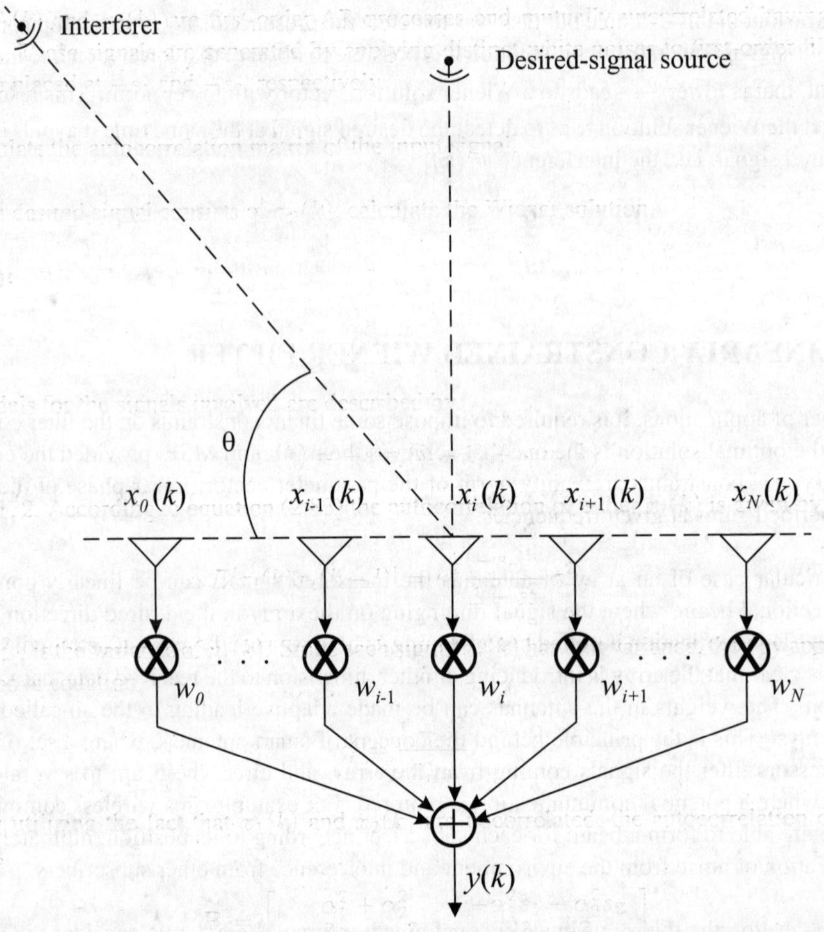

Figure 2.2 Narrowband beamformer.

where θ is the angle between the antenna array and the interferer direction of arrival, d is the distance between neighboring antennas, and c is the speed of propagation of the wave (3×10^8 m/s).

For this particular case, the LCMV filter is the one that minimizes the array output signal energy

$$\xi = E[y^2(k)] = E[\mathbf{w}^T \mathbf{x}(k)\mathbf{x}^T(k)\mathbf{w}]$$

$$\text{subject to}: \sum_{j=0}^{N} c_j w_j = f \tag{2.102}$$

where

$$\mathbf{w} = [w_0\ w_1 \ldots w_N]^T$$

$$\mathbf{x}(k) = [x_0(k)\ x_1(k) \ldots x_N(k)]^T$$

and

$$\mathbf{c} = [1\,1\ldots1]^T$$

is the constraint vector, since $\theta = 90°$. The desired gain is usually $f = 1$.

In the case the desired signal impinges the array at an angle θ with respect to the array, the incoming signal reaches the ith antenna delayed by $i\frac{d\cos\theta}{c}$ with respect to the 0th antenna [23]. Let's consider the case of a narrowband array such that all antennas detect the impinging signal with the same amplitude when measured taking into consideration their relative delays, which are multiples of $\frac{d\cos\theta}{c}$. In such a case the optimal receiver coefficients would be

$$w_i = \frac{\mathrm{e}^{\jmath\omega\tau_i}}{N+1} \tag{2.103}$$

for $i = 0, 1, \ldots, N$, in order to add coherently the delays of the desired incoming signal at a given direction θ. The impinging signal appear at the ith antenna multiplied by $\mathrm{e}^{-\jmath\omega\tau_i}$, considering the particular case of array configuration of Fig. 2.2. In this uniform linear array, the antenna locations are

$$p_i = id$$

for $i = 0, 1, \ldots, N$. Using the 0th antenna as reference, the signal will reach the array according to the following pattern

$$\begin{aligned}\tilde{\mathbf{c}} &= \mathrm{e}^{\jmath\omega t}\left[1\ \mathrm{e}^{-\jmath\omega\frac{d\cos\theta}{c}}\ \mathrm{e}^{-\jmath\omega\frac{2d\cos\theta}{c}}\ \ldots\mathrm{e}^{-\jmath\omega\frac{Nd\cos\theta}{c}}\right]^T \\ &= \mathrm{e}^{\jmath\omega t}\left[1\ \mathrm{e}^{-\jmath\frac{2\pi}{\lambda}d\cos\theta}\ \mathrm{e}^{-\jmath\frac{2\pi}{\lambda}2d\cos\theta}\ \ldots\mathrm{e}^{-\jmath\frac{2\pi}{\lambda}Nd\cos\theta}\right]^T\end{aligned} \tag{2.104}$$

where the equality $\frac{\omega}{c} = \frac{2\pi}{\lambda}$ was employed, with λ being the wavelength corresponding to the frequency ω.

By defining the variable $\psi(\omega,\theta) = \frac{2\pi}{\lambda}d\cos\theta$, we can describe the output signal of the beamformer as

$$\begin{aligned}y &= \mathrm{e}^{\jmath\omega t}\sum_{i=0}^{N} w_i \mathrm{e}^{-\jmath\psi(\omega,\theta)i} \\ &= \mathrm{e}^{\jmath\omega t}H(\omega,\theta)\end{aligned} \tag{2.105}$$

where $H(\omega,\theta)$ modifies the amplitude and phase of transmitted signal at a given frequency ω. Note that the shaping function $H(\omega,\theta)$ depends on the impinging angle.

For the sake of illustration, if the antenna separation is $d = \frac{\lambda}{2}$, $\theta = 60°$, and N is odd, then the constraint vector would be

$$\begin{aligned}\mathbf{c} &= \left[1\ \mathrm{e}^{-\jmath\frac{\pi}{2}}\ \mathrm{e}^{-\jmath\pi}\ldots\mathrm{e}^{-\jmath\frac{N\pi}{2}}\right]^T \\ &= \left[1\ -\jmath\ -1\ldots\mathrm{e}^{-\jmath\frac{N\pi}{2}}\right]^T\end{aligned} \tag{2.106}$$

Using the method of Lagrange multipliers, we can rewrite the constrained minimization problem described in equation (2.102) as

$$\xi_c = E[\mathbf{w}^T\mathbf{x}(k)\mathbf{x}^T(k)\mathbf{w}] + \lambda(\mathbf{c}^T\mathbf{w} - f) \tag{2.107}$$

The gradient of ξ_c with respect to $\mathbf{w}$ is equal to

$$\mathbf{g_w} = 2\mathbf{R}\mathbf{w} + \lambda\mathbf{c} \tag{2.108}$$

where $\mathbf{R} = E[\mathbf{x}(k)\mathbf{x}^T(k)]$. For a positive definite matrix $\mathbf{R}$, the value of $\mathbf{w}$ that satisfies $\mathbf{g_w} = \mathbf{0}$ is unique and minimizes ξ_c. Denoting $\mathbf{w}_o$ as the optimal solution, we have

$$\begin{aligned} 2\mathbf{R}\mathbf{w}_o + \lambda\mathbf{c} &= \mathbf{0} \\ 2\mathbf{c}^T\mathbf{w}_o + \lambda\mathbf{c}^T\mathbf{R}^{-1}\mathbf{c} &= 0 \\ 2f + \lambda\mathbf{c}^T\mathbf{R}^{-1}\mathbf{c} &= \mathbf{0} \end{aligned}$$

where in order to obtain the second equality, we premultiply the first equation by $\mathbf{c}^T\mathbf{R}^{-1}$. Therefore,

$$\lambda = -2(\mathbf{c}^T\mathbf{R}^{-1}\mathbf{c})^{-1}f$$

and the LCMV filter is

$$\mathbf{w}_o = \mathbf{R}^{-1}\mathbf{c}(\mathbf{c}^T\mathbf{R}^{-1}\mathbf{c})^{-1}f \tag{2.109}$$

If more constraints need to be satisfied by the filter, these can be easily incorporated in a constraint matrix and in a gain vector, such that

$$\mathbf{C}^T\mathbf{w} = \mathbf{f} \tag{2.110}$$

In this case, the LCMV filter is given by

$$\mathbf{w}_o = \mathbf{R}^{-1}\mathbf{C}(\mathbf{C}^T\mathbf{R}^{-1}\mathbf{C})^{-1}\mathbf{f} \tag{2.111}$$

If there is a desired signal, the natural objective is the minimization of the MSE, not the output energy as in the narrowband beamformer. In this case, it is straightforward to modify equation (2.107) and obtain the optimal solution

$$\mathbf{w}_o = \mathbf{R}^{-1}\mathbf{p} + \mathbf{R}^{-1}\mathbf{C}(\mathbf{C}^T\mathbf{R}^{-1}\mathbf{C})^{-1}(\mathbf{f} - \mathbf{C}^T\mathbf{R}^{-1}\mathbf{p}) \tag{2.112}$$

where $\mathbf{p} = E[d(k)\,\mathbf{x}(k)]$ see problem 20.

In the case of complex input signals and constraints the optimal solution is given by

$$\mathbf{w}_o = \mathbf{R}^{-1}\mathbf{p} + \mathbf{R}^{-1}\mathbf{C}(\mathbf{C}^H\mathbf{R}^{-1}\mathbf{C})^{-1}(\mathbf{f} - \mathbf{C}^H\mathbf{R}^{-1}\mathbf{p}) \tag{2.113}$$

where $\mathbf{C}^H\mathbf{w} = \mathbf{f}$.

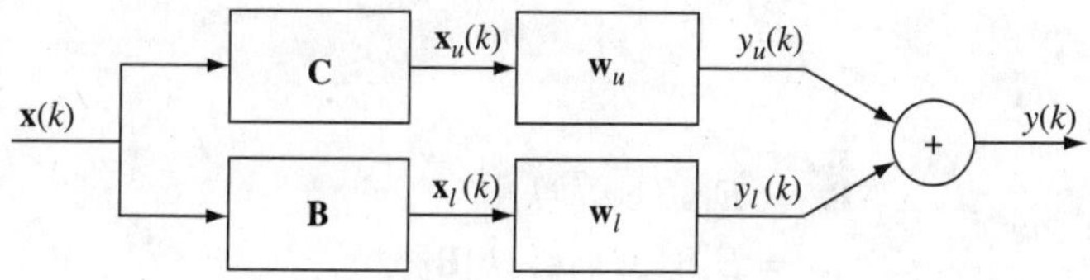

Figure 2.3 The generalized sidelobe canceller.

2.5.1 The Generalized Sidelobe Canceller

An alternative implementation to the direct-form constrained adaptive filter showed above is called the generalized sidelobe canceller (GSC) (see Fig. 2.3) [20].

For this structure the input signal vector is transformed by a matrix

$$\mathbf{T} = [\mathbf{C}\ \mathbf{B}] \tag{2.114}$$

where $\mathbf{C}$ is the constraint matrix and $\mathbf{B}$ is a *blocking matrix* that spans the null space of $\mathbf{C}$, i.e., matrix $\mathbf{B}$ satisfies

$$\mathbf{B}^T\mathbf{C} = \mathbf{0} \tag{2.115}$$

The output signal $y(k)$ shown in Fig. 2.3 is formed as

$$\begin{aligned} y(k) &= \mathbf{w}_u^T\mathbf{C}^T\mathbf{x}(k) + \mathbf{w}_l^T\mathbf{B}^T\mathbf{x}(k) \\ &= (\mathbf{C}\mathbf{w}_u + \mathbf{B}\mathbf{w}_l)^T\mathbf{x}(k) \\ &= (\mathbf{T}\mathbf{w})^T\mathbf{x}(k) \\ &= \bar{\mathbf{w}}^T\mathbf{x}(k) \end{aligned} \tag{2.116}$$

where $\mathbf{w} = [\mathbf{w}_u^T\ \mathbf{w}_l^T]^T$ and $\bar{\mathbf{w}} = \mathbf{T}\mathbf{w}$.

The linear constraints are satisfied if $\mathbf{C}^T\bar{\mathbf{w}} = \mathbf{f}$. But as $\mathbf{C}^T\mathbf{B} = \mathbf{0}$, then the condition to be satisfied becomes

$$\mathbf{C}^T\bar{\mathbf{w}} = \mathbf{C}^T\mathbf{C}\mathbf{w}_u = \mathbf{f} \tag{2.117}$$

Therefore, for the GSC structure shown in Fig. 2.3 there is a necessary condition that the upper part of the coefficient vector, $\mathbf{w}_u$, should be initialized as

$$\mathbf{w}_u = (\mathbf{C}^T\mathbf{C})^{-1}\mathbf{f} \tag{2.118}$$

Minimization of the output energy is achieved with a proper choice of $\mathbf{w}_l$. In fact, we transformed a constrained optimization problem into an unconstrained one, which in turn can be solved with the classical linear Wiener filter, i.e.,

$$\begin{aligned}\min_{\mathbf{w}_l} E[y^2(k)] &= \min_{\mathbf{w}_l} E\{[y_u(k) + \mathbf{w}_l^T \mathbf{x}_l(k)]^2\} \\ &= \mathbf{w}_{l,o} \\ &= -\mathbf{R}_l^{-1}\mathbf{p}_l \end{aligned} \tag{2.119}$$

where

$$\begin{aligned}\mathbf{R}_l &= E[\mathbf{x}_l(k)\mathbf{x}_l^T(k)] \\ &= E[\mathbf{B}^T\mathbf{x}(k)\mathbf{x}^T(k)\mathbf{B}] \\ &= \mathbf{B}^T[\mathbf{x}(k)\mathbf{x}^T(k)]\mathbf{B} \\ &= \mathbf{B}^T\mathbf{R}\mathbf{B} \end{aligned} \tag{2.120}$$

and

$$\begin{aligned}\mathbf{p}_l &= E[y_u(k)\,\mathbf{x}_l(k)] = E[\mathbf{x}_l(k)\,y_u(k)] \\ &= E[\mathbf{B}^T\mathbf{x}(k)\,\mathbf{w}_u^T\mathbf{C}^T\mathbf{x}(k)] \\ &= E[\mathbf{B}^T\mathbf{x}(k)\,\mathbf{x}^T(k)\mathbf{C}\mathbf{w}_u] \\ &= \mathbf{B}^T E[\mathbf{x}(k)\,\mathbf{x}^T(k)]\mathbf{C}\mathbf{w}_u \\ &= \mathbf{B}^T\mathbf{R}\mathbf{C}\mathbf{w}_u \\ &= \mathbf{B}^T\mathbf{R}\mathbf{C}(\mathbf{C}^T\mathbf{C})^{-1}\mathbf{f} \end{aligned} \tag{2.121}$$

where in the above derivations we utilized the results and definitions from equations (2.116) and (2.118).

Using the equations (2.120), (2.121), and (2.118), it is possible to show that

$$\mathbf{w}_{l,o} = -(\mathbf{B}^T\mathbf{R}\mathbf{B})^{-1}\mathbf{B}^T\mathbf{R}\mathbf{C}(\mathbf{C}^T\mathbf{C})^{-1}\mathbf{f} \tag{2.122}$$

Given that $\mathbf{w}_{l,o}$ is the solution to an unconstrained minimization problem of transformed quantities, any unconstrained adaptive filter can be used to estimate recursively this optimal solution. The drawback in the implementation of the GSC structure comes from the transformation of the input signal vector via a constraint matrix and a blocking matrix. Although in theory any matrix with linearly independent columns that spans the null space of $\mathbf{C}$ can be employed, in many cases the computational complexity resulting from the multiplication of $\mathbf{B}$ by $\mathbf{x}(k)$ can be prohibitive. Furthermore, if the transformation matrix $\mathbf{T}$ is not orthogonal, finite-precision effects may yield an overall unstable system. A simple solution that guarantees orthogonality in the transformation and low computational complexity can be obtained with a Householder transformation [21].

2.6 MEAN-SQUARE ERROR SURFACE

The mean-square error is a quadratic function of the parameters $\mathbf{w}$. Assuming a given fixed $\mathbf{w}$, the MSE is not a function of time and can be expressed as

$$\xi = \sigma_d^2 - 2\mathbf{w}^T\mathbf{p} + \mathbf{w}^T\mathbf{R}\mathbf{w} \tag{2.123}$$

where σ_d^2 is the variance of $d(k)$ assuming it has zero-mean. The MSE is a quadratic function of the tap weights forming a hyperparaboloid surface. The MSE surface is convex and has only positive values. For two weights, the surface is a paraboloid. Fig. 2.4 illustrates the MSE surface for a numerical example where **w** has two coefficients. If the MSE surface is intersected by a plane parallel to the **w** plane, placed at a level superior to $\xi_{\min}$, the intersection consists of an ellipse representing equal MSE contours as depicted in Fig. 2.5. Note that in this figure we showed three distinct ellipses, corresponding to different levels of MSE. The ellipses of constant MSE are all concentric.

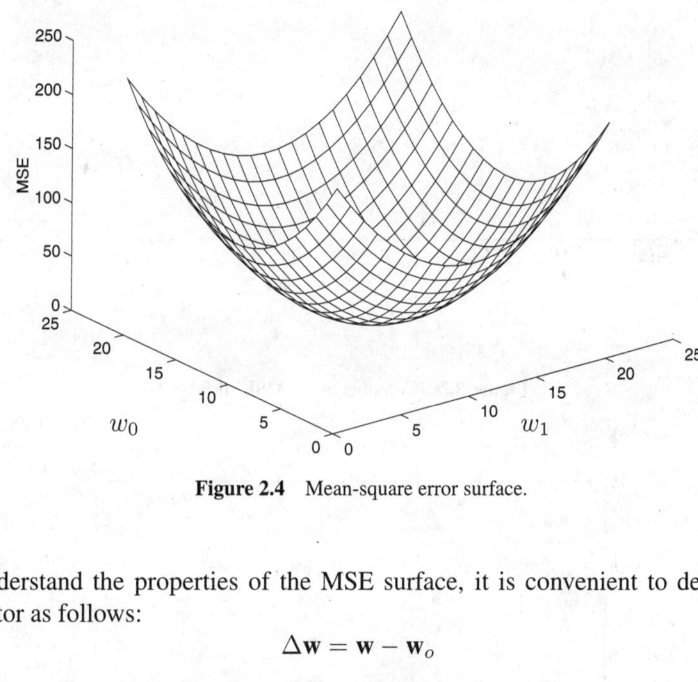

Figure 2.4 Mean-square error surface.

In order to understand the properties of the MSE surface, it is convenient to define a translated coefficient vector as follows:

$$\Delta\mathbf{w} = \mathbf{w} - \mathbf{w}_o \tag{2.124}$$

The MSE can be expressed as a function of $\Delta\mathbf{w}$ as follows:

$$\begin{aligned}
\xi &= \sigma_d^2 - \mathbf{w}_o^T\mathbf{p} + \mathbf{w}_o^T\mathbf{p} - 2\mathbf{w}^T\mathbf{p} + \mathbf{w}^T\mathbf{R}\mathbf{w} \\
&= \xi_{\min} - \Delta\mathbf{w}^T\mathbf{p} - \mathbf{w}^T\mathbf{R}\mathbf{w}_o + \mathbf{w}^T\mathbf{R}\mathbf{w} \\
&= \xi_{\min} - \Delta\mathbf{w}^T\mathbf{p} + \mathbf{w}^T\mathbf{R}\Delta\mathbf{w} \\
&= \xi_{\min} - \mathbf{w}_o^T\mathbf{R}\Delta\mathbf{w} + \mathbf{w}^T\mathbf{R}\Delta\mathbf{w} \\
&= \xi_{\min} + \Delta\mathbf{w}^T\mathbf{R}\Delta\mathbf{w}
\end{aligned} \tag{2.125}$$

where we used the results of equations (2.92) and (2.93). The corresponding error surface contours are depicted in Fig. 2.6.

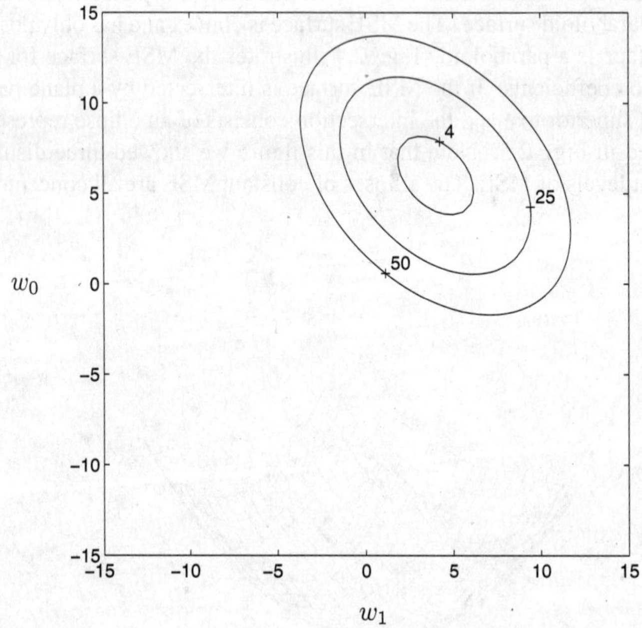

Figure 2.5 Contours of the MSE surface.

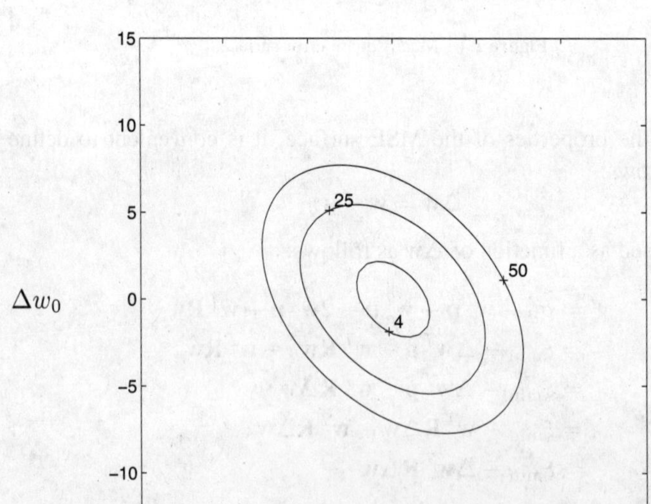

Figure 2.6 Translated contours of the MSE surface.

By employing the diagonalized form of $\mathbf{R}$, the last equation can be rewritten as follows:

$$\begin{aligned}\xi &= \xi_{\min} + \Delta\mathbf{w}^T\mathbf{Q}\mathbf{\Lambda}\mathbf{Q}^T\Delta\mathbf{w} \\ &= \xi_{\min} + \mathbf{v}^T\mathbf{\Lambda}\mathbf{v} \\ &= \xi_{\min} + \sum_{i=0}^{N}\lambda_i v_i^2 \end{aligned} \tag{2.126}$$

where $\mathbf{v} = \mathbf{Q}^T\Delta\mathbf{w}$ are the rotated parameters.

The above form for representing the MSE surface is an uncoupled form, in the sense that each component of the gradient vector of the MSE with respect to the rotated parameters is a function of a single parameter, that is

$$\mathbf{g_v}[\xi] = [2\lambda_0 v_0 \;\; 2\lambda_1 v_1 \; \ldots \; 2\lambda_N v_N]^T$$

This property means that if all v_i's are zero except one, the gradient direction coincides with the nonzero parameter axis. In other words, the rotated parameters represent the principal axes of the hyperellipse of constant MSE, as illustrated in Fig. 2.7. Note that since the rotated parameters are the result of the projection of the original parameter vector $\Delta\mathbf{w}$ on the eigenvectors $\mathbf{q}_i$ direction, it is straightforward to conclude that the eigenvectors represent the principal axes of the constant MSE hyperellipses.

The matrix of second derivatives of ξ as related to the rotated parameters is $\mathbf{\Lambda}$. We can note that the gradient will be steeper in the principal axes corresponding to larger eigenvalues. This is the direction, in the two axes case, where the ellipse is narrow.

2.7 BIAS AND CONSISTENCY

The correct interpretation of the results obtained by the adaptive-filtering algorithm requires the definitions of bias and consistency. An estimate is considered unbiased if the following condition is satisfied

$$E[\mathbf{w}(k)] = \mathbf{w}_o \tag{2.127}$$

The difference $E[\mathbf{w}(k)] - \mathbf{w}_o$ is called the bias in the parameter estimate.

An estimate is considered consistent if

$$\mathbf{w}(k) \rightarrow \mathbf{w}_o \text{ as } k \rightarrow \infty \tag{2.128}$$

Note that since $\mathbf{w}(k)$ is a random variable, it is necessary to define in which sense the limit is taken. Usually, the limit with probability one is employed. In the case of identification, a system is considered identifiable if the given parameter estimates are consistent. For a more formal treatment on this subject refer to [18].

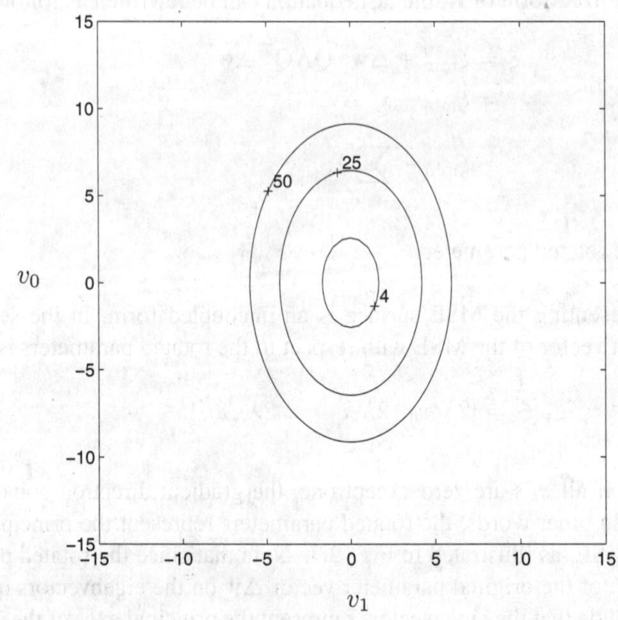

Figure 2.7 Rotated contours of the MSE surface.

2.8 NEWTON ALGORITHM

In the context of the MSE minimization discussed in the previous section, see equation (2.123), the coefficient-vector updating using the Newton method is performed as follows:

$$\mathbf{w}(k+1) = \mathbf{w}(k) - \mu\mathbf{R}^{-1}\mathbf{g_w}(k) \tag{2.129}$$

where its derivation originates from equation (1.4). Assuming the true gradient and the matrix $\mathbf{R}$ are available, the coefficient-vector updating can be expressed as

$$\mathbf{w}(k+1) = \mathbf{w}(k) - \mu\mathbf{R}^{-1}[-2\mathbf{p} + 2\mathbf{R}\mathbf{w}(k)] = (\mathbf{I} - 2\mu\mathbf{I})\mathbf{w}(k) + 2\mu\mathbf{w}_o \tag{2.130}$$

where if $\mu = 1/2$, the Wiener solution is reached in one step.

The Wiener solution can be approached using a Newton-like search algorithm, by updating the adaptive-filter coefficients as follows:

$$\mathbf{w}(k+1) = \mathbf{w}(k) - \mu\hat{\mathbf{R}}^{-1}(k)\hat{\mathbf{g}}_{\mathbf{w}}(k) \tag{2.131}$$

where $\hat{\mathbf{R}}^{-1}(k)$ is an estimate of $\mathbf{R}^{-1}$ and $\hat{\mathbf{g}}_{\mathbf{w}}(k)$ is an estimate of $\mathbf{g_w}$, both at instant k. The parameter μ is the convergence factor that regulates the convergence rate. Newton-based algorithms present, in general, fast convergence. However, the estimate of $\mathbf{R}^{-1}$ is computationally intensive and can become numerically unstable if special care is not taken. These factors made the steepest-descent-based algorithms more popular in adaptive-filtering applications.

2.9 STEEPEST-DESCENT ALGORITHM

In order to get a practical feeling of a problem that is being solved using the steepest-descent algorithm, we assume that the optimal coefficient vector, i.e., the Wiener solution, is $\mathbf{w}_o$, and that the reference signal is not corrupted by measurement noise[8].

The main objective of the present section is to study the rate of convergence, the stability, and the steady-state behavior of an adaptive filter whose coefficients are updated through the steepest-descent algorithm. It is worth mentioning that the steepest-descent method can be considered an efficient gradient-type algorithm, in the sense that it works with the true gradient vector, and not with an estimate of it. Therefore, the performance of other gradient-type algorithms can at most be close to the performance of the steepest-descent algorithm. When the objective function is the MSE, the difficult task of obtaining the matrix $\mathbf{R}$ and the vector $\mathbf{p}$ impairs the steepest-descent algorithm from being useful in adaptive-filtering applications. Its performance, however, serves as a benchmark for gradient-based algorithms.

The steepest-descent algorithm updates the coefficients in the following general form

$$\mathbf{w}(k+1) = \mathbf{w}(k) - \mu \mathbf{g_w}(k) \tag{2.132}$$

where the above expression is equivalent to equation (1.6). It is worth noting that several alternative gradient-based algorithms available replace $\mathbf{g_w}(k)$ by an estimate $\hat{\mathbf{g}}_{\mathbf{w}}(k)$, and they differ in the way the gradient vector is estimated. The true gradient expression is given in equation (2.91) and, as can be noted, it depends on the vector $\mathbf{p}$ and the matrix $\mathbf{R}$, that are usually not available.

Substituting equation (2.91) in equation (2.132), we get

$$\mathbf{w}(k+1) = \mathbf{w}(k) - 2\mu\mathbf{R}\mathbf{w}(k) + 2\mu\mathbf{p} \tag{2.133}$$

Now, some of the main properties related to the convergence behavior of the steepest-descent algorithm in stationary environment are described. First, an analysis is required to determine the influence of the convergence factor μ in the convergence behavior of the steepest-descent algorithm.

The error in the adaptive-filter coefficients when compared to the Wiener solution is defined as

$$\Delta\mathbf{w}(k) = \mathbf{w}(k) - \mathbf{w}_o \tag{2.134}$$

The steepest-descent algorithm can then be described in an alternative way, that is:

$$\begin{aligned}\Delta\mathbf{w}(k+1) &= \Delta\mathbf{w}(k) - 2\mu[\mathbf{R}\mathbf{w}(k) - \mathbf{R}\mathbf{w}_o] \\ &= \Delta\mathbf{w}(k) - 2\mu\mathbf{R}\Delta\mathbf{w}(k) \\ &= (\mathbf{I} - 2\mu\mathbf{R})\,\Delta\mathbf{w}(k)\end{aligned} \tag{2.135}$$

where the relation $\mathbf{p} = \mathbf{R}\mathbf{w}_o$ (see equation (2.92)) was employed. It can be shown from the above equation that

$$\Delta\mathbf{w}(k+1) = (\mathbf{I} - 2\mu\mathbf{R})^{k+1}\Delta\mathbf{w}(0) \tag{2.136}$$

[8]Noise added to the reference signal originated from environment and/or thermal noise.

or

$$\mathbf{w}(k+1) = \mathbf{w}_o + (\mathbf{I} - 2\mu\mathbf{R})^{k+1}[\mathbf{w}(0) - \mathbf{w}_o] \tag{2.137}$$

The equation (2.135) premultiplied by $\mathbf{Q}^T$, where $\mathbf{Q}$ is the unitary matrix that diagonalizes $\mathbf{R}$ through a similarity transformation, yields

$$\begin{aligned}
\mathbf{Q}^T \Delta\mathbf{w}(k+1) &= (\mathbf{I} - 2\mu\mathbf{Q}^T\mathbf{R}\mathbf{Q})\mathbf{Q}^T\Delta\mathbf{w}(k) \\
&= \mathbf{v}(k+1) \\
&= (\mathbf{I} - 2\mu\mathbf{\Lambda})\mathbf{v}(k) \\
&= \begin{bmatrix} 1-2\mu\lambda_0 & 0 & \cdots & 0 \\ 0 & 1-2\mu\lambda_1 & & \vdots \\ \vdots & \vdots & \ddots & \\ 0 & 0 & & 1-2\mu\lambda_N \end{bmatrix} \mathbf{v}(k)
\end{aligned} \tag{2.138}$$

In the above equation, $\mathbf{v}(k+1) = \mathbf{Q}^T\Delta\mathbf{w}(k+1)$ is the rotated coefficient-vector error. Using induction, equation (2.138) can be rewritten as

$$\begin{aligned}
\mathbf{v}(k+1) &= (\mathbf{I} - 2\mu\mathbf{\Lambda})^{k+1}\mathbf{v}(0) \\
&= \begin{bmatrix} (1-2\mu\lambda_0)^{k+1} & 0 & \cdots & 0 \\ 0 & (1-2\mu\lambda_1)^{k+1} & & \vdots \\ \vdots & \vdots & \ddots & \vdots \\ 0 & 0 & & (1-2\mu\lambda_N)^{k+1} \end{bmatrix} \mathbf{v}(0)
\end{aligned} \tag{2.139}$$

This equation shows that in order to guarantee the convergence of the coefficients, each element $1 - 2\mu\lambda_i$ must have an absolute value less than one. As a consequence, the convergence factor of the steepest-descent algorithm must be chosen in the range

$$0 < \mu < \frac{1}{\lambda_{\max}} \tag{2.140}$$

where $\lambda_{\max}$ is the largest eigenvalue of $\mathbf{R}$. In this case, all the elements of the diagonal matrix in equation (2.139) tend to zero as $k \rightarrow \infty$, resulting in $\mathbf{v}(k+1) \rightarrow 0$ for large k.

The μ value in the above range guarantees that the coefficient vector approaches the optimum coefficient vector $\mathbf{w}_o$. It should be mentioned that if matrix $\mathbf{R}$ has large eigenvalue spread, the convergence speed of the coefficients will be primarily dependent on the value of the smallest eigenvalue. Note that the slowest decaying element in equation (2.139) is given by $(1 - 2\mu\lambda_{\min})^{k+1}$.

The MSE presents a transient behavior during the adaptation process, that can be analyzed in a straightforward way if we employ the diagonalized version of $\mathbf{R}$. Recalling from equation (2.125) that

$$\xi(k) = \xi_{\min} + \Delta\mathbf{w}^T(k)\mathbf{R}\Delta\mathbf{w}(k) \tag{2.141}$$

the MSE can then be simplified as follows:

$$\begin{aligned}\xi(k) &= \xi_{\min} + \Delta\mathbf{w}^T(k)\mathbf{Q}\boldsymbol{\Lambda}\,\mathbf{Q}^T\Delta\mathbf{w}(k)\\ &= \xi_{\min} + \mathbf{v}^T(k)\boldsymbol{\Lambda}\,\mathbf{v}(k)\\ &= \xi_{\min} + \sum_{i=0}^{N}\lambda_i v_i^2(k)\end{aligned} \tag{2.142}$$

If we apply the result of equation (2.139) in equation (2.142), it can be shown that the following relation results

$$\begin{aligned}\xi(k) &= \xi_{\min} + \mathbf{v}^T(k-1)(\mathbf{I}-2\mu\boldsymbol{\Lambda})\boldsymbol{\Lambda}\,(\mathbf{I}-2\mu\boldsymbol{\Lambda})\mathbf{v}(k-1)\\ &= \xi_{\min} + \sum_{i=0}^{N}\lambda_i(1-2\mu\lambda_i)^{2k}v_i^2(0)\end{aligned} \tag{2.143}$$

The analyses presented in this section show that before the steepest-descent algorithm reaches the steady-state behavior, there is a transient period where the error is usually high and the coefficients are far from the Wiener solution. As can be seen from equation (2.139), in the case of the adaptive-filter coefficients, the convergence will follow $(N+1)$ geometric decaying curves with ratios $r_{wi} = (1-2\mu\lambda_i)$. Each of these curves can be approximated by an exponential envelope with time constant τ_{wi} as follows [5]:

$$r_{wi} = \mathrm{e}^{\frac{-1}{\tau_{wi}}} = 1 - \frac{1}{\tau_{wi}} + \frac{1}{2!\tau_{wi}^2} + \cdots \tag{2.144}$$

In general, r_{wi} is slightly smaller than one, specially in the cases of slowly decreasing modes that correspond to small values λ_i and μ. Therefore,

$$r_{wi} = (1-2\mu\lambda_i) \approx 1 - \frac{1}{\tau_{wi}} \tag{2.145}$$

then

$$\tau_{wi} \approx \frac{1}{2\mu\lambda_i}$$

for $i = 0, 1, \ldots, N$.

For the convergence of the MSE, the range of values of μ is the same to guarantee the convergence of the coefficients. In this case, due to the exponent $2k$ in equation (2.143), the geometric decaying curves have ratios given by $r_{ei} = (1-4\mu\lambda_i)$, that can be approximated by exponential envelopes with time constants given by

$$\tau_{ei} \approx \frac{1}{4\mu\lambda_i} \tag{2.146}$$

for $i = 0, 1, \ldots, N$, where it was considered that $4\mu^2\lambda_i^2 \ll 1$. In the convergence of both the error and the coefficients, the time required for the convergence depends on the ratio of the eigenvalues of the input signal. Further discussions on convergence properties that apply to gradient-type algorithms can be found in Chapter 3.

Example 2.3

The matrix **R** and the vector **p** are known for a given experimental environment:

$$\mathbf{R} = \begin{bmatrix} 1 & 0.4045 \\ 0.4045 & 1 \end{bmatrix}$$

$$\mathbf{p} = [0\ 0.2939]^T$$

$$E[d^2(k)] = 0.5$$

(a) Deduce the equation for the MSE.

(b) Choose a small value for μ, and starting the parameters at $[-1 \quad -2]^T$ plot the convergence path of the steepest-descent algorithm in the MSE surface.

(c) Repeat the previous item for the Newton algorithm starting at $[0 \ -2]^T$.

Solution:

(a) The MSE function is given by

$$\begin{aligned}\xi &= E[d^2(k)] - 2\mathbf{w}^T\mathbf{p} + \mathbf{w}^T\mathbf{R}\mathbf{w} \\ &= \sigma_d^2 - 2[w_1\ w_2]\begin{bmatrix} 0 \\ 0.2939 \end{bmatrix} + [w_1\ w_2]\begin{bmatrix} 1 & 0.4045 \\ 0.4045 & 1 \end{bmatrix}\begin{bmatrix} w_1 \\ w_2 \end{bmatrix}\end{aligned}$$

After performing the algebraic calculations, we obtain the following result

$$\xi = 0.5 + w_1^2 + w_2^2 + 0.8090w_1w_2 - 0.5878w_2$$

(b) The steepest-descent algorithm was applied to minimize the MSE using a convergence factor $\mu = 0.1/\lambda_{\max}$, where $\lambda_{\max} = 1.4045$. The convergence path of the algorithm in the MSE surface is depicted in Fig. 2.8. As can be noted, the path followed by the algorithm first approaches the main axis (eigenvector) corresponding to the smaller eigenvalue, and then follows toward the minimum in a direction increasingly aligned with this main axis.

(c) The Newton algorithm was also applied to minimize the MSE using a convergence factor $\mu = 0.1/\lambda_{\max}$. The convergence path of the Newton algorithm in the MSE surface is depicted in Fig. 2.9. The Newton algorithm follows a straight path to the minimum. □

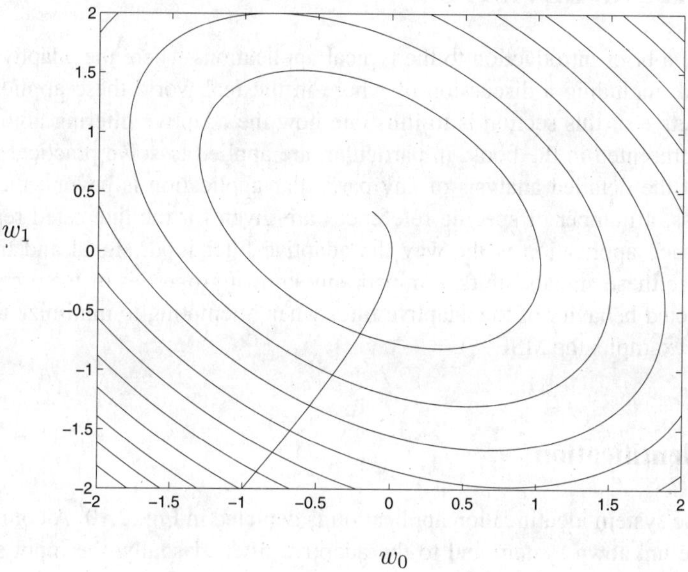

Figure 2.8 Convergence path of the steepest-descent algorithm.

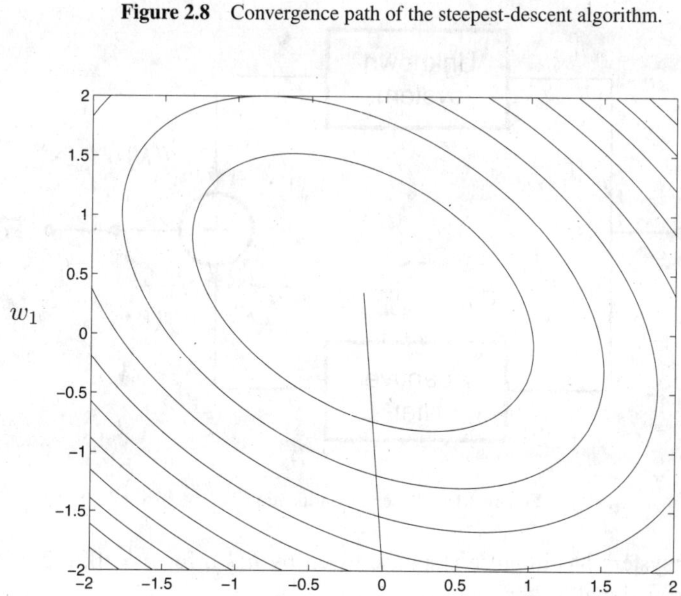

Figure 2.9 Convergence path of the Newton algorithm.

2.10 APPLICATIONS REVISITED

In this section, we give a brief introduction to the typical applications where the adaptive-filtering algorithms are required, including a discussion of where in the real world these applications are found. The main objective of this section is to illustrate how the adaptive-filtering algorithms, in general, and the ones presented in the book, in particular, are applied to solve practical problems. It should be noted that the detailed analysis of any particular application is beyond the scope of this book. Nevertheless, a number of specific references are given for the interested reader. The distinctive feature of each application is the way the adaptive filter input signal and the desired signal are chosen. Once these signals are determined, any known properties of them can be used to understand the expected behavior of the adaptive filter when attempting to minimize the chosen objective function (for example, the MSE, ξ).

2.10.1 System Identification

The typical set up of the system identification application is depicted in Fig. 2.10. A common input signal is applied to the unknown system and to the adaptive filter. Usually, the input signal is a wideband signal, in order to allow the adaptive filter to converge to a good model of the unknown system.

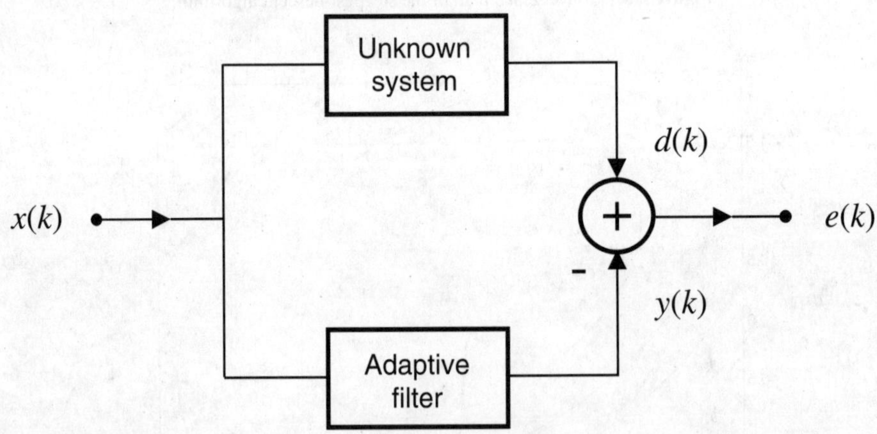

Figure 2.10 System identification.

Assume the unknown system has impulse response given by $h(k)$, for $k = 0, 1, 2, 3, \ldots, \infty$, and zero for $k < 0$. The error signal is then given by

$$\begin{aligned} e(k) &= d(k) - y(k) \\ &= \sum_{l=0}^{\infty} h(l)x(k-l) - \sum_{i=0}^{N} w_i(k)x(k-i) \end{aligned} \tag{2.147}$$

where $w_i(k)$ are the coefficients of the adaptive filter.

Assuming that $x(k)$ is a white noise, the MSE for a fixed $\mathbf{w}$ is given by

$$\begin{aligned}\xi &= E\{[\mathbf{h}^T\mathbf{x}_\infty(k) - \mathbf{w}^T\mathbf{x}_{N+1}(k)]^2\} \\ &= E\left[\mathbf{h}^T\mathbf{x}_\infty(k)\mathbf{x}_\infty^T(k)\mathbf{h} - 2\mathbf{h}^T\mathbf{x}_\infty(k)\mathbf{x}_{N+1}^T(k)\mathbf{w} + \mathbf{w}^T\mathbf{x}_{N+1}(k)\mathbf{x}_{N+1}^T(k)\mathbf{w}\right] \\ &= \sigma_x^2\sum_{i=0}^{\infty} h^2(i) - 2\sigma_x^2\mathbf{h}^T\begin{bmatrix}\mathbf{I}_{N+1} \\ \mathbf{0}_{\infty\times(N+1)}\end{bmatrix}\mathbf{w} + \mathbf{w}^T\mathbf{R}_{N+1}\mathbf{w} \end{aligned} \tag{2.148}$$

where $\mathbf{x}_\infty(k)$ and $\mathbf{x}_{N+1}(k)$ are the input signal vector with infinite and finite lengths, respectively.

By calculating the derivative of ξ with respect to the coefficients of the adaptive filter, it follows that

$$\mathbf{w}_o = \mathbf{h}_{N+1} \tag{2.149}$$

where

$$\mathbf{h}_{N+1}^T = \mathbf{h}^T\begin{bmatrix}\mathbf{I}_{N+1} \\ \mathbf{0}_{\infty\times(N+1)}\end{bmatrix} \tag{2.150}$$

If the input signal is a white noise, the best model for the unknown system is a system whose impulse response coincides with the $N+1$ first samples of the unknown system impulse response. In the cases where the impulse response of the unknown system is of finite length and the adaptive filter is of sufficient order (i.e., it has enough number of parameters), the MSE becomes zero if there is no measurement noise (or channel noise). In practical applications the measurement noise is unavoidable, and if it is uncorrelated with the input signal, the expected value of the adaptive-filter coefficients will coincide with the unknown-system impulse response samples. The output error will of course be the measurement noise. We can observe that the measurement noise introduces a variance in the estimates of the unknown system parameters.

Some real world applications of the system identification scheme include modeling of multipath communication channels [36], control systems [28], seismic exploration [37], and cancellation of echo caused by hybrids in some communication systems [38]-[42], just to mention a few.

2.10.2 Signal Enhancement

In the signal enhancement application, the reference signal consists of a desired signal $x(k)$ that is corrupted by an additive noise $n_1(k)$. The input signal of the adaptive filter is a noise signal $n_2(k)$ that is correlated with the interference signal $n_1(k)$, but uncorrelated with $x(k)$. Fig. 2.11 illustrates the configuration of the signal enhancement application. In practice, this configuration is found in acoustic echo cancellation for auditoriums [45], hearing aids, noise cancellation in hydrophones [44], cancelling of power line interference in electrocardiography [28], and in other applications. The cancelling of echo caused by the hybrid in some communication systems can also be considered a signal enhancement problem [28].

In this application, the error signal is given by

$$e(k) = x(k) + n_1(k) - \sum_{l=0}^{N} w_l n_2(k-l) = x(k) + n_1(k) - y(k) \tag{2.151}$$

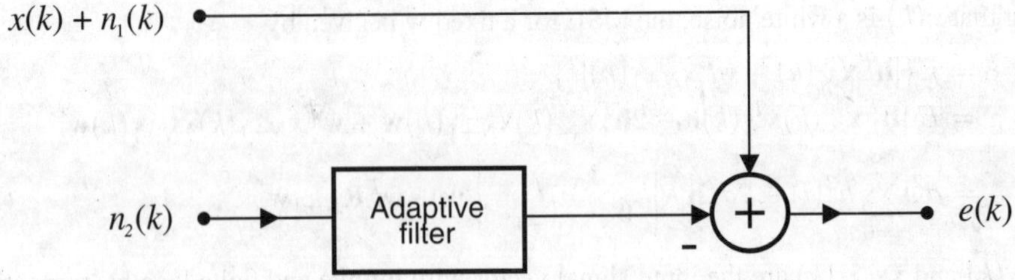

Figure 2.11 Signal enhancement ($n_1(k)$ and $n_2(k)$ are noise signals correlated to each other).

The resulting MSE is then given by

$$E[e^2(k)] = E[x^2(k)] + E\{[n_1(k) - y(k)]^2\} \tag{2.152}$$

where it was assumed that $x(k)$ is uncorrelated with $n_1(k)$ and $n_2(k)$. The above equation shows that if the adaptive filter, having $n_2(k)$ as the input signal, is able to perfectly predict the signal $n_1(k)$, the minimum MSE is given by

$$\xi_{\min} = E[x^2(k)] \tag{2.153}$$

where the error signal, in this situation, is the desired signal $x(k)$.

The effectiveness of the signal enhancement scheme depends on the high correlation between $n_1(k)$ and $n_2(k)$. In some applications, it is useful to include a delay of L samples in the reference signal or in the input signal, such that their relative delay yields a maximum cross-correlation between $y(k)$ and $n_1(k)$, reducing the MSE. This delay provides a kind of synchronization between the signals involved. An example exploring this issue will be presented in the following chapters.

2.10.3 Signal Prediction

In the signal prediction application, the adaptive-filter input consists of a delayed version of the desired signal as illustrated in Fig. 2.12. The MSE is given by

$$\xi = E\{[x(k) - \mathbf{w}^T \mathbf{x}(k-L)]^2\} \tag{2.154}$$

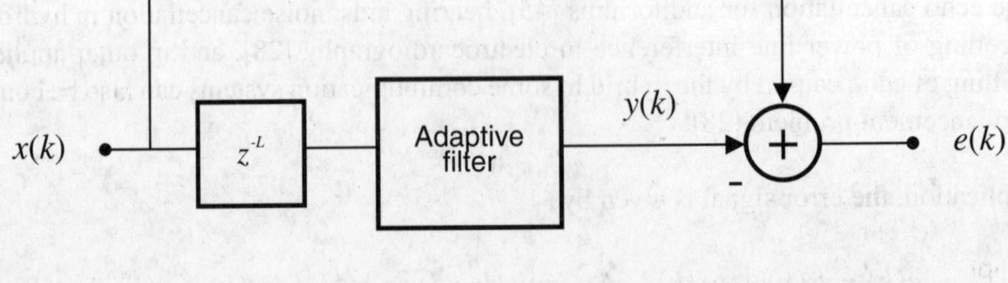

Figure 2.12 Signal prediction.

The minimization of the MSE leads to an FIR filter, whose coefficients are the elements of $\mathbf{w}$. This filter is able to predict the present sample of the input signal using as information old samples such as $x(k-L)$, $x(k-L-1), \ldots, x(k-L-N)$. The resulting FIR filter can then be considered a model for the signal $x(k)$ when the MSE is small. The minimum MSE is given by

$$\xi_{\min} = r(0) - \mathbf{w}_o^T \begin{bmatrix} r(L) \\ r(L+1) \\ \cdot \\ \cdot \\ \cdot \\ r(L+N) \end{bmatrix} \tag{2.155}$$

where $\mathbf{w}_o$ is the optimum predictor coefficient vector and $r(l) = E[x(k)x(k-l)]$ for a stationary process.

A typical predictor's application is in linear prediction coding of speech signals [43], where the predictor's task is to estimate the speech parameters. These parameters $\mathbf{w}$ are part of the coding information that is transmitted or stored along with other information inherent to the speech characteristics, such as pitch period, among others.

The adaptive signal predictor is also used for adaptive line enhancement (ALE), where the input signal is a narrowband signal (predictable) added to a wideband signal. After convergence, the predictor output will be an enhanced version of the narrowband signal.

Yet another application of the signal predictor is the suppression of narrowband interference in a wideband signal. The input signal, in this case, has the same general characteristics of the ALE. However, we are now interested in removing the narrowband interferer. For such an application, the output signal of interest is the error signal [45].

2.10.4 Channel Equalization

As can be seen from Fig. 2.13, channel equalization or inverse filtering consists of estimating a transfer function to compensate for the linear distortion caused by the channel. From another point of view, the objective is to force a prescribed dynamic behavior for the cascade of the channel (unknown system) and the adaptive filter, determined by the input signal. The first interpretation is more appropriate in communications, where the information is transmitted through dispersive channels [35], [41]. The second interpretation is appropriate for control applications, where the inverse filtering scheme generates control signals to be used in the unknown system [28].

In the ideal situation, where $n(k) = 0$ and the equalizer has sufficient order, the error signal is zero if

$$W(z)H(z) = z^{-L} \tag{2.156}$$

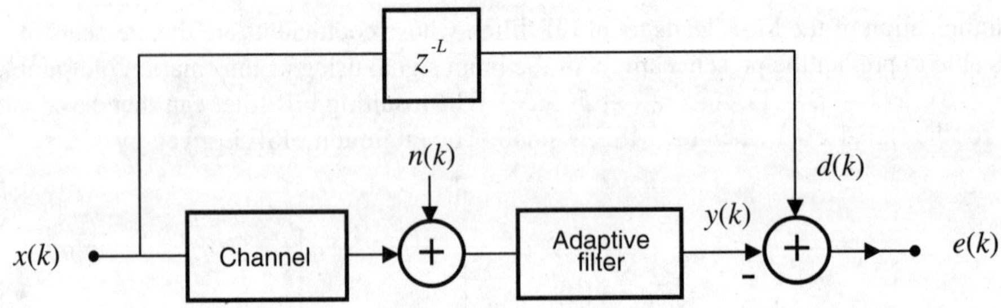

Figure 2.13 Channel equalization.

where $W(z)$ and $H(z)$ are the equalizer and unknown system transfer functions, respectively. Therefore, the ideal equalizer has the following transfer function

$$W(z) = \frac{z^{-L}}{H(z)} \tag{2.157}$$

From the above equation, we can conclude that if $H(z)$ is an IIR transfer function with nontrivial numerator and denominator polynomials, $W(z)$ will also be IIR. If $H(z)$ is an all-pole model, $W(z)$ is FIR. If $H(z)$ is an all-zero model, $W(z)$ is an all-pole transfer function.

By applying the inverse $\mathcal{Z}$-transform to equation (2.156), we can conclude that the optimal equalizer impulse response convolved with the channel impulse response produces as a result an impulse. This means that for zero additional error in the channel, the output signal $y(k)$ restores $x(k-L)$ and, therefore, one can conclude that a deconvolution process took place.

The delay in the reference signal plays an important role in the equalization process. Without the delay, the desired signal is $x(k)$, whereas the signal $y(k)$ will be mainly influenced by old samples of the input signal, since the unknown system is usually causal. As a consequence, the equalizer should also perform the task of predicting $x(k)$ simultaneously with the main task of equalizing the channel. The introduction of a delay alleviates the prediction task, leaving the equalizer free to invert the channel response. A rule of thumb for choosing the delay was proposed and analyzed in [28], where it was conjectured that the best delay should be close to half the time span of the equalizer. In practice, the reader should try different delays.

In the case the unknown system is not of minimum phase, i.e., its transfer function has zeros outside the unit circle of the $\mathcal{Z}$ plane, the optimum equalizer is either stable and noncausal, or unstable and causal. Both solutions are unacceptable. The noncausal stable solution could be better approximated by a causal FIR filter when the delay is included in the desired signal. The delay forces a time shift in the ideal impulse response of the equalizer, allowing the time span, where most of the energy is concentrated, to be in the *causal* region.

If channel noise signal is present and is uncorrelated with the channel's input signal, the error signal and $y(k)$ will be accordingly noisier. However, it should be noticed that the adaptive equalizer, in the process of reducing the MSE, disturbs the optimal solution by trying to reduce the effects of $n(k)$. Therefore, in a noisy environment the equalizer transfer function is not exactly the inverse of $H(z)$.

In practice, the noblest use of the adaptive equalizer is to compensate for the distortion caused by the transmission channel in a communication system. The main distortions caused by the channels are high attenuation and intersymbol interference (ISI). The ISI is generated when different frequency components of the transmitted signals arrive at different times at the receiver, a phenomenon caused by the nonlinear group delay of the channel [35]. For example, in a digital communication system, the time-dispersive channel extends a transmitted symbol beyond the time interval allotted to it, interfering in the past and future symbols. Under severe ISI, when short symbol space is used, the number of symbols causing ISI is large.

The channel impulse response is a time spread sequence described by $h(k)$ with the received signal being given by

$$re(k+J) = x(k)h(J) + \sum_{l=-\infty,\, l\neq k}^{k+J} x(l)h(k+J-l) + n(k+J) \tag{2.158}$$

where J denotes the channel time delay (including the sampler phase). The first term of the above equation corresponds to the desired information, the second term is the interference of the symbols sent before and after $x(k)$. The third term accounts for channel noise. Obviously only the neighboring symbols have significant influence in the second term of the above equation. The elements of the second term involving $x(l)$, for $l > k$, are called pre-cursor ISI since they are caused by components of the data signal that reach the receiver before their cursor. On the other hand, the elements involving $x(l)$, for $l < k$, are called post-cursor ISI.

In many situations, the ISI is reduced by employing an equalizer consisting of an adaptive FIR filter of appropriate length. The adaptive equalizer attempts to cancel the ISI in the presence of noise. In digital communication, a decision device is placed after the equalizer in order to identify the symbol at a given instant. The equalizer coefficients are updated in two distinct circumstances by employing different reference signals. During the equalizer training period, a previously chosen training signal is transmitted through the channel and a properly delayed version of this signal, that is prestored in the receiver end, is used as reference signal. The training signal is usually a pseudo-noise sequence long enough to allow the equalizer to compensate for the channel distortions. After convergence, the error between the adaptive-filter output and the decision device output is utilized to update the coefficients. The resulting scheme is the decision-directed adaptive equalizer. It should be mentioned that in some applications no training period is available. Usually, in this case, the decision-directed error is used all the time.

A more general equalizer scheme is the decision-feedback equalizer (DFE) illustrated in Fig. 2.14. The DFE is widely used in situations where the channel distortion is severe [35], [46]. The basic idea is to feed back, via a second FIR filter, the decisions made by the decision device that is applied to the equalized signal. The second FIR filter is preceded by a delay, otherwise there is a delay-free loop around the decision device. Assuming the decisions were correct, we are actually feeding back the symbols $x(l)$, for $l < k$, of equation (2.158). The DFE is able to cancel the post-cursor ISI for a number of past symbols (depending on the order of the FIR feedback filter), leaving more freedom for the feedforward section to take care of the remaining terms of the ISI. Some known characteristics of the DFE are [35]:

- The signals that are fed back are symbols, being noise free and allowing computational savings.
- The noise enhancement is reduced, if compared with the feedforward-only equalizer.
- Short time recovery when incorrect decisions are made.
- Reduced sensitivity to sampling phase.

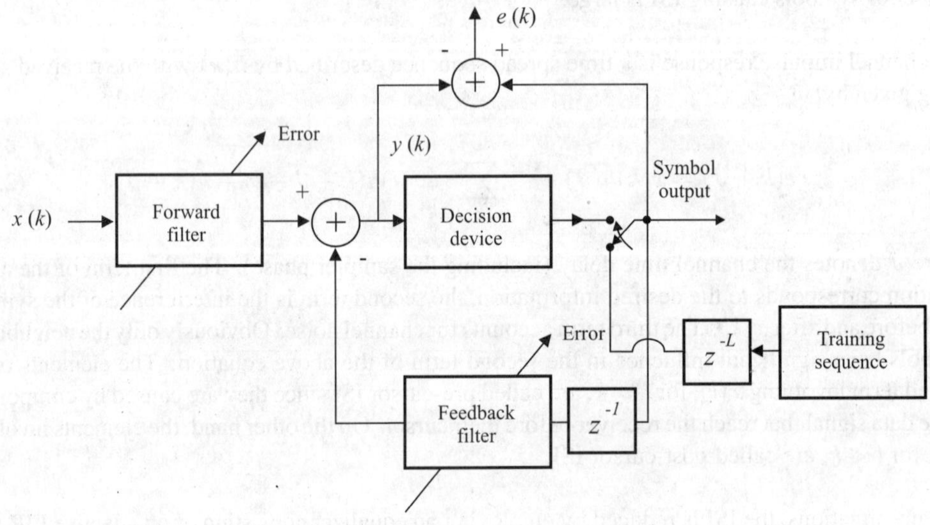

Figure 2.14 Decision-feedback equalizer.

The DFE operation starts with a training period where a known sequence is transmitted through the channel, and the same sequence is used at the receiver as the desired signal. The delay introduced in the training signal is meant to compensate for the delay the transmitted signal faces when passing through the channel. During the training period the error signal, which consists of the difference between the delayed training signal and signal $y(k)$, is minimized by adapting the coefficients of the forward and feedback filters. After this period, there is no training signal and the desired signal will consist of the decision device output signal. Assuming the decisions are correct, this *blind* way of performing the adaptation is the best solution to keep track of small changes in the channel behavior.

Example 2.4

In this example we will verify the effectiveness of the Wiener solution in environments related to the applications of noise cancellation, prediction, equalization, and identification.

(a) In a noise cancellation environment a sinusoid is corrupted by noise as follows

$$d(k) = \cos \omega_0 k + n_1(k)$$

with

$$n_1(k) = -an_1(k-1) + n(k)$$

$|a| < 1$ and $n(k)$ is a zero-mean white noise with variance $\sigma_n^2 = 1$. The input signal of the Wiener filter is described by

$$n_2(k) = -bn_2(k-1) + n(k)$$

where $|b| < 1$.

(b) In a prediction case the input signal is modeled as

$$x(k) = -ax(k-1) + n(k)$$

with $n(k)$ being a white noise with unit variance and $|a| < 1$.

(c) In an equalization problem a zero-mean white noise signal $s(k)$ with variance c is transmitted through a channel with an AR model described by

$$\hat{x}(k) = -a\hat{x}(k-1) + s(k)$$

with $|a| < 1$ and the received signal given by

$$x(k) = \hat{x}(k) + n(k)$$

whereas $n(k)$ is a zero-mean white noise with variance d and uncorrelated with $s(k)$.

(d) In a system identification problem a zero-mean white noise signal $x(k)$ with variance c is employed as the input signal to identify an AR system whose model is described by

$$v(k) = -av(k-1) + x(k)$$

where $|a| < 1$ and the desired signal is given by

$$d(k) = v(k) + n(k)$$

Repeat the problem if the system to be identified is an MA whose model is described by

$$v(k) = -ax(k-1) + x(k)$$

For all these cases describe the Wiener solution with two coefficients and comment on the results.

Solution:

Some results used in the examples are briefly reviewed. A 2×2 matrix inversion is performed as

$$\mathbf{R}^{-1} = \frac{1}{r_{11}r_{22} - r_{12}r_{21}} \begin{bmatrix} r_{22} & -r_{12} \\ -r_{21} & r_{11} \end{bmatrix}$$

where r_{ij} is the element of row i and column j of the matrix $\mathbf{R}$. For two first-order AR modeled signals $x(k)$ and $v(k)$, whose poles are respectively placed at $-a$ and $-b$ with the same white noise input with unit variance, their cross-correlations are given by[9]

$$E[x(k)v(k-l)] = \frac{(-a)^l}{1-ab}$$

for $l > 0$, and

$$E[x(k)v(k-l)] = \frac{(-b)^{-l}}{1-ab}$$

for $l < 0$, are frequently required in the following solutions.

(a)

The input signal in this case is given by $n_2(k)$, whereas the desired signal is given by $d(k)$. The elements of the correlation matrix are computed as

$$E[n_2(k)n_2(k-l)] = \frac{(-b)^{|l|}}{1-b^2}$$

The expression for the cross-correlation vector is given by

$$\begin{aligned}
\mathbf{p} &= \begin{bmatrix} E[(\cos\omega_0 k + n_1(k))n_2(k)] \\ E[(\cos\omega_0 k + n_1(k))n_2(k-1)] \end{bmatrix} \\
&= \begin{bmatrix} E[n_1(k)n_2(k)] \\ E[n_1(k)n_2(k-1)] \end{bmatrix} \\
&= \begin{bmatrix} \frac{1}{1-ab}\sigma_n^2 \\ -\frac{a}{1-ab}\sigma_n^2 \end{bmatrix} = \begin{bmatrix} \frac{1}{1-ab} \\ -\frac{a}{1-ab} \end{bmatrix}
\end{aligned}$$

where in the last expression we substituted $\sigma_n^2 = 1$.

The coefficients corresponding to the Wiener solution are given by

$$\mathbf{w}_o = \mathbf{R}^{-1}\mathbf{p} = \begin{bmatrix} 1 & b \\ b & 1 \end{bmatrix} \begin{bmatrix} \frac{1}{1-ab} \\ -\frac{a}{1-ab} \end{bmatrix} = \begin{bmatrix} 1 \\ \frac{b-a}{1-ab} \end{bmatrix}$$

The special case where $a = 0$ provides a quite illustrative solution. In this case

$$\mathbf{w}_o = \begin{bmatrix} 1 \\ b \end{bmatrix}$$

such that the error signal is given by

$$\begin{aligned}
e(k) = d(k) - y(k) &= \cos\omega_0 k + n(k) - \mathbf{w}_o^T \begin{bmatrix} n_2(k) \\ n_2(k-1) \end{bmatrix} \\
&= \cos\omega_0 k + n(k) - n_2(k) - bn_2(k-1) \\
&= \cos\omega_0 k + n(k) + bn_2(k-1) - n(k) - bn_2(k-1) = \cos\omega_0 k
\end{aligned}$$

[9]Assuming $x(k)$ and $v(k)$ are jointly WSS.

As can be observed the cosine signal is fully recovered since the Wiener filter was able to restore $n(k)$ and remove it from the desired signal.

(b)

In the prediction case the input signal is $x(k)$ and the desired signal is $x(k+L)$. Since

$$E[x(k)x(k-L)] = \frac{(-a)^{|L|}}{1-a^2}$$

the input signal correlation matrix is

$$\mathbf{R} = \begin{bmatrix} E[x^2(k)] & E[x(k)x(k-1)] \\ E[x(k)x(k-1)] & E[x^2(k-1)] \end{bmatrix}$$
$$= \begin{bmatrix} \frac{1}{1-a^2} & -\frac{a}{1-a^2} \\ -\frac{a}{1-a^2} & \frac{1}{1-a^2} \end{bmatrix}$$

Vector **p** is described by

$$\mathbf{p} = \begin{bmatrix} E[x(k+L)x(k)] \\ E[x(k+L)x(k-1)] \end{bmatrix} = \begin{bmatrix} \frac{(-a)^{|L|}}{1-a^2} \\ \frac{(-a)^{|L+1|}}{1-a^2} \end{bmatrix}$$

The expression for the optimal coefficient vector is easily derived.

$$\mathbf{w}_o = \mathbf{R}^{-1}\mathbf{p}$$
$$= (1-a^2)\begin{bmatrix} \frac{1}{1-a^2} & \frac{a}{1-a^2} \\ \frac{a}{1-a^2} & \frac{1}{1-a^2} \end{bmatrix} \begin{bmatrix} \frac{(-a)^{L}}{1-a^2} \\ \frac{(-a)^{L+1}}{1-a^2} \end{bmatrix}$$
$$= \begin{bmatrix} (-a)^L \\ 0 \end{bmatrix}$$

where in the above equation the value of L is considered positive. The predictor result tells us that an estimate $\hat{x}(k+L)$ of $x(k+L)$ can be obtained as

$$\hat{x}(k+L) = (-a)^L x(k)$$

According to our model for the signal $x(k)$, the actual value of $x(k+L)$ is

$$x(k+L) = (-a)^L x(k) + \sum_{i=0}^{L-1}(-a)^i n(k-i)$$

The results show that if $x(k)$ is an observed data at a given instant of time, the best estimate of $x(k+L)$ in terms of $x(k)$ is to average out the noise as follows

$$\hat{x}(k+L) = (-a)^L x(k) + E[\sum_{i=0}^{L-1}(-a)^i n(k-i)] = (-a)^L x(k)$$

since $E[n(k-i)] = 0$.

(c)

In this equalization problem, matrix $\mathbf{R}$ is given by

$$\mathbf{R} = \begin{bmatrix} E[x^2(k)] & E[x(k)x(k-1)] \\ E[x(k)x(k-1)] & E[x^2(k-1)] \end{bmatrix} = \begin{bmatrix} \frac{1}{1-a^2}c + d & -\frac{a}{1-a^2}c \\ -\frac{a}{1-a^2}c & \frac{1}{1-a^2}c + d \end{bmatrix}$$

By utilizing as desired signal $s(k-L)$ and recalling that it is a white noise and uncorrelated with the other signals involved in the experiment, the cross-correlation vector between the input and desired signals has the following expression

$$\mathbf{p} = \begin{bmatrix} E[x(k)s(k-L)] \\ E[x(k-1)s(k-L)] \end{bmatrix} = \begin{bmatrix} (-1)^L a^L c \\ (-1)^{L-1} a^{L-1} c \end{bmatrix}$$

The coefficients of the underlying Wiener solution are given by

$$\begin{aligned}
\mathbf{w}_o = \mathbf{R}^{-1}\mathbf{p} &= \frac{1}{\frac{c^2}{1-a^2} + 2\frac{dc}{1-a^2} + d^2} \begin{bmatrix} \frac{1}{1-a^2}c + d & \frac{a}{1-a^2}c \\ \frac{a}{1-a^2}c & \frac{1}{1-a^2}c + d \end{bmatrix} \begin{bmatrix} (-1)^L a^L c \\ (-1)^{L-1} a^{L-1} c \end{bmatrix} \\
&= \frac{(-1)^L a^L c}{\frac{c^2}{1-a^2} + 2\frac{cd}{1-a^2} + d^2} \begin{bmatrix} \frac{c}{1-a^2} + d - \frac{c}{1-a^2} \\ \frac{ac}{1-a^2} - a^{-1}d - \frac{a^{-1}c}{1-a^2} \end{bmatrix} \\
&= \frac{(-1)^L a^L c}{\frac{c^2}{1-a^2} + 2\frac{cd}{1-a^2} + d^2} \begin{bmatrix} d \\ -a^{-1}d - a^{-1}c \end{bmatrix}
\end{aligned}$$

If there is no additional noise, i.e. $d = 0$, the above result becomes

$$\mathbf{w}_o = \begin{bmatrix} 0 \\ (-1)^{L-1} a^{L-1}(1-a^2) \end{bmatrix}$$

That is, the Wiener solution is just correcting the gain of the previously received component of the input signal, namely $x(k-1)$, while not using its most recent component $x(k)$. This happens because the desired signal $s(k-L)$ at instant k has a defined correlation with any previously received symbol. On the other hand, if the signal $s(k)$ is a colored noise the Wiener filter would have a nonzero first coefficient in a noiseless environment. In case there is environmental noise, the solution tries to find a perfect balance between the desired signal modeling and the noise amplification.

(d)

In the system identification example the input signal correlation matrix is given by

$$\mathbf{R} = \begin{bmatrix} c & 0 \\ 0 & c \end{bmatrix}$$

With the desired signal $d(k)$, the cross-correlation vector is described as

$$\mathbf{p} = \begin{bmatrix} E[x(k)d(k)] \\ E[x(k-1)d(k)] \end{bmatrix} = \begin{bmatrix} c \\ -ca \end{bmatrix}$$

The coefficients of the underlying Wiener solution are given by

$$\mathbf{w}_o = \mathbf{R}^{-1}\mathbf{p} = \begin{bmatrix} \frac{1}{c} & 0 \\ 0 & \frac{1}{c} \end{bmatrix} \begin{bmatrix} c \\ -ca \end{bmatrix} = \begin{bmatrix} 1 \\ -a \end{bmatrix}$$

Note that this solution represents the best way a first-order FIR model can approximate an IIR model, since

$$W_o(z) = 1 - az^{-1}$$

and

$$\frac{1}{1+az^{-1}} = 1 - az^{-1} + a^2z^{-2} + \ldots$$

On the other hand, if the unknown model is the described FIR model such as $v(k) = -ax(k-1) + x(k)$, the Wiener solution remains the same and corresponds exactly to the unknown system model.

In all these examples, the environmental signals are considered wide-sense stationary and their statistics assumed known. In a practical situation, not only the statistics might be unknown but the environments are usually nonstationary as well. In these situations, the adaptive filters come into play since their coefficients vary with time according to measured signals from the environment.

□

2.10.5 Digital Communication System

For illustration, a general digital communication scheme over a channel consisting of a subscriber line (telephone line, for example) is shown in Fig. 2.15. In either end, the input signal is first coded and conditioned by a transmit filter. The filter shapes the pulse and limits in band the signal that is actually transmitted. The signal then crosses the hybrid to travel through a dual duplex channel. The hybrid is an impedance bridge used to transfer the transmit signal into the channel with minimal leakage to the near-end receiver. The imperfections of the hybrid cause echo that should be properly canceled.

In the channel, the signal is corrupted by white noise and crosstalk (leakage of signals being transmitted by other subscribers). After crossing the channel and the far-end hybrid, the signal is filtered by the receive filter that attenuates high-frequency noise and also acts as an antialiasing filter. Subsequently, we have a joint DFE and echo canceller, where the forward filter and echo canceller outputs are subtracted. The result after subtracting the decision feedback output is applied to the decision device. After passing through the decision device, the symbol is decoded.

Other schemes for data transmission in subscriber line exist [41]. The one shown here is for illustration purposes, having as special feature the joint equalizer and echo canceller strategy. The digital subscriber line (DSL) structure shown here has been used in integrated services digital network (ISDN) basic access, that allows a data rate of 144 Kbits/s [41]. Also, a similar scheme is employed

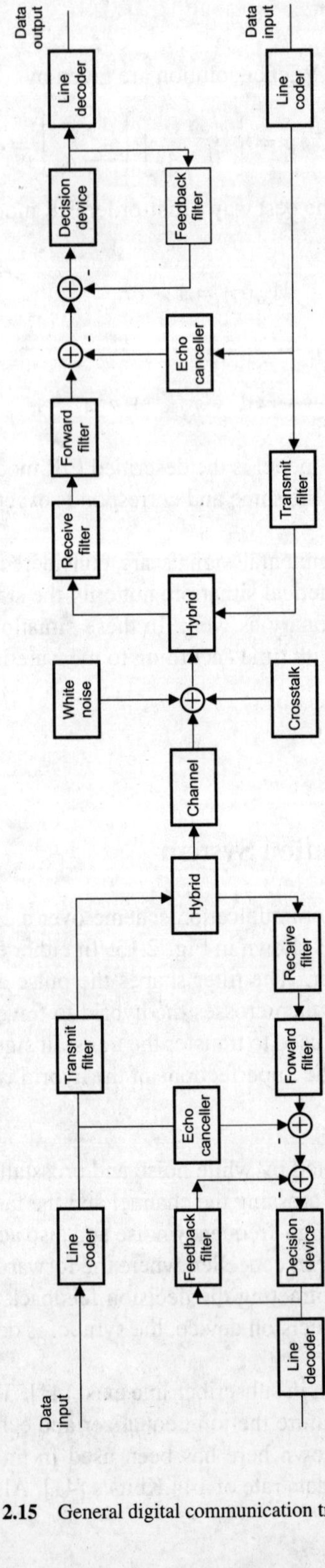

Figure 2.15 General digital communication transceiver.

in the high bit rate digital subscriber line (HDSL) [40], [47] that operates over short and conditioned loops [48], [49]. The latter system belongs to a broad class of digital subscriber line collectively known as XDSL.

2.11 CONCLUDING REMARKS

In this chapter, we described some of the concepts underlying the adaptive filtering theory. The material presented here forms the basis to understand the behavior of most adaptive-filtering algorithms in a practical implementation. The basic concept of the MSE surface searching algorithms was briefly reviewed, serving as a starting point for the development of a number of practical adaptive-filtering algorithms to be presented in the following chapters. We illustrated through several examples the expected Wiener solutions in a number of distinct situations. In addition, we presented the basic concepts of linearly-constrained Wiener filter required in array signal processing. The theory and practice of adaptive signal processing is also the main subject of some excellent books such as [27]-[34].

2.12 REFERENCES

1. D. G. Luenberger, *Introduction to Linear and Nonlinear Programming*, Addison Wesley, Reading, MA, 2nd edition, 1984.

2. R. Fletcher, *Practical Methods of Optimization*, John Wiley & Sons, New York, NY, 2nd edition, 1990.

3. A. Antoniou and W.-S. Lu, *Practical Optimization: Algorithms and Engineering Applications*, Springer, New York, NY, 2007.

4. B. Widrow and M. E. Hoff, "Adaptive switching circuits," *WESCOM Conv. Rec.*, pt. 4, pp. 96-140, 1960.

5. B. Widrow, J. M. McCool, M. G. Larimore, and C. R. Johnson, Jr., "Stationary and nonstationary learning characteristics of the LMS adaptive filters," *Proceedings of the IEEE*, vol. 64, pp. 1151-1162, Aug. 1976.

6. A. Papoulis, *Signal Analysis*, McGraw Hill, New York, NY, 1977.

7. A. V. Oppenheim, A. S. Willsky, and S. H. Nawab, *Signals and Systems*, Prentice Hall, Englewood Cliffs, NJ, 2nd edition, 1997.

8. P. S. R. Diniz, E. A. B. da Silva, and S. L. Netto, *Digital Signal Processing: System Analysis and Design*, Cambridge University Press, Cambridge, UK, 2002.

9. A. Antoniou, *Digital Signal Processing: Signals, Systems, and Filters*, McGraw Hill, New York, NY, 2005.

10. L. B. Jackson, *Digital Filters and Signal Processing*, Kluwer Academic Publishers, Norwell, MA, 3rd edition, 1996.

11. R. A. Roberts and C. T. Mullis, *Digital Signal Processing*, Addison-Wesley, Reading, MA, 1987.

12. J. G. Proakis and D. G. Manolakis, *Digital Signal Processing*, Prentice Hall, Englewood Cliffs, NJ, 4th edition, 2007.

13. T. Bose, *Digital Signal and Image Processing*, John Wiley & Sons, New York, NY, 2004.

14. A. Papoulis, *Probability, Random Variables, and Stochastic Processes*, McGraw Hill, New York, NY, 3rd edition, 1991.

15. P. Z. Peebles, Jr., *Probability, Random Variables, and Random Signal Principles*, McGraw Hill, New York, NY, 3rd edition, 1993.

16. W. A. Gardner, *Introduction to Random Processes*, McGraw Hill, New York, NY, Second edition, 1990.

17. C. R. Johnson, Jr., *Lectures on Adaptive Parameter Estimation*, Prentice Hall, Englewood Cliffs, NJ, 1988.

18. T. Söderström and P. Stoica, *System Identification*, Prentice Hall International, Hemel Hempstead, Hertfordshire, 1989.

19. G. Strang, *Linear Algebra and Its Applications*, Academic Press, New York, NY, 2nd Edition, 1980.

20. L. J. Griffiths and C. W. Jim, "An alternative approach to linearly constrained adaptive beamforming," *IEEE Trans. on Antennas and Propagation*, vol. AP-30, pp. 27-34, Jan. 1982.

21. M. L. R. de Campos, S. Werner, and J. A. Apolinário, Jr., "Constrained adaptation algorithms employing Householder transformation," *IEEE Trans. on Signal Processing*, vol.50, pp. 2187-2195, Sept. 2002.

22. D. H. Johnson and D. E. Dudgeon, *Array Signal Processing*, Prentice Hall, Englewood Cliffs, NJ, 1993.

23. H. L. Van trees, *Optimum Array Processing: Part IV of Detection, Estimation and Modulation Theory*, John Wiley & Sons, New York, NY, 2002.

24. A. Papoulis, "Predictable processes and Wold's decomposition: A review," *IEEE Trans. on Acoust., Speech, and Signal Processing*, vol. ASSP-33, pp. 933-938, Aug. 1985.

25. S. M. Kay, *Fundamentals of Statistical Signal Processing: Estimation Theory*, Prentice Hall, Englewood Cliffs, NJ, 1993.

26. S. L. Marple, Jr., *Digital Spectral Analysis*, Prentice Hall, Englewood Cliffs, NJ, 1987.

27. M. L. Honig and D. G. Messerschmitt, *Adaptive Filters: Structures, Algorithms, and Applications*, Kluwer Academic Publishers, Boston, MA, 1984.

28. B. Widrow and S. D. Stearns, *Adaptive Signal Processing*, Prentice Hall, Englewood Cliffs, NJ, 1985.

29. S. T. Alexander, *Adaptive Signal Processing*, Springer Verlag, New York, NY, 1986.

30. J. R. Treichler, C. R. Johnson, Jr., and M. G. Larimore, *Theory and Design of Adaptive Filters*, John Wiley & Sons, New York, NY, 1987.

31. M. Bellanger, *Adaptive Digital Filters and Signal Analysis*, Marcel Dekker, Inc., New York, NY, 2nd Edition, 2001.

32. P. Strobach, *Linear Prediction Theory*, Springer Verlag, New York, NY, 1990.

33. S. Haykin, *Adaptive Filter Theory*, Prentice Hall, Englewood Cliffs, NJ, 4th edition, 2002.

34. A. H. Sayed, *Fundamentals of Adaptive Filtering*, John Wiley & Sons, Hoboken, NJ, 2003.

35. S. U. Qureshi, "Adaptive Equalization," *Proceedings of the IEEE*, vol. 73, pp. 1349-1387, Sept. 1985.

36. J. G. Proakis, *Digital Communication*, McGraw Hill, New York, NY, 4th edition, 2001.

37. L. C. Wood and S. Treitel, "Seismic signal processing," *Proceedings of the IEEE*, vol. 63, pp. 649-661, Dec. 1975.

38. D. G. Messerschmitt, "Echo cancellation in speech and data transmission," *IEEE Journal on Selected Areas in Communications*, vol. SAC-2, pp. 283-296, March 1984.

39. M. L. Honig, "Echo cancellation of voiceband data signals using recursive least squares and stochastic gradient algorithms," *IEEE Trans. on Communications*, vol. COM-33, pp. 65-73, Jan. 1985.

40. S. Subramanian, D. J. Shpak, P. S. R. Diniz, and A. Antoniou, "The performance of adaptive filtering algorithms in a simulated HDSL environment," *Proc. IEEE Canadian Conf. Electrical and Computer Engineering*, Toronto, Canada, pp. TA 2.19.1-TA 2.19.5, Sept. 1992.

41. D. W. Lin, "Minimum mean-squared error echo cancellation and equalization for digital subscriber line transmission: Part I - theory and computation," *IEEE Trans. on Communications*, vol. 38, pp. 31-38, Jan. 1990.

42. D. W. Lin, "Minimum mean-squared error echo cancellation and equalization for digital subscriber line transmission: Part II - a simulation study," *IEEE Trans. on Communications*, vol. 38, pp. 39-45, Jan. 1990.

43. L. R. Rabiner and R. W. Schafer, *Digital Processing of Speech Signals*, Prentice Hall, Englewood Cliffs, NJ, 1978.

44. B. D. Van Veen and K. M. Buckley, "Beamforming: a versatile approach to spatial filtering," *IEEE Acoust., Speech, Signal Processing Magazine*, vol. 37, pp. 4-24, April 1988.

45. B. Widrow, J. R. Grover, Jr., J. M. McCool, J. Kaunitz, C. S. Williams, R. H. Hearns, J. R. Zeidler, E. Dong, Jr., and R. C. Goodlin, "Adaptive noise cancelling: Principles and applications," *Proceedings of the IEEE*, vol. 63, pp. 1692-1716, Dec. 1975.

46. M. Abdulrahman and D. D. Falconer, "Cyclostationary crosstalk suppression by decision feedback equalization on digital subscriber line," *IEEE Journal on Selected Areas in Communications*, vol. 10, pp. 640-649, April 1992.

47. H. Samueli, B. Daneshrad, R. B. Joshi, B. C. Wong, and H. T. Nicholas, III, "A 64-tap CMOS echo canceller/decision feedback equalizer for 2B1Q HDSL transceiver," *IEEE Journal on Selected Areas in Communications*, vol. 9, pp. 839-847, Aug. 1991.

48. J.-J. Werner, "The HDSL environment," *IEEE Journal on Selected Areas in Communications*, vol. 9, pp. 785-800, Aug. 1991.

49. J. W. Leichleider, "High bit rate digital subscriber lines: A review of HDSL progress," *IEEE Journal on Selected Areas in Communications*, vol. 9, pp. 769-784, Aug. 1991.

2.13 PROBLEMS

1. Suppose the input signal vector is composed by a delay line with a single input signal, compute the correlation matrix for the following input signals:

(a)

$$x(k) = \sin(\frac{\pi}{6}k) + \cos(\frac{\pi}{4}k) + n(k)$$

(b)

$$x(k) = an_1(k)\cos(\omega_0 k) + n_2(k)$$

(c)

$$x(k) = an_1(k)\sin(\omega_0 k + n_2(k))$$

(d)

$$x(k) = -a_1x(k-1) - a_2x(k-2) + n(k)$$

(e)

$$x(k) = \sum_{i=0}^{4} 0.25n(k-i)$$

(f)

$$x(k) = an(k)e^{\jmath\omega_0 k}$$

In all cases, $n(k)$, $n_1(k)$, and $n_2(k)$ are white noise processes, with zero mean and with variances σ_n^2, $\sigma_{n_1}^2$, and $\sigma_{n_2}^2$, respectively. These random signals are considered independent.

2. Consider two complex random processes represented by $x(k)$ and $y(k)$.
(a) Derive $\sigma_{xy}^2(k,l) = E[(x(k) - m_x(k))(y(l) - m_y(l))]$ as a function of $r_{xy}(k,l)$, $m_x(k)$ and $m_y(l)$.
(b) Repeat (a) if $x(k)$ and $y(k)$ are jointly WSS.
(c) Being $x(k)$ and $y(k)$ orthogonal, in which conditions are they not correlated?

3. For the correlation matrices given below, calculate their eigenvalues, eigenvectors, and conditioning numbers.
(a)

$$\mathbf{R} = \frac{1}{4}\begin{bmatrix} 4 & 3 & 2 & 1 \\ 3 & 4 & 3 & 2 \\ 2 & 3 & 4 & 3 \\ 1 & 2 & 3 & 4 \end{bmatrix}$$

(b)

$$\mathbf{R} = \begin{bmatrix} 1 & 0.95 & 0.9025 & 0.857375 \\ 0.95 & 1 & 0.95 & 0.9025 \\ 0.9025 & 0.95 & 1 & 0.95 \\ 0.857375 & 0.9025 & 0.95 & 1 \end{bmatrix}$$

(c)

$$\mathbf{R} = 50\sigma_n^2\begin{bmatrix} 1 & 0.9899 & 0.98 & 0.970 \\ 0.9899 & 1 & 0.9899 & 0.98 \\ 0.98 & 0.9899 & 1 & 0.9899 \\ 0.970 & 0.98 & 0.9899 & 1 \end{bmatrix}$$

(d)

$$\mathbf{R} = \begin{bmatrix} 1 & 0.5 & 0.25 & 0.125 \\ 0.5 & 1 & 0.5 & 0.25 \\ 0.25 & 0.5 & 1 & 0.5 \\ 0.125 & 0.25 & 0.5 & 1 \end{bmatrix}$$

4. For the correlation matrix given below, calculate its eigenvalues and eigenvectors, and form the matrix **Q**.

$$\mathbf{R} = \frac{1}{4}\begin{bmatrix} a_1 & a_2 \\ a_2 & a_1 \end{bmatrix}$$

5. The input signal of a second-order adaptive filter is described by

$$x(k) = \alpha_1 x_1(k) + \alpha_2 x_2(k)$$

where $x_1(k)$ and $x_2(k)$ are first-order AR processes and uncorrelated between themselves having both unit variance. These signals are generated by applying distinct white noises to first-order filters whose poles are placed at a and $-b$, respectively.

(a) Calculate the autocorrelation matrix of the input signal.

(b) If the desired signal consists of $\gamma x_2(k)$, calculate the Wiener solution.

6. The input signal of a first-order adaptive filter is described by

$$x(k) = \sqrt{2}x_1(k) + x_2(k) + 2x_3(k)$$

where $x_1(k)$ and $x_2(k)$ are first-order AR processes and uncorrelated between themselves having both unit variance. These signals are generated by applying distinct white noises to first-order filters whose poles are placed at -0.5 and $\frac{\sqrt{2}}{2}$, respectively. The signal $x_3(k)$ is a white noise with unit variance and uncorrelated with $x_1(k)$ and $x_2(k)$.

(a) Calculate the autocorrelation matrix of the input signal.

(b) If the desired signal consists of $\frac{1}{2}x_3(k)$, calculate the Wiener solution.

7. Repeat the previous problem if the signal $x_3(k)$ is exactly the white noise that generated $x_2(k)$.

8. In a prediction case a sinusoid is corrupted by noise as follows

$$x(k) = \cos \omega_0 k + n_1(k)$$

with

$$n_1(k) = -an_1(k-1) + n(k)$$

where $|a| < 1$. For this case describe the Wiener solution with two coefficients and comment on the results.

9. Generate the ARMA processes $x(k)$ described below. Calculate the variance of the output signal and the autocorrelation for lags 1 and 2. In all cases, $n(k)$ is zero-mean Gaussian white noise with variance 0.1.

(a)

$$x(k) = 1.9368x(k-1) - 0.9519x(k-2) + n(k) - 1.8894n(k-1) + n(k-2)$$

(b)

$$x(k) = -1.9368x(k-1) - 0.9519x(k-2) + n(k) + 1.8894n(k-1) + n(k-2)$$

Hint: For white noise generation consult for example [14], [15].

10. Generate the AR processes $x(k)$ described below. Calculate the variance of the output signal and the autocorrelation for lags 1 and 2. In all cases, $n(k)$ is zero-mean Gaussian white noise with variance 0.05.

(a)

$$x(k) = -0.8987x(k-1) - 0.9018x(k-2) + n(k)$$

(b)

$$x(k) = 0.057x(k-1) + 0.889x(k-2) + n(k)$$

11. Generate the MA processes $x(k)$ described below. Calculate the variance of the output signal and the autocovariance matrix. In all cases, $n(k)$ is zero-mean Gaussian white noise with variance 1.

(a)

$$x(k) = 0.0935n(k) + 0.3027n(k-1) + 0.4n(k-2) + 0.3027n(k-4) + 0.0935n(k-5)$$

(b)

$$x(k) = n(k) - n(k-1) + n(k-2) - n(k-4) + n(k-5)$$

(c)

$$x(k) = n(k) + 2n(k-1) + 3n(k-2) + 2n(k-4) + n(k-5)$$

12. Show that a process generated by adding two AR processes is in general an ARMA process.

13. Determine if the following processes are mean ergodic:

(a)

$$x(k) = an_1(k)\cos(\omega_0 k) + n_2(k)$$

(b)

$$x(k) = an_1(k)\sin(\omega_0 k + n_2(k))$$

(c)

$$x(k) = an(k)e^{2\jmath\omega_0 k}$$

In all cases, $n(k)$, $n_1(k)$, and $n_2(k)$ are white noise processes, with zero mean and with variances σ_n^2, $\sigma_{n_1}^2$, and $\sigma_{n_2}^2$, respectively. These random signals are considered independent.

14. Show that the minimum (maximum) value of equation (2.69) occurs when $w_i = 0$ for $i \neq j$ and λ_j is the smallest (largest) eigenvalue, respectively.

15. Suppose the matrix **R** and the vector **p** are known for a given experimental environment. Compute the Wiener solution for the following cases:

(a)

$$\mathbf{R} = \frac{1}{4}\begin{bmatrix} 4 & 3 & 2 & 1 \\ 3 & 4 & 3 & 2 \\ 2 & 3 & 4 & 3 \\ 1 & 2 & 3 & 4 \end{bmatrix}$$

$$\mathbf{p} = \begin{bmatrix} \frac{1}{2} & \frac{3}{8} & \frac{2}{8} & \frac{1}{8} \end{bmatrix}^T$$

(b)

$$\mathbf{R} = \begin{bmatrix} 1 & 0.8 & 0.64 & 0.512 \\ 0.8 & 1 & 0.8 & 0.64 \\ 0.64 & 0.8 & 1 & 0.8 \\ 0.512 & 0.64 & 0.8 & 1 \end{bmatrix}$$

$$\mathbf{p} = \frac{1}{4}[0.4096\ 0.512\ 0.64\ 0.8]^T$$

(c)

$$\mathbf{R} = \frac{1}{3}\begin{bmatrix} 3 & -2 & 1 \\ -2 & 3 & -2 \\ 1 & -2 & 3 \end{bmatrix}$$

$$\mathbf{p} = \begin{bmatrix} -2 & 1 & -\frac{1}{2} \end{bmatrix}^T$$

16. For the environments described in the previous problem, derive the updating formula for the steepest-descent method. Considering that the adaptive-filter coefficients are initially zero, calculate their values for the first ten iterations.

17. Repeat the previous problem using the Newton method.

18. Calculate the spectral decomposition for the matrices **R** of problem 15.

19. Calculate the minimum MSE for the examples of problem 15 considering that the variance of the reference signal is given by σ_d^2.

20. Derive equation (2.112).

21. Derive the constraint matrix $\mathbf{C}$ and the gain vector $\mathbf{f}$ that impose the condition of linear phase onto the linearly constrained Wiener filter.

22. Show that the optimal solutions of the LCMV filter and the GSC filter with minimum norm are equivalent and related according to $\mathbf{w}_{\text{LCMV}} = \mathbf{T}\mathbf{w}_{\text{GSC}}$, where $\mathbf{T} = [\mathbf{C}\ \mathbf{B}]$ is a full-rank transformation matrix with $\mathbf{C}^T\mathbf{B} = \mathbf{0}$ and

$$\mathbf{w}_{\text{LCMV}} = \mathbf{R}^{-1}\mathbf{C}(\mathbf{C}^T\mathbf{R}^{-1}\mathbf{C})^{-1}\mathbf{f}$$

and

$$\mathbf{w}_{\text{GSC}} = \begin{bmatrix} (\mathbf{C}^T\mathbf{C})^{-1}\mathbf{f} \\ -(\mathbf{B}^T\mathbf{R}\mathbf{B})^{-1}\mathbf{B}^T\mathbf{R}\mathbf{C}(\mathbf{C}^T\mathbf{C})^{-1}\mathbf{f} \end{bmatrix}$$

23. Calculate the time constants of the MSE and of the coefficients for the examples of problem 15 considering that the steepest-descent algorithm was employed.

24. For the examples of problem 15, describe the equations for the MSE surface.

25. Using the spectral decomposition of a Hermitian matrix show that

$$\mathbf{R}^{\frac{1}{N}} = \mathbf{Q}\mathbf{\Lambda}^{\frac{1}{N}}\mathbf{Q}^H = \sum_{i=0}^{N} \lambda_i^{\frac{1}{N}}\mathbf{q}_i\mathbf{q}_i^H$$

26. Derive the complex steepest-descent algorithm.

27. Derive the Newton algorithm for complex signals.

28. In a signal enhancement application, assume that $n_1(k) = n_2(k) * h(k)$, where $h(k)$ represents the impulse response of an unknown system. Also, assume that some small leakage of the signal $x(k)$, given by $h'(k) * x(k)$, is added to the adaptive-filter input. Analyze the consequences of this phenomenon.

29. In the equalizer application, calculate the optimal equalizer transfer function when the channel noise is present.

3

THE LEAST-MEAN-SQUARE (LMS) ALGORITHM

3.1 INTRODUCTION

The least-mean-square (LMS) is a search algorithm in which a simplification of the gradient vector computation is made possible by appropriately modifying the objective function [1]-[2]. The LMS algorithm, as well as others related to it, is widely used in various applications of adaptive filtering due to its computational simplicity [3]-[7]. The convergence characteristics of the LMS algorithm are examined in order to establish a range for the convergence factor that will guarantee stability. The convergence speed of the LMS is shown to be dependent on the eigenvalue spread of the input signal correlation matrix [2]-[6]. In this chapter, several properties of the LMS algorithm are discussed including the misadjustment in stationary and nonstationary environments [2]-[9] and tracking performance [10]-[12]. The analysis results are verified by a large number of simulation examples. Appendix B, section B.1, complements this chapter by analyzing the finite-wordlength effects in LMS algorithms.

The LMS algorithm is by far the most widely used algorithm in adaptive filtering for several reasons. The main features that attracted the use of the LMS algorithm are low computational complexity, proof of convergence in stationary environment, unbiased convergence in the mean to the Wiener solution, and stable behavior when implemented with finite-precision arithmetic. The convergence analysis of the LMS presented here utilizes the independence assumption.

3.2 THE LMS ALGORITHM

In Chapter 2 we derived the optimal solution for the parameters of the adaptive filter implemented through a linear combiner, which corresponds to the case of multiple input signals. This solution leads to the minimum mean-square error in estimating the reference signal $d(k)$. The optimal (Wiener) solution is given by

$$\mathbf{w}_o = \mathbf{R}^{-1}\mathbf{p} \tag{3.1}$$

where $\mathbf{R} = E[\mathbf{x}(k)\mathbf{x}^T(k)]$ and $\mathbf{p} = E[d(k)\mathbf{x}(k)]$, assuming that $d(k)$ and $\mathbf{x}(k)$ are jointly wide-sense stationary.

P.S.R. Diniz, *Adaptive Filtering*, DOI: 10.1007/978-0-387-68606-6_3,

If good estimates of matrix $\mathbf{R}$, denoted by $\hat{\mathbf{R}}(k)$, and of vector $\mathbf{p}$, denoted by $\hat{\mathbf{p}}(k)$, are available, a steepest-descent-based algorithm can be used to search the Wiener solution of equation (3.1) as follows:

$$\begin{aligned}\mathbf{w}(k+1) &= \mathbf{w}(k) - \mu\hat{\mathbf{g}}_{\mathbf{w}}(k) \\ &= \mathbf{w}(k) + 2\mu(\hat{\mathbf{p}}(k) - \hat{\mathbf{R}}(k)\mathbf{w}(k))\end{aligned} \tag{3.2}$$

for $k = 0, 1, 2, \ldots$, where $\hat{\mathbf{g}}_{\mathbf{w}}(k)$ represents an estimate of the gradient vector of the objective function with respect to the filter coefficients.

One possible solution is to estimate the gradient vector by employing instantaneous estimates for $\mathbf{R}$ and $\mathbf{p}$ as follows:

$$\begin{aligned}\hat{\mathbf{R}}(k) &= \mathbf{x}(k)\mathbf{x}^T(k) \\ \hat{\mathbf{p}}(k) &= d(k)\mathbf{x}(k)\end{aligned} \tag{3.3}$$

The resulting gradient estimate is given by

$$\begin{aligned}\hat{\mathbf{g}}_{\mathbf{w}}(k) &= -2d(k)\mathbf{x}(k) + 2\mathbf{x}(k)\mathbf{x}^T(k)\mathbf{w}(k) \\ &= 2\mathbf{x}(k)(-d(k) + \mathbf{x}^T(k)\mathbf{w}(k)) \\ &= -2e(k)\mathbf{x}(k)\end{aligned} \tag{3.4}$$

Note that if the objective function is replaced by the instantaneous square error $e^2(k)$, instead of the MSE, the above gradient estimate represents the true gradient vector since

$$\begin{aligned}\frac{\partial e^2(k)}{\partial \mathbf{w}} &= \left[2e(k)\frac{\partial e(k)}{\partial w_0(k)}\ 2e(k)\frac{\partial e(k)}{\partial w_1(k)}\ \ldots\ 2e(k)\frac{\partial e(k)}{\partial w_N(k)}\right]^T \\ &= -2e(k)\mathbf{x}(k) \\ &= \hat{\mathbf{g}}_{\mathbf{w}}(k)\end{aligned} \tag{3.5}$$

The resulting gradient-based algorithm is known[1] as the least-mean-square (LMS) algorithm, whose updating equation is

$$\mathbf{w}(k+1) = \mathbf{w}(k) + 2\mu e(k)\mathbf{x}(k) \tag{3.6}$$

where the convergence factor μ should be chosen in a range to guarantee convergence.

Fig. 3.1 depicts the realization of the LMS algorithm for a delay line input $\mathbf{x}(k)$. Typically, one iteration of the LMS requires $N+2$ multiplications for the filter coefficient updating and $N+1$ multiplications for the error generation. The detailed description of the LMS algorithm is shown in the table denoted as Algorithm 3.1.

It should be noted that the initialization is not necessarily performed as described in Algorithm 3.1, where the coefficients of the adaptive filter were initialized with zeros. For example, if a rough idea of the optimal coefficient value is known, these values could be used to form $\mathbf{w}(0)$ leading to a reduction in the number of iterations required to reach the neighborhood of $\mathbf{w}_o$.

[1] Because it minimizes the mean of the squared error.

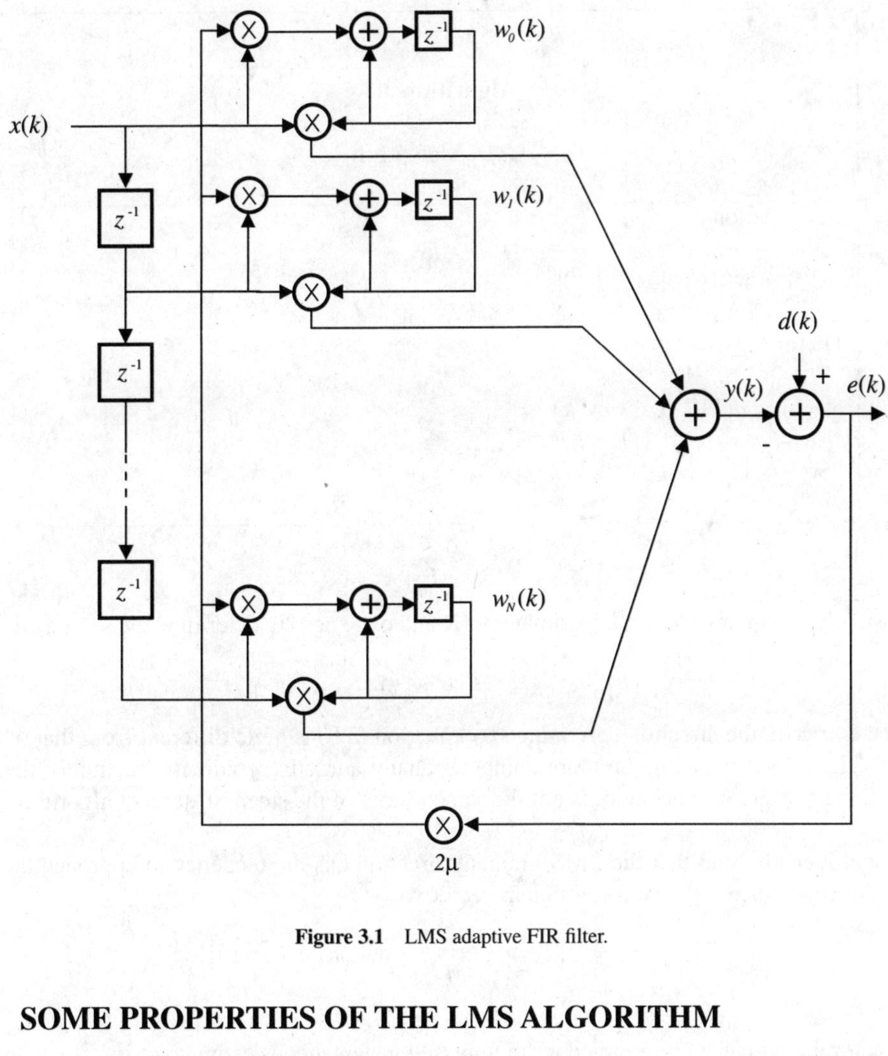

Figure 3.1 LMS adaptive FIR filter.

3.3 SOME PROPERTIES OF THE LMS ALGORITHM

In this section, the main properties related to the convergence behavior of the LMS algorithm in a stationary environment are described. The information contained here is essential to understand the influence of the convergence factor μ in various convergence aspects of the LMS algorithm.

3.3.1 Gradient Behavior

As shown in Chapter 2, see equation (2.91), the ideal gradient direction required to perform a search on the MSE surface for the optimum coefficient vector solution is

$$\begin{aligned} \mathbf{g_w}(k) &= 2\{E\left[\mathbf{x}(k)\mathbf{x}^T(k)\right]\mathbf{w}(k) - E\left[d(k)\mathbf{x}(k)\right]\} \\ &= 2[\mathbf{R}\mathbf{w}(k) - \mathbf{p}] \end{aligned} \tag{3.7}$$

Algorithm 3.1

LMS Algorithm

Initialization

$\mathbf{x}(0) = \mathbf{w}(0) = [0\,0\,\ldots\,0]^T$

Do for $k \geq 0$

$e(k) = d(k) - \mathbf{x}^T(k)\mathbf{w}(k)$

$\mathbf{w}(k+1) = \mathbf{w}(k) + 2\mu e(k)\mathbf{x}(k)$

In the LMS algorithm, instantaneous estimates of **R** and **p** are used to determine the search direction, i.e.,

$$\hat{\mathbf{g}}_{\mathbf{w}}(k) = 2\left[\mathbf{x}(k)\mathbf{x}^T(k)\mathbf{w}(k) - d(k)\mathbf{x}(k)\right] \tag{3.8}$$

As can be expected, the direction determined by equation (3.8) is quite different from that of equation (3.7). Therefore, by using the more computationally attractive gradient direction of the LMS algorithm, the convergence behavior is not the same as that of the steepest-descent algorithm.

On average, it can be said that the LMS gradient direction has the tendency to approach the ideal gradient direction since for a fixed coefficient vector **w**

$$\begin{aligned} E[\hat{\mathbf{g}}_{\mathbf{w}}(k)] &= 2\{E\left[\mathbf{x}(k)\mathbf{x}^T(k)\right]\mathbf{w} - E\left[d(k)\mathbf{x}(k)\right]\} \\ &= \mathbf{g}_{\mathbf{w}} \end{aligned} \tag{3.9}$$

hence, vector $\hat{\mathbf{g}}_{\mathbf{w}}(k)$ can be interpreted as an unbiased instantaneous estimate of $\mathbf{g}_{\mathbf{w}}$. In an ergodic environment, if, for a fixed **w** vector, $\hat{\mathbf{g}}_{\mathbf{w}}(k)$ is calculated for a large number of inputs and reference signals, the average direction tends to $\mathbf{g}_{\mathbf{w}}$, i.e.,

$$\lim_{M\to\infty} \frac{1}{M}\sum_{i=1}^{M} \hat{\mathbf{g}}_{\mathbf{w}}(k+i) \to \mathbf{g}_{\mathbf{w}} \tag{3.10}$$

3.3.2 Convergence Behavior of the Coefficient Vector

Assume that an unknown FIR filter with coefficient vector given by $\mathbf{w}_o$ is being identified by an adaptive FIR filter of the same order, employing the LMS algorithm. Measurement white noise $n(k)$ with zero mean and variance σ_n^2 is added to the output of the unknown system.

The error in the adaptive-filter coefficients as related to the ideal coefficient vector $\mathbf{w}_o$, in each iteration, is described by the $N+1$-length vector

$$\Delta\mathbf{w}(k) = \mathbf{w}(k) - \mathbf{w}_o \tag{3.11}$$

With this definition, the LMS algorithm can alternatively be described by

$$\begin{aligned}\Delta\mathbf{w}(k+1) &= \Delta\mathbf{w}(k) + 2\mu e(k)\mathbf{x}(k)\\ &= \Delta\mathbf{w}(k) + 2\mu\mathbf{x}(k)\left[\mathbf{x}^T(k)\mathbf{w}_o + n(k) - \mathbf{x}^T(k)\mathbf{w}(k)\right]\\ &= \Delta\mathbf{w}(k) + 2\mu\mathbf{x}(k)\left[e_o(k) - \mathbf{x}^T(k)\Delta\mathbf{w}(k)\right]\\ &= \left[\mathbf{I} - 2\mu\mathbf{x}(k)\mathbf{x}^T(k)\right]\Delta\mathbf{w}(k) + 2\mu e_o(k)\mathbf{x}(k)\end{aligned} \tag{3.12}$$

where $e_o(k)$ is the optimum output error given by

$$\begin{aligned}e_o(k) &= d(k) - \mathbf{w}_o^T\mathbf{x}(k)\\ &= \mathbf{w}_o^T\mathbf{x}(k) + n(k) - \mathbf{w}_o^T\mathbf{x}(k)\\ &= n(k)\end{aligned} \tag{3.13}$$

The expected error in the coefficient vector is then given by

$$E[\Delta\mathbf{w}(k+1)] = E\{[\mathbf{I} - 2\mu\mathbf{x}(k)\mathbf{x}^T(k)]\Delta\mathbf{w}(k)\} + 2\mu E[e_o(k)\mathbf{x}(k)] \tag{3.14}$$

If it is assumed that the elements of $\mathbf{x}(k)$ are statistically independent of the elements of $\Delta\mathbf{w}(k)$ and $e_o(k)$, equation (3.14) can be simplified as follows:

$$\begin{aligned}E[\Delta\mathbf{w}(k+1)] &= \{\mathbf{I} - 2\mu E[\mathbf{x}(k)\mathbf{x}^T(k)]\}E[\Delta\mathbf{w}(k)]\\ &= (\mathbf{I} - 2\mu\mathbf{R})E[\Delta\mathbf{w}(k)]\end{aligned} \tag{3.15}$$

The first assumption is justified if we assume that the deviation in the parameters is dependent on previous input signal vectors only, whereas in the second assumption we also considered that the error signal at the optimal solution is orthogonal to the elements of the input signal vector. The above expression leads to

$$E[\Delta\mathbf{w}(k+1)] = (\mathbf{I} - 2\mu\mathbf{R})^{k+1}E[\Delta\mathbf{w}(0)] \tag{3.16}$$

Equation (3.15) premultiplied by $\mathbf{Q}^T$, where $\mathbf{Q}$ is the unitary matrix that diagonalizes $\mathbf{R}$ through a similarity transformation, yields

$$\begin{aligned}E\left[\mathbf{Q}^T\Delta\mathbf{w}(k+1)\right] &= (\mathbf{I} - 2\mu\mathbf{Q}^T\mathbf{R}\mathbf{Q})E\left[\mathbf{Q}^T\Delta\mathbf{w}(k)\right]\\ &= E\left[\Delta\mathbf{w}'(k+1)\right]\\ &= (\mathbf{I} - 2\mu\mathbf{\Lambda})E\left[\Delta\mathbf{w}'(k)\right]\\ &= \begin{bmatrix} 1-2\mu\lambda_0 & 0 & \cdots & 0\\ 0 & 1-2\mu\lambda_1 & & \vdots\\ \vdots & \vdots & \ddots & \vdots\\ 0 & 0 & & 1-2\mu\lambda_N\end{bmatrix} E\left[\Delta\mathbf{w}'(k)\right]\end{aligned} \tag{3.17}$$

where $\Delta\mathbf{w}'(k+1) = \mathbf{Q}^T\Delta\mathbf{w}(k+1)$ is the rotated-coefficient error vector. The applied rotation yielded an equation where the driving matrix is diagonal, making it easier to analyze the equation's dynamic behavior. Alternatively, the above relation can be expressed as

$$\begin{aligned} E[\Delta\mathbf{w}'(k+1)] &= (\mathbf{I} - 2\mu\mathbf{\Lambda})^{k+1}E[\Delta\mathbf{w}'(0)] \\ &= \begin{bmatrix} (1-2\mu\lambda_0)^{k+1} & 0 & \cdots & 0 \\ 0 & (1-2\mu\lambda_1)^{k+1} & & \vdots \\ \vdots & \vdots & \ddots & \vdots \\ 0 & 0 & & (1-2\mu\lambda_N)^{k+1} \end{bmatrix} E[\Delta\mathbf{w}'(0)] \end{aligned} \tag{3.18}$$

This equation shows that in order to guarantee convergence of the coefficients in the mean, the convergence factor of the LMS algorithm must be chosen in the range

$$0 < \mu < \frac{1}{\lambda_{\max}} \tag{3.19}$$

where $\lambda_{\max}$ is the largest eigenvalue of $\mathbf{R}$. Values of μ in this range guarantees that all elements of the diagonal matrix in equation (3.18) tend to zero as $k \to \infty$, since $-1 < (1 - 2\mu\lambda_i) < 1$, for $i = 0, 1, \ldots, N$. As a result $E[\Delta\mathbf{w}'(k+1)]$ tends to zero for large k.

The choice of μ as above explained ensures that the mean value of the coefficient vector approaches the optimum coefficient vector $\mathbf{w}_o$. It should be mentioned that if the matrix $\mathbf{R}$ has a large eigenvalue spread, it is advisable to choose a value for μ much smaller than the upper bound. As a result, the convergence speed of the coefficients will be primarily dependent on the value of the smallest eigenvalue, responsible for the slowest mode in equation (3.18).

The key assumption for the above analysis is the so-called independence theory [4], which considers all vectors $\mathbf{x}(i)$, for $i = 0, 1, \ldots, k$, statistically independent. This assumption allowed us to consider $\Delta\mathbf{w}(k)$ independent of $\mathbf{x}(k)\mathbf{x}^T(k)$ in equation (3.14). Such an assumption, despite not being rigorously valid especially when $\mathbf{x}(k)$ consists of the elements of a delay line, leads to theoretical results that are in good agreement with the experimental results.

3.3.3 Coefficient-Error-Vector Covariance Matrix

In this subsection, we derive the expressions for the second-order statistics of the errors in the adaptive-filter coefficients. Since for large k the mean value of $\Delta\mathbf{w}(k)$ is zero, the covariance of the coefficient-error vector is defined as

$$\mathrm{cov}[\Delta\mathbf{w}(k)] = E[\Delta\mathbf{w}(k)\Delta\mathbf{w}^T(k)] = E\{[\mathbf{w}(k) - \mathbf{w}_o][\mathbf{w}(k) - \mathbf{w}_o]^T\} \tag{3.20}$$

By replacing equation (3.12) in (3.20) it follows that

$$\begin{aligned}\text{cov}[\Delta\mathbf{w}(k+1)] &= E\{\left[\mathbf{I}-2\mu\mathbf{x}(k)\mathbf{x}^T(k)\right]\Delta\mathbf{w}(k)\Delta\mathbf{w}^T(k)\left[\mathbf{I}-2\mu\mathbf{x}(k)\mathbf{x}^T(k)\right]^T\\&+[\mathbf{I}-2\mu\mathbf{x}(k)\mathbf{x}^T(k)]\Delta\mathbf{w}(k)2\mu e_o(k)\mathbf{x}^T(k)\\&+2\mu e_o(k)\mathbf{x}(k)\Delta\mathbf{w}^T(k)[\mathbf{I}-2\mu\mathbf{x}(k)\mathbf{x}^T(k)]^T\\&+4\mu^2 e_o^2(k)\mathbf{x}(k)\mathbf{x}^T(k)\}\end{aligned}\tag{3.21}$$

By considering $e_o(k)$ independent of $\Delta\mathbf{w}(k)$ and orthogonal to $\mathbf{x}(k)$, the second and third terms on the right-hand side of the above equation can be eliminated. The details of this simplification can be carried out by describing each element of the eliminated matrices explicitly. In this case,

$$\begin{aligned}\text{cov}[\Delta\mathbf{w}(k+1)] &= \text{cov}[\Delta\mathbf{w}(k)]+E[-2\mu\mathbf{x}(k)\mathbf{x}^T(k)\Delta\mathbf{w}(k)\Delta\mathbf{w}^T(k)\\&-2\mu\Delta\mathbf{w}(k)\Delta\mathbf{w}^T(k)\mathbf{x}(k)\mathbf{x}^T(k)\\&+4\mu^2\mathbf{x}(k)\mathbf{x}^T(k)\Delta\mathbf{w}(k)\Delta\mathbf{w}^T(k)\mathbf{x}(k)\mathbf{x}^T(k)\\&+4\mu^2 e_o^2(k)\mathbf{x}(k)\mathbf{x}^T(k)]\end{aligned}\tag{3.22}$$

In addition, assuming that $\Delta\mathbf{w}(k)$ and $\mathbf{x}(k)$ are independent, equation (3.22) can be rewritten as

$$\begin{aligned}\text{cov}[\Delta\mathbf{w}(k+1)] &= \text{cov}[\Delta\mathbf{w}(k)]-2\mu E[\mathbf{x}(k)\mathbf{x}^T(k)]E[\Delta\mathbf{w}(k)\Delta\mathbf{w}^T(k)]\\&-2\mu E[\Delta\mathbf{w}(k)\Delta\mathbf{w}^T(k)]E[\mathbf{x}(k)\mathbf{x}^T(k)]\\&+4\mu^2 E\left\{\mathbf{x}(k)\mathbf{x}^T(k)E[\Delta\mathbf{w}(k)\Delta\mathbf{w}^T(k)]\mathbf{x}(k)\mathbf{x}^T(k)\right\}\\&+4\mu^2 E[e_o^2(k)]E[\mathbf{x}(k)\mathbf{x}^T(k)]\\&= \text{cov}[\Delta\mathbf{w}(k)]-2\mu\mathbf{R}\,\text{cov}[\Delta\mathbf{w}(k)]\\&-2\mu\,\text{cov}[\Delta\mathbf{w}(k)]\mathbf{R}+4\mu^2\mathbf{A}+4\mu^2\sigma_n^2\mathbf{R}\end{aligned}\tag{3.23}$$

The calculation of $\mathbf{A}=E\left\{\mathbf{x}(k)\mathbf{x}^T(k)E[\Delta\mathbf{w}(k)\Delta\mathbf{w}^T(k)]\mathbf{x}(k)\mathbf{x}^T(k)\right\}$ involves fourth-order moments and the result can be obtained by expanding the matrix inside the operation $E[\cdot]$ as described in [4] and [13] for jointly Gaussian input signal samples. The result is

$$\mathbf{A}=2\mathbf{R}\,\text{cov}[\Delta\mathbf{w}(k)]\,\mathbf{R}+\mathbf{R}\,\text{tr}\{\mathbf{R}\,\text{cov}[\Delta\mathbf{w}(k)]\}\tag{3.24}$$

where $\text{tr}[\cdot]$ denotes trace of $[\cdot]$. Equation (3.23) is needed to calculate the excess mean-square error caused by the noisy estimate of the gradient employed by the LMS algorithm. As can be noted, $\text{cov}[\Delta\mathbf{w}(k+1)]$ does not tend to $\mathbf{0}$ as $k\rightarrow\infty$, due to the last term in equation (3.23) that provides an excitation in the dynamic matrix equation.

A more useful form for equation (3.23) can be obtained by premultiplying and postmultiplying it by $\mathbf{Q}^T$ and $\mathbf{Q}$ respectively, yielding

$$\begin{aligned}\mathbf{Q}^T\text{cov}[\Delta\mathbf{w}(k+1)]\mathbf{Q} &= \mathbf{Q}^T\,\text{cov}[\Delta\mathbf{w}(k)]\,\mathbf{Q}\\&-2\mu\mathbf{Q}^T\mathbf{R}\mathbf{Q}\mathbf{Q}^T\,\text{cov}[\Delta\mathbf{w}(k)]\mathbf{Q}\\&-2\mu\mathbf{Q}^T\,\text{cov}[\Delta\mathbf{w}(k)]\mathbf{Q}\mathbf{Q}^T\mathbf{R}\mathbf{Q}\\&+8\mu^2\mathbf{Q}^T\mathbf{R}\mathbf{Q}\mathbf{Q}^T\,\text{cov}[\Delta\mathbf{w}(k)]\mathbf{Q}\mathbf{Q}^T\mathbf{R}\mathbf{Q}\\&+4\mu^2\mathbf{Q}^T\mathbf{R}\mathbf{Q}\mathbf{Q}^T\,\text{tr}\{\mathbf{R}\mathbf{Q}\mathbf{Q}^T\,\text{cov}[\Delta\mathbf{w}(k)]\}\mathbf{Q}\\&+4\mu^2\sigma_n^2\mathbf{Q}^T\mathbf{R}\mathbf{Q}\end{aligned}\tag{3.25}$$

where we used the equality $\mathbf{Q}^T\mathbf{Q} = \mathbf{Q}\mathbf{Q}^T = \mathbf{I}$. Using the fact that $\mathbf{Q}^T\text{tr}[\mathbf{B}]\mathbf{Q} = \text{tr}[\mathbf{Q}^T\mathbf{B}\mathbf{Q}]\mathbf{I}$ for any $\mathbf{B}$,

$$\begin{aligned}\text{cov}[\Delta\mathbf{w}'(k+1)] &= \text{cov}[\Delta\mathbf{w}'(k)] - 2\mu\mathbf{\Lambda}\,\text{cov}[\Delta\mathbf{w}'(k)] - 2\mu\,\text{cov}[\Delta\mathbf{w}'(k)]\mathbf{\Lambda} \\ &\quad +8\mu^2\mathbf{\Lambda}\,\text{cov}[\Delta\mathbf{w}'(k)]\mathbf{\Lambda} + 4\mu^2\mathbf{\Lambda}\,\text{tr}\{\mathbf{\Lambda}\,\text{cov}[\Delta\mathbf{w}'(k)]\} + 4\mu^2\sigma_n^2\mathbf{\Lambda}\end{aligned} \tag{3.26}$$

where $\text{cov}[\Delta\mathbf{w}'(k)] = E[\mathbf{Q}^T\Delta\mathbf{w}(k)\Delta\mathbf{w}^T(k)\mathbf{Q}]$.

As will be shown in subsection 3.3.6, only the diagonal elements of $\text{cov}[\Delta\mathbf{w}'(k)]$ contribute to the excess MSE in the LMS algorithm. By defining $\mathbf{v}'(k)$ as a vector with elements consisting of the diagonal elements of $\text{cov}[\Delta\mathbf{w}'(k)]$, and $\boldsymbol{\lambda}$ as a vector consisting of the eigenvalues of $\mathbf{R}$, the following relation can be derived from the above equations

$$\begin{aligned}\mathbf{v}'(k+1) &= (\mathbf{I} - 4\mu\mathbf{\Lambda} + 8\mu^2\mathbf{\Lambda}^2 + 4\mu^2\boldsymbol{\lambda}\boldsymbol{\lambda}^T)\mathbf{v}'(k) + 4\mu^2\sigma_n^2\boldsymbol{\lambda} \\ &= \mathbf{B}\mathbf{v}'(k) + 4\mu^2\sigma_n^2\boldsymbol{\lambda}\end{aligned} \tag{3.27}$$

where the elements of $\mathbf{B}$ are given by

$$b_{ij} = \begin{cases} 1 - 4\mu\lambda_i + 8\mu^2\lambda_i^2 + 4\mu^2\lambda_i^2 & \text{for } i = j \\ 4\mu^2\lambda_i\lambda_j & \text{for } i \neq j \end{cases} \tag{3.28}$$

The value of the convergence factor μ must be chosen in a range that guarantees the convergence of $\mathbf{v}'(k)$. Since matrix $\mathbf{B}$ is symmetric, it has only real-valued eigenvalues. Also since all entries of $\mathbf{B}$ are also non-negative, the maximum among the sum of elements in any row of $\mathbf{B}$ represents an upper bound to the maximum eigenvalue of $\mathbf{B}$ and to the absolute value of any other eigenvalue, see pages 53 and 63 of [14] or the Gershgorin theorem in [15]. As a consequence, a sufficient condition to guarantee convergence is to force the sum of the elements in any row of $\mathbf{B}$ to be kept in the range $0 < \sum_{j=0}^{N} b_{ij} < 1$. Since

$$\sum_{j=0}^{N} b_{ij} = 1 - 4\mu\lambda_i + 8\mu^2\lambda_i^2 + 4\mu^2\lambda_i\sum_{j=0}^{N}\lambda_j \tag{3.29}$$

the critical values of μ are those for which the above equation approaches 1, as for any μ the expression is always positive. This will occur only if the last three terms of equation (3.29) approach zero, that is

$$-4\mu\lambda_i + 8\mu^2\lambda_i^2 + 4\mu^2\lambda_i\sum_{j=0}^{N}\lambda_j \approx 0$$

After simple manipulation the stability condition obtained is

$$0 < \mu < \frac{1}{2\lambda_{\max} + \sum_{j=0}^{N}\lambda_j} < \frac{1}{\sum_{j=0}^{N}\lambda_j} = \frac{1}{\text{tr}[\mathbf{R}]} \tag{3.30}$$

where the last and simpler expression is more widely used in practice because $\text{tr}[\mathbf{R}]$ is quite simple to estimate since it is related with the Euclidean norm squared of the input signal vector, whereas an estimate $\lambda_{\max}$ is much more difficult to obtain. It will be shown in equation (3.45) that μ controls the speed of convergence of the MSE.

The upper bound obtained for the value of μ is important from the practical point of view, because it gives us an indication of the maximum value of μ that could be used in order to achieve convergence of the coefficients. However, the reader should be advised that the given upper bound is somewhat optimistic due to the approximations and assumptions made. In most cases, the value of μ should not be chosen close to the upper bound.

3.3.4 Behavior of the Error Signal

In this subsection, the mean value of the output error in the adaptive filter is calculated, considering that the unknown system model has infinite impulse response and there is measurement noise. The error signal, when an additional measurement noise is accounted for, is given by

$$e(k) = d'(k) - \mathbf{w}^T(k)\mathbf{x}(k) + n(k) \tag{3.31}$$

where $d'(k)$ is the desired signal without measurement noise. For a given known input vector $\mathbf{x}(k)$, the expected value of the error signal is

$$\begin{aligned} E[e(k)] &= E[d'(k)] - E[\mathbf{w}^T(k)\mathbf{x}(k)] + E[n(k)] \\ &= E[d'(k)] - \mathbf{w}_o^T\mathbf{x}(k) + E[n(k)] \end{aligned} \tag{3.32}$$

where $\mathbf{w}_o$ is the optimal solution, i.e., the Wiener solution for the coefficient vector. Note that the input signal vector was assumed known in the above equation, in order to expose what can be expected if the adaptive filter converges to the optimal solution. If $d'(k)$ was generated through an infinite impulse response system, a residue error remains in the subtraction of the first two terms due to undermodeling (adaptive FIR filter with insufficient number of coefficients), i.e.,

$$E[e(k)] = E\left[\sum_{i=N+1}^{\infty} h(i)x(k-i)\right] + E[n(k)] \tag{3.33}$$

where $h(i)$, for $i = N+1, \ldots, \infty$, are the coefficients of the process that generated the part of $d'(k)$ not identified by the adaptive filter. If the input signal and $n(k)$ have zero mean, then $E[e(k)] = 0$.

3.3.5 Minimum Mean-Square Error

In this subsection, the minimum MSE is calculated for undermodeling situations and in the presence of additional noise. Let's assume again the undermodeling case where the adaptive filter has less coefficients than the unknown system in a system identification setup. In this case we can write

$$\begin{aligned} d(k) &= \mathbf{h}^T\mathbf{x}_\infty(k) + n(k) \\ &= [\mathbf{w}_o^T \; \bar{\mathbf{h}}^T] \begin{bmatrix} \mathbf{x}(k) \\ \bar{\mathbf{x}}_\infty(k) \end{bmatrix} + n(k) \end{aligned} \tag{3.34}$$

where $\mathbf{w}_o$ is a vector containing the first $N+1$ coefficients of the unknown system impulse response, $\overline{\mathbf{h}}$ contains the remaining elements of $\mathbf{h}$. The output signal of an adaptive filter with $N+1$ coefficients is given by

$$y(k) = \mathbf{w}^T(k)\mathbf{x}(k)$$

In this setup the MSE has the following expression

$$\begin{aligned}
\xi &= E\{d^2(k) - 2\mathbf{w}_o^T\mathbf{x}(k)\mathbf{w}^T(k)\mathbf{x}(k) - 2\overline{\mathbf{h}}^T\overline{\mathbf{x}}_\infty(k)\mathbf{w}^T(k)\mathbf{x}(k) \\
&\quad -2[\mathbf{w}^T(k)\mathbf{x}(k)]n(k) + [\mathbf{w}^T(k)\mathbf{x}(k)]^2\} \\
&= E\left\{d^2(k) - 2[\mathbf{w}^T(k)\ \mathbf{0}_\infty^T]\begin{bmatrix}\mathbf{x}(k)\\ \overline{\mathbf{x}}_\infty(k)\end{bmatrix}[\mathbf{w}_o^T\ \overline{\mathbf{h}}^T]\begin{bmatrix}\mathbf{x}(k)\\ \overline{\mathbf{x}}_\infty(k)\end{bmatrix}\right. \\
&\quad \left. -2[\mathbf{w}^T(k)\mathbf{x}(k)]n(k) + [\mathbf{w}^T(k)\mathbf{x}(k)]^2\right\} \\
&= E[d^2(k)] - 2[\mathbf{w}^T(k)\ \mathbf{0}_\infty^T]\mathbf{R}_\infty\begin{bmatrix}\mathbf{w}_o\\ \overline{\mathbf{h}}\end{bmatrix} + \mathbf{w}^T(k)\mathbf{R}\mathbf{w}(k)
\end{aligned} \tag{3.35}$$

where

$$\mathbf{R}_\infty = E\left\{\begin{bmatrix}\mathbf{x}(k)\\ \overline{\mathbf{x}}_\infty(k)\end{bmatrix}[\mathbf{x}^T(k)\ \overline{\mathbf{x}}_\infty^T(k)]\right\}$$

and $\mathbf{0}_\infty$ is an infinite length vector whose elements are zeros. By calculating the derivative of ξ with respect to the coefficients of the adaptive filter, it follows that (see derivations around equations (2.91) and (2.148))

$$\begin{aligned}
\hat{\mathbf{w}}_o &= \mathbf{R}^{-1}\text{trunc}\{\mathbf{p}_\infty\}_{N+1} = \mathbf{R}^{-1}\text{trunc}\left\{\mathbf{R}_\infty\begin{bmatrix}\mathbf{w}_o\\ \overline{\mathbf{h}}\end{bmatrix}\right\}_{N+1} \\
&= \mathbf{R}^{-1}\text{trunc}\{\mathbf{R}_\infty\mathbf{h}\}_{N+1}
\end{aligned} \tag{3.36}$$

where $\text{trunc}\{\mathbf{a}\}_{N+1}$ represents a vector generated by retaining the first $N+1$ elements of $\mathbf{a}$. It should be noticed that the results of equations (3.35) and (3.36) are algorithm independent.

The minimum mean-square error can be obtained from equation (3.35), when assuming the input signal is a white noise uncorrelated with the additional noise signal, that is

$$\begin{aligned}
\xi_{\min} = E[e^2(k)]_{\min} &= \sum_{i=N+1}^{\infty} h^2(i)E[x^2(k-i)] + E[n^2(k)] \\
&= \sum_{i=N+1}^{\infty} h^2(i)\sigma_x^2 + \sigma_n^2
\end{aligned} \tag{3.37}$$

This minimum error is achieved when it is assumed that the adaptive-filter multiplier coefficients are frozen at their optimum values, refer to equation (2.148) for similar discussion. In case the adaptive filter has sufficient order to model the process that generated $d(k)$, the minimum MSE that can be achieved is equal to the variance of the additional noise, given by σ_n^2. The reader should note that the effect of undermodeling discussed in this subsection generates an excess MSE with respect to σ_n^2.

3.3.6 Excess Mean-Square Error and Misadjustment

The result of the previous subsection assumes that the adaptive-filter coefficients converge to their optimal values, but in practice this is not so. Although the coefficient vector on average converges to $\mathbf{w}_o$, the instantaneous deviation $\Delta\mathbf{w}(k) = \mathbf{w}(k) - \mathbf{w}_o$, caused by the noisy gradient estimates, generates an excess MSE. The excess MSE can be quantified as described in the present subsection. The output error at instant k is given by

$$\begin{aligned} e(k) &= d(k) - \mathbf{w}_o^T\mathbf{x}(k) - \Delta\mathbf{w}^T(k)\mathbf{x}(k) \\ &= e_o(k) - \Delta\mathbf{w}^T(k)\mathbf{x}(k) \end{aligned} \tag{3.38}$$

then

$$e^2(k) = e_o^2(k) - 2e_o(k)\Delta\mathbf{w}^T(k)\mathbf{x}(k) + \Delta\mathbf{w}^T(k)\mathbf{x}(k)\mathbf{x}^T(k)\Delta\mathbf{w}(k) \tag{3.39}$$

The so-called independence theory assumes that the vectors $\mathbf{x}(k)$, for all k, are statistically independent, allowing a simple mathematical treatment for the LMS algorithm. As mentioned before, this assumption is in general not true, especially in the case where $\mathbf{x}(k)$ consists of the elements of a delay line. However, even in this case the use of the independence assumption is justified by the agreement between the analytical and the experimental results. With the independence assumption, $\Delta\mathbf{w}(k)$ can be considered independent of $\mathbf{x}(k)$, since only previous input vectors are involved in determining $\Delta\mathbf{w}(k)$. By using the assumption and applying the expected value operator to equation (3.39), we have

$$\begin{aligned} \xi(k) &= E[e^2(k)] \\ &= \xi_{\min} - 2E[\Delta\mathbf{w}^T(k)]E[e_o(k)\mathbf{x}(k)] + E[\Delta\mathbf{w}^T(k)\mathbf{x}(k)\mathbf{x}^T(k)\Delta\mathbf{w}(k)] \\ &= \xi_{\min} - 2E[\Delta\mathbf{w}^T(k)]E[e_o(k)\mathbf{x}(k)] + E\{\text{tr}[\Delta\mathbf{w}^T(k)\mathbf{x}(k)\mathbf{x}^T(k)\Delta\mathbf{w}(k)]\} \\ &= \xi_{\min} - 2E[\Delta\mathbf{w}^T(k)]E[e_o(k)\mathbf{x}(k)] + E\{\text{tr}[\mathbf{x}(k)\mathbf{x}^T(k)\Delta\mathbf{w}(k)\Delta\mathbf{w}^T(k)]\} \end{aligned} \tag{3.40}$$

where in the fourth equality we used the property $\text{tr}[\mathbf{A}\cdot\mathbf{B}] = \text{tr}[\mathbf{B}\cdot\mathbf{A}]$. The last term of the above equation can be rewritten as

$$\text{tr}\left\{E[\mathbf{x}(k)\mathbf{x}^T(k)]E[\Delta\mathbf{w}(k)\Delta\mathbf{w}^T(k)]\right\}$$

Since $\mathbf{R} = E[\mathbf{x}(k)\mathbf{x}^T(k)]$ and by the orthogonality principle $E[e_o(k)\mathbf{x}(k)] = 0$, the above equation can be simplified as follows:

$$\xi(k) = \xi_{\min} + E[\Delta\mathbf{w}^T(k)\mathbf{R}\Delta\mathbf{w}(k)] \tag{3.41}$$

The excess in the MSE is given by

$$\begin{aligned} \Delta\xi(k) &\triangleq \xi(k) - \xi_{\min} = E[\Delta\mathbf{w}^T(k)\mathbf{R}\Delta\mathbf{w}(k)] \\ &= E\{\text{tr}[\mathbf{R}\Delta\mathbf{w}(k)\Delta\mathbf{w}^T(k)]\} \\ &= \text{tr}\{E[\mathbf{R}\Delta\mathbf{w}(k)\Delta\mathbf{w}^T(k)]\} \end{aligned} \tag{3.42}$$

By using the fact that $\mathbf{Q}\mathbf{Q}^T = \mathbf{I}$, the following relation results

$$\begin{aligned} \Delta\xi(k) &= \text{tr}\left\{E[\mathbf{Q}\mathbf{Q}^T\mathbf{R}\mathbf{Q}\mathbf{Q}^T\Delta\mathbf{w}(k)\Delta\mathbf{w}^T(k)\mathbf{Q}\mathbf{Q}^T]\right\} \\ &= \text{tr}\{\mathbf{Q}\boldsymbol{\Lambda}\,\text{cov}[\Delta\mathbf{w}'(k)]\mathbf{Q}^T\} \end{aligned} \tag{3.43}$$

Therefore,

$$\Delta\xi(k) = \text{tr}\{\mathbf{\Lambda}\ \text{cov}[\Delta\mathbf{w}'(k)]\} \tag{3.44}$$

From equation (3.27), it is possible to show that

$$\Delta\xi(k) = \sum_{i=0}^{N} \lambda_i v_i'(k) = \boldsymbol{\lambda}^T \mathbf{v}'(k) \tag{3.45}$$

Since

$$v_i'(k+1) = (1 - 4\mu\lambda_i + 8\mu^2\lambda_i^2)v_i'(k) + 4\mu^2\lambda_i \sum_{j=0}^{N} \lambda_j v_j'(k) + 4\mu^2\sigma_n^2\lambda_i \tag{3.46}$$

and $v_i'(k+1) \approx v_i'(k)$ for large k, we can apply a summation operation to the above equation in order to obtain

$$\begin{aligned}\sum_{j=0}^{N} \lambda_j v_j'(k) &= \frac{\mu\sigma_n^2 \sum_{i=0}^{N} \lambda_i + 2\mu \sum_{i=0}^{N} \lambda_i^2 v_i'(k)}{1 - \mu \sum_{i=0}^{N} \lambda_i} \\ &\approx \frac{\mu\sigma_n^2 \sum_{i=0}^{N} \lambda_i}{1 - \mu \sum_{i=0}^{N} \lambda_i} \\ &= \frac{\mu\sigma_n^2 \text{tr}[\mathbf{R}]}{1 - \mu \text{tr}[\mathbf{R}]}\end{aligned} \tag{3.47}$$

where the term $2\mu \sum_{i=0}^{N} \lambda_i^2 v_i'(k)$ was considered very small as compared to the remaining terms of the numerator. This assumption is not easily justifiable, but is valid for small values of μ.

The excess mean-square error can then be expressed as

$$\xi_{\text{exc}} = \lim_{k\to\infty} \Delta\xi(k) \approx \frac{\mu\sigma_n^2 \text{tr}[\mathbf{R}]}{1 - \mu \text{tr}[\mathbf{R}]} \tag{3.48}$$

This equation, for very small μ, can be approximated by

$$\xi_{\text{exc}} \approx \mu\sigma_n^2 \text{tr}[\mathbf{R}] = \mu(N+1)\sigma_n^2\sigma_x^2 \tag{3.49}$$

where σ_x^2 is the input signal variance and σ_n^2 is the additional-noise variance.

The misadjustment M, defined as the ratio between the ξ_{exc} and the minimum MSE, is a common parameter used to compare different adaptive signal processing algorithms. For the LMS algorithm, the misadjustment is given by

$$M \triangleq \frac{\xi_{\text{exc}}}{\xi_{\text{min}}} \approx \frac{\mu \text{tr}[\mathbf{R}]}{1 - \mu \text{tr}[\mathbf{R}]} \tag{3.50}$$

3.3.7 Transient Behavior

Before the LMS algorithm reaches the steady-state behavior, a number of iterations are spent in the transient part. During this time, the adaptive-filter coefficients and the output error change from their initial values to values close to that of the corresponding optimal solution.

In the case of the adaptive-filter coefficients, the convergence in the mean will follow $(N+1)$ geometric decaying curves with ratios $r_{wi} = (1 - 2\mu\lambda_i)$. Each of these curves can be approximated by an exponential envelope with time constant τ_{wi} as follows (see equation (3.18)) [2]:

$$r_{wi} = e^{\frac{-1}{\tau_{wi}}} = 1 - \frac{1}{\tau_{wi}} + \frac{1}{2!\tau_{wi}^2} + \cdots \tag{3.51}$$

where for each iteration, the decay in the exponential envelope is equal to the decay in the original geometric curve. In general, r_{wi} is slightly smaller than one, especially for the slowly decreasing modes corresponding to small λ_i and μ. Therefore,

$$r_{wi} = (1 - 2\mu\lambda_i) \approx 1 - \frac{1}{\tau_{wi}} \tag{3.52}$$

then

$$\tau_{wi} = \frac{1}{2\mu\lambda_i}$$

for $i = 0, 1, \ldots, N$. Note that in order to guarantee convergence of the tap coefficients in the mean, μ must be chosen in the range $0 < \mu < 1/\lambda_{\max}$ (see equation (3.19)).

According to equation (3.30), for the convergence of the MSE the range of values for μ is $0 < \mu < 1/\text{tr}[\mathbf{R}]$, and the corresponding time constant can be calculated from matrix $\mathbf{B}$ in equation (3.27), by considering the terms in μ^2 small as compared to the remaining terms in matrix $\mathbf{B}$. In this case, the geometric decaying curves have ratios given by $r_{ei} = (1 - 4\mu\lambda_i)$ that can be fitted to exponential envelopes with time constants given by

$$\tau_{ei} = \frac{1}{4\mu\lambda_i} \tag{3.53}$$

for $i = 0, 1, \ldots, N$. In the convergence of both the error and the coefficients, the time required for the convergence depends on the ratio of eigenvalues of the input signal correlation matrix.

Returning to the tap coefficients case, if μ is chosen to be approximately $1/\lambda_{\max}$ the corresponding time constant for the coefficients is given by

$$\tau_{wi} \approx \frac{\lambda_{\max}}{2\lambda_i} \leq \frac{\lambda_{\max}}{2\lambda_{\min}} \tag{3.54}$$

Since the mode with the highest time constant takes longer to reach convergence, the rate of convergence is determined by the slowest mode given by $\tau_{w_{\max}} = \lambda_{\max}/(2\lambda_{\min})$. Suppose the convergence is considered achieved when the slowest mode provides an attenuation of 100, i.e.,

$$e^{\frac{-k}{\tau_{w_{\max}}}} = 0.01$$

this requires the following number of iterations in order to reach convergence:

$$k \approx 4.6 \frac{\lambda_{\max}}{2\lambda_{\min}}$$

The above situation is quite optimistic because μ was chosen to be high. As mentioned before, in practice we should choose the value of μ much smaller than the upper bound. For an eigenvalue spread approximating one, according to equation (3.30) let's choose μ smaller than $1/[(N+3)\lambda_{\max}]$.[2] In this case, the LMS algorithm will require at least

$$k \approx 4.6 \frac{(N+3)\lambda_{\max}}{2\lambda_{\min}} \approx 2.3(N+3)$$

iterations to achieve convergence in the coefficients.

The analytical results presented in this section are valid for stationary environments. The LMS algorithm can also operate in the case of nonstationary environments, as shown in the following section.

3.4 LMS ALGORITHM BEHAVIOR IN NONSTATIONARY ENVIRONMENTS

In practical situations, the environment in which the adaptive filter is embedded may be nonstationary. In these cases, the input signal autocorrelation matrix and/or the cross-correlation vector, denoted respectively by $\mathbf{R}(k)$ and $\mathbf{p}(k)$, are/is varying with time. Therefore, the optimal solution for the coefficient vector is also a time-varying vector given by $\mathbf{w}_o(k)$.

Since the optimal coefficient vector is not fixed, it is important to analyze if the LMS algorithm will be able to track changes in $\mathbf{w}_o(k)$. It is also of interest to learn how the tracking error in the coefficients given by $E[\mathbf{w}(k)] - \mathbf{w}_o(k)$ will affect the output MSE. It will be shown later that the excess MSE caused by lag in the tracking of $\mathbf{w}_o(k)$ can be separated from the excess MSE caused by the measurement noise, and therefore, without loss of generality, in the following analysis the additional noise will be considered zero.

The coefficient-vector updating in the LMS algorithm can be written in the following form

$$\begin{aligned} \mathbf{w}(k+1) &= \mathbf{w}(k) + 2\mu\mathbf{x}(k)e(k) \\ &= \mathbf{w}(k) + 2\mu\mathbf{x}(k)[d(k) - \mathbf{x}^T(k)\mathbf{w}(k)] \end{aligned} \tag{3.55}$$

Since

$$d(k) = \mathbf{x}^T(k)\mathbf{w}_o(k) \tag{3.56}$$

the coefficient updating can be expressed as follows:

$$\mathbf{w}(k+1) = \mathbf{w}(k) + 2\mu\mathbf{x}(k)[\mathbf{x}^T(k)\mathbf{w}_o(k) - \mathbf{x}^T(k)\mathbf{w}(k)] \tag{3.57}$$

[2] This choice also guarantees the convergence of the MSE.

Now assume that an ensemble of a nonstationary adaptive identification process has been built, where the input signal in each experiment is taken from the same stochastic process. The input signal is considered stationary. This assumption results in a fixed **R** matrix, and the nonstationarity is caused by the desired signal that is generated by applying the input signal to a time-varying system. With these assumptions, by using the expected value operation to the ensemble, with the coefficient updating in each experiment given by equation (3.57), and additionally assuming that $\mathbf{w}(k)$ is independent of $\mathbf{x}(k)$ yields

$$\begin{aligned} E[\mathbf{w}(k+1)] &= E[\mathbf{w}(k)] + 2\mu E[\mathbf{x}(k)\mathbf{x}^T(k)]\mathbf{w}_o(k) - 2\mu E[\mathbf{x}(k)\mathbf{x}^T(k)]E[\mathbf{w}(k)] \\ &= E[\mathbf{w}(k)] + 2\mu\mathbf{R}\{\mathbf{w}_o(k) - E[\mathbf{w}(k)]\} \end{aligned} \tag{3.58}$$

If the lag in the coefficient vector is defined by

$$\mathbf{l_w}(k) = E[\mathbf{w}(k)] - \mathbf{w}_o(k) \tag{3.59}$$

equation (3.58) can be rewritten as

$$\mathbf{l_w}(k+1) = (\mathbf{I} - 2\mu\mathbf{R})\mathbf{l_w}(k) - \mathbf{w}_o(k+1) + \mathbf{w}_o(k) \tag{3.60}$$

In order to simplify our analysis, we can premultiply the above equation by $\mathbf{Q}^T$, resulting in a decoupled set of equations given by

$$\mathbf{l}'_{\mathbf{w}}(k+1) = (\mathbf{I} - 2\mu\boldsymbol{\Lambda})\mathbf{l}'_{\mathbf{w}}(k) - \mathbf{w}'_o(k+1) + \mathbf{w}'_o(k) \tag{3.61}$$

where the vectors with superscript are the original vectors projected onto the transformed space. As can be noted, each element of the lag-error vector is determined by the following relation

$$l'_i(k+1) = (1 - 2\mu\lambda_i)l'_i(k) - w'_{oi}(k+1) + w'_{oi}(k) \tag{3.62}$$

where $l'_i(k)$ is the ith element of $\mathbf{l}'_{\mathbf{w}}(k)$. By properly interpreting the above equation, we can say that the lag is generated by applying the transformed instantaneous optimal coefficient to a first-order discrete-time *lag filter* denoted as $L''_i(z)$, i.e.,

$$L'_i(z) = -\frac{z-1}{z-1+2\mu\lambda_i}W'_{oi}(z) = L''_i(z)W'_{oi}(z) \tag{3.63}$$

The discrete-time filter transient response converges with a time constant of the exponential envelope given by

$$\tau_i = \frac{1}{2\mu\lambda_i} \tag{3.64}$$

which is of course different for each individual tap. Therefore, the tracking ability of the coefficients in the LMS algorithm is dependent on the eigenvalues of the input signal correlation matrix.

The lag in the adaptive-filter coefficients leads to an excess mean-square error. In order to calculate the excess MSE, suppose that each element of the optimal coefficient vector is modeled as a first-order Markov process. This nonstationary situation can be considered somewhat simplified as compared with some real practical situations. However, it allows a manageable mathematical analysis while retaining the essence of handling the more complicated cases. The first-order Markov process is described by

$$\mathbf{w}_o(k) = \lambda_{\mathbf{w}}\mathbf{w}_o(k-1) + \mathbf{n}_{\mathbf{w}}(k) \tag{3.65}$$

where $\mathbf{n}_{\mathbf{w}}(k)$ is a vector whose elements are zero-mean white noise processes with variance $\sigma_{\mathbf{w}}^2$, and $\lambda_{\mathbf{w}} < 1$. Note that $(1 - 2\mu\lambda_i) < \lambda_{\mathbf{w}} < 1$, for $i = 0, 1, \ldots, N$, since the optimal coefficients values must vary slower than the adaptive-filter tracking speed, i.e., $\frac{1}{2\mu\lambda_i} < \frac{1}{1-\lambda_{\mathbf{w}}}$. This model may not represent an actual system when $\lambda_{\mathbf{w}} \to 1$, since the $E[\mathbf{w}_o(k)\mathbf{w}_o^T(k)]$ will have unbounded elements if, for example, $\mathbf{n}_{\mathbf{w}}(k)$ is not exactly zero mean. A more realistic model would include a factor $(1-\lambda_{\mathbf{w}})^{\frac{p}{2}}$, for $p \geq 1$, multiplying $\mathbf{n}_{\mathbf{w}}(k)$ in order to guarantee that $E[\mathbf{w}_o(k)\mathbf{w}_o^T(k)]$ is bounded. In the following discussions, this case will not be considered since the corresponding results can be easily derived (see problem 14).

From equations (3.62) and (3.63), we can infer that the lag-error vector elements are generated by applying a first-order discrete-time system to the elements of the unknown system coefficient vector, both in the transformed space. On the other hand, the coefficients of the unknown system are generated by applying each element of the noise vector $\mathbf{n}_{\mathbf{w}}(k)$ to a first-order all-pole filter, with the pole placed at $\lambda_{\mathbf{w}}$. For the unknown coefficient vector with the above model, the lag-error vector elements can be generated by applying each element of the transformed noise vector $\mathbf{n}'_{\mathbf{w}}(k) = \mathbf{Q}^T\mathbf{n}_{\mathbf{w}}(k)$ to a discrete-time filter with transfer function

$$H_i(z) = \frac{-(z-1)z}{(z-1+2\mu\lambda_i)(z-\lambda_{\mathbf{w}})} \tag{3.66}$$

This transfer function consists of a cascade of the lag filter $L_i''(z)$ with the all-pole filter representing the first-order Markov process as illustrated in Fig. 3.2. Using the inverse $\mathcal{Z}$-transform, the variance of the elements of the vector $\mathbf{l}'_{\mathbf{w}}(k)$ can then be calculated by

$$\begin{aligned} E[l_i'^2(k)] &= \frac{1}{2\pi\jmath}\oint H_i(z)H_i(z^{-1})\sigma_{\mathbf{w}}^2 z^{-1}\,dz \\ &= \left[\frac{1}{(1-\lambda_{\mathbf{w}}-2\mu\lambda_i)(1-\lambda_{\mathbf{w}}+2\mu\lambda_i\lambda_{\mathbf{w}})}\right]\left[\frac{-\mu\lambda_i}{1-\mu\lambda_i}+\frac{1-\lambda_{\mathbf{w}}}{1+\lambda_{\mathbf{w}}}\right]\sigma_{\mathbf{w}}^2 \end{aligned} \tag{3.67}$$

If $\lambda_{\mathbf{w}}$ is considered very close to 1, it is possible to simplify the above equation as

$$E[l_i'^2(k)] \approx \frac{\sigma_{\mathbf{w}}^2}{4\mu\lambda_i(1-\mu\lambda_i)} \tag{3.68}$$

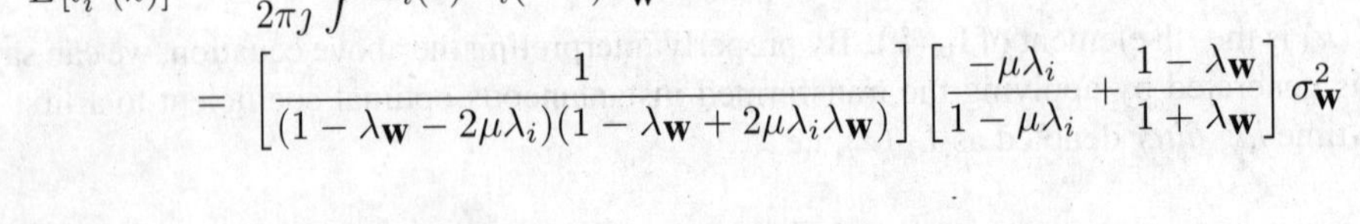

Figure 3.2 Lag model in nonstationary environment.

Any error in the coefficient vector of the adaptive filter as compared to the optimal coefficient filter generates an excess MSE (see equation (3.41)). Since the lag is one source of error in the adaptive-filter coefficients, then the excess MSE due to lag is given by

$$\begin{aligned}\xi_{\text{lag}} &= E[\mathbf{l}_{\mathbf{w}}^T(k)\mathbf{R}\mathbf{l}_{\mathbf{w}}(k)] \\ &= E\{\text{tr}[\mathbf{R}\mathbf{l}_{\mathbf{w}}(k)\mathbf{l}_{\mathbf{w}}^T(k)]\} \\ &= \text{tr}\{\mathbf{R}E[\mathbf{l}_{\mathbf{w}}(k)\mathbf{l}_{\mathbf{w}}^T(k)]\} \\ &= \text{tr}\{\mathbf{\Lambda}E[\mathbf{l}'_{\mathbf{w}}(k)\mathbf{l}'^T_{\mathbf{w}}(k)]\} \\ &= \sum_{i=0}^{N}\lambda_i E[l_i'^2(k)] \\ &\approx \frac{\sigma_{\mathbf{w}}^2}{4\mu}\sum_{i=0}^{N}\frac{1}{1-\mu\lambda_i}\end{aligned} \tag{3.69}$$

If μ is very small, the MSE due to lag tends to infinity indicating that the LMS algorithm, in this case, cannot track any change in the environment. On the other hand, for μ appropriately chosen the algorithm can track variations in the environment leading to an excess MSE. This excess MSE depends on the variance of the optimal coefficient disturbance and on the values of the input signal autocorrelation matrix eigenvalues, as indicated in equation (3.69). On the other hand, if μ is very small and $\lambda_{\mathbf{w}}$ is not very close to 1, the approximation for equation (3.67) becomes

$$E[l_i'^2(k)] \approx \frac{\sigma_{\mathbf{w}}^2}{1-\lambda_{\mathbf{w}}^2} \tag{3.70}$$

As a result the MSE due to lag is given by

$$\xi_{\text{lag}} \approx \frac{(N+1)\sigma_{\mathbf{w}}^2}{1-\lambda_{\mathbf{w}}^2} \tag{3.71}$$

It should be noticed that $\lambda_{\mathbf{w}}$ closer to 1 than the modes of the adaptive filter is the common operation region, therefore the result of equation (3.71) is not discussed further.

Now we analyze how the error due to lag interacts with the error generated by the noisy calculation of the gradient in the LMS algorithm. The overall error in the taps is given by

$$\Delta\mathbf{w}(k) = \mathbf{w}(k) - \mathbf{w}_o(k) = \{\mathbf{w}(k) - E[\mathbf{w}(k)]\} + \{E[\mathbf{w}(k)] - \mathbf{w}_o(k)\} \tag{3.72}$$

where the first error in the above equation is due to the additional noise and the second is the error due to lag. The overall excess MSE can then be expressed as

$$\begin{aligned}\xi_{\text{total}} &= E\{[\mathbf{w}(k) - \mathbf{w}_o(k)]^T\mathbf{R}[\mathbf{w}(k) - \mathbf{w}_o(k)]\} \\ &\approx E\{(\mathbf{w}(k) - E[\mathbf{w}(k)])^T\mathbf{R}(\mathbf{w}(k) - E[\mathbf{w}(k)])\} \\ &\quad + E\{(E[\mathbf{w}(k)] - \mathbf{w}_o(k))^T\mathbf{R}(E[\mathbf{w}(k)] - \mathbf{w}_o(k))\}\end{aligned} \tag{3.73}$$

since $2E\{(\mathbf{w}(k) - E[\mathbf{w}(k)])^T\mathbf{R}(E[\mathbf{w}(k)] - \mathbf{w}_o(k))\} \approx 0$, if we consider the fact that $\mathbf{w}_o(k)$ is kept fixed in each experiment of the ensemble. As a consequence, an estimate for the overall excess MSE can be obtained by adding the results of equations (3.48) and (3.69), i.e.,

$$\xi_{\text{total}} \approx \frac{\mu\sigma_n^2 \text{tr}[\mathbf{R}]}{1-\mu\text{tr}[\mathbf{R}]} + \frac{\sigma_{\mathbf{w}}^2}{4\mu}\sum_{i=0}^{N}\frac{1}{1-\mu\lambda_i} \tag{3.74}$$

If small μ is employed, the above equation can be simplified as follows:

$$\xi_{\text{total}} \approx \mu\sigma_n^2 \text{tr}[\mathbf{R}] + \frac{\sigma_{\mathbf{w}}^2}{4\mu}(N+1) \tag{3.75}$$

Differentiating the above equation with respect to μ and setting the result to zero yields an optimum value for μ given by

$$\mu_{\text{opt}} = \sqrt{\frac{(N+1)\sigma_{\mathbf{w}}^2}{4\sigma_n^2 \text{tr}[\mathbf{R}]}} \tag{3.76}$$

The μ_{opt} is supposed to lead to the minimum excess MSE. However, the user should bear in mind that the μ_{opt} can only be used if it satisfies stability conditions, and if its value can be considered small enough to validate equation (3.75). Also this value is optimum only when quantization effects are not taken into consideration, where for short-wordlength implementation the best μ should be chosen following the guidelines given in the Appendix B. It should also be mentioned that the study of the misadjustment due to nonstationarity of the environment is considerably more complicated when the input signal and the desired signal are simultaneously nonstationary [8], [10]-[17]. Therefore, the analysis presented here is only valid if the assumptions made are valid. However, the simplified analysis provides a good sample of the LMS algorithm behavior in a nonstationary environment and gives a general indication of what can be expected in more complicated situations.

The results of the analysis of the previous sections are obtained assuming that the algorithm is implemented with infinite precision[3]. However, the widespread use of adaptive-filtering algorithms in real-time requires their implementation with short wordlength, in order to meet the speed requirements. When implemented with short-wordlength precision the LMS algorithm behavior can be very different from what is expected in infinite precision. In particular, when the convergence factor μ tends to zero it is expected that the minimum mean-square error is reached in steady state; however, due to quantization effects the MSE tends to increase significantly if μ is reduced below a certain value. In fact, the algorithm can stop updating some filter coefficients if μ is not chosen appropriately. Appendix B, section B.1, presents detailed analysis of the quantization effects in the LMS algorithm.

3.5 COMPLEX LMS ALGORITHM

The LMS algorithm for complex signals, which often appear in communications applications, is derived in Appendix A. References [18]-[19] provide details related to complex differentiation required to generate algorithms working in environments with complex signals.

[3]This is an abuse of language, by infinite precision we mean very long wordlength.

By recalling that the LMS algorithm utilizes instantaneous estimates of matrix **R**, denoted by $\hat{\mathbf{R}}(k)$, and of vector **p**, denoted by $\hat{\mathbf{p}}(k)$, given by

$$\begin{aligned}\hat{\mathbf{R}}(k) &= \mathbf{x}(k)\mathbf{x}^H(k)\\ \hat{\mathbf{p}}(k) &= d^*(k)\mathbf{x}(k)\end{aligned} \tag{3.77}$$

The actual objective function being minimized is the instantaneous square error $|e(k)|^2$. According to the derivations in section A.3, the expression of the gradient estimate is

$$\hat{\mathbf{g}}_{\mathbf{w}^*}\{e(k)e^*(k)\} = -e^*(k)\mathbf{x}(k) \tag{3.78}$$

By utilizing the output error definition for the complex environment case and the instantaneous gradient expression, the updating equations for the complex LMS algorithm are described by

$$\begin{cases} e(k) = d(k) - \mathbf{w}^H(k)\mathbf{x}(k) \\ \mathbf{w}(k+1) = \mathbf{w}(k) + \mu_c e^*(k)\mathbf{x}(k) \end{cases} \tag{3.79}$$

If the convergence factor $\mu_c = 2\mu$, the expressions for the coefficient updating equation of the complex and real cases have the same form and the analysis results for the real case equally applies to the complex case[4].

An iteration of the complex LMS requires $N+2$ complex multiplications for the filter coefficient updating and $N+1$ complex multiplications for the error generation. In a non-optimized form each complex multiplication requires four real multiplications. The detailed description of the complex LMS algorithm is shown in the table denoted as Algorithm 3.2. As for any adaptive-filtering algorithm, the initialization is not necessarily performed as described in Algorithm 3.2, where the coefficients of the adaptive filter are started with zeros.

3.6 EXAMPLES

In this section, a number of examples are presented in order to illustrate the use of the LMS algorithm as well as to verify theoretical results presented in the previous sections.

3.6.1 Analytical Examples

Some analytical tools presented so far are employed to characterize two interesting types of adaptive-filtering problems. The problems are also solved with the LMS algorithm.

Example 3.1

A Gaussian white noise with unit variance colored by a filter with transfer function

[4]The missing factor 2 here originates from the term $\frac{1}{2}$ in definition of the gradient that we opted to use in order to be coherent with most literature, in actual implementation the factor 2 of the real case is usually incorporated to the μ.

Algorithm 3.2   **Complex LMS Algorithm**
Initialization $\mathbf{x}(0) = \mathbf{w}(0) = [0\,0\,\ldots\,0]^T$ Do for $k \geq 0$ $e(k) = d(k) - \mathbf{w}^H(k)\mathbf{x}(k)$ $\mathbf{w}(k+1) = \mathbf{w}(k) + \mu_c e^*(k)\mathbf{x}(k)$

$$H_{in}(z) = \frac{1}{z-0.5}$$

is transmitted through a communication channel with model given by

$$H_c(z) = \frac{1}{z+0.8}$$

and with the channel noise being Gaussian white noise with variance $\sigma_n^2 = 0.1$.

Fig. 3.3 illustrates the experimental environment. Note that $x'(k)$ is generated by first applying Gaussian white noise with variance $\sigma_{in}^2 = 1$ to a filter with transfer function $H_{in}(z)$. The result is applied to a communication channel with transfer function $H_c(z)$, and then Gaussian channel noise with variance $\sigma_n^2 = 0.1$ is added. On the other hand, $d(k)$ is generated by applying the same Gaussian noise with variance $\sigma_{in}^2 = 1$ to the filter with transfer function $H_{in}(z)$, with the result delayed by L samples.

(a) Determine the best value for the delay L.

(b) Compute the Wiener solution.

(c) Choose an appropriate value for μ and plot the convergence path for the LMS algorithm on the MSE surface.

(d) Plot the learning curves of the MSE and the filter coefficients in a single run as well as for the average of 25 runs.

Solution:

(a) In order to determine L, we will examine the behavior of the cross-correlation between the adaptive-filter input signal denoted by $x'(k)$ and the reference signal $d(k)$.

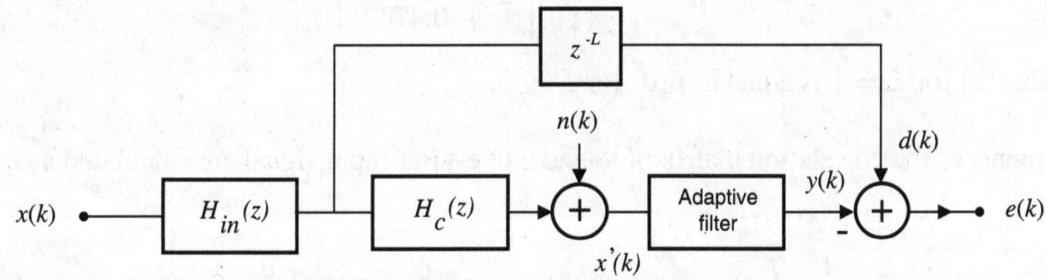

Figure 3.3 Channel equalization of Example 3.1.

The cross-correlation between $d(k)$ and $x'(k)$ is given by

$$\begin{aligned}p(i) &= E[d(k)x'(k-i)] \\ &= \frac{1}{2\pi \jmath}\oint H_{in}(z)z^{-L}z^{i}H_{in}(z^{-1})H_c(z^{-1})\sigma_{in}^2\frac{dz}{z} \\ &= \frac{1}{2\pi \jmath}\oint \frac{1}{z-0.5}z^{-L}z^{i}\frac{z}{1-0.5z}\frac{z}{1+0.8z}\sigma_{in}^2\frac{dz}{z}\end{aligned}$$

where the integration path is a counterclockwise closed contour corresponding to the unit circle.

The contour integral of the above equation can be solved through the Cauchy's residue theorem. For $L = 0$ and $L = 1$, the general solution is

$$p(0) = E[d(k)x'(k)] = \sigma_{in}^2[0.5^{-L+1}\frac{1}{0.75}\frac{1}{1.4}]$$

where in order to obtain $p(0)$, we computed the residue at the pole located at 0.5. The values of the cross-correlation for $L = 0$ and $L = 1$ are respectively

$$\begin{aligned}p(0) &= 0.47619 \\ p(0) &= 0.95238\end{aligned}$$

For $L = 2$, we have that

$$p(0) = \sigma_{in}^2[0.5^{-L+1}\frac{1}{0.75}\frac{1}{1.4} - 2] = -0.09522$$

where in this case we computed the residues at the poles located at 0.5 and at 0, respectively. For $L = 3$, we have

$$p(0) = \sigma_{in}^2[\frac{0.5^{-L+1}}{1.05} - 3.4] = 0.4095$$

From the above analysis, we see that the strongest correlation between $x'(k)$ and $d(k)$ occurs for $L = 1$. For this delay, the equalization is more effective. As a result, from the above calculations, we can obtain the elements of vector **p** as follows:

$$\mathbf{p} = \begin{bmatrix} p(0) \\ p(1) \end{bmatrix} = \begin{bmatrix} 0.9524 \\ 0.4762 \end{bmatrix}$$

Note that $p(1)$ for $L = 1$ is equal to $p(0)$ for $L = 0$.

The elements of the correlation matrix of the adaptive-filter input signal are calculated as follows:

$$\begin{aligned} r(i) &= E[x'(k)x'(k-i)] \\ &= \frac{1}{2\pi\jmath}\oint H_{in}(z)H_c(z)z^i H_{in}(z^{-1})H_c(z^{-1})\sigma_{in}^2 \frac{dz}{z} + \sigma_n^2\delta(i) \\ &= \frac{1}{2\pi\jmath}\oint \frac{1}{z-0.5}\frac{1}{z+0.8}z^i \frac{z}{1-0.5z}\frac{z}{1+0.8z}\sigma_{in}^2\frac{dz}{z} + \sigma_n^2\delta(i) \end{aligned}$$

where again the integration path is a counterclockwise closed contour corresponding to the unit circle, and $\delta(i)$ is the unitary impulse. Solving the contour integral equation, we obtain

$$\begin{aligned} r(0) &= E[x'^2(k)] \\ &= \sigma_{in}^2[\frac{1}{1.3}\frac{0.5}{0.75}\frac{1}{1.4} + \frac{-1}{1.3}\frac{-0.8}{1.4}\frac{1}{0.36}] + \sigma_n^2 = 1.6873 \end{aligned}$$

where the in order to obtain $r(0)$, we computed the residues at the poles located at 0.5 and -0.8, respectively. Similarly, we have that

$$\begin{aligned} r(1) &= E[x'(k)x'(k-1)] \\ &= \sigma_{in}^2[\frac{1}{1.3}\frac{1}{0.75}\frac{1}{1.4} + \frac{-1}{1.3}\frac{1}{1.4}\frac{1}{0.36}] = -0.7937 \end{aligned}$$

where again we computed the residues at the poles located at 0.5 and -0.8, respectively.

The correlation matrix of the adaptive-filter input signal is given by

$$\mathbf{R} = \begin{bmatrix} 1.6873 & -0.7937 \\ -0.7937 & 1.6873 \end{bmatrix}$$

(b) The coefficients corresponding to the Wiener solution are given by

$$\begin{aligned} \mathbf{w}_o &= \mathbf{R}^{-1}\mathbf{p} \\ &= 0.45106 \begin{bmatrix} 1.6873 & 0.7937 \\ 0.7937 & 1.6873 \end{bmatrix} \begin{bmatrix} 0.9524 \\ 0.4762 \end{bmatrix} \\ &= \begin{bmatrix} 0.8953 \\ 0.7034 \end{bmatrix} \end{aligned}$$

(c) The LMS algorithm is applied to minimize the MSE using a convergence factor $\mu = 1/40\text{tr}[\mathbf{R}]$, where $\text{tr}[\mathbf{R}] = 3.3746$. The value of μ is 0.0074. This small value of the convergence factor allows a smooth convergence path. The convergence path of the algorithm on the MSE surface is depicted in Fig. 3.4. As can be noted, the path followed by the LMS algorithm looks like a noisy steepest-descent path. It first approaches the main axis (eigenvector) corresponding to the smaller eigenvalue, and then follows toward the minimum in a direction increasingly aligned with this main axis.

(d) The learning curves of the MSE and the filter coefficients, in a single run are depicted in Fig. 3.5. The learning curves of the MSE and the filter coefficients, obtained by averaging the results of 25 runs, are depicted in Fig. 3.6. As can be noted, these curves are less noisy than in the single run case.

□

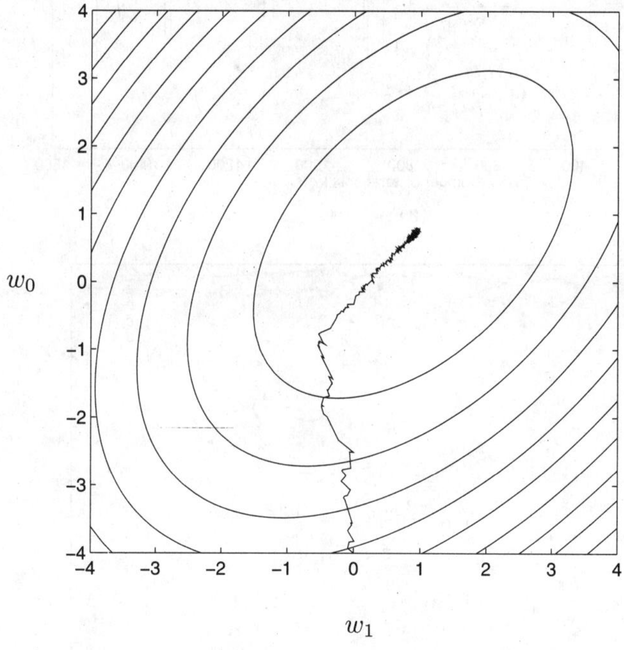

Figure 3.4 Convergence path on the MSE surface.

The adaptive-filtering problems discussed so far assumed that the signals taken from the environment were stochastic signals. Also, by assuming these signals were ergodic, we have shown that the adaptive filter is able to approach the Wiener solution by replacing the ensemble average by time averages. In conclusion, we can assume that the solution reached by the adaptive filter is based on time averages of the cross-correlations of the environment signals.

For example, if the environment signals are periodic deterministic signals, the optimal solution depends on the time average of the related cross-correlations computed over one period of the signals.

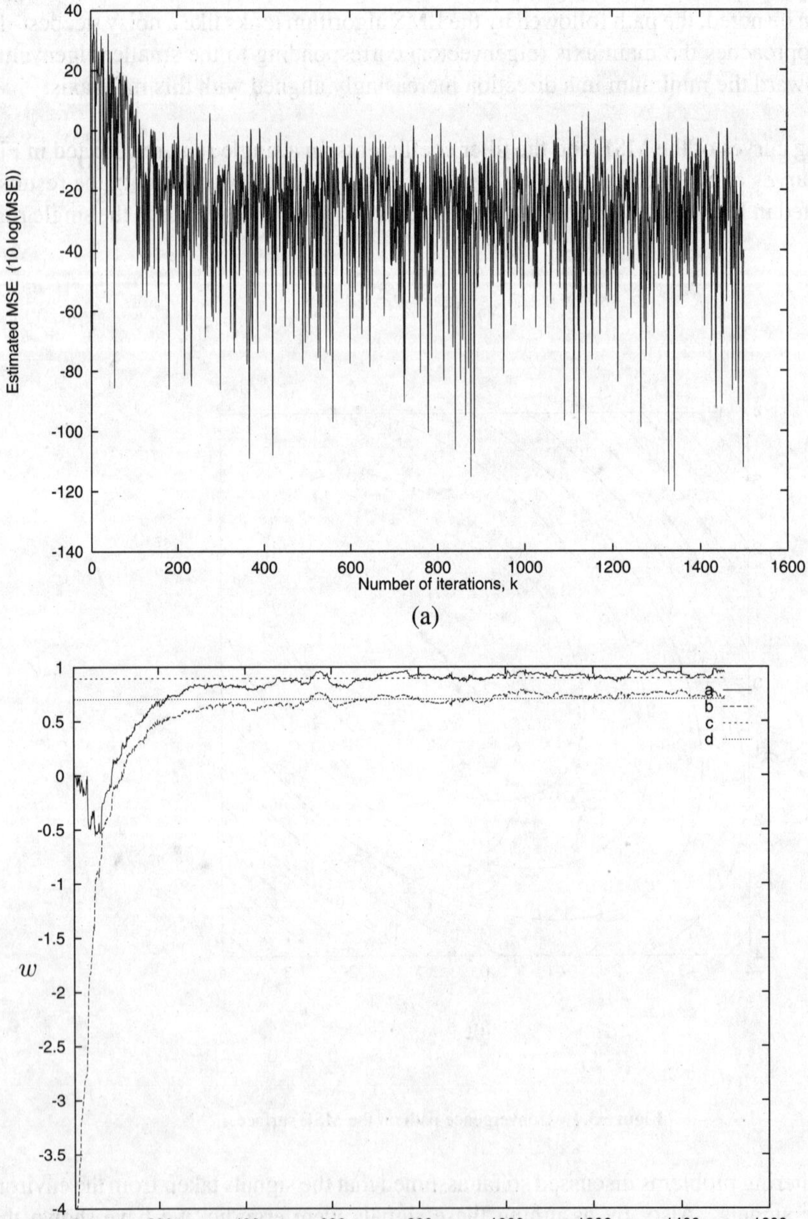

Figure 3.5 (a) Learning curve of the instantaneous squared error (b) Learning curves of the coefficients, a - first coefficient, b - second coefficient, c - optimal value for the first coefficient, d - optimal value of the second coefficient.

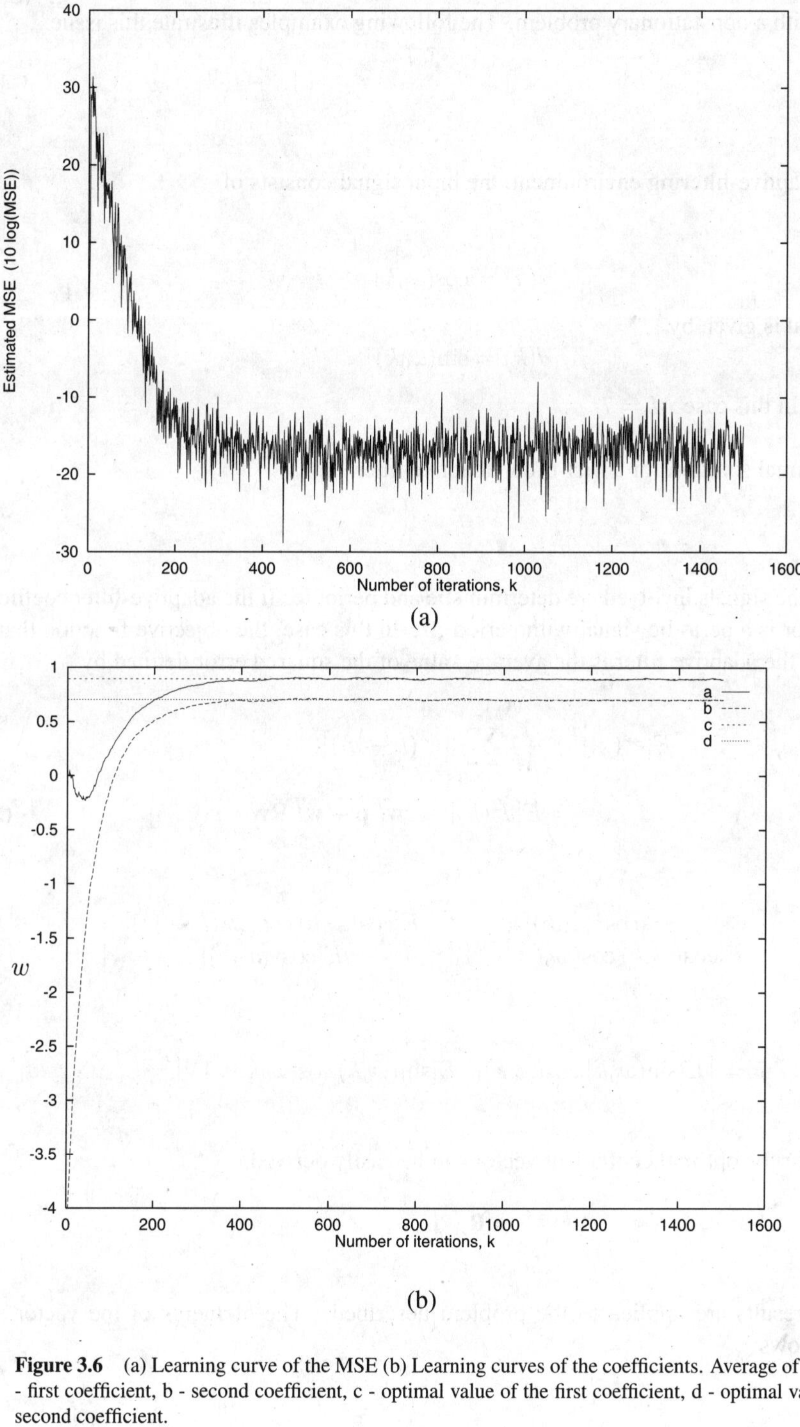

Figure 3.6 (a) Learning curve of the MSE (b) Learning curves of the coefficients. Average of 25 runs. a - first coefficient, b - second coefficient, c - optimal value of the first coefficient, d - optimal value of the second coefficient.

Note that in this case, the solution obtained using an ensemble average would be time varying since we are dealing with a nonstationary problem. The following examples illustrate this issue.

Example 3.2

Suppose in an adaptive-filtering environment, the input signal consists of

$$x(k) = \cos(\omega_0 k)$$

The desired signal is given by

$$d(k) = \sin(\omega_0 k)$$

where $\omega_0 = \frac{2\pi}{M}$. In this case $M = 7$.

Compute the optimal solution for a first-order adaptive filter.

Solution:

In this example, the signals involved are deterministic and periodic. If the adaptive-filter coefficients are fixed, the error is a periodic signal with period M. In this case, the objective function that will be minimized by the adaptive filter is the average value of the squared error defined by

$$\begin{aligned} \bar{E}[e^2(k)] &= \frac{1}{M}\sum_{m=0}^{M-1}[e^2(k-m)] \\ &= \bar{E}[d^2(k)] - 2\mathbf{w}^T\bar{\mathbf{p}} + \mathbf{w}^T\bar{\mathbf{R}}\mathbf{w} \end{aligned} \tag{3.80}$$

where

$$\bar{\mathbf{R}} = \begin{bmatrix} \bar{E}[\cos^2(\omega_0 k)] & \bar{E}[\cos(\omega_0 k)\cos(\omega_0(k-1))] \\ \bar{E}[\cos(\omega_0 k)\cos(\omega_0(k-1))] & \bar{E}[\cos^2(\omega_0 k)] \end{bmatrix}$$

and

$$\bar{\mathbf{p}} = \left[\bar{E}[\sin(\omega_0 k)\cos(\omega_0 k)] \;\; \bar{E}[\sin(\omega_0 k)\cos(\omega_0 k - 1)]\right]^T$$

The expression for the optimal coefficient vector can be easily derived.

$$\mathbf{w}_o = \bar{\mathbf{R}}^{-1}\bar{\mathbf{p}}$$

Now the above results are applied to the problem described. The elements of the vector $\bar{\mathbf{p}}$ are calculated as follows:

$$
\begin{aligned}
\bar{\mathbf{p}} &= \frac{1}{M}\sum_{m=0}^{M-1}\begin{bmatrix} d(k-m)x(k-m) \\ d(k-m)x(k-m-1) \end{bmatrix} \\
&= \frac{1}{M}\sum_{m=0}^{M-1}\begin{bmatrix} \sin(\omega_0(k-m))\cos(\omega_0(k-m)) \\ \sin(\omega_0(k-m))\cos(\omega_0(k-m-1)) \end{bmatrix} \\
&= \frac{1}{2}\begin{bmatrix} 0 \\ \sin(\omega_0) \end{bmatrix} \\
&= \begin{bmatrix} 0 \\ 0.3909 \end{bmatrix}
\end{aligned}
$$

The elements of the correlation matrix of the adaptive-filter input signal are calculated as follows:

$$
\begin{aligned}
\bar{r}(i) &= \bar{E}[x(k)x(k-i)] \\
&= \frac{1}{M}\sum_{m=0}^{M-1}[\cos(\omega_0(k-m))\cos(\omega_0(k-m-i))]
\end{aligned}
$$

where

$$
\begin{aligned}
\bar{r}(0) &= \bar{E}[\cos^2(\omega_0(k))] = 0.5 \\
\bar{r}(1) &= \bar{E}[\cos(\omega_0(k))\cos(\omega_0(k-1))] = 0.3117
\end{aligned}
$$

The correlation matrix of the adaptive-filter input signal is given by

$$
\bar{\mathbf{R}} = \begin{bmatrix} 0.5 & 0.3117 \\ 0.3117 & 0.5 \end{bmatrix}
$$

The coefficients corresponding to the optimal solution are given by

$$
\bar{\mathbf{w}}_o = \bar{\mathbf{R}}^{-1}\bar{\mathbf{p}} = \begin{bmatrix} -0.7972 \\ 1.2788 \end{bmatrix}
$$

□

Example 3.3

(a) Assume the input and desired signals are deterministic and periodic with period M. Study the LMS algorithm behavior.

(b) Choose an appropriate value for μ in the previous example and plot the convergence path for the LMS algorithm on the average error surface.

Solution:

(a) It is convenient at this point to recall the coefficient updating of the LMS algorithm

$$\mathbf{w}(k+1) = \mathbf{w}(k) + 2\mu\mathbf{x}(k)e(k) = \mathbf{w}(k) + 2\mu\mathbf{x}(k)\left[d(k) - \mathbf{x}^T(k)\mathbf{w}(k)\right]$$

This equation can be rewritten as

$$\mathbf{w}(k+1) = \left[\mathbf{I} - 2\mu\mathbf{x}(k)\mathbf{x}^T(k)\right]\mathbf{w}(k) + 2\mu d(k)\mathbf{x}(k) \tag{3.81}$$

The solution of equation (3.81), as a function of the initial values of the adaptive-filter coefficients, is given by

$$\mathbf{w}(k+1) = \prod_{i=0}^{k}\left[\mathbf{I} - 2\mu\mathbf{x}(i)\mathbf{x}^T(i)\right]\mathbf{w}(0) + \sum_{i=0}^{k}\left\{\prod_{j=i+1}^{k}\left[\mathbf{I} - 2\mu\mathbf{x}(j)\mathbf{x}^T(j)\right]2\mu d(i)\mathbf{x}(i)\right\} \tag{3.82}$$

where we define that $\prod_{j=k+1}^{k}[\cdot] = 1$ for the second product.

Assuming the value of the convergence factor μ is small enough to guarantee that the LMS algorithm will converge, the first term on the right-hand side of the above equation will vanish as $k \to \infty$. The resulting expression for the coefficient vector is given by

$$\mathbf{w}(k+1) = \sum_{i=0}^{k}\left\{\prod_{j=i+1}^{k}\left[\mathbf{I} - 2\mu\mathbf{x}(j)\mathbf{x}^T(j)\right]2\mu d(i)\mathbf{x}(i)\right\}$$

The analysis of the above solution is not straightforward. Following an alternative path based on averaging the results in a period M, we can reach conclusive results.

Let us define the average value of the adaptive-filter parameters as follows:

$$\overline{\mathbf{w}(k+1)} = \frac{1}{M}\sum_{m=0}^{M-1}\mathbf{w}(k+1-m)$$

Similar definition can be applied to the remaining parameters of the algorithm.

Considering that the signals are deterministic and periodic, we can apply the average operation to equation (3.81). The resulting equation is

$$\begin{aligned}\overline{\mathbf{w}(k+1)} &= \frac{1}{M}\sum_{m=0}^{M-1}\left[\mathbf{I} - 2\mu\mathbf{x}(k-m)\mathbf{x}^T(k-m)\right]\mathbf{w}(k-m) + \frac{1}{M}\sum_{m=0}^{M-1}2\mu d(k-m)\mathbf{x}(k-m)\\ &= \overline{\left[\mathbf{I} - 2\mu\mathbf{x}(k)\mathbf{x}^T(k)\right]\mathbf{w}(k)} + \overline{2\mu d(k)\mathbf{x}(k)}\end{aligned} \tag{3.83}$$

For large k and small μ, it is expected that the parameters converge to the neighborhood of the optimal solution. In this case, we can consider that $\overline{\mathbf{w}(k+1)} \approx \overline{\mathbf{w}(k)}$ and that the following approximation is valid

$$\overline{\mathbf{x}(k)\mathbf{x}^T(k)\mathbf{w}(k)} \approx \overline{\mathbf{x}(k)\mathbf{x}^T(k)}\ \ \overline{\mathbf{w}(k)}$$

since the parameters after convergence wander around the optimal solution. Using these approximations in (3.83), the average values of the parameters in the LMS algorithm for periodic signals are given by

$$\overline{\mathbf{w}(k)} \approx \overline{\mathbf{x}(k)\mathbf{x}^T(k)}^{-1}\ \overline{d(k)\mathbf{x}(k)} = \bar{\mathbf{R}}^{-1}\bar{\mathbf{p}}$$

(b) The LMS algorithm is applied to minimize the squared error of the problem described in Example 3.2 using a convergence factor $\mu = 1/100\text{tr}[\bar{\mathbf{R}}]$, where $\text{tr}[\bar{\mathbf{R}}] = 1$. The value of μ is 0.01. The convergence path of the algorithm on the MSE surface is depicted in Fig. 3.7. As can be verified, the parameters generated by the LMS algorithm approach the optimal solution.

□

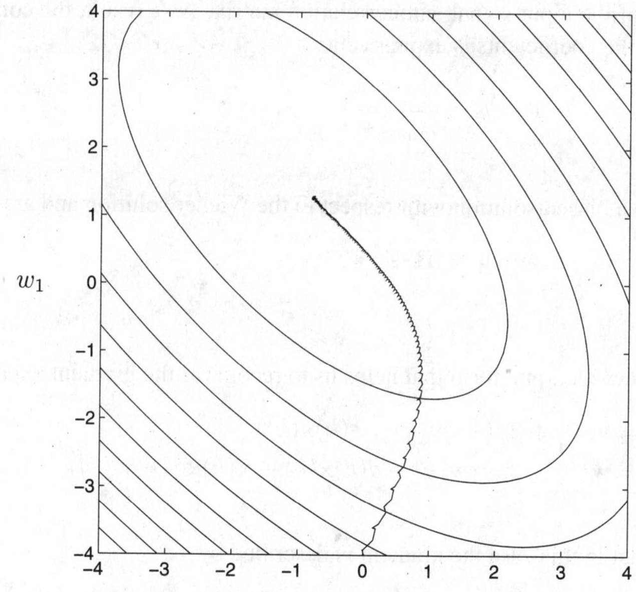

Figure 3.7 Convergence path on the MSE surface.

Example 3.4

The leaky LMS algorithm has the following updating equation

$$\mathbf{w}(k+1) = (1 - 2\mu\gamma)\mathbf{w}(k) + 2\mu e(k)\mathbf{x}(k) \tag{3.84}$$

where $0 < \gamma << 1$.

(a) Compute the range of values of μ such that the coefficients converge in average.

(b) What is the objective function this algorithm actually minimizes?

(c) What happens to the filter coefficients if the error and/or input signals become zero?

Solution:

(a) By utilizing the error expression we generate the coefficient updating equation given by

$$\mathbf{w}(k+1) = \{\mathbf{I} - 2\mu[\mathbf{x}(k)\mathbf{x}^T(k) + \gamma\mathbf{I}]\}\mathbf{w}(k) + 2\mu d(k)\mathbf{x}(k)$$

By applying the expectation operation it follows that

$$E[\mathbf{w}(k+1)] = \{\mathbf{I} - 2\mu[\mathbf{R} + \gamma\mathbf{I}]\}E[\mathbf{w}(k)] + 2\mu\mathbf{p}$$

The inclusion of γ is equivalent to add a white noise to the input signal $x(n)$, such that a value of γ is added to the eigenvalues of the input signal autocorrelation matrix. As a result, the condition for the stability in the mean for the coefficients is expressed as

$$0 < \mu < \frac{1}{\lambda_{\max} + \gamma}$$

The coefficients converge to a biased solution with respect to the Wiener solution and are given by

$$E[\mathbf{w}(k)] = [\mathbf{R} + \gamma\mathbf{I}]^{-1}\mathbf{p}$$

for $k \to \infty$.

(b) Equation (3.84) can be rewritten in a form that helps us to recognize the gradient expression.

$$\begin{aligned}\mathbf{w}(k+1) &= \mathbf{w}(k) + 2\mu(-\gamma\mathbf{w}(k) + e(k)\mathbf{x}(k)) \\ &= \mathbf{w}(k) - 2\mu(\gamma\mathbf{w}(k) - d(k)\mathbf{x}(k) + \mathbf{x}(k)\mathbf{x}^T(k)\mathbf{w}(k))\end{aligned} \tag{3.85}$$

By inspection we observe that in this case the gradient is described by

$$\mathbf{g}_{\mathbf{w}}(k) = 2\gamma\mathbf{w}(k) - 2e(k)\mathbf{x}(k) = 2\gamma\mathbf{w}(k) - 2d(k)\mathbf{x}(k) + 2\mathbf{x}(k)\mathbf{x}^T(k)\mathbf{w}(k)$$

The corresponding objective function that is indeed minimized is given by

$$\xi(k) = \{\gamma\|\mathbf{w}(k)\|^2 + e^2(k)\}$$

(c) For zero input or zero error signal after some initial iterations, the dynamic updating equation (3.84) has zero excitation. Since the eigenvalues of the transition matrix $\{\mathbf{I} - 2\mu[\mathbf{x}(k)\mathbf{x}^T(k) + \gamma\mathbf{I}]\}$ are smaller than one, then the adaptive-filter coefficients will tend to zero for large k.

□

3.6.2 System Identification Simulations

In this subsection, a system identification problem is described and solved by using the LMS algorithm. In the following chapters the same problem will be solved using other algorithms presented in the book. For the FIR adaptive filters the following identification problem is posed:

Example 3.5

An adaptive-filtering algorithm is used to identify a system with impulse response given below.

$$\mathbf{h} = [0.1\ 0.3\ 0.0\ -0.2\ -0.4\ -0.7\ -0.4\ -0.2]^T$$

Consider three cases for the input signal: colored noises with variance $\sigma_x^2 = 1$ and eigenvalue spread of their correlation matrix equal to 1.0, 20, and 80, respectively. The measurement noise is Gaussian white noise uncorrelated with the input and with variance $\sigma_n^2 = 10^{-4}$. The adaptive filter has 8 coefficients.

(a) Run the algorithm and comment on the convergence behavior in each case.

(b) Measure the misadjustment in each example and compare with the theoretical results where appropriate.

(c) Considering that fixed-point arithmetic is used, run the algorithm for a set of experiments and calculate the expected values for $||\Delta\mathbf{w}(k)_Q||^2$ and $\xi(k)_Q$ for the following case:

Additional noise: white noise with variance	$\sigma_n^2 = 0.0015$
Coefficient wordlength:	$b_c = 16$ bits
Signal wordlength:	$b_d = 16$ bits
Input signal: Gaussian white noise with variance	$\sigma_x^2 = 1.0$

(d) Repeat the previous experiment for the following cases
$b_c = 12$ bits, $b_d = 12$ bits.
$b_c = 10$ bits, $b_d = 10$ bits.

(e) Suppose the unknown system is a time-varying system whose coefficients are first-order Markov processes with $\lambda_{\mathbf{w}} = 0.99$ and $\sigma_{\mathbf{w}}^2 = 0.0015$. The initial time-varying-system multiplier coefficients are the ones above described. The input signal is Gaussian white noise with variance $\sigma_x^2 = 1.0$, and the measurement noise is also Gaussian white noise independent of the input signal and of the elements of $\mathbf{n_w}(k)$, with variance $\sigma_n^2 = 0.01$. Simulate the experiment described, measure the total excess MSE, and compare to the calculated results.

Solution:

(a) The colored input signal is generated by applying Gaussian white noise, with variance σ_v^2, to a first-order filter with transfer function

$$H(z) = \frac{z}{z-a}$$

As can be shown from the equation (2.83), the input signal correlation matrix in this case is given by

$$\mathbf{R} = \frac{\sigma_v^2}{1-a^2}\begin{bmatrix} 1 & a & \cdots & a^7 \\ a & 1 & \cdots & a^6 \\ \vdots & \vdots & \ddots & \vdots \\ a^7 & a^6 & \cdots & 1 \end{bmatrix}$$

The proper choice of the value of a, in order to obtain the desired eigenvalue spread, is not a straightforward task. Some guidelines are now discussed. For example, if the adaptive filter is of first order, the matrix **R** is two by two with eigenvalues

$$\lambda_{\max} = \frac{\sigma_v^2}{1-a^2}(1+a)$$

and

$$\lambda_{\min} = \frac{\sigma_v^2}{1-a^2}(1-a)$$

respectively. In this case, the choice of a is straightforward.

In general, it can be shown that

$$\frac{\lambda_{\max}}{\lambda_{\min}} \leq \frac{|H_{\max}(e^{\jmath\omega})|^2}{|H_{\min}(e^{\jmath\omega})|^2}$$

For a very large order adaptive filter, the eigenvalue spread approaches

$$\frac{\lambda_{\max}}{\lambda_{\min}} \approx \frac{|H_{\max}(e^{\jmath\omega})|^2}{|H_{\min}(e^{\jmath\omega})|^2} = \left\{\frac{1+a}{1-a}\right\}^2$$

where the details to reach this result can be found in page 124 of [20].

Using the above relations as guidelines, we reached the correct values of a. These values are $a = 0.6894$ and $a = 0.8702$ for eigenvalue spreads of 20 and 80, respectively.

Since the variance of the input signal should be unity, the variance of the Gaussian white noise that produces $x(k)$ should be given by

$$\sigma_v^2 = 1 - a^2$$

For the LMS algorithm, we first calculate the upper bound for μ ($\mu_{\max}$) to guarantee the algorithm stability, and run the algorithm for $\mu_{\max}$, $\mu_{\max}/5$, and $\mu_{\max}/10$.

In this example, the LMS algorithm does not converge for $\mu = \mu_{\max} \approx 0.1$. The convergence behavior for $\mu_{\max}/5$ and $\mu_{\max}/10$ are illustrated through the learning curves depicted in Fig. 3.8, where in this case the eigenvalue spread is 1. Each curve is obtained by averaging the results of 200 independent runs. As can be noticed, the reduction of the convergence factor leads to a reduction in the convergence speed. Also note that for $\mu = 0.02$ the estimated MSE is plotted only for the first 400 iterations, enough to display the convergence behavior. In all examples the tap coefficients are initialized with zero. Fig. 3.9 illustrates the learning curves for the various eigenvalue spreads, where in each case the convergence factor is $\mu_{\max}/5$. As expected the convergence rate is reduced for a high eigenvalue spread.

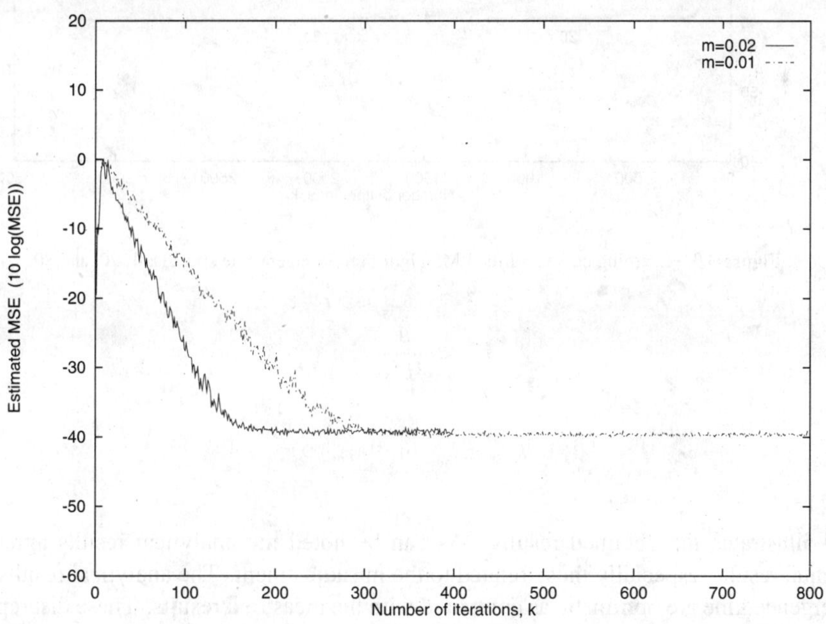

Figure 3.8 Learning curves for the LMS algorithm with convergence factors $\mu_{\max}/5$ and $\mu_{\max}/10$.

(b) The misadjustment is measured and compared with the results obtained from the following relation

$$M = \frac{\mu(N+1)\sigma_x^2}{1-\mu(N+1)\sigma_x^2}$$

Also, for the present problem we calculated the time constants τ_{wi} and τ_{ei}, and the expected number of iterations to achieve convergence using the relations

$$\tau_{wi} \approx \frac{1}{2\mu\lambda_i}$$

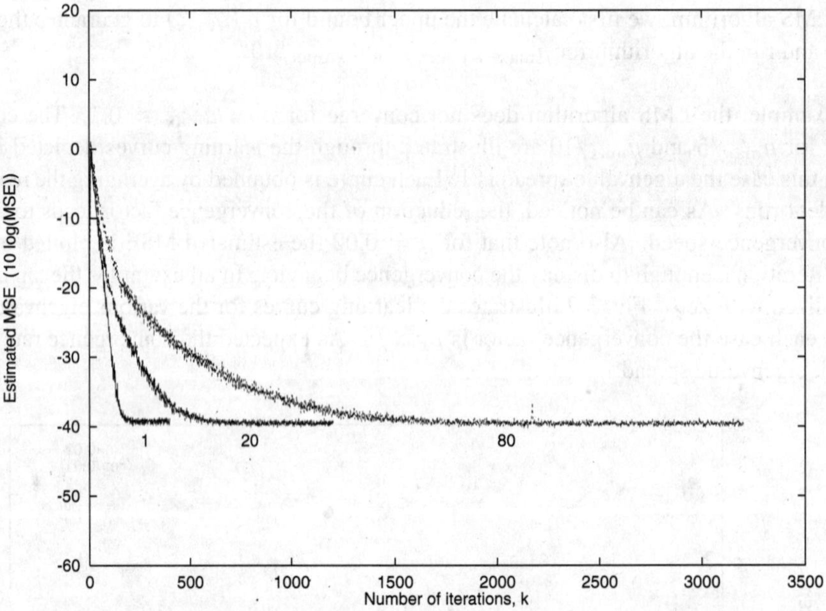

Figure 3.9 Learning curves for the LMS algorithm for eigenvalue spreads: 1, 20, and 80.

$$\tau_{ei} \approx \frac{1}{4\mu\lambda_i}$$

$$k \approx \tau_{e_{\max}} \ln(100)$$

Table 3.1 illustrates the obtained results. As can be noted the analytical results agree with the experimental results, especially those related to the misadjustment. The analytical results related to the convergence time are optimistic as compared with the measured results. These discrepancies are mainly due to the approximations in the analysis.

(c), (d) The LMS algorithm is implemented employing fixed-point arithmetic using 16, 12, and 10 bits for data and coefficient wordlengths. The chosen value of μ is 0.01. The learning curves for the MSE are depicted in Fig. 3.10. Fig. 3.11 depicts the evolution of $||\Delta\mathbf{w}(k)_Q||^2$ with the number of iterations. The experimental results show that the algorithm still works for such limited precision. In Table 3.2, we present a summary of the results obtained from simulation experiments and a comparison with the results predicted by the theory. The experimental results are obtained by averaging the results of 200 independent runs. The relations employed to calculate the theoretical results shown in Table 3.2 correspond to equations (B.26) and (B.32) derived in Appendix B. These relations are repeated here for convenience:

$$E[||\Delta\mathbf{w}(k)_Q||^2] = \frac{\mu(\sigma_n^2 + \sigma_e^2)(N+1)}{1 - \mu(N+1)\sigma_x^2} + \frac{(N+1)\sigma_{\mathbf{w}}^2}{4\mu\sigma_x^2[1 - \mu(N+1)\sigma_x^2]}$$

Table 3.1 Evaluation of the LMS Algorithm

μ	$\frac{\lambda_{max}}{\lambda_{min}}$	Misadjustment Experiment	Misadjustment Theory	$\tau_{e_{max}}$	$\tau_{w_{max}}$	Iterations
0.02	1	0.2027	0.1905	12.5	25	58
0.01280	20	0.1298	0.1141	102.5	205	473
0.01024	80	0.1045	0.0892	338.9	677.5	1561
0.01	1	0.0881	0.0870	25	50	116
0.006401	20	0.0581	0.0540	205	410	944
0.005119	80	0.0495	0.0427	677.5	1355	3121

$$\xi(k)_Q = \frac{\sigma_e^2 + \sigma_n^2}{1 - \mu(N+1)\sigma_x^2} + \frac{(N+1)\sigma_{\mathbf{w}}^2}{4\mu[1 - \mu(N+1)\sigma_x^2]}$$

The results of Table 3.2 confirm that the finite-precision implementation analysis presented is accurate.

Table 3.2 Results of the Finite Precision Implementation of the LMS Algorithm

No. of bits	$\xi(k)_Q$ Experiment	$\xi(k)_Q$ Theory	$E[\|\Delta\mathbf{w}(k)_Q\|^2]$ Experiment	$E[\|\Delta\mathbf{w}(k)_Q\|^2]$ Theory
16	$1.629\ 10^{-3}$	$1.630\ 10^{-3}$	$1.316\ 10^{-4}$	$1.304\ 10^{-4}$
12	$1.632\ 10^{-3}$	$1.631\ 10^{-3}$	$1.309\ 10^{-4}$	$1.315\ 10^{-4}$
10	$1.663\ 10^{-3}$	$1.648\ 10^{-3}$	$1.465\ 10^{-4}$	$1.477\ 10^{-4}$

(e) The performance of the LMS algorithm is also tested in the nonstationary environment above described. The excess MSE is measured and depicted in Fig. 3.12. For this example μ_{opt} is found to be greater than μ_{max}. The value of μ used in the example is 0.05. The excess MSE in steady state predicted by the relation

$$\xi_{total} \approx \frac{\mu\sigma_n^2 \mathrm{tr}[\mathbf{R}]}{1 - \mu\mathrm{tr}[\mathbf{R}]} + \frac{\sigma_{\mathbf{w}}^2}{4\mu}\sum_{i=0}^{N}\frac{1}{1 - \mu\lambda_i}$$

is 0.124, whereas the measured excess MSE in steady state is 0.118. Once more the results obtained from the analysis are accurate.

□

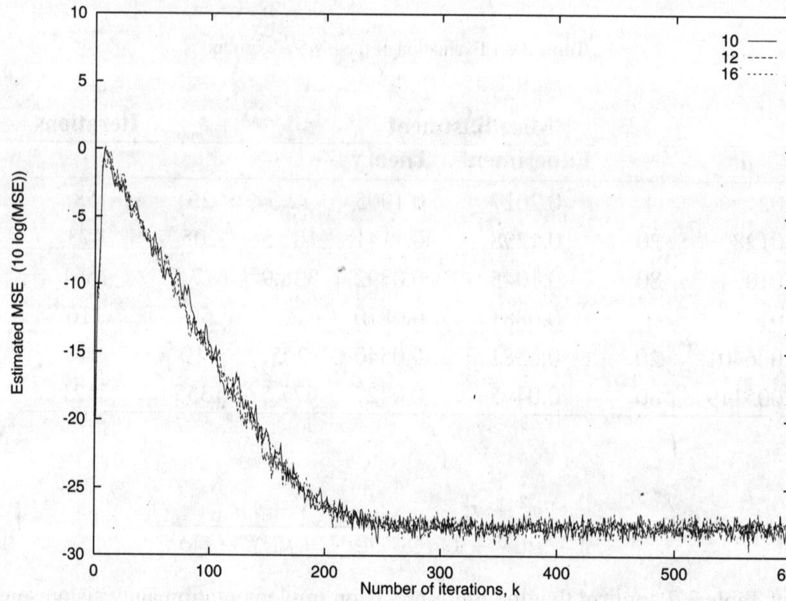

Figure 3.10 Learning curves for the LMS algorithm implemented with fixed-point arithmetic and with $\mu = 0.01$.

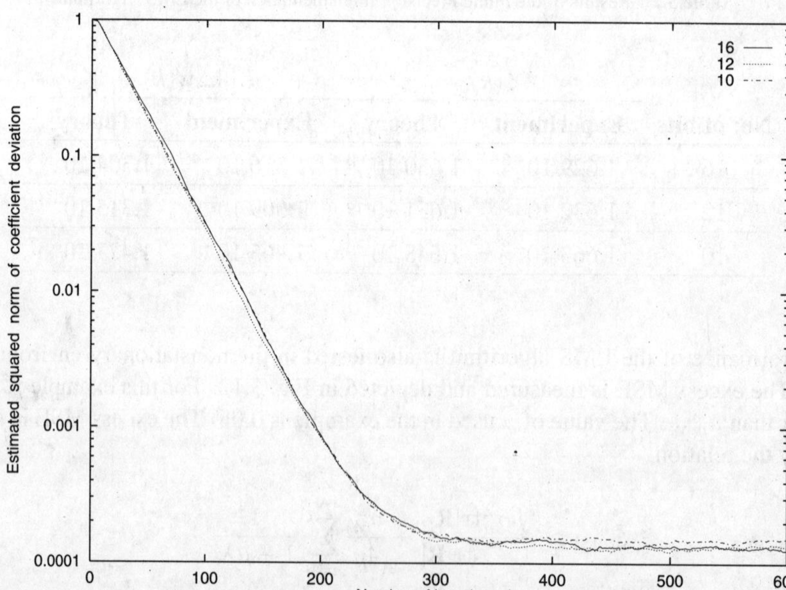

Figure 3.11 Estimate of $||\Delta\mathbf{w}(k)_Q||^2$ for the LMS algorithm implemented with fixed-point arithmetic and with $\mu = 0.01$.

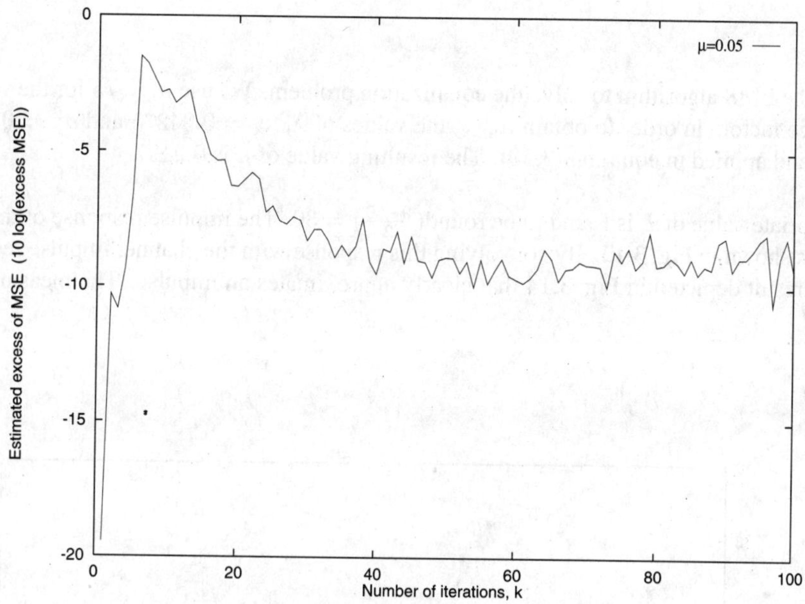

Figure 3.12 The excess MSE of the LMS algorithm in nonstationary environment, $\mu = 0.05$.

3.6.3 Channel Equalization Simulations

In this subsection an equalization example is described. This example will be used as pattern for comparison of several algorithms presented in this book.

Example 3.6

Perform the equalization of a channel with the following impulse response

$$h(k) = 0.1\ (0.5^k)$$

for $k = 0, 1, \ldots 8$. Use a known training signal that consists of independent binary samples (-1,1). An additional Gaussian white noise with variance $10^{-2.5}$ is present at the channel output.

(a) Find the impulse response of an equalizer with 50 coefficients.

(b) Convolve the equalizer impulse response at a given iteration after convergence, with the channel impulse response and comment on the result.

Solution:

We apply the LMS algorithm to solve the equalization problem. We use $\mu_{max}/5$ for the value of the convergence factor. In order to obtain μ_{max}, the values of $\lambda_{max} = 0.04275$ and $\sigma_x^2 = 0.01650$ are measured and applied in equation (3.30). The resulting value of μ is 0.2197.

The appropriate value of L is found to be round$(\frac{9+50}{2}) = 30$. The impulse response of the resulting equalizer is shown in Fig. 3.13. By convolving this response with the channel impulse response, we obtain the result depicted in Fig. 3.14 that clearly approximates an impulse. The measured MSE is 0.3492.

□

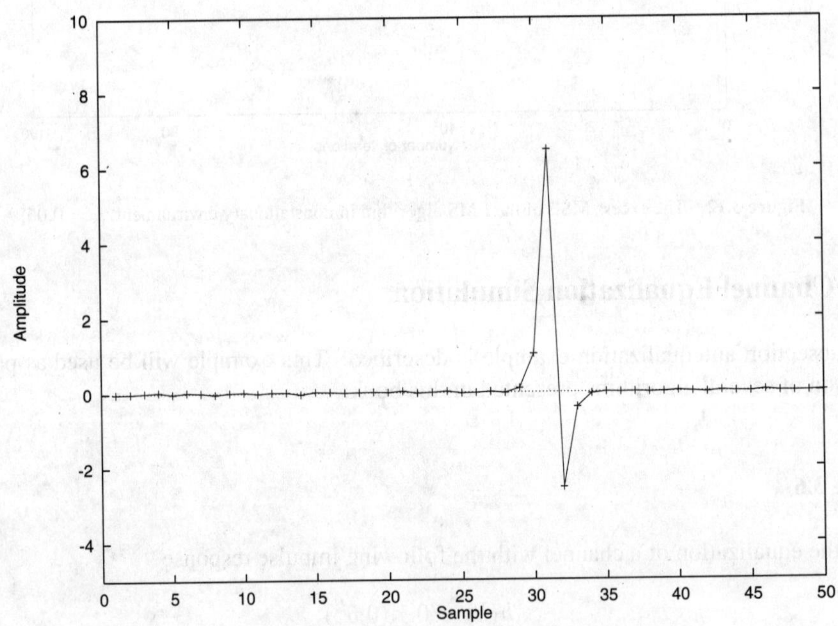

Figure 3.13 Equalizer impulse response; LMS algorithm.

3.6.4 Fast Adaptation Simulations

The exact evaluation of the learning curves of the squared error or coefficients of an adaptive filter is a difficult task. In general the solution is to run repeated simulations and average their results. For the LMS algorithm this ensemble averaging leads to results which are close to those predicted by independence theory [4], if the convergence factor is small. In fact, the independence theory is a first-order approximation in μ to the actual learning curves of $\xi(k)$ [4], [22].

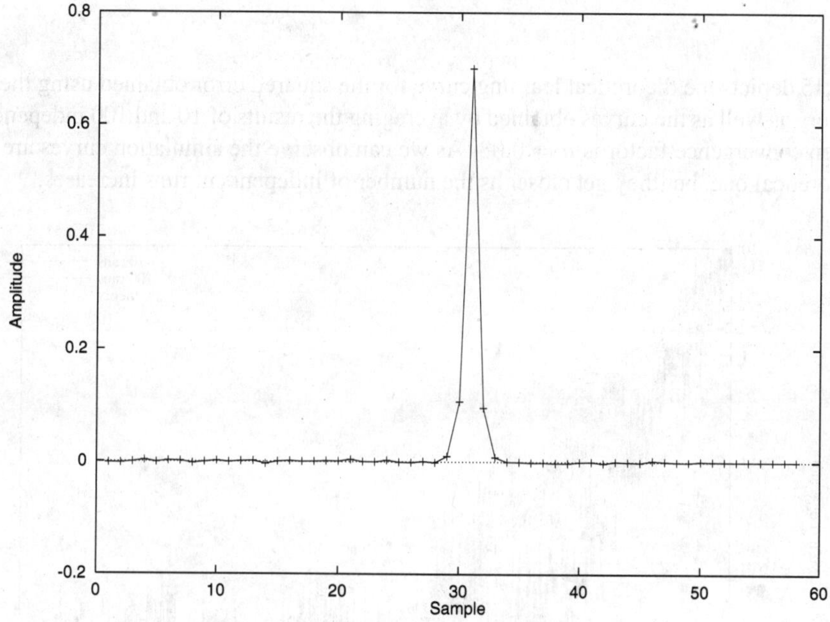

Figure 3.14 Convolution result; LMS algorithm.

However, for large μ the results from the ensemble average can be quite different from the theoretical prediction [21]. The following example explores this observation.

Example 3.7

An adaptive-filtering algorithm is used to identify a system. Consider three cases described below.

(a) The unknown system has length 10, the input signal is a stationary Gaussian noise with variance $\sigma_x^2 = 1$ and the measurement noise is Gaussian white noise uncorrelated with the input and with variance $\sigma_n^2 = 10^{-4}$.

(b) The unknown system has length 2, the input signal is a stationary uniformly distributed noise in the range -0.5 and 0.5, and there is no measurement noise.

(c) Study the behavior of the ensemble average as well as the mean square value of the coefficient error of an LMS algorithm with a single coefficient, when the input signal is a stationary uniformly distributed noise in the range $-a$ and a, and there is no measurement noise.

Solution:

(a) Fig. 3.15 depicts the theoretical learning curve for the squared error obtained using the independence theory as well as the curves obtained by averaging the results of 10 and 100 independent runs. The chosen convergence factor is $\mu = 0.08$. As we can observe the simulation curves are not close to the theoretical one, but they get closer as the number of independent runs increases.

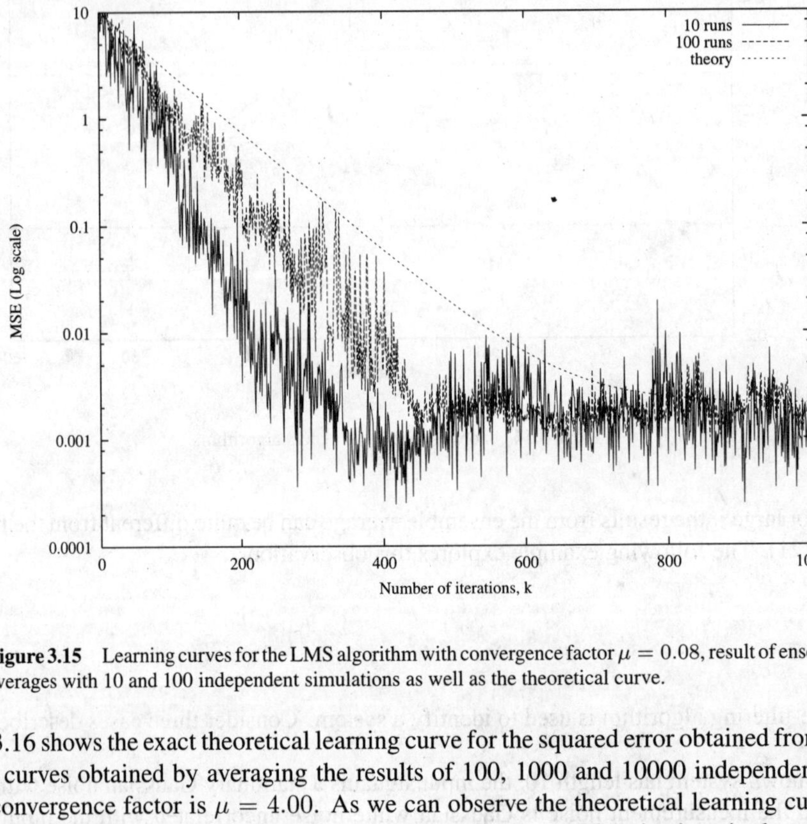

Figure 3.15 Learning curves for the LMS algorithm with convergence factor $\mu = 0.08$, result of ensemble averages with 10 and 100 independent simulations as well the theoretical curve.

(b) Fig. 3.16 shows the exact theoretical learning curve for the squared error obtained from [23] along with the curves obtained by averaging the results of 100, 1000 and 10000 independent runs. The chosen convergence factor is $\mu = 4.00$. As we can observe the theoretical learning curve diverges whereas the simulation curves converge. A closer look at this problem is given in the next item.

(c) From equation (3.12), the evolution of the squared deviation in the tap coefficient is given by

$$\Delta w^2(k+1) = \left[1 - 2\mu x^2(k)\right]^2 \Delta w^2(k)$$

where $\Delta w(0)$ is fixed, and the additional noise is zero. Note that the evolution of $\Delta w^2(k)$ is governed by the random factor $\left[1 - 2\mu x^2(k)\right]^2$. With the assumptions on the input signal these random factors form an independent, identically distributed random sequence. The above model can then be rewritten as

$$\Delta w^2(k+1) = \left\{\prod_{i=0}^{k} \left[1 - 2\mu x^2(i)\right]^2\right\} \Delta w^2(0) \tag{3.86}$$

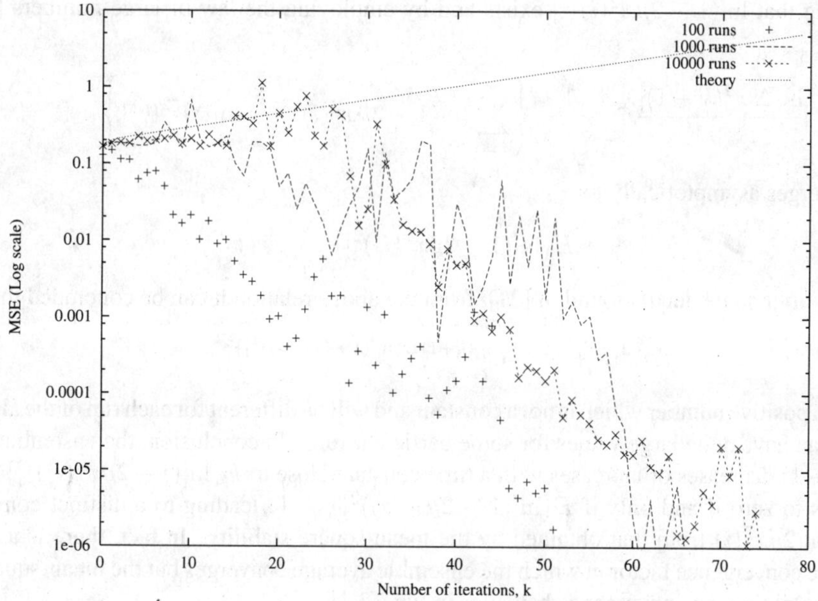

Figure 3.16 Learning curves for the LMS algorithm with convergence factor $\mu = 4.00$, result of ensemble averages with 100, 1000 and 10000 independent simulations as well as the theoretical curve.

The objective now is to study the differences between the expected value of $\Delta w^2(k+1)$ and its ensemble average. In the first case, by using the independence of the random factors in equation (3.86) we have that

$$\begin{aligned} E[\Delta w^2(k+1)] &= \left\{ \prod_{i=0}^{k} E\left[(1-2\mu x^2(i))^2\right] \right\} \Delta w^2(0) \\ &= \{E\left[(1-2\mu x^2(0))^2\right]\}^{k+1} \Delta w^2(0) \end{aligned} \tag{3.87}$$

Since the variance of the input signal is $\sigma_x^2 = \frac{a^2}{3}$ and its fourth-order moment is given by $\frac{a^4}{5}$, the above equation can be rewritten as

$$\begin{aligned} E[\Delta w^2(k+1)] &= \left\{E\left[(1-2\mu x^2(0))^2\right]\right\}^{k+1} \Delta w^2(0) \\ &= \left(1 - 4\mu\frac{a^2}{3} + 4\mu^2\frac{a^4}{5}\right)^{k+1} \Delta w^2(0) \end{aligned} \tag{3.88}$$

From the above equation we can observe that the rate of convergence of $E[\Delta w^2(k)]$ is equal to $\ln\{E\left[(1-2\mu x^2(0))^2\right]\}$.

Let's examine now how the ensemble average of $\Delta w^2(k)$ evolves, for large k and μ, by computing its logarithm as follows:

$$\ln[\Delta w^2(k+1)] = \sum_{i=0}^{k} \ln[(1-2\mu x^2(i))^2] + \ln[\Delta w^2(0)] \tag{3.89}$$

By assuming that $\ln[(1-2\mu x^2(i))^2]$ exists and by employing the law of large numbers [13], we obtain

$$\frac{\ln[\Delta w^2(k+1)]}{k+1} = \frac{1}{k+1}\left\{\sum_{i=0}^{k} \ln[(1-2\mu x^2(i))^2] + \ln[\Delta w^2(0)]\right\} \tag{3.90}$$

which converges asymptotically to

$$E\{\ln[(1-2\mu x^2(i))^2]\}$$

For large k, after some details found in [21], from the above relation it can be concluded that

$$\Delta w^2(k+1) \approx C e^{(k+1)E\{\ln[(1-2\mu x^2(i))^2]\}} \tag{3.91}$$

where C is a positive number which is not a constant and will be different for each run of the algorithm. In fact, C can have quite large values for some particular runs. In conclusion, the ensemble average of $\Delta w^2(k+1)$ decreases or increases with a time constant close to $E\{\ln[(1-2\mu x^2(i))^2]\}^{-1}$. Also it converges to zero if and only if $E\{\ln[(1-2\mu x^2(i))^2]\} < 0$, leading to a distinct convergence condition on $2\mu x^2(i)$ from that obtained by the mean-square stability. In fact, there is a range of values of the convergence factor in which the ensemble average converges but the mean-square value diverges, explaining the convergence behavior in Fig. 3.16.

Fig. 3.17 depicts the curves of $\ln\{E\left[(1-2\mu x^2(0))^2\right]\}$ (the logarithm of the rate of convergence of mean-square coefficient error) and of $E\{\ln[(1-2\mu x^2(i))^2]\}$ as a function of $2\mu x^2(i)$. For small values of $2\mu x^2(i)$ both curves are quite close, however for larger values they are somewhat different in particular at the minima of the curves which correspond to the fastest convergence rate. In addition, as the curves become further apart the convergence is faster for the ensemble average of the squared coefficient error than for the mean-square coefficient error for large k.

□

3.6.5 The Linearly Constrained LMS Algorithm

In the narrowband beamformer application discussed in section 2.5, our objective was to minimize the array output power subjecting the linear combiner coefficients to a set of constraints. Now, let us derive an adaptive version of the LCMV filter by first rewriting the linearly constrained objective function of equation (2.107) for the case of multiple constraints as

$$\begin{aligned}\xi_c &= E\left[\mathbf{w}^T\mathbf{x}(k)\mathbf{x}^T(k)\mathbf{w}\right] + \boldsymbol{\lambda}^T\left[\mathbf{C}^T\mathbf{w} - \mathbf{f}\right] \\ &= \mathbf{w}^T\mathbf{R}\mathbf{w} + \boldsymbol{\lambda}^T\left[\mathbf{C}^T\mathbf{w} - \mathbf{f}\right]\end{aligned} \tag{3.92}$$

where $\mathbf{R}$ is the input signal autocorrelation matrix, $\mathbf{C}$ is the constraint matrix, and $\boldsymbol{\lambda}$ is the vector of Lagrange multipliers.

The constrained LMS-based algorithm [24] can be derived by searching for the coefficient vector $\mathbf{w}(k+1)$ that satisfies the set of constraints and represents a small update with respect to $\mathbf{w}(k)$ in

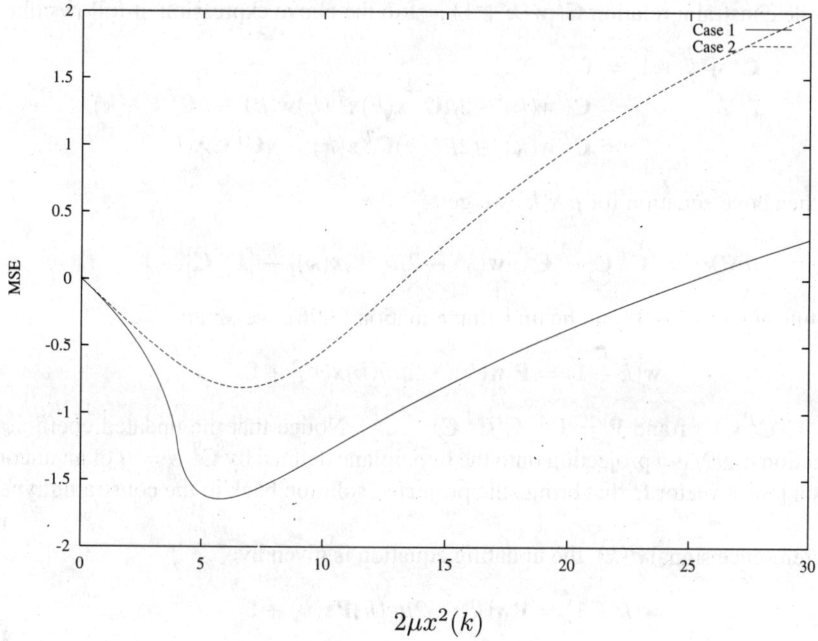

Figure 3.17 Parameters related to the rate of convergence, Case 1: $E\{[\ln[(1-2\mu x^2(i))^2]\}$, Case 2: $\ln\{E\left[(1-2\mu x^2(0))^2\right]\}$ as a function of $2\mu x^2(i)$.

the direction of the negative of the gradient (see equation (2.108)), i.e.,

$$\begin{aligned}\mathbf{w}(k+1) &= \mathbf{w}(k) - \mu\mathbf{g_w}\{\xi_c(k)\}\\ &= \mathbf{w}(k) - \mu[2\mathbf{R}(k)\mathbf{w}(k) + \mathbf{C}\boldsymbol{\lambda}(k)]\end{aligned} \tag{3.93}$$

where $\mathbf{R}(k)$ is some estimate of the input signal autocorrelation matrix at instant k, $\mathbf{C}$ is again the constraint matrix, and $\boldsymbol{\lambda}(k)$ is the $(N+1)\times 1$ vector of Lagrange multipliers.

In the particular case of the constrained LMS algorithm, matrix $\mathbf{R}(k)$ is chosen as an instantaneous rank-one estimate given by $\mathbf{x}(k)\mathbf{x}^T(k)$. In this case, we can utilize the method of Lagrange multipliers to solve the constrained minimization problem defined by

$$\begin{aligned}\xi_c(k) &= \mathbf{w}^T(k)\mathbf{x}(k)\mathbf{x}^T(k)\mathbf{w}(k) + \boldsymbol{\lambda}^T(k)\left[\mathbf{C}^T\mathbf{w}(k) - \mathbf{f}\right]\\ &= \mathbf{w}^T(k)\mathbf{x}(k)\mathbf{x}^T(k)\mathbf{w}(k) + \left[\mathbf{w}^T(k)\mathbf{C} - \mathbf{f}^T\right]\boldsymbol{\lambda}(k)\end{aligned} \tag{3.94}$$

The gradient of $\xi_c(k)$ with respect to $\mathbf{w}(k)$ is given by

$$\mathbf{g_w}\{\xi_c(k)\} = 2\mathbf{x}(k)\mathbf{x}^T(k)\mathbf{w}(k) + \mathbf{C}\boldsymbol{\lambda}(k) \tag{3.95}$$

The constrained LMS updating algorithm related to equation (3.93) becomes

$$\begin{aligned}\mathbf{w}(k+1) &= \mathbf{w}(k) - 2\mu\mathbf{x}(k)\mathbf{x}^T(k)\mathbf{w}(k) - \mu\mathbf{C}\boldsymbol{\lambda}(k)\\ &= \mathbf{w}(k) - 2\mu y(k)\mathbf{x}(k) - \mu\mathbf{C}\boldsymbol{\lambda}(k)\end{aligned} \tag{3.96}$$

If we apply the constraint relation $\mathbf{C}^T\mathbf{w}(k+1) = \mathbf{f}$ to the above expression, it follows that

$$\begin{aligned}\mathbf{C}^T\mathbf{w}(k+1) &= \mathbf{f} \\ &= \mathbf{C}^T\mathbf{w}(k) - 2\mu\mathbf{C}^T\mathbf{x}(k)\mathbf{x}^T(k)\mathbf{w}(k) - \mu\mathbf{C}^T\mathbf{C}\boldsymbol{\lambda}(k) \\ &= \mathbf{C}^T\mathbf{w}(k) - 2\mu y(k)\mathbf{C}^T\mathbf{x}(k) - \mu\mathbf{C}^T\mathbf{C}\boldsymbol{\lambda}(k)\end{aligned} \tag{3.97}$$

By solving the above equation for $\mu\boldsymbol{\lambda}(k)$ we get

$$\mu\boldsymbol{\lambda}(k) = \left[\mathbf{C}^T\mathbf{C}\right]^{-1}\mathbf{C}^T\left[\mathbf{w}(k) - 2\mu y(k)\mathbf{x}(k)\right] - \left[\mathbf{C}^T\mathbf{C}\right]^{-1}\mathbf{f} \tag{3.98}$$

If we substitute equation (3.98) in the updating equation (3.96), we obtain

$$\mathbf{w}(k+1) = \mathbf{P}[\mathbf{w}(k) - 2\mu y(k)\mathbf{x}(k)] + \mathbf{f}_c \tag{3.99}$$

where $\mathbf{f}_c = \mathbf{C}(\mathbf{C}^T\mathbf{C})^{-1}\mathbf{f}$ and $\mathbf{P} = \mathbf{I} - \mathbf{C}(\mathbf{C}^T\mathbf{C})^{-1}\mathbf{C}^T$. Notice that the updated coefficient vector given in equation (3.99) is a projection onto the hyperplane defined by $\mathbf{C}^T\mathbf{w} = \mathbf{0}$ of an unconstrained LMS solution plus a vector $\mathbf{f}_c$ that brings the projected solution back to the constraint hyperplane.

If there is a reference signal $d(k)$, the updating equation is given by

$$\mathbf{w}(k+1) = \mathbf{P}\mathbf{w}(k) + 2\mu e(k)\mathbf{P}\mathbf{x}(k) + \mathbf{f}_c \tag{3.100}$$

In the case of the constrained normalized LMS algorithm (see section 4.4), the solution satisfies $\mathbf{w}^T(k+1)\mathbf{x}(k) = d(k)$ in addition to $\mathbf{C}^T\mathbf{w}(k+1) = \mathbf{f}$ [25]. Alternative adaptation algorithms may be derived such that the solution at each iteration also satisfies a set of linear constraints [26].

For environments with complex signals and complex constraints, the updating equation is given by

$$\mathbf{w}(k+1) = \mathbf{P}\mathbf{w}(k) + \mu_c e^*(k)\mathbf{P}\mathbf{x}(k) + \mathbf{f}_c \tag{3.101}$$

where $\mathbf{C}^H\mathbf{w}(k+1) = \mathbf{f}$, $\mathbf{f}_c = \mathbf{C}(\mathbf{C}^H\mathbf{C})^{-1}\mathbf{f}$ and $\mathbf{P} = \mathbf{I} - \mathbf{C}(\mathbf{C}^H\mathbf{C})^{-1}\mathbf{C}^H$.

An efficient implementation for constrained adaptive filters was proposed in [27], which consists of applying a transformation to the input signal vector based on Householder transformation. The method can be regarded as an alternative implementation of the generalized sidelobe canceller structure, but with the advantages of always utilizing orthogonal/unitary matrices and rendering low computational complexity.

Example 3.8

An array of antennas with four elements, with inter-element spacing of 0.15 meters, receives signals from two different sources arriving at 90° and 30° of angles with respect to the axis where the antennas are placed. The desired signal impinges on the antenna at 90°. The signal of interest is a sinusoid of frequency 20MHz and the interferer signal is a sinusoid of frequency 70MHz. The sampling frequency is 2GHz.

Use the linearly-constrained LMS algorithm in order adapt the array coefficients.

Solution:

The adaptive-filter coefficients is initialized with $\mathbf{w}(0) = \mathbf{C}(\mathbf{C}^T\mathbf{C})^{-1}\mathbf{f}$. The value of μ used is 0.1. Fig. 3.18 illustrates the learning curve for the output signal. Fig. 3.19 illustrates details of the output signal in the early iterations where we can observe the presence of both sinusoid signals. In Fig. 3.20, the details of the output signal after convergence shows that mainly the desired sinusoid signal is present. The array output power response after convergence, as a function of the angle of arrival, is depicted in Fig. 3.21. From this figure, we observe the attenuation imposed by the array on signals arriving at 30° of angle, where the interference signal impinges.

□

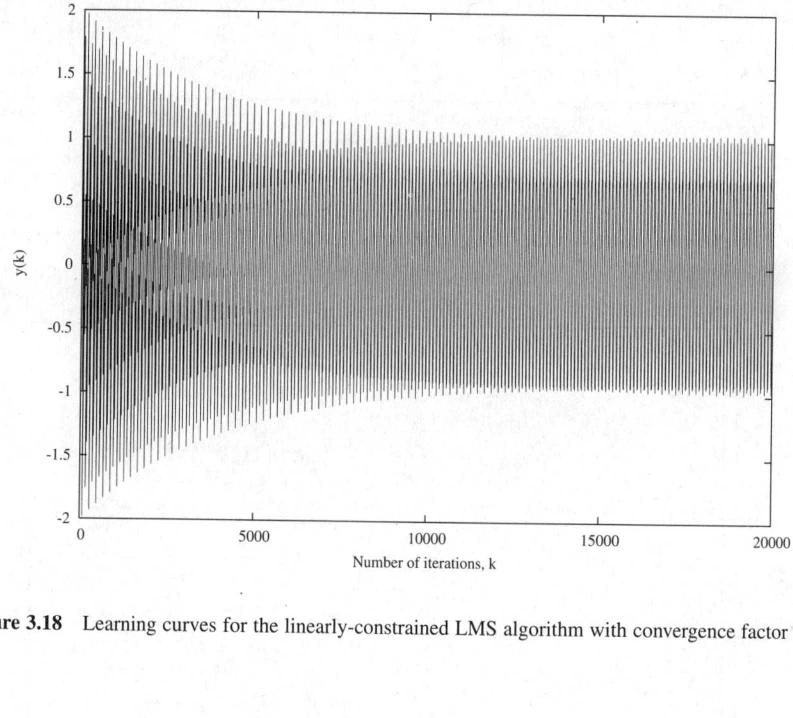

Figure 3.18 Learning curves for the linearly-constrained LMS algorithm with convergence factor $\mu = 0.1$.

3.7 CONCLUDING REMARKS

In this chapter, we studied the LMS adaptive algorithm that is certainly the most popular among the adaptive-filtering algorithms. The attractiveness of the LMS algorithm is due to its simplicity and accessible analysis under idealized conditions. As demonstrated in the present chapter, the noisy estimate of the gradient that is used in the LMS algorithm is the main source of loss in performance for stationary environments. Further discussions on the convergence behavior and on the optimality of the LMS algorithm have been reported in the open literature, see for example [28]-[34].

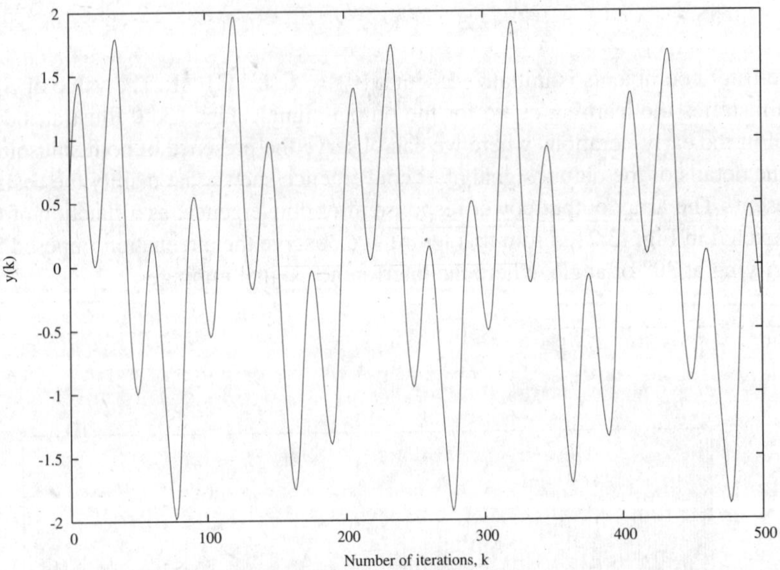

(a) Early output signal.

Figure 3.19 Learning curves for the linearly-constrained LMS algorithm with convergence factor $\mu = 0.1$; early output signal.

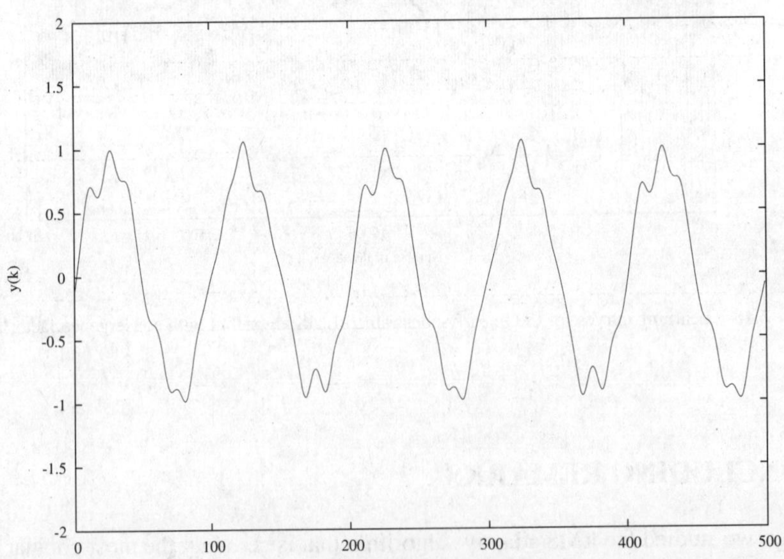

Figure 3.20 Learning curves for the linearly-constrained LMS algorithm with convergence factor $\mu = 0.1$; output signal after convergence.

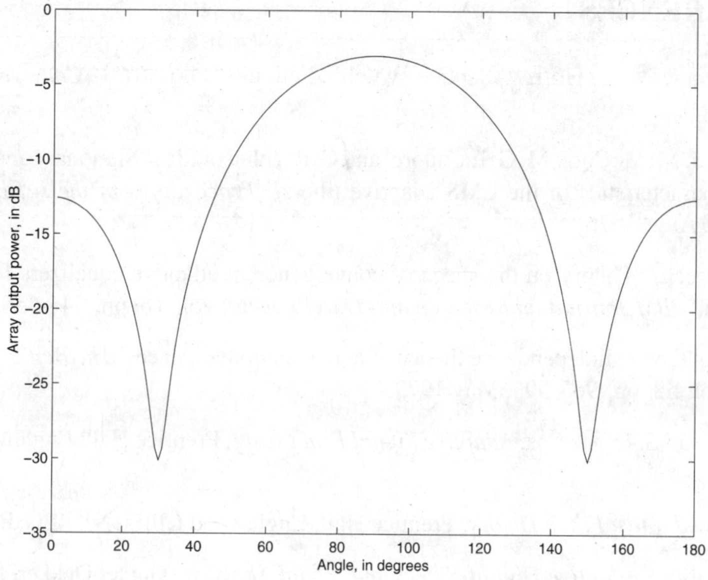

Figure 3.21 Array output power after convergence, as a function of the angle of arrival.

For nonstationary environments we showed how the algorithm behaves assuming the optimal parameter can be modeled as a first-order Markov process. The analysis allowed us to determine the conditions for adequate tracking and acceptable excess MSE. Further analysis can be found in [35].

The quantization effects on the behavior of the LMS algorithm are presented in Appendix B. The algorithm is fairly robust against quantization errors, and this is for sure one of the reasons for its choice in a number of practical applications [36]-[37].

A number of simulation examples with the LMS algorithm was presented in this chapter. The simulations included examples in system identification and equalization. Also a number of theoretical results derived in the present chapter were verified, such as the excess MSE in stationary and nonstationary environments, the finite-precision analysis etc.

3.8 REFERENCES

1. B. Widrow and M. E. Hoff, "Adaptive switching circuits," *WESCOM Conv. Rec.*, pt. 4, pp. 96-140, 1960.
2. B. Widrow, J. M. McCool, M. G. Larimore, and C. R. Johnson, Jr., "Stationary and nonstationary learning characteristics of the LMS adaptive filters," *Proceedings of the IEEE*, vol. 64, pp. 1151-1162, Aug. 1976.
3. G. Ungerboeck, "Theory on the speed of convergence in adaptive equalizers for digital communication," *IBM Journal on Research and Development*, vol. 16, pp. 546-555, Nov. 1972.
4. J. E. Mazo, "On the independence theory of equalizer convergence," *The Bell System Technical Journal*, vol. 58, pp. 963-993, May 1979.
5. B. Widrow and S. D. Stearns, *Adaptive Signal Processing*, Prentice Hall, Englewood Cliffs, NJ, 1985.
6. S. Haykin, *Adaptive Filter Theory*, Prentice Hall, Englewood Cliffs, NJ, 4th edition, 2002.
7. M. G. Bellanger, *Adaptive Digital Filters and Signal Analysis*, Marcel Dekker, Inc., New York, NY, 2nd edition, 2001.
8. D. C. Farden, "Tracking properties of adaptive signal processing algorithms," *IEEE Trans. on Acoust., Speech, and Signal Processing*, vol. ASSP-29, pp. 439-446, June 1981.
9. B. Widrow and E. Walach, "On the statistical efficiency of the LMS algorithm with nonstationary inputs," *IEEE Trans. on Information Theory*, vol. IT-30, pp. 211-221, March 1984.
10. O. Macchi, "Optimization of adaptive identification for time varying filters," *IEEE Trans. on Automatic Control*, vol. AC-31, pp. 283-287, March 1986.
11. A. Benveniste, "Design of adaptive algorithms for the tracking of time varying systems," *Int. J. Adaptive Control and Signal Processing*, vol. 1, pp. 3-29, Jan. 1987.
12. W. A. Gardner, "Nonstationary learning characteristics of the LMS algorithm," *IEEE Trans. on Circuits and Systems*, vol. CAS-34, pp. 1199-1207, Oct. 1987.
13. A. Papoulis, *Probability, Random Variables, and Stochastic Processes*, McGraw Hill, New York, NY, 3rd edition, 1991.
14. F. J. Gantmacher, *The Theory of Matrices*, Chelsea Publishing Company, vol. 2, New York, NY, 1964.
15. G. H. Golub and C. F. Van Loan, *Matrix Computations,* John Hopkins University Press, Baltimore, MD, 3rd edition, 1996.
16. V. Solo, "The limiting behavior of LMS," *IEEE Trans. on Acoust., Speech, and Signal Processing*, vol-37, pp. 1909-1922, Dec. 1989.
17. N. J. Bershad and O. M. Macchi, "Adaptive recovery of a chirped sinusoid in noise, Part 2: Performance of the LMS algorithm," *IEEE Trans. on Signal Processing*, vol. 39, pp. 595-602, March 1991.

18. D. H. Brandwood, "A complex gradient operator and its application in adaptive array theory," *IEE Proceedings Parts F and G*, vol. 130, pp. 11-16, Feb. 1983.

19. A. Hjørungnes and D. Gesbert, "Complex-valued matrix differentiation: Techniques and key results," *IEEE Trans. on Signal Processing*, vol. 55, pp. 2740-2746, June 2007.

20. D. G. Manolakis, V. K. Ingle, and S. M. Kogon, *Statistical and Adaptive Signal Processing*, McGraw Hill, New York, NY, 2000.

21. V. H. Nascimento and A. H. Sayed, "On the learning mechanism of adaptive filters," *IEEE Trans. on Signal Processing*, vol. 48, pp. 1609-1625, June 2000.

22. O. Macchi and E. Eweda, "Second-order convergence analysis of stochastic adaptive linear filter," *IEEE Trans. on Automatic Control*, vol. AC-28, pp. 76-85, March 1983.

23. S. Florian and A. Feuer, "Performance analysis of the LMS algorithm with a tapped delay line (two-dimensional case)," *IEEE Trans. on Acoust., Speech, and Signal Processing*, vol. ASSP-34, pp. 1542-1549, Dec. 1986.

24. O. L. Frost III, "An algorithm for linearly constrained adaptive array processing," *Proceedings of the IEEE*, vol. 60, pp. 926-935, Aug. 1972.

25. J. A. Apolinário Jr., S. Werner, T. I. Laakso, and P. S. R. Diniz, "Constrained normalized adaptive filtering for CDMA mobile communications," *Proc. 1998 EUSIPCO-European Signal Processing Conference*, Rhodes, Greece, pp. 2053-2056, Sept. 1998.

26. J. A. Apolinário Jr., M. L. R. de Campos, and C. P. Bernal O., "The Constrained conjugate-gradient algorithm,"*IEEE Signal Processing Letters*, vol. 7, pp. 351-354, Dec. 2000.

27. M. L. R. de Campos, S. Werner, and J. A. Apolinário, Jr., "Constrained adaptation algorithms employing Householder transformation," *IEEE Transactions on Signal Processing*, vol. 50, pp. 2187-2195, Sept. 2002.

28. A. Feuer and E. Weinstein, "Convergence analysis of LMS filters with uncorrelated Gaussian data," *IEEE Trans. on Acoust., Speech, and Signal Processing*, vol. ASSP-33, pp. 222-230, Feb. 1985.

29. D. T. Slock, "On the convergence behavior of the LMS and normalized LMS algorithms," *IEEE Trans. on Signal Processing*, vol. 40, pp. 2811-2825, Sept. 1993.

30. W. A. Sethares, D. A. Lawrence, C. R. Johnson, Jr., and R. R. Bitmead, "Parameter drift in LMS adaptive filters," *IEEE Trans. on Acoust., Speech, and Signal Processing*, vol. ASSP-34, pp. 868-878, Aug. 1986.

31. S. C. Douglas, "Exact expectation analysis of the LMS adaptive filter," *IEEE Trans. on Signal Processing*, vol. 43, pp. 2863-2871, Dec. 1995.

32. H. J. Butterweck, "Iterative analysis of the state-space weight fluctuations in LMS-type adaptive filters," *IEEE Trans. on Signal Processing*, vol. 47, pp. 2558-2561, Sept. 1999.

33. B. Hassibi, A. H. Sayed, and T. Kailath, "H^∞ optimality of the LMS Algorithm," *IEEE Trans. on Signal Processing*, vol. 44, pp. 267-280, Feb. 1996.

34. O. J. Tobias, J. C. M. Bermudez, and N. J. Bershad, "Mean weight behavior of the filtered-X LMS algorithm," *IEEE Trans. on Signal Processing*, vol. 48, pp. 1061-1075, April 2000.

35. V. Solo, "The error variance of LMS with time varying weights," *IEEE Trans. on Signal Processing*, vol. 40, pp. 803-813, April 1992.

36. S. U. Qureshi, "Adaptive Equalization," *Proceedings of the IEEE*, vol-73, pp. 1349-1387, Sept. 1985.

37. M. L. Honig, "Echo cancellation of voiceband data signals using recursive least squares and stochastic gradient algorithms," *IEEE Trans. on Communications*, vol. COM-33, pp. 65-73, Jan. 1985.

38. V. H. Nascimento and J. C. M. Bermudez, "Probability of divergence for the least-mean fourth algorithm," *IEEE Trans. on Signal Processing*, vol. 54, pp. 1376-1385, April 2006.

3.9 PROBLEMS

1. The LMS algorithm is used to predict the signal $x(k) = \cos(\pi k/3)$ using a second-order FIR filter with the first tap fixed at 1, by minimizing the mean squared value of $y(k)$. Calculate an appropriate μ, the output signal, and the filter coefficients for the first 10 iterations. Start with $\mathbf{w}^T(0) = [1\,0\,0]$.

2. The signal

$$x(k) = -0.85x(k-1) + n(k)$$

is applied to a first-order predictor, where $n(k)$ is Gaussian white noise with variance $\sigma_n^2 = 0.3$.

(a) Compute the Wiener solution.

(b) Choose an appropriate value for μ and plot the convergence path for the LMS algorithm on the MSE error surface.

(c) Plot the learning curves for the MSE and the filter coefficients in a single run as well as for the average of 25 runs.

3. Assuming it is desired to minimize the objective function $E[e^4(k)]$ utilizing a stochastic gradient type of algorithm such as the LMS. The resulting algorithm is called least-mean fourth algorithm [38]. Derive this algorithm.

4. The data-reusing LMS algorithm has the following updating equation

$$\begin{aligned}\hat{e}_l(k) &= d(k) - \hat{\mathbf{w}}_l^T(k)\mathbf{x}(k)\\ \hat{\mathbf{w}}_{l+1}(k) &= \hat{\mathbf{w}}_l(k) + 2\mu\hat{e}_l(k)\mathbf{x}(k)\end{aligned} \tag{3.102}$$

for $l = 0, 1, \ldots, L-1$, and

$$\mathbf{w}(k+1) = \hat{\mathbf{w}}_L(k) = \hat{\mathbf{w}}_{L-1}(k) + 2\mu\hat{e}_{L-1}(k)\mathbf{x}(k) \tag{3.103}$$

where $\hat{\mathbf{w}}_0(k) = \mathbf{w}(k)$.

(a) Compute the range of values of μ such that the coefficients converge in average.

(b) What is the objective function this algorithm actually minimizes?

(c) Compare its convergence speed and computational complexity with the LMS algorithm.

5. The momentum LMS algorithm has the following updating equation

$$\mathbf{w}(k+1) = \mathbf{w}(k) + 2\mu e(k)\mathbf{x}(k) + \gamma[\mathbf{w}(k) - \mathbf{w}(k-1)] \tag{3.104}$$

for $|\gamma| < 1$.

(a) Compute the range of values of μ such that the coefficients converge in average.

(b) What is the objective function this algorithm actually minimizes?

(c) Show that this algorithm can have faster convergence and higher misadjustment than the LMS algorithm.

6. An LMS algorithm can be updated in a block form. For a block of length 2 the updating equations have the following form.

$$\begin{aligned}\begin{bmatrix} e(k) \\ e(k-1) \end{bmatrix} &= \begin{bmatrix} d(k) \\ d(k-1) \end{bmatrix} - \begin{bmatrix} \mathbf{x}^T(k)\mathbf{w}(k) \\ \mathbf{x}^T(k-1)\mathbf{w}(k-1) \end{bmatrix} \\ &= \begin{bmatrix} d(k) \\ d(k-1) \end{bmatrix} - \begin{bmatrix} \mathbf{x}^T(k) \\ \mathbf{x}^T(k-1) \end{bmatrix} \mathbf{w}(k-1) \\ &\quad - \begin{bmatrix} 0 & 2\mu\mathbf{x}^T(k)\mathbf{x}(k-1) \\ 0 & 0 \end{bmatrix} \begin{bmatrix} e(k) \\ e(k-1) \end{bmatrix}\end{aligned}$$

This relation, in a more compact way, is equivalent to

$$\begin{bmatrix} e(k) \\ e(k-1) \end{bmatrix} = \begin{bmatrix} 1 & -2\mu\mathbf{x}^T(k)\mathbf{x}(k-1) \\ 0 & 1 \end{bmatrix} \left\{ \begin{bmatrix} d(k) \\ d(k-1) \end{bmatrix} - \begin{bmatrix} \mathbf{x}^T(k) \\ \mathbf{x}^T(k-1) \end{bmatrix} \mathbf{w}(k-1) \right\} \tag{3.105}$$

Derive an expression for a block of length $L+1$.

7. Use the LMS algorithm to identify a system with the transfer function given below. The input signal is a uniformly distributed white noise with variance $\sigma_x^2 = 1$, and the measurement noise is Gaussian white noise uncorrelated with the input with variance $\sigma_n^2 = 10^{-3}$. The adaptive filter has 12 coefficients.

$$H(z) = \frac{1 - z^{-12}}{1 - z^{-1}}$$

(a) Calculate the upper bound for μ ($\mu_{\max}$) to guarantee the algorithm stability.

(b) Run the algorithm for $\mu_{\max}/2$, $\mu_{\max}/10$, and $\mu_{\max}/50$. Comment on the convergence behavior in each case.

(c) Measure the misadjustment in each example and compare with the results obtained by equation (3.50).

(d) Plot the obtained FIR filter frequency response at any iteration after convergence is achieved and compare with the unknown system.

8. Repeat the previous problem using an adaptive filter with 8 coefficients and interpret the results.

9. Repeat problem 2 in case the input signal is a uniformly distributed white noise with variance $\sigma_{n_x}^2 = 0.5$ filtered by an all-pole filter given by

$$H(z) = \frac{z}{z - 0.9}$$

10. Perform the equalization of a channel with the following impulse response

$$h(k) = ku(k) - (2k - 9)u(k - 5) + (k - 9)u(k - 10)$$

Using a known training signal that consists of a binary (-1,1) random signal, generated by applying a white noise to a hard limiter (the output is 1 for positive input samples and -1 for negative). An additional Gaussian white noise with variance 10^{-2} is present at the channel output.
(a) Apply the LMS with an appropriate μ and find the impulse response of an equalizer with 100 coefficients.
(b) Convolve one of the equalizer's impulse response after convergence with the channel impulse response and comment on the result.

11. Under the assumption that the elements of $\mathbf{x}(k)$ are jointly Gaussian, show that equation (3.24) is valid.

12. In a system identification problem the input signal is generated by an autoregressive process given by

$$x(k) = -1.2x(k-1) - 0.81x(k-2) + n_x(k)$$

where $n_x(k)$ is zero-mean Gaussian white noise with variance such that $\sigma_x^2 = 1$. The unknown system is described by

$$H(z) = 1 + 0.9z^{-1} + 0.1z^{-2} + 0.2z^{-3}$$

The adaptive filter is also a third-order FIR filter, and the additional noise is zero-mean Gaussian white noise with variance $\sigma_n^2 = 0.04$. Using the LMS algorithm:
(a) Choose an appropriate μ, run an ensemble of 20 experiments, and plot the average learning curve.
(b) Plot the curve obtained using equations (3.41), (3.45), and (3.46), and compare the results.
(c) Compare the measured and theoretical values for the misadjustment.
(d) Calculate the time constants τ_{wi} and τ_{ei}, and the expected number of iterations to achieve convergence.

13. In a nonstationary environment the optimal coefficient vector is described by

$$\mathbf{w}_o(k) = -\lambda_1 \mathbf{w}_o(k-1) - \lambda_2 \mathbf{w}_o(k-2) + \mathbf{n_w}(k)$$

where $\mathbf{n_w}(k)$ is a vector whose elements are zero-mean Gaussian white processes with variance $\sigma_{\mathbf{w}}^2$. Calculate the elements of the lag-error vector.

14. Repeat the previous problem for

$$\mathbf{w}_o(k) = \lambda_w \mathbf{w}_o(k-1) + (1 - \lambda_w)\mathbf{n_w}(k)$$

15. The LMS algorithm is applied to identify a 7th-order time-varying unknown system whose coefficients are first-order Markov processes with $\lambda_{\mathbf{w}} = 0.999$ and $\sigma_{\mathbf{w}}^2 = 0.001$. The initial time-varying-system multiplier coefficients are

$$\mathbf{w}_o^T =$$
$$[0.03490 \ -0.011 \ -0.06864 \ 0.22391 \ 0.55686 \ 0.35798 \ -0.0239 \ -0.07594]$$

The input signal is Gaussian white noise with variance $\sigma_x^2 = 0.7$, and the measurement noise is also Gaussian white noise independent of the input signal and of the elements of $\mathbf{n_w}(k)$, with variance $\sigma_n^2 = 0.01$.
(a) For $\mu = 0.05$, compute the excess MSE.
(b) Repeat (a) for $\mu = 0.01$.
(c) Compute μ_{opt} and comment if it can be used.

16. Simulate the experiment described in problem 15, measure the excess MSE, and compare to the calculated results.

17. Reduce the value of $\lambda_{\mathbf{w}}$ to 0.97 in problem 15, simulate, and comment on the results.

18. Suppose a 15th-order FIR digital filter with multiplier coefficients given below is identified through an adaptive FIR filter of the same order using the LMS algorithm.
(a) Considering that fixed-point arithmetic is used, compute the expected value for $||\Delta\mathbf{w}(k)_Q||^2$ and $\xi(k)_Q$, and the probable number of iterations before the algorithm stops updating, for the following case:

Additional noise: white noise with variance	$\sigma_n^2 = 0.0015$
Coefficient wordlength:	$b_c = 16$ bits
Signal wordlength:	$b_d = 16$ bits
Input signal: Gaussian white noise with variance	$\sigma_x^2 = 0.7$
	$\mu = 0.01$

Hint: Utilize the formulas for the time constant in the LMS algorithm and equation (B.28).
(b) Simulate the experiment and plot the learning curves for the finite- and infinite-precision implementations.
(c) Compare the simulated results with those obtained through the closed form formulas.

$\mathbf{w}_o^T = [0.0219360 \quad 0.0015786 \quad -0.0602449 \quad -0.0118907 \quad 0.1375379$
$0.0574545 \quad -0.3216703 \quad -0.5287203 \quad -0.2957797 \quad 0.0002043 \quad 0.290670$
$-0.0353349 \quad -0.068210 \quad 0.0026067 \quad 0.0010333 \quad -0.0143593]$

19. Repeat the above problem for the following cases
(a) $\sigma_n^2 = 0.01$, $b_c = 12$ bits, $b_d = 12$ bits, $\sigma_x^2 = 0.7$, $\mu = 2.0\ 10^{-3}$.
(b) $\sigma_n^2 = 0.1$, $b_c = 10$ bits, $b_d = 10$ bits, $\sigma_x^2 = 0.8$, $\mu = 1.0\ 10^{-4}$.
(c) $\sigma_n^2 = 0.05$, $b_c = 14$ bits, $b_d = 14$ bits, $\sigma_x^2 = 0.8$, $\mu = 2.0\ 10^{-3}$.

20. Find the optimal value of μ (μ_{opt}) that minimizes the excess MSE given in equation (B.32), and compute for $\mu = \mu_{\text{opt}}$ the expected value of $||\Delta\mathbf{w}(k)_Q||^2$ and $\xi(k)_Q$ for the examples described in problem 19.

21. Repeat problem 18 for the case where the input signal is a first-order Markov process with $\lambda_{\mathbf{x}} = 0.95$.

22. A digital channel model can be represented by the following impulse response:

$$[-0.001 \quad -0.002 \quad 0.002 \quad 0.2 \quad 0.6 \quad 0.76 \quad 0.9 \quad 0.78 \quad 0.67 \quad 0.58$$
$$0.45 \quad 0.3 \quad 0.2 \quad 0.12 \quad 0.06 \quad 0 \quad -0.2 \quad -1 \quad -2 \quad -1 \quad 0 \quad 0.1]$$

The channel is corrupted by Gaussian white noise with power spectrum given by

$$|S(e^{\jmath\omega})|^2 = \kappa'|\omega|^{3/2}$$

where $\kappa' = 10^{-1.5}$. The training signal consists of independent binary samples (-1,1).

Design an FIR equalizer for this problem and use the LMS algorithm. Use a filter of order 50 and plot the learning curve.

23. For the previous problem, using the maximum of 51 adaptive filter coefficients, implement a DFE equalizer and compare the results with those obtained with the FIR filter. Again use the LMS algorithm.

24. Implement with fixed-point arithmetic the DFE equalizer of problem 23, using the LMS algorithm with 12 bits of wordlength for data and coefficients.

25. Use the complex LMS algorithm to equalize a channel with the transfer function given below. The input signal is a four Quadrature Amplitude Modulation (QAM)[5] signal representing a randomly generated bit stream with the signal to noise ratio $\frac{\sigma_{\tilde{x}}^2}{\sigma_n^2} = 20$ at the receiver end, that is, $\tilde{x}(k)$ is the received signal without taking into consideration the additional channel noise. The adaptive filter has 10 coefficients.

$$H(z) = (0.34 - 0.27\jmath) + (0.87 + 0.43\jmath)z^{-1} + (0.34 - 0.21\jmath)z^{-2}$$

(a) Calculate the upper bound for μ ($\mu_{\max}$) to guarantee the algorithm stability.
(b) Run the algorithm for $\mu_{\max}/2$, $\mu_{\max}/10$, and $\mu_{\max}/50$. Comment on the convergence behavior in each case.
(c) Plot the real versus imaginary parts of the received signal before and after equalization.
(d) Increase the number of coefficients to 20 and repeat the experiment in (c).

26. In a system identification problem the input signal is generated from a four QAM of the form

$$x(k) = x_{\text{re}}(k) + \jmath x_{\text{im}}(k)$$

where $x_{\text{re}}(k)$ and $x_{\text{im}}(k)$ assume values ± 1 randomly generated. The unknown system is described by

$$H(z) = 0.32 + 0.21\jmath + (-0.3 + 0.7\jmath)z^{-1} + (0.5 - 0.8\jmath)z^{-2} + (0.2 + 0.5\jmath)z^{-3}$$

The adaptive filter is also a third-order complex FIR filter, and the additional noise is zero-mean Gaussian white noise with variance $\sigma_n^2 = 0.4$. Using the complex LMS algorithm, choose an appropriate μ, run an ensemble of 20 experiments, and plot the average learning curve.

[5] The M-ary QAM constellation points are represented in by $s_i = \tilde{a}_i + \jmath\tilde{b}_i$, with $\tilde{a}_i = \pm\tilde{d}, \pm 3\tilde{d}, \ldots, \pm(\sqrt{M}-1)\tilde{d}$, and $\tilde{b}_i = \pm\tilde{d}, \pm 3\tilde{d}, \ldots, \pm(\sqrt{M}-1)\tilde{d}$. The parameter $\tilde{d}$ is represents half of the distance between two points in the constellation.

4

LMS-BASED ALGORITHMS

4.1 INTRODUCTION

There are a number of algorithms for adaptive filters which are derived from the conventional LMS algorithm discussed in the previous chapter. The objective of the alternative LMS-based algorithms is either to reduce computational complexity or convergence time. In this chapter, several LMS-based algorithms are presented and analyzed, namely, the quantized-error algorithms [1]-[11], the frequency-domain (or transform-domain) LMS algorithm [12]-[14], the normalized LMS algorithm [15], the LMS-Newton algorithm [16]-[17], and the affine projection algorithm [19]-[25]. Several algorithms that are related to the main algorithms presented in this chapter are also briefly discussed.

The quantized-error algorithms reduce the computational complexity of the LMS algorithms by representing the error signal with short wordlength or by a simple power-of-two number.

The convergence speed in the LMS-Newton algorithm is independent of the eigenvalue spread of the input signal correlation matrix. This improvement is achieved by using an estimate of the inverse of the input signal correlation matrix, leading to a substantial increase in the computational complexity.

The normalized LMS algorithm utilizes a variable convergence factor that minimizes the instantaneous error. Such a convergence factor usually reduces the convergence time but increases the misadjustment.

In the frequency-domain algorithm, a transform is applied to the input signal in order to allow the reduction of the eigenvalue spread of the transformed signal correlation matrix as compared to the eigenvalue spread of the input signal correlation matrix. The LMS algorithm applied to the better conditioned transformed signal achieves faster convergence.

The affine projection algorithm reuses old data resulting in fast convergence when the input signal is highly correlated, leading to a family of algorithms that can trade-off computational complexity with convergence speed.

P.S.R. Diniz, *Adaptive Filtering*, DOI: 10.1007/978-0-387-68606-6_4,

4.2 QUANTIZED-ERROR ALGORITHMS

The computational complexity of the LMS algorithm is mainly due to multiplications performed in the coefficient updating and in the calculation of the adaptive-filter output. In applications where the adaptive filters are required to operate in high speed, such as echo cancellation and channel equalization, it is important to minimize hardware complexity.

A first step to simplify the LMS algorithm is to apply quantization to the error signal, generating the quantized-error algorithm which updates the filter coefficients according to

$$\mathbf{w}(k+1) = \mathbf{w}(k) + 2\mu Q[e(k)]\mathbf{x}(k) \tag{4.1}$$

where $Q[\cdot]$ represents a quantization operation. The quantization function is discrete valued, bounded, and nondecreasing. The type of quantization identifies the quantized-error algorithm.

If the convergence factor μ is a power-of-two number, the coefficient updating can be implemented with simple multiplications, basically consisting of bit shifts and additions. In a number of applications, such as the echo cancellation in full-duplex data transmission [2] and equalization of channels with binary data [3], the input signal $x(k)$ is a binary signal, i.e., assumes values $+1$ and -1. In this case, the adaptive filter can be implemented without any intricate multiplication.

The quantization of the error actually implies a modification in the objective function that is minimized, denoted by $F[e(k)]$. In a general gradient-type algorithm coefficient updating is performed by

$$\mathbf{w}(k+1) = \mathbf{w}(k) - \mu\frac{\partial F[e(k)]}{\partial \mathbf{w}(k)} = \mathbf{w}(k) - \mu\frac{\partial F[e(k)]}{\partial e(k)}\frac{\partial e(k)}{\partial \mathbf{w}(k)} \tag{4.2}$$

For a linear combiner the above equation can be rewritten as

$$\mathbf{w}(k+1) = \mathbf{w}(k) + \mu\frac{\partial F[e(k)]}{\partial e(k)}\mathbf{x}(k) \tag{4.3}$$

Therefore, the objective function that is minimized in the quantized-error algorithms is such that

$$\frac{\partial F[e(k)]}{\partial e(k)} = 2Q[e(k)] \tag{4.4}$$

where $F[e(k)]$ is obtained by integrating $2Q[e(k)]$ with respect to $e(k)$. Note that the chain rule applied in equation (4.3) is not valid at the points of discontinuity of $Q[\cdot]$ where $F[e(k)]$ is not differentiable [6].

The performances of the quantized-error and LMS algorithms are obviously different. The analyses of some widely used quantized-error algorithms are presented in the following subsections.

Algorithm 4.1 Sign-Error Algorithm
Initialization $\mathbf{x}(0) = \mathbf{w}(0) = [0\,0\ldots0]^T$ Do for $k \geq 0$ $e(k) = d(k) - \mathbf{x}^T(k)\mathbf{w}(k)$ $\rho = \text{sgn}[e(k)]$ $\mathbf{w}(k+1) = \mathbf{w}(k) + 2\mu\rho\mathbf{x}(k)$

4.2.1 Sign-Error Algorithm

The simplest form for the quantization function is the sign (sgn) function defined by

$$\text{sgn}[b] = \begin{cases} 1, & b > 0 \\ 0, & b = 0 \\ -1, & b < 0 \end{cases} \tag{4.5}$$

The sign-error algorithm utilizes the sign function as the error quantizer, where the coefficient vector updating is performed by

$$\mathbf{w}(k+1) = \mathbf{w}(k) + 2\mu\,\text{sgn}[e(k)]\,\mathbf{x}(k) \tag{4.6}$$

Fig. 4.1 illustrates the realization of the sign-error algorithm for a delay line input $\mathbf{x}(k)$. If μ is a power-of-two number, one iteration of the sign-error algorithm requires $N+1$ multiplications for the error generation. The total number of additions is $2N+2$. The detailed description of the sign-error algorithm is shown in Algorithm 4.1. Obviously, the vectors $\mathbf{x}(0)$ and $\mathbf{w}(0)$ can be initialized in a different way from that described in the algorithm.

The objective function that is minimized by the sign-error algorithm is the modulus of the error multiplied by two, i.e.,

$$F[e(k)] = 2|e(k)| \tag{4.7}$$

Note that the factor two is included only to present the sign-error and LMS algorithms in a unified form. Obviously, in real implementation this factor can be merged with convergence factor μ.

Some of the properties related to the convergence behavior of the sign-error algorithm in a stationary environment are described, following the same procedure used in the previous chapter for the LMS algorithm.

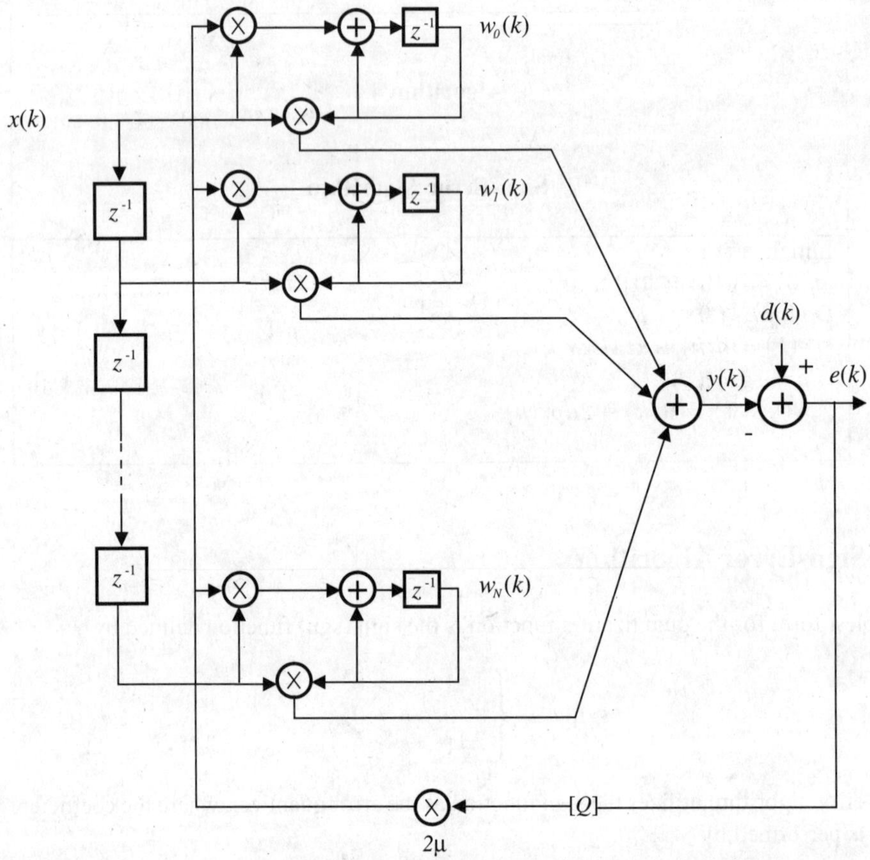

Figure 4.1 Sign-error adaptive FIR filter: $Q[e(k)] = sgn[e(k)]$.

4.2.1.1 *Steady-State Behavior of the Coefficient Vector*

The sign-error algorithm can be alternatively described by

$$\Delta\mathbf{w}(k+1) = \Delta\mathbf{w}(k) + 2\mu\,\text{sgn}[e(k)]\,\mathbf{x}(k) \tag{4.8}$$

where $\Delta\mathbf{w}(k) = \mathbf{w}(k) - \mathbf{w}_o$. The expected value of the coefficient-error vector is then given by

$$E[\Delta\mathbf{w}(k+1)] = E[\Delta\mathbf{w}(k)] + 2\mu E\{\text{sgn}[e(k)]\,\mathbf{x}(k)\} \tag{4.9}$$

The importance of the probability density function of the measurement noise $n(k)$ on the convergence of the sign-error algorithm is a noteworthy characteristic. This is due to the fact that $E\{\text{sgn}[e(k)]\,\mathbf{x}(k)\} = E\{\text{sgn}[-\Delta\mathbf{w}^T(k)\mathbf{x}(k) + n(k)]\mathbf{x}(k)\}$, where the result of the sign operation is highly dependent on the probability density function of $n(k)$. In [1], the authors present a convergence analysis of the output MSE, i.e., $E[e^2(k)]$, for different distributions of the additional noise, such as Gaussian, uniform, and binary distributions.

A closer examination of equation (4.8) indicates that even if the error signal becomes very small, the adaptive-filter coefficients will be continually updated due to the sign function applied to the error signal. Therefore, in a situation where the adaptive filter has a sufficient number of coefficients to model the desired signal, and there is no additional noise, $\Delta\mathbf{w}(k)$ will not converge to zero. In this case, $\mathbf{w}(k)$ will be convergent to a balloon centered at $\mathbf{w}_o$, when μ is appropriately chosen. The mean absolute value of $e(k)$ is also convergent to a balloon centered around zero, that means $|e(k)|$ remains smaller than the balloon radius r [6].

Recall that the desired signal without measurement noise is denoted as $d'(k)$. If it is considered that $d'(k)$ and the elements of $\mathbf{x}(k)$ are zero mean and jointly Gaussian and that the additional noise $n(k)$ is also zero mean, Gaussian, and independent of $\mathbf{x}(k)$ and $d'(k)$, the error signal will also be zero-mean Gaussian signal conditioned on $\Delta\mathbf{w}(k)$. In this case, using the results of the Price theorem described in [27] and in Papoulis [28], the following result is valid

$$E\{\text{sgn}[e(k)]\,\mathbf{x}(k)\} \approx \sqrt{\frac{2}{\pi\xi(k)}}E[\mathbf{x}(k)e(k)] \tag{4.10}$$

where $\xi(k)$ is the variance of $e(k)$ assuming the error has zero mean. The above approximation is valid for small values of μ. For large μ, $e(k)$ is dependent on $\Delta\mathbf{w}(k)$ and conditional expected value on $\Delta\mathbf{w}(k)$ should be used instead [3]-[5].

By applying equation (4.10) in equation (4.9) and by replacing $e(k)$ by $e_o(k) - \Delta\mathbf{w}^T(k)\mathbf{x}(k)$, it follows that

$$\begin{aligned} E[\Delta\mathbf{w}(k+1)] &= \left\{\mathbf{I} - 2\mu\sqrt{\frac{2}{\pi\xi(k)}}E[\mathbf{x}(k)\mathbf{x}^T(k)]\right\} E[\Delta\mathbf{w}(k)] \\ &\quad + 2\mu\sqrt{\frac{2}{\pi\xi(k)}}\;E[e_o(k)\mathbf{x}(k)] \end{aligned} \tag{4.11}$$

From the orthogonality principle we know that $E[e_o(k)\mathbf{x}(k)] = \mathbf{0}$, so that the last element of the above equation is zero. Therefore,

$$E[\Delta\mathbf{w}(k+1)] = \left[\mathbf{I} - 2\mu\sqrt{\frac{2}{\pi\xi(k)}}\mathbf{R}\right] E[\Delta\mathbf{w}(k)] \tag{4.12}$$

Following the same steps for the analysis of $E[\Delta\mathbf{w}(k)]$ in the traditional LMS algorithm, it can be shown that the coefficients of the adaptive filter implemented with the sign-error algorithm converge in the mean if the convergence factor is chosen in the range

$$0 < \mu < \frac{1}{\lambda_{\max}}\sqrt{\frac{\pi\xi(k)}{2}} \tag{4.13}$$

where $\lambda_{\max}$ is the largest eigenvalue of $\mathbf{R}$. It should be mentioned that in case $\frac{\lambda_{\max}}{\lambda_{\min}}$ is large, the convergence speed of the coefficients depends on the value of $\lambda_{\min}$ which is related to the slowest mode in equation (4.12). This conclusion can be drawn by following the same steps of the convergence analysis of the LMS algorithm, where by applying a transformation to equation (4.12) we obtain an equation similar to equation (3.17).

A more practical range for μ, avoiding the use of eigenvalue, is given by

$$0 < \mu < \frac{1}{\text{tr}[\mathbf{R}]}\sqrt{\frac{\pi\xi(k)}{2}} \tag{4.14}$$

Note that the upper bound for the value of μ requires the knowledge of the MSE, i.e., $\xi(k)$.

4.2.1.2 Coefficient-Error-Vector Covariance Matrix

The covariance of the coefficient-error vector defined as

$$\text{cov}[\Delta\mathbf{w}(k)] = E\left[(\mathbf{w}(k) - \mathbf{w}_o)(\mathbf{w}(k) - \mathbf{w}_o)^T\right] \tag{4.15}$$

is calculated by replacing equation (4.8) in equation (4.15) following the same steps used in the LMS algorithm. The resulting difference equation for $\text{cov}[\Delta\mathbf{w}(k)]$ is given by

$$\begin{aligned}\text{cov}[\Delta\mathbf{w}(k+1)] &= \text{cov}[\Delta\mathbf{w}(k)] + 2\mu E\{\text{sgn}[e(k)]\mathbf{x}(k)\Delta\mathbf{w}^T(k)\} \\ &\quad +2\mu E\{\text{sgn}[e(k)]\Delta\mathbf{w}(k)\mathbf{x}^T(k)\} + 4\mu^2\mathbf{R}\end{aligned} \tag{4.16}$$

The first term with expected value operation in the above equation can be expressed as

$$\begin{aligned}E\{\text{sgn}[e(k)]\mathbf{x}(k)\Delta\mathbf{w}^T(k)\} &= E\{\text{sgn}[e_o(k) - \Delta\mathbf{w}^T(k)\mathbf{x}(k)]\mathbf{x}(k)\Delta\mathbf{w}^T(k)\} \\ &= E\{E[\text{sgn}[e_o(k) - \Delta\mathbf{w}^T(k)\mathbf{x}(k)]\mathbf{x}(k)|\Delta\mathbf{w}(k)]\Delta\mathbf{w}^T(k)\}\end{aligned}$$

where $E[a|\Delta\mathbf{w}(k)]$ is the expected value of a conditioned on the value of $\Delta\mathbf{w}(k)$. In the first equality, $e(k)$ was replaced by the relation $d(k) - \mathbf{w}^T(k)\mathbf{x}(k) - \mathbf{w}_o^T\mathbf{x}(k) + \mathbf{w}_o^T\mathbf{x}(k) = e_o(k) - \Delta\mathbf{w}^T(k)\mathbf{x}(k)$. In the second equality, the concept of conditioned expected value was applied.

Using the Price theorem and considering that the minimum output error $e_o(k)$ is zero-mean and uncorrelated with $\mathbf{x}(k)$, the following approximations result

$$\begin{aligned}&E\{E[\text{sgn}[e_o(k) - \Delta\mathbf{w}^T(k)\mathbf{x}(k)]\mathbf{x}(k)|\Delta\mathbf{w}(k)]\Delta\mathbf{w}^T(k)\} \\ &\approx E\left\{\sqrt{\frac{2}{\pi\xi(k)}}E[e_o(k)\mathbf{x}(k) - \mathbf{x}(k)\mathbf{x}^T(k)\Delta\mathbf{w}(k)|\Delta\mathbf{w}(k)]\Delta\mathbf{w}^T(k)\right\} \\ &\approx -E\left\{\sqrt{\frac{2}{\pi\xi(k)}}\mathbf{R}\Delta\mathbf{w}(k)\Delta\mathbf{w}^T(k)\right\} \\ &= -\sqrt{\frac{2}{\pi\xi(k)}}\mathbf{R}\text{cov}[\Delta\mathbf{w}(k)]\end{aligned} \tag{4.17}$$

Following similar steps to derive the above equation, the second term with the expected value operation in equation (4.16) can be approximated as

$$E\{\text{sgn}[e(k)]\Delta\mathbf{w}(k)\mathbf{x}^T(k)\} \approx -\sqrt{\frac{2}{\pi\xi(k)}}\text{cov}[\Delta\mathbf{w}(k)]\mathbf{R} \tag{4.18}$$

Substituting equations (4.17) and (4.18) in equation (4.16), we can calculate the vector $\mathbf{v}'(k)$ consisting of diagonal elements of $\text{cov}[\Delta\mathbf{w}'(k)]$, using the same steps employed in the LMS case (see equation (3.26)). The resulting dynamic equation for $\mathbf{v}'(k)$ is given by

$$\mathbf{v}'(k+1) = \left(\mathbf{I} - 4\mu\sqrt{\frac{2}{\pi\xi(k)}}\,\mathbf{\Lambda}\right)\mathbf{v}'(k) + 4\mu^2\boldsymbol{\lambda} \tag{4.19}$$

The value of μ must be chosen in a range that guarantees the convergence of $\mathbf{v}'(k)$, which is given by

$$0 < \mu < \frac{1}{2\lambda_{\max}}\sqrt{\frac{\pi\xi(k)}{2}} \tag{4.20}$$

A more severe and practical range for μ is

$$0 < \mu < \frac{1}{2\text{tr}[\mathbf{R}]}\sqrt{\frac{\pi\xi(k)}{2}} \tag{4.21}$$

For $k \rightarrow \infty$ each element of $\mathbf{v}'(k)$ tends to

$$v_i(\infty) = \mu\sqrt{\frac{\pi\xi(\infty)}{2}} \tag{4.22}$$

4.2.1.3 *Excess Mean-Square Error and Misadjustment*

The excess MSE can be expressed as a function of the elements of $\mathbf{v}'(k)$ by

$$\Delta\xi(k) = \sum_{i=0}^{N}\lambda_i v_i(k) = \boldsymbol{\lambda}^T\mathbf{v}'(k) \tag{4.23}$$

Substituting equation (4.22) in equation (4.23) yields

$$\begin{aligned}\xi_{\text{exc}} &= \mu\sum_{i=0}^{N}\lambda_i\sqrt{\frac{\pi\xi(k)}{2}},\, k\rightarrow\infty\\ &= \mu\sum_{i=0}^{N}\lambda_i\sqrt{\pi\,\frac{\xi_{\min}+\xi_{\text{exc}}}{2}}\end{aligned} \tag{4.24}$$

since $\lim_{k\rightarrow\infty}\xi(k) = \xi_{\min} + \xi_{\text{exc}}$. Therefore,

$$\xi_{\text{exc}}^2 = \mu^2\left(\sum_{i=0}^{N}\lambda_i\right)^2\left(\frac{\pi\xi_{\min}}{2} + \frac{\pi\xi_{\text{exc}}}{2}\right) \tag{4.25}$$

There are two solutions for ξ_{exc}^2 in the above equation, where only the positive one is valid. The meaningful solution for ξ_{exc}, when μ is small, is approximately given by

$$\begin{aligned}\xi_{\text{exc}} &\approx \mu\sqrt{\frac{\pi\xi_{\min}}{2}}\sum_{i=0}^{N}\lambda_i\\ &= \mu\sqrt{\frac{\pi\xi_{\min}}{2}}\,\text{tr}[\mathbf{R}]\end{aligned} \tag{4.26}$$

By comparing the excess MSE predicted by the above equation with the corresponding equation (3.49) for the LMS algorithm, it can be concluded that both can generate the same excess MSE if μ in the sign-error algorithm is chosen such that

$$\mu = \mu_{\text{LMS}}\sqrt{\frac{2}{\pi}\xi_{\min}^{-1}} \tag{4.27}$$

The misadjustment in the sign-error algorithm is

$$M = \mu\sqrt{\frac{\pi}{2\xi_{\min}}}\ \text{tr}[\mathbf{R}] \tag{4.28}$$

Equation (4.26) would leave the impression that if there is no additional noise and there are sufficient parameters in the adaptive filter, the output MSE would converge to zero. However, when $\xi(k)$ becomes small, $||E[\Delta\mathbf{w}(k+1)]||$ in equation (4.11) can increase, since the condition of equation (4.13) will not be satisfied. This is the situation where the parameters reach the convergence balloon. In this case, from equation (4.8) we can conclude that

$$||\Delta\mathbf{w}(k+1)||^2 - ||\Delta\mathbf{w}(k)||^2 = -4\mu\ \text{sgn}[e(k)]\ e(k) + 4\mu^2||\mathbf{x}(k)||^2 \tag{4.29}$$

from where it is possible to show that a decrease in the norm of $\Delta\mathbf{w}(k)$ is obtained only when

$$|e(k)| > \mu||\mathbf{x}(k)||^2 \tag{4.30}$$

For no additional noise, first transpose the vectors in equation (4.8) and postmultiply each side by $\mathbf{x}(k)$. Next, squaring the resulting equation and applying the expected value operation on each side, the obtained result is

$$E[e^2(k+1)] = E[e^2(k)] - 4\mu E[|e(k)|\ ||\mathbf{x}(k)||^2] + 4\mu^2 E[||\mathbf{x}(k)||^4] \tag{4.31}$$

After convergence $E[e^2(k+1)] \approx E[e^2(k)]$. Also, considering that

$$E[|e(k)|\ ||\mathbf{x}(k)||^2] \approx E[|e(k)|]E[||\mathbf{x}(k)||^2]$$

and

$$\frac{E[||\mathbf{x}(k)||^4]}{E[||\mathbf{x}(k)||^2]} \approx E[||\mathbf{x}(k)||^2]$$

we conclude that

$$E[|e(k)|] \approx \mu E[||\mathbf{x}(k)||^2], k \rightarrow \infty \tag{4.32}$$

For zero-mean Gaussian $e(k)$, the following approximation is valid

$$E[|e(k)|] \approx \sqrt{\frac{2}{\pi}}\sigma_e(k), k \rightarrow \infty \tag{4.33}$$

therefore, the expected variance of $e(k)$ is

$$\sigma_e^2(k) \approx \frac{\pi}{2}\mu^2\ \text{tr}^2[\mathbf{R}], k \rightarrow \infty \tag{4.34}$$

where we used the relation $\text{tr}[\mathbf{R}] = E[||\mathbf{x}(k)||^2]$. This relation gives an estimate of the variance of the output error when no additional noise exists. As can be noted, unlike the LMS algorithm, there is an excess MSE in the sign-error algorithm caused by the nonlinear device, even when $\sigma_n^2 = 0$.

If $n(k)$ has frequently large absolute values as compared to $-\Delta\mathbf{w}^T(k)\mathbf{x}(k)$, then for most iterations $\text{sgn}[e(k)] = \text{sgn}[n(k)]$. As a result, the sign-error algorithm is fully controlled by the additional noise. In this case, the algorithm does not converge.

4.2.1.4 Transient Behavior

The ratios r_{w_i} of the geometric decaying convergence curves of the coefficients in the sign-error algorithm can be derived from equation (4.12) by employing an identical analysis of the transient behavior for the LMS algorithm. The ratios are given by

$$r_{w_i} = \left(1 - 2\mu\sqrt{\frac{2}{\pi\xi(k)}}\lambda_i\right) \tag{4.35}$$

for $i = 0, 1, \ldots, N$. If μ is chosen as suggested in equation (4.27), in order to reach the same excess MSE of the LMS algorithm, then

$$r_{w_i} = \left(1 - \frac{4}{\pi}\mu_{\text{LMS}}\sqrt{\frac{\xi_{\min}}{\xi(k)}}\lambda_i\right) \tag{4.36}$$

By recalling that r_{w_i} for the LMS algorithm is $(1 - 2\mu_{\text{LMS}}\lambda_i)$, since $\frac{2}{\pi}\sqrt{\frac{\xi_{\min}}{\xi(k)}} < 1$, it is concluded that the sign-error algorithm is slower than the LMS for the same excess MSE.

Example 4.1

Suppose in an adaptive-filtering environment that the input signal consists of

$$x(k) = \mathrm{e}^{\jmath\omega_0 k} + n(k)$$

and that the desired signal is given by

$$d(k) = \mathrm{e}^{\jmath\omega_0(k-1)}$$

where $n(k)$ is a uniformly distributed white noise with variance $\sigma_n^2 = 0.1$ and $\omega_0 = \frac{2\pi}{M}$. In this case $M = 8$.

Compute the input signal correlation matrix for a first-order adaptive filter. Calculate the value of $\mu_{\max}$ for the sign-error algorithm.

Solution:

The input signal correlation matrix for this example can be calculated as shown below:

$$\mathbf{R} = \begin{bmatrix} 1+\sigma_n^2 & \mathrm{e}^{\jmath\omega_0} \\ \mathrm{e}^{-\jmath\omega_0} & 1+\sigma_n^2 \end{bmatrix}$$

Since in this case $\text{tr}[\mathbf{R}] = 2.2$ and $\xi_{\min} = 0.1$, we have

$$\xi_{\text{exc}} \approx \mu \sqrt{\frac{\pi \xi_{\min}}{2}} \, \text{tr}[\mathbf{R}] = 0.87\mu$$

The range of values of the convergence factor is given by

$$0 < \mu < \frac{1}{2\text{tr}[\mathbf{R}]} \sqrt{\frac{\pi(\xi_{\min} + \xi_{\text{exc}})}{2}}$$

From the above expression, it is straightforward to calculate the upper bound for the convergence factor that is given by

$$\mu_{\max} \approx 0.132$$

□

4.2.2 Dual-Sign Algorithm

The dual-sign algorithm attempts to perform large corrections to the coefficient vector when the modulus of the error signal is larger than a prescribed level. The basic motivation to use the dual-sign algorithm is to avoid the slow convergence inherent to the sign-error algorithm that is caused by replacing $e(k)$ by $\text{sgn}[e(k)]$ when $|e(k)|$ is large.

The quantization function for the dual-sign algorithm is given by

$$\text{ds}[a] = \begin{cases} \gamma \, \text{sgn}[a], & |a| > \rho \\ \text{sgn}[a], & |a| \leq \rho \end{cases} \tag{4.37}$$

where $\gamma > 1$ is a power of two. The dual-sign algorithm utilizes the function above described as the error quantizer, and the coefficient updating is performed as

$$\mathbf{w}(k+1) = \mathbf{w}(k) + 2\mu \, \text{ds}[e(k)]\mathbf{x}(k) \tag{4.38}$$

The objective function that is minimized by the dual-sign algorithm is given by

$$F[e(k)] = \begin{cases} 2\gamma|e(k)| - 2\rho(\gamma - 1), & |e(k)| > \rho \\ 2|e(k)|, & |e(k)| \leq \rho \end{cases} \tag{4.39}$$

where the constant $2\rho(\gamma - 1)$ was included in the objective function to make it continuous. Obviously the gradient of $F[e(k)]$ with respect to the filter coefficients is $2\mu \, \text{ds}[e(k)]\mathbf{x}(k)$ except at points where $\text{ds}[e(k)]$ is nondifferentiable [6].

The same analysis procedure used for the sign-error algorithm can be applied to the dual-sign algorithm except for the fact that the quantization function is now different. The alternative quantization leads to particular expectations of nonlinear functions whose solutions are not presented here. The interested reader should refer to the work of Mathews [7]. The choice of γ and ρ determine the convergence behavior of the dual-sign algorithm [7], typically, a large γ tends to increase both convergence speed and excess MSE. A large ρ tends to reduce both the convergence speed and the excess MSE. If $\lim_{k\rightarrow\infty} \xi(k) \ll \rho^2$, the excess MSE of the dual-sign algorithm is approximately equal to the one given by equation (4.26) for the sign-error algorithm [7], since in this case $|e(k)|$ is usually much smaller than ρ. For a given MSE in steady state, the dual-sign algorithm is expected to converge faster than the sign-error algorithm.

4.2.3 Power-of-Two Error Algorithm

The power-of-two error algorithm applies to the error signal a quantization defined by

$$\mathrm{pe}[b] = \begin{cases} \mathrm{sgn}[b], & |b| \geq 1 \\ 2^{\mathrm{floor}[log_2|b|]}\,\mathrm{sgn}[b], & 2^{-b_d+1} \leq |b| < 1 \\ \tau\mathrm{sgn}[b], & |b| < 2^{-b_d+1} \end{cases} \tag{4.40}$$

where floor$[\cdot]$ indicates integer smaller than $[\cdot]$, b_d is the data wordlength excluding the sign bit, and τ is usually 0 or 2^{-b_d}.

The coefficient updating for the power-of-two error algorithm is given by

$$\mathbf{w}(k+1) = \mathbf{w}(k) + 2\mu\,\mathrm{pe}[e(k)]\mathbf{x}(k) \tag{4.41}$$

For $\tau = 2^{-b_d}$, the additional noise and the convergence factor can be arbitrarily small and the algorithm will not stop updating. For $\tau = 0$, when $|e(k)| < 2^{-b_d+1}$ the algorithm reaches the so-called *dead zone*, where the algorithm stops updating if $|e(k)|$ is smaller than 2^{-b_d+1} most of the time [4], [8].

A simplified and somewhat accurate analysis of this algorithm can be performed by approximating the function $\mathrm{pe}[e(k)]$ by a straight line passing through the center of each quantization step. In this case, the quantizer characteristics can be approximated by $\mathrm{pe}[e(k)] \approx \frac{2}{3}e(k)$ as illustrated in Fig. 4.2. Using this approximation, the algorithm analysis can be performed exactly in the same way as the LMS algorithm. The results for the power-of-two error algorithm can be obtained from the results for the LMS algorithm, by replacing μ by $\frac{2}{3}\mu$. It should be mentioned that such results are only approximate, and more accurate ones can be found in [8].

4.2.4 Sign-Data Algorithm

The algorithms discussed in this subsection cannot be considered as quantized error algorithms, but since they were proposed with similar motivation we decided to introduce them here. An alternative way to simplify the computational burden of the LMS algorithm is to apply quantization to the data vector $\mathbf{x}(k)$. One possible quantization scheme is to apply the sign function to the input signals, giving rise to the sign-data algorithm whose coefficient updating is performed as

$$\mathbf{w}(k+1) = \mathbf{w}(k) + 2\mu e(k)\,\mathrm{sgn}[\mathbf{x}(k)] \tag{4.42}$$

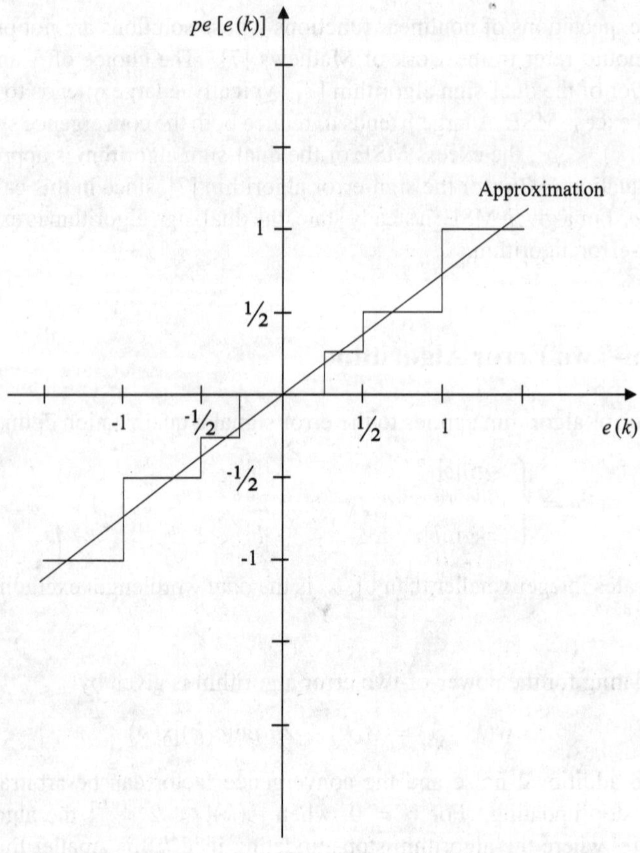

Figure 4.2 Transfer characteristic of a quantizer with 3 bits and $\tau = 0$.

where the sign operation is applied to each element of the input vector.

The quantization of the data vector can lead to a decrease in the convergence speed, and possible divergence. In the LMS algorithm, the average gradient direction follows the true gradient direction (or steepest-descent direction), whereas in the sign-data algorithm only a discrete set of directions can be followed. The limitation in the gradient direction followed by the sign-data algorithm may cause updates that result in frequent increase in the squared error, leading to instability. Therefore, it is relatively easy to find inputs that would lead to the convergence of the LMS algorithm and to the divergence of the sign-data algorithm [6], [9]. It should be mentioned, however, that the sign-data algorithm is stable for Gaussian inputs, and, as such, has been found useful in certain applications.

Another related algorithm is the sign-sign algorithm that has very simple implementation. The coefficient updating in this case is given by

$$\mathbf{w}(k+1) = \mathbf{w}(k) + 2\mu\ \mathrm{sgn}[e(k)]\ \mathrm{sgn}[\mathbf{x}(k)] \tag{4.43}$$

The sign-sign algorithm also presents the limitations related to the quantized-data algorithm.

4.3 THE LMS-NEWTON ALGORITHM

In this section, the LMS-Newton algorithm incorporating estimates of the second-order statistics of the environment signals is introduced. The objective of the algorithm is to avoid the slow convergence of the LMS algorithm when the input signal is highly correlated. The improvement in the convergence rate is achieved at the expense of an increased computational complexity.

Nonrecursive realization of the adaptive filter leads to an MSE surface that is a quadratic function of the filter coefficients. For the direct-form FIR structure, the MSE can be described by

$$\xi(k+1) = \xi(k) + \mathbf{g_w}^T(k)\left[\mathbf{w}(k+1) - \mathbf{w}(k)\right] + \left[\mathbf{w}(k+1) - \mathbf{w}(k)\right]^T \mathbf{R}\left[\mathbf{w}(k+1) - \mathbf{w}(k)\right] \tag{4.44}$$

$\xi(k)$ represents the MSE when the adaptive-filter coefficients are fixed at $\mathbf{w}(k)$ and $\mathbf{g_w}(k) = -2\mathbf{p} + 2\mathbf{R}\mathbf{w}(k)$ is the gradient vector of the MSE surface as related to the filter coefficients at $\mathbf{w}(k)$. The MSE is minimized at the instant $k+1$ if

$$\mathbf{w}(k+1) = \mathbf{w}(k) - \frac{1}{2}\mathbf{R}^{-1}\mathbf{g_w}(k) \tag{4.45}$$

This equation is the updating formula of the Newton method. Note that in the ideal case, where matrix $\mathbf{R}$ and gradient vector $\mathbf{g_w}(k)$ are known precisely, $\mathbf{w}(k+1) = \mathbf{R}^{-1}\mathbf{p} = \mathbf{w}_o$. Therefore, the Newton method converges to the optimal solution in a single iteration, as expected for a quadratic objective function.

In practice, only estimates of the autocorrelation matrix $\mathbf{R}$ and of the gradient vector are available. These estimates can be applied to the Newton updating formula in order to derive a Newton-like method given by

$$\mathbf{w}(k+1) = \mathbf{w}(k) - \mu\hat{\mathbf{R}}^{-1}(k)\hat{\mathbf{g}}_{\mathbf{w}}(k) \tag{4.46}$$

The convergence factor μ is introduced so that the algorithm can be protected from divergence, originated by the use of noisy estimates of $\mathbf{R}$ and $\mathbf{g_w}(k)$.

For stationary input signals, an unbiased estimate of $\mathbf{R}$ is

$$\begin{aligned}\hat{\mathbf{R}}(k) &= \frac{1}{k+1}\sum_{i=0}^{k}\mathbf{x}(i)\mathbf{x}^T(i)\\ &= \frac{k}{k+1}\hat{\mathbf{R}}(k-1) + \frac{1}{k+1}\mathbf{x}(k)\mathbf{x}^T(k)\end{aligned} \tag{4.47}$$

since

$$\begin{aligned}E[\hat{\mathbf{R}}(k)] &= \frac{1}{k+1}\sum_{i=0}^{k}E[\mathbf{x}(i)\mathbf{x}^T(i)]\\ &= \mathbf{R}\end{aligned} \tag{4.48}$$

However, this is not a practical estimate for $\mathbf{R}$, since for large k any change on the input signal statistics would be disregarded due to the infinite memory of the estimation algorithm.

Another form to estimate the autocorrelation matrix can be generated by employing a weighted summation as follows:

$$\begin{aligned}\hat{\mathbf{R}}(k) &= \alpha\mathbf{x}(k)\mathbf{x}^T(k) + (1-\alpha)\hat{\mathbf{R}}(k-1)\\ &= \alpha\mathbf{x}(k)\mathbf{x}^T(k) + \alpha\sum_{i=0}^{k-1}(1-\alpha)^{k-i}\mathbf{x}(i)\mathbf{x}^T(i)\end{aligned} \tag{4.49}$$

where in practice, α is a small factor chosen in the range $0 < \alpha \leq 0.1$. This range of values of α allows a good balance between the present and past input signal information. By taking the expected value on both sides of the above equation and assuming that $k \to \infty$, it follows that

$$\begin{aligned}E[\hat{\mathbf{R}}(k)] &= \alpha\sum_{i=0}^{k}(1-\alpha)^{k-i}E[\mathbf{x}(i)\mathbf{x}^T(i)]\\ &= \mathbf{R} \qquad k \to \infty\end{aligned} \tag{4.50}$$

Therefore, the estimate of $\mathbf{R}$ of equation (4.49) is unbiased.

In order to avoid inverting $\hat{\mathbf{R}}(k)$, which is required by the Newton-like algorithm, we can use the so-called matrix inversion lemma given by

$$[\mathbf{A}+\mathbf{BCD}]^{-1} = \mathbf{A}^{-1} - \mathbf{A}^{-1}\mathbf{B}[\mathbf{DA}^{-1}\mathbf{B}+\mathbf{C}^{-1}]^{-1}\mathbf{DA}^{-1} \tag{4.51}$$

where $\mathbf{A}$, $\mathbf{B}$, $\mathbf{C}$ and $\mathbf{D}$ are matrices of appropriate dimensions, and $\mathbf{A}$ and $\mathbf{C}$ are nonsingular. The above relation can be proved by simply showing that the result of premultiplying the expression on the right-hand side by $\mathbf{A}+\mathbf{BCD}$ is the identity matrix (see problem 21). If we choose $\mathbf{A} = (1-\alpha)$ $\hat{\mathbf{R}}(k-1)$, $\mathbf{B} = \mathbf{D}^T = \mathbf{x}(k)$, and $\mathbf{C} = \alpha$, it can be shown that

$$\hat{\mathbf{R}}^{-1}(k) = \frac{1}{1-\alpha}\left[\hat{\mathbf{R}}^{-1}(k-1) - \frac{\hat{\mathbf{R}}^{-1}(k-1)\mathbf{x}(k)\mathbf{x}^T(k)\hat{\mathbf{R}}^{-1}(k-1)}{\frac{1-\alpha}{\alpha}+\mathbf{x}^T(k)\hat{\mathbf{R}}^{-1}(k-1)\mathbf{x}(k)}\right] \tag{4.52}$$

The resulting equation to calculate $\hat{\mathbf{R}}^{-1}(k)$ is less complex to update (of order N^2 multiplications) than the direct inversion of $\hat{\mathbf{R}}(k)$ at every iteration (of order N^3 multiplications).

If the estimate for the gradient vector used in the LMS algorithm is applied in equation (4.46), the following coefficient updating formula for the LMS-Newton algorithm results

$$\mathbf{w}(k+1) = \mathbf{w}(k) + 2\,\mu\, e(k)\,\hat{\mathbf{R}}^{-1}(k)\mathbf{x}(k) \tag{4.53}$$

The complete LMS-Newton algorithm is outlined in Algorithm 4.2. It should be noticed that alternative initialization procedures to the one presented in Algorithm 4.2 are possible.

As previously mentioned, the LMS gradient direction has the tendency to approach the ideal gradient direction. Similarly, the vector resulting from the multiplication of $\hat{\mathbf{R}}^{-1}(k)$ to the LMS gradient

Algorithm 4.2

LMS-Newton Algorithm

Initialization

$\hat{\mathbf{R}}^{-1}(-1) = \delta\mathbf{I}$ (δ a small positive constant)

$\mathbf{w}(0) = \mathbf{x}(-1) = [0\,0\ldots 0]^T$

Do for $k \geq 0$

$e(k) = d(k) - \mathbf{x}^T(k)\mathbf{w}(k)$

$\hat{\mathbf{R}}^{-1}(k) = \frac{1}{1-\alpha}\left[\hat{\mathbf{R}}^{-1}(k-1) - \frac{\hat{\mathbf{R}}^{-1}(k-1)\mathbf{x}(k)\mathbf{x}^T(k)\hat{\mathbf{R}}^{-1}(k-1)}{\frac{1-\alpha}{\alpha}+\mathbf{x}^T(k)\hat{\mathbf{R}}^{-1}(k-1)\mathbf{x}(k)}\right]$

$\mathbf{w}(k+1) = \mathbf{w}(k) + 2\,\mu\, e(k)\,\hat{\mathbf{R}}^{-1}(k)\mathbf{x}(k)$

direction tends to approach the Newton direction. Therefore, we can conclude that the LMS-Newton algorithm converges in a more straightforward path to the minimum of the MSE surface. It can also be shown that the convergence characteristics of the algorithm is independent of the eigenvalue spread of **R**.

The LMS-Newton algorithm is mathematically identical to the recursive least-squares (RLS) algorithm if the forgetting factor (λ) in the latter is chosen such that $2\mu = \alpha = 1 - \lambda$ [39]. Since a complete discussion of the RLS algorithm is given later, no further discussion of the LMS-Newton algorithm is included here.

4.4 THE NORMALIZED LMS ALGORITHM

If one wishes to increase the convergence speed of the LMS algorithm without using estimates of the input signal correlation matrix, a variable convergence factor is a natural solution. The normalized LMS algorithm usually converges faster than the LMS algorithm, since it utilizes a variable convergence factor aiming at the minimization of the instantaneous output error.

The updating equation of the LMS algorithm can employ a variable convergence factor μ_k in order to improve the convergence rate. In this case, the updating formula is expressed as

$$\mathbf{w}(k+1) = \mathbf{w}(k) + 2\mu_k e(k)\mathbf{x}(k) = \mathbf{w}(k) + \Delta\tilde{\mathbf{w}}(k) \tag{4.54}$$

where μ_k must be chosen with the objective of achieving a faster convergence. A possible strategy is to reduce the instantaneous squared error as much as possible. The motivation behind this strategy is that the instantaneous squared error is a good and simple estimate of the MSE.

The instantaneous squared error is given by

$$e^2(k) = d^2(k) + \mathbf{w}^T(k)\mathbf{x}(k)\mathbf{x}^T(k)\mathbf{w}(k) - 2d(k)\mathbf{w}^T(k)\mathbf{x}(k) \tag{4.55}$$

If a change given by $\tilde{\mathbf{w}}(k) = \mathbf{w}(k) + \Delta\tilde{\mathbf{w}}(k)$ is performed in the weight vector, the corresponding squared error can be shown to be

$$\tilde{e}^2(k) = e^2(k) + 2\Delta\tilde{\mathbf{w}}^T(k)\mathbf{x}(k)\mathbf{x}^T(k)\mathbf{w}(k) + \Delta\tilde{\mathbf{w}}^T(k)\mathbf{x}(k)\mathbf{x}^T(k)\Delta\tilde{\mathbf{w}}(k) - 2d(k)\Delta\tilde{\mathbf{w}}^T(k)\mathbf{x}(k) \tag{4.56}$$

It then follows that

$$\begin{aligned}\Delta e^2(k) &\triangleq \tilde{e}^2(k) - e^2(k)\\ &= -2\Delta\tilde{\mathbf{w}}^T(k)\mathbf{x}(k)e(k) + \Delta\tilde{\mathbf{w}}^T(k)\mathbf{x}(k)\mathbf{x}^T(k)\Delta\tilde{\mathbf{w}}(k)\end{aligned} \tag{4.57}$$

In order to increase the convergence rate, the objective is to make $\Delta e^2(k)$ negative and minimum by appropriately choosing μ_k.

By replacing $\Delta\tilde{\mathbf{w}}(k) = 2\mu_k e(k)\mathbf{x}(k)$ in equation (4.57), it follows that

$$\Delta e^2(k) = -4\mu_k e^2(k)\mathbf{x}^T(k)\mathbf{x}(k) + 4\mu_k^2 e^2(k)[\mathbf{x}^T(k)\mathbf{x}(k)]^2 \tag{4.58}$$

The value of μ_k such that $\frac{\partial \Delta e^2(k)}{\partial \mu_k} = 0$ is given by

$$\mu_k = \frac{1}{2\mathbf{x}^T(k)\mathbf{x}(k)} \tag{4.59}$$

This value of μ_k leads to a negative value of $\Delta e^2(k)$, and, therefore, it corresponds to a minimum point of $\Delta e^2(k)$.

Using this variable convergence factor, the updating equation for the LMS algorithm is then given by

$$\mathbf{w}(k+1) = \mathbf{w}(k) + \frac{e(k)\mathbf{x}(k)}{\mathbf{x}^T(k)\mathbf{x}(k)} \tag{4.60}$$

Usually a fixed convergence factor μ_n is introduced in the updating formula in order to control the misadjustment, since all the derivations are based on instantaneous values of the squared errors and not on the MSE. Also a parameter γ should be included, in order to avoid large step sizes when $\mathbf{x}^T(k)\mathbf{x}(k)$ becomes small. The coefficient updating equation is then given by

$$\mathbf{w}(k+1) = \mathbf{w}(k) + \frac{\mu_n}{\gamma + \mathbf{x}^T(k)\mathbf{x}(k)}\, e(k)\,\mathbf{x}(k) \tag{4.61}$$

The resulting algorithm is called the normalized LMS algorithm, and is summarized in Algorithm 4.3.

Algorithm 4.3

The Normalized LMS Algorithm

Initialization
$\mathbf{x}(0) = \hat{\mathbf{w}}(0) = [0\,0\ldots0]^T$
choose μ_n in the range $0 < \mu_n \leq 2$
$\gamma =$ small constant
Do for $k \geq 0$
$e(k) = d(k) - \mathbf{x}^T(k)\mathbf{w}(k)$
$\mathbf{w}(k+1) = \mathbf{w}(k) + \frac{\mu_n}{\gamma+\mathbf{x}^T(k)\mathbf{x}(k)}\, e(k)\, \mathbf{x}(k)$

The range of values of μ_n to guarantee stability can be derived by first considering that $E[\mathbf{x}^T(k)\mathbf{x}(k)] = \text{tr}[\mathbf{R}]$ and that

$$E\left[\frac{e(k)\mathbf{x}(k)}{\mathbf{x}^T(k)\mathbf{x}(k)}\right] \approx \frac{E[e(k)\mathbf{x}(k)]}{E[\mathbf{x}^T(k)\mathbf{x}(k)]}$$

Next, consider that the average value of the convergence factor actually applied to the LMS direction $2e(k)\mathbf{x}(k)$ is $\frac{\mu_n}{2\,\text{tr}[\mathbf{R}]}$. Finally, by comparing the updating formula of the standard LMS algorithm with that of the normalized LMS algorithm, the desired upper bound result follows:

$$0 < \mu = \frac{\mu_n}{2\,\text{tr}[\mathbf{R}]} < \frac{1}{\text{tr}[\mathbf{R}]} \tag{4.62}$$

or $0 < \mu_n < 2$.

4.5 THE TRANSFORM-DOMAIN LMS ALGORITHM

The transform-domain LMS algorithm is another technique to increase the convergence speed of the LMS algorithm when the input signal is highly correlated. The basic idea behind this methodology is to modify the input signal to be applied to the adaptive filter such that the conditioning number of the corresponding correlation matrix is improved.

In the transform-domain LMS algorithm, the input signal vector $\mathbf{x}(k)$ is transformed in a more convenient vector $\mathbf{s}(k)$, by applying an orthonormal (or unitary) transform [10]-[12], i.e.,

$$\mathbf{s}(k) = \mathbf{T}\mathbf{x}(k) \tag{4.63}$$

where $\mathbf{T}\mathbf{T}^T = \mathbf{I}$. The MSE surface related to the direct-form implementation of the FIR adaptive filter can be described by

$$\xi(k) = \xi_{\min} + \Delta\mathbf{w}^T(k)\mathbf{R}\Delta\mathbf{w}(k) \tag{4.64}$$

where $\Delta\mathbf{w}(k) = \mathbf{w}(k) - \mathbf{w}_o$. In the transform-domain case, the MSE surface becomes

$$\begin{aligned}\xi(k) &= \xi_{\min} + \Delta\hat{\mathbf{w}}^T(k)E[\mathbf{s}(k)\mathbf{s}^T(k)]\Delta\hat{\mathbf{w}}(k) \\ &= \xi_{\min} + \Delta\hat{\mathbf{w}}^T(k)\mathbf{TRT}^T\Delta\hat{\mathbf{w}}(k)\end{aligned} \tag{4.65}$$

where $\hat{\mathbf{w}}(k)$ represents the adaptive coefficients of the transform-domain filter. Fig. 4.3 depicts the transform-domain adaptive filter.

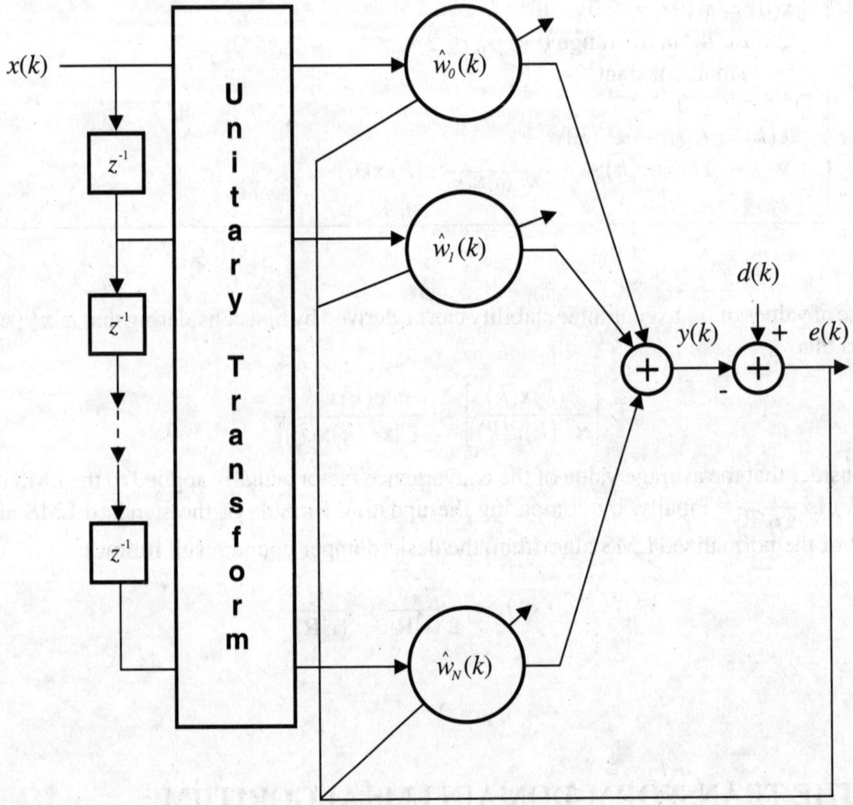

Figure 4.3 Transform-domain adaptive filter.

The effect of applying the transformation matrix $\mathbf{T}$ to the input signal is to rotate the error surface as illustrated in the numerical examples of Figs. 4.4 and 4.5. It can be noticed that the eccentricity of the MSE surface remains unchanged by the application of the transformation, and, therefore, the eigenvalue spread is unaffected by the transformation. As a consequence, no improvement in the convergence rate is expected to occur. However, if in addition each element of the transform output is power normalized, the distance between the points where the equal-error contours (given by the ellipses) meet the coefficient axes ($\Delta\hat{w}_0$ and $\Delta\hat{w}_1$) and the origin (point 0×0) are equalized. As a result, a reduction in the eigenvalue spread is expected, especially when the coefficient axes are almost aligned with the principal axes of the ellipses. Fig. 4.6 illustrates the effect of power normalization. The perfect alignment and power normalization means that the error surface will become a hyperparaboloid spheric, with the eigenvalue spread becoming equal to one. Alternatively, it means that the transform was able to turn the elements of the vector $\mathbf{s}(k)$ uncorrelated. Fig. 4.7 shows another error surface which after properly rotated and normalized is transformed into the error surface of Fig. 4.8.

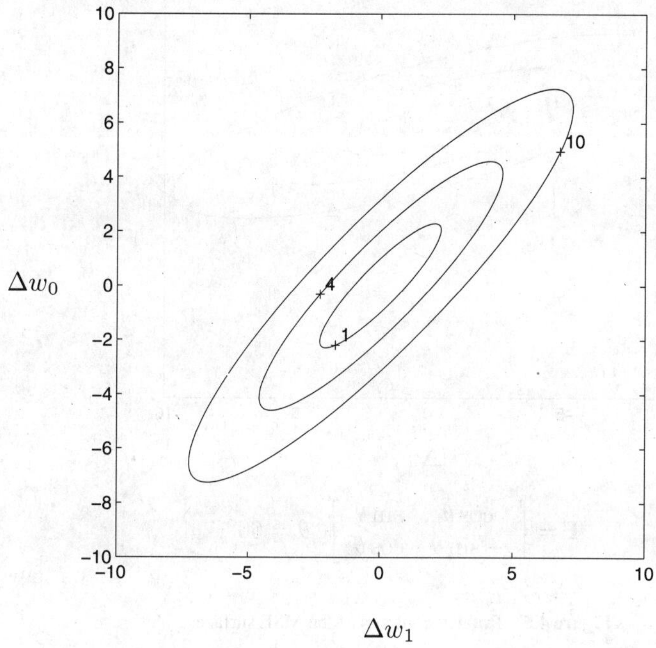

$$\mathbf{R} = \begin{bmatrix} 1 & -0.9 \\ -0.9 & 1 \end{bmatrix}$$

Figure 4.4 Contours of the original MSE surface.

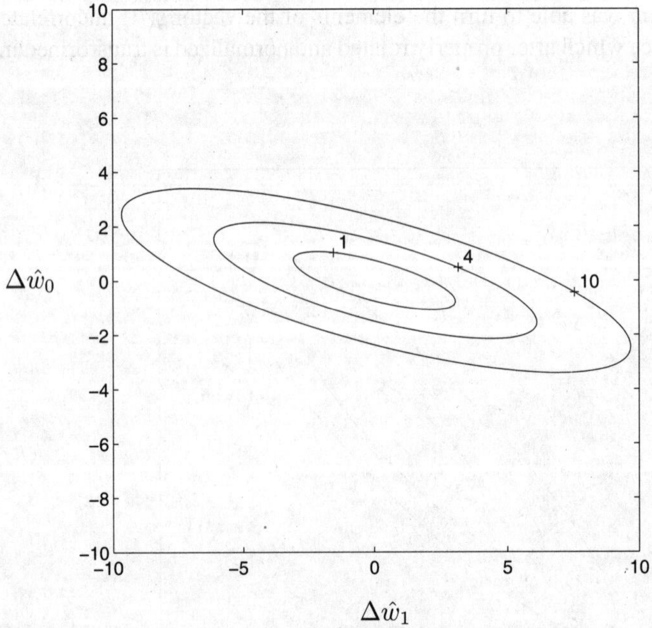

$$\mathbf{T} = \begin{bmatrix} \cos\theta & \sin\theta \\ -\sin\theta & \cos\theta \end{bmatrix} \quad \theta = 60^\circ$$

Figure 4.5 Rotated contours of the MSE surface.

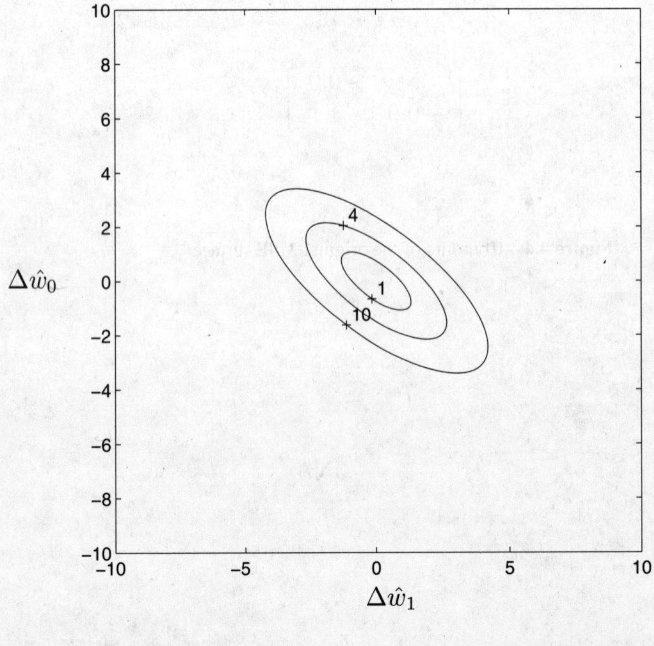

Figure 4.6 Contours of the power normalized MSE surface.

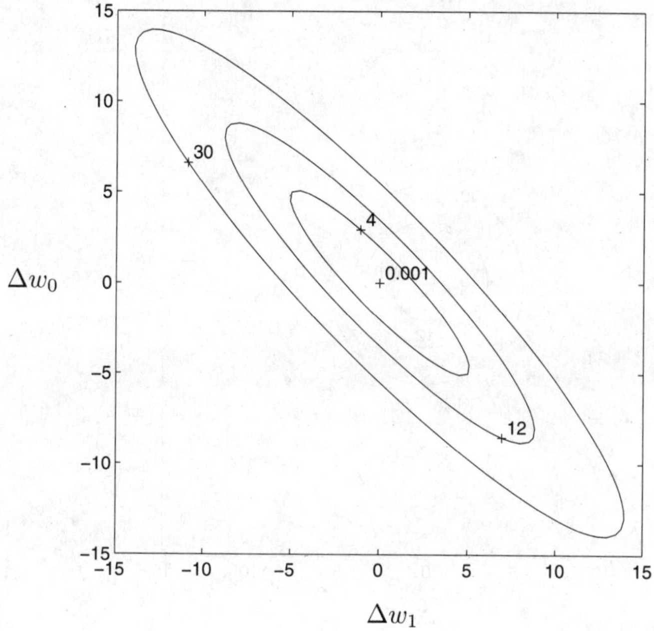

$$\mathbf{R} = \begin{bmatrix} 1 & 0.92 \\ 0.92 & 1 \end{bmatrix}$$

Figure 4.7 Contours of the original MSE surface.

The autocorrelation matrix related to the transform-domain filter is given by

$$\mathbf{R}_s = \mathbf{T}\mathbf{R}\mathbf{T}^T \tag{4.66}$$

therefore if the elements of $\mathbf{s}(k)$ are uncorrelated, matrix $\mathbf{R}_s$ is diagonal, meaning that the application of the transformation matrix was able to diagonalize the autocorrelation matrix $\mathbf{R}$. It can then be concluded that $\mathbf{T}^T$, in this case, corresponds to a matrix whose columns consist of the orthonormal eigenvectors of $\mathbf{R}$. The resulting transformation matrix corresponds to the Karhunen-Loève Transform (KLT)[26].

The normalization of $\mathbf{s}(k)$ and subsequent application of the LMS algorithm would lead to a transform-domain algorithm with the limitation that the solution would be independent of the input signal power. An alternative solution, without this limitation, is to apply the normalized LMS algorithm to update the coefficients of the transform-domain algorithm. We can give an interpretation for the good performance of this solution. Assuming the transform was efficient in the rotation of the MSE surface, the variable convergence factor is large in the update of the coefficients corresponding to low signal power. On the other hand, the convergence factor is small if the corresponding transform

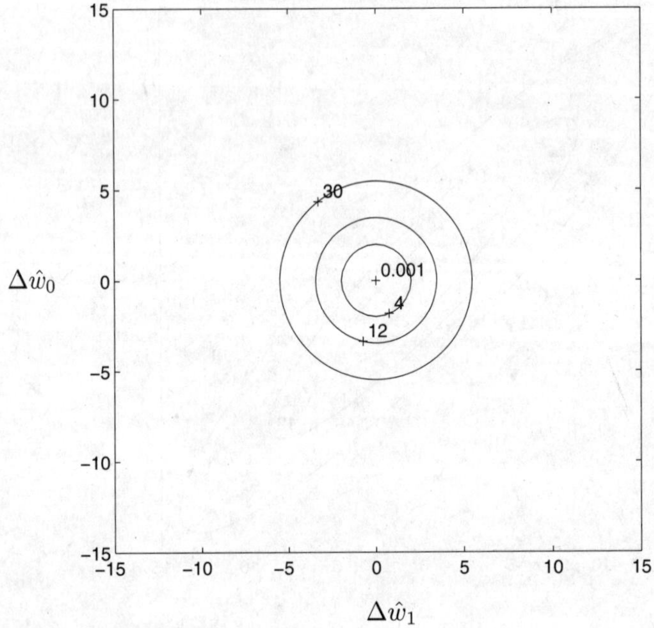

$$\mathbf{TR} = \begin{bmatrix} 1 & 0 \\ 0 & 1 \end{bmatrix}$$

Figure 4.8 Contours of the rotated and power normalized MSE surface.

output power is high. Specifically, the signals $s_i(k)$ are normalized by their power denoted by $\sigma_i^2(k)$ only when applied in the updating formula. The coefficient update equation in this case is

$$\hat{w}_i(k+1) = \hat{w}_i(k) + \frac{2\mu}{\gamma + \sigma_i^2(k)} e(k) s_i(k) \tag{4.67}$$

where $\sigma_i^2(k) = \alpha s_i^2(k) + (1-\alpha)\sigma_i^2(k-1)$, α is a small factor chosen in the range $0 < \alpha \leq 0.1$, and γ is also a small constant to avoid that the second term of the update equation becomes too large when $\sigma_i^2(k)$ is small.

In matrix form the above updating equation can be rewritten as

$$\hat{\mathbf{w}}(k+1) = \hat{\mathbf{w}}(k) + 2\mu e(k) \boldsymbol{\Sigma}^{-2}(k) \mathbf{s}(k) \tag{4.68}$$

where $\boldsymbol{\Sigma}^{-2}(k)$ is a diagonal matrix containing as elements the inverse of the power estimates of the elements of $\mathbf{s}(k)$ added to γ.

It can be shown that if μ is chosen appropriately, the adaptive-filter coefficients converge to

$$\hat{\mathbf{w}}_o = \mathbf{R}_s{}^{-1}\mathbf{p}_s \tag{4.69}$$

where $\mathbf{R}_s = \mathbf{TRT}^T$ and $\mathbf{p}_s = \mathbf{Tp}$. As a consequence, the optimum coefficient vector is

$$\hat{\mathbf{w}}_o = (\mathbf{TRT}^T)^{-1}\mathbf{Tp} = \mathbf{TR}^{-1}\mathbf{p} = \mathbf{Tw}_o \tag{4.70}$$

The convergence speed of the coefficient vector $\hat{\mathbf{w}}(k)$ is determined by the eigenvalue spread of $\mathbf{\Sigma}^{-2}(k)\mathbf{R}_s$.

The requirement on the transformation matrix is that it should be invertible. If the matrix $\mathbf{T}$ is not square (number of columns larger than rows), the space spanned by the polynomials formed with the rows of $\mathbf{T}$ will be of dimension $N+1$, but these polynomials are of order larger than N. This subspace does not contain the complete space of polynomials of order N. In general, except for very specific desired signals, the entire space of Nth-order polynomials would be required. For an invertible matrix $\mathbf{T}$ there is a one-to-one correspondence between the solutions obtained by the LMS and transform-domain LMS algorithms. Although the transformation matrix is not required to be unitary, it appears that no advantages are obtained by using nonunitary transforms [13].

The best unitary transform for the transform-domain adaptive filter is the KLT. However, since the KLT is a function of the input signal, it cannot be efficiently computed in real time. An alternative is to choose a unitary transform that is close to the KLT of the particular input signal. By close is meant that both transforms perform nearly the same rotation of the MSE surface. In any situation, the choice of an appropriate transform is not an easy task. Some guidelines can be given, such as: i) Since the KLT of a real signal is real, the chosen transform should be real for real input signals; ii) For speech signals the discrete-time cosine transform (DCT) is a good approximation for the KLT [26]; iii) Transforms with fast algorithms should be given special attention.

A number of real transforms such as DCT, discrete-time Hartley transform, and others, are available [26]. Most of them have fast algorithms or can be implemented in recursive frequency-domain format. In particular, the outputs of the DCT are given by

$$s_0(k) = \frac{1}{\sqrt{N+1}}\sum_{l=0}^{N} x(k-l) \tag{4.71}$$

and

$$s_i(k) = \sqrt{\frac{2}{N+1}}\sum_{l=0}^{N} x(k-l)\cos\left[\pi i\frac{(2l+1)}{2(N+1)}\right] \tag{4.72}$$

From Fig. 4.3, we observe that the delay line and the unitary transform form a single-input and multiple-output preprocessing filter. In case the unitary transform is the DCT, the transfer function from the input to the outputs of the DCT preprocessing filter can be described in a recursive format as follows:

$$T_i(z) = \frac{k_0}{N+1}\cos\tau_i\,\frac{[z^{N+1}-(-1)^i](z-1)}{z^N[z^2-(2\cos 2\tau_i)z+1]} \tag{4.73}$$

Algorithm 4.4

The Transform-Domain LMS Algorithm

Initialization
$\mathbf{x}(0) = \hat{\mathbf{w}}(0) = [0\,0\ldots 0]^T$
$\gamma =$ small constant
$0 < \alpha \leq 0.1$
Do for each $x(k)$ and $d(k)$ given for $k \geq 0$
$\mathbf{s}(k) = \mathbf{T}\mathbf{x}(k)$
$e(k) = d(k) - \mathbf{s}^T(k)\hat{\mathbf{w}}(k)$
$\hat{\mathbf{w}}(k+1) = \hat{\mathbf{w}}(k) + 2\,\mu\, e(k)\,\mathbf{\Sigma}^{-2}(k)\mathbf{s}(k)$

where

$$k_0 = \begin{cases} \sqrt{2} & if \quad i = 0 \\ 2 & if \quad i = 1, ..., N \end{cases}$$

and $\tau_i = \frac{\pi i}{2(N+1)}$. The derivation details are not given here, since they are beyond the scope of this text.

For complex input signals, the discrete-time Fourier transform (DFT) is a natural choice due to its efficient implementations.

Although no general procedure is available to choose the best transform when the input signal is not known *a priori*, the decorrelation performed by the transform, followed by the power normalization, is sufficient to reduce the eigenvalue spread for a broad (not all) class of input signals. Therefore, the transform-domain LMS algorithms are expected to converge faster than the standard LMS algorithm in most applications [13].

The complete transform-domain LMS algorithm is outlined on Algorithm 4.4.

Example 4.2

Repeat the equalization problem of example 3.1 of the previous chapter using the transform-domain LMS algorithm.

(a) Compute the Wiener solution.

(b) Choose an appropriate value for μ and plot the convergence path for the transform-domain LMS algorithm on the MSE surface.

Solution:

(a) In this example, the correlation matrix of the adaptive-filter input signal is given by

$$\mathbf{R} = \begin{bmatrix} 1.6873 & -0.7937 \\ -0.7937 & 1.6873 \end{bmatrix}$$

and the cross-correlation vector $\mathbf{p}$ is

$$\mathbf{p} = \begin{bmatrix} 0.9524 \\ 0.4762 \end{bmatrix}$$

For square matrix $\mathbf{R}$ of dimension 2, the transformation matrix corresponding to the cosine transform is given by

$$\mathbf{T} = \begin{bmatrix} \frac{\sqrt{2}}{2} & \frac{\sqrt{2}}{2} \\ \frac{\sqrt{2}}{2} & -\frac{\sqrt{2}}{2} \end{bmatrix}$$

For this filter order, the above transformation matrix coincides with the KLT.

The coefficients corresponding to the Wiener solution of the transform-domain filter are given by

$$\begin{aligned} \hat{\mathbf{w}}_o &= (\mathbf{TRT}^T)^{-1}\mathbf{Tp} \\ &= \begin{bmatrix} \frac{1}{0.8936} & 0 \\ 0 & \frac{1}{2.4810} \end{bmatrix} \begin{bmatrix} 1.0102 \\ 0.3367 \end{bmatrix} \\ &= \begin{bmatrix} 1.1305 \\ 0.1357 \end{bmatrix} \end{aligned}$$

(b) The transform-domain LMS algorithm is applied to minimize the MSE using a small convergence factor $\mu = 1/300$, in order to obtain a smoothly converging curve. The convergence path of the algorithm in the MSE surface is depicted in Fig. 4.9. As can be noted, the transformation aligned the coefficient axes with the main axes of the ellipses belonging to the error surface. The reader should notice that the algorithm follows an almost straight path to the minimum and that the effect of the eigenvalue spread is compensated by the power normalization. The convergence in this case is faster than for the LMS case.

□

From the transform-domain LMS algorithm point of view, we can consider that the LMS-Newton algorithm attempts to utilize an estimate of the KLT through $\hat{\mathbf{R}}^{-1}(k)$. On the other hand, the normalized LMS algorithm utilizes an identity transform with an instantaneous estimate of the input signal power given by $\mathbf{x}^T(k)\mathbf{x}(k)$.

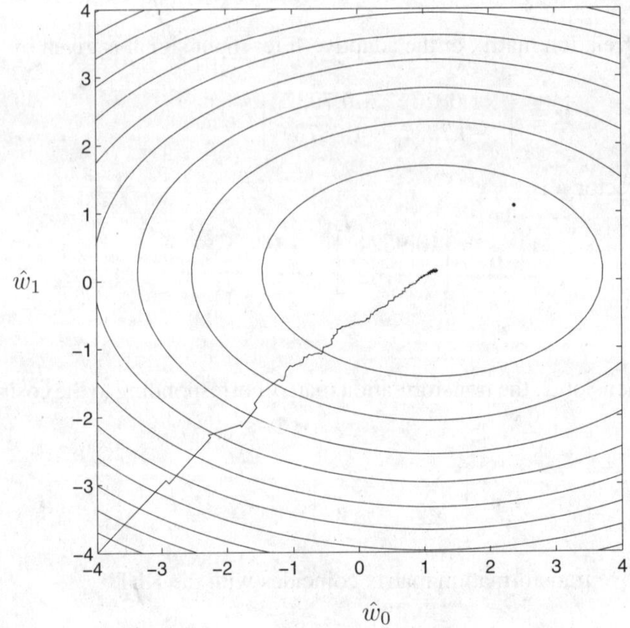

Figure 4.9 Convergence path of the transform-domain adaptive filter.

4.6 THE AFFINE PROJECTION ALGORITHM

There are situations where it is possible to recycle the old data signal in order to improve the convergence of the adaptive-filtering algorithms. Data-reusing algorithms [18], [19]-[24] are considered an alternative to increase the speed of convergence in adaptive-filtering algorithms in situations where the input signal is correlated. The penalty to be paid by data reusing is increased algorithm misadjustment, and as usual a trade-off between final misadjustment and convergence speed is achieved through the introduction of a convergence factor.

Let's assume we keep the last $L+1$ input signal vectors in a matrix as follows:

$$\begin{aligned}\mathbf{X}_{\text{ap}}(k) &= \begin{bmatrix} x(k) & x(k-1) & \cdots & x(k-L+1) & x(k-L) \\ x(k-1) & x(k-2) & \cdots & x(k-L) & x(k-L-1) \\ \vdots & \vdots & \ddots & \vdots & \vdots \\ x(k-N) & x(k-N-1) & \cdots & x(k-L-N+1) & x(k-L-N) \end{bmatrix} \\ &= [\mathbf{x}(k)\, \mathbf{x}(k-1) \ldots \mathbf{x}(k-L)] \end{aligned} \tag{4.74}$$

We can also define some vectors representing the partial reusing results at a given iteration k, such as the adaptive-filter output, the desired signal, and the error vectors.

These vectors are

$$\mathbf{y}_{\mathrm{ap}}(k) = \mathbf{X}_{\mathrm{ap}}^T(k)\mathbf{w}(k) = \begin{bmatrix} y_{\mathrm{ap},0}(k) \\ y_{\mathrm{ap},1}(k) \\ \vdots \\ y_{\mathrm{ap},L}(k) \end{bmatrix} \tag{4.75}$$

$$\mathbf{d}_{\mathrm{ap}}(k) = \begin{bmatrix} d(k) \\ d(k-1) \\ \vdots \\ d(k-L) \end{bmatrix} \tag{4.76}$$

$$\mathbf{e}_{\mathrm{ap}}(k) = \begin{bmatrix} e_{\mathrm{ap},0}(k) \\ e_{\mathrm{ap},1}(k) \\ \vdots \\ e_{\mathrm{ap},L}(k) \end{bmatrix} = \begin{bmatrix} d(k) - y_{\mathrm{ap},0}(k) \\ d(k-1) - y_{\mathrm{ap},1}(k) \\ \vdots \\ d(k-L) - y_{\mathrm{ap},L}(k) \end{bmatrix} = \mathbf{d}_{\mathrm{ap}}(k) - \mathbf{y}_{\mathrm{ap}}(k) \tag{4.77}$$

The objective of the affine projection algorithm is to minimize

$$\begin{aligned} &\frac{1}{2}\|\mathbf{w}(k+1) - \mathbf{w}(k)\|^2 \\ &\text{subject to :} \\ &\mathbf{d}_{\mathrm{ap}}(k) - \mathbf{X}_{\mathrm{ap}}^T(k)\mathbf{w}(k+1) = \mathbf{0} \end{aligned} \tag{4.78}$$

The affine projection algorithm maintains the next coefficient vector $\mathbf{w}(k+1)$ as close as possible to the current one[1] $\mathbf{w}(k)$, while forcing the *a posteriori*[2] error to be zero.

Using the method of Lagrange multipliers to turn the constrained minimization into an unconstrained one, the unconstrained function to be minimized is

$$F[\mathbf{w}(k+1)] = \frac{1}{2}\|\mathbf{w}(k+1) - \mathbf{w}(k)\|^2 + \boldsymbol{\lambda}_{\mathrm{ap}}^T(k)[\mathbf{d}(k) - \mathbf{X}_{\mathrm{ap}}^T(k)\mathbf{w}(k+1)] \tag{4.79}$$

where $\boldsymbol{\lambda}_{\mathrm{ap}}(k)$ is an $(L+1) \times 1$ vector of Lagrange multipliers. The above expression can be rewritten as

$$\begin{aligned} F[\mathbf{w}(k+1)] &= \frac{1}{2}\left[\mathbf{w}(k+1) - \mathbf{w}(k)\right]^T\left[\mathbf{w}(k+1) - \mathbf{w}(k)\right] \\ &\quad + \left[\mathbf{d}^T(k) - \mathbf{w}^T(k+1)\mathbf{X}_{\mathrm{ap}}(k)\right]\boldsymbol{\lambda}_{\mathrm{ap}}(k) \end{aligned} \tag{4.80}$$

The gradient of $F[\mathbf{w}(k+1)]$ with respect to $\mathbf{w}(k+1)$ is given by

$$\mathbf{g}_{\mathbf{w}}\{F[\mathbf{w}(k+1)]\} = \frac{1}{2}\left[2\mathbf{w}(k+1) - 2\mathbf{w}(k)\right] - \mathbf{X}_{\mathrm{ap}}(k)\boldsymbol{\lambda}_{\mathrm{ap}}(k) \tag{4.81}$$

---

[1]This procedure is known as minimal distance principle.

[2]The *a posteriori* error is the one computed with the current available data (up to instant k) using the already updated coefficient vector $\mathbf{w}(k+1)$.

Algorithm 4.5   **The Affine Projection Algorithm**
Initialization $\mathbf{x}(0) = \mathbf{w}(0) = [0\,0\ldots0]^T$ choose μ in the range $0 < \mu \leq 2$ $\gamma =$ small constant   Do for $k \geq 0$ $\mathbf{e}_{\text{ap}}(k) = \mathbf{d}_{\text{ap}}(k) - \mathbf{X}_{\text{ap}}^T(k)\mathbf{w}(k)$ $\mathbf{w}(k+1) = \mathbf{w}(k) + \mu\mathbf{X}_{\text{ap}}(k)\left(\mathbf{X}_{\text{ap}}^T(k)\mathbf{X}_{\text{ap}}(k) + \gamma\mathbf{I}\right)^{-1}\mathbf{e}_{\text{ap}}(k)$

After setting the gradient of $F[\mathbf{w}(k+1)]$ with respect to $\mathbf{w}(k+1)$ equal to zero, we get

$$\mathbf{w}(k+1) = \mathbf{w}(k) + \mathbf{X}_{\text{ap}}(k)\boldsymbol{\lambda}_{\text{ap}}(k) \tag{4.82}$$

If we substitute equation (4.82) in the constraint relation of equation (4.78), we obtain

$$\mathbf{X}_{\text{ap}}^T(k)\mathbf{X}_{\text{ap}}(k)\boldsymbol{\lambda}_{\text{ap}}(k) = \mathbf{d}_{\text{ap}}(k) - \mathbf{X}_{\text{ap}}^T(k)\mathbf{w}(k) = \mathbf{e}_{\text{ap}}(k) \tag{4.83}$$

The update equation is now given by equation (4.82) with $\boldsymbol{\lambda}_{\text{ap}}(k)$ being the solution of equation (4.83), i.e.,

$$\mathbf{w}(k+1) = \mathbf{w}(k) + \mathbf{X}_{\text{ap}}(k)\left(\mathbf{X}_{\text{ap}}^T(k)\mathbf{X}_{\text{ap}}(k)\right)^{-1}\mathbf{e}_{\text{ap}}(k) \tag{4.84}$$

The above algorithm corresponds to the conventional affine projection algorithm [19] with unity convergence factor. A trade-off between final misadjustment and convergence speed is achieved through the introduction of a convergence factor as follows

$$\mathbf{w}(k+1) = \mathbf{w}(k) + \mu\mathbf{X}_{\text{ap}}(k)\left(\mathbf{X}_{\text{ap}}^T(k)\mathbf{X}_{\text{ap}}(k)\right)^{-1}\mathbf{e}_{\text{ap}}(k) \tag{4.85}$$

Note that with the convergence factor the *a posteriori* error is no longer zero. In fact, when measurement noise is present in the environment, zeroing the *a posteriori* error is not a good idea since we are forcing the adaptive filter to compensate for the effect of a noise signal which is uncorrelated with the adaptive-filter input signal. The result is a high misadjustment when the convergence factor is one. The description of the affine projection algorithm is given in Algorithm 4.5, where an identity matrix multiplied by a small constant was added to the matrix $\mathbf{X}_{\text{ap}}^T(k)\mathbf{X}_{\text{ap}}(k)$ in order to avoid numerical problems in the matrix inversion. The order of the matrix to be inverted depends on the number of data vectors being reused.

Let's define the hyperplane $\mathcal{S}(k)$ as follows

$$\mathcal{S}(k) = \{\mathbf{w}(k+1) \in \mathcal{R}^{N+1} : d(k) - \mathbf{w}^T(k+1)\mathbf{x}(k) = 0\} \tag{4.86}$$

It is noticed that the *a posteriori* error over this hyperplane is zero, that is, given the current input data stored in the vector $\mathbf{x}(k)$ the coefficients are updated to a point where the error computed with the coefficients updated is zero. This definition allows an insightful geometric interpretation for the affine projection algorithm.

In the affine projection algorithm the coefficients are computed such that they belong to an $L+1$-dimensional subspace $\in \mathbb{R}^{N+1}$, where $\mathbb{R}$ represents the set of real numbers, spanned by the $L+1$ columns of $\mathbf{X}_{\text{ap}}(k)$. The objective of having $L+1$ *a posteriori* errors equal to zero has infinity number of solutions, such that any solution on $\mathcal{S}(k)$ can be added to a coefficient vector lying on $\mathcal{S}^{\perp}(k)$. By also minimizing $\frac{1}{2}\|\mathbf{w}(k+1) - \mathbf{w}(k)\|^2$ specifies a solution with minimum disturbance. The matrix $\mathbf{X}_{\text{ap}}(k)\left(\mathbf{X}_{\text{ap}}^T(k)\mathbf{X}_{\text{ap}}(k)\right)^{-1}\mathbf{X}_{\text{ap}}^T(k)$ represents an orthogonal projection operator on the $L+1$-dimensional subspace of $\mathbb{R}^{N+1}$ spanned by the $L+1$ columns of $\mathbf{X}_{\text{ap}}(k)$. This projection matrix has $L+1$ eigenvalues equal to 1 and $N-L$ eigenvalues of value 0. On the other hand, the matrix $\mathbf{I} - \mu\mathbf{X}_{\text{ap}}(k)\left(\mathbf{X}_{\text{ap}}^T(k)\mathbf{X}_{\text{ap}}(k)\right)^{-1}\mathbf{X}_{\text{ap}}^T(k)$ has $L+1$ eigenvalues equal to 1 and $N-L$ eigenvalues of value $1-\mu$.

When $L=0$ and $L=1$ the affine projection algorithm has the normalized LMS and binormalized LMS algorithms [21] as special cases, respectively. In the binormalized case the matrix inversion has closed form solution. Fig. 4.10 illustrates the updating of the coefficient vector for a two-dimensional problem for the LMS algorithm, for the normalized LMS algorithm, for the normalized LMS algorithm with a single data reuse[3], and the binormalized LMS algorithm. Here we assume that the coefficients are originally at $\tilde{\mathbf{w}}$ when the new data vector $\mathbf{x}(k)$ becomes available and $\mathbf{x}(k-1)$ is still stored, and this scenario is used to illustrate the coefficient updating of related algorithms. In addition, it is assumed an environment with no additional noise and a system identification with sufficient order, where the LMS algorithm utilizes a small convergence factor whereas the remaining algorithms use unit convergence factor. The conventional LMS algorithm takes a step towards $\mathcal{S}(k)$ yielding a solution $\mathbf{w}(k+1)$, anywhere between points 1 and 3 in Fig. 4.10, that is closer to $\mathcal{S}(k)$ than $\tilde{\mathbf{w}}$. The NLMS algorithm with unit convergence factor performs a line search in the direction of $\mathbf{x}(k)$ to yield in a single step the solution $\mathbf{w}(k+1)$, represented by point 3 in Fig. 4.10, which belongs to $\mathcal{S}(k)$. A single reuse of the previous data using normalized LMS algorithm would lead to point 4. The binormalized LMS algorithm, which corresponds to an affine projection algorithm with two projections, yields the solution that belongs to $\mathcal{S}(k-1)$ and $\mathcal{S}(k)$, represented by point 5 in Fig. 4.10. As an illustration, it is possible to observe in Fig. 4.11 that by repeatedly re-utilizing the data vectors $\mathbf{x}(k)$ and $\mathbf{x}(k-1)$ to update the coefficients with the normalized LMS algorithm would reach point 5 in a zig-zag pattern after an infinite number of iterations. This approach is known as Kaczmarz method [22].

For a noise-free environment and sufficient-order identification problem, the optimal solution $\mathbf{w}_o$ is at the intersection of $L+1$ hyperplanes constructed with linearly independent input signal vectors. The affine projection algorithm with unit convergence factor updates the coefficient to the intersection. Fig. 4.12 illustrates the coefficient updating for a three-dimensional problem for the normalized

[3]In this algorithm the updating is performed in two steps: $\hat{\mathbf{w}}(k) = \mathbf{w}(k) + \frac{e(k)\mathbf{x}(k)}{\mathbf{x}^T(k)\mathbf{x}(k)}$ and $\mathbf{w}(k+1) = \hat{\mathbf{w}}(k) + \frac{\hat{e}(k-1)\mathbf{x}(k-1)}{\mathbf{x}^T(k-1)\mathbf{x}(k-1)}$, where in the latter case $\hat{e}(k-1)$ is computed with the previous data $d(k-1)$ and $\mathbf{x}(k-1)$ using the coefficients $\hat{\mathbf{w}}(k)$.

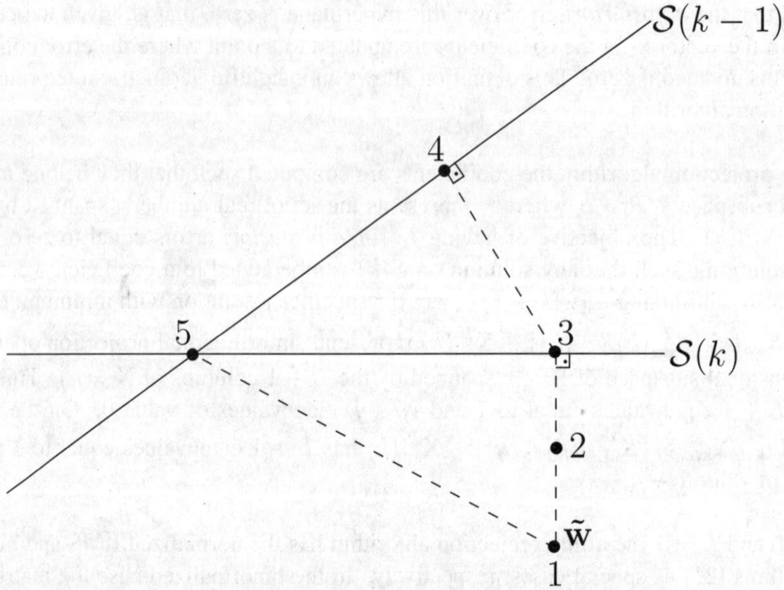

Figure 4.10 Coefficient vector updating for the normalized LMS algorithm and binormalized LMS algorithm.

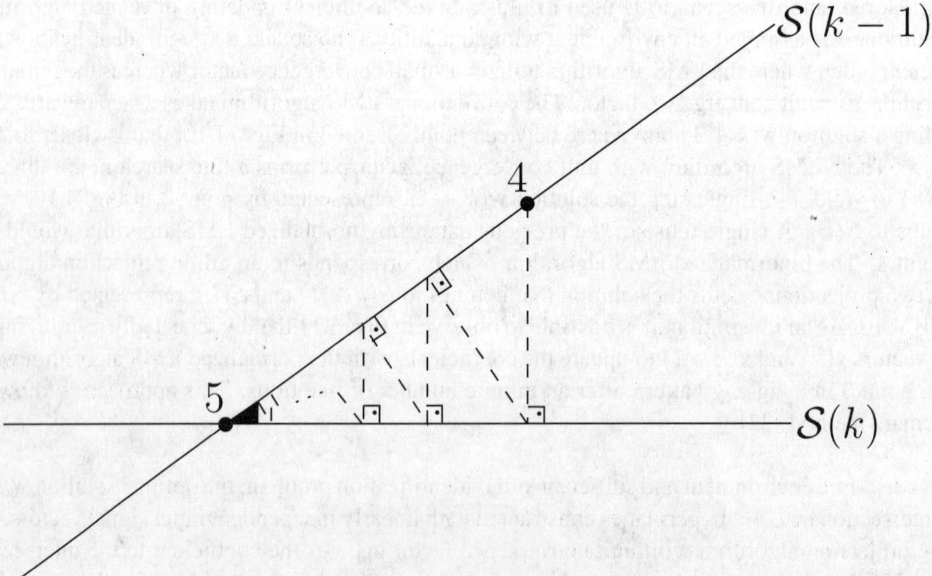

Figure 4.11 Multiple data reuse for the normalized LMS algorithm.

and binormalized LMS algorithms. It can be observed in Fig. 4.12 that $\mathbf{x}(k)$ and, consequently, $\mathbf{g_w}[e^2(k)]$ are orthogonal to the hyperplane $\mathcal{S}(k)$. Similarly, $\mathbf{x}(k-1)$ is orthogonal to the hyperplane $\mathcal{S}(k-1)$. The normalized LMS algorithm moves the coefficients from point 1 to point 2, whereas the binormalized LMS algorithm updates the coefficients to point 3 at the intersection of the two hyperplanes.

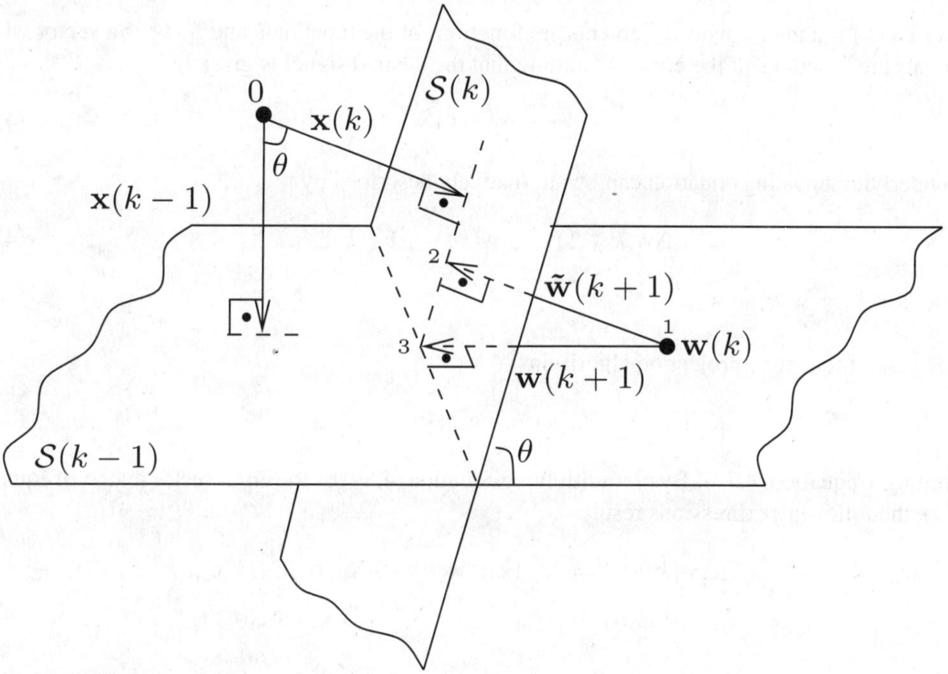

Figure 4.12 Three-dimensional coefficient vector updating for the normalized LMS algorithm and binormalized LMS algorithm.

The affine projection algorithm combines data reusing, orthogonal projections of L consecutive gradient directions, and normalization in order to achieve faster convergence than many other LMS-based algorithms. At each iteration, the affine projection algorithm yields the solution $\mathbf{w}(k+1)$ which is at the intersection of hyperplanes $\mathcal{S}(k), \mathcal{S}(k-1), \ldots, \mathcal{S}(k-L)$ and is as close as possible to $\mathbf{w}(k)$. The computational complexity of the affine projection algorithm is related to the number of data vectors being reused which ultimately determines the order of the matrix to be inverted. Some fast versions of the algorithm can be found in [20], [25]. It is also possible to reduce computations by employing data-selective strategies as will be discussed in Chapter 6.

4.6.1 Misadjustment in the Affine Projection Algorithm

The analysis of the affine projection algorithm is somewhat more involved than some of the LMS-based algorithms. The following framework provides an alternative analysis approach utilizing the concept of energy conservation [44]-[48]. This framework has been widely used in recent literature to analyze several adaptive-filtering algorithms [48]. In particular, the approach is very useful to analyze the behavior of the affine projection algorithm in a rather simple manner [47].

A general adaptive-filtering algorithm utilizes the following coefficient updating form

$$\mathbf{w}(k+1) = \mathbf{w}(k) - \mu \mathbf{F_x}(k)\mathbf{f_e}(k) \tag{4.87}$$

where $\mathbf{F_x}(k)$ is a matrix whose elements are functions of the input data and $\mathbf{f_e}(k)$ is a vector whose elements are functions of the error. Assuming that the desired signal is given by

$$d(k) = \mathbf{w}_o^T \mathbf{x}(k) + n(k) \tag{4.88}$$

the underlying updating equation can be alternatively described by

$$\Delta\mathbf{w}(k+1) = \Delta\mathbf{w}(k) - \mu \mathbf{F_x}(k)\mathbf{f_e}(k) \tag{4.89}$$

where $\Delta\mathbf{w}(k) = \mathbf{w}(k) - \mathbf{w}_o$.

In the case of the affine projection algorithm

$$\mathbf{f_e}(k) = -\mathbf{e}_{\text{ap}}(k) \tag{4.90}$$

according to equation (4.77). By premultiplying equation (4.89) by the input vector matrix of equation (4.74), the following expressions result

$$\begin{aligned}\mathbf{X}_{\text{ap}}^T(k)\Delta\mathbf{w}(k+1) &= \mathbf{X}_{\text{ap}}^T(k)\Delta\mathbf{w}(k) + \mu\mathbf{X}_{\text{ap}}^T(k)\mathbf{F_x}(k)\mathbf{e}_{\text{ap}}(k)\\ -\tilde{\varepsilon}_{\text{ap}}(k) &= -\tilde{\mathbf{e}}_{\text{ap}}(k) + \mu\mathbf{X}_{\text{ap}}^T(k)\mathbf{F_x}(k)\mathbf{e}_{\text{ap}}(k)\end{aligned} \tag{4.91}$$

where

$$\tilde{\varepsilon}_{\text{ap}}(k) = -\mathbf{X}_{\text{ap}}^T(k)\Delta\mathbf{w}(k+1) \tag{4.92}$$

is the noiseless *a posteriori* error vector and

$$\tilde{\mathbf{e}}_{\text{ap}}(k) = -\mathbf{X}_{\text{ap}}^T(k)\Delta\mathbf{w}(k) = \mathbf{e}_{\text{ap}}(k) - \mathbf{n}_{\text{ap}}(k) \tag{4.93}$$

is the noiseless *a priori* error vector with

$$\mathbf{n}_{\text{ap}}(k) = \begin{bmatrix} n(k) \\ n(k-1) \\ \vdots \\ n(k-L) \end{bmatrix}$$

being the standard noise vector.

For the regularized affine projection algorithm

$$\mathbf{F_x}(k) = \mathbf{X}_{\text{ap}}(k)\left(\mathbf{X}_{\text{ap}}^T(k)\mathbf{X}_{\text{ap}}(k) + \gamma\mathbf{I}\right)^{-1}$$

where the matrix $\gamma\mathbf{I}$ is added to the matrix to be inverted in order to avoid numerical problems in the inversion operation in the cases $\mathbf{X}_{\text{ap}}^T(k)\mathbf{X}_{\text{ap}}(k)$ is ill conditioned.

By solving equation (4.91), we get

$$\frac{1}{\mu}\left(\mathbf{X}_{\text{ap}}^T(k)\mathbf{X}_{\text{ap}}(k)\right)^{-1}\left(\tilde{\mathbf{e}}_{\text{ap}}(k) - \tilde{\varepsilon}_{\text{ap}}(k)\right) = \left(\mathbf{X}_{\text{ap}}^T(k)\mathbf{X}_{\text{ap}}(k) + \gamma\mathbf{I}\right)^{-1}\mathbf{e}_{\text{ap}}(k)$$

If we replace the above equation in

$$\Delta\mathbf{w}(k+1) = \Delta\mathbf{w}(k) + \mu\mathbf{X}_{\text{ap}}(k)\left(\mathbf{X}_{\text{ap}}^T(k)\mathbf{X}_{\text{ap}}(k) + \gamma\mathbf{I}\right)^{-1}\mathbf{e}_{\text{ap}}(k) \tag{4.94}$$

which corresponds to equation (4.89) for the affine projection case, it is possible to deduce that

$$\begin{aligned}\Delta\mathbf{w}(k+1) - \mathbf{X}_{\text{ap}}(k)\left(\mathbf{X}_{\text{ap}}^T(k)\mathbf{X}_{\text{ap}}(k)\right)^{-1}\tilde{\mathbf{e}}_{\text{ap}}(k) &= \\ \Delta\mathbf{w}(k) - \mathbf{X}_{\text{ap}}(k)\left(\mathbf{X}_{\text{ap}}^T(k)\mathbf{X}_{\text{ap}}(k)\right)^{-1}\tilde{\varepsilon}_{\text{ap}}(k)&\end{aligned} \tag{4.95}$$

From the above equation it is possible to prove that

$$\begin{aligned}E\left[\|\Delta\mathbf{w}(k+1)\|^2\right] + E\left[\tilde{\mathbf{e}}_{\text{ap}}^T(k)\left(\mathbf{X}_{\text{ap}}^T(k)\mathbf{X}_{\text{ap}}(k)\right)^{-1}\tilde{\mathbf{e}}_{\text{ap}}(k)\right] &= \\ E\left[\|\Delta\mathbf{w}(k)\|^2\right] + E\left[\tilde{\varepsilon}_{\text{ap}}^T(k)\left(\mathbf{X}_{\text{ap}}^T(k)\mathbf{X}_{\text{ap}}(k)\right)^{-1}\tilde{\varepsilon}_{\text{ap}}(k)\right]&\end{aligned} \tag{4.96}$$

Proof:

One can now calculate the Euclidean norm of both sides of equation (4.95)

$$\begin{aligned}&\left[\Delta\mathbf{w}(k+1) - \mathbf{X}_{\text{ap}}(k)\left(\mathbf{X}_{\text{ap}}^T(k)\mathbf{X}_{\text{ap}}(k)\right)^{-1}\tilde{\mathbf{e}}_{\text{ap}}(k)\right]^T \\ \times\ &\left[\Delta\mathbf{w}(k+1) - \mathbf{X}_{\text{ap}}(k)\left(\mathbf{X}_{\text{ap}}^T(k)\mathbf{X}_{\text{ap}}(k)\right)^{-1}\tilde{\mathbf{e}}_{\text{ap}}(k)\right] = \\ &\left[\Delta\mathbf{w}(k) - \mathbf{X}_{\text{ap}}(k)\left(\mathbf{X}_{\text{ap}}^T(k)\mathbf{X}_{\text{ap}}(k)\right)^{-1}\tilde{\varepsilon}_{\text{ap}}(k)\right]^T \\ \times\ &\left[\Delta\mathbf{w}(k) - \mathbf{X}_{\text{ap}}(k)\left(\mathbf{X}_{\text{ap}}^T(k)\mathbf{X}_{\text{ap}}(k)\right)^{-1}\tilde{\varepsilon}_{\text{ap}}(k)\right]\end{aligned}$$

By performing the inner products one by one, the above equation becomes

$$\begin{aligned}&\Delta\mathbf{w}^T(k+1)\Delta\mathbf{w}(k+1) - \Delta\mathbf{w}^T(k+1)\mathbf{X}_{\text{ap}}(k)\left(\mathbf{X}_{\text{ap}}^T(k)\mathbf{X}_{\text{ap}}(k)\right)^{-1}\tilde{\mathbf{e}}_{\text{ap}}(k) \\ -\ &\left[\mathbf{X}_{\text{ap}}(k)\left(\mathbf{X}_{\text{ap}}^T(k)\mathbf{X}_{\text{ap}}(k)\right)^{-1}\tilde{\mathbf{e}}_{\text{ap}}(k)\right]^T\Delta\mathbf{w}(k+1) \\ +\ &\left[\mathbf{X}_{\text{ap}}(k)\left(\mathbf{X}_{\text{ap}}^T(k)\mathbf{X}_{\text{ap}}(k)\right)^{-1}\tilde{\mathbf{e}}_{\text{ap}}(k)\right]^T\left[\mathbf{X}_{\text{ap}}(k)\left(\mathbf{X}_{\text{ap}}^T(k)\mathbf{X}_{\text{ap}}(k)\right)^{-1}\tilde{\mathbf{e}}_{\text{ap}}(k)\right] = \\ &\Delta\mathbf{w}^T(k)\Delta\mathbf{w}(k) - \Delta\mathbf{w}^T(k)\mathbf{X}_{\text{ap}}(k)\left(\mathbf{X}_{\text{ap}}^T(k)\mathbf{X}_{\text{ap}}(k)\right)^{-1}\tilde{\varepsilon}_{\text{ap}}(k) \\ -\ &\left[\mathbf{X}_{\text{ap}}(k)\left(\mathbf{X}_{\text{ap}}^T(k)\mathbf{X}_{\text{ap}}(k)\right)^{-1}\tilde{\varepsilon}_{\text{ap}}(k)\right]^T\Delta\mathbf{w}(k) \\ +\ &\left[\mathbf{X}_{\text{ap}}(k)\left(\mathbf{X}_{\text{ap}}^T(k)\mathbf{X}_{\text{ap}}(k)\right)^{-1}\tilde{\varepsilon}_{\text{ap}}(k)\right]^T\left[\mathbf{X}_{\text{ap}}(k)\left(\mathbf{X}_{\text{ap}}^T(k)\mathbf{X}_{\text{ap}}(k)\right)^{-1}\tilde{\varepsilon}_{\text{ap}}(k)\right]\end{aligned}$$

Since $\tilde{\varepsilon}_{\text{ap}}(k) = -\mathbf{X}_{\text{ap}}^T(k)\Delta\mathbf{w}(k+1)$ and $\tilde{\mathbf{e}}_{\text{ap}}(k) = -\mathbf{X}_{\text{ap}}^T(k)\Delta\mathbf{w}(k)$

$$
\begin{aligned}
& \|\Delta\mathbf{w}(k+1)\|^2 + \tilde{\varepsilon}_{\text{ap}}^T(k)\left(\mathbf{X}_{\text{ap}}^T(k)\mathbf{X}_{\text{ap}}(k)\right)^{-1}\tilde{\mathbf{e}}_{\text{ap}}(k) \\
+ \quad & \tilde{\mathbf{e}}_{\text{ap}}^T(k)\left(\mathbf{X}_{\text{ap}}^T(k)\mathbf{X}_{\text{ap}}(k)\right)^{-1}\tilde{\varepsilon}_{\text{ap}}(k) + \tilde{\mathbf{e}}_{\text{ap}}^T(k)\left(\mathbf{X}_{\text{ap}}^T(k)\mathbf{X}_{\text{ap}}(k)\right)^{-1}\tilde{\mathbf{e}}_{\text{ap}}(k) = \\
& \|\Delta\mathbf{w}(k)\|^2 + \tilde{\mathbf{e}}_{\text{ap}}^T(k)\left(\mathbf{X}_{\text{ap}}^T(k)\mathbf{X}_{\text{ap}}(k)\right)^{-1}\tilde{\varepsilon}_{\text{ap}}(k) \\
+ \quad & \tilde{\varepsilon}_{\text{ap}}^T(k)\left(\mathbf{X}_{\text{ap}}^T(k)\mathbf{X}_{\text{ap}}(k)\right)^{-1}\tilde{\mathbf{e}}_{\text{ap}}(k) + \tilde{\varepsilon}_{\text{ap}}^T(k)\left(\mathbf{X}_{\text{ap}}^T(k)\mathbf{X}_{\text{ap}}(k)\right)^{-1}\tilde{\varepsilon}_{\text{ap}}(k)
\end{aligned}
$$

By removing the equal terms on both sides of the last equation the following equality holds

$$
\begin{aligned}
& \|\Delta\mathbf{w}(k+1)\|^2 + \tilde{\mathbf{e}}_{\text{ap}}^T(k)\left(\mathbf{X}_{\text{ap}}^T(k)\mathbf{X}_{\text{ap}}(k)\right)^{-1}\tilde{\mathbf{e}}_{\text{ap}}(k) = \\
& \|\Delta\mathbf{w}(k)\|^2 + \tilde{\varepsilon}_{\text{ap}}^T(k)\left(\mathbf{X}_{\text{ap}}^T(k)\mathbf{X}_{\text{ap}}(k)\right)^{-1}\tilde{\varepsilon}_{\text{ap}}(k)
\end{aligned} \tag{4.97}
$$

As can be observed no approximations were utilized so far. Now by applying the expected value operation on both sides of the above equation, the expression of equation (4.96) holds.

□

If it is assumed that the algorithm has converged, that is, the coefficients remain in average unchanged, then $E\left[\|\Delta\mathbf{w}(k+1)\|^2\right] = E\left[\|\Delta\mathbf{w}(k)\|^2\right]$. As a result the following equality holds in the steady state.

$$
E\left[\tilde{\mathbf{e}}_{\text{ap}}^T(k)\left(\mathbf{X}_{\text{ap}}^T(k)\mathbf{X}_{\text{ap}}(k)\right)^{-1}\tilde{\mathbf{e}}_{\text{ap}}(k)\right] = E\left[\tilde{\varepsilon}_{\text{ap}}^T(k)\left(\mathbf{X}_{\text{ap}}^T(k)\mathbf{X}_{\text{ap}}(k)\right)^{-1}\tilde{\varepsilon}_{\text{ap}}(k)\right] \tag{4.98}
$$

In the above expression it is useful to remove the dependence on the *a posteriori* error, what can be achieved by applying equation (4.91) to the affine projection algorithm case.

$$
\tilde{\varepsilon}_{\text{ap}}(k) = \tilde{\mathbf{e}}_{\text{ap}}(k) - \mu\mathbf{X}_{\text{ap}}^T(k)\mathbf{X}_{\text{ap}}(k)\left(\mathbf{X}_{\text{ap}}^T(k)\mathbf{X}_{\text{ap}}(k) + \gamma\mathbf{I}\right)^{-1}\mathbf{e}_{\text{ap}}(k) \tag{4.99}
$$

By substituting equation (4.99) in equation (4.98) we get

$$
\begin{aligned}
& E\left[\tilde{\mathbf{e}}_{\text{ap}}^T(k)\left(\mathbf{X}_{\text{ap}}^T(k)\mathbf{X}_{\text{ap}}(k)\right)^{-1}\tilde{\mathbf{e}}_{\text{ap}}(k)\right] = E\left[\tilde{\mathbf{e}}_{\text{ap}}^T(k)\left(\mathbf{X}_{\text{ap}}^T(k)\mathbf{X}_{\text{ap}}(k)\right)^{-1}\tilde{\mathbf{e}}_{\text{ap}}(k)\right. \\
- \quad & \mu\tilde{\mathbf{e}}_{\text{ap}}^T(k)\left(\mathbf{X}_{\text{ap}}^T(k)\mathbf{X}_{\text{ap}}(k) + \gamma\mathbf{I}\right)^{-1}\mathbf{e}_{\text{ap}}(k) - \mu\mathbf{e}_{\text{ap}}^T(k)\left(\mathbf{X}_{\text{ap}}^T(k)\mathbf{X}_{\text{ap}}(k) + \gamma\mathbf{I}\right)^{-1}\tilde{\mathbf{e}}_{\text{ap}}(k) \\
+ \quad & \left.\mu^2\mathbf{e}_{\text{ap}}^T(k)\left(\mathbf{X}_{\text{ap}}^T(k)\mathbf{X}_{\text{ap}}(k) + \gamma\mathbf{I}\right)^{-1}\mathbf{X}_{\text{ap}}^T(k)\mathbf{X}_{\text{ap}}(k)\left(\mathbf{X}_{\text{ap}}^T(k)\mathbf{X}_{\text{ap}}(k) + \gamma\mathbf{I}\right)^{-1}\mathbf{e}_{\text{ap}}(k)\right]
\end{aligned} \tag{4.100}
$$

The above expression can be simplified as

$$
\mu^2 E\Big[\mathbf{e}_{\text{ap}}^T(k)\hat{\mathbf{S}}_{\text{ap}}(k)\hat{\mathbf{R}}_{\text{ap}}(k)\hat{\mathbf{S}}_{\text{ap}}(k)\mathbf{e}_{\text{ap}}(k)\Big] = \mu E\Big[\tilde{\mathbf{e}}_{\text{ap}}^T(k)\hat{\mathbf{S}}_{\text{ap}}(k)\mathbf{e}_{\text{ap}}(k) + \mathbf{e}_{\text{ap}}^T(k)\hat{\mathbf{S}}_{\text{ap}}(k)\tilde{\mathbf{e}}_{\text{ap}}(k)\Big] \tag{4.101}
$$

where the following definitions are employed to simplify the discussion

$$\hat{\mathbf{R}}_{\text{ap}}(k) = \mathbf{X}_{\text{ap}}^T(k)\mathbf{X}_{\text{ap}}(k)$$
$$\hat{\mathbf{S}}_{\text{ap}}(k) = \left(\mathbf{X}_{\text{ap}}^T(k)\mathbf{X}_{\text{ap}}(k) + \gamma\mathbf{I}\right)^{-1} \tag{4.102}$$

By rescuing the definition of the error squared of equation (3.39) and applying the expected value operator we obtain

$$\xi(k) = E[e^2(k)] = E[n^2(k)] - 2E[n(k)\Delta\mathbf{w}^T(k)\mathbf{x}(k)] + E[\Delta\mathbf{w}^T(k)\mathbf{x}(k)\mathbf{x}^T(k)\Delta\mathbf{w}(k)] \tag{4.103}$$

If the coefficients have weak dependency of the additional noise and applying the orthogonality principle, we can simplify the above expression as follows

$$\xi(k) = \sigma_n^2 + E[\Delta\mathbf{w}^T(k)\mathbf{x}(k)\mathbf{x}^T(k)\Delta\mathbf{w}(k)]$$
$$= \sigma_n^2 + E[\tilde{e}_{\text{ap},0}^2(k)] \tag{4.104}$$

where $\tilde{e}_{\text{ap},0}(k)$ is the first element of vector $\tilde{\mathbf{e}}_{\text{ap}}(k)$.

In order to compute the excess of mean-square error we can remove the value of $E[\tilde{e}_{\text{ap},0}^2(k)]$ from equation (4.101). Since our aim is to compute $E[\tilde{e}_{\text{ap},0}^2(k)]$, we can substitute equation (4.93) in equation (4.101) in order to get rid of $\mathbf{e}_{\text{ap}}(k)$. The resulting expression is given by

$$E\left[\mu(\tilde{\mathbf{e}}_{\text{ap}}(k) + \mathbf{n}_{\text{ap}}(k))^T\hat{\mathbf{S}}_{\text{ap}}(k)\hat{\mathbf{R}}_{\text{ap}}(k)\hat{\mathbf{S}}_{\text{ap}}(k)(\tilde{\mathbf{e}}_{\text{ap}}(k) + \mathbf{n}_{\text{ap}}(k))\right] =$$
$$E\left[\tilde{\mathbf{e}}_{\text{ap}}^T(k)\hat{\mathbf{S}}_{\text{ap}}(k)(\tilde{\mathbf{e}}_{\text{ap}}(k) + \mathbf{n}_{\text{ap}}(k)) + (\tilde{\mathbf{e}}_{\text{ap}}(k) + \mathbf{n}_{\text{ap}}(k))^T\hat{\mathbf{S}}_{\text{ap}}(k)\tilde{\mathbf{e}}_{\text{ap}}(k)\right] \tag{4.105}$$

By considering the noise white and statistically independent of the input signal, the above relation can be further simplified as

$$\mu E\left[\tilde{\mathbf{e}}_{\text{ap}}^T(k)\hat{\mathbf{S}}_{\text{ap}}(k)\hat{\mathbf{R}}_{\text{ap}}(k)\hat{\mathbf{S}}_{\text{ap}}(k)\tilde{\mathbf{e}}_{\text{ap}}(k) + \mathbf{n}_{\text{ap}}^T(k)\hat{\mathbf{S}}_{\text{ap}}(k)\hat{\mathbf{R}}_{\text{ap}}(k)\hat{\mathbf{S}}_{\text{ap}}(k)\mathbf{n}_{\text{ap}}(k)\right] =$$
$$2E\left[\tilde{\mathbf{e}}_{\text{ap}}^T(k)\hat{\mathbf{S}}_{\text{ap}}(k)\tilde{\mathbf{e}}_{\text{ap}}(k)\right] \tag{4.106}$$

The above expression, after some rearrangements, can be rewritten as

$$2E\left\{\text{tr}[\tilde{\mathbf{e}}_{\text{ap}}(k)\tilde{\mathbf{e}}_{\text{ap}}^T(k)\hat{\mathbf{S}}_{\text{ap}}(k)]\right\} - \mu E\left\{\text{tr}[\tilde{\mathbf{e}}_{\text{ap}}(k)\tilde{\mathbf{e}}_{\text{ap}}^T(k)\hat{\mathbf{S}}_{\text{ap}}(k)\hat{\mathbf{R}}_{\text{ap}}(k)\hat{\mathbf{S}}_{\text{ap}}(k)]\right\} =$$
$$\mu E\left\{\text{tr}[\mathbf{n}_{\text{ap}}(k)\mathbf{n}_{\text{ap}}^T(k)\hat{\mathbf{S}}_{\text{ap}}(k)\hat{\mathbf{R}}_{\text{ap}}(k)\hat{\mathbf{S}}_{\text{ap}}(k)]\right\} \tag{4.107}$$

where we used the property $\text{tr}[\mathbf{A}\cdot\mathbf{B}] = \text{tr}[\mathbf{B}\cdot\mathbf{A}]$.

In addition, if matrix $\hat{\mathbf{R}}_{\text{ap}}(k)$ is invertible it can be noticed that

$$\hat{\mathbf{S}}_{\text{ap}}(k) = \left[\hat{\mathbf{R}}_{\text{ap}}(k) + \gamma\mathbf{I}\right]^{-1}$$
$$= \hat{\mathbf{R}}_{\text{ap}}^{-1}(k)\left[\mathbf{I} - \gamma\hat{\mathbf{R}}_{\text{ap}}^{-1}(k) + \gamma^2\hat{\mathbf{R}}_{\text{ap}}^{-2}(k) - \gamma^3\hat{\mathbf{R}}_{\text{ap}}^{-3}(k) + \cdots\right]$$
$$\approx \hat{\mathbf{R}}_{\text{ap}}^{-1}(k)\left[\mathbf{I} - \gamma\hat{\mathbf{R}}_{\text{ap}}^{-1}(k)\right] \approx \hat{\mathbf{R}}_{\text{ap}}^{-1}(k) \tag{4.108}$$

where the last two relations are valid for $\gamma \ll 1$.

By assuming that the matrix $\hat{\mathbf{S}}_{\text{ap}}(k)$ is statistically independent of the noiseless *a priori* error after convergence, and of the noise, the equation (4.107) can be rewritten as

$$\begin{aligned}&2\text{tr}\left\{E[\tilde{\mathbf{e}}_{\text{ap}}(k)\tilde{\mathbf{e}}_{\text{ap}}^T(k)]E[\hat{\mathbf{S}}_{\text{ap}}(k)]\right\} - \mu\text{tr}\left\{E[\tilde{\mathbf{e}}_{\text{ap}}(k)\tilde{\mathbf{e}}_{\text{ap}}^T(k)]E[\hat{\mathbf{S}}_{\text{ap}}(k)]\right\}\\&+\gamma\mu\text{tr}\left\{E[\tilde{\mathbf{e}}_{\text{ap}}(k)\tilde{\mathbf{e}}_{\text{ap}}^T(k)]\right\} = \mu\text{tr}\left\{E[\mathbf{n}_{\text{ap}}(k)\mathbf{n}_{\text{ap}}^T(k)]E[\hat{\mathbf{S}}_{\text{ap}}(k)]\right\} - \gamma\mu\text{tr}\left\{E[\mathbf{n}_{\text{ap}}(k)\mathbf{n}_{\text{ap}}^T(k)]\right\}\end{aligned} \tag{4.109}$$

This equation can be further simplified by assuming the noise is white[4] and γ is small leading to the following expression

$$(2-\mu)\text{tr}\{E[\tilde{\mathbf{e}}_{\text{ap}}(k)\tilde{\mathbf{e}}_{\text{ap}}^T(k)]E[\hat{\mathbf{S}}_{\text{ap}}(k)]\} = \mu\sigma_n^2\text{tr}\{E[\hat{\mathbf{S}}_{\text{ap}}(k)]\} \tag{4.110}$$

Our task now is to compute $E[\tilde{\mathbf{e}}_{\text{ap}}(k)\tilde{\mathbf{e}}_{\text{ap}}^T(k)]$ where we will assume in the process that this matrix is diagonal dominant whose final result has the following form

$$E[\tilde{\mathbf{e}}_{\text{ap}}(k)\tilde{\mathbf{e}}_{\text{ap}}^T(k)] = \mathbf{A}E[\tilde{e}_{\text{ap},0}^2(k)] + \mu^2\mathbf{B}\sigma_n^2$$

Proof:

The i-th rows of equations (4.92) and (4.93) are given by

$$\tilde{\varepsilon}_{\text{ap},i}(k) = -\mathbf{x}^T(k-i)\Delta\mathbf{w}(k+1) \tag{4.111}$$

and

$$\tilde{e}_{\text{ap},i}(k) = -\mathbf{x}^T(k-i)\Delta\mathbf{w}(k) = e_{\text{ap},i}(k) - n(k-i) \tag{4.112}$$

for $i = 0, \ldots, L$. Using in equation (4.91) the fact that $\mathbf{X}_{\text{ap}}^T(k)\mathbf{F}_{\mathbf{X}}(k) \approx \mathbf{I}$ for small γ, then

$$-\tilde{\boldsymbol{\varepsilon}}_{\text{ap}}(k) = -\tilde{\mathbf{e}}_{\text{ap}}(k) + \mu\mathbf{e}_{\text{ap}}(k) \tag{4.113}$$

By properly utilizing in equations (4.111) and (4.112) the i-th row of equation (4.91), we obtain

$$\begin{aligned}\tilde{\varepsilon}_{\text{ap},i}(k) &= -\mathbf{x}^T(k-i)\Delta\mathbf{w}(k+1)\\&= (1-\mu)\tilde{e}_{\text{ap},i}(k) - \mu n(k-i)\\&= (1-\mu)\mathbf{x}^T(k-i)\Delta\mathbf{w}(k) - \mu n(k-i)\end{aligned} \tag{4.114}$$

Squaring the above equation, assuming the coefficients are weakly dependent on the noise which is in turn white noise, and following closely the procedure to derive equation (4.96) from equation (4.95), we get

$$E[(\mathbf{x}^T(k-i)\Delta\mathbf{w}(k+1))^2] = (1-\mu)^2E[(\mathbf{x}^T(k-i)\Delta\mathbf{w}(k))^2] + \mu^2\sigma_n^2 \tag{4.115}$$

[4] In this case, $E[\mathbf{n}_{\text{ap}}(k)\mathbf{n}_{\text{ap}}^T(k)] = \sigma_n^2\mathbf{I}$.

The above expression relates the squared values of the *a posteriori* and *a priori* errors. However, the same kind of relation holds for the previous time instant, that is

$$E[(\mathbf{x}^T(k-i-1)\Delta\mathbf{w}(k))^2] = (1-\mu)^2 E[(\mathbf{x}^T(k-i-1)\Delta\mathbf{w}(k-1))^2] + \mu^2\sigma_n^2$$

or

$$E[\tilde{e}^2_{\mathrm{ap},i+1}(k)] = (1-\mu)^2 E[\tilde{e}^2_{\mathrm{ap},i}(k-1)] + \mu^2\sigma_n^2 \tag{4.116}$$

Note that for $i = 0$ this term corresponds to the second diagonal element of the matrix $E[\tilde{\mathbf{e}}_{\mathrm{ap}}(k)\tilde{\mathbf{e}}^T_{\mathrm{ap}}(k)]$. Specifically we can compute $E[\tilde{e}^2_{\mathrm{ap},1}(k)]$ as

$$\begin{aligned} E[(\mathbf{x}^T(k-1)\Delta\mathbf{w}(k))^2] &= E[\tilde{e}^2_{\mathrm{ap},1}(k)] \\ &= (1-\mu)^2 E[(\mathbf{x}^T(k-1)\Delta\mathbf{w}(k-1))^2] + \mu^2\sigma_n^2 \\ &= (1-\mu)^2 E[\tilde{e}^2_{\mathrm{ap},0}(k-1)] + \mu^2\sigma_n^2 \end{aligned} \tag{4.117}$$

For $i = 1$ equation (4.116) becomes

$$\begin{aligned} E[(\mathbf{x}^T(k-2)\Delta\mathbf{w}(k))^2] &= E[\tilde{e}^2_{\mathrm{ap},2}(k)] \\ &= (1-\mu)^2 E[(\mathbf{x}^T(k-2)\Delta\mathbf{w}(k-1))^2] + \mu^2\sigma_n^2 \\ &= (1-\mu)^2 E[\tilde{e}^2_{\mathrm{ap},1}(k-1)] + \mu^2\sigma_n^2 \end{aligned} \tag{4.118}$$

By substituting equation (4.117) in the above equation it follows that

$$E[\tilde{e}^2_{\mathrm{ap},2}(k)] = (1-\mu)^4 E[\tilde{e}^2_{\mathrm{ap},0}(k-2)] + [1+(1-\mu)^2]\mu^2\sigma_n^2 \tag{4.119}$$

By induction one can prove that

$$E[\tilde{e}^2_{\mathrm{ap},i+1}(k)] = (1-\mu)^{2(i+1)} E[\tilde{e}^2_{\mathrm{ap},0}(k-i-1)] + \left[1+\sum_{l=1}^{i}(1-\mu)^{2l}\right]\mu^2\sigma_n^2 \tag{4.120}$$

By assuming that $E[\tilde{e}^2_{\mathrm{ap},0}(k)] \approx E[\tilde{e}^2_{\mathrm{ap},0}(k-i)]$ for $i = 0, \ldots, L$, then

$$E[\tilde{\mathbf{e}}_{\mathrm{ap}}(k)\tilde{\mathbf{e}}^T_{\mathrm{ap}}(k)] = \mathbf{A}E[\tilde{e}^2_{\mathrm{ap},0}(k)] + \mu^2\mathbf{B}\sigma_n^2 \tag{4.121}$$

with

$$\mathbf{A} = \begin{bmatrix} 1 & & & & \\ & (1-\mu)^2 & & \mathbf{0} & \\ & & (1-\mu)^4 & & \\ & \mathbf{0} & & \ddots & \\ & & & & (1-\mu)^{2L} \end{bmatrix}$$

$$\mathbf{B} = \begin{bmatrix} 0 & & & & & & \\ & 1 & & & \mathbf{0} & & \\ & & 1+(1-\mu)^2 & & & & \\ & & & \ddots & & & \\ & \mathbf{0} & & & 1+\sum_{l=1}^{i}(1-\mu)^{2l} & & \\ & & & & & \ddots & \\ & & & & & & 1+\sum_{l=1}^{L-1}(1-\mu)^{2l} \end{bmatrix}$$

where it was also considered that the above matrix $E[\tilde{\mathbf{e}}_{ap}(k)\tilde{\mathbf{e}}_{ap}^T(k)]$ was diagonal dominant, as it is usually the case in practice. Note from the above relation that the convergence factor μ should be chosen in the range $0 < \mu < 2$, so that the elements of the noiseless *a priori* error remain bounded for any value of L.

□

We have available all the quantities required to calculate the excess of MSE in the affine projection algorithm. Specifically, we can substitute the result of equation (4.121) in equation (4.110) obtaining

$$(2-\mu)\left[E[\tilde{e}_{ap,0}^2(k)]\text{tr}\{\mathbf{A}E[\hat{\mathbf{S}}_{ap}(k)]\} + \mu^2\sigma_n^2\text{tr}\{\mathbf{B}E[\hat{\mathbf{S}}_{ap}(k)]\}\right] = \mu\sigma_n^2\text{tr}\{E[\hat{\mathbf{S}}_{ap}(k)]\} \quad (4.122)$$

The second term on the left-hand side can be neglected in case the signal to noise ratio is high. For small μ this term also becomes substantially smaller than the term on the right-hand side. For μ close to one the referred terms become comparable only for large L, when the misadjustment becomes less sensitive to L. In the following discussions we will not consider the term multiplied by μ^2.

Assuming the diagonal elements of $E[\hat{\mathbf{S}}_{ap}(k)]$ are equal and the matrix $\mathbf{A}$ multiplying it on the left-hand side is a diagonal matrix, after a few manipulations it is possible to deduce that

$$\begin{aligned} E[\tilde{e}_{ap,0}^2(k)] &= \frac{\mu}{2-\mu}\sigma_n^2\frac{\text{tr}\{E[\hat{\mathbf{S}}_{ap}(k)]\}}{\text{tr}\{\mathbf{A}E[\hat{\mathbf{S}}_{ap}(k)]\}} \\ &= \frac{(L+1)\mu}{2-\mu}\frac{1-(1-\mu)^2}{1-(1-\mu)^{2(L+1)}}\sigma_n^2 \end{aligned} \quad (4.123)$$

Therefore, the misadjustment for the affine projection algorithm is given by

$$M = \frac{(L+1)\mu}{2-\mu}\frac{1-(1-\mu)^2}{1-(1-\mu)^{2(L+1)}} \quad (4.124)$$

For large L and small $1-\mu$, this equation can be approximated by

$$M = \frac{(L+1)\mu}{(2-\mu)} \quad (4.125)$$

In [23], by considering a simplified model for the input signal vector consisting of vectors with discrete angular orientation and the independence assumption, an expression for the misadjustment of the affine projection algorithm was derived, that is

$$M = \frac{\mu}{2-\mu}E\left[\frac{1}{\|\mathbf{x}(k)\|^2}\right]\text{tr}[\mathbf{R}] \quad (4.126)$$

which is independent of L. It is observed in the experiments that higher number of reuses leads to higher misadjustment, as indicated in equation (4.125). The equivalent expression of (4.126) using the derivations presented here would lead to

$$M = \frac{(L+1)\mu}{2-\mu}E\left[\frac{1}{\|\mathbf{x}(k)\|^2}\right]\text{tr}[\mathbf{R}] \quad (4.127)$$

which can obtained from equation (4.123) by considering that

$$\text{tr}\{E[\hat{\mathbf{S}}_{\text{ap}}(k)]\} \approx (L+1)E\left[\frac{1}{\|\mathbf{x}(k)\|^2}\right]$$

and

$$\frac{1}{\text{tr}\{\mathbf{A}E[\hat{\mathbf{S}}_{\text{ap}}(k)]\}} \approx \text{tr}[\mathbf{R}]$$

for μ close to one.

4.6.2 Behavior in Nonstationary Environments

In a nonstationary environment the error in the coefficients is described by the following vector

$$\Delta\mathbf{w}(k+1) = \mathbf{w}(k+1) - \mathbf{w}_o(k+1) \tag{4.128}$$

where $\mathbf{w}_o(k+1)$ is the optimal time-varying vector. For this case, equation (4.95) becomes

$$\Delta\mathbf{w}(k+1) = \Delta\hat{\mathbf{w}}(k) + \mu\mathbf{X}_{\text{ap}}(k)\left(\mathbf{X}_{\text{ap}}^T(k)\mathbf{X}_{\text{ap}}(k) + \gamma\mathbf{I}\right)^{-1}\mathbf{e}_{\text{ap}}(k) \tag{4.129}$$

where $\Delta\hat{\mathbf{w}}(k) = \mathbf{w}(k) - \mathbf{w}_o(k+1)$. By premultiplying the above expression by $\mathbf{X}_{\text{ap}}^T(k)$ it follows that

$$\begin{aligned}\mathbf{X}_{\text{ap}}^T(k)\Delta\mathbf{w}(k+1) &= \mathbf{X}_{\text{ap}}^T(k)\Delta\hat{\mathbf{w}}(k) + \mu\mathbf{X}_{\text{ap}}^T(k)\mathbf{X}_{\text{ap}}(k)\left(\mathbf{X}_{\text{ap}}^T(k)\mathbf{X}_{\text{ap}}(k) + \gamma\mathbf{I}\right)^{-1}\mathbf{e}_{\text{ap}}(k)\\ -\tilde{\varepsilon}_{\text{ap}}(k) &= -\tilde{\mathbf{e}}_{\text{ap}}(k) + \mu\mathbf{X}_{\text{ap}}^T(k)\mu\mathbf{X}_{\text{ap}}(k)\left(\mathbf{X}_{\text{ap}}^T(k)\mathbf{X}_{\text{ap}}(k) + \gamma\mathbf{I}\right)^{-1}\mathbf{e}_{\text{ap}}(k)\end{aligned} \tag{4.130}$$

By solving the equation (4.130), it is possible to show that

$$\frac{1}{\mu}\left(\mathbf{X}_{\text{ap}}^T(k)\mathbf{X}_{\text{ap}}(k)\right)^{-1}\left[\tilde{\mathbf{e}}_{\text{ap}}(k) - \tilde{\varepsilon}_{\text{ap}}(k)\right] = \left(\mathbf{X}_{\text{ap}}^T(k)\mathbf{X}_{\text{ap}}(k) + \gamma\mathbf{I}\right)^{-1}\mathbf{e}_{\text{ap}}(k) \tag{4.131}$$

Following the same procedure to derive equation (4.95), we can now substitute equation (4.131) in equation (4.129) in order to deduce that

$$\begin{aligned}&\Delta\mathbf{w}(k+1) - \mathbf{X}_{\text{ap}}(k)\left(\mathbf{X}_{\text{ap}}^T(k)\mathbf{X}_{\text{ap}}(k)\right)^{-1}\tilde{\mathbf{e}}_{\text{ap}}(k)\\ &= \Delta\hat{\mathbf{w}}(k) - \mathbf{X}_{\text{ap}}(k)\left(\mathbf{X}_{\text{ap}}^T(k)\mathbf{X}_{\text{ap}}(k)\right)^{-1}\tilde{\varepsilon}_{\text{ap}}(k)\end{aligned} \tag{4.132}$$

By computing the energy on both sides of this equation as previously performed in equation (4.96), it is possible to show that

$$\begin{aligned}&E\left[\|\Delta\mathbf{w}(k+1)\|^2\right] + E\left[\tilde{\mathbf{e}}_{\text{ap}}^T(k)\left(\mathbf{X}_{\text{ap}}^T(k)\mathbf{X}_{\text{ap}}(k)\right)^{-1}\tilde{\mathbf{e}}_{\text{ap}}(k)\right]\\ &= E\left[\|\Delta\hat{\mathbf{w}}(k)\|^2\right] + E\left[\tilde{\varepsilon}_{\text{ap}}^T(k)\left(\mathbf{X}_{\text{ap}}^T(k)\mathbf{X}_{\text{ap}}(k)\right)^{-1}\tilde{\varepsilon}_{\text{ap}}(k)\right]\\ &= E\left[\|\Delta\mathbf{w}(k) + \Delta\mathbf{w}_o(k+1)\|^2\right] + E\left[\tilde{\varepsilon}_{\text{ap}}^T(k)\left(\mathbf{X}_{\text{ap}}^T(k)\mathbf{X}_{\text{ap}}(k)\right)^{-1}\tilde{\varepsilon}_{\text{ap}}(k)\right]\\ &\approx E\left[\|\Delta\mathbf{w}(k)\|^2\right] + E\left[\|\Delta\mathbf{w}_o(k+1)\|^2\right] + E\left[\tilde{\varepsilon}_{\text{ap}}^T(k)\left(\mathbf{X}_{\text{ap}}^T(k)\mathbf{X}_{\text{ap}}(k)\right)^{-1}\tilde{\varepsilon}_{\text{ap}}(k)\right]\end{aligned} \tag{4.133}$$

where $\Delta\mathbf{w}_o(k+1) = \mathbf{w}_o(k) - \mathbf{w}_o(k+1)$, and in the last equality we have assumed that $E\left[\Delta\mathbf{w}^T(k)\Delta\mathbf{w}_o(k+1)\right] \approx 0$. This assumption is valid for simple models for the time-varying behavior of the unknown system, such as random walk model [28][5]. We will adopt this assumption in order to simplify our analysis.

The time-varying characteristic of the unknown system leads to an excess mean-square error. As before, in order to calculate the excess MSE we assume that each element of the optimal coefficient vector is modeled as a first-order Markov process. As previously mentioned, this nonstationary environment can be considered somewhat simplified, but allows a manageable mathematical analysis. The first-order Markov process is described by

$$\mathbf{w}_o(k) = \lambda_{\mathbf{w}}\mathbf{w}_o(k-1) + \kappa_{\mathbf{w}}\mathbf{n}_{\mathbf{w}}(k) \tag{4.134}$$

where $\mathbf{n}_{\mathbf{w}}(k)$ is a vector whose elements are zero-mean white noise processes with variance $\sigma_{\mathbf{w}}^2$, and $\lambda_{\mathbf{w}} < 1$. If $\kappa_{\mathbf{w}} = 1$ this model may not represent a real system when $\lambda_{\mathbf{w}} \rightarrow 1$, since the $E[\mathbf{w}_o(k)\mathbf{w}_o^T(k)]$ will have unbounded elements if, for example, $\mathbf{n}_{\mathbf{w}}(k)$ is not exactly zero mean. A better model utilizes a factor $\kappa_{\mathbf{w}} = (1-\lambda_{\mathbf{w}})^{\frac{p}{2}}$, for $p \geq 1$, multiplying $\mathbf{n}_{\mathbf{w}}(k)$ in order to guarantee that $E[\mathbf{w}_o(k)\mathbf{w}_o^T(k)]$ is bounded.

In our derivations of the excess of MSE, the covariance of $\Delta\mathbf{w}_o(k+1) = \mathbf{w}_o(k) - \mathbf{w}_o(k+1)$ is required. That is

$$\begin{aligned}
\text{cov}[\Delta\mathbf{w}_o(k+1)] &= E\left[(\mathbf{w}_o(k+1) - \mathbf{w}_o(k))(\mathbf{w}_o(k+1) - \mathbf{w}_o(k))^T\right] \\
&= E\left[(\lambda_{\mathbf{w}}\mathbf{w}_o(k) + \kappa_{\mathbf{w}}\mathbf{n}_{\mathbf{w}}(k) - \mathbf{w}_o(k))(\lambda_{\mathbf{w}}\mathbf{w}_o(k) + \kappa_{\mathbf{w}}\mathbf{n}_{\mathbf{w}}(k) - \mathbf{w}_o(k))^T\right] \\
&= E\left\{[(\lambda_{\mathbf{w}} - 1)\mathbf{w}_o(k) + \kappa_{\mathbf{w}}\mathbf{n}_{\mathbf{w}}(k)][(\lambda_{\mathbf{w}} - 1)\mathbf{w}_o(k) + \kappa_{\mathbf{w}}\mathbf{n}_{\mathbf{w}}(k)]^T\right\}
\end{aligned} \tag{4.135}$$

Since each element of $\mathbf{n}_{\mathbf{w}}(k)$ is a zero-mean white noise process with variance $\sigma_{\mathbf{w}}^2$, and $\lambda_{\mathbf{w}} < 1$, by applying the result of equation (2.82), it follows that

$$\begin{aligned}
\text{cov}[\Delta\mathbf{w}_o(k+1)] &= \kappa_{\mathbf{w}}^2\sigma_{\mathbf{w}}^2\frac{(1-\lambda_{\mathbf{w}})^2}{1-\lambda_{\mathbf{w}}^2}\mathbf{I} + \kappa_{\mathbf{w}}^2\sigma_{\mathbf{w}}^2\mathbf{I} \\
&= \kappa_{\mathbf{w}}^2\left[\frac{1-\lambda_{\mathbf{w}}}{1+\lambda_{\mathbf{w}}} + 1\right]\sigma_{\mathbf{w}}^2\mathbf{I}
\end{aligned} \tag{4.136}$$

By employing this result, we can compute

$$E\left[\|\Delta\mathbf{w}_o(k+1)\|^2\right] = \text{tr}\{\text{cov}[\Delta\mathbf{w}_o(k+1)]\} = (N+1)\left[\frac{2\kappa_{\mathbf{w}}^2}{1+\lambda_{\mathbf{w}}}\right]\sigma_{\mathbf{w}}^2 \tag{4.137}$$

We are now in a position to solve equation (4.133) utilizing the result of equation (4.137). Again by assuming that the algorithm has converged, that is, the Euclidean norm of the coefficients increment

[5] In this model the coefficients change according to $\mathbf{w}_o(k) = \mathbf{w}_o(k-1) + \mathbf{n}_{\mathbf{w}}(k)$.

remains in average unchanged, then $E\left[\|\Delta\mathbf{w}(k+1)\|^2\right] = E\left[\|\Delta\mathbf{w}(k)\|^2\right]$. As a result, equation (4.133) can be rewritten as

$$\begin{aligned} E\left[\tilde{\mathbf{e}}_{\text{ap}}^T(k)\left(\mathbf{X}_{\text{ap}}^T(k)\mathbf{X}_{\text{ap}}(k)\right)^{-1}\tilde{\mathbf{e}}_{\text{ap}}(k)\right] &= E\left[\tilde{\varepsilon}_{\text{ap}}^T(k)\left(\mathbf{X}_{\text{ap}}^T(k)\mathbf{X}_{\text{ap}}(k)\right)^{-1}\tilde{\varepsilon}_{\text{ap}}(k)\right] \\ &\quad +(N+1)\left[\frac{2\kappa_{\mathbf{w}}^2}{1+\lambda_{\mathbf{w}}}\right]\sigma_{\mathbf{w}}^2 \end{aligned} \tag{4.138}$$

Leading to the equivalent of equation (4.101) as follows

$$\begin{aligned} \mu^2 E\left[\mathbf{e}_{\text{ap}}^T(k)\hat{\mathbf{S}}_{\text{ap}}(k)\hat{\mathbf{R}}_{\text{ap}}(k)\hat{\mathbf{S}}_{\text{ap}}(k)\mathbf{e}_{\text{ap}}(k)\right] &= \mu E\left[\tilde{\mathbf{e}}_{\text{ap}}^T(k)\hat{\mathbf{S}}_{\text{ap}}(k)\mathbf{e}_{\text{ap}}(k) + \mathbf{e}_{\text{ap}}^T(k)\hat{\mathbf{S}}_{\text{ap}}(k)\tilde{\mathbf{e}}_{\text{ap}}(k)\right] \\ &\quad +(N+1)\left[\frac{2\kappa_{\mathbf{w}}^2}{1+\lambda_{\mathbf{w}}}\right]\sigma_{\mathbf{w}}^2 \end{aligned} \tag{4.139}$$

By solving this equation following precisely the same procedure as equation (4.101) was solved, we can derive the excess of MSE only due to the time-varying unknown system.

$$\xi_{\text{lag}} = \frac{N+1}{\mu(2-\mu)}\left[\frac{2\kappa_{\mathbf{w}}^2}{1+\lambda_{\mathbf{w}}}\right]\sigma_{\mathbf{w}}^2 \tag{4.140}$$

By taking into consideration the additional noise and the time-varying parameters to be estimated, the overall excess of MSE is given by

$$\begin{aligned} \xi_{\text{exc}} &= \frac{(L+1)\mu}{2-\mu}\frac{1-(1-\mu)^2}{1-(1-\mu)^{2(L+1)}}\sigma_n^2 + \frac{N+1}{\mu(2-\mu)}\left[\frac{2\kappa_{\mathbf{w}}^2}{1+\lambda_{\mathbf{w}}}\right]\sigma_{\mathbf{w}}^2 \\ &= \frac{1}{2-\mu}\left\{(L+1)\mu\frac{1-(1-\mu)^2}{1-(1-\mu)^{2(L+1)}}\sigma_n^2 + \frac{N+1}{\mu}\left[\frac{2\kappa_{\mathbf{w}}^2}{1+\lambda_{\mathbf{w}}}\right]\sigma_{\mathbf{w}}^2\right\} \end{aligned} \tag{4.141}$$

If $\kappa_{\mathbf{w}} = 1$, L is large, and $|1-\mu| < 1$, the above expression becomes simpler

$$\xi_{\text{exc}} = \frac{1}{2-\mu}\left\{(L+1)\mu\sigma_n^2 + \frac{2(N+1)}{\mu(1+\lambda_{\mathbf{w}})}\sigma_{\mathbf{w}}^2\right\} \tag{4.142}$$

As can be observed, the contribution due to the lag is inversely proportional to the value of μ. This is an expected result since for small values of μ an adaptive-filtering algorithm will face difficulties in tracking the variations in the unknown system.

4.6.3 Transient Behavior

This subsection presents some considerations related to the behavior of the affine projection algorithm during the transient. In order to achieve this goal we start by removing the dependence of equation (4.96) on the noiseless *a posteriori* error through equation (4.99), very much like previously performed in the derivations of equations (4.100) and (4.101). The resulting expression is

$$\begin{aligned} E\left[\|\Delta\mathbf{w}(k+1)\|^2\right] &= E\left[\|\Delta\mathbf{w}(k)\|^2\right] + \mu^2 E\left[\mathbf{e}_{\text{ap}}^T(k)\hat{\mathbf{S}}_{\text{ap}}(k)\hat{\mathbf{R}}_{\text{ap}}(k)\hat{\mathbf{S}}_{\text{ap}}(k)\mathbf{e}_{\text{ap}}(k)\right] \\ &\quad -\mu E\left[\tilde{\mathbf{e}}_{\text{ap}}^T(k)\hat{\mathbf{S}}_{\text{ap}}(k)\mathbf{e}_{\text{ap}}(k) + \mathbf{e}_{\text{ap}}^T(k)\hat{\mathbf{S}}_{\text{ap}}(k)\tilde{\mathbf{e}}_{\text{ap}}(k)\right] \end{aligned} \tag{4.143}$$

Since from equation (4.93)

$$\begin{aligned}\mathbf{e}_{\text{ap}}(k) &= \tilde{\mathbf{e}}_{\text{ap}}(k) + \mathbf{n}_{\text{ap}}(k) \\ &= -\mathbf{X}_{\text{ap}}^T(k)\Delta\mathbf{w}(k) + \mathbf{n}_{\text{ap}}(k)\end{aligned}$$

the above expression (4.143) can be rewritten as

$$\begin{aligned}&E\left[\|\Delta\mathbf{w}(k+1)\|^2\right] = E\left[\|\Delta\mathbf{w}(k)\|^2\right] \\ &+\mu^2 E\left[\left(-\Delta\mathbf{w}^T(k)\mathbf{X}_{\text{ap}}(k) + \mathbf{n}_{\text{ap}}^T(k)\right)\hat{\mathbf{S}}_{\text{ap}}(k)\hat{\mathbf{R}}_{\text{ap}}(k)\hat{\mathbf{S}}_{\text{ap}}(k)\left(-\mathbf{X}_{\text{ap}}^T(k)\Delta\mathbf{w}(k) + \mathbf{n}_{\text{ap}}(k)\right)\right] \\ &-\mu E\left[\left(-\Delta\mathbf{w}^T(k)\mathbf{X}_{\text{ap}}(k)\right)\hat{\mathbf{S}}_{\text{ap}}(k)\left(-\mathbf{X}_{\text{ap}}^T(k)\Delta\mathbf{w}(k) + \mathbf{n}_{\text{ap}}(k)\right)\right. \\ &\left.+\left(-\Delta\mathbf{w}^T(k)\mathbf{X}_{\text{ap}}(k) + \mathbf{n}_{\text{ap}}^T(k)\right)\hat{\mathbf{S}}_{\text{ap}}(k)\left(-\mathbf{X}_{\text{ap}}^T(k)\Delta\mathbf{w}(k)\right)\right] \end{aligned} \tag{4.144}$$

By considering the noise white and uncorrelated with the other quantities of this recursion, the above equation can be simplified to

$$\begin{aligned}E\left[\|\Delta\mathbf{w}(k+1)\|^2\right] &= E\left[\|\Delta\mathbf{w}(k)\|^2\right] - 2\mu E\left[\Delta\mathbf{w}^T(k)\mathbf{X}_{\text{ap}}(k)\hat{\mathbf{S}}_{\text{ap}}(k)\mathbf{X}_{\text{ap}}^T(k)\Delta\mathbf{w}(k)\right] \\ &+\mu^2 E\left[\Delta\mathbf{w}^T(k)\mathbf{X}_{\text{ap}}(k)\hat{\mathbf{S}}_{\text{ap}}(k)\hat{\mathbf{R}}_{\text{ap}}(k)\hat{\mathbf{S}}_{\text{ap}}(k)\mathbf{X}_{\text{ap}}^T(k)\Delta\mathbf{w}(k)\right] \\ &+\mu^2 E\left[\mathbf{n}_{\text{ap}}^T(k)\hat{\mathbf{S}}_{\text{ap}}(k)\hat{\mathbf{R}}_{\text{ap}}(k)\hat{\mathbf{S}}_{\text{ap}}(k)\mathbf{n}_{\text{ap}}(k)\right]\end{aligned} \tag{4.145}$$

By applying the property that $\text{tr}[\mathbf{AB}] = \text{tr}[\mathbf{BA}]$, this relation is equivalent to

$$\begin{aligned}\text{tr}\{\text{cov}[\Delta\mathbf{w}(k+1)]\} &= \text{tr}\left[\text{cov}[\Delta\mathbf{w}(k)]\right] - 2\mu\text{tr}\left\{E\left[\mathbf{X}_{\text{ap}}(k)\hat{\mathbf{S}}_{\text{ap}}(k)\mathbf{X}_{\text{ap}}^T(k)\Delta\mathbf{w}(k)\Delta\mathbf{w}^T(k)\right]\right\} \\ &+\mu^2\text{tr}\left\{E\left[\mathbf{X}_{\text{ap}}(k)\hat{\mathbf{S}}_{\text{ap}}(k)\hat{\mathbf{R}}_{\text{ap}}(k)\hat{\mathbf{S}}_{\text{ap}}(k)\mathbf{X}_{\text{ap}}^T(k)\Delta\mathbf{w}(k)\Delta\mathbf{w}^T(k)\right]\right\} \\ &+\mu^2\text{tr}\left\{E\left[\hat{\mathbf{S}}_{\text{ap}}(k)\hat{\mathbf{R}}_{\text{ap}}(k)\hat{\mathbf{S}}_{\text{ap}}(k)\right]E\left[\mathbf{n}_{\text{ap}}(k)\mathbf{n}_{\text{ap}}^T(k)\right]\right\}\end{aligned} \tag{4.146}$$

By assuming that the $\Delta\mathbf{w}(k+1)$ is independent of the data and the noise is white, it follows that

$$\begin{aligned}\text{tr}\{\text{cov}[\Delta\mathbf{w}(k+1)]\} &= \text{tr}\left\{\left[\mathbf{I} - E\left(2\mu\mathbf{X}_{\text{ap}}(k)\hat{\mathbf{S}}_{\text{ap}}(k)\mathbf{X}_{\text{ap}}^T(k)\right.\right.\right. \\ &\left.\left.\left.+\ \mu^2\mathbf{X}_{\text{ap}}(k)\hat{\mathbf{S}}_{\text{ap}}(k)\hat{\mathbf{R}}_{\text{ap}}(k)\hat{\mathbf{S}}_{\text{ap}}(k)\mathbf{X}_{\text{ap}}^T(k)\right)\right]\text{cov}[\Delta\mathbf{w}(k)]\right\} \\ &+\mu^2\sigma_n^2\text{tr}\left\{E\left[\hat{\mathbf{S}}_{\text{ap}}(k)\hat{\mathbf{R}}_{\text{ap}}(k)\hat{\mathbf{S}}_{\text{ap}}(k)\right]\right\}\end{aligned} \tag{4.147}$$

Now by recalling that

$$\hat{\mathbf{S}}_{\text{ap}}(k) \approx \hat{\mathbf{R}}_{\text{ap}}^{-1}(k)\left[\mathbf{I} - \gamma\hat{\mathbf{R}}_{\text{ap}}^{-1}(k)\right]$$

and by utilizing the unitary matrix $\mathbf{Q}$, that in the present discussion diagonalizes $E[\mathbf{X}_{\text{ap}}(k)\hat{\mathbf{S}}_{\text{ap}}(k)\mathbf{X}_{\text{ap}}^T(k)]$, the following relation is valid

$$\begin{aligned}\text{tr}\{\text{cov}[\Delta\mathbf{w}(k+1)]\mathbf{QQ}^T\} &= \text{tr}\left\{\mathbf{QQ}^T\left[\mathbf{I} - E\left(2\mu\mathbf{X}_{\text{ap}}(k)\hat{\mathbf{S}}_{\text{ap}}(k)\mathbf{X}_{\text{ap}}^T(k)\right.\right.\right. \\ &\left.\left.\left.+\ (1-\gamma)\mu^2\mathbf{X}_{\text{ap}}(k)\hat{\mathbf{S}}_{\text{ap}}(k)\mathbf{X}_{\text{ap}}^T(k)\right)\right]\mathbf{QQ}^T\text{cov}[\Delta\mathbf{w}(k)]\mathbf{QQ}^T\right\} \\ &+(1-\gamma)\mu^2\sigma_n^2\text{tr}\left\{E\left[\hat{\mathbf{S}}_{\text{ap}}(k)\right]\right\}\end{aligned} \tag{4.148}$$

Again by applying the property that $\text{tr}[\mathbf{AB}] = \text{tr}[\mathbf{BA}]$ and assuming γ small, it follows that

$$\begin{aligned}\text{tr}\{\mathbf{Q}^T\text{cov}[\Delta\mathbf{w}(k+1)]\mathbf{Q}\} &= \text{tr}\Big\{\mathbf{Q}\Big[\mathbf{I}-\mathbf{Q}^T E\Big(2\mu\mathbf{X}_{\text{ap}}(k)\hat{\mathbf{S}}_{\text{ap}}(k)\mathbf{X}_{\text{ap}}^T(k)\\ &\quad+\mu^2\mathbf{X}_{\text{ap}}(k)\hat{\mathbf{S}}_{\text{ap}}(k)\mathbf{X}_{\text{ap}}^T(k)\Big)\mathbf{Q}\Big]\mathbf{Q}^T\text{cov}[\Delta\mathbf{w}(k)]\mathbf{Q}\mathbf{Q}^T\Big\}\\ &\quad+\mu^2\sigma_n^2\text{tr}\Big\{E\Big[\hat{\mathbf{S}}_{\text{ap}}(k)\Big]\Big\} \end{aligned} \tag{4.149}$$

By defining

$$\Delta\mathbf{w}'(k+1) = \Delta\mathbf{w}(k+1)\mathbf{Q}$$

equation (4.149) can be rewritten as

$$\begin{aligned}\text{tr}\{\text{cov}[\Delta\mathbf{w}'(k+1)]\} &= \text{tr}\Big\{\mathbf{Q}^T\mathbf{Q}\Big[\mathbf{I}-\mathbf{Q}^T E\Big(-2\mu\mathbf{X}_{\text{ap}}(k)\hat{\mathbf{S}}_{\text{ap}}(k)\mathbf{X}_{\text{ap}}^T(k)\\ &\quad+\mu^2\mathbf{X}_{\text{ap}}(k)\hat{\mathbf{S}}_{\text{ap}}(k)\mathbf{X}_{\text{ap}}^T(k)\Big)\mathbf{Q}\Big]\text{cov}[\Delta\mathbf{w}'(k)]\Big\}\\ &\quad+\mu^2\sigma_n^2\text{tr}\Big\{E\Big[\hat{\mathbf{S}}_{\text{ap}}(k)\Big]\Big\}\\ &= \text{tr}\Big\{\Big[\mathbf{I}-2\mu\hat{\mathbf{\Lambda}}+\mu^2\hat{\mathbf{\Lambda}}\Big]\text{cov}[\Delta\mathbf{w}'(k)]\Big\}+\mu^2\sigma_n^2\text{tr}\Big\{E\Big[\hat{\mathbf{S}}_{\text{ap}}(k)\Big]\Big\} \end{aligned} \tag{4.150}$$

where $\hat{\mathbf{\Lambda}}$ is a diagonal matrix whose elements are the eigenvalues of $E[\mathbf{X}_{\text{ap}}(k)\hat{\mathbf{S}}_{\text{ap}}(k)\mathbf{X}_{\text{ap}}^T(k)]$, denoted as $\hat{\lambda}_i$, for $i = 0, \ldots, N$.

By using the likely assumption that $\text{cov}[\Delta\mathbf{w}'(k+1)]$ and $\hat{\mathbf{S}}_{\text{ap}}(k)$ are diagonal dominant, we can disregard the trace operator in the above equation and observe that the geometric decaying curves have ratios $r_{\text{cov}[\Delta\mathbf{w}(k)]} = (1 - 2\mu\hat{\lambda}_i + \mu^2\hat{\lambda}_i)$. As a result, according to the considerations in the derivation of equation (3.52), it is possible to infer that the convergence time constant is given by

$$\begin{aligned}\tau_{ei} &= \tau_{\text{cov}[\Delta\mathbf{w}(k)]}\\ &= \frac{1}{\mu\hat{\lambda}_i}\frac{1}{2-\mu}\end{aligned} \tag{4.151}$$

since the error squared depends on the convergence of the diagonal elements of the covariance matrix of the coefficient-error vector, see discussions around equation (3.53). As can be observed, the time constants for error convergence are dependent on the inverse of the eigenvalues of $E[\mathbf{X}_{\text{ap}}(k)\hat{\mathbf{S}}_{\text{ap}}(k)\mathbf{X}_{\text{ap}}^T(k)]$. However, since μ is not constrained by these eigenvalues, the speed of convergence is expected to be higher than for the LMS algorithm, particularly in situations where the eigenvalue spread of the input signal is high. Simulation results confirm the improved performance of the affine projection algorithm.

4.6.4 Complex Affine Projection Algorithm

Using the method of Lagrange multipliers to transform the constrained minimization into an unconstrained one, the unconstrained function to be minimized is

$$F[\mathbf{w}(k+1)] = \frac{1}{2}\|\mathbf{w}(k+1)-\mathbf{w}(k)\|^2 + \text{re}\Big\{\boldsymbol{\lambda}_{\text{ap}}^T(k)[\mathbf{d}_{\text{ap}}(k) - \mathbf{X}_{\text{ap}}^T(k)\mathbf{w}^*(k+1)]\Big\} \tag{4.152}$$

where $\boldsymbol{\lambda}_{\rm ap}(k)$ is a complex $(L+1) \times 1$ vector of Lagrange multipliers, and the real part operator is required in order to turn the overall objective function real valued. The above expression can be rewritten as

$$\begin{aligned} F[\mathbf{w}(k+1)] &= \frac{1}{2}[\mathbf{w}(k+1) - \mathbf{w}(k)]^H [\mathbf{w}(k+1) - \mathbf{w}(k)] \\ &+ \frac{1}{2}\boldsymbol{\lambda}_{\rm ap}^H(k)\left[\mathbf{d}_{\rm ap}^*(k) - \mathbf{X}_{\rm ap}^H(k)\mathbf{w}(k+1)\right] \\ &+ \frac{1}{2}\boldsymbol{\lambda}_{\rm ap}^T(k)\left[\mathbf{d}_{\rm ap}(k) - \mathbf{X}_{\rm ap}^T(k)\mathbf{w}^*(k+1)\right] \end{aligned} \tag{4.153}$$

The gradient of $F[\mathbf{w}(k+1)]$ with respect to $\mathbf{w}^*(k+1)$ is given by[6]

$$\frac{\partial F[\mathbf{w}(k+1)]}{\partial \mathbf{w}^*(k+1)} = \mathbf{g}_{\mathbf{w}^*}\{F[\mathbf{w}(k+1)]\} = \frac{1}{2}[\mathbf{w}(k+1) - \mathbf{w}(k)] - \frac{1}{2}\mathbf{X}_{\rm ap}(k)\boldsymbol{\lambda}_{\rm ap}(k) \tag{4.154}$$

After setting the gradient of $F[\mathbf{w}(k+1)]$ with respect to $\mathbf{w}^*(k+1)$ equal to zero, the expression below follows

$$\mathbf{w}(k+1) = \mathbf{w}(k) + \mathbf{X}_{\rm ap}(k)\boldsymbol{\lambda}_{\rm ap}(k) \tag{4.155}$$

By replacing equation (4.155) in the constraint relation $\mathbf{d}_{\rm ap}^*(k) - \mathbf{X}_{\rm ap}^H(k)\mathbf{w}(k+1) = \mathbf{0}$, we generate the expression

$$\mathbf{X}_{\rm ap}^H(k)\mathbf{X}_{\rm ap}(k)\boldsymbol{\lambda}_{\rm ap}(k) = \mathbf{d}_{\rm ap}^*(k) - \mathbf{X}_{\rm ap}^H(k)\mathbf{w}(k) = \mathbf{e}_{\rm ap}^*(k) \tag{4.156}$$

The update equation is now given by equation (4.155) with $\boldsymbol{\lambda}_{\rm ap}(k)$ being the solution of equation (4.156), i.e.,

$$\mathbf{w}(k+1) = \mathbf{w}(k) + \mathbf{X}_{\rm ap}(k)\left(\mathbf{X}_{\rm ap}^H(k)\mathbf{X}_{\rm ap}(k)\right)^{-1}\mathbf{e}_{\rm ap}^*(k) \tag{4.157}$$

This updating equation corresponds to the complex affine projection algorithm with unity convergence factor. As common practice, we introduce a convergence factor in order to trade-off final misadjustment and convergence speed as follows

$$\mathbf{w}(k+1) = \mathbf{w}(k) + \mu\mathbf{X}_{\rm ap}(k)\left(\mathbf{X}_{\rm ap}^H(k)\mathbf{X}_{\rm ap}(k)\right)^{-1}\mathbf{e}_{\rm ap}^*(k) \tag{4.158}$$

The description of the complex affine projection algorithm is given in Algorithm 4.6, where as before a regularization is introduced through an identity matrix multiplied by a small constant added to the matrix $\mathbf{X}_{\rm ap}^H(k)\mathbf{X}_{\rm ap}(k)$ in order to avoid numerical problems in the matrix inversion.

4.7 SIMULATION EXAMPLES

In this section, some adaptive-filtering problems are described and solved by using some of the algorithms presented in this chapter.

[6] The reader should recall that when computing the gradient with respect to $\mathbf{w}^*(k+1)$, $\mathbf{w}(k+1)$ is treated as a constant.

Algorithm 4.6

Complex Affine Projection Algorithm

Initialization
$\mathbf{x}(0) = \mathbf{w}(0) = [0\,0 \ldots 0]^T$
choose μ in the range $0 < \mu \leq 2$
$\gamma =$ small constant
Do for $k \geq 0$
$\mathbf{e}_{\text{ap}}^*(k) = \mathbf{d}_{\text{ap}}^*(k) - \mathbf{X}_{\text{ap}}^H(k)\mathbf{w}(k)$
$\mathbf{w}(k+1) = \mathbf{w}(k) + \mu \mathbf{X}_{\text{ap}}(k)\left(\mathbf{X}_{\text{ap}}^H(k)\mathbf{X}_{\text{ap}}(k) + \gamma \mathbf{I}\right)^{-1} \mathbf{e}_{\text{ap}}^*(k)$

Example 4.3: Transform-Domain LMS Algorithm

Use the transform-domain LMS algorithm to identify the system described in example of subsection 3.6.2. The transform is the DCT.

Solution:

All the results presented here for the transform-domain LMS algorithm are obtained by averaging the results of 200 independent runs.

We run the algorithm with a value of $\mu = 0.01$, with $\alpha = 0.05$ and $\gamma = 10^{-6}$. With this value of μ, the misadjustment of the transform-domain LMS algorithm is about the same as that of the LMS algorithm with $\mu = 0.02$. In Fig. 4.13, the learning curves for the eigenvalue spreads 20 and 80 are illustrated. First note that the convergence speed is about the same for different eigenvalue spreads, showing the effectiveness of the rotation performed by the transform in this case. If we compare these curves with those of Fig. 3.9 for the LMS algorithm, we conclude that the transform-domain LMS algorithm has better performance than the LMS algorithm for high eigenvalue spread. For an eigenvalue spread equal to 20, the transform-domain LMS algorithm requires around 200 iterations to achieve convergence, whereas the LMS requires at least 500 iterations. This improvement is achieved without increasing the misadjustment as can be verified by comparing the results of Tables 3.1 and 4.1.

The reader should bear in mind that the improvements in convergence of the transform-domain LMS algorithm can be achieved only if the transformation is effective. In this example, since the input signal is colored using a first-order all-pole filter, the cosine transform is known to be effective because it approximates the KLT.

The finite-precision implementation of the transform-domain LMS algorithm presents similar performance to that of the LMS algorithm, as can be verified by comparing the results of Tables 3.2 and 4.2. An eigenvalue spread of one is used in this example. The value of μ is 0.01, while the remaining parameter values are $\gamma = 2^{-b_d}$ and $\alpha = 0.05$. The value of μ in this case is chosen the same as for the LMS algorithm.

□

Table 4.1 Evaluation of the Transform-Domain LMS Algorithm

$\frac{\lambda_{max}}{\lambda_{min}}$	**Misadjustment**
1	0.2027
20	0.2037
80	0.2093

Table 4.2 Results of the Finite-Precision Implementation of the Transform-Domain LMS Algorithm

	$\xi(k)_Q$	$E[\|\Delta\mathbf{w}(k)_Q\|^2]$
No of bits	**Experiment**	**Experiment**
16	$1.627\ 10^{-3}$	$1.313\ 10^{-4}$
12	$1.640\ 10^{-3}$	$1.409\ 10^{-4}$
10	$1.648\ 10^{-3}$	$1.536\ 10^{-4}$

Example 4.4: Affine Projection Algorithm

An adaptive-filtering algorithm is used to identify the system described in example of subsection 3.6.2 using the affine projection algorithm using $L = 0$, $L = 1$ and $L = 4$. Do not consider the finite-precision case.

Solution:

Fig. 4.14 depicts the estimate of the MSE learning curve of the affine projection algorithm for the case of eigenvalue spread equal to 1, obtained by averaging the results of 200 independent runs. As can be noticed by increasing L the algorithm becomes faster. The chosen convergence factor is $\mu = 0.4$, and the measured misadjustments are $M = 0.32$ for $L = 0$, $M = 0.67$ for $L = 1$, and $M = 2.05$ for $L = 4$. In all cases $\gamma = 0$ is utilized, and for $L = 1$ in the first iteration we start with $L = 0$, whereas for $L = 4$ in the first four iterations we employ $L = 0, 1, 2,$ and 3, respectively. If we consider that the term $E\left[\frac{1}{\|\mathbf{x}(k)\|^2}\right] \approx \frac{1}{(N+1)\sigma_x^2}$, the expected misadjustment according to equation (4.126) is $M = 0.25$, which is somewhat close to the measured ones considering the above approximation as well as the approximations in the derivation of the theoretical formula.

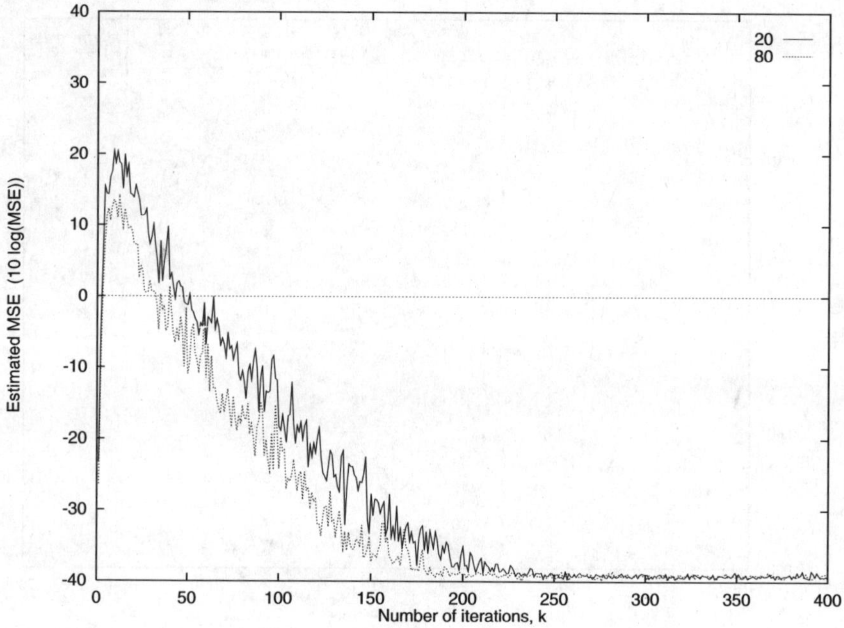

Figure 4.13 Learning curves for the transform-domain LMS algorithm for eigenvalue spreads: 20 and 80.

Fig. 4.15 depicts the average of the squared error obtained from 200 independent runs for the case of eigenvalue spread equal to 80. Again we verify that by increasing L the algorithm becomes faster. The chosen convergence factor is also $\mu = 0.4$, and the measured misadjustments for three values of the eigenvalue spread are listed in Table 4.3. It can be observed that higher eigenvalue spreads do not increase the misadjustment substantially.

Table 4.3 Evaluation of the Affine Projection Algorithm, $\mu = 0.4$

$\frac{\lambda_{max}}{\lambda_{min}}$	**Misadjustment, $L = 0$**		**Misadjustment, $L = 1$**		**Misadjustment, $L = 4$**	
	Experiment	**Theory**	**Experiment**	**Theory**	**Experiment**	**Theory**
1	0.32	0.25	0.67	0.37	2.05	0.81
20	0.35	0.25	0.69	0.37	2.29	0.81
80	0.37	0.25	0.72	0.37	2.43	0.81

In Fig. 4.16, it is shown the effect of using different values for the convergence factor, when $L = 1$ and the eigenvalue spread is equal to 1. For $\mu = 0.2$ the misadjustment is $M = 0.30$, for $\mu = 0.4$ the misadjustment is $M = 0.67$, and for $\mu = 1$ the misadjustment is $M = 1.56$.

□

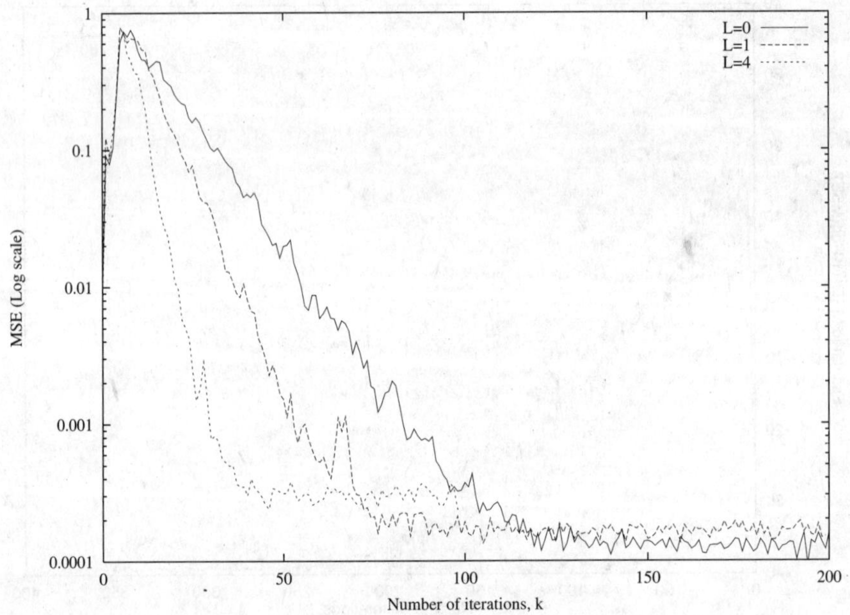

Figure 4.14 Learning curves for the affine projection algorithms for $L = 0$, $L = 1$, and $L = 4$, eigenvalue spread equal 1.

4.7.1 Signal Enhancement Simulation

In this subsection, a signal enhancement simulation environment is described. This example will also be employed in some of the following chapters.

In a signal enhancement problem, the reference signal is

$$r(k) = \sin(0.2\pi k) + n_r(k)$$

where $n_r(k)$ is zero-mean Gaussian white noise with variance $\sigma^2_{n_r} = 10$. The input signal is given by $n_r(k)$ passed through a filter with the following transfer function

$$H(z) = \frac{0.4}{z^2 - 1.36z + 0.79}$$

The adaptive filter is a 20th-order FIR filter. In all examples, a delay $L = 10$ is applied to the reference signal.

Example 4.5: Quantized-Error and Normalized LMS Algorithms

Using the sign-error, power-of-two error with $b_d = 12$, and normalized LMS algorithms:
(a) Choose an appropriate μ in each case and run an ensemble of 50 experiments. Plot the average

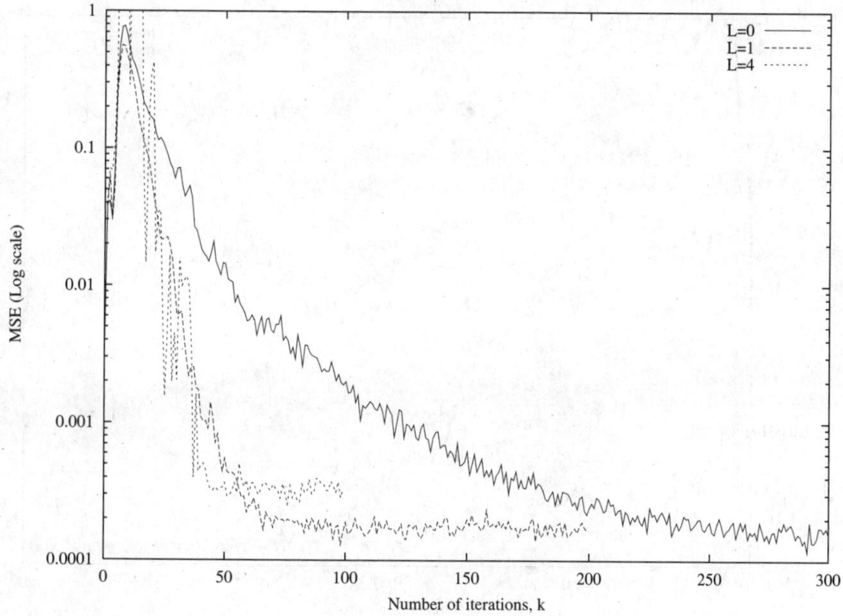

Figure 4.15 Learning curves for the affine projection algorithms for $L = 0$, $L = 1$, and $L = 4$, eigenvalue spread equal 80.

learning curve.
(b) Plot the output errors and comment on the results.

Solution:

The maximum value of μ for the LMS algorithm in this example is 0.005. The value of μ for both the sign-error and power-of-two LMS algorithms is chosen 0.001. The coefficients of the adaptive filter are initialized with zero. For the normalized LMS algorithm $\mu_n = 0.4$ and $\gamma = 10^{-6}$ are used. Fig. 4.17 depicts the learning curves for the three algorithms. The results show that the sign-error and power-of-two error algorithms present similar convergence speed, whereas the normalized LMS algorithm is faster to converge. The reader should notice that the MSE after convergence is not small since we are dealing with an example where the signal to noise ratio is low.

The DFT with 128 points of the input signal is shown in Fig. 4.18 where the presence of the sinusoid cannot be noted. In the same figure are shown the DFT of the error and the error signal itself, for the experiment using the normalized LMS algorithm. In the cases of DFT, the result presented is the magnitude of the DFT outputs. As can be verified, the output error tends to produce a signal with the same period of the sinusoid after convergence and the DFT shows clearly the presence of the sinusoid. The other two algorithms lead to similar results.

□

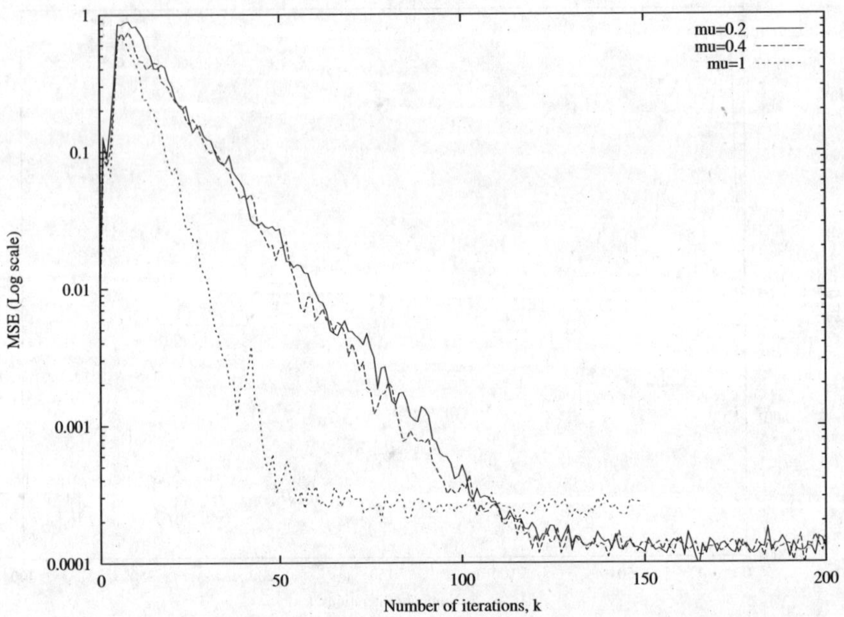

Figure 4.16 Learning curves for the affine projection algorithms for $\mu = 0.2$, $\mu = 0.4$, and $\mu = 1$.

4.7.2 Signal Prediction Simulation

In this subsection a signal prediction simulation environment is described. This example will also be used in some of the following chapters.

In a prediction problem the input signal is

$$x(k) = -\sqrt{2}\ \sin(0.2\pi k) + \sqrt{2}\ \sin(0.05\pi k) + n_x(k)$$

where $n_x(k)$ is zero-mean Gaussian white noise with variance $\sigma_{n_x}^2 = 1$. The adaptive filter is a fourth-order FIR filter.

(a) Run an ensemble of 50 experiments and plot the average learning curve.
(b) Determine the zeros of the resulting FIR filter and comment on the results.

Example 4.6: Quantized-Error and Normalized LMS Algorithms

We solve the above problem using the sign-error, power-of-two error with $b_d = 12$, and normalized LMS algorithms.

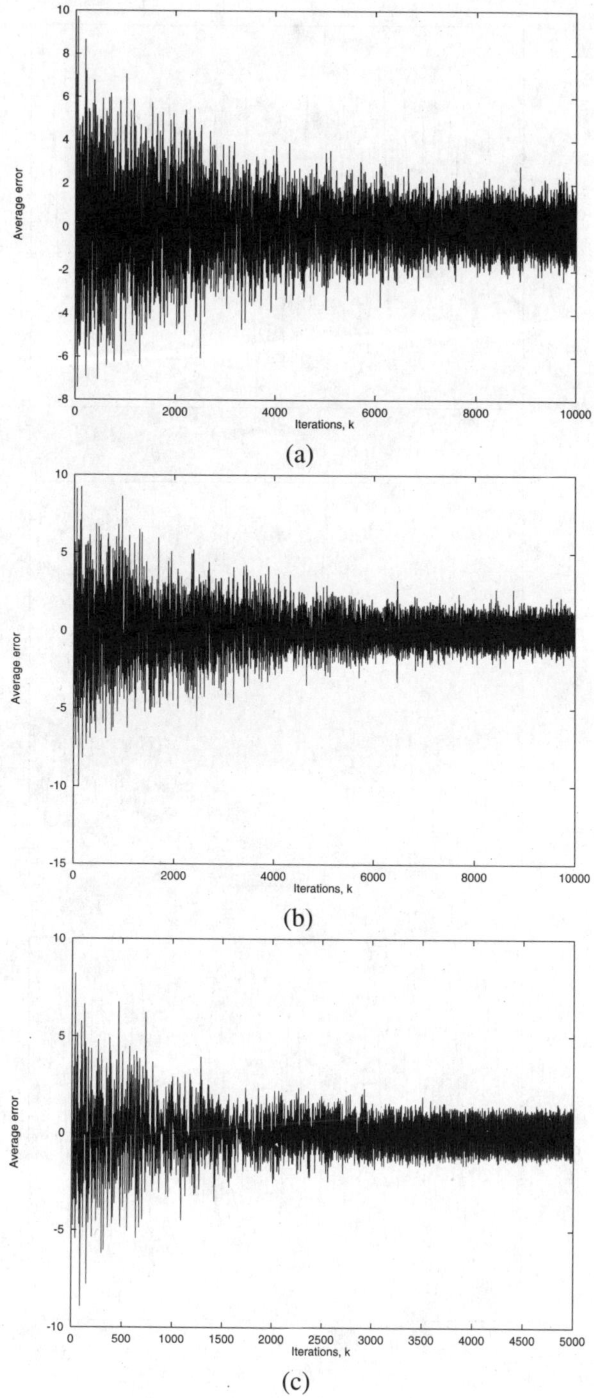

Figure 4.17 Learning curves for the (a) Sign-error, (b) Power-of-two, and (c) Normalized LMS algorithms.

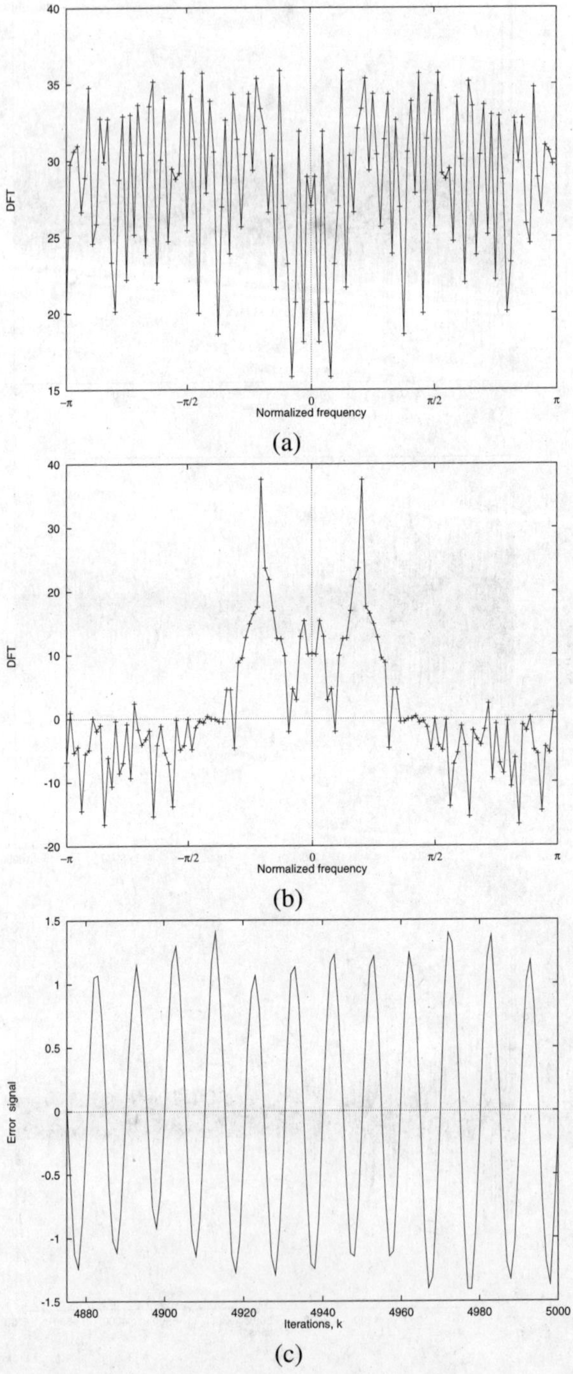

Figure 4.18 (a) DFT of the input signal, (b) DFT of the error signal, (c) The output error for the normalized LMS algorithm.

Solution:

In the first step, each algorithm is tested in order to determine experimentally the maximum value of μ in which the convergence is achieved. The choice of the convergence factor is $\mu_{\max}/5$ for each algorithm. The chosen values of μ for the sign-error and power-of-two LMS algorithms are 0.0028 and 0.0044, respectively. For the normalized LMS algorithm, $\mu_n = 0.4$ and $\gamma = 10^{-6}$ are used. The coefficients of the adaptive filter are initialized with zero. In all cases, we notice a strong attenuation of the predictor response around the frequencies of the two sinusoids. See, for example, the response depicted in Fig. 4.19 obtained by running the power-of-two LMS algorithm. The learning curves for the three algorithms are depicted in Fig. 4.20. The zeros of the transfer function from the input to the output error are calculated for the power-of-two algorithm:

$$-0.3939;\ -0.2351 \pm \jmath 0.3876;\ -0.6766 \pm \jmath 0.3422$$

Notice that the predictor tends to place its zeros at low frequencies, in order to attenuate the two low-frequency sinusoids.

In the experiments, we notice that for a given additional noise, smaller convergence factor leads to higher attenuation at the sinusoid frequencies. This is an expected result since the excess MSE is smaller. Another observation is that the attenuation also grows as the signal to noise ratio is reduced, again due to the smaller MSE.

□

4.8 CONCLUDING REMARKS

In this chapter, a number of adaptive-filtering algorithms were presented derived from the LMS algorithm. There were two basic directions followed in the derivation of the algorithms: one direction was to search for simpler algorithms from the computational point of view, and the other was to sophisticate the LMS algorithm looking for improvements in performance. The simplified algorithms lead to low-power, low-complexity and/or high-speed integrated circuit implementations [29], at a cost of increasing the misadjustment and/or of losing convergence speed among other things [30]. The simplified algorithms discussed here were the quantized-error algorithms.

We also introduced the LMS-Newton algorithm, whose performance is independent of the eigenvalue spread of the input signal correlation matrix. This algorithm is related to the RLS algorithm which will be discussed in the following chapter, although some distinctive features exist between them [39]. Newton-type algorithms with reduced computational complexity are also known [40]-[41], and the main characteristic of this class of algorithms is to reduce the computation involving the inverse of the estimate of **R**.

In the normalized LMS algorithm, the straightforward objective was to find the step size that minimizes the instantaneous output error. There are many papers dealing with the analysis [31]-[33] and applications [34] of the normalized LMS algorithm. The idea of using variable step size in the LMS

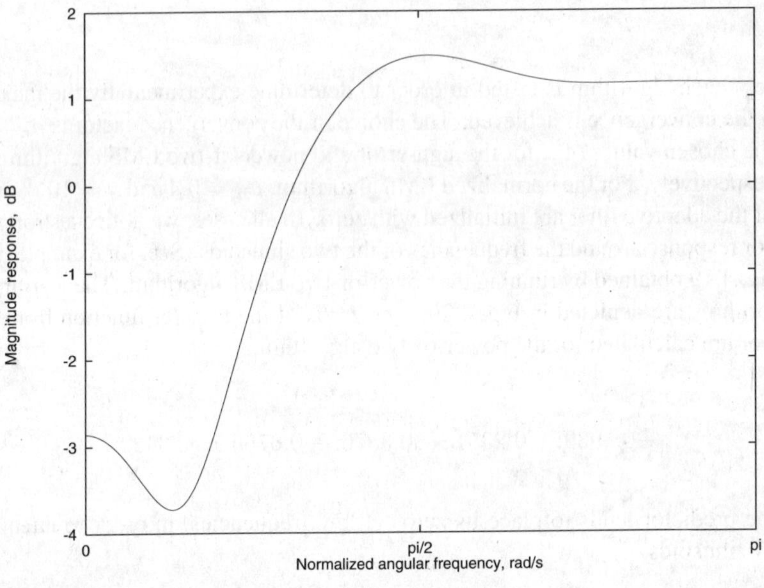

Figure 4.19 Magnitude response of the FIR adaptive filter at a given iteration after convergence using the power-of-two LMS algorithm.

and normalized LMS algorithms can lead to a number of interesting algorithms [35]-[37], that in some cases are very efficient in tracking nonstationary environments [38].

The transform-domain LMS algorithm aimed at reducing the eigenvalue spread of the input signal correlation matrix. Several frequency-domain adaptive algorithms, which are related in some sense to the transform-domain LMS algorithm, have been investigated in the recent years [42]. Such algorithms exploit the whitening property associated with the normalized transform-domain LMS algorithm, and most of them update the coefficients at a rate lower than the input sampling rate. One of the resulting structures, presented in [43], can be interpreted as a direct generalization of the transform-domain LMS algorithm and is called generalized adaptive subband decomposition structure. Such structure consists of a small-size fixed transform, which is applied to the input sequence, followed by sparse adaptive subfilters which are updated at the input rate. In high-order adaptive-filtering problems, the use of this structure with appropriately chosen transform-size and sparsity factor can lead to significant convergence rate improvement for colored input signals when compared to the standard LMS algorithm. The convergence rate improvement is achieved without the need for large transform sizes. Other algorithms to deal with high-order adaptive filters are discussed in Chapter 12.

The affine projection algorithm is very appealing in applications requiring a trade-off between convergence speed and computational complexity. Although the algorithms in the affine projection family might have high misadjustment, their combination with deterministic objective functions leading to data selective updating results in computationally efficient algorithms with low misadjustment and high convergence speed [24], as will be discussed in Chapter 6. Several simulation examples involving the LMS-based algorithms were presented in this chapter. These examples aid the reader to understand what are the main practical characteristics of the LMS-based algorithms.

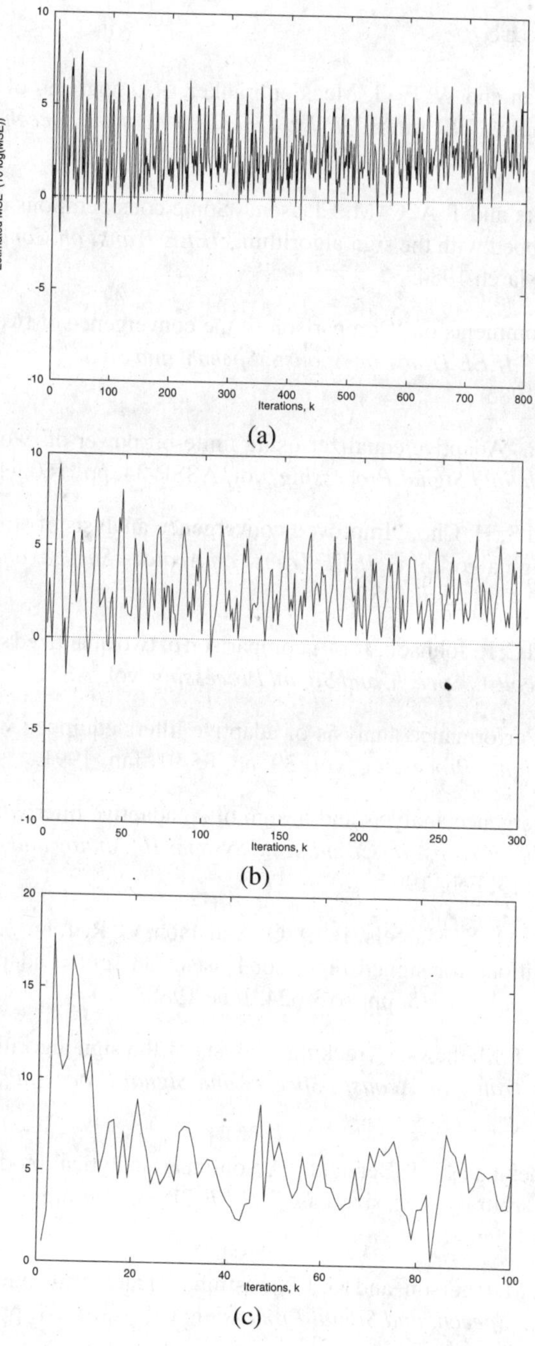

Figure 4.20 Learning curves for the (a) Sign-error, (b) Power-of-two, and (c) Normalized LMS algorithms.

4.9 REFERENCES

1. T. A. C. M. Claasen and W. F. G. Mecklenbräuker, "Comparison of the convergence of two algorithms for adaptive FIR filters," *IEEE Trans. on Acoust., Speech, and Signal Processing*, vol. ASSP-29, pp. 670-678, June 1981.

2. N. A. M. Verhoeckx and T. A. C. M. Claasen, "Some considerations on the design of adaptive digital filters equipped with the sign algorithm," *IEEE Trans. on Communications*, vol. COM-32, pp. 258-266, March 1984.

3. N. J. Bershad, "Comments on 'Comparison of the convergence of two algorithms for adaptive FIR digital filters'," *IEEE Trans. on Acoust., Speech, and Signal Processing*, vol. ASSP-33, pp. 1604-1606, Dec. 1985.

4. P. Xue and B. Liu, "Adaptive equalizer using finite-bit power-of-two quantizer," *IEEE Trans. on Acoust., Speech, and Signal Processing*, vol. ASSP-34, pp. 1603-1611, Dec. 1986.

5. V. J. Mathews and S. H. Cho, "Improved convergence analysis of stochastic gradient adaptive filters using the sign algorithm," *IEEE Trans. on Acoust., Speech, and Signal Processing*, vol. ASSP-35, pp. 450-454, April 1987.

6. W. A. Sethares and C. R. Johnson, Jr., "A comparison of two quantized state adaptive algorithms," *IEEE Trans. on Acoust., Speech, and Signal Processing*, vol. ASSP-37, pp. 138-143, Jan. 1989.

7. V. J. Mathews, "Performance analysis of adaptive filters equipped with dual sign algorithm," *IEEE Trans. on Signal Processing*, vol. 39, pp. 85-91, Jan. 1991.

8. E. Eweda, "Convergence analysis and design of an adaptive filter with finite-bit power-of-two quantizer error," *IEEE Trans. on Circuits and Systems II : Analog and Digital Signal Processing*, vol. 39, pp. 113-115, Feb. 1992.

9. W. A. Sethares, I. M. X. Mareels, B. D. O. Anderson, C. R. Johnson, Jr., and R. R. Bitmead, "Excitation conditions for signed regressor least mean square adaptation," *IEEE Trans. on Circuits and Systems*, vol. 35, pp. 613-624, June 1988.

10. S. H. Cho and V. J. Mathews, "Tracking analysis of the sign algorithm in nonstationary environments," *IEEE Trans. on Acoust., Speech, and Signal Processing*, vol. 38, pp. 2046-2057, Dec. 1990.

11. J. C. M. Bermudez and N. J. Bershad, "A nonlinear analytical model for the quantized LMS algorithm: The arbitrary step size case," *IEEE Trans. on Signal Processing*, vol. 44, pp. 1175-1183, May 1996.

12. S. S. Narayan, A. M. Peterson, and M. J. Narasimha, "Transform domain LMS algorithm," *IEEE Trans. on Acoust., Speech, and Signal Processing*, vol. ASSP-31, pp. 609-615, June 1983.

13. D. F. Marshall, W. K. Jenkins, and J. J. Murphy, "The use of orthogonal transform for improving performance of adaptive filters, " *IEEE Trans. on Circuits and Systems*, vol. 36, pp. 474-484, April 1989.

14. J. C. Lee and C. K. Un, "Performance of transform-domain LMS adaptive digital filters," *IEEE Trans. on Acoust., Speech, and Signal Processing*, vol. ASSP-34, pp. 499-510, June 1986.

15. F. F. Yassa, "Optimality in the choice of convergence factor for gradient based adaptive algorithms," *IEEE Trans. on Acoust., Speech, and Signal Processing*, vol. ASSP-35, pp. 48-59, Jan. 1987.

16. B. Widrow and S. D. Stearns, *Adaptive Signal Processing*, Prentice Hall, Englewood Cliffs, NJ, 1985.

17. P. S. R. Diniz and L. W. Biscainho, "Optimal variable step size for the LMS/Newton algorithm with application to subband adaptive filtering," *IEEE Trans. on Signal Processing*, vol. SP-40, pp. 2825-2829, Nov. 1992.

18. S. Roy and J. J. Shynk, "Analysis of the data-reusing LMS algorithm," *Proc. Midwest Symposium on Circuits and Systems*, Urbana, IL, pp. 1127-1130, Aug. 1989.

19. K. Ozeki and T. Umeda, "An adaptive filtering algorithm using an orthogonal projection to an affine subspace and its properties," *Electronics and Communications in Japan*, vol. 67-A, pp. 19-27, 1984.

20. S. L. Gay and S. Tavathia, "The fast affine projection algorithm," *Proc. IEEE Int. Conf. on Acoust., Speech, and Signal Processing*, Detroit, MI, pp. 3023-3026, May 1995.

21. J. A. Apolinário, M. L. R. de Campos, and P. S. R. Diniz, "The binormalized data-reusing LMS algorithm," *IEEE Trans. on Signal Processing*, vol. 48, pp. 3235-3242, Nov. 2000.

22. R. A. Soni, K. A. Gallivan, and W. K. Jenkins, "Low-complexity data-reusing methods in adaptive filtering," *IEEE Trans. on Signal Processing*, vol. 52, pp. 394-405, Feb. 2004.

23. S. G. Sankaran and A. A. (Louis) Beex, "Convergence behavior of affine projection algorithms," *IEEE Trans. on Signal Processing*, vol. 48, pp. 1086-1096, April 2000.

24. S. Werner and P. S. R. Diniz, "Set-membership affine projection algorithm," *IEEE Signal Processing Letters*, vol. 8, pp. 231-235, Aug 2001.

25. G.-O. Glentis, K. Berberidis, and S. Theodoridis, "Efficient least squares adaptive algorithms for FIR transversal filtering," *IEEE Signal Processing Magazine*, vol. 16, pp. 13-41, July 1999.

26. N. S. Jayant and P. Noll, *Digital Coding of Waveforms: Principles and Applications to Speech and Video*, Prentice Hall, Englewood Cliffs, NJ, 1984.

27. R. Price, "A useful theorem for nonlinear devices having Gaussian inputs," *IRE Trans. on Information Theory*, vol. IT-4, pp. 69-72, June 1958.

28. A. Papoulis, *Probability, Random Variables, and Stochastic Processes*, 3rd edition, McGraw-Hill, New York, NY, 1991.

29. H. Samueli, B. Daneshrad, R. B. Joshi, B. C. Wong, and H. T. Nicholas, III, "A 64-tap CMOS echo canceller/decision feedback equalizer for 2B1Q HDSL ," *IEEE Journal on Selected Areas in Communications*, vol. 9, pp. 839-847, Aug. 1991.

30. C. R. Johnson, Jr., *Lectures on Adaptive Parameter Estimation*, Prentice Hall, Englewood Cliffs, NJ, 1988.

31. D. T. Slock, "On the convergence behavior of the LMS and normalized LMS algorithms," *IEEE Trans. on Signal Processing*, vol. 40, pp. 2811-2825, Sept. 1993.

32. N. J. Bershad, "Analysis of the normalized LMS algorithm with Gaussian inputs," *IEEE Trans. on Acoust., Speech, and Signal Processing*, vol. ASSP-34, pp. 793-806, Aug. 1986.

33. M. Tarrab and A. Feuer, "Convergence and performance analysis of the normalized LMS algorithm with uncorrelated Gaussian data," *IEEE Trans. on Information Theory*, vol. IT-34, pp. 680-691, July 1988.

34. J. F. Doherty, "An adaptive algorithm for stable decision-feedback filtering," *IEEE Trans. on Circuits and Systems–II: Analog and Digital Signal Processing*, vol. 40, pp. 1-8, Jan. 1993.

35. W. B. Mikhael, F. H. Fu, L. G. Kazovsky, G. S. Kang, and L. J. Fransen, "Adaptive filter with individual adaptation of parameters," *IEEE Trans. on Circuits and Systems*, vol. 33, pp. 677-686, July 1986.

36. R. W. Harris, D. M. Chabries, and F. A. Bishop, "A variable step (VS) adaptive filter algorithm," *IEEE Trans. on Acoust., Speech, and Signal Processing*, vol. ASSP-34, pp. 309-316, April 1986.

37. C. S. Modlin and J. M. Cioffi, "A fast decision feedback LMS algorithm using multiple step sizes," *Proc. IEEE Inter. Conf. on Communications*, New Orleans, pp. 1201-1205, May 1994.

38. S. D. Peters and A. Antoniou, "Environment estimation for enhanced NLMS adaptation," *Proc. IEEE Pac. Rim Conf. on Comm., Comp. and Sig. Proc.*, Victoria, Canada, pp. 342-345, May 1993.

39. P. S. R. Diniz, M. L. R. de Campos, and A. Antoniou, "Analysis of LMS-Newton adaptive filtering algorithms with variable convergence factor," *IEEE Trans. on Signal Processing*, vol. 43, pp. 617-627, March 1995.

40. D. F. Marshall and W. K. Jenkins, "A fast quasi-Newton adaptive filtering algorithm," *IEEE Trans. on Signal Processing*, vol. 40, pp. 1652-1662, July 1993.

41. G. V. Moustakides and S. Theodoridis, "Fast Newton transversal filters - A new class of adaptive estimation algorithm," *IEEE Trans. on Signal Processing*, vol. 39, pp. 2184-2193, Oct. 1991.

42. J. J. Shynk, "Frequency-domain and multirate adaptive filtering," *IEEE Signal Processing Magazine*, vol. 9, pp. 15-37, Jan. 1992.

43. M. R. Petraglia and S. K. Mitra, "Adaptive FIR filter structure based on the generalized subband decomposition of FIR filters," *IEEE Transactions on Circuits and Systems II: Analog and Digital Signal Processing*, vol. 40, pp. 354-362, June 1993.

44. A. H. Sayed and M. Rupp, "Error-energy bounds for adaptive gradient algorithms," *IEEE Trans. on Signal Processing*, vol. 44, pp. 1982-1989, August 1996.

45. N. R. Yousef and A. H. Sayed, "A unified approach to the steady-state and tracking analyses of adaptive filters," *IEEE Trans. on Signal Processing*, vol. 49, pp. 314-324, Feb. 2001.

46. T. Y. Al-Naffouri and A. H. Sayed, "Transient analysis of adaptive filters with error nonlinearities," *IEEE Trans. on Signal Processing*, vol. 51, pp. 653-663, March 2003.

47. H.-C. Shin and A. H. Sayed, "Mean-square performance of a family of affine projection algorithms," *IEEE Trans. on Signal Processing*, vol. 52, pp. 90-102, Jan. 2004.

48. A. H. Sayed, *Fundamentals of Adaptive Filtering*, John Wiley & Sons, Hoboken, NJ, 2003.

49. M. L. R. de Campos and A. Antoniou, "A new quasi-Newton adaptive filtering algorithm," *IEEE Trans. on Circuits and Systems–II: Analog and Digital Signal Processing*, vol. 44, pp. 924-934, Nov. 1997.

4.10 PROBLEMS

1. From equation (4.16) derive the difference equation for $\mathbf{v}'(k)$ given by equation (4.19).
2. Prove the validity of equation (4.27).
3. The sign-error algorithm is used to predict the signal $x(k) = \sin(\pi k/3)$ using a second-order FIR filter with the first tap fixed at 1, by minimizing the mean square value of $y(k)$. This is an alternative way to interpret how the predictor works. Calculate an appropriate μ, the output signal $y(k)$, and the filter coefficients for the first 10 iterations. Start with $\mathbf{w}^T(0) = [1\ 0\ 0]$.
4. Derive an LMS-Newton algorithm leading to zero *a posteriori* error.
5. Derive the updating equations of the affine projection algorithm, for $L = 1$.
6. Use the sign-error algorithm to identify a system with the transfer function given below. The input signal is a uniformly distributed white noise with variance $\sigma_x^2 = 1$, and the measurement noise is Gaussian white noise uncorrelated with the input with variance $\sigma_n^2 = 10^{-3}$. The adaptive filter has 12 coefficients.

$$H(z) = \frac{1 - z^{-12}}{1 + z^{-1}}$$

   (a) Calculate the upper bound for μ ($\mu_{\max}$) to guarantee the algorithm stability.
   (b) Run the algorithm for $\mu_{\max}/2$, $\mu_{\max}/5$, and $\mu_{\max}/10$. Comment on the convergence behavior in each case.
   (c) Measure the misadjustment in each example and compare with the results obtained by equation (4.28).
   (d) Plot the obtained FIR filter frequency response at any iteration after convergence is achieved and compare with the unknown system.
7. Repeat the previous problem using an adaptive filter with 8 coefficients and interpret the results.

8. Repeat problem 6 when the input signal is a uniformly distributed white noise with variance $\sigma_{n_x}^2 = 0.5$, filtered by an all-pole filter given by

$$H(z) = \frac{z}{z - 0.9}$$

9. In problem 6, consider that the additional noise has the following variances (a) $\sigma_n^2 = 0$, (b) $\sigma_n^2 = 1$. Comment on the results obtained in each case.

10. Perform the equalization of a channel with the following impulse response

$$h(k) = ku(k) - (2k - 9)u(k - 5) + (k - 9)u(k - 10)$$

using a known training signal consisting of a binary (-1,1) random signal. An additional Gaussian white noise with variance 10^{-2} is present at the channel output.
(a) Apply the sign-error with an appropriate μ and find the impulse response of an equalizer with 15 coefficients.
(b) Convolve the equalizer impulse response at an iteration after convergence, with the channel impulse response and comment on the result.

11. In a system identification problem, the input signal is generated by an autoregressive process given by

$$x(k) = -1.2x(k - 1) - 0.81x(k - 2) + n_x(k)$$

where $n_x(k)$ is zero-mean Gaussian white noise with variance such that $\sigma_x^2 = 1$. The unknown system is described by

$$H(z) = 1 + 0.9z^{-1} + 0.1z^{-2} + 0.2z^{-3}$$

The adaptive filter is also a third-order FIR filter. Using the sign-error algorithm:
(a) Choose an appropriate μ, run an ensemble of 20 experiments, and plot the average learning curve.
(b) Measure the excess MSE and compare the results with the theoretical value.

12. In the previous problem, calculate the time constant τ_{wi} and the expected number of iterations to achieve convergence.

13. The sign-error algorithm is applied to identify a 7th-order time-varying unknown system whose coefficients are first-order Markov processes with $\lambda_{\mathbf{w}} = 0.999$ and $\sigma_{\mathbf{w}}^2 = 0.001$. The initial time-varying system multiplier coefficients are

$$\mathbf{w}_o^T = [0.03490 \quad -0.011 \quad -0.06864 \;\; 0.22391 \;\; 0.55686 \;\; 0.35798 \quad -0.0239 \quad -0.07594]$$

The input signal is Gaussian white noise with variance $\sigma_x^2 = 0.7$, and the measurement noise is also Gaussian white noise independent of the input signal and of the elements of $\mathbf{n_w}(k)$, with variance $\sigma_n^2 = 0.01$.
For $\mu = 0.01$, simulate the experiment described and measure the excess MSE.

14. Reduce the value of $\lambda_{\mathbf{w}}$ to 0.95 in problem 13, simulate, and comment on the results.

15. Suppose a 15th-order FIR digital filter with multiplier coefficients given below, is identified through an adaptive FIR filter of the same order using the sign-error algorithm. Use fixed-point arithmetic and run simulations for the following case.
    Additional noise: white noise with variance $\sigma_n^2 = 0.0015$
    Coefficient wordlength: $b_c = 16$ bits
    Signal wordlength: $b_d = 16$ bits
    Input signal: Gaussian white noise with variance $\sigma_x^2 = 0.7$
    $\mu = 0.01$

    $\mathbf{w}_o^T$ = [0.0219360 0.0015786 – 0.0602449 – 0.0118907 0.1375379 0.0574545 – 0.3216703 – 0.5287203 – 0.2957797 0.0002043 0.290670 −0.0353349 – 0.068210 0.0026067 0.0010333 – 0.0143593]

    Plot the learning curves of the estimates of $E[||\Delta\mathbf{w}(k)_Q||^2]$ and $\xi(k)_Q$ obtained through 25 independent runs, for the finite- and infinite-precision implementations.

16. Repeat the above problem for the following cases
    (a) $\sigma_n^2 = 0.01$, $b_c = 12$ bits, $b_d = 12$ bits, $\sigma_x^2 = 0.7$, $\mu = 10^{-4}$.
    (b) $\sigma_n^2 = 0.1$, $b_c = 10$ bits, $b_d = 10$ bits, $\sigma_x^2 = 0.8$, $\mu = 2.0\ 10^{-5}$.
    (c) $\sigma_n^2 = 0.05$, $b_c = 14$ bits, $b_d = 16$ bits, $\sigma_x^2 = 0.8$, $\mu = 3.5\ 10^{-4}$.

17. Repeat problem 15 for the case where the input signal is a first-order Markov process with $\lambda_{\mathbf{x}} = 0.95$.

18. Repeat problem 6 for the dual-sign algorithm given $\gamma = 16$ and $\rho = 1$, and comment on the results.

19. Repeat problem 6 for the power-of-two error algorithm given $b_d = 6$ and $\tau = 2^{-b_d}$, and comment on the results.

20. Repeat problem 6 for the sign-data and sign-sign algorithms and compare the results.

21. Show the validity of the matrix inversion lemma defined in equation (4.51).

22. For the setup described in problem 8, choose an appropriate μ and run the LMS-Newton algorithm.
    (a) Measure the misadjustment.
    (b) Plot the frequency response of the FIR filter obtained after convergence is achieved and compare with the unknown system.

23. Repeat problem 8 using the normalized LMS algorithm.

24. Repeat problem 8 using the transform-domain LMS algorithm with DCT. Compare the results with those obtained with the standard LMS algorithm.

25. Repeat problem 8 using the affine projection algorithm.

26. Repeat problem 8 using the transform-domain LMS algorithm with DCT.

27. For the input signal described in problem 8, derive the autocorrelation matrix of order one (2 × 2). Apply the DCT and the normalization to $\mathbf{R}$ in order to generate $\hat{\mathbf{R}} = \Sigma^{-2}\mathbf{T}\mathbf{R}\mathbf{T}^T$. Compare the eigenvalue spreads of $\mathbf{R}$ and $\hat{\mathbf{R}}$.

28. Repeat the previous problem for **R** with dimension 3 by 3.

29. Use the complex affine projection algorithm with $L = 3$ to equalize a channel with the transfer function given below. The input signal is a four QAM signal representing a randomly generated bit stream with the signal to noise ratio $\frac{\sigma_{\tilde{x}}^2}{\sigma_n^2} = 20$ at the receiver end, that is, $\tilde{x}(k)$ is the received signal without taking into consideration the additional channel noise. The adaptive filter has 10 coefficients.

$$H(z) = (0.34 - 0.27\jmath) + (0.87 + 0.43\jmath)z^{-1} + (0.34 - 0.21\jmath)z^{-2}$$

(a) Run the algorithm for $\mu = 0.1$, $\mu = 0.4$, and $\mu = 0.8$. Comment on the convergence behavior in each case.
(b) Plot the real versus imaginary parts of the received signal before and after equalization.
(c) Increase the number of coefficients to 20 and repeat the experiment in (b).

30. Repeat the problem 29 for the case of the normalized LMS algorithm.

31. In a system identification problem the input signal is generated from a four QAM of the form

$$x(k) = x_{\text{re}}(k) + \jmath x_{\text{im}}(k)$$

where $x_{\text{re}}(k)$ and $x_{\text{im}}(k)$ assume values ± 1 randomly generated. The unknown system is described by

$$H(z) = 0.32 + 0.21\jmath + (-0.3 + 0.7\jmath)z^{-1} + (0.5 - 0.8\jmath)z^{-2} + (0.2 + 0.5\jmath)z^{-3}$$

The adaptive filter is also a third-order complex FIR filter, and the additional noise is composed of zero-mean Gaussian white noises in the real and imaginary parts with variance $\sigma_n^2 = 0.4$. Using the complex affine projection algorithm with $L = 1$, choose an appropriate μ, run an ensemble of 20 experiments, and plot the average learning curve.

32. Repeat the problem 31 utilizing the affine projection algorithm with $L = 4$.

33. Derive a complex transform-domain LMS algorithm for the case the transformation matrix is the DFT.

34. The Quasi-Newton algorithm first proposed in [49] is described by the following set of equations

$$\begin{aligned}
e(k) &= d(k) - \mathbf{w}^T(k)\mathbf{x}(k) \\
\mu(k) &= \frac{1}{2\mathbf{x}^T(k)\hat{\mathbf{R}}^{-1}(k)\mathbf{x}(k)} \\
\mathbf{w}(k+1) &= \mathbf{w}(k) + 2\,\mu(k)\,e(k)\,\hat{\mathbf{R}}^{-1}(k)\mathbf{x}(k) \\
\hat{\mathbf{R}}^{-1}(k+1) &= \hat{\mathbf{R}}^{-1}(k) - 2\mu(k)\,(1 - \mu(k))\,\hat{\mathbf{R}}^{-1}(k)\mathbf{x}(k)\mathbf{x}^T(k)\hat{\mathbf{R}}^{-1}(k)
\end{aligned} \tag{4.159}$$

(a) Apply this algorithm as well as the binormalized LMS algorithm to identify the system

$$H(z) = 1 + z^{-1} + z^{-2}$$

when the additional noise is a uniformly distributed white noise with variance $\sigma_n^2 = 0.01$, and the input signal is a Gaussian noise with unit variance filtered by an all-pole filter given by

$$G(z) = \frac{0.19z}{z - 0.9}$$

Through simulations, compare the convergence speed of the two algorithms when their misadjustments are approximately the same. The later condition can be met by choosing the μ in the binormalized LMS algorithm appropriately.

5

CONVENTIONAL RLS ADAPTIVE FILTER

5.1 INTRODUCTION

Least-squares algorithms aim at the minimization of the sum of the squares of the difference between the desired signal and the model filter output [1]-[2]. When new samples of the incoming signals are received at every iteration, the solution for the least-squares problem can be computed in recursive form resulting in the recursive least-squares (RLS) algorithms. The conventional version of these algorithms will be the topic of the present chapter.

The RLS algorithms are known to pursue fast convergence even when the eigenvalue spread of the input signal correlation matrix is large. These algorithms have excellent performance when working in time-varying environments. All these advantages come with the cost of an increased computational complexity and some stability problems, which are not as critical in LMS-based algorithms [3]-[4].

Several properties related to the RLS algorithms are discussed including misadjustment, tracking behavior, which are verified through a number of simulation results.

Appendix C deals with the quantization effects in the conventional RLS algorithm. Appendix D provides an introduction to Kalman filters whose special case can be related to the RLS algorithms.

5.2 THE RECURSIVE LEAST-SQUARES ALGORITHM

The objective here is to choose the coefficients of the adaptive filter such that the output signal $y(k)$, during the period of observation, will match the desired signal as closely as possible in the least-squares sense. The minimization process requires the information of the input signal available so far. Also, the objective function we seek to minimize is deterministic.

The generic FIR adaptive filter realized in the direct form is shown in Fig. 5.1. The input signal information vector at a given instant k is given by

$$\mathbf{x}(k) = [x(k)\, x(k-1) \ldots x(k-N)]^T \tag{5.1}$$

P.S.R. Diniz, *Adaptive Filtering*, DOI: 10.1007/978-0-387-68606-6_5,

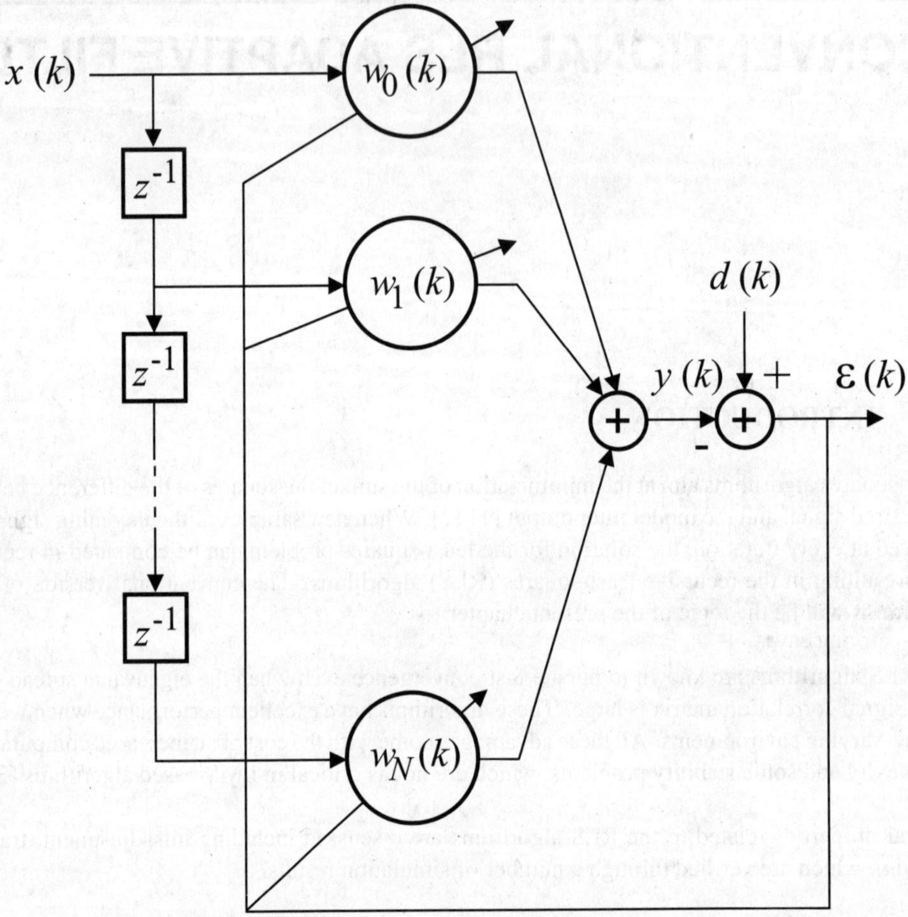

Figure 5.1 Adaptive FIR filter.

where N is the order of the filter. The coefficients $w_j(k)$, for $j = 0, 1, \ldots, N$, are adapted aiming at the minimization of a given objective function. In the case of least-squares algorithms, the objective function is deterministic and is given by

$$\begin{aligned}\xi^d(k) &= \sum_{i=0}^{k} \lambda^{k-i} \varepsilon^2(i) \\ &= \sum_{i=0}^{k} \lambda^{k-i} \left[d(i) - \mathbf{x}^T(i)\mathbf{w}(k)\right]^2\end{aligned} \tag{5.2}$$

where $\mathbf{w}(k) = [w_o(k)\, w_1(k) \ldots w_N(k)]^T$ is the adaptive-filter coefficient vector and $\varepsilon(i)$ is the *a posteriori* output error[1] at instant i. The parameter λ is an exponential weighting factor that should be chosen in the range $0 \ll \lambda \leq 1$. This parameter is also called forgetting factor since the information of the distant past has an increasingly negligible effect on the coefficient updating.

It should be noticed that in the development of the LMS and LMS-based algorithms we utilized the *a priori* error. In the RLS algorithms $\varepsilon(k)$ is used to denote the *a posteriori* error whereas $e(k)$ denotes the *a priori* error. The *a posteriori* error will be our first choice in the development of the RLS-based algorithms.

As can be noted, each error consists of the difference between the desired signal and the filter output, using the most recent coefficients $\mathbf{w}(k)$. By differentiating $\xi^d(k)$ with respect to $\mathbf{w}(k)$, it follows that

$$\frac{\partial \xi^d(k)}{\partial \mathbf{w}(k)} = -2\sum_{i=0}^{k} \lambda^{k-i}\mathbf{x}(i)[d(i) - \mathbf{x}^T(i)\mathbf{w}(k)] \tag{5.3}$$

By equating the result to zero, it is possible to find the optimal vector $\mathbf{w}(k)$ that minimizes the least-squares error, through the following relation:

$$-\sum_{i=0}^{k} \lambda^{k-i}\mathbf{x}(i)\mathbf{x}^T(i)\mathbf{w}(k) + \sum_{i=0}^{k} \lambda^{k-i}\mathbf{x}(i)d(i) = \begin{bmatrix} 0 \\ 0 \\ \vdots \\ 0 \end{bmatrix}$$

The resulting expression for the optimal coefficient vector $\mathbf{w}(k)$ is given by

$$\begin{aligned}\mathbf{w}(k) &= \left[\sum_{i=0}^{k} \lambda^{k-i}\mathbf{x}(i)\mathbf{x}^T(i)\right]^{-1} \sum_{i=0}^{k} \lambda^{k-i}\mathbf{x}(i)d(i) \\ &= \mathbf{R}_D^{-1}(k)\mathbf{p}_D(k)\end{aligned} \tag{5.4}$$

where $\mathbf{R}_D(k)$ and $\mathbf{p}_D(k)$ are called the deterministic correlation matrix of the input signal and the deterministic cross-correlation vector between the input and desired signals, respectively.

In equation (5.4) it was assumed that $\mathbf{R}_D(k)$ is nonsingular. However, if $\mathbf{R}_D(k)$ is singular a generalized inverse [1] should be used instead in order to obtain a solution for $\mathbf{w}(k)$ that minimizes $\xi^d(k)$. Since we are assuming that in most practical applications the input signal has persistence of excitation, the cases requiring generalized inverse are not discussed here. It should be mentioned that if the input signal is considered to be zero for $k < 0$ then $\mathbf{R}_D(k)$ will always be singular for $k < N$, i.e., during the initialization period. During this period, the optimal value of the coefficients can be calculated for example by the backsubstitution algorithm to be presented in subsection 9.2.1.

The straightforward computation of the inverse of $\mathbf{R}_D(k)$ results in an algorithm with computational complexity of $O[N^3]$. In the conventional RLS algorithm the computation of the inverse matrix is

[1] The *a posteriori* error is computed after the coefficient vector is updated, and taking into consideration the most recent input data vector $\mathbf{x}(k)$.

avoided through the use of the matrix inversion lemma [1], first presented in the previous chapter for the LMS-Newton algorithm. Using the matrix inversion lemma, see equation (4.51), the inverse of the deterministic correlation matrix can then be calculated in the following form

$$\mathbf{S}_D(k) = \mathbf{R}_D^{-1}(k) = \frac{1}{\lambda}\left[\mathbf{S}_D(k-1) - \frac{\mathbf{S}_D(k-1)\mathbf{x}(k)\mathbf{x}^T(k)\mathbf{S}_D(k-1)}{\lambda + \mathbf{x}^T(k)\mathbf{S}_D(k-1)\mathbf{x}(k)}\right] \tag{5.5}$$

The complete conventional RLS algorithm is described in Algorithm 5.1.

Algorithm 5.1 **Conventional RLS Algorithm**
Initialization $\mathbf{S}_D(-1) = \delta\mathbf{I}$ where δ can be the inverse of the input signal power estimate $\mathbf{p}_D(-1) = \mathbf{x}(-1) = [0\,0\ldots 0]^T$ Do for $k \geq 0$: $\mathbf{S}_D(k) = \frac{1}{\lambda}[\mathbf{S}_D(k-1) - \frac{\mathbf{S}_D(k-1)\mathbf{x}(k)\mathbf{x}^T(k)\mathbf{S}_D(k-1)}{\lambda+\mathbf{x}^T(k)\mathbf{S}_D(k-1)\mathbf{x}(k)}]$ $\mathbf{p}_D(k) = \lambda\mathbf{p}_D(k-1) + d(k)\mathbf{x}(k)$ $\mathbf{w}(k) = \mathbf{S}_D(k)\mathbf{p}_D(k)$ If necessary compute $y(k) = \mathbf{w}^T(k)\mathbf{x}(k)$ $\varepsilon(k) = d(k) - y(k)$

An alternative way to describe the conventional RLS algorithm can be obtained if equation (5.4) is rewritten in the following form

$$\left[\sum_{i=0}^{k}\lambda^{k-i}\mathbf{x}(i)\mathbf{x}^T(i)\right]\mathbf{w}(k) = \lambda\left[\sum_{i=0}^{k-1}\lambda^{k-i-1}\mathbf{x}(i)d(i)\right] + \mathbf{x}(k)d(k) \tag{5.6}$$

By considering that $\mathbf{R}_D(k-1)\mathbf{w}(k-1) = \mathbf{p}_D(k-1)$, it follows that

$$\begin{aligned}\left[\sum_{i=0}^{k}\lambda^{k-i}\mathbf{x}(i)\mathbf{x}^T(i)\right]\mathbf{w}(k) &= \lambda\mathbf{p}_D(k-1) + \mathbf{x}(k)d(k)\\ &= \lambda\mathbf{R}_D(k-1)\mathbf{w}(k-1) + \mathbf{x}(k)d(k)\\ &= \left[\sum_{i=0}^{k}\lambda^{k-i}\mathbf{x}(i)\mathbf{x}^T(i) - \mathbf{x}(k)\mathbf{x}^T(k)\right]\mathbf{w}(k-1) + \mathbf{x}(k)d(k)\end{aligned} \tag{5.7}$$

where in the last equality the matrix $\mathbf{x}(k)\mathbf{x}^T(k)$ was added and subtracted inside square bracket on the right side of equation (5.7). Now, define the *a priori* error as

$$e(k) = d(k) - \mathbf{x}^T(k)\mathbf{w}(k-1) \tag{5.8}$$

By expressing $d(k)$ as a function of the *a priori* error and replacing the result in equation (5.7), after few manipulations, it can be shown that

$$\mathbf{w}(k) = \mathbf{w}(k-1) + e(k)\mathbf{S}_D(k)\mathbf{x}(k) \tag{5.9}$$

With equation (5.9), it is straightforward to generate an alternative conventional RLS algorithm as shown in Algorithm 5.2.

Algorithm 5.2

Alternative RLS Algorithm

Initialization

$\mathbf{S}_D(-1) = \delta\mathbf{I}$

where δ can be the inverse of an estimate of the input signal power

$\mathbf{x}(-1) = \mathbf{w}(-1) = [0\,0\ldots0]^T$

Do for $k \geq 0$

$e(k) = d(k) - \mathbf{x}^T(k)\mathbf{w}(k-1)$

$\boldsymbol{\psi}(k) = \mathbf{S}_D(k-1)\mathbf{x}(k)$

$\mathbf{S}_D(k) = \frac{1}{\lambda}[\mathbf{S}_D(k-1) - \frac{\boldsymbol{\psi}(k)\boldsymbol{\psi}^T(k)}{\lambda+\boldsymbol{\psi}^T(k)\mathbf{x}(k)}]$

$\mathbf{w}(k) = \mathbf{w}(k-1) + e(k)\mathbf{S}_D(k)\mathbf{x}(k)$

If necessary compute

$y(k) = \mathbf{w}^T(k)\mathbf{x}(k)$

$\varepsilon(k) = d(k) - y(k)$

In Algorithm 5.2, $\boldsymbol{\psi}(k)$ is an auxiliary vector required to reduce the computational burden defined by

$$\boldsymbol{\psi}(k) = \mathbf{S}_D(k-1)\mathbf{x}(k) \tag{5.10}$$

Further reduction in the number of divisions is possible if an additional auxiliary vector is used, defined as

$$\boldsymbol{\phi}(k) = \frac{\boldsymbol{\psi}(k)}{\lambda + \boldsymbol{\psi}^T(k)\mathbf{x}(k)} \tag{5.11}$$

This vector can be used to update $\mathbf{S}_D(k)$ as follows:

$$\mathbf{S}_D(k) = \frac{1}{\lambda}\left[\mathbf{S}_D(k-1) - \boldsymbol{\psi}(k)\boldsymbol{\phi}^T(k)\right] \tag{5.12}$$

As will be discussed, the above relation can lead to stability problems in the RLS algorithm.

5.3 PROPERTIES OF THE LEAST-SQUARES SOLUTION

In this section, some properties related to the least-squares solution are discussed in order to give some insight to the algorithm behavior in several situations to be discussed later on.

5.3.1 Orthogonality Principle

Define the matrices $\mathbf{X}(k)$ and $\mathbf{d}(k)$ that contain all the information about the input signal vector $\mathbf{x}(k)$ and the desired signal vector $d(k)$ as follows:

$$\mathbf{X}(k) = \begin{bmatrix} x(k) & \lambda^{1/2}x(k-1) & \cdots & \lambda^{(k-1)/2}x(1) & \lambda^{k/2}x(0) \\ x(k-1) & \lambda^{1/2}x(k-2) & \cdots & \lambda^{(k-1)/2}x(0) & 0 \\ \vdots & \vdots & & \vdots & \vdots \\ x(k-N) & \lambda^{1/2}x(k-N-1) & \cdots & 0 & 0 \end{bmatrix}$$
$$= [\mathbf{x}(k)\ \lambda^{1/2}\mathbf{x}(k-1) \ldots \lambda^{k/2}\mathbf{x}(0)] \tag{5.13}$$
$$\mathbf{d}(k) = [d(k)\ \lambda^{1/2}d(k-1) \ldots \lambda^{k/2}d(0)]^T \tag{5.14}$$

where $\mathbf{X}(k)$ is $(N+1) \times (k+1)$ and $\mathbf{d}(k)$ is $(k+1) \times 1$.

By using the matrices above defined it is possible to replace the least-squares solution of equation (5.4) by the following relation

$$\mathbf{X}(k)\mathbf{X}^T(k)\mathbf{w}(k) = \mathbf{X}(k)\mathbf{d}(k) \tag{5.15}$$

The product $\mathbf{X}^T(k)\mathbf{w}(k)$ forms a vector including all the adaptive-filter outputs when the coefficients are given by $\mathbf{w}(k)$. This vector corresponds to an estimate of $\mathbf{d}(k)$. Hence, defining

$$\mathbf{y}(k) = \mathbf{X}^T(k)\mathbf{w}(k) = [y(k)\ \lambda^{1/2}y(k-1) \ldots \lambda^{k/2}y(0)]^T \tag{5.16}$$

it follows from equation (5.15) that

$$\mathbf{X}(k)\mathbf{X}^T(k)\mathbf{w}(k) - \mathbf{X}(k)\mathbf{d}(k) = \mathbf{X}(k)[\mathbf{y}(k) - \mathbf{d}(k)] = \mathbf{0} \tag{5.17}$$

This relation means that the weighted-error vector given by

$$\boldsymbol{\varepsilon}(k) = \begin{bmatrix} \varepsilon(k) \\ \lambda^{1/2}\varepsilon(k-1) \\ \vdots \\ \lambda^{k/2}\varepsilon(0) \end{bmatrix} = \mathbf{d}(k) - \mathbf{y}(k) \tag{5.18}$$

is in the null space of $\mathbf{X}(k)$, i.e., the weighted-error vector is orthogonal to all row vectors of $\mathbf{X}(k)$. This justifies the fact that (5.15) is often called normal equation. A geometrical interpretation can easily be given for a least-squares problem solution with a single coefficient filter.

Example 5.1

Suppose that $\lambda = 1$ and that the following signals are involved in the least-squares problem

$$\mathbf{d}(1) = \begin{bmatrix} 0.5 \\ 1.5 \end{bmatrix} \quad \mathbf{X}(1) = [1 \ \ -2]$$

The optimal coefficient is given by

$$\begin{aligned} \mathbf{X}(1)\mathbf{X}^T(1)\mathbf{w}(1) &= [1 \ \ -2]\begin{bmatrix} 1 \\ -2 \end{bmatrix} \mathbf{w}(1) \\ &= \mathbf{X}(1)\mathbf{d}(1) \\ &= [1 \ \ -2]\begin{bmatrix} 0.5 \\ 1.5 \end{bmatrix} \end{aligned}$$

After performing the calculations the result is

$$\mathbf{w}(1) = -\frac{1}{2}$$

The output of the adaptive filter with coefficient given by $\mathbf{w}(1)$ is

$$\mathbf{y}(1) = \begin{bmatrix} -\frac{1}{2} \\ 1 \end{bmatrix}$$

Note that

$$\begin{aligned} \mathbf{X}(1)[\mathbf{y}(1) - \mathbf{d}(1)] &= [1 \ \ -2]\begin{bmatrix} -1 \\ -0.5 \end{bmatrix} \\ &= 0 \end{aligned}$$

Fig. 5.2 illustrates the fact that $\mathbf{y}(1)$ is the projection of $\mathbf{d}(1)$ in the $\mathbf{X}(1)$ direction. In the general case we can say that the vector $\mathbf{y}(k)$ is the projection of $\mathbf{d}(k)$ onto the subspace spanned by the rows of $\mathbf{X}(k)$.

□

5.3.2 Relation Between Least-Squares and Wiener Solutions

When $\lambda = 1$ the matrix $\frac{1}{k+1}\mathbf{R}_D(k)$ for large k is a consistent estimate of the input signal autocorrelation matrix $\mathbf{R}$, if the process from which the input signal was taken is ergodic. The same observation is valid for the vector $\frac{1}{k+1}\mathbf{p}_D(k)$ related to $\mathbf{p}$ if the desired signal is also ergodic. In this case,

$$\mathbf{R} = \lim_{k\to\infty} \frac{1}{k+1}\sum_{i=0}^{k} \mathbf{x}(i)\mathbf{x}^T(i) = \lim_{k\to\infty} \frac{1}{k+1}\mathbf{R}_D(k) \tag{5.19}$$

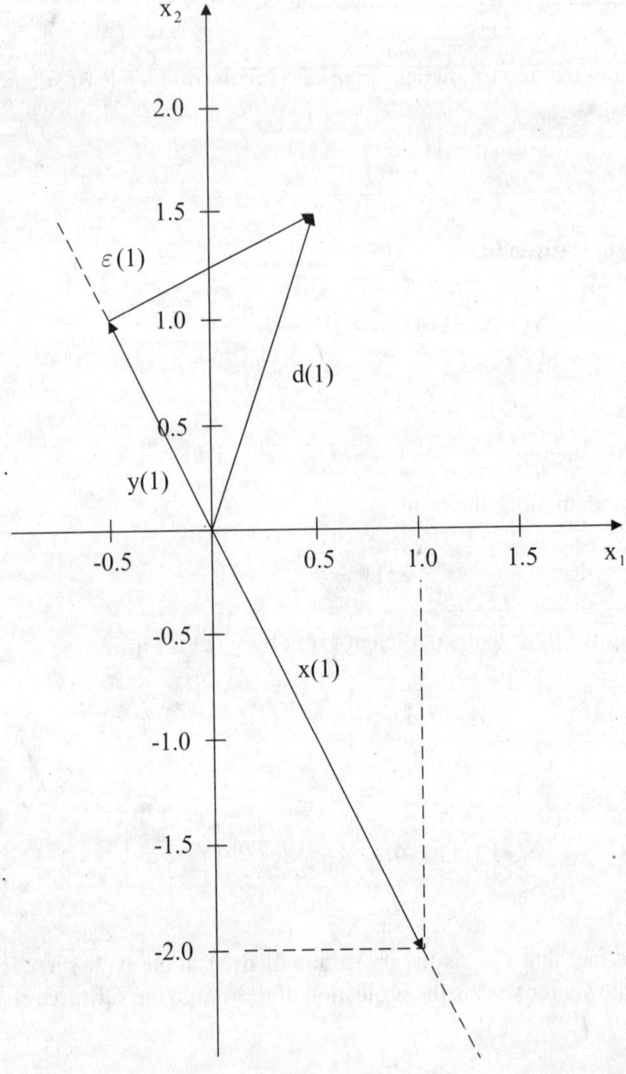

Figure 5.2 Geometric interpretation of least-squares solution.

and

$$\mathbf{p} = \lim_{k\to\infty} \frac{1}{k+1} \sum_{i=0}^{k} \mathbf{x}(i)d(i) = \lim_{k\to\infty} \frac{1}{k+1} \mathbf{p}_D(k) \tag{5.20}$$

It can then be shown that

$$\mathbf{w}(k) = \mathbf{R}_D^{-1}(k)\mathbf{p}_D(k) = \mathbf{R}^{-1}\mathbf{p} = \mathbf{w}_o \tag{5.21}$$

when k tends to infinity. This result indicates that the least-squares solution tends to the Wiener solution if the signals involved are ergodic and stationary. The stationarity requirement is due to the fact that the estimate of $\mathbf{R}$ given by equation (5.19) is not sensitive to any changes in $\mathbf{R}$ for large values of k. If the input signal is nonstationary $\mathbf{R}_D(k)$ is a biased estimate for $\mathbf{R}$. Note that in this case $\mathbf{R}$ is time varying.

5.3.3 Influence of the Deterministic Autocorrelation Initialization

The initialization of $\mathbf{S}_D(-1) = \delta\mathbf{I}$ causes a bias in the coefficients estimated by the adaptive filter. Suppose that the initial value given to $\mathbf{R}_D(k)$ is taken into account in the actual RLS solution as follows:

$$\begin{aligned}\sum_{i=-1}^{k} \lambda^{k-i}\mathbf{x}(i)\mathbf{x}^T(i)\mathbf{w}(k) &= [\sum_{i=0}^{k} \lambda^{k-i}\mathbf{x}(i)\mathbf{x}^T(i) + \frac{\lambda^{k+1}}{\delta}\mathbf{I}]\mathbf{w}(k) \\ &= \mathbf{p}_D(k) \end{aligned} \tag{5.22}$$

By recognizing that the deterministic autocorrelation matrix leading to an unbiased solution does not include the initialization matrix, we now examine the influence of this matrix. By multiplying $\mathbf{S}_D(k) = \mathbf{R}_D^{-1}(k)$ on both sides of equation (5.22), and by considering $k \to \infty$, it can be concluded that

$$\mathbf{w}(k) + \frac{\lambda^{k+1}}{\delta}\mathbf{S}_D(k)\mathbf{w}(k) = \mathbf{w}_o \tag{5.23}$$

where $\mathbf{w}_o$ is the optimal solution for the RLS algorithm.

The bias caused by the initialization of $\mathbf{S}_D(k)$ is approximately

$$\mathbf{w}(k) - \mathbf{w}_o \approx -\frac{\lambda^{k+1}}{\delta}\mathbf{S}_D(k)\mathbf{w}_o \tag{5.24}$$

For $\lambda < 1$, it is straightforward to conclude that the bias tends to zero as k tends to infinity. On the other hand, when $\lambda = 1$ the elements of $\mathbf{S}_D(k)$ get smaller when the number of iterations increase, as a consequence this matrix approaches a null matrix for large k.

The RLS algorithm would reach the optimum solution for the coefficients after $N+1$ iterations if no measurement noise is present, and the influence of the initialization matrix $\mathbf{S}_D(-1)$ is negligible at this point. This result follows from the fact that after $N+1$ iterations, the input signal vector has enough information to allow the adaptive algorithm to identify the coefficients of the unknown system. In other words, enough information means the tap delay line is filled with information of the input signal.

5.3.4 Steady-State Behavior of the Coefficient Vector

In order to understand better the steady-state behavior of the adaptive-filter coefficients, suppose that an FIR filter with coefficients given by $\mathbf{w}_o$ is being identified by an adaptive FIR filter of the same order employing an LS algorithm. Also assume that a measurement noise signal $n(k)$ is added to the desired signal before the error signal is calculated as follows:

$$d(k) = \mathbf{w}_o^T\mathbf{x}(k) + n(k) \tag{5.25}$$

where the additional noise is considered to be a white noise with zero mean and variance given by σ_n^2.

Given the adaptive-filter input vectors $\mathbf{x}(k)$, for $k = 0, 1, \ldots$, we are interested in calculating the average values of the adaptive-filter coefficients $w_i(k)$, for $i = 0, 1, \ldots, N$. The desired result is the following equality valid for $k \geq N$.

$$\begin{aligned} E[\mathbf{w}(k)] &= E\left\{\left[\mathbf{X}(k)\mathbf{X}^T(k)\right]^{-1}\mathbf{X}(k)\mathbf{d}(k)\right\} \\ &= E\left\{\left[\mathbf{X}(k)\mathbf{X}^T(k)\right]^{-1}\mathbf{X}(k)[\mathbf{X}^T(k)\mathbf{w}_o + \mathbf{n}(k)]\right\} \\ &= E\left\{\left[\mathbf{X}(k)\mathbf{X}^T(k)\right]^{-1}\mathbf{X}(k)\mathbf{X}^T(k)\mathbf{w}_o\right\} = \mathbf{w}_o \end{aligned} \tag{5.26}$$

where $\mathbf{n}(k) = [n(k)\ \lambda^{1/2}n(k-1)\ \lambda n(k-2)\ \ldots\ \lambda^{k/2}n(0)]^T$ is the noise vector, whose elements were considered orthogonal to the input signal. The above equation shows that the estimate given by the LS algorithm is an unbiased estimate when $\lambda \leq 1$.

A more accurate analysis reveals the behavior of the adaptive-filter coefficients during the transient period. The error in the filter coefficients can be described by the following $(N+1) \times 1$ vector

$$\Delta\mathbf{w}(k) = \mathbf{w}(k) - \mathbf{w}_o \tag{5.27}$$

It follows from equation (5.7) that

$$\mathbf{R}_D(k)\mathbf{w}(k) = \lambda\mathbf{R}_D(k-1)\mathbf{w}(k-1) + \mathbf{x}(k)d(k) \tag{5.28}$$

Defining the minimum output error as

$$e_o(k) = d(k) - \mathbf{x}^T(k)\mathbf{w}_o \tag{5.29}$$

and replacing $d(k)$ in equation (5.28), it can be deduced that

$$\mathbf{R}_D(k)\Delta\mathbf{w}(k) = \lambda\mathbf{R}_D(k-1)\Delta\mathbf{w}(k-1) + \mathbf{x}(k)e_o(k) \tag{5.30}$$

where the following relation was used

$$\mathbf{R}_D(k) = \lambda\mathbf{R}_D(k-1) + \mathbf{x}(k)\mathbf{x}^T(k) \tag{5.31}$$

The solution of equation (5.30) is given by

$$\Delta\mathbf{w}(k) = \lambda^{k+1}\mathbf{S}_D(k)\mathbf{R}_D(-1)\Delta\mathbf{w}(-1) + \mathbf{S}_D(k)\sum_{i=0}^{k}\lambda^{k-i}\mathbf{x}(i)e_o(i) \tag{5.32}$$

By replacing $\mathbf{R}_D(-1)$ by $\frac{1}{\delta}\,\mathbf{I}$ and taking the expected value of the resulting equation, it follows that

$$E[\Delta\mathbf{w}(k)] = \frac{\lambda^{k+1}}{\delta}E[\mathbf{S}_D(k)]\Delta\mathbf{w}(-1) + E[\mathbf{S}_D(k)\sum_{i=0}^{k}\lambda^{k-i}\mathbf{x}(i)e_o(i)] \tag{5.33}$$

Since $\mathbf{S}_D(k)$ is dependent on all past input signal vectors, becoming relatively invariant when the number of iterations increase, the contribution of any individual $\mathbf{x}(i)$ can be considered negligible. Also, due to the orthogonality principle, $e_o(i)$ can also be considered uncorrelated to all elements of $\mathbf{x}(i)$. This means that the last vector in equation (5.33) cannot have large element values. On the other hand, the first vector in equation (5.33) can have large element values only during the initial convergence, since as $k \to \infty$, $\lambda^{k+1} \to 0$ and $\mathbf{S}_D(k)$ is expected to have a nonincreasing behavior, i.e., $\mathbf{R}_D(k)$ is assumed to remain positive definite as $k \to \infty$ and the input signal power does not become too small. The above discussion leads to the conclusion that the adaptive-filter coefficients tend to the optimal values in $\mathbf{w}_o$ almost independently from the eigenvalue spread of the input signal correlation matrix.

If we consider the spectral decomposition of the matrix $E[\mathbf{S}_D(k)]$ (see equation (2.65)), the dependency on the eigenvalues of $\mathbf{R}$ can be easily accounted for in the simple case of $\lambda = 1$. Applying the expected value operator to the relation of equation (5.19), we can infer that

$$E[\mathbf{S}_D(k)] \approx \frac{\mathbf{R}^{-1}}{(k+1)} \tag{5.34}$$

for large k. Now consider the slowest decaying mode of the spectral decomposition of $E[\mathbf{S}_D(k)]$ given by

$$\mathbf{S}_{D_{\max}} = \frac{\mathbf{q}_{\min}\mathbf{q}_{\min}^T}{(k+1)\lambda_{\min}} \tag{5.35}$$

where $\lambda_{\min}$ is the smallest eigenvalue of $\mathbf{R}$ and $\mathbf{q}_{\min}$ is the corresponding eigenvector. Applying this result to equation (5.33), with $\lambda = 1$, we can conclude that the value of the minimum eigenvalue affects the convergence of the filter coefficients only in the first few iterations, because the term $k+1$ in the denominator reduces the values of the elements of $\mathbf{S}_{D_{\max}}$.

Further interesting properties of the coefficients generated by the LS algorithm are:

- The estimated coefficients are the best linear unbiased solution to the identification problem [1], in the sense that no other unbiased solution generated by alternative approaches has lower variance.
- If the additive noise is normally distributed the LS solution reaches the Cramer-Rao lower bound, resulting in a minimum-variance unbiased solution [1]. The Cramer-Rao lower bound establishes a lower bound to the coefficient-error-vector covariance matrix for any unbiased estimator of the optimal parameter vector $\mathbf{w}_o$.

5.3.5 Coefficient-Error-Vector Covariance Matrix

So far, we have shown that the estimation parameters in the vector $\mathbf{w}(k)$ converge on average to their optimal value of the vector $\mathbf{w}_o$. However, it is essential to analyze the coefficient-error-vector covariance matrix in order to determine how good is the obtained solution, in the sense that we are measuring how far the parameters wander around the optimal solution.

Using the same convergence assumption of the last section, it will be shown here that for $\lambda = 1$ the coefficient-error-vector covariance matrix is given by

$$\text{cov}[\Delta\mathbf{w}(k)] = E\left[(\mathbf{w}(k) - \mathbf{w}_o)(\mathbf{w}(k) - \mathbf{w}_o)^T\right] = \sigma_n^2 E[\mathbf{S}_D(k)] \tag{5.36}$$

Proof:

First note that by using equations (5.4) and (5.15), the following relations are verified

$$\mathbf{w}(k) - \mathbf{w}_o = \mathbf{S}_D(k)\mathbf{p}_D(k) - \mathbf{S}_D(k)\mathbf{S}_D^{-1}(k)\mathbf{w}_o \tag{5.37}$$

$$= \left[\mathbf{X}(k)\mathbf{X}^T(k)\right]^{-1}\mathbf{X}(k)\left[\mathbf{d}(k) - \mathbf{X}^T(k)\mathbf{w}_o\right] \tag{5.38}$$

$$= \left[\mathbf{X}(k)\mathbf{X}^T(k)\right]^{-1}\mathbf{X}(k)\mathbf{n}(k) \tag{5.39}$$

where $\mathbf{n}(k) = [n(k)\ \lambda^{1/2}n(k-1)\ \lambda n(k-2)\ \ldots\ \lambda^{k/2}n(0)]^T$.

Applying the last equation to the covariance of the coefficient-error-vector it follows that

$$\begin{aligned}\text{cov}[\Delta\mathbf{w}(k)] &= E\left\{\left[\mathbf{X}(k)\mathbf{X}^T(k)\right]^{-1}\mathbf{X}(k)E[\mathbf{n}(k)\mathbf{n}^T(k)]\mathbf{X}^T(k)\left[\mathbf{X}(k)\mathbf{X}^T(k)\right]^{-1}\right\} \\ &= E\left\{\sigma_n^2\mathbf{S}_D(k)\mathbf{X}(k)\Lambda\mathbf{X}^T(k)\mathbf{S}_D(k)\right\}\end{aligned}$$

where

$$\Lambda = \begin{bmatrix} 1 & & & & \\ & \lambda & & \mathbf{0} & \\ & & \lambda^2 & & \\ & \mathbf{0} & & \ddots & \\ & & & & \lambda^k \end{bmatrix}$$

For $\lambda = 1$, $\Lambda = \mathbf{I}$, it follows that

$$\begin{aligned}\text{cov}[\Delta\mathbf{w}(k)] &= E\left[\sigma_n^2\mathbf{S}_D(k)\mathbf{X}(k)\mathbf{X}^T(k)\mathbf{S}_D(k)\right] \\ &= E\left[\sigma_n^2\mathbf{S}_D(k)\mathbf{R}_D(k)\mathbf{S}_D(k)\right] \\ &= \sigma_n^2 E\left[\mathbf{S}_D(k)\right]\end{aligned}$$

□

Therefore, when $\lambda = 1$, the coefficient-error-vector covariance matrix tends to decrease its norm as time progresses since $\mathbf{S}_D(k)$ is also norm decreasing. The variance of the additional noise $n(k)$ influences directly the norm of the covariance matrix.

5.3.6 Behavior of the Error Signal

It is important to understand how the error signal behaves in the RLS algorithm. When a measurement noise is present in the adaptive-filtering process, the *a priori* error signal is given by

$$e(k) = d'(k) - \mathbf{w}^T(k-1)\mathbf{x}(k) + n(k) \tag{5.40}$$

where $d'(k) = \mathbf{w}_o^T\mathbf{x}(k)$ is the desired signal without measurement noise.

Again if the input signal is considered known (conditional expectation), then

$$\begin{aligned} E[e(k)] &= E[d'(k)] - E[\mathbf{w}^T(k-1)]\mathbf{x}(k) + E[n(k)] \\ &= E[\mathbf{w}_o^T - \mathbf{w}_o^T]\mathbf{x}(k) + E[n(k)] \\ &= E[n(k)] \end{aligned} \tag{5.41}$$

assuming that the adaptive-filter order is sufficient to model perfectly the desired signal.

From equation (5.41), it can be concluded that if the noise signal has zero mean then

$$E[e(k)] = 0$$

It is also important to assess the minimum mean value of the squared error that is reachable using an RLS algorithm. The minimum mean-square error (MSE) in the presence of external uncorrelated noise is given by

$$\xi_{\min} = E[e^2(k)] = E[e_o^2(k)] = E[n^2(k)] = \sigma_n^2 \tag{5.42}$$

where it is assumed that the adaptive-filter multiplier coefficients were frozen at their optimum values and that the number of coefficients of the adaptive filter is sufficient to model the desired signal. In the conditions described the *a priori* error corresponds to the minimum output error as defined in equation (5.29). It should be noted, however, that if the additive noise is correlated with the input and the desired signals, a more complicated expression for the MSE results, accounting for the referred correlation.

When employing the *a posteriori* error the value of minimum MSE, denoted by $\xi_{\min,p}$, differs from the corresponding value related to the *a priori* error. First note that by using equation (5.39), the following relation is verified

$$\Delta\mathbf{w}(k) = \mathbf{S}_D(k)\mathbf{X}(k)\mathbf{n}(k) \tag{5.43}$$

When a measurement noise is present in the adaptive-filtering process, the *a posteriori* error signal is given by

$$\varepsilon(k) = d'(k) - \mathbf{w}^T(k)\mathbf{x}(k) + n(k) = -\Delta\mathbf{w}^T(k)\mathbf{x}(k) + e_o(k) \tag{5.44}$$

The expression for the MSE related to the *a posteriori* error is then given by

$$\begin{aligned} \xi(k) &= E[\varepsilon^2(k)] \\ &= E[e_o^2(k)] - 2E[\mathbf{x}^T(k)\Delta\mathbf{w}(k)e_o(k)] + E[\Delta\mathbf{w}^T(k)\mathbf{x}(k)\mathbf{x}^T(k)\Delta\mathbf{w}(k)] \end{aligned} \tag{5.45}$$

By replacing the expression (5.43) in equation (5.45) above, the following relations follow

$$\begin{aligned}
\xi(k) =& E[e_o^2(k)] - 2E[\mathbf{x}^T(k)\mathbf{S}_D(k)\mathbf{X}(k)\mathbf{n}(k)e_o(k)] + E[\Delta\mathbf{w}^T(k)\mathbf{x}(k)\mathbf{x}^T(k)\Delta\mathbf{w}(k)] \\
=& \sigma_n^2 - 2E[\mathbf{x}^T(k)\mathbf{S}_D(k)\mathbf{X}(k)] \begin{bmatrix} \sigma_n^2 \\ 0 \\ \vdots \\ 0 \end{bmatrix} + E[\Delta\mathbf{w}^T(k)\mathbf{x}(k)\mathbf{x}^T(k)\Delta\mathbf{w}(k)] \\
=& \sigma_n^2 - 2E[\mathbf{x}^T(k)\mathbf{S}_D(k)\mathbf{x}(k)]\sigma_n^2 + E[\Delta\mathbf{w}^T(k)\mathbf{x}(k)\mathbf{x}^T(k)\Delta\mathbf{w}(k)] \\
=& \xi_{\text{min,p}} + E[\Delta\mathbf{w}^T(k)\mathbf{x}(k)\mathbf{x}^T(k)\Delta\mathbf{w}(k)]
\end{aligned} \tag{5.46}$$

where in the second equality it was considered that the additional noise is uncorrelated with the input signal and that $e_o(k) = n(k)$. This equality occurs when the adaptive filter has sufficient order to identify the unknown system.

Note that $\xi_{\text{min,p}}$ related to the *a posteriori* error in equation (5.46) is not the same as minimum MSE of the *a priori* error, denoted in this book by ξ_{min}. The last term, that is $E[\Delta\mathbf{w}^T(k)\mathbf{x}(k)\mathbf{x}^T(k)\Delta\mathbf{w}(k)]$, in equation (5.46) determines the excess MSE of the RLS algorithm.

It is possible to verify that the following expressions for $\xi_{\text{min,p}}$ are accurate approximations

$$\begin{aligned}
\xi_{\text{min,p}} &= \{1 - 2E[\mathbf{x}^T(k)\mathbf{S}_D(k)\mathbf{x}(k)]\}\sigma_n^2 \\
&= \left\{1 - 2\text{tr}\left[E\left(\mathbf{S}_D(k)\mathbf{x}(k)\mathbf{x}^T(k)\right)\right]\right\}\sigma_n^2 \\
&= \left\{1 - 2\text{tr}\left[\frac{1-\lambda}{1-\lambda^{k+1}}\mathbf{I}\right]\right\}\sigma_n^2 \\
&= \left\{1 - 2(N+1)\left[\frac{1-\lambda}{1-\lambda^{k+1}}\right]\right\}\sigma_n^2 \\
&= \left\{1 - 2(N+1)\left[\frac{1}{1+\lambda+\lambda^2+\ldots+\lambda^k}\right]\right\}\sigma_n^2
\end{aligned} \tag{5.47}$$

In the above expression, it is considered that $\mathbf{S}_D(k)$ is slowly varying as compared to $\mathbf{x}(k)$ when $\lambda \rightarrow 1$, such that

$$E[\mathbf{S}_D(k)\mathbf{x}(k)\mathbf{x}^T(k)] \approx E\left[\mathbf{S}_D(k)\right]E\left[\mathbf{x}(k)\mathbf{x}^T(k)\right]$$

and that by using equation (5.55)

$$E\left[\mathbf{S}_D(k)\mathbf{x}(k)\mathbf{x}^T(k)\right] \approx \frac{1-\lambda}{1-\lambda^{k+1}}\mathbf{I}$$

Equation (5.47) applies to the case where $\lambda < 1$, and as can be observed from the term multiplying $N+1$ there is a transient for small k which dies away when the number of iterations increases[2]. If we fit the decrease in the term multiplying $N+1$ at each iteration to an exponential envelop, the

[2]The expression for $\xi_{\text{min,p}}$ can be negative, however, $\xi(k)$ is always non negative.

time constant will be $\frac{1}{\lambda^{k+1}}$. Unlike the LMS algorithm, this time constant is time varying and is not related to the eigenvalue spread of the input signal correlation matrix.

Example 5.2

Repeat the equalization problem of example 3.1 using the RLS algorithm.

(a) Using $\lambda = 0.99$, run the algorithm and save matrix $\mathbf{S}_D(k)$ at iteration 500 and compare with the inverse of the input signal correlation matrix.

(b) Plot the convergence path for the RLS algorithm on the MSE surface.

Solution:

(a) The inverse of matrix **R**, as computed in the example 3.1, is given by

$$\mathbf{R}^{-1} = 0.45106 \begin{bmatrix} 1.6873 & 0.7937 \\ 0.7937 & 1.6873 \end{bmatrix} = \begin{bmatrix} 0.7611 & 0.3580 \\ 0.3580 & 0.7611 \end{bmatrix}$$

The initialization matrix $\mathbf{S}_D(-1)$ is a diagonal matrix with the diagonal elements equal to 0.1. The matrix $\mathbf{S}_D(k)$ at the 500th iteration, obtained by averaging the results of 30 experiments, is

$$\mathbf{S}_D(500) = \begin{bmatrix} 0.0078 & 0.0037 \\ 0.0037 & 0.0078 \end{bmatrix}$$

Also, the obtained values of the deterministic cross-correlation vector is

$$\mathbf{p}_D(500) = \begin{bmatrix} 95.05 \\ 46.21 \end{bmatrix}$$

Now, we divide each element of the matrix $\mathbf{R}^{-1}$ by

$$\frac{1-\lambda^{k+1}}{1-\lambda} = 99.34$$

since in a stationary environment $E[\mathbf{S}_D(k)] = \frac{1-\lambda}{1-\lambda^{k+1}}\mathbf{R}^{-1}$, see equation (5.55) for a formal proof.

The resulting matrix is

$$\frac{1}{99.34}\mathbf{R}^{-1} = \begin{bmatrix} 0.0077 & 0.0036 \\ 0.0036 & 0.0077 \end{bmatrix}$$

As can be noted the values of the elements of the above matrix are close to the average values of the corresponding elements of matrix $\mathbf{S}_D(500)$.

Similarly, if we multiply the cross-correlation vector $\mathbf{p}$ by 99.34, the result is

$$99.34\mathbf{p} = \begin{bmatrix} 94.61 \\ 47.31 \end{bmatrix}$$

The values of the elements of this vector are also close to the corresponding elements of $\mathbf{p}_D(500)$.

(b) The convergence path of the RLS algorithm on the MSE surface is depicted in Fig. 5.3. The reader should notice that the RLS algorithm approaches the minimum using large steps when the coefficients of the adaptive filter are far away from the optimum solution.

□

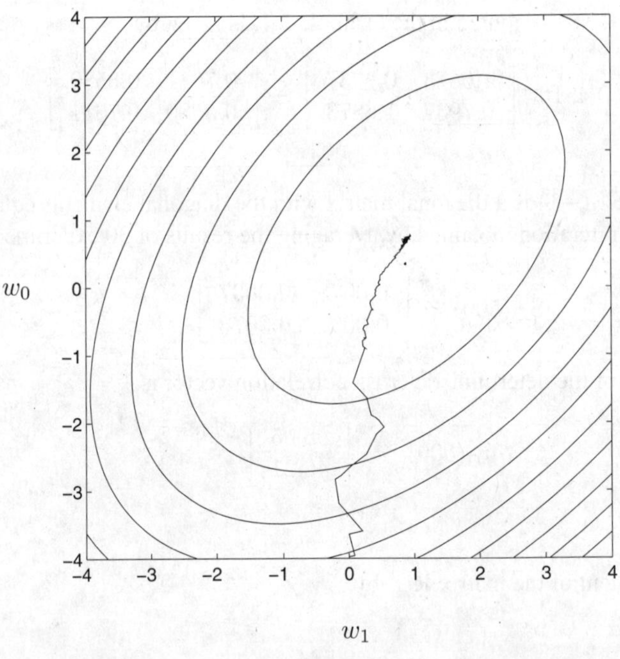

Figure 5.3 Convergence path of the RLS adaptive filter.

5.3.7 Excess Mean-Square Error and Misadjustment

In a practical implementation of the recursive least-squares algorithm, the best estimation for the unknown parameter vector is given by $\mathbf{w}(k)$, whose expected value is $\mathbf{w}_o$. However, there is always an excess MSE at the output caused by the error in the coefficient estimation, namely $\Delta\mathbf{w}(k) =$

$\mathbf{w}(k) - \mathbf{w}_o$. The mean-square error is (see equation (5.46))

$$\begin{aligned}\xi(k) &= \xi_{\min,p} + E\{[\mathbf{w}(k) - \mathbf{w}_o]^T \mathbf{x}(k)\mathbf{x}^T(k)[\mathbf{w}(k) - \mathbf{w}_o]\} \\ &= \xi_{\min,p} + E[\Delta\mathbf{w}^T(k)\mathbf{x}(k)\mathbf{x}^T(k)\Delta\mathbf{w}(k)]\end{aligned} \quad (5.48)$$

Now considering that $\Delta w_j(k)$, for $j = 0, 1, \ldots, N$, are random variables with zero mean and independent of $\mathbf{x}(k)$, the MSE can be calculated as follows

$$\begin{aligned}\xi(k) &= \xi_{\min,p} + E[\Delta\mathbf{w}^T(k)\mathbf{R}\Delta\mathbf{w}(k)] \\ &= \xi_{\min,p} + E\{\text{tr}[\mathbf{R}\Delta\mathbf{w}(k)\Delta\mathbf{w}^T(k)]\} \\ &= \xi_{\min,p} + \text{tr}\{\mathbf{R}E[\Delta\mathbf{w}(k)\Delta\mathbf{w}^T(k)]\} \\ &= \xi_{\min,p} + \text{tr}\{\mathbf{R}\text{cov}[\Delta\mathbf{w}(k)]\}\end{aligned} \quad (5.49)$$

On a number of occasions it is interesting to consider the analysis for $\lambda = 1$ separated from that for $\lambda < 1$.

Excess MSE for $\lambda = 1$

By applying in equation (5.49) the results of equations (5.36) and (5.19), and considering that

$$\xi_{\min,p} = \left(1 - 2\frac{N+1}{k+1}\right)\xi_{\min} = \left(1 - 2\frac{N+1}{k+1}\right)\sigma_n^2$$

for $\lambda = 1$ (see equations (5.42) and (5.47)), we can infer that

$$\begin{aligned}\xi(k) &= \left[1 - 2\frac{N+1}{k+1}\right]\sigma_n^2 + \text{tr}\left\{\mathbf{R}E[\mathbf{S}_D(k)]\right\}\sigma_n^2 \\ &= \left[1 - 2\frac{N+1}{k+1} + \text{tr}(\mathbf{R}\frac{\mathbf{R}^{-1}}{k+1})\right]\sigma_n^2 \text{ for } k \to \infty \\ &= \left(1 - 2\frac{N+1}{k+1} + \frac{N+1}{k+1}\right)\sigma_n^2 \text{ for } k \to \infty \\ &= \left(1 - \frac{N+1}{k+1}\right)\sigma_n^2 \text{ for } k \to \infty\end{aligned}$$

As can be noted the minimum MSE can be reached only after the algorithm has operated on a number of samples larger than the filter order.

Excess MSE for $\lambda < 1$

Again assuming that the mean-square error surface is quadratic as considered in equation (5.48), the expected excess in the MSE is then defined by

$$\Delta\xi(k) = E[\Delta\mathbf{w}^T(k)\mathbf{R}\Delta\mathbf{w}(k)] \quad (5.50)$$

The objective now is to calculate and analyze the excess MSE when $\lambda < 1$. From equation (5.30) one can show that

$$\Delta\mathbf{w}(k) = \lambda\mathbf{S}_D(k)\mathbf{R}_D(k-1)\Delta\mathbf{w}(k-1) + \mathbf{S}_D(k)\mathbf{x}(k)e_o(k) \tag{5.51}$$

By applying equation (5.51) to (5.50), it follows that

$$E[\Delta\mathbf{w}^T(k)\mathbf{R}\Delta\mathbf{w}(k)] = \rho_1 + \rho_2 + \rho_3 + \rho_4 \tag{5.52}$$

where

$$\begin{aligned}
\rho_1 &= \lambda^2 E[\Delta\mathbf{w}^T(k-1)\mathbf{R}_D(k-1)\mathbf{S}_D(k)\mathbf{R}\mathbf{S}_D(k)\mathbf{R}_D(k-1)\Delta\mathbf{w}(k-1)]\\
\rho_2 &= \lambda E[\Delta\mathbf{w}^T(k-1)\mathbf{R}_D(k-1)\mathbf{S}_D(k)\mathbf{R}\mathbf{S}_D(k)\mathbf{x}(k)e_o(k)]\\
\rho_3 &= \lambda E[\mathbf{x}^T(k)\mathbf{S}_D(k)\mathbf{R}\mathbf{S}_D(k)\mathbf{R}_D(k-1)\Delta\mathbf{w}(k-1)e_o(k)]\\
\rho_4 &= E[\mathbf{x}^T(k)\mathbf{S}_D(k)\mathbf{R}\mathbf{S}_D(k)\mathbf{x}(k)e_o^2(k)]
\end{aligned}$$

Now each term in equation (5.52) will be evaluated separately.

1- Evaluation of ρ_1

First note that as $k \to \infty$, it can be assumed that $\mathbf{R}_D(k) \approx \mathbf{R}_D(k-1)$, then

$$\rho_1 \approx \lambda^2 E[\Delta\mathbf{w}^T(k-1)\mathbf{R}\Delta\mathbf{w}(k-1)] \tag{5.53}$$

2- Evaluation of ρ_2

Since each element of $\mathbf{R}_D(k)$ is given by

$$r_{d,ij}(k) = \sum_{l=0}^{k} \lambda^{k-l} x(l-i)x(l-j) \tag{5.54}$$

for $0 \leq i,\ j \leq N$. Therefore,

$$E[r_{d,ij}(k)] = \sum_{l=0}^{k} \lambda^{k-l} E[x(l-i)x(l-j)]$$

If $x(k)$ is stationary, $r(i-j) = E[x(l-i)x(l-j)]$ is independent of the value l, then

$$E[r_{d,ij}(k)] = r(i-j)\frac{1-\lambda^{k+1}}{1-\lambda} \approx \frac{r(i-j)}{1-\lambda} \tag{5.55}$$

Equation (5.55) allows the conclusion that

$$E[\mathbf{R}_D(k)] \approx \frac{1}{1-\lambda}E[\mathbf{x}(k)\mathbf{x}^T(k)] = \frac{1}{1-\lambda}\mathbf{R} \tag{5.56}$$

In each step, it can be considered that

$$\mathbf{R}_D(k) = \frac{1}{1-\lambda}\mathbf{R} + \Delta\mathbf{R}(k) \tag{5.57}$$

where $\Delta\mathbf{R}(k)$ is a symmetric error matrix with zero-mean stochastic entries that are independent of the input signal. From equations (5.56) and (5.57), it can be concluded that

$$\mathbf{S}_D(k)\mathbf{R} \approx (1-\lambda)\left[\mathbf{I} - (1-\lambda)\mathbf{R}^{-1}\Delta\mathbf{R}(k)\right] \tag{5.58}$$

where in the last relation $\mathbf{S}_D(k)\Delta\mathbf{R}(k)$ was considered approximately equal to

$$(1-\lambda)\mathbf{R}^{-1}\Delta\mathbf{R}(k)$$

by using equation (5.56) and disregarding second-order errors.

In the long run, it is known that $E[\mathbf{S}_D(k)\mathbf{R}] = (1-\lambda)\mathbf{I}$, that means the second term inside the square bracket in equation (5.58) is a measure of the perturbation caused by $\Delta\mathbf{R}(k)$ in the product $\mathbf{S}_D(k)\mathbf{R}$. Denoting the perturbation by $\Delta\mathbf{I}(k)$, that is

$$\Delta\mathbf{I}(k) = (1-\lambda)\mathbf{R}^{-1}\Delta\mathbf{R}(k) \tag{5.59}$$

it can be concluded that

$$\begin{aligned}\rho_2 &\approx \lambda(1-\lambda)E\left\{\Delta\mathbf{w}^T(k-1)[\mathbf{I} - \Delta\mathbf{I}^T(k)]\mathbf{x}(k)e_o(k)\right\}\\ &\approx \lambda(1-\lambda)E[\Delta\mathbf{w}^T(k-1)]E[\mathbf{x}(k)e_o(k)] = 0\end{aligned} \tag{5.60}$$

where it was considered that $\Delta\mathbf{w}^T(k-1)$ is independent of $\mathbf{x}(k)$ and $e_o(k)$, $\Delta\mathbf{I}(k)$ was also considered an independent error matrix with zero mean, and finally we used the fact that $\mathbf{x}(k)$ and $e_o(k)$ are orthogonal.

3- Following a similar approach it can be shown that

$$\begin{aligned}\rho_3 &\approx \lambda(1-\lambda)E\left\{\mathbf{x}^T(k)[\mathbf{I} - \Delta\mathbf{I}(k)]\Delta\mathbf{w}(k-1)e_o(k)\right\}\\ &\approx \lambda(1-\lambda)E[\mathbf{x}^T(k)e_o(k)]E[\Delta\mathbf{w}(k-1)] = 0\end{aligned} \tag{5.61}$$

4- Evaluation of ρ_4

$$\begin{aligned}\rho_4 &= E[\mathbf{x}^T(k)\mathbf{S}_D(k)\mathbf{R}\mathbf{S}_D(k)\mathbf{R}\mathbf{R}^{-1}\mathbf{x}(k)e_o^2(k)]\\ &\approx (1-\lambda)^2E\left\{\mathbf{x}^T(k)[\mathbf{I} - \Delta\mathbf{I}(k)]^2\mathbf{R}^{-1}\mathbf{x}(k)\right\}\xi_{\min}\end{aligned} \tag{5.62}$$

where equations (5.58) and (5.29) were used and $e_o(k)$ was considered independent of $\mathbf{x}(k)$ and $\Delta\mathbf{I}(k)$. By using the property that

$$E\left\{\mathbf{x}^T(k)[\mathbf{I} - \Delta\mathbf{I}(k)]^2\mathbf{R}^{-1}\mathbf{x}(k)\right\} = \mathrm{tr}E\left\{[\mathbf{I} - \Delta\mathbf{I}(k)]^2\mathbf{R}^{-1}\mathbf{x}(k)\mathbf{x}^T(k)\right\}$$

and recalling that $\Delta\mathbf{I}(k)$ has zero mean and is independent of $\mathbf{x}(k)$, then equation (5.62) is simplified to

$$\rho_4 = (1-\lambda)^2\mathrm{tr}\{\mathbf{I} + E[\Delta\mathbf{I}^2(k)]\}\xi_{\min} \tag{5.63}$$

where tr[·] means trace of [·], and we utilized the fact that $E\{\mathbf{R}^{-1}\mathbf{x}(k)\mathbf{x}^T(k)\} = \mathbf{I}$.

By using equations (5.53), (5.60), and (5.63), it follows that

$$E[\Delta\mathbf{w}^T(k)\mathbf{R}\Delta\mathbf{w}(k)] = \lambda^2 E[\Delta\mathbf{w}^T(k-1)\mathbf{R}\Delta\mathbf{w}(k-1)] + (1-\lambda)^2 \text{tr}\{\mathbf{I} + E[\Delta\mathbf{I}^2(k)]\}\xi_{\text{min}} \tag{5.64}$$

Asymptotically, the solution of the above equation is

$$\xi_{\text{exc}} = \frac{1-\lambda}{1+\lambda}\text{tr}\{\mathbf{I} + E[\Delta\mathbf{I}^2(k)]\}\xi_{\text{min}} \tag{5.65}$$

Note that the term given by $E[\Delta\mathbf{I}^2(k)]$ is not easy to estimate and is dependent on fourth-order statistics of the input signal. However, in specific situations, it is possible to compute an approximate estimate for this matrix. In steady state, it can be considered for white noise input signal that only the diagonal elements of $\mathbf{R}$ and $\Delta\mathbf{R}$ are important to the generation of excess MSE. Even when the input signal is not white, this diagonal dominance can be considered a reasonable approximation in most of the cases. From the definition of $\Delta\mathbf{I}(k)$ in equation (5.59), it follows that

$$E[\Delta\mathbf{I}_{ii}^2(k)] = (1-\lambda)^2 \frac{E[\Delta r_{ii}^2(k)]}{[\sigma_x^2]^2} \tag{5.66}$$

where σ_x^2 is variance of $x(k)$. By calculating $\Delta\mathbf{R}(k) - \lambda\Delta\mathbf{R}(k-1)$ using equation (5.57), we show that

$$\Delta r_{ii}(k) = \lambda\Delta r_{ii}(k-1) + x(k-i)x(k-i) - r_{ii} \tag{5.67}$$

Squaring the above equation, applying the expectation operation, and using the independence between $\Delta r_{ii}(k)$ and $x(k)$, it follows that

$$E[\Delta r_{ii}^2(k)] = \lambda^2 E[\Delta r_{ii}^2(k-1)] + E\left\{[x(k-i)x(k-i) - r_{ii}]^2\right\} \tag{5.68}$$

Therefore, asymptotically

$$E[\Delta r_{ii}^2(k)] = \frac{1}{1-\lambda^2}\sigma^2_{x^2(k-i)} = \frac{1}{1-\lambda^2}\sigma^2_{x^2} \tag{5.69}$$

By substituting equation (5.69) in (5.66), it becomes

$$E[\Delta\mathbf{I}_{ii}^2(k)] = \frac{1-\lambda}{1+\lambda}\frac{\sigma^2_{x^2}}{(\sigma_x^2)^2} = \frac{1-\lambda}{1+\lambda}\mathcal{K} \tag{5.70}$$

where $\mathcal{K} = \frac{\sigma^2_{x^2}}{(\sigma_x^2)^2}$ is dependent on input signal statistics. For Gaussian signals, $\mathcal{K} = 2$ [7].

Returning to our main objective, the excess MSE can then be described as

$$\xi_{\text{exc}} = (N+1)\frac{1-\lambda}{1+\lambda}\left(1 + \frac{1-\lambda}{1+\lambda}\mathcal{K}\right)\xi_{\text{min}} \tag{5.71}$$

If λ is approximately one and $\mathcal{K}$ is not very large then

$$\xi_{\text{exc}} = (N+1)\frac{1-\lambda}{1+\lambda}\xi_{\min} \tag{5.72}$$

this expression can be reached through a simpler analysis [6]. However, the more complete derivation shown here can give more insight to the interpretation of the results obtained by using the RLS algorithm, mainly when λ is not very close to one.

The misadjustment formula can be deduced from equation (5.71)

$$M = \frac{\xi_{\text{exc}}}{\xi_{\min}} = (N+1)\frac{1-\lambda}{1+\lambda}\left(1+\frac{1-\lambda}{1+\lambda}\mathcal{K}\right) \tag{5.73}$$

As can be noted, the decrease of λ from one brings a fourth-order statistics term into the picture by increasing the misadjustment. Then, the fast adaptation of the RLS algorithm, that corresponds to smaller λ, brings a noisier steady-state response. Therefore, when working in a stationary environment the best choice for λ would be one, if the excess MSE in the steady state is considered high for other values of λ. However, other problems such as instability due to quantization noise are prone to occur when $\lambda = 1$.

5.4 BEHAVIOR IN NONSTATIONARY ENVIRONMENTS

In cases where the input signal and/or the desired signal are nonstationary, the optimal values of the coefficients are time variant and described by $\mathbf{w}_o(k)$. That means the autocorrelation matrix $\mathbf{R}(k)$ and/or the cross-correlation vector $\mathbf{p}(k)$ are time variant. For example, typically in a system identification application the autocorrelation matrix $\mathbf{R}(k)$ is time invariant while the cross-correlation matrix $\mathbf{p}(k)$ is time variant, because in this case the designer can choose the input signal. On the other hand, in equalization, prediction, and signal enhancement applications both the input and the desired signal are nonstationary leading to time-varying matrices $\mathbf{R}(k)$ and $\mathbf{p}(k)$.

The objective in the present section is to analyze how close the RLS algorithm is able to track the time-varying solution $\mathbf{w}_o(k)$. Also, it is of interest to learn how the tracking error in $\mathbf{w}(k)$ affects the output MSE [7]. Here, the effects of the measurement noise are not considered, since only the nonstationary effects are desired. Also, both effects on the MSE can be added since, in general, they are independent.

Recall from equations (5.8) and (5.9) that

$$\mathbf{w}(k) = \mathbf{w}(k-1) + \mathbf{S}_D(k)\mathbf{x}(k)[d(k) - \mathbf{x}^T(k)\mathbf{w}(k-1)] \tag{5.74}$$

and

$$d(k) = \mathbf{x}^T(k)\mathbf{w}_o(k-1) + e'_o(k) \tag{5.75}$$

The error signal $e'_o(k)$ is the minimum error at iteration k being generated by the nonstationarity of the environment. One can replace equation (5.75) in (5.74) in order to obtain the following relation

$$\mathbf{w}(k) = \mathbf{w}(k-1) + \mathbf{S}_D(k)\mathbf{x}(k)\mathbf{x}^T(k)[\mathbf{w}_o(k-1) - \mathbf{w}(k-1)] + \mathbf{S}_D(k)\mathbf{x}(k)e'_o(k) \tag{5.76}$$

By taking the expected value of equation (5.76), considering that $\mathbf{x}(k)$ and $e'_o(k)$ are approximately orthogonal, and that $\mathbf{w}(k-1)$ is independent of $\mathbf{x}(k)$, then

$$E[\mathbf{w}(k)] = E[\mathbf{w}(k-1)] + E[\mathbf{S}_D(k)\mathbf{x}(k)\mathbf{x}^T(k)]\left\{\mathbf{w}_o(k-1) - E[\mathbf{w}(k-1)]\right\} \tag{5.77}$$

It is now needed to compute $E[\mathbf{S}_D(k)\mathbf{x}(k)\mathbf{x}^T(k)]$ in the case of nonstationary input signal. From equations (5.54) and (5.56), one can show that

$$\mathbf{R}_D(k) = \sum_{l=0}^{k} \lambda^{k-l}\mathbf{R}(l) + \Delta\mathbf{R}(k) \tag{5.78}$$

since $E[\mathbf{R}_D(k)] = \sum_{l=0}^{k} \lambda^{k-l}\mathbf{R}(l)$. The matrix $\Delta\mathbf{R}(k)$ is again considered a symmetric error matrix with zero-mean stochastic entries that are independent of the input signal.

If the environment is considered to be varying at a slower pace than the memory of the adaptive RLS algorithm, then

$$\mathbf{R}_D(k) \approx \frac{1}{1-\lambda}\mathbf{R}(k) + \Delta\mathbf{R}(k) \tag{5.79}$$

Considering that $(1-\lambda)||\mathbf{R}^{-1}(k)\Delta\mathbf{R}(k)|| < 1$ and using the same procedure to deduce equation (5.58), we obtain

$$\mathbf{S}_D(k) \approx (1-\lambda)\mathbf{R}^{-1}(k) - (1-\lambda)^2\mathbf{R}^{-1}(k)\Delta\mathbf{R}(k)\mathbf{R}^{-1}(k) \tag{5.80}$$

it then follows that

$$\begin{aligned} E[\mathbf{w}(k)] = {} & E[\mathbf{w}(k-1)] + \{(1-\lambda)E[\mathbf{R}^{-1}(k)\mathbf{x}(k)\mathbf{x}^T(k)] \\ & - (1-\lambda)^2 E[\mathbf{R}^{-1}(k)\Delta\mathbf{R}(k)\mathbf{R}^{-1}(k)\mathbf{x}(k)\mathbf{x}^T(k)]\}\left\{\mathbf{w}_o(k-1) - E[\mathbf{w}(k-1)]\right\} \\ \approx {} & E[\mathbf{w}(k-1)] + (1-\lambda)\left\{\mathbf{w}_o(k-1) - E[\mathbf{w}(k-1)]\right\} \end{aligned} \tag{5.81}$$

where it was considered that $\Delta\mathbf{R}(k)$ is independent of $\mathbf{x}(k)$ and has zero expected value.

Now defining the lag-error vector in the coefficients as

$$\mathbf{l_w}(k) = E[\mathbf{w}(k)] - \mathbf{w}_o(k) \tag{5.82}$$

From equation (5.81), it can be concluded that

$$\mathbf{l_w}(k) = \lambda\mathbf{l_w}(k-1) - \mathbf{w}_o(k) + \mathbf{w}_o(k-1) \tag{5.83}$$

Equation (5.83) is equivalent to say that the lag is generated by applying the optimal instantaneous value $\mathbf{w}_o(k)$ through a first-order discrete-time filter as follows:

$$L_i(z) = -\frac{z-1}{z-\lambda}W_{oi}(z) \tag{5.84}$$

The discrete-time filter transient response converges with a time constant given by

$$\tau = \frac{1}{1-\lambda} \tag{5.85}$$

The time constant is of course the same for each individual coefficient. Note that the tracking ability of the coefficients in the RLS algorithm is independent of the eigenvalues of the input signal correlation matrix.

The lag in the coefficients leads to an excess MSE. In order to calculate the MSE suppose that the optimal coefficient values are first-order Markov processes described by

$$\mathbf{w}_o(k) = \lambda_{\mathbf{w}}\mathbf{w}_o(k-1) + \mathbf{n}_{\mathbf{w}}(k) \tag{5.86}$$

where $\mathbf{n}_{\mathbf{w}}(k)$ is a vector whose elements are zero-mean white noise processes with variance $\sigma_{\mathbf{w}}^2$, and $\lambda_{\mathbf{w}} < 1$. Note that $\lambda < \lambda_{\mathbf{w}} < 1$, since the optimal coefficients values must vary slower than the filter tracking speed, that means $\frac{1}{1-\lambda} < \frac{1}{1-\lambda_{\mathbf{w}}}$.

The excess MSE due to lag is then given by (see the derivations around equation (3.41))

$$\begin{aligned}
\xi_{\text{lag}} &= E[\mathbf{l}_{\mathbf{w}}^T(k)\mathbf{R}\mathbf{l}_{\mathbf{w}}(k)] \\
&= E\left\{\text{tr}[\mathbf{R}\mathbf{l}_{\mathbf{w}}(k)\mathbf{l}_{\mathbf{w}}^T(k)]\right\} \\
&= \text{tr}\left\{\mathbf{R}E[\mathbf{l}_{\mathbf{w}}(k)\mathbf{l}_{\mathbf{w}}^T(k)]\right\} \\
&= \text{tr}\left\{\mathbf{\Lambda}E[\mathbf{l}'_{\mathbf{w}}(k)\mathbf{l}'^T_{\mathbf{w}}(k)]\right\} \\
&= \sum_{i=0}^{N} \lambda_i E[l_i'^2(k)]
\end{aligned} \tag{5.87}$$

For $\lambda_{\mathbf{w}}$ not close to one, it is a bit more complicated to deduce the excess MSE due to lag than for $\lambda_{\mathbf{w}} \approx 1$. However, the effort is worth it because the resulting expression is more accurate. From equation (5.84), we can see that the lag-error vector elements are generated by applying a first-order discrete-time system to the elements of the unknown system coefficient vector. On the other hand, the coefficients of the unknown system are generated by applying each element of the noise vector $\mathbf{n}_{\mathbf{w}}(k)$ to a first-order all-pole filter, with the pole placed at $\lambda_{\mathbf{w}}$. For the unknown coefficient vector with the above model, the lag-error vector elements can be generated by applying the elements of the noise vector $\mathbf{n}_{\mathbf{w}}(k)$ to a discrete-time filter with transfer function

$$H(z) = \frac{-(z-1)z}{(z-\lambda)(z-\lambda_{\mathbf{w}})} \tag{5.88}$$

This transfer function consists of a cascade of the lag filter with the all-pole filter representing the first-order Markov process. The solution for the variance of the lag terms l_i can be computed through the inverse $\mathcal{Z}$-transform as follows:

$$E[l_i'^2(k)] = \frac{1}{2\pi \jmath}\oint H(z)H(z^{-1})\sigma_{\mathbf{w}}^2 z^{-1}\, dz \tag{5.89}$$

The above integral can be solved using the residue theorem as previously shown in the LMS algorithm case.

Using the solution for the variance of the lag terms of equation (5.89) for values of $\lambda_{\mathbf{w}} < 1$, and substituting the result in the last term of equation (5.87) it can be shown that

$$\begin{aligned}\xi_{\text{lag}} &\approx \frac{\text{tr}[\mathbf{R}]\sigma_{\mathbf{w}}^2}{\lambda_{\mathbf{w}}(1+\lambda^2) - \lambda(1+\lambda_{\mathbf{w}}^2)}\left(\frac{1-\lambda}{1+\lambda} - \frac{1-\lambda_{\mathbf{w}}}{1+\lambda_{\mathbf{w}}}\right) \\ &= \frac{(N+1)\sigma_{\mathbf{w}}^2\sigma_x^2}{\lambda_{\mathbf{w}}(1+\lambda^2) - \lambda(1+\lambda_{\mathbf{w}}^2)}\left(\frac{1-\lambda}{1+\lambda} - \frac{1-\lambda_{\mathbf{w}}}{1+\lambda_{\mathbf{w}}}\right) \end{aligned} \tag{5.90}$$

where it was used the fact that $\text{tr}[\mathbf{R}] = \sum_{i=0}^{N} \lambda_i = (N+1)\sigma_x^2$, for a tap delay line. It should be noticed that assumptions such as the correlation matrix $\mathbf{R}$ being diagonal and the input signal being white noise were not required in this derivation.

If $\lambda = 1$ and $\lambda_{\mathbf{w}} \approx 1$, the MSE due to lag tends to infinity indicating that the RLS algorithm in this case cannot track any change in the environment. On the other hand, for $\lambda < 1$ the algorithm can track variations in the environment, leading to an excess MSE that depends on the variance of the optimal coefficient disturbance and on the input signal variance.

For $\lambda_{\mathbf{w}} = 1$ and $\lambda \approx 1$, it is possible to rewrite equation (5.90) as

$$\xi_{\text{lag}} \approx (N+1)\frac{\sigma_{\mathbf{w}}^2}{2(1-\lambda)}\sigma_x^2 \tag{5.91}$$

The total excess MSE accounting for the lag and finite memory is given by

$$\xi_{\text{total}} \approx (N+1)\left[\frac{1-\lambda}{1+\lambda}\xi_{\min} + \frac{\sigma_{\mathbf{w}}^2\sigma_x^2}{2(1-\lambda)}\right] \tag{5.92}$$

By differentiating the above equation with respect to λ and setting the result to zero, an optimum value for λ can be found that yields minimum excess MSE.

$$\lambda_{\text{opt}} = \frac{1 - \frac{\sigma_{\mathbf{w}}\sigma_x}{2\sigma_n}}{1 + \frac{\sigma_{\mathbf{w}}\sigma_x}{2\sigma_n}} \tag{5.93}$$

In the above equation we used $\sigma_n = \sqrt{\xi_{\min}}$. Note that the optimal value of λ does not depend on the adaptive-filter order N, and can be used when it falls in an acceptable range of values for λ. Also, this value is optimum only when quantization effects are not important and the first-order Markov model (with $\lambda_{\mathbf{w}} \approx 1$) is a good approximation for the nonstationarity of the desired signal.

When implemented with finite-precision arithmetic, the conventional RLS algorithm behavior can differ significantly from what is expected under infinite precision. A series of inconvenient effects can show up in the practical implementation of the conventional RLS algorithm, such as divergence and freezing in the updating of the adaptive-filter coefficients. Appendix C, presents a detailed analysis of the finite-wordlength effects in the RLS algorithm.

5.5 COMPLEX RLS ALGORITHM

In the complex data case the RLS objective function is given by

$$\xi^d(k) = \sum_{i=0}^{k} \lambda^{k-i}|\varepsilon(i)|^2 = \sum_{i=0}^{k} \lambda^{k-i}|d(i) - \mathbf{w}^H(i)\mathbf{x}(k)|^2$$
$$= \sum_{i=0}^{k} \lambda^{k-i} \left[d(i) - \mathbf{w}^H(i)\mathbf{x}(k)\right] \left[d^*(i) - \mathbf{w}^T(i)\mathbf{x}^*(k)\right] \tag{5.94}$$

Differentiating $\xi^d(k)$ with respect to the complex coefficient $\mathbf{w}^*(k)$ leads to[3]

$$\frac{\partial \xi^d(k)}{\partial \mathbf{w}^*(k)} = -\sum_{i=0}^{k} \lambda^{k-i}\mathbf{x}(i)[d^*(i) - \mathbf{w}^T(i)\mathbf{x}^*(k)] \tag{5.95}$$

The optimal vector $\mathbf{w}(k)$ that minimizes the least-squares error, is computed by equating the above equation to zero that is

$$-\sum_{i=0}^{k} \lambda^{k-i}\mathbf{x}(i)\mathbf{x}^H(i)\mathbf{w}(k) + \sum_{i=0}^{k} \lambda^{k-i}\mathbf{x}(i)d^*(i) = \begin{bmatrix} 0 \\ 0 \\ \vdots \\ 0 \end{bmatrix}$$

leading to the following expression

$$\mathbf{w}(k) = \left[\sum_{i=0}^{k} \lambda^{k-i}\mathbf{x}(i)\mathbf{x}^H(i)\right]^{-1} \sum_{i=0}^{k} \lambda^{k-i}\mathbf{x}(i)d^*(i)$$
$$= \mathbf{R}_D^{-1}(k)\mathbf{p}_D(k) \tag{5.96}$$

The matrix inversion lemma to the case of complex data is given by

$$\mathbf{S}_D(k) = \mathbf{R}_D^{-1}(k) = \frac{1}{\lambda}\left[\mathbf{S}_D(k-1) - \frac{\mathbf{S}_D(k-1)\mathbf{x}(k)\mathbf{x}^H(k)\mathbf{S}_D(k-1)}{\lambda + \mathbf{x}^H(k)\mathbf{S}_D(k-1)\mathbf{x}(k)}\right] \tag{5.97}$$

The complete conventional RLS algorithm is described in Algorithm 5.3.

An alternative complex RLS algorithm has an updating equation described by

$$\mathbf{w}(k) = \mathbf{w}(k-1) + e^*(k)\mathbf{S}_D(k)\mathbf{x}(k) \tag{5.98}$$

where

$$e(k) = d(k) - \mathbf{w}^H(k-1)\mathbf{x}(k) \tag{5.99}$$

With equation (5.98), it is straightforward to generate an alternative conventional RLS algorithm as shown in Algorithm 5.4.

---

[3]Again the reader should recall that when computing the gradient with respect to $\mathbf{w}^*(k)$, $\mathbf{w}(k)$ is treated as a constant.

Algorithm 5.3

Conventional Complex RLS Algorithm

Initialization

$\mathbf{S}_D(-1) = \delta\mathbf{I}$

where δ can be the inverse of the input signal power estimate

$\mathbf{p}_D(-1) = \mathbf{x}(-1) = [0\,0\ldots0]^T$

Do for $k \geq 0$:

$\mathbf{S}_D(k) = \frac{1}{\lambda}[\mathbf{S}_D(k-1) - \frac{\mathbf{S}_D(k-1)\mathbf{x}(k)\mathbf{x}^H(k)\mathbf{S}_D(k-1)}{\lambda+\mathbf{x}^H(k)\mathbf{S}_D(k-1)\mathbf{x}(k)}]$

$\mathbf{p}_D(k) = \lambda\mathbf{p}_D(k-1) + d^*(k)\mathbf{x}(k)$

$\mathbf{w}(k) = \mathbf{S}_D(k)\mathbf{p}_D(k)$

If necessary compute

$y(k) = \mathbf{w}^H(k)\mathbf{x}(k)$

$\varepsilon(k) = d(k) - y(k)$

Algorithm 5.4

Alternative Complex RLS Algorithm

Initialization

$\mathbf{S}_D(-1) = \delta\mathbf{I}$

where δ can be the inverse of an estimate of the input signal power

$\mathbf{x}(-1) = \mathbf{w}(-1) = [0\,0\ldots0]^T$

Do for $k \geq 0$

$e(k) = d(k) - \mathbf{w}^H(k-1)\mathbf{x}(k)$

$\boldsymbol{\psi}(k) = \mathbf{S}_D(k-1)\mathbf{x}(k)$

$\mathbf{S}_D(k) = \frac{1}{\lambda}[\mathbf{S}_D(k-1) - \frac{\boldsymbol{\psi}(k)\boldsymbol{\psi}^H(k)}{\lambda+\boldsymbol{\psi}^H(k)\mathbf{x}(k)}]$

$\mathbf{w}(k) = \mathbf{w}(k-1) + e^*(k)\mathbf{S}_D(k)\mathbf{x}(k)$

If necessary compute

$y(k) = \mathbf{w}^H(k)\mathbf{x}(k)$

$\varepsilon(k) = d(k) - y(k)$

5.6 SIMULATION EXAMPLES

In this section, some adaptive-filtering problems described in the last two chapters are solved using the conventional RLS algorithm presented in this chapter.

Example 5.3: System Identification Simulations

The conventional RLS algorithm is employed in the identification of the system described in the subsection 3.6.2. The forgetting factor is chosen as $\lambda = 0.99$.

Solution:

In the first test, we address the sensitivity of the RLS algorithm to the eigenvalue spread of the input signal correlation matrix. The measured simulation results are obtained by ensemble averaging 200 independent runs. The learning curves of the mean-squared *a priori* error are depicted in Fig. 5.4, for different values of the eigenvalue spread. Also, the measured misadjustment in each example is given in Table 5.1. From these results, we conclude that the RLS algorithm is insensitive to the eigenvalue spread. It is worth mentioning at this point that the convergence speed of the RLS algorithm is affected by the choice of λ, since a smaller value of λ leads to faster convergence while increasing the misadjustment in stationary environment. Table 5.1 shows the misadjustment predicted by theory, calculated using the relation repeated below. As can be seen from this table the analytical results agree with those obtained through simulations.

$$M = (N+1)\frac{1-\lambda}{1+\lambda}(1+\frac{1-\lambda}{1+\lambda}\mathcal{K})$$

Table 5.1 Evaluation of the RLS Algorithm

	Misadjustment	
$\frac{\lambda_{\max}}{\lambda_{\min}}$	**Experiment**	**Theory**
1	0.04211	0.04020
20	0.04211	0.04020
80	0.04547	0.04020

The conventional RLS algorithm is implemented with finite-precision arithmetic, using fixed-point representation with 16, 12, and 10 bits, respectively. The results presented are measured before any sign of instability is noticed. Table 5.2 summarizes the results of the finite-precision implementation of the conventional RLS algorithm. Note that in most cases there is a close agreement between the measurement results and those predicted by the equations given below. These equations correspond to equations (C.37) and (C.48) derived in Appendix C.

$$E[||\Delta\mathbf{w}(k)_Q||^2]\approx\frac{(1-\lambda)(N+1)}{2\lambda}\frac{\sigma_n^2+\sigma_e^2}{\sigma_x^2}+\frac{(N+1)\sigma_{\mathbf{w}}^2}{2\lambda(1-\lambda)}$$

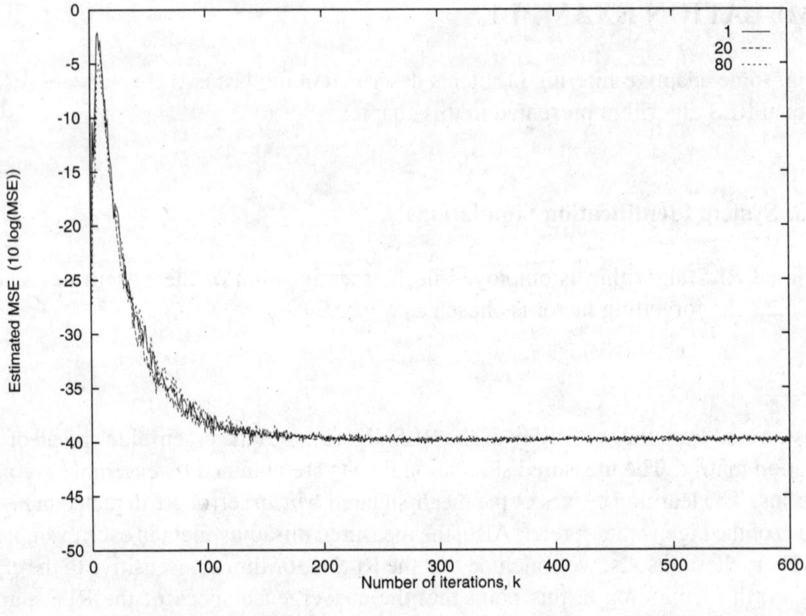

Figure 5.4 Learning curves for RLS algorithm for eigenvalue spreads: 1, 20, and 80; $\lambda = 0.99$.

$$\xi(k)_Q \approx \xi_{\min} + \sigma_e^2 + \frac{(N+1)\sigma_{\mathbf{w}}^2 \sigma_x^2}{2\lambda(1-\lambda)}$$

For the simulations with 12 and 10 bits, the discrepancy between the measured and theoretical estimates of $E[||\Delta\mathbf{w}(k)_Q||^2]$ are caused by the freezing of some coefficients.

If the results presented here are compared with the results presented in Table 3.2 for the LMS, we notice that both the LMS and the RLS algorithms performed well in the finite-precision implementation. The reader should bear in mind that the conventional RLS algorithm requires an expensive strategy to keep the deterministic correlation matrix positive definite, as discussed in Appendix C.

Table 5.2 Results of the Finite-Precision Implementation of the RLS Algorithm

	$\xi(k)_Q$		$E[\|\|\Delta\mathbf{w}(k)_Q\|\|^2]$	
No. of bits	**Experiment**	**Theory**	**Experiment**	**Theory**
16	$1.566\ 10^{-3}$	$1.500\ 10^{-3}$	$6.013\ 10^{-5}$	$6.061\ 10^{-5}$
12	$1.522\ 10^{-3}$	$1.502\ 10^{-3}$	$3.128\ 10^{-5}$	$6.261\ 10^{-5}$
10	$1.566\ 10^{-3}$	$1.532\ 10^{-3}$	$6.979\ 10^{-5}$	$9.272\ 10^{-5}$

The simulations related to the experiment described for nonstationary environments are also performed. From the simulations we measure the total excess MSE, and then compare the results to those obtained with the expression below.

$$\xi_{\text{exc}} \approx (N+1)\frac{1-\lambda}{1+\lambda}(1+\frac{1-\lambda}{1+\lambda}\mathcal{K})\xi_{\min} + \frac{(N+1)\sigma_{\mathbf{w}}^2\sigma_x^2}{\lambda_{\mathbf{w}}(1+\lambda^2)-\lambda(1+\lambda_{\mathbf{w}}^2)}(\frac{1-\lambda}{1+\lambda}-\frac{1-\lambda_{\mathbf{w}}}{1+\lambda_{\mathbf{w}}})$$

An attempt to use the optimal value of λ is made. The predicted optimal value, in this case, is too small and as a consequence $\lambda = 0.99$ is used. The measured excess MSE is 0.0254, whereas the theoretical value predicted by the above equation is 0.0418. Note that the theoretical result is not as accurate as all the previous cases discussed so far, due to a number of approximations used in the analysis. However, the above equation provides a good indication of what is expected in the practical implementation. By choosing a smaller value for λ a better tracking performance is obtained, a situation where the above equation is not as accurate.

□

Example 5.4: Signal Enhancement Simulations

We solved the same signal enhancement problem described in the subsection 4.7.1 with the conventional RLS and LMS algorithms.

Solution:

For the LMS algorithm, the convergence factor is chosen $\mu_{\max}/5$. The resulting value for μ in the LMS case is 0.001, whereas $\lambda = 1.0$ is used for the RLS algorithm. The learning curves for the algorithms are shown in Fig. 5.5, where we can verify the faster convergence of the RLS algorithm. By plotting the output errors after convergence, we noted the large variance of the MSE for both algorithms. This result is due to the small signal to noise ratio, in this case. Fig. 5.6 depicts the output error and its DFT with 128 points for the RLS algorithm. In both cases, we can clearly detect the presence of the sinusoid.

□

5.7 CONCLUDING REMARKS

In this chapter, we introduced the conventional RLS algorithm and discussed various aspects related to its performance behavior. Much of the results obtained herein through mathematical analysis are valid for the whole class of RLS algorithms to be presented in the following chapters, except for the finite-precision analysis since that depends on the form the internal calculations of each algorithm are performed. The analysis presented here is far from being complete. However, the main aspects of the conventional RLS have been addressed, such as: convergence behavior and tracking capabilities.

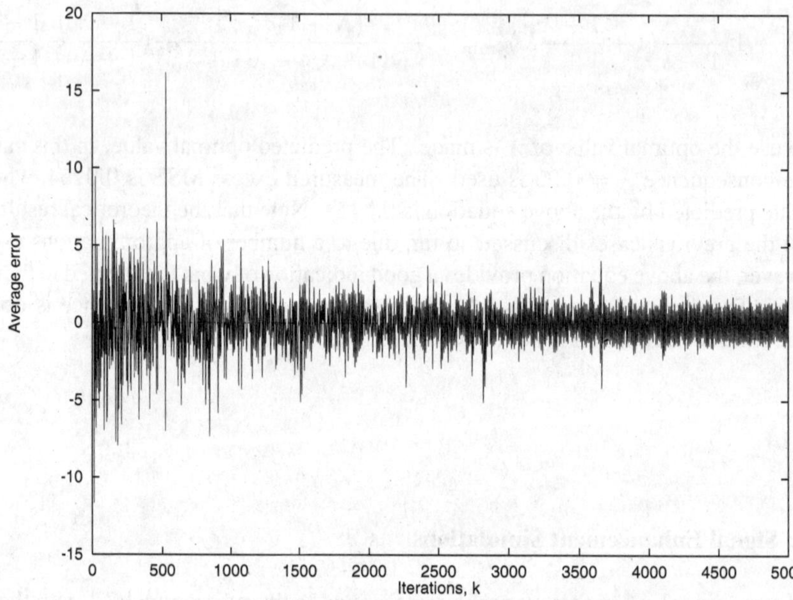

(a)

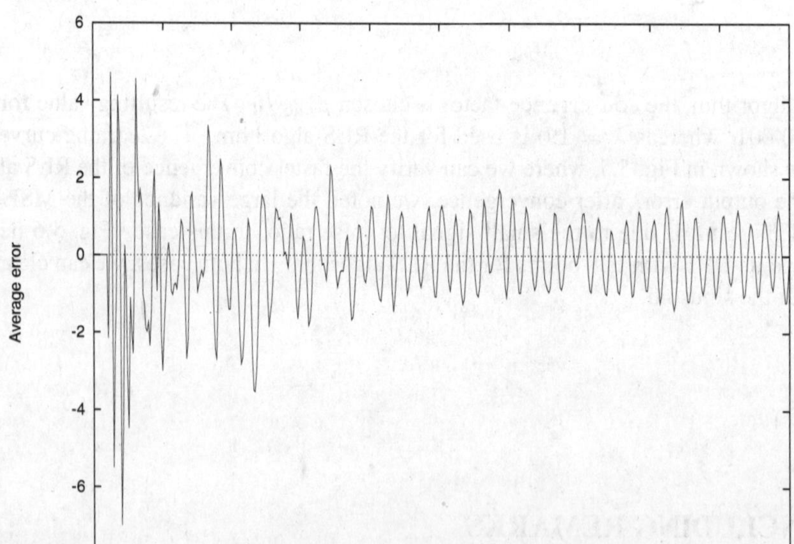

(b)

Figure 5.5 Learning curves for the (a) LMS and (b) RLS algorithms.

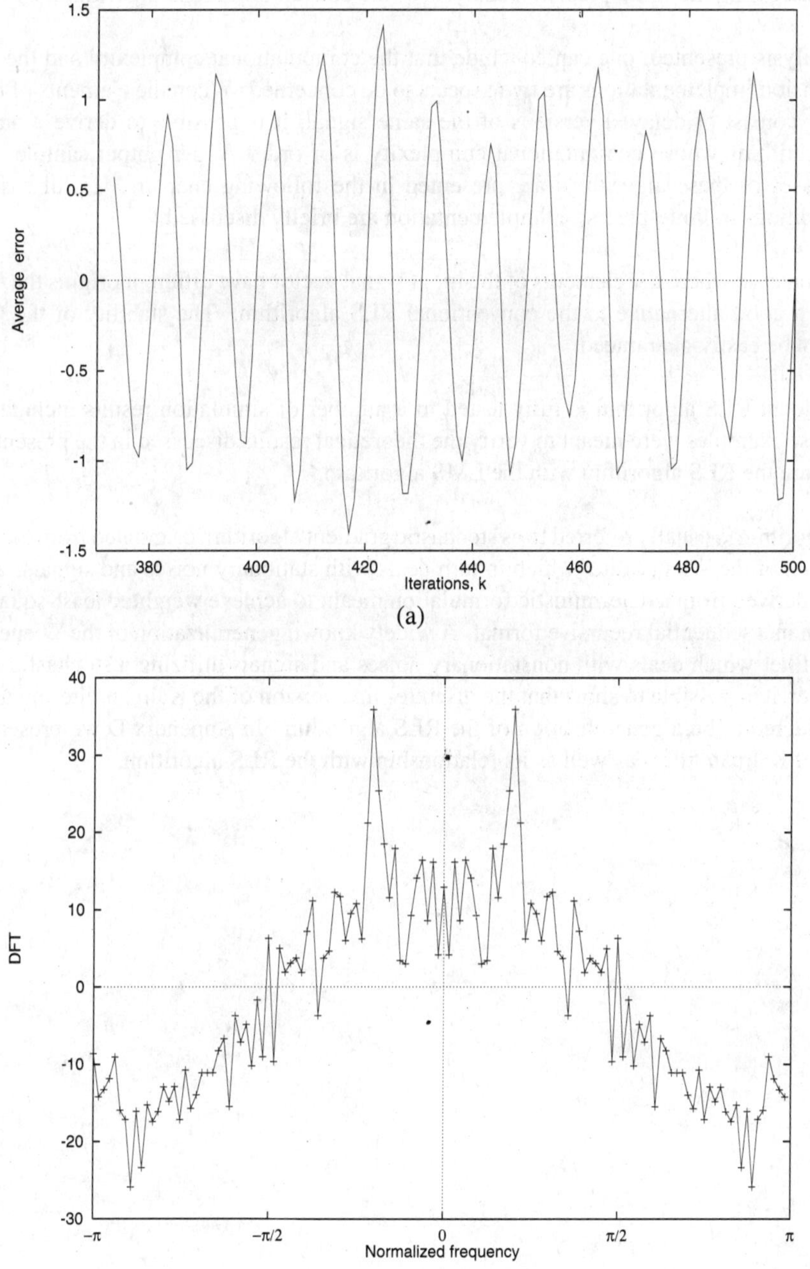

Figure 5.6 (a) Output error for the RLS algorithm and (b) DFT of the output error.

The interested reader should consult [9]-[11] for some further results. Appendix C complements this chapter by addressing the finite-precision analysis of the conventional RLS algorithm.

From the analysis presented, one can conclude that the computational complexity and the stability in finite-precision implementations are two aspects to be concerned. When the elements of the input signal vector consist of delayed versions of the same signal, it is possible to derive a number of fast RLS algorithms whose computational complexity is of order N per output sample. Several different classes of these algorithms are presented in the following chapters. In all cases, their stability conditions in finite-precision implementation are briefly discussed.

For the general case where the elements of the input signal vector have different origins the QR-RLS algorithm is a good alternative to the conventional RLS algorithm. The stability of the QR-RLS algorithm can be easily guaranteed.

The conventional RLS algorithm is fully tested in a number of simulation results included in this chapter. These examples were meant to verify the theoretical results discussed in the present chapter and to compare the RLS algorithm with the LMS algorithm.

The LMS algorithm is usually referred to as stochastic gradient algorithm originated from the stochastic formulation of the Wiener filter which in turn deals with stationary noises and signals. The RLS algorithm is derived from a deterministic formulation meant to achieve weighted least-squares error minimization in a sequential recursive format. A widely known generalization of the Wiener filter is the Kalman filter which deals with nonstationary noises and signals utilizing a stochastic formulation. However, it is possible to show that the discrete-time version of the Kalman filtering algorithm can be considered to be a generalization of the RLS algorithm. In Appendix D we present a brief description of Kalman filters as well as its relationship with the RLS algorithm.

5.8 REFERENCES

1. G. C. Goodwin and R. L. Payne, *Dynamic System Identification: Experiment Design and Data Analysis*, Academic Press, New York, NY, 1977.

2. S. Haykin, *Adaptive Filter Theory*, Prentice Hall, Englewood Cliffs, NJ, 4th edition, 2002.

3. S. H. Ardalan, "Floating-point analysis of recursive least-squares and least-mean squares adaptive filters," *IEEE Trans. on Circuits and Systems*, vol. CAS-33, pp. 1192-1208, Dec. 1986.

4. J. M. Cioffi, "Limited precision effects in adaptive filtering," *IEEE Trans. on Circuits and Systems*, vol. CAS-34, pp. 821-833, July 1987.

5. R. S. Medaugh and L. J. Griffiths, "A comparison of two linear predictors," *Proc. IEEE Intern. Conf. on Acoust., Speech, Signal Processing*, Atlanta, GA, pp. 293-296, April 1981.

6. F. Ling and J. G. Proakis, "Nonstationary learning characteristics of least squares adaptive estimation algorithms," *Proc. IEEE Intern. Conf. on Acoust., Speech, Signal Processing*, San Diego, CA, pp. 30.3.1.-30.3.4, March 1984.

7. E. Eleftheriou and D. D. Falconer, "Tracking properties and steady-state performance of RLS adaptive filter algorithms," *IEEE Trans. on Acoust., Speech, and Signal Processing*, vol. ASSP-34, pp. 1097-1110, Oct. 1986.

8. J. M. Cioffi and T. Kailath, "Fast recursive-least-squares transversal filters for adaptive filtering," *IEEE Trans. on Acoust., Speech, and Signal Processing*, vol. ASSP-32, pp. 304-337, April 1984.

9. S. Ardalan, "On the sensitivity of transversal RLS algorithms to random perturbations in the filter coefficients," *IEEE Trans. on Acoust., Speech, and Signal Processing*, vol. 36, pp. 1781-1783, Nov. 1988.

10. C. R. Johnson, Jr., *Lectures on Adaptive Parameter Estimation*, Prentice Hall, Englewood Cliffs, NJ, 1988.

11. O. M. Macchi and N. J. Bershad, "Adaptive recovery of a chirped sinusoid in noise, Part 1: performance of the RLS algorithm," *IEEE Trans. on Signal Processing*, vol. 39, pp. 583-594, March 1991.

5.9 PROBLEMS

1. The RLS algorithm is used to predict the signal $x(k) = \cos \frac{\pi k}{3}$ using a second-order FIR filter with the first tap fixed at 1. Given $\lambda = 0.98$, calculate the output signal $y(k)$ and the tap coefficients for the first 10 iterations. Note that we aim the minimization of $E[y^2(k)]$.

   Start with $\mathbf{w}^T(-1) = [1\ 0\ 0]$ and $\delta = 100$.

2. Show that the solution in equation (5.4) is a minimum point.

3. Show that $\mathbf{S}_D(k)$ approaches a null matrix for large k, when $\lambda = 1$.

4. Suppose that the measurement noise $n(k)$ is a random signal with zero-mean and the probability density with normal distribution. In a sufficient-order identification of an FIR system with optimal coefficients given by $\mathbf{w}_o$, show that the least-squares solution with $\lambda = 1$ is also normally distributed with mean $\mathbf{w}_o$ and covariance $E[\mathbf{S}_D(k)\sigma_n^2]$.

5. Prove that equation (5.42) is valid. What is the result when $n(k)$ has zero mean and is correlated to the input signal $x(k)$?

   Hint: You can use the relation $E[e^2(k)] = E[e(k)]^2 + \sigma^2[e(k)]$, where $\sigma^2[\cdot]$ means variance of $[\cdot]$.

6. Consider that the additive noise $n(k)$ is uncorrelated with the input and the desired signals and is also a nonwhite noise with autocorrelation matrix $\mathbf{R}_n$. Determine the transfer function of a prewhitening filter that applied to $d'(k) + n(k)$ and $x(k)$ generates the optimum least-squares solution $\mathbf{w}_o = \mathbf{R}^{-1}\mathbf{p}$ for $k \rightarrow \infty$.

7. Show that if the additive noise is uncorrelated with $d'(k)$ and $x(k)$, and nonwhite, the least-squares algorithm will converge asymptotically to the optimal solution.

8. In problem 4, when $n(k)$ is correlated to $x(k)$, is $\mathbf{w}_o$ still the optimal solution? If not, what is the optimal solution?

9. Show that in the RLS algorithm the following relation is true

$$\xi^d(k) = \lambda \xi^d(k-1) + \varepsilon(k)e(k)$$

   where $e(k)$ is the *a priori* error as defined in equation (5.8).

10. Prove the validity of the approximation in equation (5.80).

11. Demonstrate that the updating formula for the complex RLS algorithm is given by equation (5.98).

12. Show that for an input signal with diagonal dominant correlation matrix $\mathbf{R}$ the following approximation related to equations (C.28) and (C.32) is valid.

$$E\{\mathbf{N}_{\mathbf{S}_D}(k)\mathbf{x}(k)\mathbf{x}^T(k)\text{cov}[\Delta\mathbf{w}(k-1)_Q]\mathbf{x}(k)\mathbf{x}^T(k)\mathbf{N}_{\mathbf{S}_D}(k)\} \approx \sigma^2_{\mathbf{S}_D}\sigma_x^4\text{tr}\,\{\text{cov}[\Delta\mathbf{w}(k-1)_Q]\}\,\mathbf{I}$$

13. Derive the equations (C.35), (C.36), and (C.37).

14. The conventional RLS algorithm is applied to identify a 7th-order time-varying unknown system whose coefficients are first-order Markov processes with $\lambda_{\mathbf{w}} = 0.999$ and $\sigma^2_{\mathbf{w}} = 0.033$. The initial time-varying system multiplier coefficients are

$$\mathbf{w}_o^T = [0.03490\ -0.01100\ -0.06864\ 0.22391\ 0.55686\ 0.35798\ -0.02390\ -0.07594]$$

   The input signal is Gaussian white noise with variance $\sigma_x^2 = 1$ and the measurement noise is also Gaussian white noise independent of the input signal and of the elements of $\mathbf{n}_{\mathbf{w}}(k)$, with variance $\sigma_n^2 = 0.01$.

(a) For $\lambda = 0.97$, compute the excess MSE.

(b) Repeat (a) for $\lambda = \lambda_{\text{opt}}$.

(c) Simulate the experiment described, measure the excess MSE, and compare to the calculated results.

15. Reduce the value of $\lambda_{\mathbf{w}}$ to 0.97 in problem 14, simulate, and comment on the results.

16. Suppose a 15th-order FIR digital filter with multiplier coefficients given below is identified through an adaptive FIR filter of the same order using the conventional RLS algorithm. Consider that fixed-point arithmetic is used.

Additional noise : white noise with variance	$\sigma_n^2 = 0.0015$
Coefficient wordlength:	$b_c = 16$ bits
Signal wordlength:	$b_d = 16$ bits
Input signal: Gaussian white noise with variance	$\sigma_x^2 = 0.7$
	$\lambda = \lambda_{\text{opt}}$

$\mathbf{w}_o^T = [0.0219360 \ \ 0.0015786 \ \ -0.0602449 - 0.0118907 \ \ 0.1375379$
$0.0574545 \ \ -0.3216703 \ \ -0.5287203 \ \ -0.2957797 \ \ 0.0002043 \ \ 0.290670$
$-0.0353349 \ \ -0.0068210 \ \ 0.0026067 \ \ 0.0010333 \ \ -0.0143593]$

(a) Compute the expected value for $||\Delta\mathbf{w}(k)_Q||^2$ and $\xi(k)_Q$ for the described case.

(b) Simulate the identification example described and compare the simulated results with those obtained through the closed form formulas.

(c) Plot the learning curves for the finite- and infinite-precision implementations. Also, plot $E[||\Delta\mathbf{w}(k)||^2]$ versus k in both cases.

17. Repeat the above problem for the following cases

(a) $\sigma_n^2 = 0.01, b_c = 9$ bits, $b_d = 9$ bits, $\sigma_x^2 = 0.7, \lambda = \lambda_{\text{opt}}$.

(b) $\sigma_n^2 = 0.1, b_c = 10$ bits, $b_d = 10$ bits, $\sigma_x^2 = 0.8, \lambda = \lambda_{\text{opt}}$.

(c) $\sigma_n^2 = 0.05, b_c = 8$ bits, $b_d = 16$ bits, $\sigma_x^2 = 0.8, \lambda = \lambda_{\text{opt}}$.

18. In problem 17, compute (do not simulate) $E[||\Delta\mathbf{w}(k)_Q||^2], \xi(k)_Q$, and the probable number of iterations before the algorithm stop updating for $\lambda = 1, \lambda = 0.98, \lambda = 0.96$, and $\lambda = \lambda_{\text{opt}}$.

19. Repeat problem 16 for the case where the input signal is a first-order Markov process with $\lambda_{\mathbf{x}} = 0.95$.

20. A digital channel model can be represented by the following impulse response:

$$[-0.001 \ \ -0.002 \ 0.002 \ 0.2 \ 0.6 \ 0.76 \ 0.9 \ 0.78 \ 0.67 \ 0.58$$
$$0.45 \ 0.3 \ 0.2 \ 0.12 \ 0.06 \ 0 \ \ -0.2 \ \ -1 \ \ -2 \ \ -1 \ 0 \ 0.1]$$

The channel is corrupted by Gaussian noise with power spectrum given by

$$|S(e^{\jmath\omega})|^2 = \kappa'|\omega|^{3/2}$$

where $\kappa' = 10^{-1.5}$. The training signal consists of independent binary samples (-1,1).

Design an FIR equalizer for this problem and use the RLS algorithm. Use a filter of order 50 and plot the learning curve.

21. For the previous problem, using the maximum of 51 adaptive-filter coefficients, implement a DFE equalizer and compare the results with those obtained with the FIR equalizer. Again use the RLS algorithm.

22. Use the complex RLS algorithm to equalize a channel with the transfer function given below. The input signal is a four QAM signal representing a randomly generated bit stream with the signal to noise ratio $\frac{\sigma_{\tilde{x}}^2}{\sigma_n^2} = 20$ at the receiver end, that is, $\tilde{x}(k)$ is the received signal without taking into consideration the additional channel noise. The adaptive filter has 10 coefficients.

$$H(z) = (0.34 - 0.27\jmath) + (0.87 + 0.43\jmath)z^{-1} + (0.34 - 0.21\jmath)z^{-2}$$

(a) Use an appropriate value for λ in the range $0.95 - 0.99$, run the algorithm and comment on the convergence behavior.
(b) Plot the real versus imaginary parts of the received signal before and after equalization.
(c) Increase the number of coefficients to 20 and repeat the experiment in (b).

23. In a system identification problem the input signal is generated from a four QAM of the form

$$x(k) = x_{\text{re}}(k) + \jmath x_{\text{im}}(k)$$

where $x_{\text{re}}(k)$ and $x_{\text{im}}(k)$ assume values ± 1 randomly generated. The unknown system is described by

$$H(z) = 0.5 + 0.2\jmath + (-0.1 + 0.4\jmath)z^{-1} + (0.2 - 0.4\jmath)z^{-2} + (0.2 + 0.7\jmath)z^{-3}$$

The adaptive filter is also a third-order complex FIR filter, and the additional noise is zero-mean Gaussian white noise with variance $\sigma_n^2 = 0.3$. Using the complex RLS algorithm run an ensemble of 20 experiments, and plot the average learning curve.

24. Apply the Kalman filter to equalize the system

$$H(z) = \frac{0.19z}{z - 0.9}$$

when the additional noise is a uniformly distributed white noise with variance $\sigma_n^2 = 0.1$, and the input signal to the channel is a Gaussian noise with unit variance.

6

DATA-SELECTIVE ADAPTIVE FILTERING

6.1 INTRODUCTION

The families of adaptive-filtering algorithms introduced so far present a tradeoff between speed of convergence and the misadjustment after the transient. These characteristics are easily observable in stationary environments. In general fast converging algorithms tend to be very dynamic, a feature not necessarily advantageous after convergence in a stationary environment. In this chapter, an alternative formulation to govern the updating of the adaptive-filter coefficients is introduced. The basic assumption is that the additional noise is considered bounded, and the bound is either known or can be estimated [1]. The key strategy of the formulation is to find a *feasibility set*[1] such that the bounded error specification is met for any member of this set. As a result, the *set-membership filtering* (SMF) is aimed at estimating the feasibility set itself or a member of this set [2].

As a byproduct, the SMF allows the reduction of computational complexity in adaptive filtering, since the filter coefficients are updated only when the output estimation error is higher than the pre-determined upper bound [2]-[3].

Set-membership adaptive filters employ a deterministic objective function related to a bounded error constraint on the filter output, such that the updates belong to a set of feasible solutions. The objective function resembles the prescribed specifications of non-adaptive digital filter design. In the latter, any filter whose amplitude ripples in some frequency bands are smaller than given bounds is an acceptable solution. The main difference is that in the SMF the considered bound applies to the time-domain output error. As compared with their competing algorithms such as the normalized LMS, affine projection, and RLS algorithms [4]-[11], the SMF algorithms lead to reduced computational complexity primarily due to data-selective updates.

Usually the set-membership algorithms perform updates more frequently during the early iterations in stationary environments. As such, if these updates entail more computational complexity than available, some alternative solution is required. A possible strategy to maintain some control on the

[1]This set is defined as the set of filter coefficients leading to output errors whose moduli fall below a prescribed upper bound.

P.S.R. Diniz, *Adaptive Filtering*, DOI: 10.1007/978-0-387-68606-6_6,

amount of computational resources is to adopt partial update, where only a subset of the adaptive-filter coefficients are updated at each iteration. The resulting algorithms are collectively called partial-update (PU) algorithms [14]-[19].

This chapter presents several alternative set-membership algorithms which are closely related to the normalized LMS algorithm [7], the binormalized data-reusing LMS algorithm (here denoted as SM-BNLMS) [10], and the affine projection (SM-AP) [11] algorithm. In addition, this chapter describes the set-membership affine projection algorithm with partial update in some detail. The family of algorithms described in this chapter leads to more flexible management of the computational resources, in comparison with the algorithms presented in the previous chapters.

6.2 SET-MEMBERSHIP FILTERING

The SMF concept is a framework applicable to adaptive-filtering problems that are linear in parameters. The adaptive-filter output is given by

$$y(k) = \mathbf{w}^T \mathbf{x}(k) \tag{6.1}$$

where $\mathbf{x}(k) = [x_0(k)\ x_1(k)\ \ldots x_N(k)]^T$ is the input signal vector, and $\mathbf{w} = [w_0\ w_1 \ldots w_N]^T$ is the parameter vector.

Considering a desired signal sequence $d(k)$ and a sequence of input vectors $\mathbf{x}(k)$, both for for $k = 0, 1, 2, \ldots, \infty$, the estimation error sequence $e(k)$ is calculated as

$$e(k) = d(k) - \mathbf{w}^T \mathbf{x}(k) \tag{6.2}$$

also for $k = 0, 1, 2, \ldots, \infty$. The vectors $\mathbf{x}(k)$ and $\mathbf{w} \in \mathbb{R}^{N+1}$, where $\mathbb{R}$ represents the set of real numbers, whereas $y(k)$ and $e(k)$ represent the adaptive-filter output signal and output error, respectively. The objective of the SMF is to design $\mathbf{w}$ such that the magnitude of estimation output error is upper bounded by a prescribed quantity $\bar{\gamma}$. If the value of $\bar{\gamma}$ is properly chosen there are several valid estimates for $\mathbf{w}$. In summary, any filter parameter leading to a magnitude of the output estimation error smaller than a deterministic threshold is an acceptable solution. From the bounded error constraint results a set of filters rather than a single estimate. If $\bar{\gamma}$ is chosen too small there might be no solution.

Assuming that $\bar{\mathcal{S}}$ denotes the set of all possible input-desired data pairs $(\mathbf{x}, d)$ of interest, it is possible to define Θ as the set of all possible vectors $\mathbf{w}$ leading to output errors whose magnitudes are bounded by $\bar{\gamma}$ whenever $(\mathbf{x}, d) \in \bar{\mathcal{S}}$. The set Θ, called *feasibility set*, is given by

$$\Theta = \bigcap_{(\mathbf{x},d)\in\bar{\mathcal{S}}} \{\mathbf{w} \in \mathbb{R}^{N+1} : |d - \mathbf{w}^T\mathbf{x}| \leq \bar{\gamma}\} \tag{6.3}$$

Now let's consider the practical case where only measured data are available. Given a set of data pairs $\{\mathbf{x}(i), d(i)\}$, for $i = 0, 1, \ldots, k$, let's define $\mathcal{H}(k)$ as the set containing all vectors $\mathbf{w}$ such that the associated output error at time instant k is upper bounded in magnitude by $\bar{\gamma}$. That is,

$$\mathcal{H}(k) = \{\mathbf{w} \in \mathbb{R}^{N+1} : |d(k) - \mathbf{w}^T\mathbf{x}(k)| \leq \bar{\gamma}\} \tag{6.4}$$

The set $\mathcal{H}(k)$ is usually referred to as the *constraint set*. The boundaries of $\mathcal{H}(k)$ are hyperplanes. For the two-dimensional case, where the coefficient vector has two elements, $\mathcal{H}(k)$ comprises the region between the lines where $d(k) - \mathbf{w}^T\mathbf{x}(k) = \pm\bar{\gamma}$ as depicted in Fig. 6.1. For more dimensions, $\mathcal{H}(k)$ represents the region between two parallel hyperplanes in the parameter space $\mathbf{w}$.

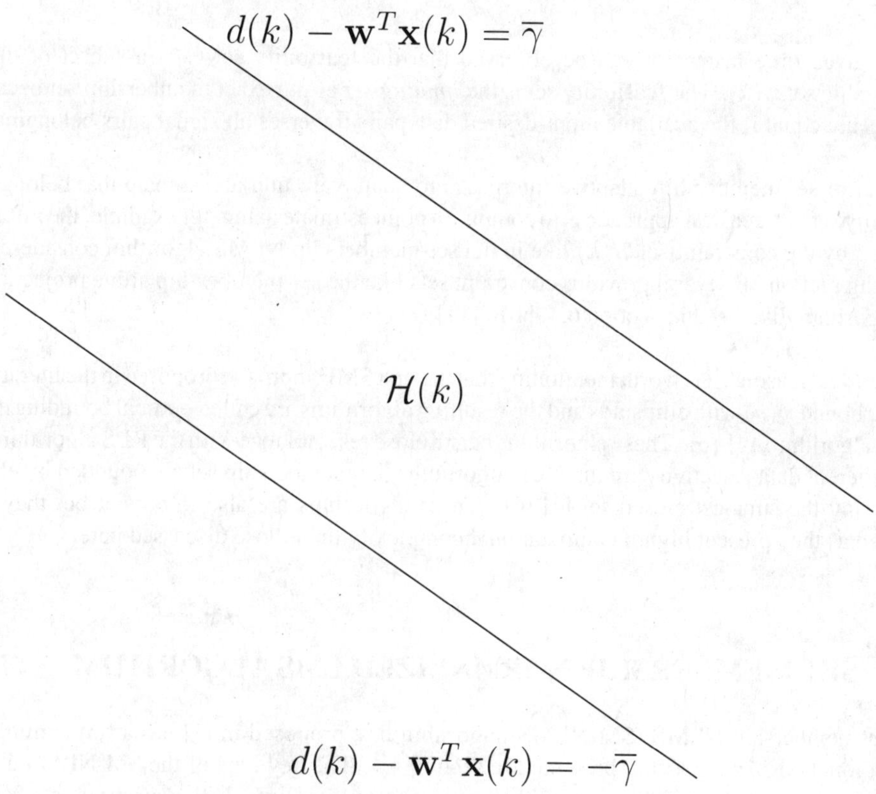

Figure 6.1 Constraint set in $\mathbf{w}$ plane for a two-dimension example.

Since for each data pair there is an associated constraint set, the intersection of the constraint sets over all the available time instants $i = 0, 1, \ldots, k$, is called the *exact membership set* $\psi(k)$, formally defined as

$$\psi(k) = \bigcap_{i=0}^{k} \mathcal{H}(i) \tag{6.5}$$

The set $\psi(k)$ represents a polygon in the parameter space whose location is one of the main objectives of the set-membership filtering.

For a set of data pairs including substantial innovation, the polygon in $\mathbf{w}$, $\psi(k)$, should become small. This property usually occurs after a large number of iterations k, when most likely $\psi(k) = \psi(k-1)$ since $\psi(k-1)$ is entirely contained in the constraint set $\mathcal{H}(k)$ as depicted in Fig. 6.2.a. In this case, the adaptive-filter coefficients do not need updating because the current membership set is totally inside the constraint set, resulting in a selection of update which is data dependent. The selective updating of the set-membership filtering brings about opportunities for power and computational savings, so crucial in devices such as mobile terminals. On the other hand, in the early iterations it is highly possible that the constraint set reduces the size of the membership-set polygon as illustrates Fig. 6.2.b.

At any given time instant, it can be observed that the feasibility set Θ is a subset of the exact membership set $\psi(k)$. The feasibility set is the *limiting set* of the exact membership set because the two sets are equal if the available input-desired data pairs traverses all signal pairs belonging to $\bar{\mathcal{S}}$.

The goal of set-membership adaptive filtering is to adaptively find an estimate that belongs to the feasibility set. The easiest approach is to compute a point estimate using, for example, the information provided by the constraint set $\mathcal{H}(k)$ like in the set-membership NLMS algorithm considered in the following section, or several previous constraint sets like the set-membership affine projection (SM-AP) algorithm discussed in section 6.4, [10], [11].

For historical reasons, it is worth mentioning that the first SMF approach proposed in the literature tries to outer bound $\psi(k)$ with ellipsoids and the resulting algorithms are called optimal bounding ellipsoid (OBE) algorithms [4]-[6]. These algorithms bear a close resemblance with the RLS algorithm [2] and have inherent data selectivity. In the OBE algorithms the membership set are bounded by ellipsoids comprising the smallest closed set [4]-[6]. These algorithms are also important but they are not included as they present higher computational complexity than those discussed here.

6.3 SET-MEMBERSHIP NORMALIZED LMS ALGORITHM

The set-membership NLMS (SM-NLMS) algorithm first proposed in [7] has a form similar to the conventional NLMS algorithm presented in section 4.4. The key idea of the SM-NLMS algorithm is to perform a test to verify if the previous estimate $\mathbf{w}(k)$ lies outside the constraint set $\mathcal{H}(k)$, i.e., $|d(k) - \mathbf{w}^T(k)\mathbf{x}(k)| > \bar{\gamma}$. If the modulus of the error signal is greater than the specified bound, the new estimate $\mathbf{w}(k+1)$ will be updated to the closest boundary of $\mathcal{H}(k)$ at a minimum distance, i.e., the SM-NLMS minimizes $\|\mathbf{w}(k+1) - \mathbf{w}(k)\|^2$ subjected to $\mathbf{w}(k+1) \in \mathcal{H}(k)$ [12]. The updating is performed by an orthogonal projection of the previous estimate onto the closest boundary of $\mathcal{H}(k)$. Fig. 6.3 illustrates the updating procedure of the SM-NLMS algorithm.

In order to derive the update equations, first consider the *a priori* error $e(k)$ given by

$$e(k) = d(k) - \mathbf{w}^T(k)\mathbf{x}(k) \tag{6.6}$$

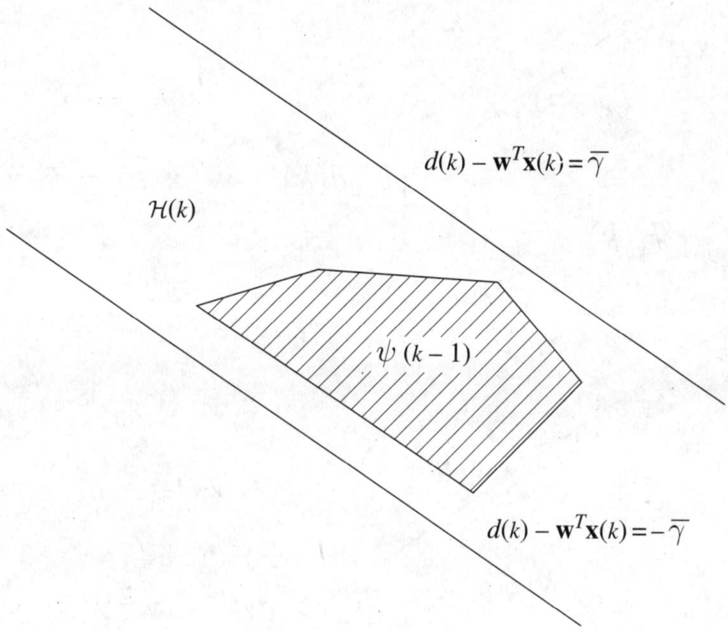

(a) Exact membership set, $\psi\,(k-1)$, contained in the constraint set, $\psi\,(k-1) \subset \mathcal{H}(k)$.

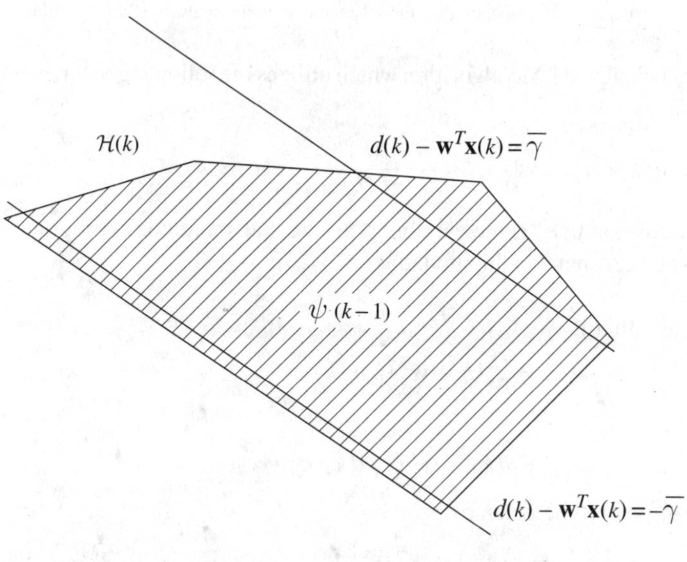

(b) Exact membership set, $\psi\,(k-1)$, not contained in the constraint set, $\psi\,(k-1) \not\subseteq \mathcal{H}(k)$.

Figure 6.2 Exact membership set $\psi(k)$ and its possible intersection with the constraint set $\mathcal{H}(k)$.

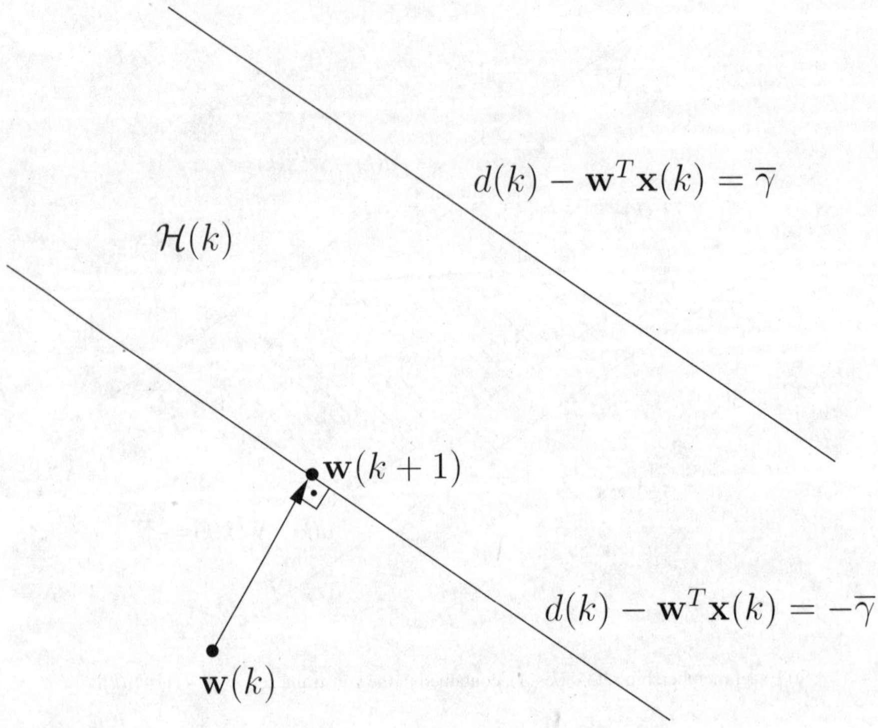

Figure 6.3 Coefficient vector updating for the set-membership normalized LMS algorithm.

then, let's start with the normalized LMS algorithm which utilizes the following recursion for updating $\mathbf{w}(k)$

$$\mathbf{w}(k+1) = \mathbf{w}(k) + \frac{\mu(k)}{\gamma + \mathbf{x}^T(k)\mathbf{x}(k)}\, e(k)\,\mathbf{x}(k) \tag{6.7}$$

where in the present discussion $\mu(k)$ is the variable step size that should be appropriately chosen in order to satisfy the desired set-membership updating.

The update should occur either if

$$e(k) = d(k) - \mathbf{w}^T(k)\mathbf{x}(k) > \bar{\gamma}$$

or

$$e(k) = d(k) - \mathbf{w}^T(k)\mathbf{x}(k) < -\bar{\gamma}$$

and the *a posteriori error* should be given by

$$\begin{aligned}\varepsilon(k) &= d(k) - \mathbf{w}^T(k+1)\mathbf{x}(k) = \pm\bar{\gamma}\\ &= d(k) - \mathbf{w}^T(k)\mathbf{x}(k) - \frac{\mu(k)}{\gamma + \mathbf{x}^T(k)\mathbf{x}(k)}\, e(k)\,\mathbf{x}^T(k)\mathbf{x}(k)\\ &= e(k) - \frac{\mu(k)}{\gamma + \mathbf{x}^T(k)\mathbf{x}(k)}\, e(k)\,\mathbf{x}^T(k)\mathbf{x}(k)\end{aligned} \tag{6.8}$$

where $\varepsilon(k)$ becomes equal to $\pm\bar{\gamma}$ because the coefficients are updated to the closest boundary of $\mathcal{H}(k)$. Since γ, whose only task is regularization, is a small constant it can be disregarded leading to the following equality

$$\varepsilon(k) = e(k)[1 - \mu(k)] = \pm\bar{\gamma} \tag{6.9}$$

The above equation leads to

$$1 - \mu(k) = \pm\frac{\bar{\gamma}}{e(k)} \tag{6.10}$$

where the plus (+) sign applies for the case when $e(k) > 0$ and the minus (−) sign applies for the case where $e(k) < 0$. Therefore, by inspection we conclude that the variable step size, $\mu(k)$, is given by

$$\mu(k) = \begin{cases} 1 - \frac{\bar{\gamma}}{|e(k)|} & \text{if } |e(k)| > \bar{\gamma} \\ 0 & \text{otherwise} \end{cases} \tag{6.11}$$

The updating equations (6.6), (6.11), and (6.7) are quite similar to those of the NLMS algorithm except for the variable step size $\mu(k)$. The SM-NLMS algorithm is outlined in Algorithm 6.1. As a rule of thumb, the value of $\bar{\gamma}$ is chosen around $\sqrt{5}\sigma_n$, where σ_n^2 is the variance of the additional noise, some further discussion in this matter is found in section 6.7 [7], [25].

The reader should recall that the NLMS algorithm minimizes $\|\mathbf{w}(k+1) - \mathbf{w}(k)\|^2$ subjected to the constraint that $\mathbf{w}^T(k+1)\mathbf{x}(k) = d(k)$, as such it is a particular case of the SM-NLMS algorithm by choosing the bound $\bar{\gamma} = 0$. It should be noticed that by using a step size $\mu(k) = 1$ in the SM-NLMS whenever $\mathbf{w}(k) \notin \mathcal{H}(k)$, one performs a valid update since the hyperplane with zero *a posteriori* error lies in $\mathcal{H}(k)$. In this case, the resulting algorithm does not minimize the Euclidean distance $\|\mathbf{w}(k+1) - \mathbf{w}(k)\|^2$ since the *a posteriori* error is zero and less than $\bar{\gamma}$.

6.4 SET-MEMBERSHIP AFFINE PROJECTION ALGORITHM

The exact membership set $\psi(k)$ previously defined in equation (6.5) suggests the use of more constraint-sets in the update [11]. This section generalizes the concept of set-membership in order to conceive algorithms whose updates belong to the past $L+1$ constraint sets. In order to achieve our goal, it is convenient to express $\psi(k)$ as

$$\psi(k) = \left(\bigcap_{i=0}^{k-L-1} \mathcal{H}(i)\right)\left(\bigcap_{j=k-L}^{k} \mathcal{H}(j)\right) = \psi^{k-L-1}(k)\bigcap\psi^{L+1}(k) \tag{6.12}$$

where $\psi^{L+1}(k)$ represents the intersection of the $L+1$ last constraint sets, whereas $\psi^{k-L-1}(k)$ is the intersection of the first $k-L$ constraint sets. The aim of this derivation is to conceive an algorithm whose coefficient update belongs to the last $L+1$ constraint-sets, i.e., $\mathbf{w}(k+1) \in \psi^{L+1}(k)$.

Just like in the original affine projection algorithm of section 4.6, we can retain the last $L+1$ input signal vectors in a matrix as follows:

$$\mathbf{X}_{\text{ap}}(k) = [\mathbf{x}(k)\,\mathbf{x}(k-1)\ldots\mathbf{x}(k-L)] \tag{6.13}$$

Algorithm 6.1

The Set-Membership Normalized LMS Algorithm

Initialization

$\mathbf{x}(0) = \mathbf{w}(0) = [0 \ldots 0]^T$

choose $\bar{\gamma}$ around $\sqrt{5}\sigma_n$

$\gamma =$ small constant

Do for $k \geq 0$

$e(k) = d(k) - \mathbf{x}^T(k)\mathbf{w}(k)$

$\mu(k) = \begin{cases} 1 - \frac{\bar{\gamma}}{|e(k)|} & \text{if } |e(k)| > \bar{\gamma} \\ 0 & \text{otherwise} \end{cases}$

$\mathbf{w}(k+1) = \mathbf{w}(k) + \frac{\mu(k)}{\gamma + \mathbf{x}^T(k)\mathbf{x}(k)} e(k)\,\mathbf{x}(k)$

where $\mathbf{X}_{\text{ap}}(k) \in \mathbb{R}^{(N+1)\times(L+1)}$ contains the corresponding retained inputs, with $\mathbf{x}(k)$ being the input-signal vector

$$\mathbf{x}(k) = [x(k)\ x(k-1)\ \ldots\ x(k-N)]^T \tag{6.14}$$

The vectors representing the data considered at a given iteration k, such as the desired signal and error vectors are given by

$$\mathbf{d}_{\text{ap}}(k) = \begin{bmatrix} d(k) \\ d(k-1) \\ \vdots \\ d(k-L) \end{bmatrix} \tag{6.15}$$

$$\mathbf{e}_{\text{ap}}(k) = \begin{bmatrix} e_{\text{ap},0}(k) \\ e_{\text{ap},1}(k) \\ \vdots \\ e_{\text{ap},L}(k) \end{bmatrix} \tag{6.16}$$

where $\mathbf{d}_{\text{ap}}(k) \in \mathbb{R}^{(L+1)\times 1}$ contains the desired outputs from the $L+1$ last time instants.

Consider that $\mathcal{S}(k-i+1)$ denotes the hyperplane which contains all vectors $\mathbf{w}$ such that $d(k-i+1) - \mathbf{w}^T\mathbf{x}(k-i+1) = \bar{\gamma}_i(k)$, for $i = 1, \ldots, L+1$, where the parameters $\bar{\gamma}_i(k)$ represent the bound constraint to be satisfied by the error magnitudes after coefficient updating. Some particular choices for the parameters $\bar{\gamma}_i(k)$ are discussed later on, for now any choice satisfying the bound constraint is valid. That is, if all $\bar{\gamma}_i(k)$ are chosen such that $|\bar{\gamma}_i(k)| \leq \bar{\gamma}$ then $\mathcal{S}(k-i+1) \in \mathcal{H}(k-i+1)$, for $i = 1, \ldots, L+1$. Vector $\bar{\boldsymbol{\gamma}}(k) \in \mathbb{R}^{(L+1)\times 1}$ specifies the point in $\psi^{L+1}(k)$, where

$$\bar{\boldsymbol{\gamma}}(k) = [\bar{\gamma}_1(k)\ \bar{\gamma}_2(k)\ \ldots\ \bar{\gamma}_{L+1}(k)]^T \tag{6.17}$$

The objective function to be minimized in the set-membership affine projection (SM-AP) algorithm can now be stated. Perform a coefficient update whenever $\mathbf{w}(k) \notin \psi^{L+1}(k)$ in such a way that[2]

$$\min \|\mathbf{w}(k+1) - \mathbf{w}(k)\|^2 \tag{6.18}$$

subject to:

$$\mathbf{d}_{\text{ap}}(k) - \mathbf{X}_{\text{ap}}^T(k)\mathbf{w}(k+1) = \bar{\boldsymbol{\gamma}}(k) \tag{6.19}$$

where the constraint can be rewritten as $d(k-i+1) - \mathbf{x}^T(k-i+1)\mathbf{w}(k+1) = \bar{\gamma}_i(k)$, for $i = 1, \ldots, L+1$. Fig. 6.4 illustrates a typical coefficient update related to the SM-AP algorithm for the case with two coefficients, $L = 1$ and $|\bar{\gamma}_i(k)| < |\bar{\gamma}|$, such that $\mathbf{w}(k+1)$ is not placed at the border of $\mathcal{H}(k)$.

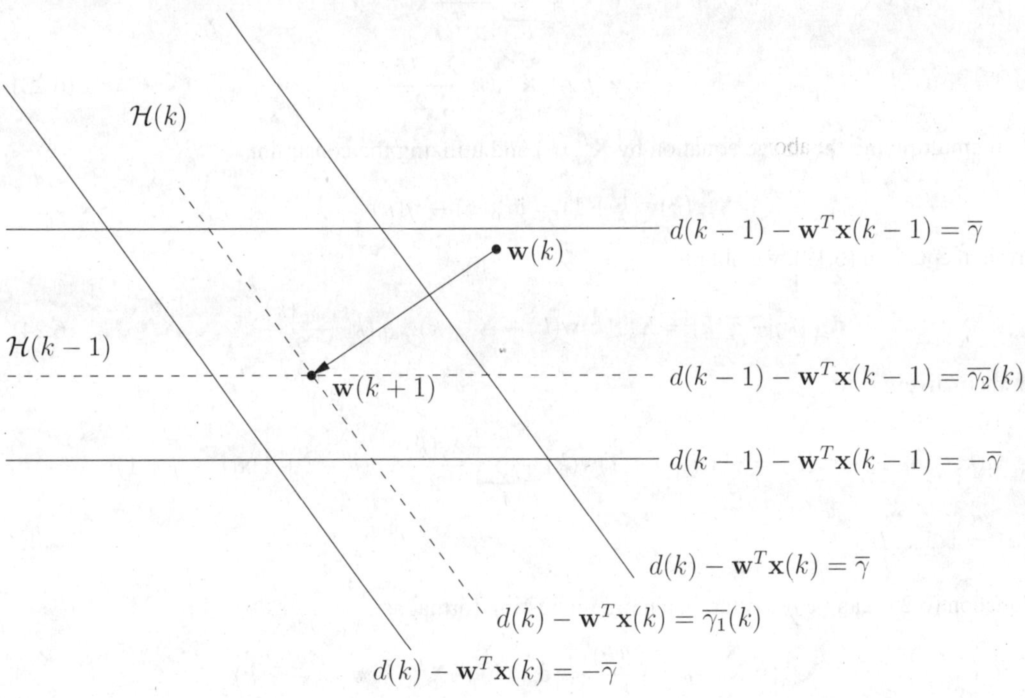

Figure 6.4 SM-AP algorithm coefficient update.

Using the method of Lagrange multipliers [12], the unconstrained function to be minimized is

$$F[\mathbf{w}(k+1)] = \|\mathbf{w}(k+1) - \mathbf{w}(k)\|^2 + \boldsymbol{\lambda}_{\text{ap}}^T(k)[\mathbf{d}_{\text{ap}}(k) - \mathbf{X}_{\text{ap}}^T(k)\mathbf{w}(k+1) - \bar{\boldsymbol{\gamma}}(k)] \tag{6.20}$$

where the vector of Lagrange multipliers, $\boldsymbol{\lambda}_{\text{ap}}(k) \in \mathbb{R}^{(L+1)\times 1}$, is given by

$$\boldsymbol{\lambda}_{\text{ap}}(k) = [\lambda_{\text{ap},1}(k)\ \lambda_{\text{ap},2}(k) \ldots\ \lambda_{\text{ap},L+1}(k)]^T \tag{6.21}$$

[2]The reader should note that in earlier definition of the objective function related to the affine projection algorithm a constant $\frac{1}{2}$ was multiplied to the norm to be minimized. This constant is not relevant and is only used when it simplifies the algorithm derivation.

such that the constraints can be rewritten in the above equation as follows

$$\begin{aligned} F[\mathbf{w}(k+1)] &= \|\mathbf{w}(k+1) - \mathbf{w}(k)\|^2 \\ &+ \sum_{i=1}^{L+1} \lambda_{\text{ap},i}(k)[d(k-i+1) - \mathbf{x}^T(k-i+1)\mathbf{w}(k+1) - \bar{\gamma}_i(k)] \end{aligned} \tag{6.22}$$

We solve the minimization problem of equation (6.18) by first setting the gradient of the function $F[\mathbf{w}(k+1)]$ with respect to $\mathbf{w}(k+1)$ equal to zero, in order to derive the following equation

$$\begin{aligned} \mathbf{w}(k+1) &= \mathbf{w}(k) + \sum_{i=1}^{L+1} \frac{\lambda_i(k)}{2}\mathbf{x}(k-i+1) \\ &= \mathbf{w}(k) + \mathbf{X}_{\text{ap}}(k)\frac{\boldsymbol{\lambda}_{\text{ap}}(k)}{2} \end{aligned} \tag{6.23}$$

By premultiplying the above equation by $\mathbf{X}_{\text{ap}}^T(k)$ and utilizing the constraints

$$\mathbf{X}_{\text{ap}}^T(k)\mathbf{w}(k+1) = \mathbf{d}_{\text{ap}}(k) - \bar{\boldsymbol{\gamma}}(k)$$

given in equation (6.19), we obtain

$$\mathbf{d}_{\text{ap}}(k) - \bar{\boldsymbol{\gamma}}(k) = \mathbf{X}_{\text{ap}}^T(k)\mathbf{w}(k) + \mathbf{X}_{\text{ap}}^T(k)\mathbf{X}_{\text{ap}}(k)\frac{\boldsymbol{\lambda}_{\text{ap}}(k)}{2} \tag{6.24}$$

or alternatively

$$d(k-i+1) - \bar{\gamma}_i(k) = \mathbf{x}^T(k-i+1)\mathbf{w}(k) + \sum_{j=1}^{L+1} \frac{\lambda_j(k)}{2}\mathbf{x}^T(k-i+1)\mathbf{x}(k-j+1)$$

for $i = 1, \ldots, L+1$.

Equation (6.24) can be rewritten in a more interesting format as

$$\begin{aligned} \mathbf{X}_{\text{ap}}^T(k)\mathbf{X}_{\text{ap}}(k)\frac{\boldsymbol{\lambda}_{\text{ap}}(k)}{2} &= \mathbf{d}_{\text{ap}}(k) - \mathbf{X}_{\text{ap}}^T(k)\mathbf{w}(k) - \bar{\boldsymbol{\gamma}}(k) \\ &= \mathbf{e}_{\text{ap}}(k) - \bar{\boldsymbol{\gamma}}(k) \end{aligned} \tag{6.25}$$

leading to

$$\frac{\boldsymbol{\lambda}_{\text{ap}}(k)}{2} = \left[\mathbf{X}_{\text{ap}}^T(k)\mathbf{X}_{\text{ap}}(k)\right]^{-1}[\mathbf{e}_{\text{ap}}(k) - \bar{\boldsymbol{\gamma}}(k)] \tag{6.26}$$

It is now possible to derive the updating equation by starting from equation (6.23) with $\boldsymbol{\lambda}_{\text{ap}}(k)$ being given by equation (6.26), i.e.,

$$\mathbf{w}(k+1) = \begin{cases} \mathbf{w}(k) + \mathbf{X}_{\text{ap}}(k)\left[\mathbf{X}_{\text{ap}}^T(k)\mathbf{X}_{\text{ap}}(k)\right]^{-1}[\mathbf{e}_{\text{ap}}(k) - \bar{\boldsymbol{\gamma}}(k)] & \text{if } |e(k)| > \bar{\gamma} \\ \mathbf{w}(k) & \text{otherwise} \end{cases} \tag{6.27}$$

Algorithm 6.2

The Set-Membership Affine Projection Algorithm

Initialization
$\mathbf{x}(0) = \mathbf{w}(0) = [0 \ \ldots \ 0]^T$
choose $\bar{\gamma}$ around $\sqrt{5}\sigma_n$
$\gamma =$ small constant
Do for $k \geq 0$
$\mathbf{e}_{\text{ap}}(k) = \mathbf{d}_{\text{ap}}(k) - \mathbf{X}_{\text{ap}}^T(k)\mathbf{w}(k)$

$$\mathbf{w}(k+1) = \begin{cases} \mathbf{w}(k) + \mathbf{X}_{\text{ap}}(k)\left[\mathbf{X}_{\text{ap}}^T(k)\mathbf{X}_{\text{ap}}(k) + \gamma\mathbf{I}\right]^{-1}\left[\mathbf{e}_{\text{ap}}(k) - \bar{\boldsymbol{\gamma}}(k)\right] & \text{if } |e(k)| > \bar{\gamma} \\ \mathbf{w}(k) & \text{otherwise} \end{cases}$$

where

$$\mathbf{e}_{\text{ap}}(k) = [e(k) \ \varepsilon(k-1) \ \ldots \ \varepsilon(k-L)]^T \tag{6.28}$$

with $\varepsilon(k-i) = d(k-i) - \mathbf{x}^T(k-i)\mathbf{w}(k)$ denoting the *a posteriori* error calculated with the data pair of iteration $k-i$ using the coefficients of iteration k. Algorithm 6.2 describes in detail the general form of the SM-AP algorithm.

Several properties related to the SM-AP algorithm are straightforward to infer.

- For time instants $k < L+1$, i.e., during initialization, we can only assume knowledge of $\mathcal{H}(i)$ for $i = 0, 1, \ldots, k$. As a result, if an update is needed when $k < L+1$, the algorithm is used with the only $k+1$ constraint sets available.
- In order to verify if an update $\mathbf{w}(k+1)$ is required, we only have to check if $\mathbf{w}(k) \notin \mathcal{H}(k)$ since due to previous updates $\mathbf{w}(k) \in \mathcal{H}(k-i+1)$ holds for $i = 2, \ldots, L+1$.
- By choosing the bound $\bar{\gamma} = 0$, it is possible to verify that the algorithm becomes the conventional AP algorithm with unity step-size.

6.4.1 A Trivial Choice for Vector $\bar{\boldsymbol{\gamma}}(k)$

In the above discussions no specific choice for the parameters $\bar{\gamma}_i(k)$ has been discussed except for the requirement that the adaptive-filter coefficients should be in $\mathcal{H}(k-i+1)$, meaning that $|\bar{\gamma}_i(k)| \leq \bar{\gamma}$. There is infinite number of possible choices for $\bar{\gamma}_i(k)$, each leading to a different update.

The most trivial choice would be $\bar{\boldsymbol{\gamma}}(k) = \mathbf{0}$, i.e., to force the *a posteriori* errors to be zero at the last $L+1$ time instants. If we replace $\bar{\boldsymbol{\gamma}}(k) = \mathbf{0}$ in equation (6.24) and solving for $\boldsymbol{\lambda}_{\text{ap}}(k)$ the following recursions result

$$\frac{\boldsymbol{\lambda}_{\text{ap}}(k)}{2} = \left(\mathbf{X}_{\text{ap}}^T(k)\mathbf{X}_{\text{ap}}(k)\right)^{-1}\mathbf{e}_{\text{ap}}(k) \tag{6.29}$$

The update recursion is given by

$$\mathbf{w}(k+1) = \begin{cases} \mathbf{w}(k) + \mathbf{X}_{\text{ap}}(k)\left(\mathbf{X}_{\text{ap}}^T(k)\mathbf{X}_{\text{ap}}(k)\right)^{-1}\mathbf{e}_{\text{ap}}(k) & \text{if } |e(k)| > \bar{\gamma} \\ \mathbf{w}(k) & \text{otherwise} \end{cases} \tag{6.30}$$

The above updating equation is identical to the conventional affine-projection (AP) algorithm with unity step size whenever an update takes place, that is, $\mathbf{w}(k) \notin \mathcal{H}(k)$. However, owing to the data selectivity, the SM-AP algorithm leads to considerable reduction in complexity as compared with the conventional AP algorithm. Fig. 6.5 depicts a typical coefficient update, where for illustration purposes $\mathbf{w}(k)$ does not lie in the zero *a posteriori* hyperplane belonging to $\mathcal{H}(k-1)$.

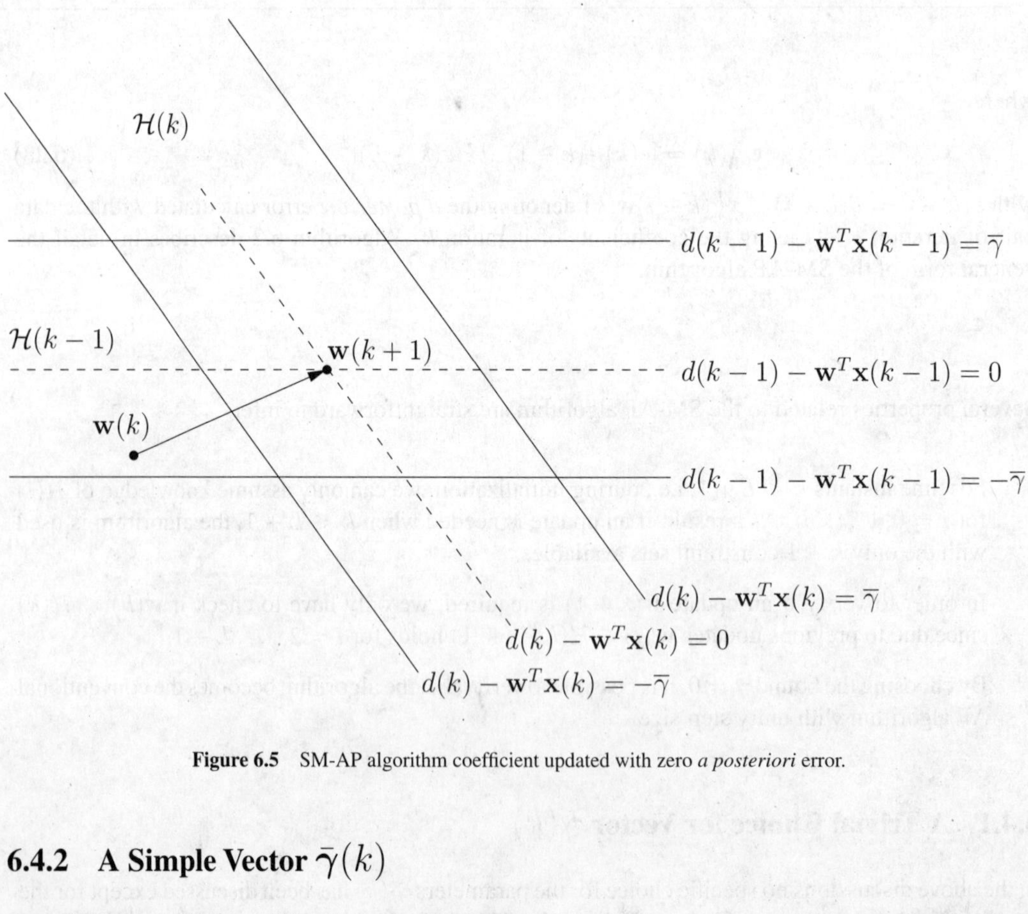

Figure 6.5 SM-AP algorithm coefficient updated with zero *a posteriori* error.

6.4.2 A Simple Vector $\bar{\boldsymbol{\gamma}}(k)$

Any choice for $\bar{\gamma}_i(k)$ is valid as long as they correspond to points represented by the adaptive-filter coefficients in $\mathcal{H}(k-i+1)$, i.e., $|\bar{\gamma}_i(k)| \leq \bar{\gamma}$. One can exploit this freedom in order to make the resulting algorithm more suitable for a target application. A particularly simple SM-AP version is obtained if $\bar{\gamma}_i(k)$ for $i \neq 1$ corresponds to the *a posteriori* error $\varepsilon(k-i+1) = d(k-i+1) - \mathbf{w}^T(k)\mathbf{x}(k-i+1)$ and $\bar{\gamma}_1(k) = e(k)/|e(k)|$. Since the coefficients were updated considering previous data pairs then at this point it is true that $\mathbf{w}(k) \in \mathcal{H}(k-i+1)$, i.e., $|\varepsilon(k-i+1)| = |d(k-i+1) - \mathbf{x}^T(k-i+1)\mathbf{w}(k)| \leq \bar{\gamma}$, for $i = 2, \ldots, L+1$. Therefore, by choosing $\bar{\gamma}_i(k) = \varepsilon(k-i+1)$, for $i \neq 1$, all the elements on the right-hand side of equation (6.24) become zero, except for first element.

It is only left now the choice of the constraint value $\bar{\gamma}_1(k)$, that can be selected as in the SM-NLMS algorithm where $\bar{\gamma}_1(k)$ is such that the solution lies at the nearest boundary of $\mathcal{H}(k)$, i.e.,

$$\bar{\gamma}_1(k) = \bar{\gamma}\frac{e(k)}{|e(k)|} \tag{6.31}$$

Such choices utilized in equation (6.25) leads to

$$\mathbf{X}_{\mathrm{ap}}^T(k)\mathbf{X}_{\mathrm{ap}}(k)\frac{\boldsymbol{\lambda}_{\mathrm{ap}}(k)}{2} = \mu(k)e(k)\mathbf{u}_1 \tag{6.32}$$

where $\mu(k) = 1 - \frac{\bar{\gamma}}{|e(k)|}$ and $\mathbf{u}_1 = [1\ 0\ \ldots\ 0]^T$.

The resulting update equation is then given by

$$\mathbf{w}(k+1) = \mathbf{w}(k) + \mathbf{X}_{\mathrm{ap}}(k)\left[\mathbf{X}_{\mathrm{ap}}^T(k)\mathbf{X}_{\mathrm{ap}}(k)\right]^{-1}\mu(k)e(k)\mathbf{u}_1 \tag{6.33}$$

where

$$e(k) = d(k) - \mathbf{w}^T(k)\mathbf{x}(k) \tag{6.34}$$

$$\mu(k) = \begin{cases} 1 - \frac{\bar{\gamma}}{|e(k)|} & \text{if } |e(k)| > \bar{\gamma} \\ 0 & \text{otherwise} \end{cases} \tag{6.35}$$

This algorithm minimizes the Euclidean distance $\|\mathbf{w}(k+1) - \mathbf{w}(k)\|^2$ subject to the constraint $\mathbf{w}(k+1) \in \psi^{L+1}(k)$ such that the *a posteriori* errors at iteration $k-i$, $\varepsilon(k-i)$, are kept constant for $i = 2, \ldots, L+1$. Fig. 6.6 illustrates a typical coefficient updating for the simplified SM-AP algorithm where it is observed that the *a posteriori* error related to previous data remains unaltered.

The simplified SM-AP algorithm given by equation (6.33) will perform an update if and only if $\mathbf{w}(k) \notin \mathcal{H}(k)$, or $e(k) > \bar{\gamma}$. The step-by-step description of the simplified SM-AP algorithm is presented in Algorithm 6.3.

6.4.3 Reducing the Complexity in the Simplified SM-AP Algorithm

In the updating expression of equation (6.33) vector $\mathbf{u}_1$ has a special form which can be exploited in order to reduce the computational complexity.

The inverse matrix in equation (6.33) can be partitioned as

$$\begin{aligned}\left[\mathbf{X}_{\mathrm{ap}}^T(k)\mathbf{X}_{\mathrm{ap}}(k)\right]^{-1} &= \left\{\left[\mathbf{x}(k)\ \tilde{\mathbf{X}}_{\mathrm{ap}}(k)\right]^T\left[\mathbf{x}(k)\ \tilde{\mathbf{X}}_{\mathrm{ap}}(k)\right]\right\}^{-1} \\ &= \begin{bmatrix} a & \mathbf{b}^T \\ \mathbf{b} & \mathbf{C}\end{bmatrix}\end{aligned} \tag{6.36}$$

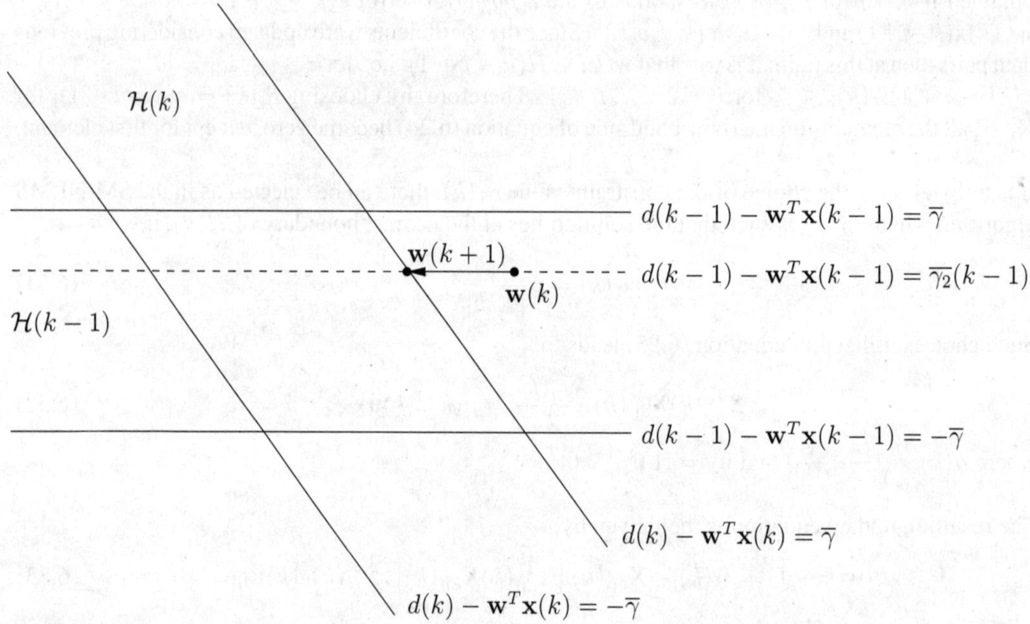

Figure 6.6 Simplified SM-AP algorithm coefficient update with constant *a posteriori* error.

Algorithm 6.3

The Simplified Set-Membership Affine Projection Algorithm

Initialization

$\mathbf{x}(0) = \mathbf{w}(0) = [0 \ \ldots \ 0]^T$

choose $\bar{\gamma}$ around $\sqrt{5}\sigma_n$

$\gamma =$ small constant

Do for $k \geq 0$

$\mathbf{e}_{\text{ap}}(k) = \mathbf{d}_{\text{ap}}(k) - \mathbf{X}^T_{\text{ap}}(k)\mathbf{w}(k)$

$$\mu(k) = \begin{cases} 1 - \frac{\bar{\gamma}}{|e(k)|} & \text{if } |e(k)| > \bar{\gamma} \\ 0 & \text{otherwise} \end{cases}$$

$$\mathbf{w}(k+1) = \mathbf{w}(k) + \mathbf{X}_{\text{ap}}(k)\left[\mathbf{X}^T_{\text{ap}}(k)\mathbf{X}_{\text{ap}}(k) + \gamma\mathbf{I}\right]^{-1}\mu(k)e(k)\mathbf{u}_1$$

where

$$a = \left[\varphi^T(k)\varphi(k)\right]^{-1} \tag{6.37}$$

$$\mathbf{b} = -\left[\tilde{\mathbf{X}}_{\text{ap}}^T(k)\tilde{\mathbf{X}}_{\text{ap}}(k)\right]^{-1} \tilde{\mathbf{X}}_{\text{ap}}^T(k)\mathbf{x}(k)a \tag{6.38}$$

with $\varphi(k)$ defined as

$$\varphi(k) = \mathbf{x}(k) - \tilde{\mathbf{X}}_{\text{ap}}(k)\left[\tilde{\mathbf{X}}_{\text{ap}}^T(k)\tilde{\mathbf{X}}_{\text{ap}}(k)\right]^{-1} \tilde{\mathbf{X}}_{\text{ap}}^T(k)\mathbf{x}(k) \tag{6.39}$$

where the vector $\varphi(k) \in \mathbb{R}^{(N+1)\times 1}$, see problem 12.

As a result,

$$\begin{aligned}
\mathbf{X}_{\text{ap}}(k)\left[\mathbf{X}_{\text{ap}}^T(k)\mathbf{X}_{\text{ap}}(k)\right]^{-1}\mathbf{u}_1 &= \left[\mathbf{x}(k)\;\tilde{\mathbf{X}}_{\text{ap}}(k)\right]\begin{bmatrix} a \\ \mathbf{b} \end{bmatrix} \\
&= \left[\mathbf{x}(k)\;\tilde{\mathbf{X}}_{\text{ap}}(k)\right]\begin{bmatrix} 1 \\ \frac{\mathbf{b}}{a} \end{bmatrix} a \\
&= \left[\mathbf{x}(k) - \left[\tilde{\mathbf{X}}_{\text{ap}}^T(k)\tilde{\mathbf{X}}_{\text{ap}}(k)\right]^{-1}\tilde{\mathbf{X}}_{\text{ap}}^T(k)\mathbf{x}(k)\right] a \\
&= \varphi(k)\left[\varphi^T(k)\varphi(k)\right]^{-1}
\end{aligned} \tag{6.40}$$

where the last equality follows from equations (6.37) and (6.39).

An efficient expression for the coefficient update is obtained using the partition in equation (6.36), that is

$$\mathbf{w}(k+1) = \mathbf{w}(k) + \frac{\varphi(k)}{\varphi^T(k)\varphi(k)}e(k) \tag{6.41}$$

where $\varphi(k)$ is defined as in equation (6.39). This representation of the SM-AP algorithm is computationally attractive since it utilizes matrices with lower dimensions than those presented in equation (6.33), specifically matrix $\left[\tilde{\mathbf{X}}_{\text{ap}}^T(k)\tilde{\mathbf{X}}_{\text{ap}}(k)\right]$ in equation (6.39) has dimension $L \times L$ whereas matrix $\left[\mathbf{X}_{\text{ap}}^T(k)\mathbf{X}_{\text{ap}}(k)\right]$ in equation (6.33) has dimension $(L+1)\times(L+1)$. The number of reuses L is in most of the cases chosen in the range $0 \leq L \leq 5$, therefore the strategy for reducing the computational burden of the inversion brings about significant benefit.

6.5 SET-MEMBERSHIP BINORMALIZED LMS ALGORITHMS

In the SM-AP algorithm the computational complexity is directly related to the number of data reuses. The main component of the computational burden is the information matrix inversion. Since the SM-NLMS algorithm only considers the constraint set $\mathcal{H}(k)$ in its update, it has low complexity per update whereas its convergence speed follows the pattern of the NLMS algorithm. Both algorithms have their convergence speed governed by the eigenvalue spread of the input-signal correlation matrix. In order to alleviate this drawback while keeping the implementation complexity as low as possible, an attractive particular solution for the SM-AP algorithm is the set-membership binormalized LMS (SM-BNLMS) algorithm. Two algorithms are derived requiring that the solution belongs to the constraint sets at time instants k and $k-1$, i.e., $\mathbf{w}(k+1) \in \mathcal{H}(k) \cap \mathcal{H}(k-1)$, which are general cases of the binormalized LMS algorithm [13]. The SM-BNLMS algorithms can be seen as extensions of the SM-NLMS algorithm that use two consecutive constraint sets for each update, and also as special cases of the SM-AP algorithms.

Let's assume $\mathcal{S}(k-i+1)$, for $i=1,2$, denote the hyperplanes which contain all vectors $\mathbf{w}$ such that $d(k-i+1) - \mathbf{w}^T\mathbf{x}(k-i+1) = \bar{\gamma}_i(k)$, where $\bar{\gamma}_i(k)$ are the values of the bound constraints that should be met in order to validate a given estimate. Specifically, if $\bar{\gamma}_i(k)$, for $i=1,2$, are chosen such that $|\bar{\gamma}_i(k)| \leq \bar{\gamma}$, then $\mathcal{S}(k-i+1) \in \mathcal{H}(k-i+1)$ [10].

Whenever $\mathbf{w}(k) \notin \mathcal{H}(k) \cap \mathcal{H}(k-1)$, we can propose an objective function such as

$$\begin{aligned}
&\min \|\mathbf{w}(k+1) - \mathbf{w}(k)\|^2 \\
&\text{subject to:} \\
&d(k) - \mathbf{x}^T(k)\mathbf{w}(k+1) = \bar{\gamma}_1(k) \\
&d(k-1) - \mathbf{x}^T(k-1)\mathbf{w}(k+1) = \bar{\gamma}_2(k)
\end{aligned} \tag{6.42}$$

where the pair of thresholds $(\bar{\gamma}_1(k), \bar{\gamma}_2(k))$ specifies the point in $\mathcal{H}(k) \cap \mathcal{H}(k-1)$ where the final parameter estimate will be placed. The previously shown Fig. 6.4 illustrates how the coefficients are updated to prescribed *a posteriori* errors determined by $(\bar{\gamma}_1(k), \bar{\gamma}_2(k))$.

In principle there is a need to verify if an update according to equation (6.42) is required, where such an update can be skip if $\mathbf{w}(k) \in \mathcal{H}(k) \cap \mathcal{H}(k-1)$. There are ways of keeping $\mathbf{w}(k+1) \in \mathcal{H}(k-1)$ whenever an update is required, that is, whenever $\mathbf{w}(k) \notin \mathcal{H}(k)$. This type of solution is discussed further in subsection 6.5.2. At any rate, we can solve the general constrained minimization problem of equation (6.42) for the binormalized case by applying Lagrange multiplier method, resulting in the following unconstrained objective function

$$\begin{aligned}
F[\mathbf{w}(k+1)] &= \|\mathbf{w}(k+1) - \mathbf{w}(k)\|^2 + \lambda_1(k)[d(k) - \mathbf{x}^T(k)\mathbf{w}(k+1) - \bar{\gamma}_1(k)] \\
&\quad + \lambda_2(k)[d(k-1) - \mathbf{x}^T(k-1)\mathbf{w}(k+1) - \bar{\gamma}_2(k)]
\end{aligned} \tag{6.43}$$

By computing the gradient of equation (6.43) with respect to $\mathbf{w}(k+1)$, setting the result to zero, we get

$$\begin{aligned}
\mathbf{w}(k+1) &= \mathbf{w}(k) + \mathbf{X}_{\text{ap}}(k)\frac{\boldsymbol{\lambda}_{\text{ap}}(k)}{2} \\
&= \mathbf{w}(k) + [\mathbf{x}(k)\,\mathbf{x}(k-1)]\begin{bmatrix} \frac{\lambda_1(k)}{2} \\ \frac{\lambda_2(k)}{2} \end{bmatrix}
\end{aligned} \tag{6.44}$$

where this expression is the specialized form of equation (6.23) to the binormalized case.

The Lagrange multipliers are obtained by replacing equation (6.44) in the constraints of equation (6.42) such that

$$\begin{bmatrix} \mathbf{x}^T(k) \\ \mathbf{x}^T(k-1) \end{bmatrix} [\mathbf{x}(k)\ \mathbf{x}(k-1)] \begin{bmatrix} \frac{\lambda_1(k)}{2} \\ \frac{\lambda_2(k)}{2} \end{bmatrix} = \begin{bmatrix} d(k) \\ d(k-1) \end{bmatrix} - \begin{bmatrix} \mathbf{x}^T(k) \\ \mathbf{x}^T(k-1) \end{bmatrix} \mathbf{w}(k) - \bar{\gamma}(k)$$

$$= \begin{bmatrix} e(k) \\ \varepsilon(k-1) \end{bmatrix} - \begin{bmatrix} \bar{\gamma}_1(k) \\ \bar{\gamma}_2(k) \end{bmatrix} \tag{6.45}$$

By solving the above equation we obtain

$$\frac{\lambda_1(k)}{2} = \frac{[e(k) - \bar{\gamma}_1(k)]\,\|\mathbf{x}(k-1)\|^2 - [\varepsilon(k-1) - \bar{\gamma}_2(k)]\,\mathbf{x}^T(k)\mathbf{x}(k-1)}{\|\mathbf{x}(k)\|^2\|\mathbf{x}(k-1)\|^2 - [\mathbf{x}^T(k-1)\mathbf{x}(k)]^2} \tag{6.46}$$

$$\frac{\lambda_2(k)}{2} = \frac{[\varepsilon(k-1) - \bar{\gamma}_2(k)]\,\|\mathbf{x}(k)\|^2 - [e(k) - \bar{\gamma}_1(k)]\,\mathbf{x}^T(k-1)\mathbf{x}(k)}{\|\mathbf{x}(k)\|^2\|\mathbf{x}(k-1)\|^2 - [\mathbf{x}^T(k-1)\mathbf{x}(k)]^2} \tag{6.47}$$

where the errors in the above equations are the *a priori* error at iteration k, defined as $e(k) = d(k) - \mathbf{w}^T(k)\mathbf{x}(k)$, and the *a posteriori* error at iteration $k-1$, defined as $\varepsilon(k-1) = d(k-1) - \mathbf{w}^T(k)\mathbf{x}(k-1)$.

The expression for the coefficient update of the SM-BNLMS algorithm is then given by

$$\mathbf{w}(k+1) = \begin{cases} \mathbf{w}(k) + \frac{\lambda_1(k)}{2}\mathbf{x}(k) + \frac{\lambda_2(k)}{2}\mathbf{x}(k-1) & \text{if } |e(k)| > \bar{\gamma} \\ \mathbf{w}(k) & \text{otherwise} \end{cases} \tag{6.48}$$

Some special forms of the SM-BNLMS algorithm are following discussed.

6.5.1 SM-BNLMS Algorithm 1

The first form of the SM-BNLMS algorithm is derived by employing two steps, where in each step we minimize the Euclidean distance between the old filter coefficients and the new update, subjected to the constraint that the new update lies in constraint set $\mathcal{H}(k)$. Then, we test if the new update belongs in the previous constraint set $\mathcal{H}(k-1)$ and if not a new update takes place. Basically, the SM-BNLMS algorithm 1 performs a step according to the SM-NLMS algorithm and if the solution belongs to both constraint sets $\mathcal{H}(k)$ and $\mathcal{H}(k-1)$ no further update is required. If the initial step moves the solution away from $\mathcal{H}(k-1)$, then a second update is performed in order to place the solution at the intersection of $\mathcal{H}(k)$ and $\mathcal{H}(k-1)$ at a minimum distance from $\mathbf{w}(k)$. Fig. 6.7 illustrates the coefficient updates according to the situations discussed so far. As desired, the SM-BNLMS algorithm 1 minimizes $\|\mathbf{w}(k+1) - \mathbf{w}(k)\|^2$ subject to the constraint that $\mathbf{w}(k+1) \in \mathcal{H}(k) \cap \mathcal{H}(k-1)$.

The updating equation for the SM-BNLMS algorithm 1 can be derived by first performing an orthogonal projection of $\mathbf{w}(k)$ onto the nearest boundary of $\mathcal{H}(k)$ just like in the SM-NLMS algorithm

$$\hat{\mathbf{w}}(k) = \mathbf{w}(k) + \mu(k)\frac{e(k)\mathbf{x}(k)}{\|\mathbf{x}(k)\|^2} \tag{6.49}$$

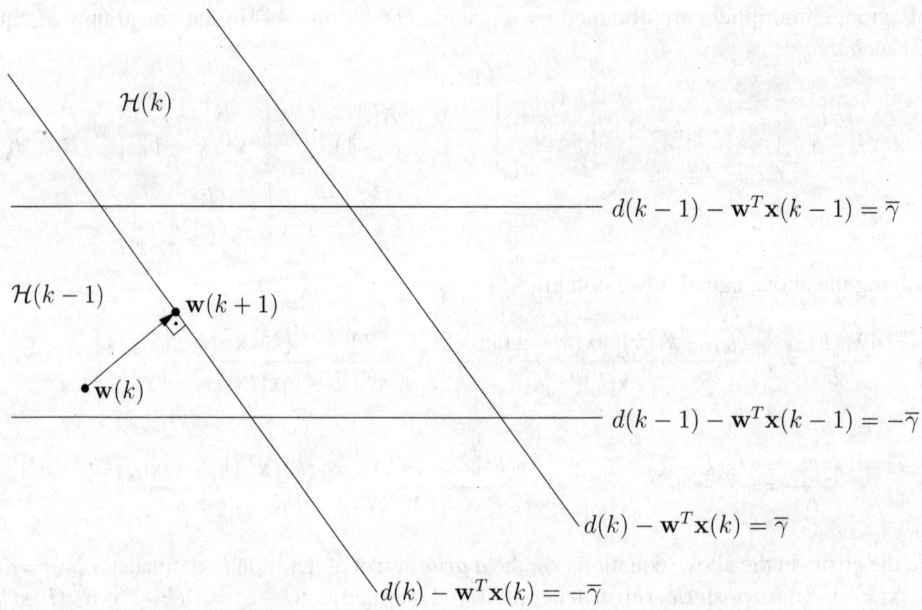

(a) $\mathbf{w}(k+1) \in \mathcal{H}(k) \cap \mathcal{H}(k-1)$, no further update.

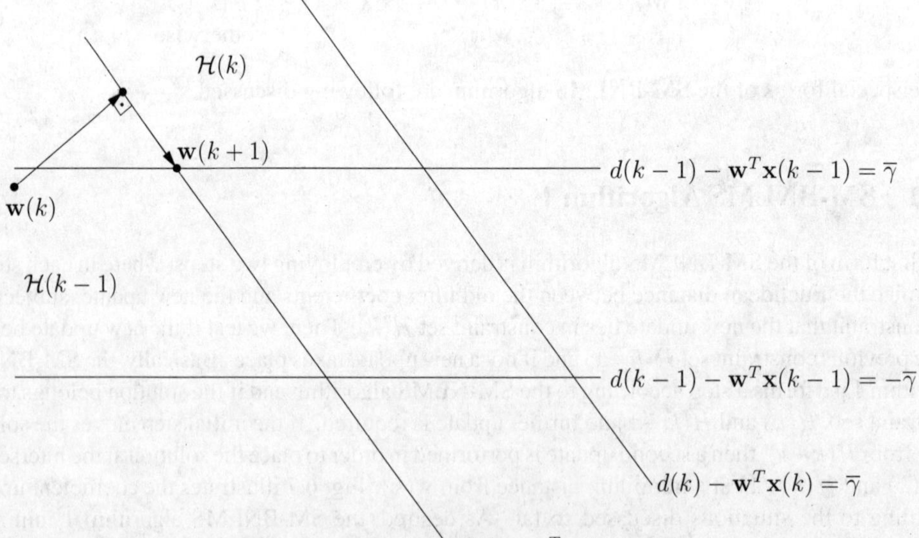

(b) $\mathbf{w}(k+1) \notin \mathcal{H}(k-1)$, further update.

Figure 6.7 Possible coefficient updates for the SM-BNLMS algorithm 1.

where $\mu(k)$ is the variable convergence factor given by equation (6.11) and $e(k)$ is the *a priori* output error defined in equation (6.6). If $\hat{\mathbf{w}}(k) \in \mathcal{H}(k-1)$, i.e., $|d(k-1) - \hat{\mathbf{w}}^T(k)\mathbf{x}(k-1)| \leq \bar{\gamma}$, no further update is required, therefore $\mathbf{w}(k+1) = \hat{\mathbf{w}}(k)$. On the other hand, if $\hat{\mathbf{w}}(k) \notin \mathcal{H}(k-1)$ a second step is necessary in order to move the solution to the intersection of $\mathcal{H}(k)$ and $\mathcal{H}(k-1)$ at a minimum distance. This second step is performed in the orthogonal direction with respect to the first step, namely $\mathbf{x}^{\perp}(k)$. The resulting second updating is then performed in the following form

$$\mathbf{w}(k+1) = \hat{\mathbf{w}}(k) + \hat{\mu}(k)\frac{\varepsilon(k-1)\mathbf{x}^{\perp}(k)}{\|\mathbf{x}^{\perp}(k)\|^2} \tag{6.50}$$

where

$$\mathbf{x}^{\perp}(k) = \left(\mathbf{I} - \frac{\mathbf{x}(k)\mathbf{x}^T(k)}{\|\mathbf{x}(k)\|^2}\right)\mathbf{x}(k-1) \tag{6.51}$$

$$\varepsilon(k-1) = d(k-1) - \hat{\mathbf{w}}^T(k)\mathbf{x}(k-1) \tag{6.52}$$

$$\hat{\mu}(k) = 1 - \frac{\bar{\gamma}}{|\varepsilon(k-1)|} \tag{6.53}$$

Algorithm 6.4 describes in detail the SM-BNLMS algorithm 1, where we utilized an explicit form for $\mathbf{x}^{\perp}(k)$, see problem 2. It is straightforward to observe that if the bound of the estimation error is chosen to be zero, i.e., $\bar{\gamma} = 0$, the updating equations of the SM-BNLMS algorithm 1 coincide with those of binormalized LMS algorithm with unity step-size [13].

In the SM-BNLMS algorithm 1 if the constraint sets $\mathcal{H}(k)$ and $\mathcal{H}(k-1)$ are parallel, the denominator term $\|\mathbf{x}^{\perp}(k)\|^2$ is zero, since this term is given by

$$\|\mathbf{x}^{\perp}(k)\|^2 = \|\mathbf{x}(k-1)\|^2 - \frac{[\mathbf{x}^T(k-1)\mathbf{x}(k)]^2}{\|\mathbf{x}(k)\|^2}$$

As a result the second step of equation (6.50) is not performed to avoid division by zero.

6.5.2 SM-BNLMS Algorithm 2

The SM-BNLMS algorithm 2 reduces the computational complexity per update even further by avoiding the intermediate constraint check required by the SM-BNLMS algorithm 1. A smart idea to avoid extra computation is, at instant k, to maintain the value of the *a posteriori* error $\varepsilon(k-1)$, which utilizes the data from instant $k-1$, equal to the constraint threshold, that is by choosing $\bar{\gamma}_2(k) = \varepsilon(k-1)$. Since the previous coefficient estimate $\mathbf{w}(k) \in \mathcal{H}(k-1)$, then it is a fact that $\varepsilon(k-1) \leq \bar{\gamma}$. Therefore by choosing $\bar{\gamma}_2(k) = \varepsilon(k-1)$ then $\bar{\gamma}_2(k) \leq \bar{\gamma}$. On the other hand, if we choose $\bar{\gamma}_1(k)$ such that the update lies on the closest boundary of $\mathcal{H}(k)$, i.e., $\bar{\gamma}_1(k) = \bar{\gamma}\text{sign}[e(k)]$, the new coefficient estimate $\mathbf{w}(k+1)$ lies on the nearest boundary of $\mathcal{H}(k)$ such that the *a posteriori* error at iteration $k-1$, $\varepsilon(k-1)$, is kept constant. By specializing the updating equation of the general SM-BNLMS algorithm to the SM-BNLMS algorithm 2 case, we have

$$\mathbf{w}(k+1) = \mathbf{w}(k) + \frac{\lambda_1'(k)}{2}\mathbf{x}(k) + \frac{\lambda_2'(k)}{2}\mathbf{x}(k-1) \tag{6.54}$$

Algorithm 6.4

The Set-Membership Binormalized LMS Algorithm 1

Initialization

$\mathbf{x}(0) = \mathbf{w}(0) = [0...0]^T$

choose $\bar{\gamma}$ around $\sqrt{5}\sigma_n$

$\gamma =$ small constant

Do for $k \geq 0$

$e(k) = d(k) - \mathbf{x}^T(k)\mathbf{w}(k)$

$\mu(k) = \begin{cases} 1 - \frac{\bar{\gamma}}{|e(k)|} & \text{if } |e(k)| > \bar{\gamma} \\ 0 & \text{otherwise} \end{cases}$

$\hat{\mathbf{w}}(k) = \mathbf{w}(k) + \mu(k)\frac{e(k)\mathbf{X}(k)}{\gamma + \|\mathbf{X}(k)\|^2}$

$\varepsilon(k-1) = d(k-1) - \hat{\mathbf{w}}^T(k)\mathbf{x}(k-1)$

$\hat{\mu}(k) = \begin{cases} 1 - \frac{\bar{\gamma}}{|\varepsilon(k-1)|}, & \text{if } |e(k)| > \bar{\gamma} \text{ and } |\varepsilon(k-1)| > \bar{\gamma} \\ 0 & \text{otherwise} \end{cases}$

$\frac{\lambda_1(k)}{2} = -\frac{\hat{\mu}(k)\varepsilon(k-1)\mathbf{X}^T(k-1)\mathbf{X}(k)}{\gamma + \|\mathbf{X}(k)\|^2\|\mathbf{X}(k-1)\|^2 - [\mathbf{X}^T(k-1)\mathbf{X}(k)]^2}$

$\frac{\lambda_2(k)}{2} = \frac{\hat{\mu}(k)\varepsilon(k-1)\|\mathbf{X}(k)\|^2}{\gamma + \|\mathbf{X}(k)\|^2\|\mathbf{X}(k-1)\|^2 - [\mathbf{X}^T(k-1)\mathbf{X}(k)]^2}$

$\mathbf{w}(k+1) = \hat{\mathbf{w}}(k) + \frac{\lambda_1(k)}{2}\mathbf{x}(k) + \frac{\lambda_2(k)}{2}\mathbf{x}(k-1)$

where

$$\frac{\lambda_1'(k)}{2} = \frac{\mu(k)e(k)\|\mathbf{x}(k-1)\|^2}{\|\mathbf{x}(k)\|^2\|\mathbf{x}(k-1)\|^2 - [\mathbf{x}^T(k-1)\mathbf{x}(k)]^2} \tag{6.55}$$

$$\frac{\lambda_2'(k)}{2} = -\frac{\mu(k)e(k)\mathbf{x}^T(k-1)\mathbf{x}(k)}{\|\mathbf{x}(k)\|^2\|\mathbf{x}(k-1)\|^2 - [\mathbf{x}^T(k-1)\mathbf{x}(k)]^2} \tag{6.56}$$

$$\mu(k) = \begin{cases} 1 - \frac{\bar{\gamma}}{|e(k)|}, & \text{if } |e(k)| > \bar{\gamma} \\ 0 & \text{otherwise} \end{cases} \tag{6.57}$$

Fig. 6.6 depicts the update procedure of the SM-BNLMS algorithm 2, whereas Algorithm 6.5 describes it stepwise.

In the SM-BNLMS algorithm 2 if the constraint sets $\mathcal{H}(k)$ and $\mathcal{H}(k-1)$ are parallel, the denominators of the $\lambda_i'(k)$, for $i = 1, 2$ are zero. In this case, in order to avoid division by zero a regularization factor, as in equation (6.7), is employed instead.

Algorithm 6.5

The Set-Membership Binormalized LMS Algorithm 2

Initialization

$\mathbf{x}(0) = \mathbf{w}(0) = [0...0]^T$

choose $\bar{\gamma}$ around $\sqrt{5}\sigma_n$

$\gamma =$ small constant

Do for $k \geq 0$

$e(k) = d(k) - \mathbf{x}^T(k)\mathbf{w}(k)$

$$\mu(k) = \begin{cases} 1 - \frac{\bar{\gamma}}{|e(k)|} & \text{if } |e(k)| > \bar{\gamma} \\ 0 & \text{otherwise} \end{cases}$$

$$\frac{\lambda_1'(k)}{2} = \frac{\mu(k)e(k)\|\mathbf{X}(k-1)\|^2}{\gamma + \|\mathbf{X}(k)\|^2\|\mathbf{X}(k-1)\|^2 - [\mathbf{X}^T(k-1)\mathbf{X}(k)]^2}$$

$$\frac{\lambda_2'(k)}{2} = \frac{\mu(k)e(k)\mathbf{X}^T(k-1)\mathbf{X}(k)}{\gamma + \|\mathbf{X}(k)\|^2\|\mathbf{X}(k-1)\|^2 - [\mathbf{X}^T(k-1)\mathbf{X}(k)]^2}$$

$$\mathbf{w}(k+1) = \mathbf{w}(k) + \frac{\lambda_1'(k)}{2}\mathbf{x}(k) + \frac{\lambda_2'(k)}{2}\mathbf{x}(k-1)$$

6.6 COMPUTATIONAL COMPLEXITY

A brief comparison of the computational complexity among some algorithms presented in this chapter is appropriate at this point. The figure of merit considered is the number of multiplications, additions, and divisions, where it is assumed that the implementation minimizes the number of divisions, multiplications, and additions in that order. Table 6.1 lists the computational complexity for several algorithms, where in the case of the SM-BNLMS algorithm 1 there are two entries since the update complexity is related to the number of steps a given update requires. Two steps are required if after the first step $\hat{\mathbf{w}}(k) \notin \mathcal{H}(k-1)$. The SM-BNLMS algorithm 2 has fix complexity whenever an update occurs whereas for the SM-BNLMS algorithm 1 the complexity depends not only on when an update occurs but also how often the second step takes place. As expected the two versions of the SM-BNLMS algorithm lead to a small increase in computational complexity when compared with the SM-NLMS algorithm. On the other hand, the former algorithms usually require less updates and converge faster than the SM-NLMS algorithm.

The computational complexity reduction is essential in applications where the filter order is high and the resources are limited. Therefore, special care should be taken to exploit opportunities to reduce the computational burden, for example, assuming the value of $\|\mathbf{x}(k-1)\|^2$ at iteration k is unknown. If $\|\mathbf{x}(k-1)\|^2$ is known, we can compute $\|\mathbf{x}(k)\|^2$ using only two additional multiplications through $\|\mathbf{x}(k)\|^2 = \|\mathbf{x}(k-1)\|^2 + x^2(k) - x^2(k-N)$, also in case the value of $x^2(k-N)$ is prestored then only one multiplication is required. This strategy has been considered when evaluating the multiplication and addition counts of the SM-BNLMS algorithms. If update occurs at two successive time instants, $\|\mathbf{x}(k-1)\|^2$ and $\mathbf{x}^T(k-1)\mathbf{x}(k-2)$ have already been computed in the previous update, as a result,

Table 6.1 Computational Complexity in Set-membership Algorithms, † The Numbers for the RLS Apply to the Particular Implementation of Algorithm 5.2

Algorithm	Multiplication	Addition	Division
LMS	$2N+3$	$2N+2$	0
NLMS	$2N+3$	$2N+5$	1
SM-NLMS	$2N+4$	$2N+6$	1
SM-BNLMS 1 (1 step)	$3N+4$	$3N+7$	1
SM-BNLMS 1 (2 steps)	$5N+13$	$5N+16$	2
SM-BNLMS 2	$3N+11$	$3N+10$	1
RLS†	$3N^2+11N+8$	$3N^2+7N+4$	1

the number of multiplications and additions in such updates can be further reduced by approximately $N+1$ for the SM-NLMS algorithm and $2N+2$ for the SM-BNLMS algorithms 1 and 2, depending on the implementation. Finally, note that if one continuously computes $\|\mathbf{x}(k)\|^2$ and $\mathbf{x}^T(k)\mathbf{x}(k-1)$, regardless if an update is required or not, the SM-BNLMS algorithm 2 is always more efficient than SM-BNLMS algorithm 1.

6.7 TIME-VARYING $\bar{\gamma}$

In this section, an automatic way to choose $\bar{\gamma}$ is presented in order to avoid overbounding and underbounding of such a crucial parameter. In case $\bar{\gamma}$ is chosen too small the feasibility set might become null, whereas if the threshold parameter is chosen too big the resulting estimate might be meaningless and inconsistent [24].

Let's first consider the case of channel equalization application such as that of Fig. 2.13. In a typical multiuser communication environment the noise signal vector can be composed as follows [32]

$$\mathbf{n}(k) = \mathbf{n}_{\mathrm{n}}(k) + \mathbf{n}_{\mathrm{ISI}}(k) + \mathbf{n}_{\mathrm{MAI}}(k) \tag{6.58}$$

where $\mathbf{n}(k) = [n(k)\ \ n(k-1)\ \ldots\ n(k-N)]^T$, and

- $\mathbf{n}_{\mathrm{n}}(k)$ represents the contribution of the environment noise.
- $\mathbf{n}_{\mathrm{ISI}}(k)$ is the contribution of the intersymbol interference (ISI) originated when the transmitted signal crosses a channel with memory, in other words, whenever multiple paths are perceived by the receiver.
- $\mathbf{n}_{\mathrm{MAI}}(k)$ accounts for the multi-access interference (MAI), that is, the signals from other users that reach the receiver.

At the equalizer output, the disturbance due to noise can be accounted for as follows

$$y_{\text{n}}(k) = \mathbf{w}^T(k)\mathbf{n}(k) \tag{6.59}$$

where $\mathbf{w}^T(k)$ is the equalizer coefficient vector and $y_{\text{n}}(k)$ is the noise signal vector filtered by the equalizer. As a result, the equalizer output $y(k)$ is described by

$$y(k) = y_{\bar{\text{n}}}(k) + y_{\text{n}}(k) \tag{6.60}$$

with $y_{\bar{\text{n}}}(k)$ representing the equalized signal when there is no noise at the adaptive-filter input.

The average power of the disturbance, for a given equalizer with parameters $\mathbf{w}(k)$, can be calculated as

$$\sigma^2_{y_{\text{n}}}(k) = E[y^2_{\text{n}}(k)] = \mathbf{w}^T(k)E[\mathbf{n}(k)\mathbf{n}^T(k)]\mathbf{w}(k) = \|\mathbf{w}(k)\|^2\sigma^2_{\text{n}}(k) \tag{6.61}$$

Assuming there is an estimate of $\sigma^2_{y_{\text{n}}}(k)$ denoted as $\hat{\sigma}^2_{y_{\text{n}}}(k) = \|\mathbf{w}(k)\|^2\hat{\sigma}^2_{\text{n}}(k)$ we can generate a time-varying threshold parameter as follows

$$\bar{\gamma}(k+1) = \alpha\bar{\gamma}(k) + (1-\alpha)\sqrt{\beta\|\mathbf{w}(k)\|^2\hat{\sigma}^2_{\text{n}}(k)} \tag{6.62}$$

where α is a forgetting factor and β is a constant to be set. As justified in [25], a range of values for β leading to a good compromise between misadjustment and speed of convergence is $4 \leq \beta \leq 5$.

In equalization environments the best way to estimate $\sigma^2_{\text{n}}(k)$ is to remove the effect of the detected symbols from $x(k)$ in order to get a rough estimate of $n(k)$ [26]-[28], and from this estimate compute

$$\hat{\sigma}^2_{\text{n}}(k+1) = \alpha\hat{\sigma}^2_{\text{n}}(k) + (1-\alpha)\hat{n}^2(k) \tag{6.63}$$

where again α is a forgetting factor. Fig. 6.8 illustrates how the environment noise can be typically estimated in a general equalizer setup.

For system identification environment as depicted in Fig. 2.10, an estimate of the additional noise plus an eventual effect of undermodeling can be calculated from the output error itself. If the input signal and the additional noise are considered white noise and uncorrelated, see equation (2.148) for details, the MSE can be calculated as

$$\begin{aligned}\xi &= E[e^2(k)] \\ &= E\{[\mathbf{h}^T\mathbf{x}_\infty(k) - \mathbf{w}^T\mathbf{x}_{N+1}(k)]^2 + n^2(k)\} \\ &= \sigma_x^2\sum_{i=N+1}^{\infty} h^2(i) + \sigma_n^2\end{aligned} \tag{6.64}$$

where $\mathbf{x}_\infty(k)$ and $\mathbf{x}_{N+1}(k)$ are the input signal vector with infinite and finite lengths, respectively. Likewise the equalization setup, a time-varying threshold parameter for the system identification application is given by

$$\bar{\gamma}(k+1) = \alpha\bar{\gamma}(k) + (1-\alpha)\sqrt{\beta\hat{\sigma}^2_{\text{n}}(k)} \tag{6.65}$$

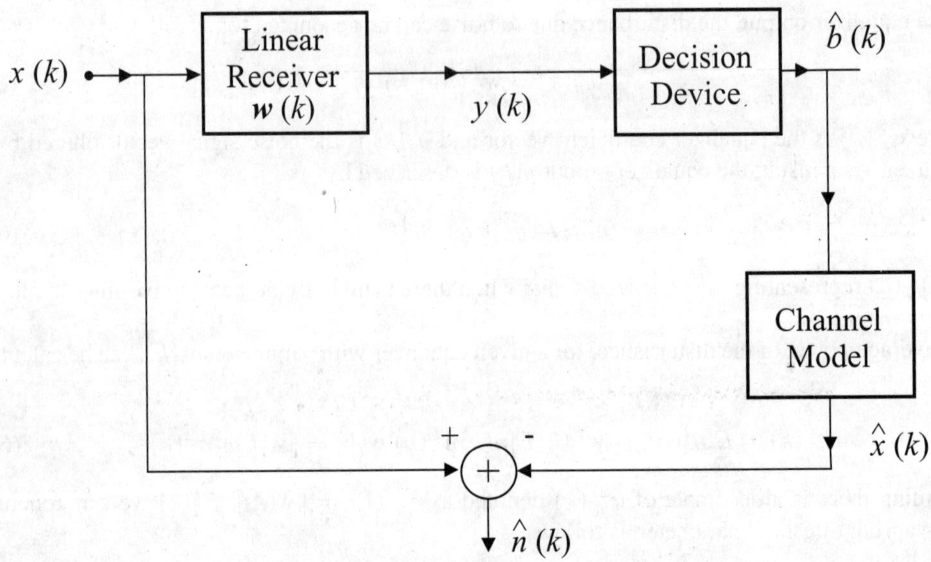

Figure 6.8 Environment noise estimation.

where for this case

$$\hat{\sigma}_{\mathrm{n}}^{2}(k+1) = \alpha\hat{\sigma}_{\mathrm{n}}^{2}(k) + (1-\alpha)e^{2}(k) \tag{6.66}$$

In [25] some analytical expressions are developed in order to provide values for $\bar{\gamma}(k)$ such that the some prescribed updating rate are nearly satisfied after the algorithm has reached convergence.

6.8 PARTIAL-UPDATE ADAPTIVE FILTERING

In several applications the number of coefficients to be updated might be prohibitive, therefore some strategies to control the computational complexity is desirable. In some cases like in acoustics echo cancellation, which might use a few thousands of adaptive coefficients, the convergence would entail a large number of iterations, calling for more sophisticated updating algorithms which are inherently more computationally intensive. A good compromise might be to update only part of the filter coefficients at each iteration instant, generating a family of algorithms called partial-update (PU) algorithms. The most widely known PU algorithm in the literature is the normalized LMS with partial update [14]-[19].

In this section special emphasis is given to the set-membership partial-update affine projection (SM-PUAP) algorithms. The combination of the partial-update with set-membership allows the updating of a selected set of coefficients whenever an update is needed. The resulting algorithms capitalize not only from the sparse updating related to the set-membership framework but also from the partial update of the coefficients, reducing the average computational complexity. It is expected that the SM-PUAP algorithms have comparable performance to that of SM-AP algorithms and affine projection algorithms with partial-update whereas computational complexity is reduced with respect to both updating schemes.

Two versions of the SM-PUAP algorithm are discussed:

- Fix partial update, where a constant number of coefficients is updated whenever required.
- Variable partial update, where the number of coefficients to be updated vary up to a maximum prescribed number.

In the partial update adaptation strategy, the main objective is to perform updates in $\bar{M}$ out of the $N+1$ adaptive-filter coefficients. The $\bar{M}$ coefficients to be updated at time instant k are selected through an index set $\mathcal{I}_{\bar{M}}(k) = \{i_0(k) \ \ldots \ i_{\bar{M}-1}(k)\}$ where the indexes $\{i_j(k)\}_{j=0}^{\bar{M}-1}$ are chosen from the set $\{0 \ 1 \ \ldots \ N\}$ representing the available coefficients to be updated. The partition of the $N+1$ coefficients into mutually exclusive subsets, each with $\bar{M}$ elements, plays a key role in the performance and in the effectiveness of the partial-update strategy. As a result, $\mathcal{I}_{\bar{M}}(k)$ varies with the iteration index k such that the $\bar{M}$ coefficients to be updated can change according to the iteration. The choice of which $\bar{M}$ coefficients should be updated is related to the objective function considered in the algorithm derivation.

As already known, in the SM-AP algorithms the new coefficient vector can be obtained as the vector $\mathbf{w}(k+1)$ that minimizes the Euclidean distance $\|\mathbf{w}(k+1) - \mathbf{w}(k)\|^2$, subject to the constraint that the moduli of *a posteriori* errors fall below certain prescribed threshold. The same idea can be used in order to derive the SM-PUAP algorithm, specifically the vector $\mathbf{w}(k+1)$ is chosen by minimizing the Euclidean distance $\|\mathbf{w}(k+1) - \mathbf{w}(k)\|^2$ subject to the constraint $\mathbf{w}(k+1) \in \mathcal{H}(k)$ in such a way that only $\bar{M}$ coefficients are updated. If $\mathbf{w}(k) \in \mathcal{H}(k)$, there is no update and the Euclidean distance is zero.

The objective function to be minimized in the set-membership partial-update affine projection (SM-PUAP) algorithm is following described. A coefficient update is performed whenever $\mathbf{w}(k) \notin \psi^{L+1}(k)$ such that

$$\min \|\mathbf{w}(k+1) - \mathbf{w}(k)\|^2 \tag{6.67}$$

subject to:

$$\begin{aligned} &\mathbf{d}_{\text{ap}}(k) - \mathbf{X}_{\text{ap}}^T(k)\mathbf{w}(k+1) = \bar{\gamma}(k) \\ &\tilde{\mathbf{C}}_{\mathcal{I}_{\bar{M}(k)}} [\mathbf{w}(k+1) - \mathbf{w}(k)] = \mathbf{0} \end{aligned} \tag{6.68}$$

where $\bar{\gamma}(k)$ is a vector determining a point within the constraint set $\mathcal{H}(k)$, such that $|\bar{\gamma}_i(k)| \leq \bar{\gamma}$, for $i = 0, 1, \ldots, L$. The matrix $\tilde{\mathbf{C}}_{\mathcal{I}_{\bar{M}(k)}} = \mathbf{I} - \mathbf{C}_{\mathcal{I}_{\bar{M}}(k)}$ is a complementary matrix of $\mathbf{C}_{\mathcal{I}_{\bar{M}}(k)}$ enforcing $\tilde{\mathbf{C}}_{\mathcal{I}_{\bar{M}}(k)}\mathbf{w}(k+1) = \tilde{\mathbf{C}}_{\mathcal{I}_{\bar{M}}(k)}\mathbf{w}(k)$, such that only $\bar{M}$ coefficients are updated. A possible choice for $\bar{\gamma}_0(k)$ is such that the updated vector belongs to the closest bounding hyperplane in $\mathcal{H}(k)$, i.e., $\bar{\gamma}_0(k) = \bar{\gamma}\, e(k)/|e(k)|$. On the other hand, some alternative choices $|\bar{\gamma}_i(k)| \leq \bar{\gamma}$, for $i = 1, 2, \ldots, L$, had been discussed. The matrix $\mathbf{C}_{\mathcal{I}_{\bar{M}}(k)}$ is a diagonal matrix that determines the coefficients to be updated at instant k, if an update is required. This matrix has $\bar{M}$ nonzero elements equal to one placed at positions indicated by $\mathcal{I}_{\bar{M}}(k)$.

Applying the method of Lagrange multipliers gives the recursive updating rule

$$\mathbf{w}(k+1) = \mathbf{w}(k) + \mathbf{C}_{\mathcal{I}_{\bar{M}}(k)}\mathbf{X}_{\text{ap}}(k)\left[\mathbf{X}_{\text{ap}}^{T}(k)\mathbf{C}_{\mathcal{I}_{\bar{M}}(k)}\mathbf{X}_{\text{ap}}(k)\right]^{-1}\left[\mathbf{e}_{\text{ap}}(k) - \bar{\boldsymbol{\gamma}}(k)\right] \tag{6.69}$$

The updating equation of the SM-PUAP algorithm is given by

$$\mathbf{w}(k+1) = \begin{cases} \mathbf{w}(k) + \mathbf{C}_{\mathcal{I}_{\bar{M}}(k)}\mathbf{X}_{\text{ap}}(k)\left[\mathbf{X}_{\text{ap}}^{T}(k)\mathbf{C}_{\mathcal{I}_{\bar{M}}(k)}\mathbf{X}_{\text{ap}}(k)\right]^{-1}\left[\mathbf{e}_{\text{ap}}(k) - \bar{\boldsymbol{\gamma}}(k)\right] & \text{if } |e(k)| > \bar{\gamma} \\ \mathbf{w}(k) & \text{otherwise} \end{cases} \tag{6.70}$$

As can be noticed from equation (6.70), for a fixed value of $\|\mathbf{e}_{\text{ap}}(k) - \bar{\boldsymbol{\gamma}}(k)\|^2$, the Euclidean distance between two consecutive coefficient vectors is minimized if $\|\mathbf{X}_{\text{ap}}^{T}(k)\mathbf{C}_{\mathcal{I}_{\bar{M}}(k)}\mathbf{X}_{\text{ap}}(k)\|$ is maximized. As a result, a natural choice for the $\bar{M}$ coefficients to be updated are those that will be multiplied by the elements of $\mathbf{X}_{\text{ap}}(k)$ with the largest norm.

Like in the case of the SM-AP algorithm of equation (6.33), it is straightforward to derive a simplified version of the SM-PUAP algorithm, whose update equation is given by

$$\mathbf{w}(k+1) = \mathbf{w}(k) + \mathbf{C}_{\mathcal{I}_{\bar{M}}(k)}\mathbf{X}_{\text{ap}}(k)\left[\mathbf{X}_{\text{ap}}^{T}(k)\mathbf{C}_{\mathcal{I}_{\bar{M}}(k)}\mathbf{X}_{\text{ap}}(k)\right]^{-1}\mu(k)e(k)\mathbf{u}_1 \tag{6.71}$$

where

$$e(k) = d(k) - \mathbf{w}^{T}(k)\mathbf{x}(k) \tag{6.72}$$

$$\mu(k) = \begin{cases} 1 - \frac{\bar{\gamma}}{|e(k)|} & \text{if } |e(k)| > \bar{\gamma} \\ 0 & \text{otherwise} \end{cases} \tag{6.73}$$

This algorithm also minimizes the Euclidean distance $\|\mathbf{w}(k+1) - \mathbf{w}(k)\|^2$ subject to the constraint $\mathbf{w}(k+1) \in \psi^{L+1}(k)$ maintaining the values of the *a posteriori* errors, $\varepsilon(k-i)$, at iteration $k-i$. Note that $\mu(k)$ starts with high values, becomes small when the error reduces, and reaches zero whenever moduli of the errors become smaller than the threshold. An interesting choice for the index set $\mathcal{I}_{\bar{M}}(k)$ specifying the coefficients to be updated is the $\bar{M}$ coefficients leading to the maximum value of $\|\mathbf{X}_{\text{ap}}^{T}(k)\mathbf{C}_{\mathcal{I}_{\bar{M}}(k)}\mathbf{X}_{\text{ap}}(k)\|$. Algorithm 6.6 describes in detail the simplified version of the SM-PUAP algorithm.

6.8.1 Set-Membership Partial-Update NLMS Algorithm

The simplest form of the SM-PUAP algorithm is the set-membership partial-update NLMS (SM-PUNLMS) algorithm. The updating equation of the SM-PUNLMS algorithm follows directly from

Algorithm 6.6

The Simplified Set-Membership Partial-Update Affine Projection Algorithm

Initialization

$\mathbf{x}(0) = \mathbf{w}(0) = [0 \ \ldots \ 0]^T$

choose $\bar{\gamma}$ around $\sqrt{5}\sigma_n$

$\gamma =$ small constant

Do for $k \geq 0$

$\mathbf{e}_{\text{ap}}(k) = \mathbf{d}_{\text{ap}}(k) - \mathbf{X}_{\text{ap}}^T(k)\mathbf{w}(k)$

$\mu(k) = \begin{cases} 1 - \frac{\bar{\gamma}}{|e(k)|} & \text{if } |e(k)| > \bar{\gamma} \\ 0 & \text{otherwise} \end{cases}$

$\mathbf{w}(k+1) = \mathbf{w}(k) + \mathbf{C}_{\mathcal{I}_{\bar{M}}(k)}\mathbf{X}_{\text{ap}}(k)\left[\mathbf{X}_{\text{ap}}^T(k)\mathbf{C}_{\mathcal{I}_{\bar{M}}(k)}\mathbf{X}_{\text{ap}}(k) + \gamma\mathbf{I}\right]^{-1}\mu(k)e(k)\mathbf{u}_1$

equation (6.71) and is given by

$$\mathbf{w}(k+1) = \mathbf{w}(k) + \mu(k)\frac{e(k)\mathbf{C}_{\mathcal{I}_{\bar{M}}(k)}\mathbf{x}(k)}{\|\mathbf{C}_{\mathcal{I}_{\bar{M}}(k)}\mathbf{x}(k)\|^2} \tag{6.74}$$

where

$$\mu(k) = \begin{cases} 1 - \frac{\bar{\gamma}}{|e(k)|} & \text{if } \mathbf{w}(k) \notin \mathcal{H}(k), \text{ i.e., if } |e(k)| > \bar{\gamma} \\ 0 & \text{otherwise} \end{cases} \tag{6.75}$$

In [17], a number of properties and an interesting geometrical interpretation of the SM-PU-NLMS algorithm update are provided, some of these results are discussed here. Fig. 6.9 depicts the situation where one coefficient updates out of three, i.e., $\bar{M} = 1$ and $N + 1 = 3$. As can be observed, the element $x_2(k)$ is the largest in magnitude among the elements of $\mathbf{x}(k)$, therefore a natural choice for $\mathbf{C}_{\mathcal{I}_{\bar{M}_2}(k)}$ is a diagonal matrix whose diagonal elements are $[0 \ 1 \ 0]$. The solution denoted by $\mathbf{w}_{\text{SM-NLMS}}$ is obtained by an orthogonal projection starting from $\mathbf{w}(k)$ onto the closest boundary of the constraint set $\mathcal{H}(k)$. The angle denoted by θ shown in Fig. 6.9 is the angle between the direction of update $\mathbf{C}_{\mathcal{I}_{\bar{M}_2}(k)}\mathbf{x}(k) = [0 \ x_2(k) \ 0]^T$ and the input vector $\mathbf{x}(k)$. When $\bar{M}$ coefficients are updated, the general expression for the cosine of θ in $\mathbb{R}^{N+1}$ is given by the relation

$$\cos\theta = \frac{\|\mathbf{C}_{\mathcal{I}_{\bar{M}}(k)}\mathbf{x}(k)\|}{\|\mathbf{x}(k)\|} \tag{6.76}$$

whereas for the case in discussion, the particular expression for the cosine is

$$\cos\theta = \frac{|x_2(k)|}{\sqrt{|x_1(k)|^2 + |x_2(k)|^2 + |x_3(k)|^2}}$$

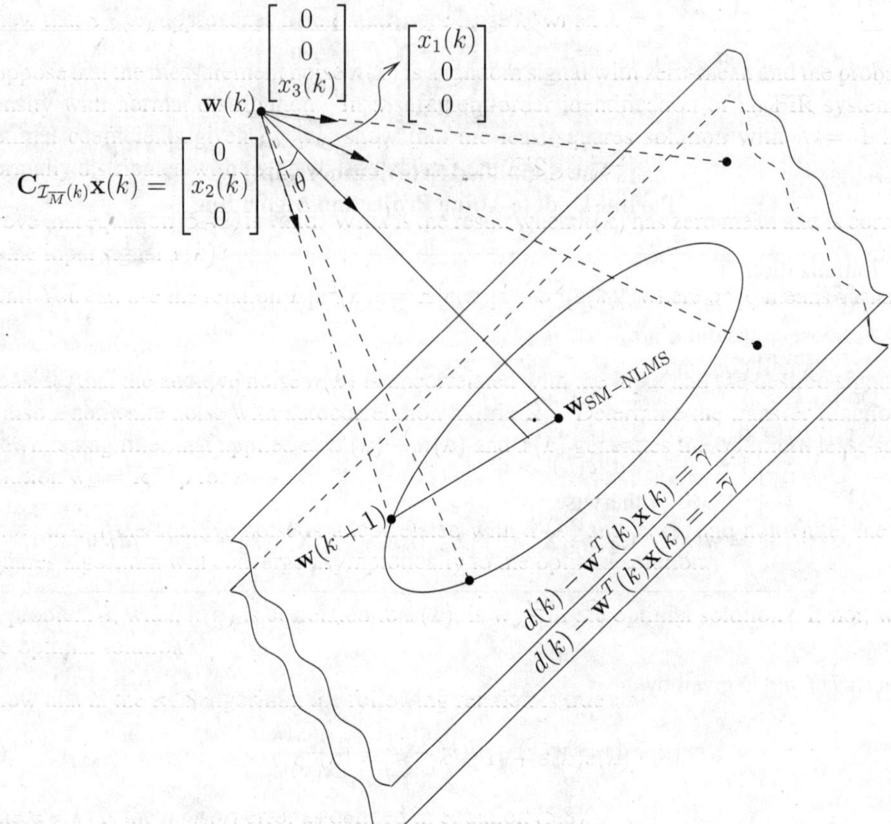

Figure 6.9 Partial coefficient update for $\mathbb{R}^3$ and $\bar{M} = 1$, where $|x_2(k)| > |x_1(k)| > |x_3(k)|$.

The SM-PUNLMS algorithm may face convergence problem whenever trying to find a solution in the constraint set. If the number of coefficients to be updated is small, $\cos\theta$ might become small according to equation (6.76), with θ becoming close to $\frac{\pi}{2}$, as can be observed in Fig. 6.10. As a result, the solution in the constraint set will depart from the SM-NLMS solution, and will give rise to stability problems.

A possible solution is to increase $\bar{M}$ up to the point where the solution provided by the SM-PUNLMS algorithm reaches a prescribed closer distance of SM-NLMS or NLMS solutions. Unfortunately this solution does not impose an upper bound on the value of $\bar{M}$, and it is highly probable that during initial iterations $\bar{M}$ would be close to the overall number of filter coefficients $N+1$. On the other hand, it is desirable that $\bar{M} \ll N+1$ in order to make the partial update effective in reducing the computational complexity.

Let's first define as $\bar{M}_{\max}$ the maximum number of coefficients that can be updated at any given iteration. It is now required to derive a strategy to control the number of coefficients to be updated while keeping a bound on the norm of the update. If $\|\mathbf{C}_{\mathcal{I}_{\bar{M}}(k)}\mathbf{x}(k)\|^2 = \|\mathbf{x}(k)\|^2$, it is straightforward

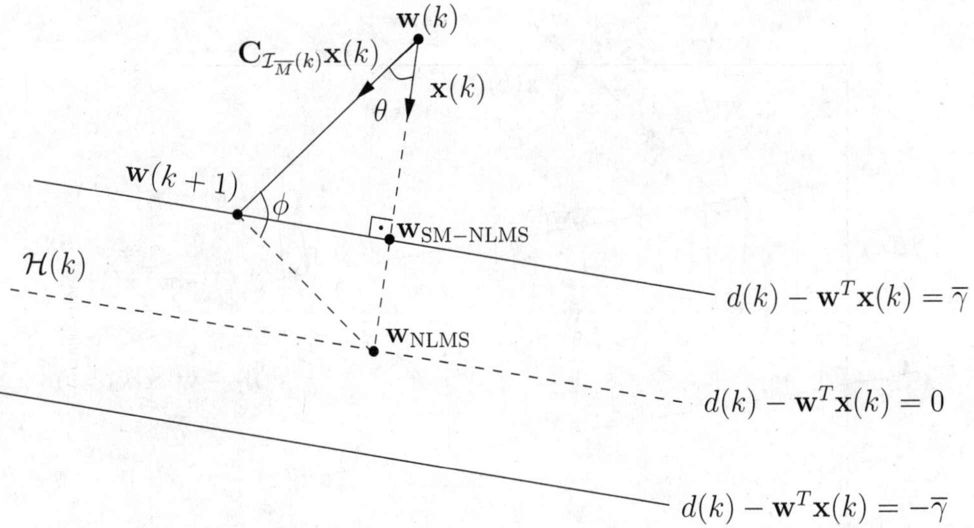

Figure 6.10 Projection in partial-update algorithms.

to verify that the angle ϕ is equal to $\frac{\pi}{2}$ and $\mathbf{w}_{\text{NLMS}} - \mathbf{w}(k+1)$ represents the projection of $\mathbf{w}_{\text{NLMS}} - \mathbf{w}(k)$ into $\mathbf{C}_{\mathcal{I}_{\bar{M}}(k)}\mathbf{x}(k)$. For angle $\phi < \frac{\pi}{2}$ the norm of the updating term might become large in order to meet the error modulus requirement, placing the partial solution far way from $\mathbf{w}_{\text{NLMS}}$ and $\mathbf{w}_{\text{SM-NLMS}}$. Indeed, whenever $\phi \geq \frac{\pi}{2}$ the norm of the updating term becomes smaller than the one required to turn the *a posteriori* error equal to zero (the one reaching $\mathbf{w}_{\text{NLMS}}$). Then, an alternative solution is to increase the number of coefficients to update up to the condition that $\|\mathbf{C}_{\mathcal{I}_{\bar{M}}(k)}\mathbf{x}(k)\|^2 \geq \mu(k)\|\mathbf{x}(k)\|^2$, for $\mu(k) = 1 - \bar{\gamma}/|e(k)|$, or $\bar{M} = \bar{M}_{\max}$. This strategy will keep the angle ϕ lower bounded by $\frac{\pi}{2}$. If $\bar{M} = \bar{M}_{\max}$, increase the threshold $\bar{\gamma}$ temporarily at the kth iteration to

$$\bar{\gamma}(k) = \frac{(\|\mathbf{x}(k)\|^2 - \|\mathbf{C}_{\mathcal{I}_{\bar{M}(k)}}\mathbf{x}(k)\|^2)}{\|\mathbf{x}(k)\|^2}|e(k)| \tag{6.77}$$

Fig. 6.11 shows that this strategy temporarily expands the constraint set in order to allow a feasible solution in the case where the required number of coefficients to meet a solution in the constraint set exceeds $\bar{M}_{\max}$, at a given iteration.

Another possible strategy for the partial update is to choose the set of coefficients to be updated in a random manner [19] utilizing randomly partitions of the $N+1$ coefficients consisting of mutually exclusive subsets of $\bar{M}$ elements, each determined by the index set $\mathcal{I}_{\bar{M}}(k) = \{i_0(k) \ldots i_{\bar{M}-1}(k)\}$, as previously defined. This solution avoids the possible instability of the partial-update LMS algorithm originated by the choice of subsets in a deterministic manner, since in the latter case it is usually possible to generate input signals in which the algorithm fails to converge.

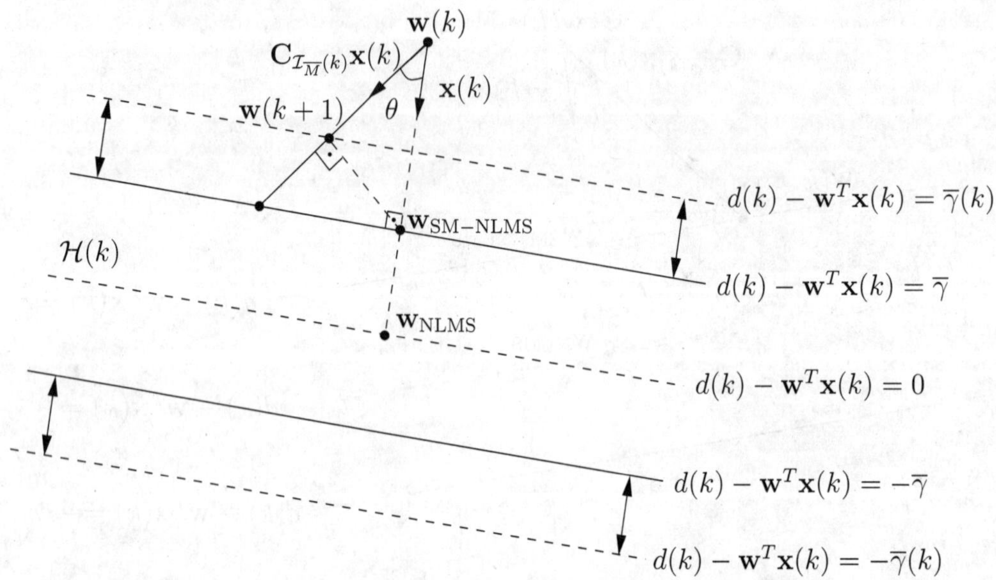

Figure 6.11 Variable constraint set $\mathcal{H}(k)$ with threshold $\overline{\gamma}(k)$.

6.9 SIMULATION EXAMPLES

In this section, some adaptive-filtering problems are described and solved by using some of the algorithms presented in this chapter.

Example 6.1: SM-AP Algorithms, system identification simulation

An adaptive-filtering algorithm is used to identify the system described in the example of subsection 3.6.2 using the following SM-AP algorithms:

- Set-membership affine projection using $L = 0$, $L = 1$ and $L = 4$.
- Set-membership partial-update affine projection with $\bar{M} = 5$, using $L = 0$, $L = 1$ and $L = 2$ and only for the eigenvalue spread of 20.

Do not consider the finite-precision case.

Solution:

All the results presented here for the affine projection and the SM-AP algorithms are obtained by averaging the results of 200 independent runs. We first run the affine projection algorithm with a value of $\mu = 0.18$, with $\gamma = 10^{-6}$. With this value of μ, the misadjustment of the affine projection algorithm is about the same as that of the LMS algorithm with $\mu = 0.0128$ and eigenvalue spread of the input signal autocorrelation matrix of 20, see Table 3.1. Fig. 6.12 illustrates the learning curves for the eigenvalue spread 80 and distinct values of L. As expected the convergence speed and the misadjustment increase with the value of L.

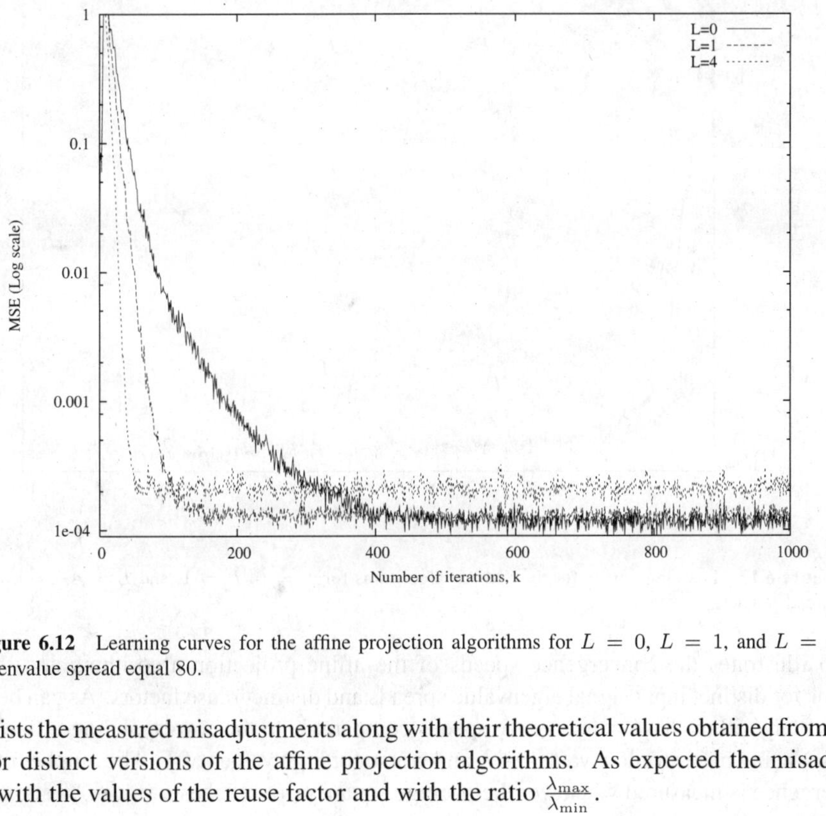

Figure 6.12 Learning curves for the affine projection algorithms for $L = 0$, $L = 1$, and $L = 4$, eigenvalue spread equal 80.

Table 6.2 lists the measured misadjustments along with their theoretical values obtained from equation (4.125) for distinct versions of the affine projection algorithms. As expected the misadjustment increases with the values of the reuse factor and with the ratio $\frac{\lambda_{\max}}{\lambda_{\min}}$.

Table 6.2 Evaluation of the Affine Projection Algorithm, $\mu = 0.18$

$\frac{\lambda_{\max}}{\lambda_{\min}}$	**Misadjustment, $L = 0$**		**Misadjustment, $L = 1$**		**Misadjustment, $L = 4$**	
	Experiment	**Theory**	**Experiment**	**Theory**	**Experiment**	**Theory**
1	0.1275	0.0989	0.2665	0.1978	0.9554	0.4945
20	0.1458	0.0989	0.2951	0.1978	1.0881	0.4945
80	0.1708	0.0989	0.3157	0.1978	1.2091	0.4945

Figs. 6.13, 6.14 and 6.15 depict the the learning curves for the simplified SM-AP algorithm for the eigenvalue spreads 1, 20 and 80, respectively. In each figure, distinct values of L are tested and the value of $\bar{\gamma}$ is $\sqrt{5}\sigma_n$. As can be observed, the convergence speed and the misadjustment increase with

the value of L. As will be discussed, a reduction in the misadjustment is achieved at the expense of mild increase in number of iterations for convergence.

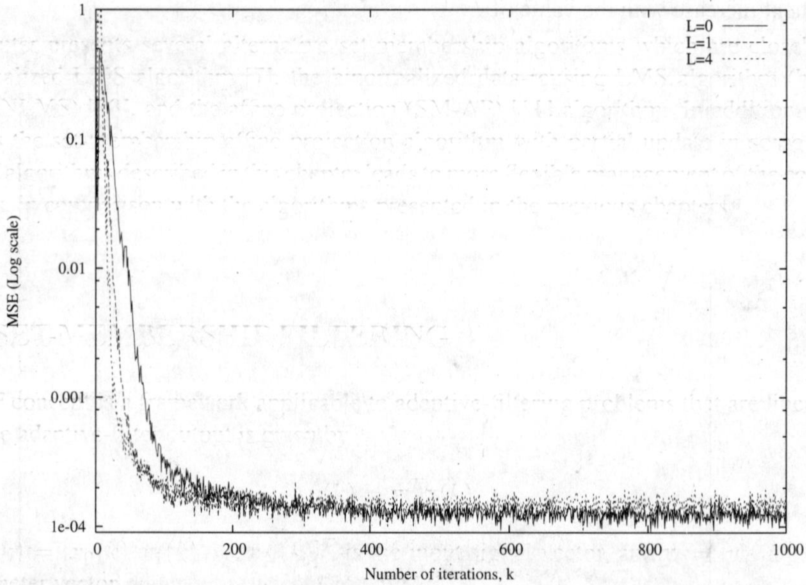

Figure 6.13 Learning curves for the SM-AP algorithms for $L = 0$, $L = 1$, and $L = 4$, eigenvalue spread equal 1.

Table 6.3 illustrates the convergence speeds of the affine projection algorithms and the SM-AP algorithms for distinct input signal eigenvalue spreads and distinct reuse factors. As can be observed, the SM-AP algorithms have convergence speeds comparable to the corresponding affine projection algorithms, being better for low values of L and worse for high values of L. The number of iterations for convergence is measured whenever the average square error reaches a value 5% above the noise floor.

Table 6.3 Convergence Speed of the Affine Projection and SM-AP Algorithms

	Convergence speed					
$\frac{\lambda_{max}}{\lambda_{min}}$	$L = 0$		$L = 1$		$L = 4$	
	AP	SM-AP	AP	SM-AP	AP	SM-AP
1	316	227	213	225	143	201
20	465	344	195	227	137	200
80	644	468	197	229	135	200

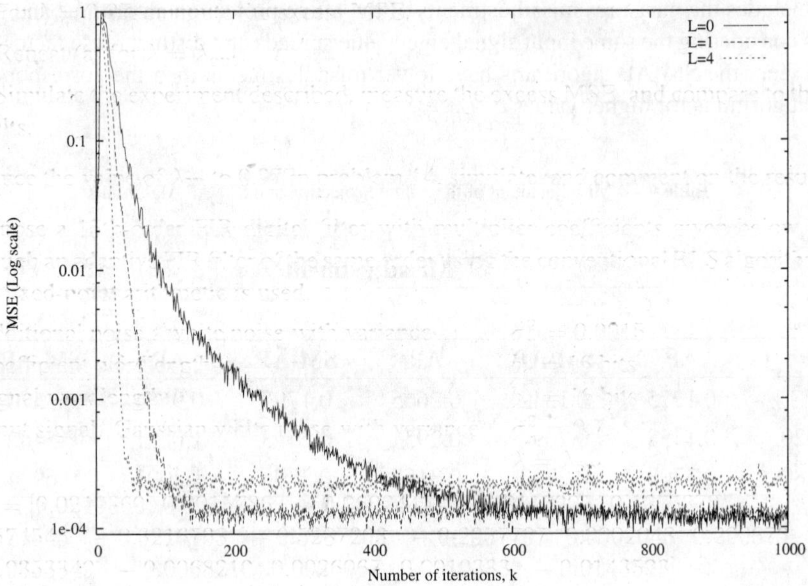

Figure 6.14 Learning curves for the SM-AP algorithms for $L = 0$, $L = 1$, and $L = 4$, eigenvalue spread equal 20.

Figure 6.15 Learning curves for the SM-AP algorithms for $L = 0$, $L = 1$, and $L = 4$, eigenvalue spread equal 80.

Table 6.4 includes the measures misadjustments of the affine projection algorithms and the SM-AP algorithms considering the same input signal eigenvalue spreads and distinct reuse factors as before. As can be seen, the SM-AP algorithms have lower misadjustments than the corresponding affine projection algorithms for higher values of L.

Table 6.4 Misadjustment of the Affine Projection and SM-AP Algorithms

	Misadjustment					
$\frac{\lambda_{\max}}{\lambda_{\min}}$	$L=0$		$L=1$		$L=4$	
	AP	**SM-AP**	AP	**SM-AP**	AP	**SM-AP**
1	0.1275	0.1542	0.2665	0.1797	0.9554	0.3570
20	0.1458	0.2094	0.2951	0.2793	1.0881	0.5462
80	0.1708	0.2723	0.3157	0.3895	1.2091	0.6934

The SM-PUAP algorithm was set to update only 5 coefficients per iteration. For the SM-PUAP algorithm the learning curves are depicted in Fig. 6.16 for distinct values of L. The values of $\bar{\gamma}$ for $L = 0, 1$, and 2 are $\sqrt{5\sigma_n^2}$, $\sqrt{7\sigma_n^2}$, and $\sqrt{17\sigma_n^2}$, respectively. The corresponding measured misadjustments were 0.1979, 0.3137, and 0.8189. An efficient algorithm for the best selection of the updating coefficients in the partial-updating affine projection algorithm is an open problem, although some approximate solutions exist [21]. The choice of the coefficients to be updated relies on a low complexity procedure to sort out the $\bar{M}$ columns of $\mathbf{X}_{\text{ap}}^T(k)$ consisting of choosing the ones whose Euclidean norm have higher values.

□

6.9.1 Echo Cancellation Environment

The elimination of echo signals in communication networks and in hands-free communication environment are challenging problems in which adaptive filtering plays a major role [29]-[30].

The network echo, also known as line echo, is caused by the hybrid transformer whose main task is to convert the two-wire loop connection between the end user and the central office into a four-wire circuit. In the two-wire case, the signal in both directions traverses the two wires, whereas in the four-wires the signals in each direction are separated. Fig. 6.17 illustrates a very simplified long-distance telephone system where the location of the echo canceller is also included. The four-wire circuit exists only in long-distance connections and the delay included in Fig. 6.17 accounts for the traveling time of the signal from one hybrid to the other. Usually the far-end hybrid leaks back to the phone its own transmitted signal giving rise to the echo. If the echo delay is over 100 ms, its effect in the conversation is very disturbing. The early solution comprised of echo suppressor, whose aim was removing the echo from the talker by cutting off the outgoing hybrid port whenever an incoming signal is detected. This approach works well for low round trip delays, but for large delays an adaptive echo canceller is more effective.

Echo cancellers are also used in acoustics echo cancellation problems where its task is to model the transfer function from the loudspeaker to the microphone in a given room. This application is more challenging than the network echo cancellation since the echo path impulse response is much longer, usually well above 500 taps, and changes quite rapidly. As depicted in Figs. 6.17 and 6.18, the echo cancellation problems in networks and in acoustics are closely related, with the latter requiring more sophisticated algorithms such as the subband adaptive filters of Chapter 12.

For both applications two measures of performance are the echo return loss (ERL) and the the echo return loss enhancement (ERLE). The ERL is ratio of the returned-echo power and the input-signal power, measuring the attenuation faced by the signal in the echo path. The ERL, measured in dB, is defined as

$$\text{ERL} = -10\log\frac{\sigma_d^2}{\sigma_x^2} = -10\log\frac{E[d^2(k)]}{E[x^2(k)]} \tag{6.78}$$

The ERLE measures the reduction in the echo obtained by utilizing the echo canceller, that is

$$\text{ERLE} = -10\log\frac{\sigma_e^2}{\sigma_d^2} = -10\log\frac{E[e^2(k)]}{E[d^2(k)]} \tag{6.79}$$

For simulation purposes we will utilize the models recommended by the International Telecommunication Union (ITU) in the ITU-T recommendation G.168 for digital network echo cancellers

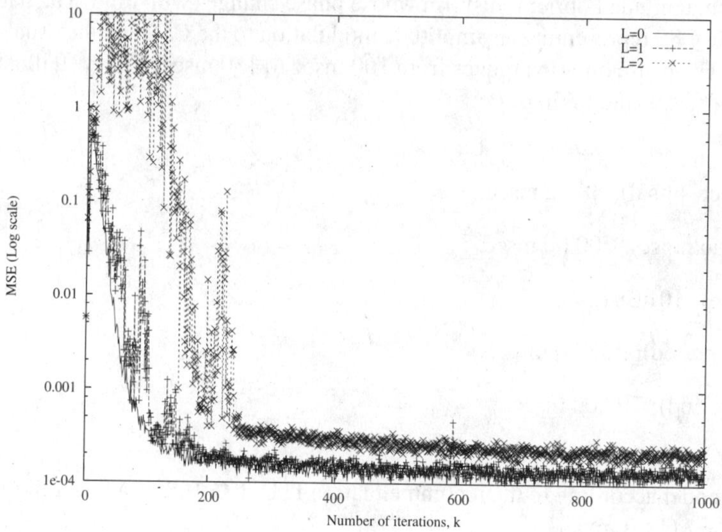

Figure 6.16 Learning curves for the SM-PUAP algorithms for $L = 0$, $L = 1$, and $L = 2$, eigenvalue spread equal 20.

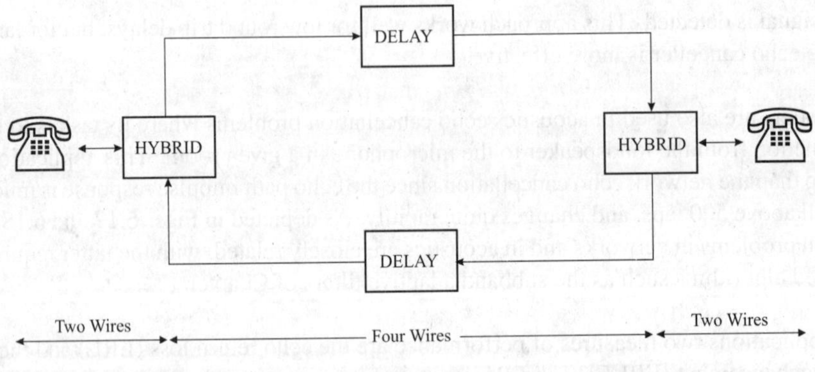

Figure 6.17 Two-wire to four-wire conversion in long-distance telephone calls.

[31]. The main focus is to highlight the typical artificial input signals and echo path models utilized to test the performance of the echo canceller algorithms. The echo cancellers should be disabled during signaling transmission periods, however, no mention is given here to this and many other practical issues described in the recommendation related to the actual implementation of the echo canceller, see [31] for details.

The tests recommended by the standard ITU-T G.168 utilize particular signals such as noise, tones, facsimile signals and a set of composite source signals (CSS). In our simulations we apply the CSS input signal as input to the echo cancellers. The CSS simulates speech characteristics in single talk and double talk enabling a performance test for echo cancellers for speech signals. The CSS consists of speech signal, non speech signal and pauses. The speech signal activates the speech detectors and has approximately 50 msec of duration. The speech signal is followed by a pseudo-noise signal having constant magnitude Fourier transform whose phase changes with time. The pause is the third component of the CSS representing an amplitude modulation to the CSS and the usual pauses during a conversation. The pause duration ranges from 100 msec to 150 msec. Fig. 6.19 illustrates the CSS for single talk. The specific timings are:

- Tvst (Speech signal): 48.62 msec.
- Tpn (Pseudo noise): 200.00 msec.
- Tpst (Pause): 101.38 msec.
- Tst1 (Half period): 350.00 msec.
- Tst (Full period): 700.00 msec.

The echo path model according to the recommendation ITU-T G.168, is a linear digital filter whose impulse response $h(k)$ is given by

$$h(k) = (K_i 10^{-\mathrm{ERL}/20}) m_i(k - \delta) \tag{6.80}$$

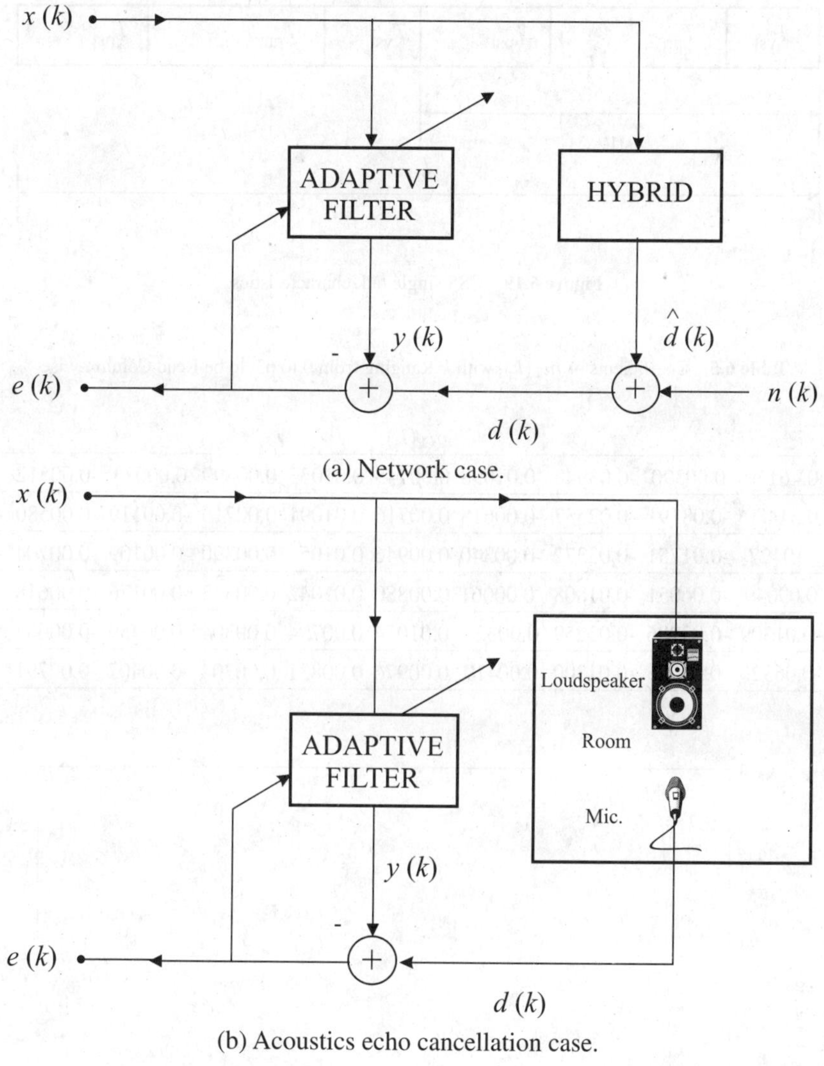

Figure 6.18 Echo cancellation setups.

where ERL is the echo return loss defined in equation (6.78) and $h(k)$ consists of a delayed and attenuated version of any sequence sorted from $m_i(k), i = 1, 2, ..., 8$, for the channel models 1 to 8. These models represent channels whose origins range from hybrid simulation models to measured responses on telephone networks. The constants K_i are determined by the input signal used in the test [31] and are different for distinct echo path models.

Just for illustration Table 6.5 shows the sequence $m_1(k)$ composing the echo-path impulse response. In this case, for CSS type input signal, the scaling signal should be signal $K_1 = 1.39$ and the minimum value of the ERL to be used in the test is 6 dB. The resulting echo-path impulse response is depicted in Fig. 6.20. For the other cases, please refer to [31].

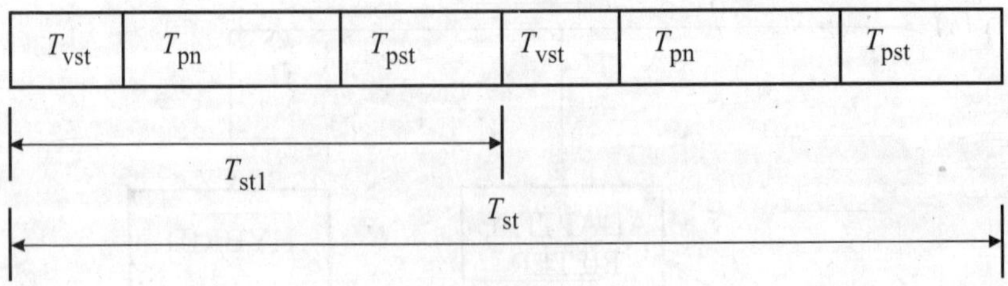

Figure 6.19 CSS single talk characteristics.

Table 6.5 Coefficients of $m_1(k)$, with k Ranging from 0 to 63, to be Read Column-wise

$m_1(k)$										
-0.00436	0.46150	0.00390	-0.03948	-0.01098	0.00745	0.01033	0.00899	0.00073	-0.00512	-0.00772
-0.00829	0.34480	-0.08191	-0.02557	-0.00618	0.00716	0.01091	0.00716	-0.00119	-0.00580	-0.00820
-0.02797	-10427	-0.01751	-0.03372	-0.00340	0.00946	0.01053	0.00390	-0.00109	-0.00704	-0.00839
-0.04208	0.09049	-0.06051	-0.01808	-0.00061	0.00880	0.01042	0.00313	-0.00176	-0.00618	-0.00724
-0.17968	-0.01309	-0.03796	-0.02259	0.00323	0.01014	0.00794	0.00304	-0.00359	-0.00685	
-0.11215	-0.06320	-0.04055	-0.01300	0.00419	0.00976	0.00831	0.00304	-0.00407	-0.00791	

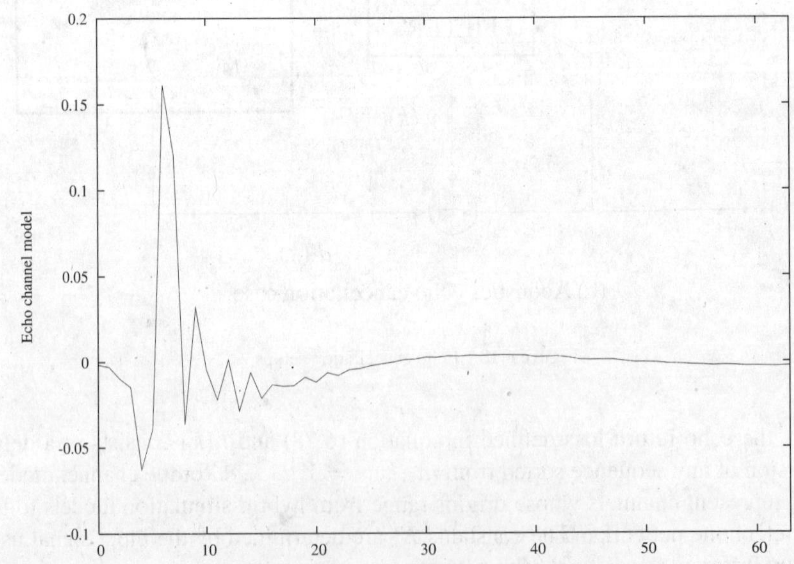

Figure 6.20 Echo-path impulse response.

Example 6.2: Echo cancellation simulations

For the algorithms pointed below, run simulations for an echo cancellation experiment consisting of 50 independent runs describing the average performance for single talk input signal for one of the eight channel models described in [31], specifically the one described in Table 6.5 with an ERL = 12dBs. List the resulting ERLE in dB for each algorithm as well as their respective number of iterations to convergence, measured whenever the average of the last 100 error signals is 10% above the error in steady state. Utilize echo cancellers with sufficient order.

- Normalized LMS algorithm.
- RLS algorithm.
- SM-NLMS algorithm.
- The simplified SM-AP algorithm with $L = 0, 1, 4$.
- The SM-PUAP algorithm with $L = 0, 1, 4$, and $\bar{M} = \text{floor}[\frac{2(N+1)}{3}]$ where floor$[\cdot]$ indicates the largest integer smaller than $[\cdot]$.

For channel model 1, depict the learning curves for the simplified SM-AP and the SM-PUAP algorithms.

Solution:

The numbers and figures presented in this example are result of averaging fifty independent runs. The normalized LMS algorithm utilizes a value of $\mu = 0.5$, with the value of the regularization parameter of $\gamma = 10^{-6}$. The forgetting factor of the RLS algorithm is $\lambda = 0.99$. These values of μ and λ were chosen after some simulation trials indicating favorable performances of the corresponding algorithms. In the SM-AP algorithms distinct values of L are tested and the value of $\bar{\gamma}$ is 0.0002.

Fig. 6.21 depicts the CSS signal utilized in this example. Fig. 6.22 illustrates the error signal for the simplified SM-AP algorithm with $L = 0, 1, 4$, where it can be observed that the error reduces faster for the case with $L = 4$ since the algorithm is more sophisticate, even though the convergence speeds for $L = 1$ and $L = 4$ are quite similar. Fig. 6.23 shows that with the SM-PUAP algorithm the convergence speed is not substantially reduced, showing that the partial updating strategy is very useful. A low complexity way to choose the elements to be updated was to sort out the $\bar{M}$ columns of $\mathbf{X}_{\text{ap}}^T(k)$ whose Euclidean norm have higher values. The SM-PUAP algorithm was set to update only $\frac{2}{3}$ of the coefficients.

Table 6.6 lists the relevant parameters in the echo cancellation environment, namely the ERLE in dB for each algorithm as well as their respective convergence speed. As can be seen in Table 6.6, the algorithms SM-NLMS ($L = 0$), SM-AP, and SM-PUAP require less updates than the remaining algorithms compared. The fastest converging algorithm is the SM-AP ($L = 4$) but it requires the highest computational complexity among the set-membership algorithms. The algorithms SM-AP and SM-PUAP, with $L = 4$, are faster converging than the RLS while requiring much less updates and

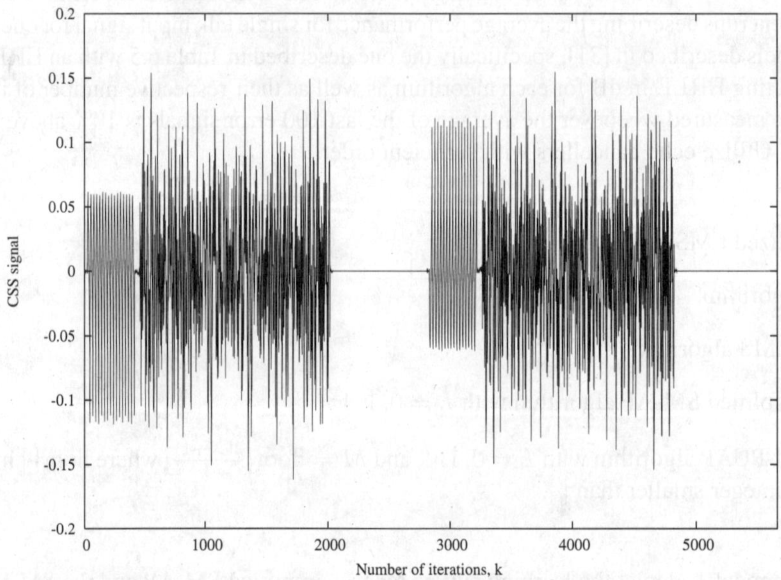

Figure 6.21 CSS signal.

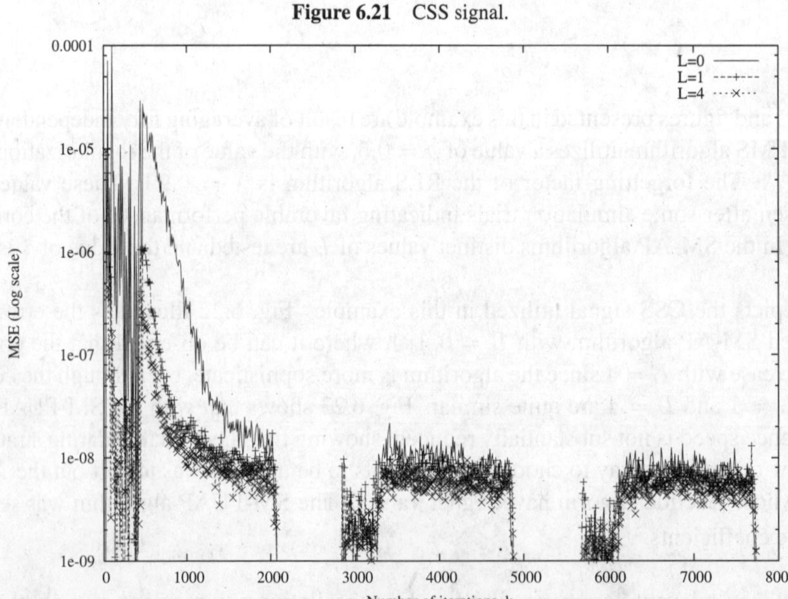

Figure 6.22 Learning curves for the simplified SM-AP algorithm with $L = 0, 1, 4$.

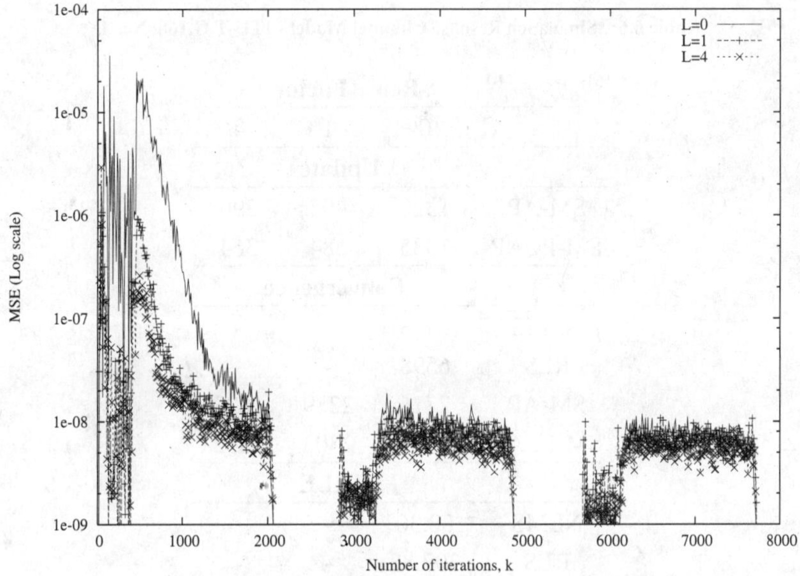

Figure 6.23 Learning curves for the SM-PUAP algorithm $L = 0, 1, 4$.

computations. On the other hand the RLS algorithm lead to much higher ERLE than the remaining algorithms followed by the NLMS. The SM-NLMS and NLMS algorithms have less computations but are slow converging as compared to the remaining SM-AP algorithms of this example.

□

6.9.2 Wireless Channel Environment

A typical continuous-time model for mobile communication channels is described by [33]

$$\tilde{h}(t) = \sum_{i=0}^{I} \sqrt{p_i} a_i(t) b_i(t - \tau_i) \tag{6.81}$$

where t is the time variable, p_i represents the power of the ith tap weight of the FIR model, $a_i(t)$ is complex Gaussian process that multiplies the corresponding transmitted symbol denoted by $b_i(t-\tau_i)$, and τ_i accounts for the relative delay that the receiver detects the ith replica of the transmitted symbols.

The power spectrum of $a_i(t)$ is responsible for the rate of fading of the ith replica (or reflection) of the transmitted symbols. This spectrum is also known as Doppler spectrum. The complete model requires the specification of the Doppler spectrum of the tap weights, denoted by $R_a(f)$ with f being the analog frequency, the delays τ_i, as well as the powers p_i, for $i = 0, \ldots, I$.

Table 6.6 Simulation Results: **Channel Model - ITU-T G.168, No. 1**

	Reuse Factor L		
	0	1	4
	Updates		
SM-AP	1320	497	290
SM-PUAP	1335	584	364
	Convergence		
NLMS	8423	-	-
RLS	6598	-	-
SM-AP	2716	2289	1832
SM-PUAP	2725	2303	1832
	ERLE		
NLMS	80.30	-	-
RLS	307.83	-	-
SM-AP	42.96	43.00	43.62
SM-PUAP	43.87	42.72	43.42

The process $a_i(t)$ is the result of a cluster of multipath components that cannot be resolved by the receiver, arriving within a range of delays[3]. Usually for outdoor environments of mobile communication systems, the model for the Doppler power spectrum is represented by the Jakes model [32] given by

$$R_a(f) = \begin{cases} \frac{1}{\pi f_{\mathrm{D}}} \frac{1}{\sqrt{1-(\frac{f}{f_{\mathrm{D}}})^2}} & \text{for} \quad |f| \leq f_{\mathrm{D}} \\ 0 & \text{for} \quad |f| > f_{\mathrm{D}} \end{cases}$$

where $f_{\mathrm{D}} = \frac{v}{\lambda_s} = \frac{v f_o}{c}$ is the maximum Doppler frequency shift, λ_s is the carrier wavelength, v is the mobile velocity in m/s, c is the speed of light (3.00×10^8 m/s), and f_o is the carrier central frequency.

If we assume that the input signal is complex and bandlimited to a bandwidth around BW, the received signal can be generated by filtering the input signal through a tapped delay line whose tap coefficients are given by $\sqrt{p_i} a_i(t)$ and the delay elements correspond to $T = \frac{1}{\mathrm{BW}}$ [33].

As an illustration, Table 6.7 lists the parameters of test channel models for an outdoor to indoor environment to be utilized in simulations. These models originate from a standard described in [34] for the Universal Mobile Telecommunications System (UMTS). In Table 6.7, the delays are relative

[3] $\tau_i - \frac{1}{2\mathrm{BW}} < \tau < \tau_i + \frac{1}{2\mathrm{BW}}$ with BW denoting the bandwidth of the transmitted signal.

to the first tap whereas the power is relative to the strongest tap. The Doppler power spectrum applies to each tap.

Let's consider for illustration a typical case of UMTS where the chip duration is 260.04 nano seconds (ns) for a transmission rate of 3.84 Mc/s (Mega chips per second). In the case the time difference between two multipath components is at least $260.04 = \frac{1}{3.84}$ ns, it is possible for the receiver to separate them. For example according to Table 6.7, in a digital simulation environment where the input signal is sampled at chip rate for channel B, it will be possible to verify the presence of the multipath signals of the taps at approximately

- 1 chip from the reference for tap 2.
- 3 chips from the reference for tap 3.
- 5 chips from the reference for tap 4.
- 9 chips from the reference for tap 5.
- 14 chips from the reference for tap 6.

where it was taken into consideration that the relative delays in the table represent the time where the energy of the continuous-time reflection reaches its maximum.

Table 6.7 Channel Model Parameters: Outdoor to Indoor Test Environment with Jakes Doppler Spectrum

Tap	Channel A		Channel B	
	Relative Delay ns	Average Power dB	Relative Delay ns	Average Power dB
1	0	0	0	0
2	110	-9.7	200	-0.9
3	190	-19.2	800	-4.9
4	410	-22.8	1200	-8.0
5	–	–	2300	-7.8
6	–	–	3700	-23.9

The coefficients of a time-varying channel including the Doppler effects can be generated as depicted in Fig. 6.24, where $\bar{n}(t)$ is a Gaussian noise source and the Doppler filter is an approximation of $H_{\mathrm{D}}(f) = \sqrt{R_a(f)}$. Fig. 6.25 shows an efficient way to generate the coefficients of the channel model [36]-[37], where from two real-valued Gaussian sources with $\bar{N}+1$ points we calculate their symmetrical FFT spectrum [38]. Then we multiply the FFT outputs by $H_{\mathrm{D}}(f_{\bar{m}})$ where $f_{\bar{m}} = \bar{m}\frac{2f_{\mathrm{D}}}{\bar{N}+1}$ for $\bar{m} = 0, 1, \ldots, \bar{N}$, and the resulting vector is applied as input to an $\bar{N}+1$ length IFFT. The quadrature and in-phase results are squared at each point in time, added, with the result square rooted. Finally, an $\bar{N}+1$-length time series is generated. In an actual simulation environment the

Gaussian noise is generated with around α_1 samples per period of the maximum Doppler frequency, that is $\frac{1}{\alpha_1 f_{\rm D}}$, therefore the sampling rate of the channel coefficients is around $\alpha_1 f_{\rm D}$ with α_1 being an integer usually chosen in the range 5 to 12. As can be noticed the coefficients of the channel model are generated from the Jakes model of the Doppler effect. However, the system simulation takes place at much higher frequency rate denoted as $f_{\rm sim}$. As a result, an interpolation operation given by $L_{\rm sim} = \text{floor}[\frac{f_{\rm sim}}{\alpha_1 f_{\rm D}}]$ should be applied to the coefficients of the channel model.

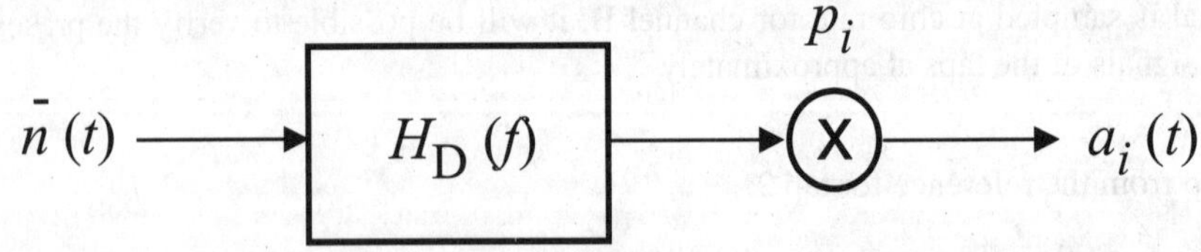

Figure 6.24 Generation of multipath coefficient including Doppler effect.

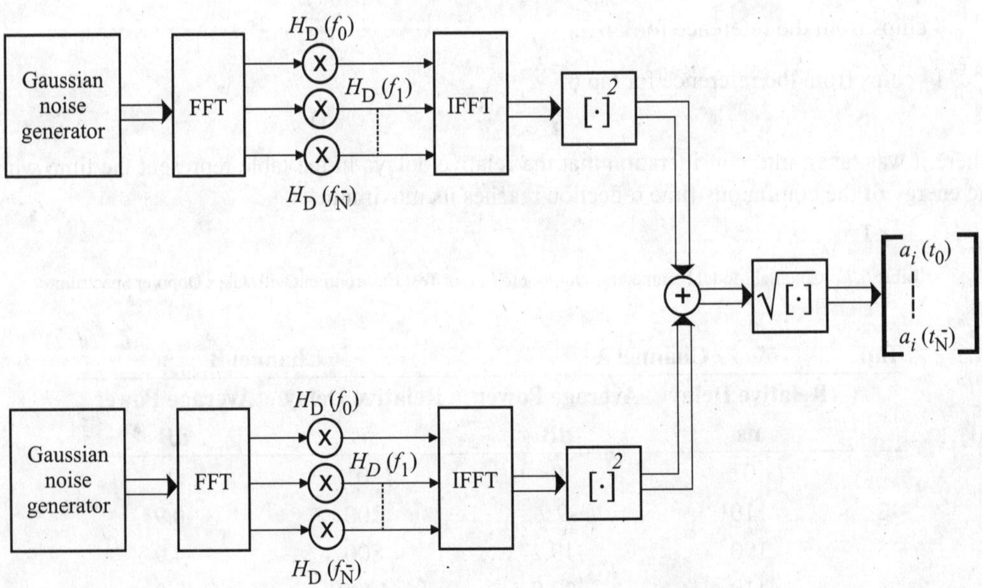

Figure 6.25 Simulation setup for Jakes model.

Example 6.3: CDMA receiver simulations

Consider a downlink connection of a synchronous direct-sequence code-division multiple access (DS-CDMA) system with J users, $G+1$ chips per symbol and a channel with $I+1$ paths. Assume the user receiver is moving at $v = 30.00$m/s and the carrier frequency is at $f_o = 1.0$GHz. We consider a simple model for the channel inspired by the UMTS test model above described. The channel model should be generated at a simulation sampling rate of at least $f_{\rm sim} = \alpha_2 \times \frac{1}{T} = \alpha_2 \text{BW}$ samples per second, with α_2 being normally an integer ranging from 5 to 12. It is worth emphasizing again that the channel coefficients will be generated at a much lower rate than the simulation sampling rate. As a result, some standard interpolation technique [38] should be used in order to match the channel model generation rate with simulation sampling rate.

Consider that the chip rate of the CDMA system is 0.5 Mc/s (Mega chips per second) and that we utilize $\alpha_2 = 10$ samples per chip to simulate the system. As such, the CDMA system simulation has sampling rate equal to 5 Msamples/s. In this case the interpolation factor applied to the chip level signal should be $L_{\text{interp}} = \text{floor}[\frac{5\text{Mc/s}}{\frac{1}{T}}] = \text{floor}[\frac{5\text{Mc/s}}{\text{BW}}]$, where floor$[\cdot]$ indicates the largest integer smaller than $[\cdot]$. The sampling frequency that the channel model should be generated is then given by

$$f_{\text{sim}} \approx \text{BW}L_{\text{interp}} \approx \alpha_1 f_{\text{D}} L_{\text{sim}}$$

in Msamples/s. In this particular discussion, assuming the input signal sampling rate equal to the chip rate the interpolation factor L_{interp} is equal to 10. Note that in the above discussion we satisfy the sampling theorem by utilizing the complex envelope, that is the complex lowpass equivalent, of the transmitted signal. This means the sampling rate is equal to the channel bandwidth.

Assuming the channel model as described is constant during each symbol interval, and that the input signal vector $\mathbf{x}(k)$ is given by[4]

$$\begin{aligned} \mathbf{x}(k) &= \sum_{j=1}^{J} A_j b_j(k)\mathbf{C}_j\mathbf{h}(k) + \mathbf{n}(k) \\ &= \sum_{j=1}^{J} A_j b_j(k)\mathbf{C}_j\mathbf{h}(k) + \mathbf{n}_{\text{n}}(k) + \mathbf{n}_{\text{ISI}}(k) \end{aligned} \tag{6.82}$$

where $\mathbf{x}(k)$ is an $(N+1 = G+I+1) \times 1$ vector and $\mathbf{n}_{\text{n}}(k)$ is defined in equation (6.58). We consider that $\mathbf{n}_{\text{n}}(k)$ is a complex Gaussian noise vector with $E[\mathbf{n}_{\text{n}}(k)\mathbf{n}_{\text{n}}^H(k)] = \sigma_n^2\mathbf{I}$. The symbols $b_j(k)$ are four QAM given by $\frac{\sqrt{2}}{2}\{\pm 1 \pm \jmath 1\}$, where the amplitude of user j is A_j. The channel vector is $\mathbf{h}(k) = [h_0(k) \ \ldots \ h_I(k)]^T$ and the $(N+1) \times (I+1)$ convolution matrix $\mathbf{C}_k$ contains one-chip shifted versions of the signature sequence for user j given by $\mathbf{s}_j = [s_{j,0} \ \ s_{j,1} \ldots s_{j,G}]^T$. Matrix $\mathbf{C}_k$ has the following format

$$\mathbf{C}_j = \begin{bmatrix} s_{j,0} & 0 & 0 & \cdots & 0 \\ s_{j,1} & s_{j,0} & 0 & \cdots & 0 \\ s_{j,2} & s_{j,1} & s_{j,0} & \cdots & 0 \\ \vdots & \vdots & \ddots & \vdots & \vdots \\ 0 & s_{j,G} & s_{j,G-1} & \cdots & s_{j,G-I} \\ \vdots & \vdots & \ddots & \vdots & \vdots \\ 0 & 0 & \cdots & 0 & s_{j,G} \end{bmatrix} \tag{6.83}$$

This example aims to access the bit error rate (BER) performance of some adaptive-filtering algorithms such as:

[4]In an actual implementation $\mathbf{x}(k)$ originates from the received signal after filtering it through a chip-pulse matched filter and then sampled at chip rate.

- Normalized LMS algorithm.
- RLS algorithm.
- SM-NLMS algorithm.
- The simplified SM-AP algorithm with $L = 4$ and time-varying $\bar{\gamma}(k)$.
- The SM-PUAP algorithm with $L = 1$, and $\bar{M} = \text{floor}[\frac{(N+1)}{1.8}]$ where floor$[\cdot]$ indicates the largest integer smaller than $[\cdot]$.

The receiver of the DS-CDMA system simulation setup is depicted in Fig. 6.26 where we utilize as spreading sequences the Gold sequences of length $G + 1 = 7$ listed in Table 6.26 [35]. The Gold sequences are not orthogonal to each other leaving some multi-access interference from the other users in the CDMA system on the information of the user of interest, even in synchronous transmission case.

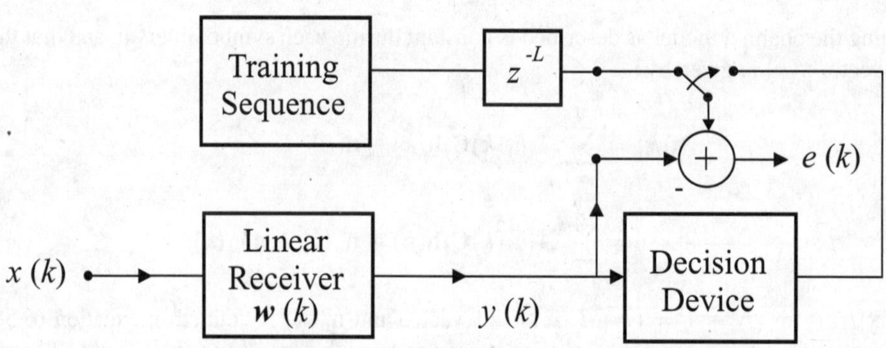

Figure 6.26 Simulations setup for DS-CDMA example.

Table 6.8 Length 7 Gold Sequences

Sequences	Gold Sequences						
s_1	1	0	0	1	0	1	1
s_2	1	1	1	0	1	0	0
s_3	0	1	1	1	1	1	1
s_4	1	1	1	0	0	0	1
s_5	1	0	1	0	1	1	0
s_6	0	0	0	0	1	0	1
s_7	1	1	0	1	1	0	0
s_8	0	0	1	1	0	0	0
s_9	0	1	0	0	0	1	0

All users are synchronized such that all their information face the same channel with three paths with relative powers given by 0, -0.9 and -4.9 dB, respectively. The relative delays between the paths are determined by a uniformly distributed random variable whose outcome is mapped to integers in the range 1 to 4, where these integers represent the number of chips.

The system starts with 5 users where all the 4 interferers have transmission powers 3 dB below the desired user power level. The corresponding signal to noise ratio, defined as the ratio between the desired user symbol energy per bit and the environment noise, is given by $E_b/N_0 = 20$ dB. The quantity $N_0/2$ corresponds to power spectral density of the noise for positive and negative frequencies, that is N_0 is average noise power per bandwidth where the noise is measured at the receiver input. At 2000 symbols, an interferer with the same power as the desired user power enters the system, whereas 2 interferers with the same level of power disconnect. This dynamic behavior aims at addressing, for this particular example, if some noticeable disturbance to the receiver performance originates from user access and disconnection from the system.

Plot the evolution of the estimated of the noise and ISI powers as compared with the actual interference power.

Solution:

For this example we measure the results by averaging the outcomes from 50 independent runs. In the case of the normalized LMS algorithm the value of μ is 0.3, whereas the regularization parameter value is $\gamma = 10^{-6}$. The RLS algorithm is implemented with $\lambda = 0.97$. Again these values of μ and λ were chosen after some simulation trials. The SM-AP algorithm uses $L = 4$ and variable $\bar{\gamma}$, whereas the SM-PUAP algorithm uses $L = 1$.

For a better view of the results the channel was allowed to change completely in an interval of 50 symbols, equivalent to 450 chips. Fig. 6.27 depicts the first 450 samples of the learning curves for the algorithms compared in this example, whereas Fig. 6.28 shows the behavior of these algorithms in the long run. In both figures the channel changes are noticeable every 350 chips, where the first change occurs at around 370 chips due to the channel plus spreading delays. As can be observed, the NLMS, RLS, SM-NLMS and the SM-AP algorithms were able to track the changes in the channel to some extent. However, as shown in Fig. 6.29, the simplified SM-PUAP algorithm with $L = 1$ using variable $\bar{\gamma}$ has very competitive performance since it is among the fastest converging in such nonstationary environment. Very similar result is obtained with the simplified SM-AP algorithm which has higher computational cost. All the algorithms did not show any noticeable distinct behavior after the entrance and exit of users in the system, mainly due to the fact that the channel model changes at every 50 symbols was the main source of changes.
Fig. 6.30 plots the evolution of the estimated noise and ISI powers as compared with the actual interference power. The estimated curve was obtained using equation (6.63) with $\alpha = 0.96$. As can be observed, the estimated average power of the interferences follows closely its actual value for this particular example, except at iteration 2000 when the interference estimate takes a few iterations to track the true interference power. The accurate estimate of the disturbances turns the SM-AP algorithms very attractive since virtually no environment dependent parameter is required to achieve good overall performance.

□

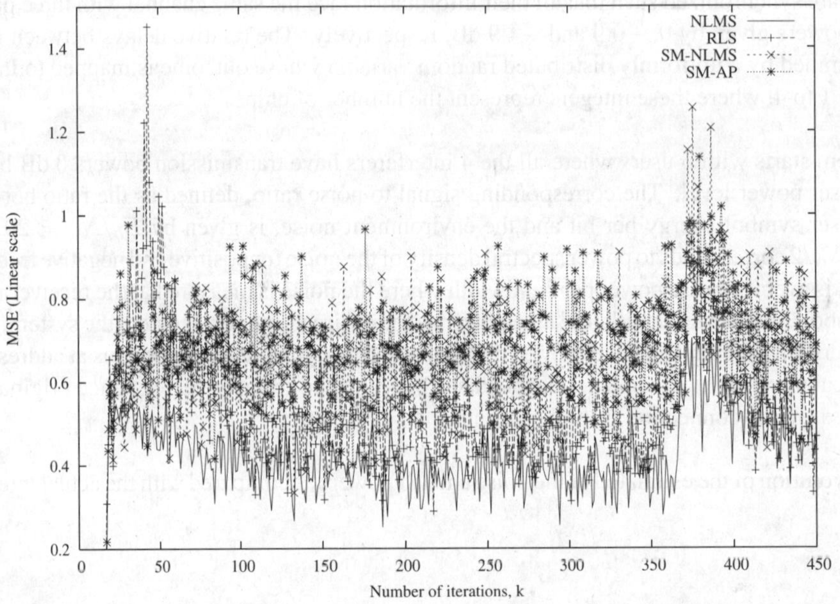

Figure 6.27 Learning curves for the NLMS, RLS, SM-NLMS, and SM-AP algorithms; 250 iterations.

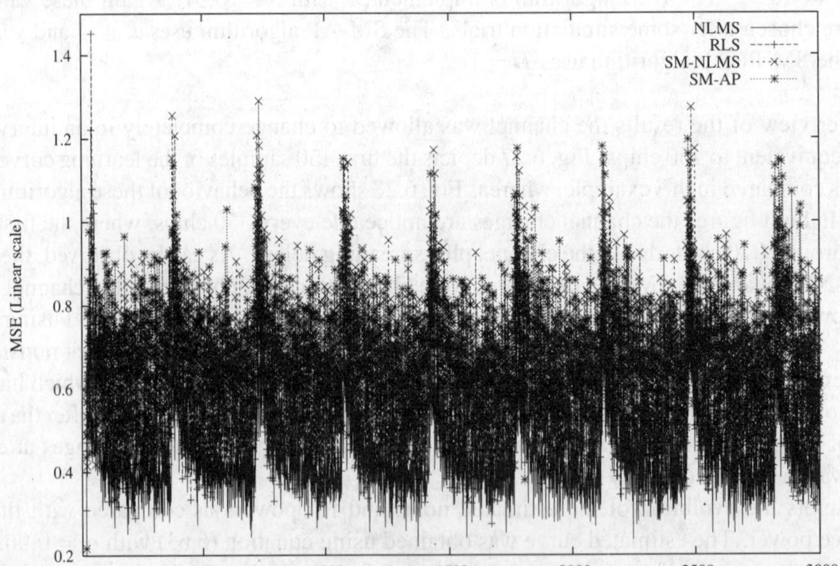

Figure 6.28 Learning curves for the NLMS, RLS, SM-NLMS, and SM-AP algorithms.

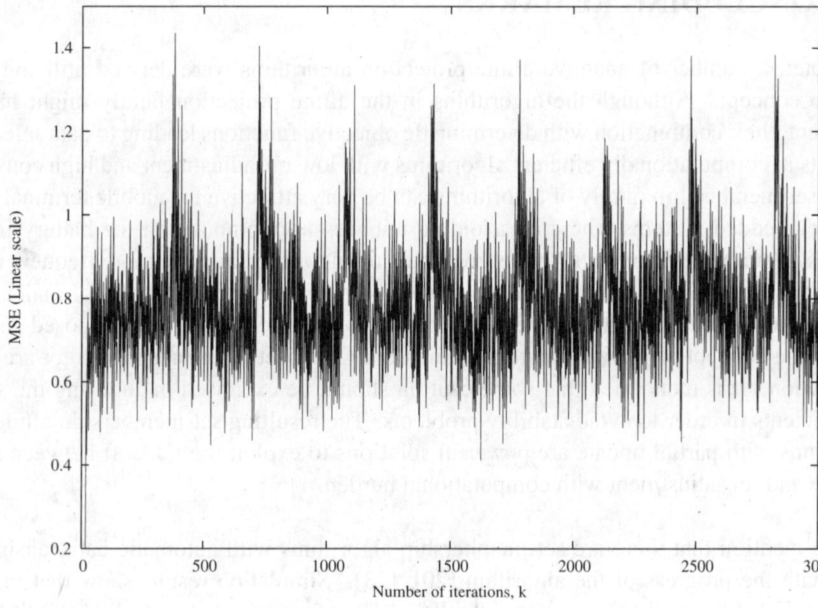

Figure 6.29 Learning curve SM-PUAP algorithms, $L = 1$.

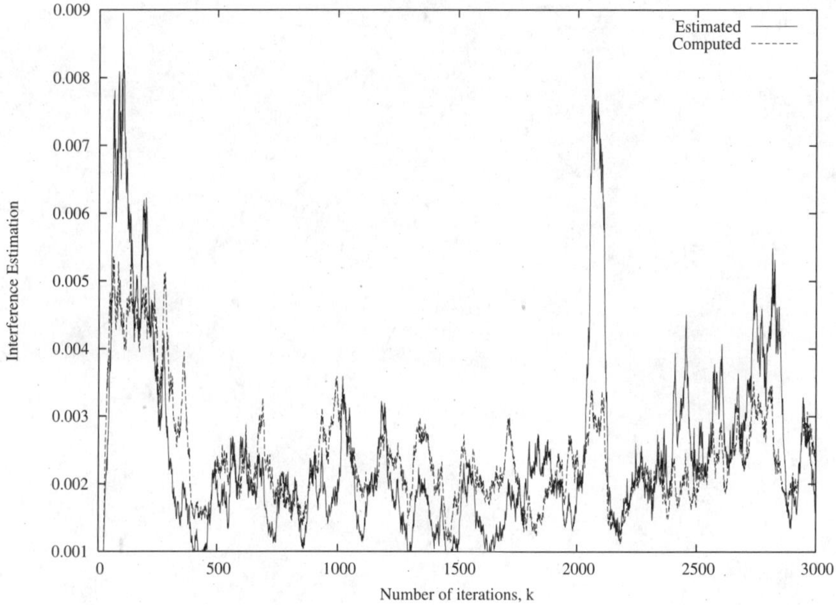

Figure 6.30 Interference estimation.

6.10 CONCLUDING REMARKS

In this chapter, a number of adaptive affine projection algorithms were derived utilizing the set-membership concept. Although the algorithms in the affine projection family might have high misadjustment, their combination with deterministic objective functions leading to data selective updating results in computationally efficient algorithms with low misadjustment and high convergence speed. The set-membership family of algorithms can be very attractive for mobile terminals, sensor arrays and embedded systems where by avoiding unnecessary computation the battery life is increased. In stationary environments, the set-membership algorithms require more frequent updating during the early iterations, as a consequence, if the computational complexity is of major concern some strategy to reduce even further the computation count is required. The proposed solution is to introduce the concept of partial update, in which only a subset of the adaptive filter are updated in each iteration. It is mentioned that some caution should be exercised in choosing the selection of the coefficients in order to avoid stability problems. The resulting set-membership affine projection algorithms with partial update are powerful solutions to exploit the tradeoff between speed of convergence and misadjustment with computational burden.

It should be mention that there are set-membership algorithms with automatic data-reusing factor according with the progress of the algorithm [20], [23]. Simulation results show that in most of iterations the SM-AP algorithm requires a small number of reuses, that is in the limit, it becomes the SM-NLMS or the SM-BNLMS algorithms. The set-membership technique can also be applied to generate constrained affine projection algorithms with low computational complexity as proposed in [39].

6.11 REFERENCES

1. F. C. Schweppe, "Recursive state estimate: Unknown but bounded errors and system inputs," *IEEE Trans. on Automatic Control*, vol. 13, pp. 22-28, Feb. 1968.
2. E. Fogel and Y.-F. Huang, "On the value of information in system identification - bounded noise case," *Automatica*, vol. 18, pp. 229-238, March 1982.
3. J. R. Deller, "Set-membership identification in digital signal processing," *IEEE Acoust., Speech, Signal Processing Magazine*, vol. 6, pp. 4-20, Oct. 1989.
4. J. R. Deller, M. Nayeri, and S. F. Odeh, "Least-squares identification with error bounds for real-time signal processing and control," *Proceedings of IEEE*, vol. 81, pp. 815-849, June 1993.
5. S. Dasgupta and Y.-F. Huang, "Asymptotically convergent modified recursive least-squares with data dependent updating and forgetting factor for systems with bounded noise," *IEEE Trans. on Information Theory*, vol. 33, pp. 383-392, May 1987.
6. A. K. Rao and Y.-F. Huang, "Tracking characteristics of an OBE parameter-estimation algorithm," *IEEE Trans. on Signal Processing*, vol. 41, pp. 1140-1148, March 1993.
7. S. Gollamudi, S. Nagaraj, S. Kapoor, and Y.-F. Huang, "Set-membership filtering and a set-membership normalized LMS algorithm with an adaptive step size," *IEEE Signal Processing Letters*, vol. 5, pp. 111-114, May 1998.
8. S. Gollamudi, S. Nagaraj, S. Kapoor, and Y.-F. Huang, "Set-membership adaptive equalization and updator-shared implementation for multiple channel communications systems," *IEEE Trans. on Signal Processing*, vol. 46, pp. 2372-2384, Sept. 1998.
9. S. Nagaraj, S. Gollamudi, S. Kapoor, and Y.-F. Huang, "BEACON: An adaptive set-membership filtering technique with sparse updates," *IEEE Trans. on Signal Processing*, vol. 47, pp. 2928-2941, Nov. 1999.
10. P. S. R. Diniz and S. Werner, "Set-membership binormalized data reusing LMS algorithms," *IEEE Trans. on Signal Processing*, vol. 51, pp. 124-134, Jan. 2003.
11. S. Werner and P. S. R. Diniz, "Set-membership affine projection algorithm," *IEEE Signal Processing Letters*, vol. 8, pp. 231-235, Aug. 2001.
12. A. Antoniou and W.-S. Lu, *Practical Optimization: Algorithms and Engineering Applications*, Springer, New York, NY, 2007.
13. J. A. Apolinário, M. L. R. de Campos, and P. S. R. Diniz, "The binormalized data-reusing LMS algorithm," *IEEE Trans. on Signal Processing*, vol. 48, pp. 3235-3242, Nov. 2000.
14. S. C. Douglas, "Adaptive filters employing partial updating," *IEEE Trans. on Circuits and Systems II: Analog and Digital Signal Processing*, vol. 44, pp. 209-216, March 1997.
15. T. Aboulnasr and K. Mayyas, "Complexity reduction of the NLMS algorithm via selective coefficient updating," *IEEE Trans. on Signal Processing*, vol. 47, pp. 1421-1427, May 1999.

16. S. C. Douğançay and O. Tanrikulu, "Adaptive Filtering algorithms with selective partial updating," *IEEE Trans. on Circuits and Systems II: Analog and Digital Signal Processing*, vol. 48, pp. 762-769, Aug. 2001.

17. S. Werner, M. L. R. de Campos, and P. S. R. Diniz, "Partial-update NLMS algorithm with data-selective updating," *IEEE Trans. on Signal Processing*, vol. 52, pp. 938-949, April 2004.

18. S. Werner, M. L. R. de Campos, and P. S. R. Diniz, "Mean-squared analysis of the partial-update NLMS algorithm," *Brazilian Telecommunications Journal - SBrT*, vol. 18, pp. 77-85, June 2003.

19. M. Godavarti and A. O. Hero, III, "Partial update LMS algorithms," *IEEE Trans. on Signal Processing*, vol. 53, pp. 2384-2399, July 2005.

20. S. Werner, J. A. Apolinário Jr., and P. S. R. Diniz, "Set-membership proportionate affine projection algorithms," *EURASIP- Journal on Audio, Speech, and Music Processing*, vol. 2007, pp. 1-10, Article ID 34242, 2007.

21. S. C. Douğançay and O. Tanrikulu, "Selective-partial-update NLMS and affine projection algorithms for acoustic echo cancellation," *Proc. IEEE Intern. Conf. on Acoust. Speech and Signal Processing*, Istanbul, Turkey, pp. 448-451, June 2000.

22. P. S. R. Diniz, R. P. Braga, and S. Werner, "Set-membership affine projection algorithm for echo cancellation," *Proc. IEEE Intern. Symposium on Circuits and Systems*, Island of Kos, Greece, pp. 405-408, May 2006.

23. S. Werner, P. S. R. Diniz, and J. E. W. Moreira, "Set-membership affine projection algorithm with variable data-reuse factor," *Proc. IEEE Intern. Symposium on Circuits and Systems*, Island of Kos, Greece, pp. 261-264, May 2006.

24. L. Guo and Y.-F. Huang, "Set-membership adaptive filtering with parameter-dependent error bound tuning," *Proc. IEEE Intern. Conf. on Acoust. Speech and Signal Processing*, Philadelphia, PA, pp. IV-369 - IV-372, May 2005.

25. J. F. Galdino, J. A. Apolinário, Jr., and M. L. R. de Campos, "A set-membership NLMS algorithm with time-varying error bound," *Proc. IEEE Intern. Symposium on Circuits and Systems*, Island of Kos, Greece, pp. 277-280, May 2006.

26. R. C. de Lamare and P. S. R. Diniz, "Set-membership adaptive algorithms based on time-varying error bounds for DS-CDMA systems," *Proc. IEEE Intern. Symposium on Circuits and Systems*, Island of Kos, Greece, pp. 3466-3469, May 2006.

27. R. C. de Lamare and P. S. R. Diniz, "Blind constrained set-membership algorithms with time-varying bounds for CDMA interference suppression," *Proc. IEEE Intern. Conf. on Acoust. Speech and Signal Processing*, Toulouse, France, pp. IV-617 - IV-620, May 2006.

28. R. C. de Lamare and P. S. R. Diniz, "Set-membership adaptive algorithms based on time-varying error bounds and their application to interference suppression," *IEEE International Telecommunications Symposium*, Fortaleza, Brazil, pp. 563-567, Sept. 2006.

29. E. Hänsler and G. Schmidt, *Acoustics Echo and Noise Control: A Practical Approach*, John Wiley & Sons, Hoboken, NJ, 2004.

30. J. Benesty, T. Gänsler, D. R. Morgan, M. M. Sondhi, and S. L. Gay, *Advances in Network and Acoustics Echo Cancellation*, Springer-Verlag, Berlin, Germany, 2001.

31. International Telecommunication Union, *ITU-T G.168 Standard for Digital Network Echo Cancellers*, June 2002.

32. J. G. Proakis, *Digital Communications*, McGraw Hill, New York, NY, 4th edition, 2001.

33. M. C. Jeruchim, P. Balaban, and K. San Shanmugan, *Simulation of Communication Systems: Modeling Methodology and Techniques*, Kluwer Academic Publishers, Norwell, MA, 2nd edition, 2000.

34. European Telecommunications Standards Institute, *UMTS, technical report TR 101 112 V3.2.0*, 1998-04.

35. E. H. Dinan and B. Jabbari, "Spreading codes for direct sequence CDMA and wideband CDMA cellular networks," *IEEE Communications Magazine*, vol. 36, pp. 48-58, Sep. 1998.

36. J. I. Smith, "Computer generated multipath fading simulation for mobile radio," *IEEE Trans. on Vehicular Technology*, vol. 24, pp. 39-40, Aug. 1975.

37. T. S. Rappaport, *Wireless Communications: Principles and Practice*, Prentice Hall, Englewood Cliffs, NJ, 2nd edition, 2002.

38. P. S. R. Diniz, E. A. B. da Silva, and S. L. Netto, *Digital Signal Processing: System Analysis and Design*, Cambridge University Press, Cambridge, UK, 2002.

39. S. Werner, J. A. Apolinário Jr., M. L. R. de Campos, and P. S. R. Diniz, "Low-complexity constrained affine-projection algorithms," *IEEE Trans. on Signal Processing*, vol. 53, pp. 4545-4555, Dec. 2005.

6.12 PROBLEMS

1. In a system identification application the unknown system has transfer function given by

$$H(z) = \frac{1}{2} - \frac{1}{2}z^{-1}$$

whereas the input signal is a binary $(-1, 1)$ random signal, and the additional noise is generated via $(-\frac{1}{4}, \frac{1}{4})$ by tossing a fair coin. Evaluate by hand the first 10 iterations SM-NLMS algorithm.

2. Show that the updating equation (6.50) is equivalent to the second coefficient updating of Algorithm 6.4.

3. Repeat problem 1 for the SM-BNLMS algorithm 1.

4. Repeat problem 1 for the SM-BNLMS algorithm 2.

5. Perform the equalization of a channel with the following impulse response

$$h(k) = ku(k) - (2k - 9)u(k - 5) + (k - 9)u(k - 10)$$

using a known training signal consisting of a binary (-1,1) random signal. An additional Gaussian white noise with variance 10^{-2} is present at the channel output.
(a) Apply the SM-NLMS algorithm with an appropriate $\bar{\gamma}$ and find the impulse response of an equalizer with 15 coefficients.
(b) Convolve the equalizer impulse response at an iteration after convergence, with the channel impulse response and comment on the result.

6. In a system identification problem, the input signal is generated by an autoregressive process given by

$$x(k) = -1.2x(k - 1) - 0.81x(k - 2) + n_x(k)$$

where $n_x(k)$ is zero-mean Gaussian white noise with variance such that $\sigma_x^2 = 1$. The unknown system is described by

$$H(z) = 1 + 0.9z^{-1} + 0.1z^{-2} + 0.2z^{-3}$$

The adaptive filter is also a third-order FIR filter, and the additional noise is a zero-mean Gaussian noise with variance given by $\sigma_n^2 = 0.001$.
Using the SM-BNLMS algorithm:
(a) Choose an appropriate $\bar{\gamma}$, run an ensemble of 20 experiments, and plot the average learning curve.
(b) Measure the excess MSE.

7. Derive the complex versions of the SM-BNLMS algorithms 1 and 2 to equalize a channel with the transfer function given below. The input signal is a four QAM signal representing a randomly generated bit stream with the signal to noise ratio $\frac{\sigma_{\tilde{x}}^2}{\sigma_n^2} = 20$ at the receiver end, that is, $\tilde{x}(k)$ is the received signal without taking into consideration the additional channel noise. The adaptive filter has 10 coefficients.

$$H(z) = (0.34 - 0.27\jmath) + (0.87 + 0.43\jmath)z^{-1} + (0.34 - 0.21\jmath)z^{-2}$$

(a) Run the algorithm for $\mu = 0.1$, $\mu = 0.4$, and $\mu = 0.8$. Comment on the convergence behavior in each case.
(b) Plot the real versus imaginary parts of the received signal before and after equalization.
(c) Increase the number of coefficients to 20 and repeat the experiment in (b).

8. In a system identification problem, the input signal is generated from a four QAM of the form

$$x(k) = x_{\text{re}}(k) + \jmath x_{\text{im}}(k)$$

where $x_{\text{re}}(k)$ and $x_{\text{im}}(k)$ assume values ± 1 randomly generated. The unknown system is described by

$$H(z) = 0.32 + 0.21\jmath + (-0.3 + 0.7\jmath)z^{-1} + (0.5 - 0.8\jmath)z^{-2} + (0.2 + 0.5\jmath)z^{-3}$$

Algorithm 6.7

The Complex Set-Membership Affine Projection Algorithm

Initialization

$\mathbf{x}(0) = \mathbf{w}(0) = [0 \ \ldots \ 0]^T$

choose $\bar{\gamma}$ around $\sqrt{5}\sigma_n$

$\gamma =$ small constant

Do for $k \geq 0$

$\mathbf{e}^*_{\text{ap}}(k) = \mathbf{d}^*_{\text{ap}}(k) - \mathbf{X}^H_{\text{ap}}(k)\mathbf{w}(k)$

$\mu(k) = \begin{cases} 1 - \frac{\bar{\gamma}}{|e(k)|} & \text{if } |e(k)| > \bar{\gamma} \\ 0 & \text{otherwise} \end{cases}$

$\mathbf{w}(k+1) = \mathbf{w}(k) + \mathbf{X}_{\text{ap}}(k)\left[\mathbf{X}^H_{\text{ap}}(k)\mathbf{X}_{\text{ap}}(k) + \gamma\mathbf{I}\right]^{-1}\mu(k)e^*(k)\mathbf{u}_1$

The adaptive filter is also a third-order complex FIR filter, and the additional noise is zero-mean Gaussian white noise with variance $\sigma_n^2 = 0.04$. Derive and use the complex set-membership normalized LMS algorithm, choose an appropriate $\bar{\gamma}$, run an ensemble of 20 experiments, and plot the average learning curve.

9. Repeat problem 8 utilizing the complex version of SM-AP algorithm, detailed in Algorithm 6.7 provided, with $L = 4$.

10. The double threshold SM-AP algorithm can be derived for applications such as echo cancellation where there is no interest in reducing the error signal power beyond certain level. Derive an SM-AP algorithm by choosing the vector $\bar{\boldsymbol{\gamma}}(k)$ in such a way that the echo canceller does not reduce the output error power below the power of the far-end signal. Instead of using as threshold a single value of $\bar{\gamma}$, the proposed algorithm uses a range for the acceptable output error value between $\bar{\gamma}_1$ and $\bar{\gamma}_2$, where $\bar{\gamma}_1 > \bar{\gamma}_2$, as depicted in Fig. 6.31.

11. In applications where the parameters to be estimated are dominated by few dominant coefficients, that is, they are sparse, it is often desirable to employ a proportionate adaptation strategy where weights are assigned to parameter components proportional to their magnitude [20]. The updating equation of the proportionate SM-AP algorithm is given by

$$\mathbf{w}(k+1) = \mathbf{w}(k) + \mathbf{P}(k)\mathbf{X}_{\text{ap}}(k)\left[\mathbf{X}^T_{\text{ap}}(k)\mathbf{X}_{\text{ap}}(k)\right]^{-1}\left[\mathbf{e}_{\text{ap}}(k) - \bar{\boldsymbol{\gamma}}(k)\right] \tag{6.84}$$

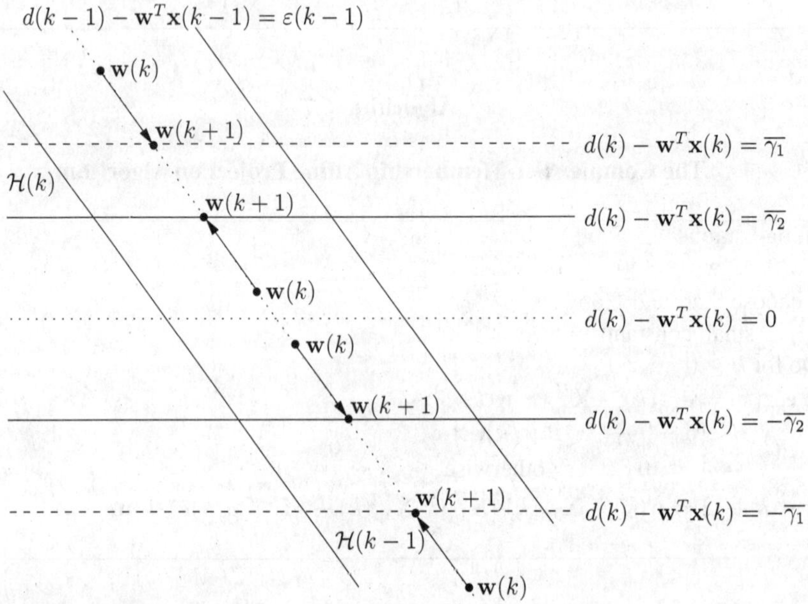

Figure 6.31 SM-AP algorithm with double threshold.

where

$$\mathbf{P}(k) = \mu(k) \begin{bmatrix} p_0(k) & 0 & \cdots & 0 \\ 0 & p_1(k) & & \vdots \\ \vdots & 0 & & \vdots \\ & \vdots & & 0 \\ 0 & 0 & \cdots & p_N(k) \end{bmatrix}$$

$$\mu(k) = \begin{cases} 1 - \frac{\bar{\gamma}}{|e(k)|} & \text{if } |e(k)| > \bar{\gamma} \\ 0 & \text{otherwise} \end{cases}$$

and

$$p_i(k) = \frac{1 - \kappa\mu(k)}{N+1} + \frac{\kappa\mu(k)|w|_i(k)}{\sum_{i=0}^{N} |w|_i(k)}$$

Use the proportionate adaption algorithm identify a system whose impulse response is given below.

$$h(k) = [1 \ 0 \ 0 \ 0.5 \ 0 \ 2]$$

The input signal is a uniformly distributed white noise with variance $\sigma_x^2 = 1$, and the measurement noise is Gaussian white noise uncorrelated with the input with variance $\sigma_n^2 = 5.25 \ 10^{-3}$. The adaptive filter has 6 coefficients.

(a) Use $\kappa = 0.5$, experiment some values of $\bar{\gamma}$ and discuss the results.
(b) Plot the obtained FIR filter impulse response at any iteration after convergence is achieved

and compare with the unknown system.
(c) Compare the best solution with that obtained by the corresponding SM-AP algorithm.

12. Prove from equations (6.36) to (6.39) that

$$\left\{ \left[\mathbf{x}(k)\, \tilde{\mathbf{X}}_{\text{ap}}(k) \right]^T \left[\mathbf{x}(k)\, \tilde{\mathbf{X}}_{\text{ap}}(k) \right] \right\} \begin{bmatrix} a \\ \mathbf{b} \end{bmatrix} = \begin{bmatrix} 1 \\ 0 \\ \vdots \\ 0 \end{bmatrix}$$

13. In SM-PUAP algorithm only $N + 1 - \bar{M}$ coefficients are updated at a given iteration. Exploit this fact to derive a reduced complexity algorithm by generalizing the procedure used to derive equations (6.36) to (6.41).

14. Identify a typical Channel A model for wireless environment described in Table 6.7 with the SM-BNLMS algorithm 2, using as input signal a Gaussian white noise and such that the signal to noise ratio at the receiver end is 10dB's. Determine though simulations the approximate number of training symbols to achieve a good channel estimation of sufficient order.

7

ADAPTIVE LATTICE-BASED RLS ALGORITHMS

7.1 INTRODUCTION

There are a large number of algorithms that solve the least-squares problem in a recursive form. In particular, the algorithms based on the lattice realization are very attractive because they allow modular implementation and require a reduced number of arithmetic operations (of order N) [1]-[7]. As a consequence, the lattice recursive least-squares (LRLS) algorithms are considered fast implementations of the RLS problem.

The LRLS algorithms are derived by solving the forward and backward linear prediction problems simultaneously. The lattice-based formulation provides the prediction and the general adaptive filter (joint-process estimation) solutions of all intermediate orders from 1 to N simultaneously. Consequently, the order of the adaptive filter can be increased or decreased without affecting the lower order solutions. This property allows the user to activate or deactivate sections of the lattice realization in real time according to performance requirements.

Unlike the RLS algorithm previously discussed, which requires only time-recursive equations, the lattice RLS algorithms use time-update and order-update equations. A key feature of the LRLS algorithms is that the prediction process discloses the properties (or the model) of the input signal. The internal signals of the prediction part retain in a sense nonredundant information of the input signal that can be utilized in a decoupled form in the following processing. This mechanism is inherently built in the lattice algorithm derivations.

The performance of the LRLS algorithms when implemented with infinite-precision arithmetic is identical to that of any other RLS algorithm. However, in finite-precision implementation each algorithm will perform differently.

In this chapter, several forms of the LRLS algorithm are presented. First, the standard LRLS algorithm based on *a posteriori* errors is presented, followed by the normalized version. The algorithms with error feedback are also derived. Finally, the LRLS algorithm based on *a priori* errors is described.

P.S.R. Diniz, *Adaptive Filtering*, DOI: 10.1007/978-0-387-68606-6_7,

7.2 RECURSIVE LEAST-SQUARES PREDICTION

The solutions of the RLS forward and backward prediction problems are essential to derive the order-updating equations inherent to the LRLS algorithms. In both cases, the results are derived following the same derivation procedure as in the conventional RLS algorithm, since the only distinct feature of the prediction problems is the definition of the reference signal $d(k)$. For example, in the forward prediction case we have $d(k) = x(k)$ whereas the input signal vector has the sample $x(k-1)$ as the most recent data. For the backward prediction case $d(k) = x(k-i-1)$, where the index i defines the sample in the past which we wish to predict, and the input signal vector has $x(k)$ as the most recent data. In this section, these solutions are studied and the results demonstrate how information can be exchanged between the forward and backward predictor solutions.

7.2.1 Forward Prediction Problem

The objective of the forward prediction is to predict a future sample of a given input sequence using the currently available information of the sequence. For example, one can try to predict the value of $x(k)$ using past samples $x(k-1), x(k-2)\ldots$, through an FIR prediction filter with $i+1$ coefficients as

$$y_f(k, i+1) = \mathbf{w}_f^T(k, i+1)\mathbf{x}(k-1, i+1) \tag{7.1}$$

where $y_f(k, i+1)$ is the predictor output signal,

$$\mathbf{w}_f(k, i+1) = [w_{f0}(k)\ w_{f1}(k) \ldots w_{fi}(k)]^T$$

is the FIR forward prediction coefficient vector, and

$$\mathbf{x}(k-1, i+1) = [x(k-1)\ x(k-2) \ldots x(k-i-1)]^T$$

is the available input signal vector. The second variable included in the vectors of equation (7.1) is to indicate the vector dimension, since it is required in the order-updating equations of the LRLS algorithm. This second variable will be included where needed in the present chapter.

The instantaneous *a posteriori* forward prediction error is given by

$$\varepsilon_f(k, i+1) = x(k) - \mathbf{w}_f^T(k, i+1)\mathbf{x}(k-1, i+1) \tag{7.2}$$

For the RLS formulation of the forward prediction problem, define the weighted forward prediction error vector as

$$\boldsymbol{\varepsilon}_f(k, i+1) = \hat{\mathbf{x}}(k) - \mathbf{X}^T(k-1, i+1)\mathbf{w}_f(k, i+1) \tag{7.3}$$

where

$$\hat{\mathbf{x}}(k) = [x(k)\ \lambda^{1/2}x(k-1)\ \lambda x(k-2) \ldots \lambda^{k/2}x(0)]^T$$

$$\boldsymbol{\varepsilon}_f(k,i+1) = [\varepsilon_f(k,i+1)\ \lambda^{1/2}\varepsilon_f(k-1,i+1)\ \lambda\varepsilon_f(k-2,i+1)\ldots\lambda^{k/2}\varepsilon_f(0,i+1)]^T$$

and

$$\mathbf{X}(k-1,i+1) = \begin{bmatrix} x(k-1) & \lambda^{1/2}x(k-2) & \cdots & \lambda^{(k-2)/2}x(1) & \lambda^{(k-1)/2}x(0) & 0 \\ x(k-2) & \lambda^{1/2}x(k-3) & \cdots & \lambda^{(k-2)/2}x(0) & 0 & 0 \\ \vdots & \vdots & & \vdots & \vdots & \vdots \\ x(k-i-1) & \lambda^{1/2}x(k-i-2) & \cdots & 0 & 0 & 0 \end{bmatrix}$$

It is straightforward to show that $\boldsymbol{\varepsilon}_f(k,i+1)$ can be rewritten as

$$\boldsymbol{\varepsilon}_f(k,i+1) = \mathbf{X}^T(k,i+2)\begin{bmatrix} 1 \\ -\mathbf{w}_f(k,i+1) \end{bmatrix} \tag{7.4}$$

The objective function that we want to minimize in the least-squares sense is the forward prediction error given by

$$\begin{aligned} \xi_f^d(k,i+1) &= \boldsymbol{\varepsilon}_f^T(k,i+1)\boldsymbol{\varepsilon}_f(k,i+1) \\ &= \sum_{l=0}^{k}\lambda^{k-l}\varepsilon_f^2(l,i+1) \\ &= \sum_{l=0}^{k}\lambda^{k-l}[x(l)-\mathbf{x}^T(l-1,i+1)\mathbf{w}_f(k,i+1)]^2 \end{aligned} \tag{7.5}$$

By differentiating $\xi_f^d(k,i+1)$ with respect to $\mathbf{w}_f(k,i+1)$ and equating the result to zero, we can find the optimum coefficient vector that minimizes the objective function, namely,

$$\begin{aligned} \mathbf{w}_f(k,i+1) &= \left[\sum_{l=0}^{k}\lambda^{k-l}\mathbf{x}(l-1,i+1)\mathbf{x}^T(l-1,i+1)\right]^{-1}\sum_{l=0}^{k}\lambda^{k-l}\mathbf{x}(l-1,i+1)x(l) \\ &= [\mathbf{X}(k-1,i+1)\mathbf{X}^T(k-1,i+1)]^{-1}\mathbf{X}(k-1,i+1)\hat{\mathbf{x}}(k) \\ &= \mathbf{R}_{Df}^{-1}(k-1,i+1)\mathbf{p}_{Df}(k,i+1) \end{aligned} \tag{7.6}$$

where $\mathbf{R}_{Df}(k-1,i+1)$ is equal to the deterministic correlation matrix $\mathbf{R}_D(k-1)$ of order $i+1$ and $\mathbf{p}_{Df}(k,i+1)$ is the deterministic cross-correlation vector between $x(l)$ and $\mathbf{x}(l-1,i+1)$.

The exponentially weighted sum of squared errors can be written as (see equation (7.5)):

$$\begin{aligned}
\xi_f^d(k,i+1) &= \sum_{l=0}^{k} \lambda^{k-l} \left\{ x^2(l) - 2x(l)\mathbf{x}^T(l-1,i+1)\mathbf{w}_f(k,i+1) \right. \\
&\quad \left. + \left[\mathbf{x}^T(l-1,i+1)\mathbf{w}_f(k,i+1)\right]^2 \right\} \\
&= \sum_{l=0}^{k} \lambda^{k-l} \left[x^2(l) - x(l)\mathbf{x}^T(l-1,i+1)\mathbf{w}_f(k,i+1)\right] \\
&\quad + \sum_{l=0}^{k} \lambda^{k-l} \left[-x(l) + \mathbf{x}^T(l-1,i+1)\mathbf{w}_f(k,i+1)\right] \mathbf{x}^T(l-1,i+1)\mathbf{w}_f(k,i+1) \\
&= \sum_{l=0}^{k} \lambda^{k-l} x(l) \left[x(l) - \mathbf{x}^T(l-1,i+1)\mathbf{w}_f(k,i+1)\right] \\
&\quad + \left[\textstyle\sum_{l=0}^{k} - \lambda^{k-l} x(l)\mathbf{x}^T(l-1,i+1) \right. \\
&\quad \left. + \mathbf{w}_f^T(k,i+1) \textstyle\sum_{l=0}^{k} \lambda^{k-l}\mathbf{x}(l-1,i+1)\mathbf{x}^T(l-1,i+1)\right] \mathbf{w}_f(k,i+1)
\end{aligned} \tag{7.7}$$

If we replace equation (7.6) in the second term of the last relation above, it can be shown by using the fact that $\mathbf{R}_D(k-1)$ is symmetric that this term is zero. Therefore, the minimum value of $\xi_f^d(k,i+1)$[1] is given by

$$\begin{aligned}
\xi_{f_{\min}}^d(k,i+1) &= \sum_{l=0}^{k} \lambda^{k-l} x(l)[x(l) - \mathbf{x}^T(l-1,i+1)\mathbf{w}_f(k,i+1)] \\
&= \sum_{l=0}^{k} \lambda^{k-l} x^2(l) - \mathbf{p}_{Df}^T(k,i+1)\mathbf{w}_f(k,i+1) \\
&= \sigma_f^2(k) - \mathbf{w}_f^T(k,i+1)\mathbf{p}_{Df}(k,i+1)
\end{aligned} \tag{7.8}$$

By combining equation (7.6) for $\mathbf{w}_f(k,i)$ and equation (7.8) for $\xi_{f_{\min}}^d(k,i+1)$ the following matrix equation can be obtained

$$\begin{bmatrix} \sigma_f^2(k) & \mathbf{p}_{Df}^T(k,i+1) \\ \mathbf{p}_{Df}(k,i+1) & \mathbf{R}_{Df}(k-1,i+1) \end{bmatrix} \begin{bmatrix} 1 \\ -\mathbf{w}_f(k,i+1) \end{bmatrix} = \begin{bmatrix} \xi_{f_{\min}}^d(k,i+1) \\ \mathbf{0} \end{bmatrix} \tag{7.9}$$

[1] Notice that no special notation was previously used for the minimum value of the RLS objective function. However, when deriving the lattice algorithms, this definition is necessary.

Since $\sigma_f^2(k) = \sum_{l=0}^{k} \lambda^{k-l} x^2(l)$ and $\mathbf{p}_{Df}(k, i+1) = \sum_{l=0}^{k} \lambda^{k-l} \mathbf{x}(l-1, i+1)x(l)$, it is possible to conclude that the leftmost term of equation (7.9) can be rewritten as

$$\begin{aligned}
&\begin{bmatrix} \sum_{l=0}^{k} \lambda^{k-l} x^2(l) & \sum_{l=0}^{k} \lambda^{k-l} \mathbf{x}^T(l-1, i+1)x(l) \\ \sum_{l=0}^{k} \lambda^{k-l} \mathbf{x}(l-1, i+1)x(l) & \sum_{l=0}^{k} \lambda^{k-l} \mathbf{x}(l-1, i+1)\mathbf{x}^T(l-1, i+1) \end{bmatrix} \\
&= \sum_{l=0}^{k} \lambda^{k-l} \begin{bmatrix} x(l) \\ \mathbf{x}(l-1, i+1) \end{bmatrix} [x(l) \;\; \mathbf{x}^T(l-1, i+1)] \\
&= \mathbf{R}_D(k, i+2)
\end{aligned} \tag{7.10}$$

Therefore,

$$\mathbf{R}_D(k, i+2) \begin{bmatrix} 1 \\ -\mathbf{w}_f(k, i+1) \end{bmatrix} = \begin{bmatrix} \xi^d_{f_{\min}}(k, i+1) \\ \mathbf{0} \end{bmatrix}$$

where $\mathbf{R}_D(k, i+2)$ corresponds to $\mathbf{R}_D(k)$ used in the previous chapter with dimension $i+2$. The above equation relates the deterministic correlation matrix of order $i+2$ to the minimum least-squares forward prediction error. The appropriate partitioning of matrix $\mathbf{R}_D(k, i+2)$ enables the derivation of the order-updating equation for the predictor tap coefficients, as will be discussed later.

7.2.2 Backward Prediction Problem

The objective of the backward predictor is to generate an estimate of a past sample of a given input sequence using the currently available information of the sequence. For example, sample $x(k-i-1)$ can be estimated from $\mathbf{x}(k, i+1)$, through an FIR backward prediction filter with $i+1$ coefficients as

$$y_b(k, i+1) = \mathbf{w}_b^T(k, i+1)\mathbf{x}(k, i+1) \tag{7.11}$$

where $y_b(k, i+1)$ is the backward predictor output signal, and

$$\mathbf{w}_b^T(k, i+1) = [w_{b0}(k) \; w_{b1}(k) \ldots w_{bi}(k)]^T$$

is the FIR backward prediction coefficient vector.

The instantaneous *a posteriori* backward prediction error is given by

$$\varepsilon_b(k, i+1) = x(k-i-1) - \mathbf{w}_b^T(k, i+1)\mathbf{x}(k, i+1) \tag{7.12}$$

The weighted backward prediction error vector is defined as

$$\boldsymbol{\varepsilon}_b(k, i+1) = \hat{\mathbf{x}}(k-i-1) - \mathbf{X}^T(k, i+1)\mathbf{w}_b(k, i+1) \tag{7.13}$$

where

$$\hat{\mathbf{x}}(k-i-1) = [x(k-i-1) \;\; \lambda^{1/2} x(k-i-2) \ldots \lambda^{(k-i-1)/2} x(0) \;\; 0 \ldots 0]^T$$

$$\boldsymbol{\varepsilon}_b(k, i+1) = [\varepsilon_b(k, i+1) \;\; \lambda^{1/2} \varepsilon_b(k-1, i+1) \ldots \lambda^{k/2} \varepsilon_b(0, i+1)]^T$$

and

$$\mathbf{X}(k,i+1) = \begin{bmatrix} x(k) & \lambda^{1/2}x(k-1) & \cdots & \lambda^{(k-1)/2}x(1) & \lambda^{(k)/2}x(0) \\ x(k-1) & \lambda^{1/2}x(k-2) & \cdots & \lambda^{(k-2)/2}x(0) & 0 \\ \vdots & \vdots & & \vdots & \vdots \\ x(k-i) & \lambda^{1/2}x(k-i-1) & \cdots & 0 \cdots & 0 \end{bmatrix}$$

The error vector can be rewritten as

$$\boldsymbol{\varepsilon}_b(k,i+1) = \mathbf{X}^T(k,i+2)\begin{bmatrix} -\mathbf{w}_b(k,i+1) \\ 1 \end{bmatrix} \tag{7.14}$$

The objective function to be minimized in the backward prediction problem is given by

$$\begin{aligned}\xi_b^d(k,i+1) &= \boldsymbol{\varepsilon}_b^T(k,i+1)\boldsymbol{\varepsilon}_b(k,i+1) \\ &= \sum_{l=0}^{k}\lambda^{k-l}\varepsilon_b^2(l,i+1) \\ &= \sum_{l=0}^{k}\lambda^{k-l}[x(l-i-1)-\mathbf{x}^T(l,i+1)\mathbf{w}_b(k,i+1)]^2 \end{aligned} \tag{7.15}$$

The optimal solution for the coefficient vector is

$$\begin{aligned}\mathbf{w}_b(k,i+1) &= \left[\sum_{l=0}^{k}\lambda^{k-l}\mathbf{x}(l,i+1)\mathbf{x}^T(l,i+1)\right]^{-1}\sum_{l=0}^{k}\lambda^{k-l}\mathbf{x}(l,i+1)x(l-i-1) \\ &= [\mathbf{X}(k,i+1)\mathbf{X}^T(k,i+1)]^{-1}\mathbf{X}(k,i+1)\hat{\mathbf{x}}(k-i-1) \\ &= \mathbf{R}_{Db}^{-1}(k,i+1)\mathbf{p}_{Db}(k,i+1) \end{aligned} \tag{7.16}$$

where $\mathbf{R}_{Db}(k,i+1)$ is equal to the deterministic correlation matrix $\mathbf{R}_D(k)$ of order $i+1$, and $\mathbf{p}_{Db}(k,i+1)$ is the deterministic cross-correlation vector between $x(l-i-1)$ and $\mathbf{x}(l,i+1)$.

Using the same procedure to derive the minimum least-squares solution in the RLS problem, it can be shown that the minimum value of $\xi_b^d(k)$ is given by

$$\begin{aligned}\xi_{b_{\min}}^d(k,i+1) &= \sum_{l=0}^{k}\lambda^{k-l}x(l-i-1)[x(l-i-1)-\mathbf{x}^T(l,i+1)\mathbf{w}_b(k,i+1)] \\ &= \sum_{l=0}^{k}\lambda^{k-l}x^2(l-i-1)-\mathbf{p}_{Db}^T(k,i+1)\mathbf{w}_b(k,i+1) \\ &= \sigma_b^2(k)-\mathbf{w}_b^T(k,i+1)\mathbf{p}_{Db}(k,i+1) \end{aligned} \tag{7.17}$$

By combining equations (7.16) and (7.17), we obtain

$$\begin{aligned}
&\begin{bmatrix} \mathbf{R}_{Db}(k,i+1) & \mathbf{p}_{Db}(k,i+1) \\ \mathbf{p}_{Db}^T(k,i+1) & \sigma_b^2(k) \end{bmatrix} \begin{bmatrix} -\mathbf{w}_b(k,i+1) \\ 1 \end{bmatrix} \\
&= \begin{bmatrix} \sum_{l=0}^{k} \lambda^{k-l}\mathbf{x}(l,i+1)\mathbf{x}^T(l,i+1) & \sum_{l=0}^{k} \lambda^{k-l}\mathbf{x}(l,i+1)x(l-i-1) \\ \sum_{l=0}^{k} \lambda^{k-l}\mathbf{x}^T(l,i+1)x(l-i-1) & \sum_{l=0}^{k} \lambda^{k-l}x^2(l-i-1) \end{bmatrix} \\
&\quad \cdot \begin{bmatrix} -\mathbf{w}_b(k,i+1) \\ 1 \end{bmatrix} \\
&= \mathbf{R}_D(k,i+2) \begin{bmatrix} -\mathbf{w}_b(k,i+1) \\ 1 \end{bmatrix} \\
&= \begin{bmatrix} \mathbf{0} \\ \xi_{b_{\min}}^d(k,i+1) \end{bmatrix}
\end{aligned} \tag{7.18}$$

where $\mathbf{R}_D(k,i+2)$ is equal to $\mathbf{R}_D(k)$ of dimension $i+2$. The above equation relates the deterministic correlation matrix of order $i+1$ to the minimum least-squares backward prediction error. This equation is important in the derivation of the order-updating equation for the backward predictor tap coefficients. This issue is discussed in the following section.

7.3 ORDER-UPDATING EQUATIONS

The objective of this section is to derive the order-updating equations for the forward and backward prediction errors. These equations are the starting point to generate the lattice realization.

7.3.1 A New Parameter $\delta(k,i)$

Using the results of equations (7.9) and (7.10), and the decomposition of $\mathbf{R}_D(k,i+2)$ given in equation (7.18), we can show that

$$\begin{aligned}
\mathbf{R}_D(k,i+2) \begin{bmatrix} 1 \\ -\mathbf{w}_f(k,i) \\ 0 \end{bmatrix} &= \begin{bmatrix} \mathbf{R}_D(k,i+1) & \mathbf{p}_{Db}(k,i+1) \\ \mathbf{p}_{Db}^T(k,i+1) & \sigma_b^2(k) \end{bmatrix} \begin{bmatrix} 1 \\ -\mathbf{w}_f(k,i) \\ 0 \end{bmatrix} \\
&= \begin{bmatrix} \xi_{f_{\min}}^d(k,i) \\ \mathbf{0} \\ \mathbf{p}_{Db}^T(k,i+1) \begin{bmatrix} 1 \\ -\mathbf{w}_f(k,i) \end{bmatrix} \end{bmatrix} \\
&= \begin{bmatrix} \xi_{f_{\min}}^d(k,i) \\ \mathbf{0} \\ \delta_f(k,i) \end{bmatrix}
\end{aligned} \tag{7.19}$$

where relation (7.9) was employed in the second equality. From the last element relation of the above vector and the definition of $\mathbf{p}_{Db}(k, i+1)$, we obtain

$$\begin{aligned}\delta_f(k,i) &= \sum_{l=0}^{k} \lambda^{k-l} x(l)x(l-i-1) - \sum_{l=0}^{k} \lambda^{k-l} x(l-i-1)\mathbf{x}^T(l-1,i)\mathbf{w}_f(k,i) \\ &= \sum_{l=0}^{k} \lambda^{k-l} x(l)x(l-i-1) - \sum_{l=0}^{k} \lambda^{k-l} x(l-i-1)y_f(l,i) \\ &= \sum_{l=0}^{k} \lambda^{k-l} \varepsilon_f(l,i)x(l-i-1)\end{aligned}$$

and $y_f(l,i) = \mathbf{x}^T(l-1,i)\mathbf{w}_f(k,i)$ is the output of a forward prediction filter of order $i-1$. Note that the parameter $\delta_f(k,i)$ can be interpreted as the deterministic cross-correlation between the forward prediction error $\varepsilon_f(l,i)$ with the coefficients fixed at $\mathbf{w}_f(k,i)$ and the desired signal of the backward predictor filter $x(l-i-1)$.

Similarly, using the results of equations (7.17) and (7.18) it can be shown that

$$\begin{aligned}\mathbf{R}_D(k,i+2)\begin{bmatrix} 0 \\ -\mathbf{w}_b(k-1,i) \\ 1 \end{bmatrix} &= \begin{bmatrix} \sigma_f^2(k) & \mathbf{p}_{Df}^T(k,i+1) \\ \mathbf{p}_{Df}(k,i+1) & \mathbf{R}_D(k-1,i+1) \end{bmatrix}\begin{bmatrix} 0 \\ -\mathbf{w}_b(k-1,i) \\ 1 \end{bmatrix} \\ &= \begin{bmatrix} \mathbf{p}_{Df}^T(k,i+1)\begin{bmatrix} -\mathbf{w}_b(k-1,i) \\ 1 \end{bmatrix} \\ \mathbf{0} \\ \xi_{b_{\min}}^d(k-1,i) \end{bmatrix} \\ &= \begin{bmatrix} \delta_b(k,i) \\ \mathbf{0} \\ \xi_{b_{\min}}^d(k-1,i) \end{bmatrix}\end{aligned} \tag{7.20}$$

where in the second equality we applied the result of equation (7.18), and

$$\begin{aligned}\delta_b(k,i) &= \sum_{l=0}^{k} \lambda^{k-l} x(l-i-1)x(l) - \sum_{l=0}^{k} \lambda^{k-l} x(l)\mathbf{x}^T(l-1,i)\mathbf{w}_b(k-1,i) \\ &= \sum_{l=0}^{k} \lambda^{k-l} x(l-i-1)x(l) - \sum_{l=0}^{k} \lambda^{k-l} x(l)y_b(l-1,i) \\ &= \sum_{l=0}^{k} \lambda^{k-l} \varepsilon_b(l-1,i)x(l)\end{aligned}$$

where $y_b(l-1,i) = \mathbf{x}^T(l-1,i)\mathbf{w}_b(k-1,i)$ is the output of a backward prediction filter of order $i-1$ with the input data of instant $l-1$, when the coefficients of the predictor are $\mathbf{w}_b(k-1,i)$. The parameter $\delta_b(k,i)$ can be interpreted as the deterministic cross-correlation between the backward prediction error $\varepsilon_b(l-1,i)$ and the desired signal of the forward predictor filter $x(l)$.

In equations (7.19) and (7.20) two new parameters were defined, namely $\delta_f(k,i)$ and $\delta_b(k,i)$. In the following derivations we will show that these parameters are equal. If $\mathbf{R}_D(k,i+2)$ is premultiplied by $[0 \quad -\mathbf{w}_b^T(k-1,i) \quad 1]$ and postmultiplied by $[1 \quad -\mathbf{w}_f(k,i) \quad 0]^T$, it can be shown that

$$[0 \quad -\mathbf{w}_b^T(k-1,i) \quad 1]\ \mathbf{R}_D(k,i+2)\begin{bmatrix} 1 \\ -\mathbf{w}_f(k,i) \\ 0 \end{bmatrix} = \delta_f(k,i) \tag{7.21}$$

By transposing the first and last terms of equation (7.20) the following relation is obtained

$$[0 \quad -\mathbf{w}_b^T(k-1,i) \quad 1]\ \mathbf{R}_D(k,i+2) = [\delta_b(k,i) \quad \mathbf{0}^T \quad \xi^d_{b_{\min}}(k-1,i)] \tag{7.22}$$

By substituting this result in equation (7.21), we obtain

$$[\delta_b(k,i) \quad \mathbf{0}^T \quad \xi^d_{b_{\min}}(k-1,i)]\begin{bmatrix} 1 \\ -\mathbf{w}_f(k,i) \\ 0 \end{bmatrix} = \delta_b(k,i) \tag{7.23}$$

Therefore, from equations (7.21) and (7.23) we conclude that

$$\delta_f(k,i) = \delta_b(k,i) = \delta(k,i) \tag{7.24}$$

In effect, the deterministic cross-correlations between $\varepsilon_f(l,i)$ and $x(l-i-1)$ and between $\varepsilon_b(l-1,i)$ and $x(l)$ are equal.

7.3.2 Order Updating of $\xi^d_{b_{\min}}(k,i)$ and $\mathbf{w}_b(k,i)$

The order updating of the minimum LS error and the tap coefficients for the backward predictor can be deduced by multiplying equation (7.19) by the scalar $\delta(k,i)/\xi^d_{f_{\min}}(k,i)$, i.e.,

$$\frac{\delta(k,i)}{\xi^d_{f_{\min}}(k,i)}\mathbf{R}_D(k,i+2)\begin{bmatrix} 1 \\ -\mathbf{w}_f(k,i) \\ 0 \end{bmatrix} = \begin{bmatrix} \delta(k,i) \\ \mathbf{0} \\ \frac{\delta^2(k,i)}{\xi^d_{f_{\min}}(k,i)} \end{bmatrix} \tag{7.25}$$

Subtracting equation (7.20) from this result yields

$$\mathbf{R}_D(k,i+2)\begin{bmatrix} \frac{\delta(k,i)}{\xi^d_{f_{\min}}(k,i)} \\ -\mathbf{w}_f(k,i)\frac{\delta(k,i)}{\xi^d_{f_{\min}}(k,i)} + \mathbf{w}_b(k-1,i) \\ -1 \end{bmatrix} = \begin{bmatrix} \mathbf{0} \\ -\xi^d_{b_{\min}}(k-1,i) + \frac{\delta^2(k,i)}{\xi^d_{f_{\min}}(k,i)} \end{bmatrix} \tag{7.26}$$

Comparing equations (7.18) and (7.26), we conclude that

$$\xi^d_{b_{\min}}(k, i+1) = \xi^d_{b_{\min}}(k-1, i) - \frac{\delta^2(k, i)}{\xi^d_{f_{\min}}(k, i)} \tag{7.27}$$

and

$$\mathbf{w}_b(k, i+1) = \begin{bmatrix} 0 \\ \mathbf{w}_b(k-1, i) \end{bmatrix} - \frac{\delta(k, i)}{\xi^d_{f_{\min}}(k, i)} \begin{bmatrix} -1 \\ \mathbf{w}_f(k, i) \end{bmatrix} \tag{7.28}$$

7.3.3 Order Updating of $\xi^d_{f_{\min}}(k, i)$ and $\mathbf{w}_f(k, i)$

Similarly, by multiplying equation (7.20) by $\delta(k, i)/\xi^d_{b_{\min}}(k-1, i)$, we get

$$\frac{\delta(k, i)}{\xi^d_{b_{\min}}(k-1, i)} \mathbf{R}_D(k, i+2) \begin{bmatrix} 0 \\ -\mathbf{w}_b(k-1, i) \\ 1 \end{bmatrix} = \begin{bmatrix} \frac{\delta^2(k,i)}{\xi^d_{b_{\min}}(k-1,i)} \\ \mathbf{0} \\ \delta(k, i) \end{bmatrix} \tag{7.29}$$

Subtracting equation (7.29) from equation (7.19), it follows that

$$\mathbf{R}_D(k, i+2) \begin{bmatrix} 1 \\ \frac{\delta(k,i)}{\xi^d_{b_{\min}}(k-1,i)} \mathbf{w}_b(k-1, i) - \mathbf{w}_f(k, i) \\ -\frac{\delta(k,i)}{\xi^d_{b_{\min}}(k-1,i)} \end{bmatrix} = \begin{bmatrix} \xi^d_{f_{\min}}(k, i) - \frac{\delta^2(k,i)}{\xi^d_{b_{\min}}(k-1,i)} \\ \mathbf{0} \end{bmatrix} \tag{7.30}$$

Comparing this equation with equation (7.9), we conclude that

$$\xi^d_{f_{\min}}(k, i+1) = \xi^d_{f_{\min}}(k, i) - \frac{\delta^2(k, i)}{\xi^d_{b_{\min}}(k-1, i)} \tag{7.31}$$

and

$$\mathbf{w}_f(k, i+1) = \begin{bmatrix} \mathbf{w}_f(k, i) \\ 0 \end{bmatrix} - \frac{\delta(k, i)}{\xi^d_{b_{\min}}(k-1, i)} \begin{bmatrix} \mathbf{w}_b(k-1, i) \\ -1 \end{bmatrix} \tag{7.32}$$

7.3.4 Order Updating of Prediction Errors

The order updating of the *a posteriori* forward and backward prediction errors can be derived as described below. From the definition of *a posteriori* forward error, we have

$$\begin{aligned} \varepsilon_f(k, i+1) &= \mathbf{x}^T(k, i+2) \begin{bmatrix} 1 \\ -\mathbf{w}_f(k, i+1) \end{bmatrix} \\ &= \mathbf{x}^T(k, i+2) \begin{bmatrix} 1 \\ -\mathbf{w}_f(k, i) \\ 0 \end{bmatrix} + \frac{\delta(k, i)}{\xi^d_{b_{\min}}(k-1, i)} \mathbf{x}^T(k, i+2) \begin{bmatrix} 0 \\ \mathbf{w}_b(k-1, i) \\ -1 \end{bmatrix} \\ &= \varepsilon_f(k, i) - \kappa_f(k, i)\varepsilon_b(k-1, i) \end{aligned} \tag{7.33}$$

where in the second equality we employed the order-updating equation (7.32) for the forward prediction coefficients. The coefficient $\kappa_f(k,i) = \frac{\delta(k,i)}{\xi^d_{b_{\min}}(k-1,i)}$ is the so-called forward reflection coefficient.

The order updating of the *a posteriori* backward prediction error is obtained by using equation (7.28) as

$$\begin{aligned}
\varepsilon_b(k,i+1) &= \mathbf{x}^T(k,i+2)\begin{bmatrix} -\mathbf{w}_b(k,i+1) \\ 1 \end{bmatrix} \\
&= \mathbf{x}^T(k,i+2)\begin{bmatrix} 0 \\ -\mathbf{w}_b(k-1,i) \\ 1 \end{bmatrix} + \frac{\delta(k,i)}{\xi^d_{f_{\min}}(k,i)}\mathbf{x}^T(k,i+2)\begin{bmatrix} -1 \\ \mathbf{w}_f(k,i) \\ 0 \end{bmatrix} \\
&= \varepsilon_b(k-1,i) - \kappa_b(k,i)\varepsilon_f(k,i)
\end{aligned} \tag{7.34}$$

where we employed the order-updating equation for the backward prediction coefficients (7.28) in the second equality. The coefficient $\kappa_b(k,i) = \frac{\delta(k,i)}{\xi^d_{f_{\min}}(k,i)}$ is the backward reflection coefficient.

Equations (7.33) and (7.34) above can be implemented with a lattice section as illustrated in Fig. 7.1.a. An order-increasing lattice-based forward and backward predictor can be constructed as illustrated in Fig. 7.1.b. The coefficients $\kappa_b(k,i)$ and $\kappa_f(k,i)$ are often called reflection coefficients of the lattice realization.

In the first section of the lattice, the forward and backward prediction errors are equal to the input signal itself since no prediction is performed before the first lattice section; therefore

$$\varepsilon_b(k,0) = \varepsilon_f(k,0) = x(k) \tag{7.35}$$

and

$$\xi^d_{f_{\min}}(k,0) = \xi^d_{b_{\min}}(k,0) = \sum_{l=0}^{k}\lambda^{k-l}x^2(l) = x^2(k) + \lambda\xi^d_{f_{\min}}(k-1,0) \tag{7.36}$$

A closer look at equations (7.9) and (7.18) leads to the conclusion that the backward and forward predictors utilize the same information matrix $\mathbf{R}_D(k,i+2)$. This result was key in deriving the expressions for the *a posteriori* forward and backward prediction errors of equations (7.33) and (7.34). Of particular note, these expressions can be shown to be independent of the predictor tap coefficients. This result will be proved in the following section, which will present an updating formula for $\delta(k,i)$ that is not directly dependent on $\mathbf{w}_f(k,i)$ and $\mathbf{w}_b(k-1,i)$.

Now that all order-updating equations are available, it is necessary to derive the time-updating equations to allow the adaptation of the lattice predictor coefficients.

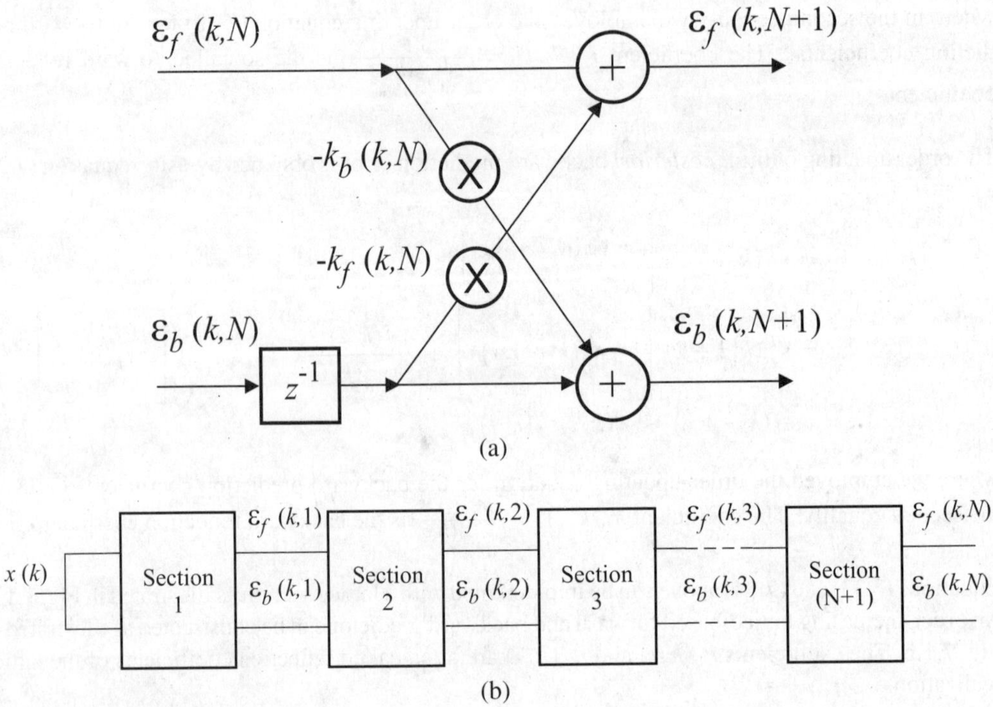

Figure 7.1 Least-squares lattice-based predictor.

7.4 TIME-UPDATING EQUATIONS

The time-updating equations are required to deal with the new incoming data that becomes available. Recall that up to this point in this text we have studied adaptive-filtering algorithms utilizing the new incoming data as soon as it becomes available. In this section, the time-updating equations for the internal quantities of the lattice algorithm are derived.

7.4.1 Time Updating for Prediction Coefficients

From equation (7.6), the time updating of the forward prediction filter coefficients is given by

$$\begin{aligned}\mathbf{w}_f(k,i) &= \mathbf{S}_D(k-1,i)\mathbf{p}_{Df}(k,i)\\ &= \mathbf{R}_D^{-1}(k-1,i)\mathbf{p}_{Df}(k,i)\end{aligned} \tag{7.37}$$

This is the standard expression for the computation of the optimal coefficient vector leading to the minimization of the LS objective function and adapted to the forward prediction case.

The updating formula of $\mathbf{S}_D(k,i)$ based on the matrix inversion lemma derived in the previous chapter (see Algorithm 5.2) for the conventional RLS algorithm can be used in equation (7.37). The resulting equation is given by

$$\begin{aligned}
\mathbf{w}_f(k,i) &= \frac{1}{\lambda}\left[\mathbf{S}_D(k-2,i) - \frac{\boldsymbol{\psi}(k-1,i)\boldsymbol{\psi}^T(k-1,i)}{\lambda+\boldsymbol{\psi}^T(k-1,i)\mathbf{x}(k-1,i)}\right]\mathbf{p}_{Df}(k,i)\\
&= \frac{1}{\lambda}\left[\mathbf{S}_D(k-2,i) - \frac{\boldsymbol{\psi}(k-1,i)\mathbf{x}^T(k-1,i)\mathbf{S}_D(k-2,i)}{\lambda+\boldsymbol{\psi}^T(k-1,i)\mathbf{x}(k-1,i)}\right]\\
&\quad\cdot[\lambda\mathbf{p}_{Df}(k-1,i)+x(k)\mathbf{x}(k-1,i)]\\
&= \mathbf{w}_f(k-1,i) - \frac{\boldsymbol{\psi}(k-1,i)\mathbf{x}^T(k-1,i)\mathbf{w}_f(k-1,i)}{\lambda+\boldsymbol{\psi}^T(k-1,i)\mathbf{x}(k-1,i)} + \frac{x(k)}{\lambda}\mathbf{c}
\end{aligned} \tag{7.38}$$

where in the we have applied the time-recursive updating formula of $\mathbf{p}_{Df}(k,i)$ in the second equality, and we have replaced $\mathbf{S}_D(k-2,i)\mathbf{p}_{Df}(k-1,i)$ by $\mathbf{w}_f(k-1,i)$ in the second term of the final expression. Vector $\mathbf{c}$ is given by

$$\begin{aligned}
\mathbf{c} &= \mathbf{S}_D(k-2,i)\mathbf{x}(k-1,i) - \frac{\boldsymbol{\psi}(k-1,i)\mathbf{x}^T(k-1,i)\mathbf{S}_D(k-2,i)\mathbf{x}(k-1,i)}{\lambda+\boldsymbol{\psi}^T(k-1,i)\mathbf{x}(k-1,i)}\\
&= \frac{\lambda\mathbf{S}_D(k-2,i)\mathbf{x}(k-1,i)}{\lambda+\boldsymbol{\psi}^T(k-1,i)\mathbf{x}(k-1,i)}
\end{aligned}$$

It is convenient at this point to recall that $\boldsymbol{\psi}(k-1,i) = \mathbf{S}_D(k-2,i)\mathbf{x}(k-1,i)$ (see equation (5.10)).

The last term in equation (7.38) can be simplified if we apply the refined definition based on equation (5.11)

$$\boldsymbol{\phi}(k-1,i) = \frac{\boldsymbol{\psi}(k-1,i)}{\lambda+\boldsymbol{\psi}^T(k-1,i)\mathbf{x}(k-1,i)} \tag{7.39}$$

where $\boldsymbol{\phi}(k-1,i)$ now includes the order index i. Using this definition in the second and third terms of the last expression of equation (7.38), it can be shown that

$$\begin{aligned}
\mathbf{w}_f(k,i) &= \mathbf{w}_f(k-1,i) + \boldsymbol{\phi}(k-1,i)[x(k) - \mathbf{w}_f^T(k-1,i)\mathbf{x}(k-1,i)]\\
&= \mathbf{w}_f(k-1,i) + \boldsymbol{\phi}(k-1,i)e_f(k,i)
\end{aligned} \tag{7.40}$$

where $e_f(k,i)$ is the *a priori* forward prediction error of a predictor of order $i-1$[2], so-called because it utilizes the tap coefficients of the previous instant $k-1$.

Following similar steps to those used to derive equation (7.40), we can show that the time updating for the backward predictor filter is given by

$$\begin{aligned}
\mathbf{w}_b(k,i) &= \frac{1}{\lambda}\left[\mathbf{S}_D(k-1,i) - \frac{\boldsymbol{\psi}(k,i)\boldsymbol{\psi}^T(k,i)}{\lambda+\boldsymbol{\psi}^T(k,i)\mathbf{x}(k,i)}\right][\lambda\mathbf{p}_{Db}(k-1,i)+\mathbf{x}(k,i)x(k-i)]\\
&= \mathbf{w}_b(k-1,i) - \boldsymbol{\phi}(k,i)\mathbf{x}^T(k,i)\mathbf{w}_b(k-1,i) + \boldsymbol{\phi}(k,i)x(k-i)\\
&= \mathbf{w}_b(k-1,i) + \boldsymbol{\phi}(k,i)e_b(k,i)
\end{aligned} \tag{7.41}$$

where $e_b(k,i)$ is the *a priori* backward prediction error of a predictor filter of order $i-1$.

[2]The predictor filter is of order $i-1$ whereas the predictor including the desired signal is of order i.

7.4.2 Time Updating for $\delta(k, i)$

From the computational point of view, it would be interesting to compute the prediction errors without explicitly using the predictor's tap coefficients, because working with these coefficients requires the use of inner products. In order to achieve this, a time-updating expression for $\delta(k, i)$ is derived. A byproduct of this derivation is the introduction of a new parameter, namely $\gamma(k, i)$, that is shown to be a conversion factor between *a priori* and *a posteriori* errors.

From the definition in equation (7.19), we have

$$\delta(k, i) = \mathbf{p}_{Db}^T(k, i+1) \begin{bmatrix} 1 \\ -\mathbf{w}_f(k, i) \end{bmatrix} \tag{7.42}$$

where $\mathbf{p}_{Db}(k, i+1)$ can be expressed in recursive form as

$$\begin{aligned} \mathbf{p}_{Db}(k, i+1) &= \sum_{l=0}^{k} \lambda^{k-l} \mathbf{x}(l, i+1) x(l-i-1) \\ &= \mathbf{x}(k, i+1) x(k-i-1) + \lambda \mathbf{p}_{Db}(k-1, i+1) \end{aligned} \tag{7.43}$$

Substituting equations (7.40) and (7.43) in equation (7.42), we get

$$\begin{aligned} \delta(k, i) &= [x(k-i-1)\mathbf{x}^T(k, i+1) + \lambda \mathbf{p}_{Db}^T(k-1, i+1)] \\ &\quad \cdot \begin{bmatrix} 1 \\ -\mathbf{w}_f(k-1, i) - \boldsymbol{\phi}(k-1, i) e_f(k, i) \end{bmatrix} \\ &= \lambda \delta(k-1, i) + \lambda \mathbf{p}_{Db}^T(k-1, i+1) \begin{bmatrix} 0 \\ -\boldsymbol{\phi}(k-1, i) e_f(k, i) \end{bmatrix} \\ &\quad + x(k-i-1)\mathbf{x}^T(k, i+1) \begin{bmatrix} 1 \\ -\mathbf{w}_f(k-1, i) \end{bmatrix} \\ &\quad + x(k-i-1)\mathbf{x}^T(k, i+1) \begin{bmatrix} 0 \\ -\boldsymbol{\phi}(k-1, i) e_f(k, i) \end{bmatrix} \end{aligned} \tag{7.44}$$

where the equality of equation (7.42) for the order index $i-1$ was used to obtain the first term of the last equality.

We now derive two relations which are essential to obtain a time-updating equation for $\delta(k, i)$. The resulting equation is efficient from the computational point of view. From the definitions of $\boldsymbol{\phi}(k-1, i)$ and $\boldsymbol{\psi}(k-1, i)$, (see equation (7.39) and the comments after equation (7.38) respectively) it can be shown that

$$
\begin{aligned}
\mathbf{p}_{Db}^T(k-1,i+1)\begin{bmatrix} 0 \\ \boldsymbol{\phi}(k-1,i) \end{bmatrix} &= \mathbf{p}_{Db}^T(k-2,i)\boldsymbol{\phi}(k-1,i) \\
&= \frac{\mathbf{p}_{Db}^T(k-2,i)\boldsymbol{\psi}(k-1,i)}{\lambda+\boldsymbol{\psi}^T(k-1,i)\mathbf{x}(k-1,i)} \\
&= \frac{\mathbf{p}_{Db}^T(k-2,i)\mathbf{S}_D(k-2,i)\mathbf{x}(k-1,i)}{\lambda+\boldsymbol{\psi}^T(k-1,i)\mathbf{x}(k-1,i)} \\
&= \frac{\mathbf{w}_b^T(k-2,i)\mathbf{x}(k-1,i)}{\lambda+\boldsymbol{\psi}^T(k-1,i)\mathbf{x}(k-1,i)} \\
&= -\frac{e_b(k-1,i)-x(k-i-1)}{\lambda+\boldsymbol{\psi}^T(k-1,i)\mathbf{x}(k-1,i)}
\end{aligned} \tag{7.45}
$$

Now using equation (7.39) it is possible to obtain the relation

$$
\begin{aligned}
\mathbf{x}^T(k,i+1)\begin{bmatrix} 0 \\ \boldsymbol{\phi}(k-1,i) \end{bmatrix} &= \frac{\mathbf{x}^T(k-1,i)\mathbf{S}_D(k-2,i)\mathbf{x}(k-1,i)}{\lambda+\boldsymbol{\psi}^T(k-1,i)\mathbf{x}(k-1,i)} \\
&= \frac{\boldsymbol{\psi}^T(k-1,i)\mathbf{x}(k-1,i)}{\lambda+\boldsymbol{\psi}^T(k-1,i)\mathbf{x}(k-1,i)}
\end{aligned} \tag{7.46}
$$

If we recall that the *a priori* forward prediction error can be computed in the form

$$
\mathbf{x}^T(k,i+1)\begin{bmatrix} 1 \\ -\mathbf{w}_f(k-1,i) \end{bmatrix} = e_f(k,i)
$$

and by substituting equations (7.45) and (7.46) into equation (7.44), after some straightforward manipulations, we obtain the following time-updating equation for $\delta(k,i)$

$$
\begin{aligned}
\delta(k,i) &= \lambda\delta(k-1,i) + \frac{\lambda e_b(k-1,i)e_f(k,i)}{\lambda+\boldsymbol{\psi}^T(k-1,i)\mathbf{x}(k-1,i)} \\
&= \lambda\delta(k-1,i) + \gamma(k-1,i)e_b(k-1,i)e_f(k,i)
\end{aligned} \tag{7.47}
$$

where

$$
\begin{aligned}
\gamma(k-1,i) &= \frac{\lambda}{\lambda+\boldsymbol{\psi}^T(k-1,i)\mathbf{x}(k-1,i)} \\
&= 1-\boldsymbol{\phi}^T(k-1,i)\mathbf{x}(k-1,i)
\end{aligned} \tag{7.48}
$$

The last relation follows from the definition of $\boldsymbol{\phi}(k-1,i)$ in equation (7.39). Parameter $\gamma(k-1,i)$ plays a key role in the relation between the *a posteriori* and *a priori* prediction errors, as will be demonstrated below.

In order to allow the derivation of a lattice-based algorithm utilizing only *a posteriori* errors, the relationship between the *a priori* and *a posteriori* errors is now derived. The *a posteriori* forward prediction error is related to the *a priori* forward prediction error as

$$\begin{aligned}\varepsilon_f(k,i) &= x(k) - \mathbf{w}_f^T(k,i)\mathbf{x}(k-1,i)\\ &= x(k) - \mathbf{w}_f^T(k-1,i)\mathbf{x}(k-1,i) - \boldsymbol{\phi}^T(k-1,i)\mathbf{x}(k-1,i)e_f(k,i)\\ &= e_f(k,i)[1-\boldsymbol{\phi}^T(k-1,i)\mathbf{x}(k-1,i)]\\ &= e_f(k,i)\gamma(k-1,i) \qquad (7.49)\end{aligned}$$

Similarly, the relationship between *a posteriori* and *a priori* backward prediction errors can be expressed as

$$\begin{aligned}\varepsilon_b(k,i) &= x(k-i) - \mathbf{w}_b^T(k,i)\mathbf{x}(k,i)\\ &= x(k-i) - \mathbf{w}_b^T(k-1,i)\mathbf{x}(k,i) - \boldsymbol{\phi}^T(k,i)\mathbf{x}(k,i)e_b(k,i)\\ &= e_b(k,i)[1-\boldsymbol{\phi}^T(k,i)\mathbf{x}(k,i)]\\ &= e_b(k,i)\gamma(k,i) \qquad (7.50)\end{aligned}$$

Parameter $\gamma(k,i)$ is often called a conversion factor between *a priori* and *a posteriori* errors.

Using equations (7.49) and (7.50), equation (7.47) can be expressed as

$$\delta(k,i) = \lambda\delta(k-1,i) + \frac{\varepsilon_b(k-1,i)\varepsilon_f(k,i)}{\gamma(k-1,i)} \qquad (7.51)$$

As a general rule each variable of the lattice-based algorithms requires an order-updating equation. Therefore, an order-updating equation for $\gamma(k,i)$ is necessary. This is the objective of the derivations in the following subsection.

7.4.3 Order Updating for $\gamma(k,i)$

Variable $\gamma(k-1,i)$ is defined by

$$\gamma(k-1,i) = 1 - \boldsymbol{\phi}^T(k-1,i)\mathbf{x}(k-1,i)$$

where $\boldsymbol{\phi}(k-1,i) = \mathbf{S}_D(k-1,i)\mathbf{x}(k-1,i)$. The relation for $\boldsymbol{\phi}(k-1,i)$ can be obtained by replacing $\mathbf{S}_D(k-1,i)$ by the expression derived by the matrix inversion lemma of equation (5.5) and verifying that the resulting simplified expression leads to equation (7.39). By multiplying the expression $\boldsymbol{\phi}(k-1,i) = \mathbf{S}_D(k-1,i)\mathbf{x}(k-1,i)$ by $\mathbf{R}_D(k-1,i)$ on both sides, we obtain the following relation

$$\mathbf{R}_D(k-1,i)\boldsymbol{\phi}(k-1,i) = \mathbf{x}(k-1,i) \qquad (7.52)$$

With this equation, we will be able to derive an order-updating equation for $\boldsymbol{\phi}(k-1,i)$ with the aid of an appropriate partitioning of $\mathbf{R}_D(k-1,i)$.

By partitioning matrix $\mathbf{R}_D(k-1,i)$ as in equation (7.19), we get

$$\mathbf{R}_D(k-1,i)\begin{bmatrix}\boldsymbol{\phi}(k-1,i-1)\\ 0\end{bmatrix} = \begin{bmatrix}\mathbf{R}_D(k-1,i-1) & \mathbf{p}_{Db}(k-1,i-1)\\ \mathbf{p}_{Db}^T(k-1,i-1) & \sigma_b^2(k-1)\end{bmatrix} \cdot \begin{bmatrix}\boldsymbol{\phi}(k-1,i-1)\\ 0\end{bmatrix}$$

$$= \begin{bmatrix}\mathbf{R}_{Db}(k-1,i-1)\boldsymbol{\phi}(k-1,i-1)\\ \mathbf{p}_{Db}^T(k-1,i-1)\boldsymbol{\phi}(k-1,i-1)\end{bmatrix}$$

We can proceed by replacing $\boldsymbol{\phi}(k-1,i-1)$ using equation (7.52) in the last element of the above vector, that is,

$$\begin{aligned}\mathbf{R}_D(k-1,i)\begin{bmatrix}\boldsymbol{\phi}(k-1,i-1)\\ 0\end{bmatrix} &= \begin{bmatrix}\mathbf{R}_{Db}(k-1,i-1)\boldsymbol{\phi}(k-1,i-1)\\ \mathbf{p}_{Db}^T(k-1,i-1)\mathbf{S}_{Db}(k-1,i-1)\mathbf{x}(k-1,i-1)\end{bmatrix}\\ &= \begin{bmatrix}\mathbf{R}_{Db}(k-1,i-1)\boldsymbol{\phi}(k-1,i-1)\\ \mathbf{w}_b^T(k-1,i-1)\mathbf{x}(k-1,i-1)\end{bmatrix}\\ &= \begin{bmatrix}\mathbf{x}(k-1,i-1)\\ x(k-i)-\varepsilon_b(k-1,i-1)\end{bmatrix}\\ &= \mathbf{x}(k-1,i)-\begin{bmatrix}\mathbf{0}\\ \varepsilon_b(k-1,i-1)\end{bmatrix}\end{aligned} \tag{7.53}$$

By multiplying the above equation by $\mathbf{S}_D(k-1,i)$, we have

$$\begin{bmatrix}\boldsymbol{\phi}(k-1,i-1)\\ 0\end{bmatrix} = \boldsymbol{\phi}(k-1,i)-\mathbf{S}_D(k-1,i)\begin{bmatrix}\mathbf{0}\\ \varepsilon_b(k-1,i-1)\end{bmatrix} \tag{7.54}$$

If we replace the above relation in the definition of the conversion factor, we deduce

$$\begin{aligned}\gamma(k-1,i) &= 1-\boldsymbol{\phi}^T(k-1,i)\mathbf{x}(k-1,i)\\ &= \gamma(k-1,i-1)-[\mathbf{0}^T\ \varepsilon_b(k-1,i)]^T\mathbf{S}_D(k-1,i)\mathbf{x}(k-1,i)\end{aligned} \tag{7.55}$$

This equation can be expressed into a more useful form by using a partitioned version of $\mathbf{S}_D(k-1,i)$ given by

$$\begin{aligned}\mathbf{S}_D(k-1,i) &= \begin{bmatrix}0 & \mathbf{0}^T\\ \mathbf{0} & \mathbf{S}_D(k-2,i-1)\end{bmatrix}\\ &+\frac{1}{\xi^d_{f_{\min}}(k-1,i-1)}\begin{bmatrix}1\\ -\mathbf{w}_f(k-1,i-1)\end{bmatrix}[1\ \ -\mathbf{w}_f^T(k-1,i-1)]\end{aligned} \tag{7.56}$$

The proof of validity of the above expression follows.

Proof:

The partitioned expression of $\mathbf{R}_D(k-1,i)$ is

$$\mathbf{R}_D(k-1,i) = \begin{bmatrix} 0 & \mathbf{0}^T \\ \mathbf{0} & \mathbf{R}_D(k-2,i-1) \end{bmatrix} + \begin{bmatrix} \sigma_f^2(k-1) & \mathbf{p}_{Df}^T(k-1,i-1) \\ \mathbf{p}_{Df}(k-1,i-1) & \mathbf{0}_{i-1,i-1} \end{bmatrix} \tag{7.57}$$

By assuming equation (7.56) is valid and premultiplying it by $\mathbf{R}_D(k-1,i)$ as in equation (7.57), it follows that

$$\begin{aligned}
\mathbf{R}_D(k-1,i)\mathbf{S}_D(k-1,i) &= \begin{bmatrix} 0 & \mathbf{0}^T \\ \mathbf{0} & \mathbf{I}_{i-1,i-1} \end{bmatrix} + \begin{bmatrix} 0 & \mathbf{p}_{Df}^T(k-1,i-1)\mathbf{S}_D(k-2,i-1) \\ \mathbf{0} & \mathbf{0}^T \end{bmatrix} \\
&\quad + \frac{1}{\xi_{f_{\min}}^d(k-1,i-1)}\mathbf{R}_D(k-1,i) \\
&\quad \cdot \begin{bmatrix} 1 \\ -\mathbf{w}_f(k-1,i-1) \end{bmatrix} [1 \ \ -\mathbf{w}_f^T(k-1,i-1)] \\
&= \begin{bmatrix} 0 & \mathbf{0}^T \\ \mathbf{0} & \mathbf{I}_{i-1,i-1} \end{bmatrix} + \begin{bmatrix} 0 & \mathbf{w}_f^T(k-1,i-1) \\ \mathbf{0} & \mathbf{0}_{i-2,i-2} \end{bmatrix} \\
&\quad + \frac{1}{\xi_{f_{\min}}^d(k-1,i-1)} \begin{bmatrix} \xi_{f_{\min}}^d(k-1,i-1) \\ \mathbf{0} \end{bmatrix} \\
&\quad \cdot [1 \ \ -\mathbf{w}_f^T(k-1,i-1)] \\
&= \begin{bmatrix} 0 & \mathbf{w}_f^T(k-1,i-1) \\ \mathbf{0} & \mathbf{I}_{i-1,i-1} \end{bmatrix} + \begin{bmatrix} 1 & -\mathbf{w}_f^T(k-1,i-1) \\ \mathbf{0} & \mathbf{0}_{i-1,i} \end{bmatrix} = \mathbf{I}_{i,i}
\end{aligned}$$

proving the validity of equation (7.56).

□

By applying equation (7.56) in equation (7.55), we obtain

$$\begin{aligned}
\gamma(k,i+1) &= 1 - \boldsymbol{\phi}^T(k,i+1)\mathbf{x}(k,i+1) \\
&= \gamma(k-1,i) - \frac{\varepsilon_f^2(k,i)}{\xi_{f_{\min}}^d(k,i)}
\end{aligned} \tag{7.58}$$

Following a similar method to that used in deriving equation (7.56), it can be shown that

$$\begin{aligned}
\mathbf{S}_D(k-1,i) &= \begin{bmatrix} \mathbf{S}_D(k-1,i-1) & \mathbf{0}_{i-1} \\ \mathbf{0}_{i-1}^T & 0 \end{bmatrix} \\
&\quad + \frac{1}{\xi_{b_{\min}}^d(k-1,i-1)} \begin{bmatrix} -\mathbf{w}_b(k-1,i-1) \\ 1 \end{bmatrix} [-\mathbf{w}_b^T(k-1,i-1) \ 1]
\end{aligned} \tag{7.59}$$

Now by replacing the above equation in equation (7.55), we can show that

$$\begin{aligned}\gamma(k-1,i) &= \gamma(k-1,i-1) - \frac{\varepsilon_b(k-1,i-1)}{\xi^d_{b_{\min}}(k-1,i-1)}\left[-\mathbf{w}_b^T(k-1,i-1)\ 1\right]\mathbf{x}(k-1,i)\\ &= \gamma(k-1,i-1) - \frac{\varepsilon_b^2(k-1,i-1)}{\xi^d_{b_{\min}}(k-1,i-1)}\end{aligned} \tag{7.60}$$

The last equation completes the set of relations required to solve the backward and forward prediction problems. In the following section, the modeling of a reference signal (joint-processor estimation) is discussed.

7.5 JOINT-PROCESS ESTIMATION

In the previous sections, we considered only the forward and backward prediction problems and explored some common features in their solutions. In a more general situation, the goal is to predict the behavior of one process represented by $d(k)$ through measurements of a related process contained in $\mathbf{x}(k, i+1)$. Therefore, it is important to derive an adaptive lattice-based realization to match a desired signal $d(k)$ through the minimization of the weighted squared error function given by

$$\begin{aligned}\xi^d(k,i+1) &= \sum_{l=0}^{k} \lambda^{k-l}\varepsilon^2(l,i+1)\\ &= \sum_{l=0}^{k} \lambda^{k-l}[d(l) - \mathbf{w}^T(k,i+1)\mathbf{x}(l,i+1)]^2\end{aligned} \tag{7.61}$$

where $y(k,i+1) = \mathbf{w}^T(k,i+1)\mathbf{x}(k,i+1)$ is the adaptive-filter output signal and $\varepsilon(l,i+1)$ is the *a posteriori* error at a given instant l if the adaptive-filter coefficients were fixed at $\mathbf{w}(k,i+1)$. The minimization procedure of $\xi^d(k,i+1)$ is often called joint-process estimation.

The prediction lattice realization generates the forward and backward prediction errors and requires some feedforward coefficients to allow the minimization of $\xi^d(k,i+1)$. In fact, the lattice predictor in this case works as a signal processing building block which improves the quality of the signals (in the sense of reducing the eigenvalue spread of the autocorrelation matrix) that are inputs to the output taps. The question is where should the taps be placed. We give some statistical arguments for this choice here. First, we repeat, for convenience, the expression of the backward prediction error:

$$\varepsilon_b(k,i+1) = \mathbf{x}^T(k,i+2)\begin{bmatrix} -\mathbf{w}_b(k,i+1) \\ 1 \end{bmatrix}$$

From the orthogonality property of the RLS algorithm, for $k \to \infty$, we can infer that

$$E[\varepsilon_b(k,i+1)x(k-l)] = 0$$

for $l = 0, 1, \ldots, i$. From this equation, it is possible to show that

$$E[\varepsilon_b(k,i+1)\mathbf{x}^T(k,i+1)] = \mathbf{0}^T$$

If we postmultiply the above equation by $[-\mathbf{w}_b(k,i)\ 1]^T$, we obtain

$$E\left\{\varepsilon_b(k,i+1)\mathbf{x}^T(k,i+1)\begin{bmatrix}-\mathbf{w}_b(k,i)\\ 1\end{bmatrix}\right\} = E[\varepsilon_b(k,i+1)\varepsilon_b(k,i)] = 0$$

This result shows that backward prediction errors of consecutive orders are uncorrelated. Using similar arguments one can show that $E[\varepsilon_b(k,i+1)\varepsilon_b(k,l)] = 0$, for $l = 0, 1, \ldots, i$.

In problem 4, it is shown that backward prediction errors are uncorrelated with each other in the sense of time averaging and, as a consequence, should be naturally chosen as inputs to the output taps. The objective function can now be written as

$$\begin{aligned}\xi^d(k,i+1) &= \sum_{l=0}^{k}\lambda^{k-l}\varepsilon^2(l,i+1)\\ &= \sum_{l=0}^{k}\lambda^{k-l}[d(l) - \hat{\boldsymbol{\varepsilon}}_b^T(k,i+1)\mathbf{v}(l,i+1)]^2\end{aligned} \tag{7.62}$$

where $\hat{\boldsymbol{\varepsilon}}_b^T(k,i+1) = [\varepsilon_b(k,0)\ \varepsilon_b(k,1)\ldots\varepsilon_b(k,i)]$ is the backward prediction error vector and $\mathbf{v}^T(k,i+1) = [v_0(k)\ v_1(k)\ldots v_i(k)]$ is the feedforward coefficient vector.

The main objective of the present section is to derive a time-updating formula for the output tap coefficients. From equations (7.61) and (7.62), it is obvious that the lattice realization generates the optimal estimation by using a parameterization different from that related to the direct-form realization. We can derive the updating equations for the elements of the forward coefficient vector using the order-updating equation for the tap coefficients of the direct-form realization. Employing equation (7.59), the equivalent optimal solution with the direct-form realization can be expressed as

$$\begin{aligned}\mathbf{w}(k,i+1) &= \mathbf{S}_D(k,i+1)\mathbf{p}_D(k,i+1)\\ &= \begin{bmatrix}\mathbf{S}_D(k,i) & \mathbf{0}_i\\ \mathbf{0}_i^T & 0\end{bmatrix}\mathbf{p}_D(k,i+1)\\ &\quad + \frac{1}{\xi^d_{b_{\min}}(k,i)}\begin{bmatrix}-\mathbf{w}_b(k,i)\\ 1\end{bmatrix}[-\mathbf{w}_b^T(k,i)\ 1]\mathbf{p}_D(k,i+1)\\ &= \begin{bmatrix}\mathbf{w}(k,i)\\ 0\end{bmatrix} + \frac{\delta_D(k,i)}{\xi^d_{b_{\min}}(k,i)}\begin{bmatrix}-\mathbf{w}_b(k,i)\\ 1\end{bmatrix}\end{aligned} \tag{7.63}$$

where

$$\begin{aligned}\delta_D(k,i) &= [-\mathbf{w}_b^T(k,i)\ 1]\mathbf{p}_D(k,i+1)\\ &= -\mathbf{w}_b^T(k,i)\sum_{l=0}^{k}\lambda^{k-l}\mathbf{x}(l,i)d(l) + \sum_{l=0}^{k}\lambda^{k-l}x(l-i)d(l)\\ &= \sum_{l=0}^{k}\lambda^{k-l}\varepsilon_b(l,i)d(l)\end{aligned}$$

and

$$\mathbf{p}_D(k, i+1) = \sum_{l=0}^{k} \lambda^{k-l}\mathbf{x}(l, i+1)d(l)$$

Since

$$\mathbf{p}_D(k, i+1) = \lambda\mathbf{p}_D(k-1, i+1) + d(k)\mathbf{x}(k, i+1)$$

and

$$\mathbf{w}_b(k, i) = \mathbf{w}_b(k-1, i) + \boldsymbol{\phi}(k, i)e_b(k, i)$$

see equation (7.41), by following the same steps we used to deduce the time update of $\delta(k, i)$ in equation (7.47), we can show that

$$\delta_D(k, i) = \lambda\delta_D(k-1, i) + \frac{\varepsilon(k, i)\varepsilon_b(k, i)}{\gamma(k, i)} \tag{7.64}$$

By calculating the output signal of the joint-process estimator using the order-updating equation (7.63) for the direct-form realization, we can show that

$$\mathbf{w}^T(k, i+1)\mathbf{x}(k, i+1) = [\mathbf{w}^T(k, i)\ 0]\mathbf{x}(k, i+1) + \frac{\delta_D(k, i)}{\xi^d_{b_{\min}}(k, i)}[-\mathbf{w}^T_b(k, i)\ 1]\mathbf{x}(k, i+1) \tag{7.65}$$

This equation can be rewritten as

$$y(k, i+1) = y(k, i) + \frac{\delta_D(k, i)}{\xi^d_{b_{\min}}(k, i)}\varepsilon_b(k, i) \tag{7.66}$$

where it can now be noticed that the joint-predictor output $y(k, i+1)$ is a function of the backward prediction error $\varepsilon_b(k, i)$. This was the motivation for using the decomposition of $\mathbf{S}_D(k, i+1)$ given by equation (7.59) in equation (7.63).

The feedforward multiplier coefficients can be identified as

$$v_i(k) = \frac{\delta_D(k, i)}{\xi^d_{b_{\min}}(k, i)} \tag{7.67}$$

and the *a posteriori* output error of the adaptive filter of order i from 1 to N are obtained simultaneously, where

$$\varepsilon(k, i+1) = \varepsilon(k, i) - v_i(k)\varepsilon_b(k, i) \tag{7.68}$$

The above result was derived by subtracting $d(k)$ from both sides of equation (7.66). The resulting lattice realization is depicted in Fig. 7.2.

We now have available all the relations required to generate the lattice recursive least-squares adaptive-filtering algorithm based on *a posteriori* estimation errors. The algorithm is described in Algorithm 7.1, which highlights in boxes the terms that should be saved in order to avoid repeated computation.

Algorithm 7.1

Lattice RLS Algorithm
Based on *A Posteriori* Errors

Initialization

Do for $i = 0, 1 \ldots, N$
$\delta(-1,i) = \delta_D(-1,i) = 0$ (assuming $x(k) = 0$ for $k < 0$)
$\xi^d_{b_{\min}}(-1,i) = \xi^d_{f_{\min}}(-1,i) = \epsilon$ (a small positive constant)
$\gamma(-1,i) = 1$
$\varepsilon_b(-1,i) = 0$
End

Do for $k \geq 0$
$\gamma(k,0) = 1$
$\varepsilon_b(k,0) = \varepsilon_f(k,0) = x(k)$ (7.35)
$\xi^d_{b_{\min}}(k,0) = \xi^d_{f_{\min}}(k,0) = x^2(k) + \lambda\xi^d_{f_{\min}}(k-1,0)$ (7.36)
$\varepsilon(k,0) = d(k)$

Do for $i = 0, 1 \ldots, N$

$$\delta(k,i) = \lambda\delta(k-1,i) + \boxed{\frac{\varepsilon_b(k-1,i)}{\gamma(k-1,i)}}\varepsilon_f(k,i) \quad (7.51)$$

$$\gamma(k,i+1) = \gamma(k,i) - \frac{\varepsilon_b^2(k,i)}{\xi^d_{b_{\min}}(k,i)} \quad (7.60)$$

$$\kappa_b(k,i) = \frac{\delta(k,i)}{\xi^d_{f_{\min}}(k,i)}$$

$$\kappa_f(k,i) = \frac{\delta(k,i)}{\xi^d_{b_{\min}}(k-1,i)}$$

$$\varepsilon_b(k,i+1) = \varepsilon_b(k-1,i) - \kappa_b(k,i)\varepsilon_f(k,i) \quad (7.34)$$

$$\varepsilon_f(k,i+1) = \varepsilon_f(k,i) - \kappa_f(k,i)\varepsilon_b(k-1,i) \quad (7.33)$$

$$\xi^d_{b_{\min}}(k,i+1) = \xi^d_{b_{\min}}(k-1,i) - \delta(k,i)\kappa_b(k,i) \quad (7.27)$$

$$\xi^d_{f_{\min}}(k,i+1) = \xi^d_{f_{\min}}(k,i) - \delta(k,i)\kappa_f(k,i) \quad (7.31)$$

Feedforward Filtering

$$\delta_D(k,i) = \lambda\delta_D(k-1,i) + \boxed{\frac{\varepsilon_b(k,i)}{\gamma(k,i)}}\varepsilon(k,i) \quad (7.64)$$

$$v_i(k) = \frac{\delta_D(k,i)}{\xi^d_{b_{\min}}(k,i)} \quad (7.67)$$

$$\varepsilon(k,i+1) = \varepsilon(k,i) - v_i(k)\varepsilon_b(k,i) \quad (7.68)$$

End
End

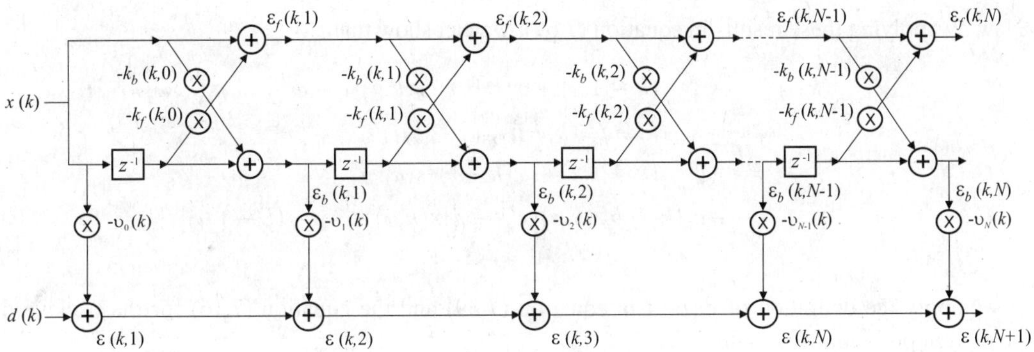

Figure 7.2 Joint-process estimation lattice realization.

7.6 TIME RECURSIONS OF THE LEAST-SQUARES ERROR

In this section, we provide a set of relations for the time updating of the minimum LS error of the prediction problems. These relations allow the derivation of two important equations involving the ratio of conversion factor of consecutive order prediction problems, namely $\frac{\gamma(k-1,i+1)}{\gamma(k-1,i)}$ and $\frac{\gamma(k,i+1)}{\gamma(k-1,i)}$. The results provided in this section are required for the derivation of some alternative lattice algorithms such as the error feedback, as well as for the fast RLS algorithms of Chapter 8.

By replacing each term in the definition of the minimum weighted least-squares error for the backward prediction problem by their time-updating equation, we have (see equations (7.16), (7.17))

$$\begin{aligned}
\xi^d_{b_{\min}}(k,i) &= \sigma_b^2(k) - \mathbf{w}_b^T(k,i)\mathbf{p}_{Db}(k,i)\\
&= \sigma_b^2(k) - \left[\mathbf{w}_b^T(k-1,i) + e_b(k,i)\boldsymbol{\phi}^T(k,i)\right]\left[\lambda\mathbf{p}_{Db}(k-1,i) + x(k-i)\mathbf{x}(k,i)\right]\\
&= \sigma_b^2(k) - \lambda\mathbf{w}_b^T(k-1,i)\mathbf{p}_{Db}(k-1,i) - x(k-i)\mathbf{w}_b^T(k-1,i)\mathbf{x}(k,i)\\
&\quad - \lambda e_b(k,i)\boldsymbol{\phi}^T(k,i)\mathbf{p}_{Db}(k-1,i) - e_b(k,i)\boldsymbol{\phi}^T(k,i)\mathbf{x}(k,i)x(k-i)\\
&= x^2(k-i) + \lambda\sigma_b^2(k-1) - \lambda\mathbf{w}_b^T(k-1,i)\mathbf{p}_{Db}(k-1,i)\\
&\quad -x(k-i)\mathbf{w}_b^T(k-1,i)\mathbf{x}(k,i) - \lambda e_b(k,i)\boldsymbol{\phi}^T(k,i)\mathbf{p}_{Db}(k-1,i)\\
&\quad -e_b(k,i)\boldsymbol{\phi}^T(k,i)\mathbf{x}(k,i)x(k-i)
\end{aligned} \tag{7.69}$$

By combining the second and third terms, we get

$$\lambda[\sigma_b^2(k-1) - \mathbf{w}_b^T(k-1,i)\mathbf{p}_{Db}(k-1,i)] = \lambda\xi^d_{b_{\min}}(k-1,i)$$

Similarly, by combining the first, fourth and sixth terms, we obtain

$$\begin{aligned}
&x(k-i)[x(k-i) - \mathbf{w}_b^T(k-1,i)\mathbf{x}(k,i) - e_b(k,i)\boldsymbol{\phi}^T(k,i)\mathbf{x}(k,i)]\\
&= x(k-i)[e_b(k,i) - e_b(k,i)\boldsymbol{\phi}^T(k,i)\mathbf{x}(k,i)]\\
&= x(k-i)e_b(k,i)[1 - \boldsymbol{\phi}^T(k,i)\mathbf{x}(k,i)]
\end{aligned}$$

Now by applying these results in equation (7.69), we can show that

$$\begin{aligned}\xi^d_{b_{\min}}(k,i) &= \lambda\xi^d_{b_{\min}}(k-1,i) + x(k-i)e_b(k,i)[1-\boldsymbol{\phi}^T(k,i)\mathbf{x}(k,i)] \\ &\quad -\lambda e_b(k,i)\boldsymbol{\phi}^T(k,i)\mathbf{p}_{Db}(k-1,i) \\ &= \lambda\xi^d_{b_{\min}}(k-1,i) + x(k-i)e_b(k,i) \\ &\quad -e_b(k,i)\boldsymbol{\phi}^T(k,i)[x(k-i)\mathbf{x}(k,i) + \lambda\mathbf{p}_{Db}(k-1,i)]\end{aligned}$$

If we apply the definition of $\boldsymbol{\phi}(k,i)$ in equation (7.39) and the equation (7.16) for the backward prediction problem, we obtain

$$\begin{aligned}\xi^d_{b_{\min}}(k,i) &= \lambda\xi^d_{b_{\min}}(k-1,i) + x(k-i)e_b(k,i) - e_b(k,i)\boldsymbol{\phi}^T(k,i)\mathbf{p}_{Db}(k,i) \\ &= \lambda\xi^d_{b_{\min}}(k-1,i) + x(k-i)e_b(k,i) - e_b(k,i)\mathbf{x}^T(k,i)\mathbf{S}_D(k-1,i)\mathbf{p}_{Db}(k,i) \\ &= \lambda\xi^d_{b_{\min}}(k-1,i) + e_b(k,i)[x(k-i) - \mathbf{w}_b^T(k,i)\mathbf{x}(k,i)] \\ &= \lambda\xi^d_{b_{\min}}(k-1,i) + e_b(k,i)\varepsilon_b(k,i) \\ &= \lambda\xi^d_{b_{\min}}(k-1,i) + \frac{\varepsilon_b^2(k,i)}{\gamma(k,i)}\end{aligned} \tag{7.70}$$

Following similar steps to those used to obtain the above equation, we can show that

$$\xi^d_{f_{\min}}(k,i) = \lambda\xi^d_{f_{\min}}(k-1,i) + \frac{\varepsilon_f^2(k,i)}{\gamma(k-1,i)} \tag{7.71}$$

From the last two equations, we can easily infer the relations that are useful in deriving alternative lattice-based algorithms, namely the normalized and error-feedback algorithms. These relations are

$$\begin{aligned}\frac{\lambda\xi^d_{b_{\min}}(k-2,i)}{\xi^d_{b_{\min}}(k-1,i)} &= 1 - \frac{\varepsilon_b^2(k-1,i)}{\gamma(k-1,i)\xi^d_{b_{\min}}(k-1,i)} \\ &= \frac{\gamma(k-1,i+1)}{\gamma(k-1,i)}\end{aligned} \tag{7.72}$$

and

$$\begin{aligned}\frac{\lambda\xi^d_{f_{\min}}(k-1,i)}{\xi^d_{f_{\min}}(k,i)} &= 1 - \frac{\varepsilon_f^2(k,i)}{\gamma(k-1,i)\xi^d_{f_{\min}}(k,i)} \\ &= \frac{\gamma(k,i+1)}{\gamma(k-1,i)}\end{aligned} \tag{7.73}$$

where equations (7.60) and (7.58), respectively, were used in the derivation of the right-hand-side expressions of the above equations.

7.7 NORMALIZED LATTICE RLS ALGORITHM

An alternative form of the lattice RLS algorithm can be obtained by applying a judicious normalization to the internal variables of the algorithm, keeping their magnitude bounded by one. This normalized lattice is specially suitable for fixed-point arithmetic implementation. Also, this algorithm requires fewer recursions and variables than the unnormalized lattices, i.e., only three equations per prediction section per time sample.

7.7.1 Basic Order Recursions

A natural way to normalize the backward and forward prediction errors is to divide them by the square root of the corresponding weighted least-squares error. However, it will be shown that a wiser strategy leads to a reduction in the number of recursions. At the same time, we must think of a way to normalize variable $\delta(k,i)$. In the process of normalizing $\varepsilon_f(k,i)$, $\varepsilon_b(k,i)$, and $\delta(k,i)$, we can reduce the number of equations by eliminating the conversion variable $\gamma(k,i+1)$. Note that $\gamma(k,i+1)$ is originally normalized. These goals can be reached if the normalization of $\delta(k,i)$ is performed as

$$\overline{\delta}(k,i) = \frac{\delta(k,i)}{\sqrt{\xi^d_{f_{\min}}(k,i)\xi^d_{b_{\min}}(k-1,i)}} \tag{7.74}$$

By noting that the conversion variable $\gamma(k-1,i)$ divides the product $\varepsilon_f(k,i)\varepsilon_b(k-1,i)$ in the time-updating formula (7.51), we can devise a way to perform the normalization of the prediction errors leading to its elimination. The appropriate normalization of the forward and backward estimation errors are, respectively, performed as

$$\overline{\varepsilon}_f(k,i) = \frac{\varepsilon_f(k,i)}{\sqrt{\gamma(k-1,i)\xi^d_{f_{\min}}(k,i)}} \tag{7.75}$$

$$\overline{\varepsilon}_b(k,i) = \frac{\varepsilon_b(k,i)}{\sqrt{\gamma(k,i)\xi^d_{b_{\min}}(k,i)}} \tag{7.76}$$

where the terms $\sqrt{\xi^d_{f_{\min}}(k,i)}$ and $\sqrt{\xi^d_{b_{\min}}(k,i)}$ perform the power normalization whereas $\sqrt{\gamma(k-1,i)}$ and $\sqrt{\gamma(k,i)}$ perform the so-called angle normalization, since $\gamma(k,i)$ is related to the angle between the spaces spanned by $\mathbf{x}(k-1,i)$ and $\mathbf{x}(k,i)$.

From the above equations and equation (7.51), we can show that

$$\begin{aligned}\overline{\delta}(k,i)\sqrt{\xi^d_{f_{\min}}(k,i)\xi^d_{b_{\min}}(k-1,i)} &= \lambda\overline{\delta}(k-1,i)\sqrt{\xi^d_{f_{\min}}(k-1,i)\xi^d_{b_{\min}}(k-2,i)}\\ &\quad +\overline{\varepsilon}_b(k-1,i)\overline{\varepsilon}_f(k,i)\sqrt{\xi^d_{f_{\min}}(k,i)\xi^d_{b_{\min}}(k-1,i)}\end{aligned} \tag{7.77}$$

Therefore,

$$\overline{\delta}(k,i) = \lambda\overline{\delta}(k-1,i)\sqrt{\frac{\xi^d_{f_{\min}}(k-1,i)\xi^d_{b_{\min}}(k-2,i)}{\xi^d_{f_{\min}}(k,i)\xi^d_{b_{\min}}(k-1,i)}} + \overline{\varepsilon}_b(k-1,i)\overline{\varepsilon}_f(k,i) \tag{7.78}$$

We now show that the term under the square root in the above equation can be expressed in terms of the normalized errors by using equations (7.72), (7.73), (7.75), and (7.76), that is,

$$\begin{aligned}\frac{\lambda\xi^d_{b_{\min}}(k-2,i)}{\xi^d_{b_{\min}}(k-1,i)} &= \frac{\gamma(k-1,i+1)}{\gamma(k-1,i)} \\ &= 1 - \frac{\varepsilon^2_b(k-1,i)}{\gamma(k-1,i)\xi^d_{b_{\min}}(k-1,i)} \\ &= 1 - \overline{\varepsilon}^2_b(k-1,i)\end{aligned} \tag{7.79}$$

and

$$\begin{aligned}\frac{\lambda\xi^d_{f_{\min}}(k-1,i)}{\xi^d_{f_{\min}}(k,i)} &= \frac{\gamma(k,i+1)}{\gamma(k-1,i)} \\ &= 1 - \frac{\varepsilon^2_f(k,i)}{\gamma(k-1,i)\xi^d_{f_{\min}}(k,i)} \\ &= 1 - \overline{\varepsilon}^2_f(k,i)\end{aligned} \tag{7.80}$$

Substituting the last two equations into equation (7.78), we can show that

$$\overline{\delta}(k,i) = \overline{\delta}(k-1,i)\sqrt{(1-\overline{\varepsilon}^2_b(k-1,i))(1-\overline{\varepsilon}^2_f(k,i))} + \overline{\varepsilon}_b(k-1,i)\overline{\varepsilon}_f(k,i) \tag{7.81}$$

Following a similar procedure used to derive the time-updating equation for $\overline{\delta}(k,i)$, one can derive the order-updating equation of the normalized forward and backward prediction errors. In the case of the forward prediction error, the following order-updating relation results:

$$\overline{\varepsilon}_f(k,i+1) = \left[\overline{\varepsilon}_f(k,i) - \overline{\delta}(k,i)\overline{\varepsilon}_b(k-1,i)\right]\sqrt{\frac{\xi^d_{f_{\min}}(k,i)}{\xi^d_{f_{\min}}(k,i+1)}}\sqrt{\frac{\gamma(k-1,i)}{\gamma(k-1,i+1)}} \tag{7.82}$$

Here again, we can express the functions under the square roots in terms of normalized variables. Using equations (7.31), (7.74), and (7.77), it can be shown that

$$\overline{\varepsilon}_f(k,i+1) = \frac{\overline{\varepsilon}_f(k,i) - \overline{\delta}(k,i)\overline{\varepsilon}_b(k-1,i)}{\sqrt{1-\overline{\delta}^2(k,i)}\sqrt{1-\overline{\varepsilon}^2_b(k-1,i)}} \tag{7.83}$$

If the same steps to derive $\bar{\varepsilon}_f(k, i+1)$ are followed, we can derive the order-updating equation for the backward prediction error as

$$\bar{\varepsilon}_b(k, i+1) = \left[\bar{\varepsilon}_b(k-1, i) - \bar{\delta}(k, i)\bar{\varepsilon}_f(k, i)\right] \sqrt{\frac{\xi^d_{b_{\min}}(k-1, i)}{\xi^d_{b_{\min}}(k, i+1)}} \sqrt{\frac{\gamma(k-1, i)}{\gamma(k, i+1)}}$$
$$= \frac{\bar{\varepsilon}_b(k-1, i) - \bar{\delta}(k, i)\bar{\varepsilon}_f(k, i)}{\sqrt{1 - \bar{\delta}^2(k, i)}\sqrt{1 - \bar{\varepsilon}^2_f(k, i)}} \tag{7.84}$$

7.7.2 Feedforward Filtering

The procedure to generate the joint-processor estimator is repeated here, using normalized variables. Define

$$\bar{\delta}_D(k, i) = \frac{\delta_D(k, i)}{\sqrt{\xi^d_{\min}(k, i)\xi^d_{b_{\min}}(k, i)}} \tag{7.85}$$

and

$$\bar{\varepsilon}(k, i) = \frac{\varepsilon(k, i)}{\sqrt{\gamma(k, i)\xi^d_{\min}(k, i)}} \tag{7.86}$$

Using a similar approach to that used to derive equation (7.31), one can show that

$$\xi^d_{\min}(k, i+1) = \xi^d_{\min}(k, i) - \frac{\delta^2_D(k, i)}{\xi^d_{b_{\min}}(k, i)} \tag{7.87}$$

The procedure used to derive the order-updating equations for the normalized prediction errors and the parameter $\bar{\delta}(k, i)$ can be followed to derive the equivalent parameters in the joint-process estimation case. For the *a posteriori* output error the following equation results

$$\bar{\varepsilon}(k, i+1) = \sqrt{\frac{\gamma(k, i)}{\gamma(k, i+1)}} \sqrt{\frac{\xi^d_{\min}(k, i)}{\xi^d_{\min}(k, i+1)}} \left[\bar{\varepsilon}(k, i) - \bar{\delta}_D(k, i)\bar{\varepsilon}_b(k, i)\right]$$
$$= \frac{1}{\sqrt{1 - \bar{\varepsilon}^2_b(k, i)}} \frac{1}{\sqrt{1 - \bar{\delta}^2_D(k, i)}} \left[\bar{\varepsilon}(k, i) - \bar{\delta}_D(k, i)\bar{\varepsilon}_b(k, i)\right] \tag{7.88}$$

The order-updating equation of $\bar{\delta}_D(k, i)$ is (see equation (7.78))

$$\bar{\delta}_D(k, i) = \sqrt{\frac{\lambda^2 \xi^d_{\min}(k-1, i)\xi^d_{b_{\min}}(k-1, i)}{\xi^d_{\min}(k, i)\xi^d_{b_{\min}}(k, i)}} \bar{\delta}_D(k-1, i) + \bar{\varepsilon}(k, i)\bar{\varepsilon}_b(k, i)$$
$$= \sqrt{(1 - \bar{\varepsilon}^2_b(k, i))(1 - \bar{\varepsilon}^2(k, i))}\bar{\delta}_D(k-1, i) + \bar{\varepsilon}(k, i)\bar{\varepsilon}_b(k, i) \tag{7.89}$$

where we used the fact that

$$\frac{\lambda \xi_{\min}^{d}(k-1,i)}{\xi_{\min}^{d}(k,i)} = 1 - \bar{\varepsilon}^{2}(k,i) \tag{7.90}$$

The normalized lattice RLS algorithm based on *a posteriori* errors is described in Algorithm 7.2.

Notice that in the updating formulas of the normalized errors, the terms involving the square root operation could be conveniently implemented through separate multiplier coefficients, namely $\eta_f(k,i)$, $\eta_b(k,i)$, and $\eta_D(k,i)$. In this way, one can perform the order updating by calculating the numerator first and proceeding with a single multiplication. These coefficients are given by

$$\eta_f(k,i+1) = \frac{1}{\sqrt{1-\bar{\delta}^{2}(k,i)}\sqrt{1-\bar{\varepsilon}_b^{2}(k-1,i)}} \tag{7.91}$$

$$\eta_b(k,i+1) = \frac{1}{\sqrt{1-\bar{\delta}^{2}(k,i)}\sqrt{1-\bar{\varepsilon}_f^{2}(k,i)}} \tag{7.92}$$

$$\eta_D(k,i+1) = \frac{1}{\sqrt{1-\bar{\varepsilon}_b^{2}(k,i)}\sqrt{1-\bar{\delta}_D^{2}(k,i)}} \tag{7.93}$$

With these multipliers it is straightforward to obtain the structure for the joint-processor estimator depicted in Fig. 7.3.

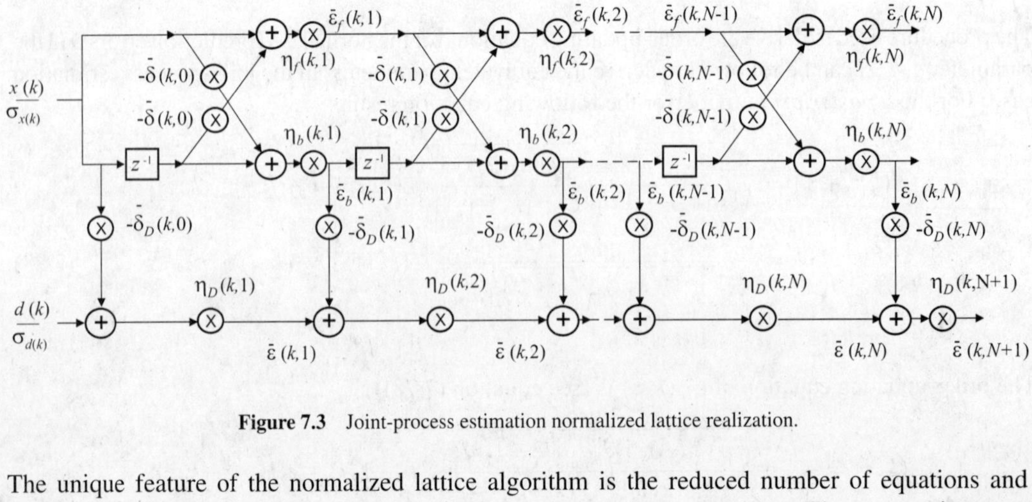

Figure 7.3 Joint-process estimation normalized lattice realization.

The unique feature of the normalized lattice algorithm is the reduced number of equations and variables at the expense of employing a number of square root operations. These operations can be costly to implement in most types of hardware architectures. Another interesting feature of the

Algorithm 7.2

Normalized Lattice RLS Algorithm
Based on *A Posteriori* Error

Initialization

Do for $i = 0, 1 \ldots, N$
$\overline{\delta}(-1, i) = 0$ (assuming $x(k) = d(k) = 0$ for $k < 0$)
$\overline{\delta}_D(-1, i) = 0$
$\overline{\varepsilon}_b(-1, i) = 0$
End

$\sigma_x^2(-1) = \sigma_d^2(-1) = \epsilon$ (ϵ small positive constant)

Do for $k \geq 0$
$\sigma_x^2(k) = \lambda\sigma_x^2(k-1) + x^2(k)$ (Input signal energy)
$\sigma_d^2(k) = \lambda\sigma_d^2(k-1) + d^2(k)$ (Reference signal energy)
$\overline{\varepsilon}_b(k, 0) = \overline{\varepsilon}_f(k, 0) = x(k)/\sigma_x(k)$
$\overline{\varepsilon}(k, 0) = d(k)/\sigma_d(k)$

Do for $i = 0, 1 \ldots, N$

$$\overline{\delta}(k, i) = \overline{\delta}(k-1, i)\sqrt{(1 - \overline{\varepsilon}_b^2(k-1, i))(1 - \overline{\varepsilon}_f^2(k, i))} + \overline{\varepsilon}_b(k-1, i)\overline{\varepsilon}_f(k, i) \quad (7.81)$$

$$\overline{\varepsilon}_b(k, i+1) = \frac{\overline{\varepsilon}_b(k-1, i) - \overline{\delta}(k, i)\overline{\varepsilon}_f(k, i)}{\sqrt{(1 - \overline{\delta}^2(k, i))(1 - \overline{\varepsilon}_f^2(k, i))}} \quad (7.84)$$

$$\overline{\varepsilon}_f(k, i+1) = \frac{\overline{\varepsilon}_f(k, i) - \overline{\delta}(k, i)\overline{\varepsilon}_b(k-1, i)}{\sqrt{(1 - \overline{\delta}^2(k, i))(1 - \overline{\varepsilon}_b^2(k-1, i))}} \quad (7.83)$$

Feedforward Filter

$$\overline{\delta}_D(k, i) = \overline{\delta}_D(k-1, i)\sqrt{(1 - \overline{\varepsilon}_b^2(k, i))(1 - \overline{\varepsilon}^2(k, i))} + \overline{\varepsilon}(k, i)\overline{\varepsilon}_b(k, i) \quad (7.89)$$

$$\overline{\varepsilon}(k, i+1) = \frac{1}{\sqrt{(1 - \overline{\varepsilon}_b^2(k, i))(1 - \overline{\delta}_D^2(k, i))}}\left[\overline{\varepsilon}(k, i) - \overline{\delta}_D(k, i)\overline{\varepsilon}_b(k, i)\right] \quad (7.88)$$

End

End

normalized lattice algorithm is that the forgetting factor λ does not appear in the internal updating equations; it appears only in the calculation of the energy of the input and reference signals. This property may be advantageous from the computational point of view in situations where there is a need to vary the value of λ. On the other hand, since all internal variables are normalized, the actual amplitude of the error signals and other quantities do not match those in other lattice structures. In fact, from the normalized lattice structure one can only effectively extract the shape of the frequency model the structure identifies, since the mapping between the parameters of normalized and non normalized structures is computationally intensive.

7.8 ERROR-FEEDBACK LATTICE RLS ALGORITHM

The reflection coefficients of the lattice algorithm have so far been updated in an indirect way, without time recursions. This section describes an alternative method of updating the reflection coefficients using time updating. These updating equations are recursive in nature and are often called direct updating, since the updating equations used for $\kappa_b(k,i)$ and $\kappa_f(k,i)$ in Algorithm 7.1 are dependent exclusively on quantities other than past reflection coefficients. Algorithms employing the recursive time updating are called error-feedback lattice RLS algorithms. These algorithms have better numerical properties than their indirect updating counterparts [3].

7.8.1 Recursive Formulas for the Reflection Coefficients

The derivation of a direct updating equation for $\kappa_f(k,i)$ starts by replacing $\delta(k,i)$ by its time-updating equation (7.51)

$$\begin{aligned}\kappa_f(k,i) &= \frac{\delta(k,i)}{\xi^d_{b_{\min}}(k-1,i)} \\ &= \frac{\lambda\delta(k-1,i)}{\xi^d_{b_{\min}}(k-1,i)} + \frac{\varepsilon_b(k-1,i)\varepsilon_f(k,i)}{\gamma(k-1,i)\xi^d_{b_{\min}}(k-1,i)}\end{aligned}$$

By multiplying and dividing the first term by $\xi^d_{b_{\min}}(k-2,i)$ and next using equation (7.72) in the first and second terms, we obtain

$$\begin{aligned}\kappa_f(k,i) &= \frac{\delta(k-1,i)}{\xi^d_{b_{\min}}(k-2,i)}\frac{\lambda\xi^d_{b_{\min}}(k-2,i)}{\xi^d_{b_{\min}}(k-1,i)} + \frac{\varepsilon_b(k-1,i)\varepsilon_f(k,i)}{\gamma(k-1,i)\xi^d_{b_{\min}}(k-1,i)} \\ &= \kappa_f(k-1,i)\frac{\gamma(k-1,i+1)}{\gamma(k-1,i)} + \frac{\varepsilon_b(k-1,i)\varepsilon_f(k,i)\gamma(k-1,i+1)}{\gamma^2(k-1,i)\lambda\xi^d_{b_{\min}}(k-2,i)} \\ &= \frac{\gamma(k-1,i+1)}{\gamma(k-1,i)}\left[\kappa_f(k-1,i) + \frac{\varepsilon_b(k-1,i)\varepsilon_f(k,i)}{\gamma(k-1,i)\lambda\xi^d_{b_{\min}}(k-2,i)}\right] \end{aligned} \quad (7.94)$$

Similarly, using equations (7.51) and (7.73), it is straightforward to show that

$$\kappa_b(k,i) = \frac{\gamma(k,i+1)}{\gamma(k-1,i)}\left[\kappa_b(k-1,i) + \frac{\varepsilon_b(k-1,i)\varepsilon_f(k,i)}{\gamma(k-1,i)\lambda\xi^d_{f_{\min}}(k-1,i)}\right] \tag{7.95}$$

The feedforward coefficients can also be time updated in a recursive form, by appropriately combining equations (7.64), (7.67), and (7.72). The time-recursive updating equation for $v_i(k)$ is

$$v_i(k) = \frac{\gamma(k,i+1)}{\gamma(k,i)}\left[v_i(k-1) + \frac{\varepsilon(k,i)\varepsilon_b(k,i)}{\gamma(k,i)\lambda\xi^d_{b_{\min}}(k-1,i)}\right] \tag{7.96}$$

The error-feedback LRLS algorithm described in Algorithm 7.3 employs the equations (7.94), (7.95), and (7.96). This algorithm is directly derived from Algorithm 7.1.

Alternative *a posteriori* LRLS algorithms can be obtained if we replace equations (7.27) and (7.31) by (7.70) and (7.72) in Algorithms 7.1 and 7.3, respectively. These modifications as well as possible others do not change the behavior of the LRLS algorithm when implemented with infinite precision (long wordlength). However, differences exist in computational complexity and in the effects of quantization error propagation.

7.9 LATTICE RLS ALGORITHM BASED ON *A PRIORI* ERRORS

The lattice algorithms presented so far are based on *a posteriori* errors; however alternative algorithms based on *a priori* errors exist and one of them is derived in this section.

The time updating of the quantity $\delta(k,i)$ as a function of the *a priori* errors was previously derived (see equation (7.47)) and is repeated here for convenience.

$$\delta(k,i) = \lambda\delta(k-1,i) + \gamma(k-1,i)e_b(k-1,i)e_f(k,i) \tag{7.97}$$

The time updating of the forward prediction *a priori* error can be obtained by using equation (7.32) as

$$\begin{aligned}
e_f(k,i+1) &= \mathbf{x}^T(k,i+2)\begin{bmatrix}1\\ -\mathbf{w}_f(k-1,i+1)\end{bmatrix}\\
&= \mathbf{x}^T(k,i+2)\begin{bmatrix}1\\ -\mathbf{w}_f(k-1,i)\\ 0\end{bmatrix} + \frac{\delta(k-1,i)}{\xi^d_{b_{\min}}(k-2,i)}\mathbf{x}^T(k,i+2)\begin{bmatrix}0\\ \mathbf{w}_b(k-2,i)\\ -1\end{bmatrix}\\
&= e_f(k,i) - \frac{\delta(k-1,i)}{\xi^d_{b_{\min}}(k-2,i)}e_b(k-1,i)\\
&= e_f(k,i) - \kappa_f(k-1,i)e_b(k-1,i)
\end{aligned} \tag{7.98}$$

Algorithm 7.3

Error-Feedback LRLS Algorithm
Based on *A Posteriori* Errors

Initialization

Do for $i = 0, 1 \ldots, N$

$\kappa_b(-1,i) = \kappa_f(-1,i) = v_i(-1) = \delta(-1,i) = 0, \gamma(-1,i) = 1$

$\xi^d_{b_{\min}}(-2,i) = \xi^d_{b_{\min}}(-1,i) = \xi^d_{f_{\min}}(-1,i) = \epsilon$ (a small positive constant)

$\varepsilon_b(-1,i) = 0$

End

Do for $k \geq 0$

$\gamma(k,0) = 1$

$$\varepsilon_b(k,0) = \varepsilon_f(k,0) = x(k) \quad (7.35)$$

$$\xi^d_{f_{\min}}(k,0) = \xi^d_{b_{\min}}(k,0) = x^2(k) + \lambda \xi^d_{f_{\min}}(k-1,0) \quad (7.36)$$

$\varepsilon(k,0) = d(k)$

Do for $i = 0, 1 \ldots, N$

$$\delta(k,i) = \lambda\delta(k-1,i) + \boxed{\frac{\varepsilon_b(k-1,i)\varepsilon_f(k,i)}{\gamma(k-1,i)}} \quad (7.51)$$

$$\gamma(k,i+1) = \gamma(k,i) - \frac{\varepsilon_b^2(k,i)}{\xi^d_{b_{\min}}(k,i)} \quad (7.60)$$

$$\kappa_f(k,i) = \boxed{\frac{\gamma(k-1,i+1)}{\gamma(k-1,i)}}\left[\kappa_f(k-1,i) + \boxed{\frac{\varepsilon_b(k-1,i)\varepsilon_f(k,i)}{\gamma(k-1,i)}}\frac{1}{\lambda\xi^d_{b_{\min}}(k-2,i)}\right] \quad (7.94)$$

$$\kappa_b(k,i) = \frac{\gamma(k,i+1)}{\gamma(k-1,i)}\left[\kappa_b(k-1,i) + \boxed{\frac{\varepsilon_b(k-1,i)\varepsilon_f(k,i)}{\gamma(k-1,i)}}\frac{1}{\lambda\xi^d_{f_{\min}}(k-1,i)}\right] \quad (7.95)$$

$$\varepsilon_b(k,i+1) = \varepsilon_b(k-1,i) - \kappa_b(k,i)\varepsilon_f(k,i) \quad (7.34)$$

$$\varepsilon_f(k,i+1) = \varepsilon_f(k,i) - \kappa_f(k,i)\varepsilon_b(k-1,i) \quad (7.33)$$

$$\xi^d_{f_{\min}}(k,i+1) = \xi^d_{f_{\min}}(k,i) - \frac{\delta^2(k,i)}{\xi^d_{b_{\min}}(k-1,i)} \quad (7.31)$$

$$\xi^d_{b_{\min}}(k,i+1) = \xi^d_{b_{\min}}(k-1,i) - \frac{\delta^2(k,i)}{\xi^d_{f_{\min}}(k,i)} \quad (7.27)$$

Feedforward Filtering

$$v_i(k) = \boxed{\frac{\gamma(k,i+1)}{\gamma(k,i)}}\left[v_i(k-1) + \frac{\varepsilon(k,i)\varepsilon_b(k,i)}{\gamma(k,i)\lambda\xi^d_{b_{\min}}(k-1,i)}\right] \quad (7.96)$$

$$\varepsilon(k,i+1) = \varepsilon(k,i) - v_i(k)\varepsilon_b(k,i) \quad (7.68)$$

End

End

With equation (7.28), we can generate the time-updating equation of the backward prediction *a priori* error as

$$\begin{aligned} e_b(k, i+1) &= \mathbf{x}^T(k, i+2) \begin{bmatrix} 0 \\ -\mathbf{w}_b(k-2, i) \\ 1 \end{bmatrix} - \frac{\delta(k-1, i)}{\xi^d_{f_{\min}}(k-1, i)} \mathbf{x}^T(k, i+2) \begin{bmatrix} -1 \\ \mathbf{w}_f(k-1, i) \\ 0 \end{bmatrix} \\ &= e_b(k-1, i) - \frac{\delta(k-1, i)}{\xi^d_{f_{\min}}(k-1, i)} e_f(k, i) \\ &= e_b(k-1, i) - \kappa_b(k-1, i) e_f(k, i) \end{aligned} \quad (7.99)$$

The order updating of $\gamma(k-1, i)$ can be derived by employing the relations of equations (7.50) and (7.60). The result is

$$\gamma(k-1, i+1) = \gamma(k-1, i) - \frac{\gamma^2(k-1, i) e_b^2(k-1, i)}{\xi^d_{b_{\min}}(k-1, i)} \quad (7.100)$$

The updating of the feedforward coefficients of the lattice realization based on *a priori* errors is performed by the following equations

$$\delta_D(k, i) = \lambda \delta_D(k-1, i) + \gamma(k, i) e_b(k, i) e(k, i) \quad (7.101)$$

$$e(k, i+1) = e(k, i) - v_i(k-1) e_b(k, i) \quad (7.102)$$

$$v_i(k-1) = \frac{\delta_D(k-1, i)}{\xi^d_{b_{\min}}(k-1, i)} \quad (7.103)$$

The derivations are omitted since they follow the same steps of the predictor equations.

An LRLS algorithm based on *a priori* errors is described in Algorithm 7.4. The normalized and error-feedback versions of the LRLS algorithm based on *a priori* errors also exist and their derivations are left as problems.

7.10 QUANTIZATION EFFECTS

A major issue related to the implementation of adaptive filters is their behavior when implemented with finite-precision arithmetic. In particular, the roundoff errors arising from the quantization of the internal quantities of an algorithm propagate internally and can even cause instability. The numerical stability and accuracy are algorithm dependent. In this section, we summarize some of the results obtained in the literature related to the LRLS algorithms [3], [7]-[8].

One of the first attempts to study the numerical accuracy of the lattice algorithms was reported in [7]. Special attention was given to the normalized lattice RLS algorithm, since this algorithm is suitable for fixed-point arithmetic implementation, due to its internal normalization. In this study, it was

Algorithm 7.4

LRLS Algorithm
Based on *A Priori* Errors

Initialization
Do for $i = 0, 1 \ldots, N$
$\delta(-1,i) = \delta_D(-1,i) = 0$ (assuming $x(k) = 0$ for $k < 0$)
$\gamma(-1,i) = 1$
$\xi^d_{b_{\min}}(-1,i) = \xi^d_{f_{\min}}(-1,i) = \epsilon$ (a small positive constant)
$e_b(-1,i) = 0$
$\kappa_f(-1,i) = \kappa_b(-1,i) = 0$
End

Do for $k \geq 0$
$\gamma(k,0) = 1$
$e_b(k,0) = e_f(k,0) = x(k)$
$\xi^d_{f_{\min}}(k,0) = \xi^d_{b_{\min}}(k,0) = x^2(k) + \lambda \xi^d_{f_{\min}}(k-1,0)$
$e(k,0) = d(k)$
Do for $i = 0, 1 \ldots, N$

$$\delta(k,i) = \lambda\delta(k-1,i) + \gamma(k-1,i)e_b(k-1,i)e_f(k,i) \tag{7.47}$$

$$\gamma(k,i+1) = \gamma(k,i) - \frac{\boxed{\gamma^2(k,i)e_b^2(k,i)}}{\xi^d_{b_{\min}}(k,i)} \tag{7.100}$$

$$e_b(k,i+1) = e_b(k-1,i) - \kappa_b(k-1,i)e_f(k,i) \tag{7.99}$$

$$e_f(k,i+1) = e_f(k,i) - \kappa_f(k-1,i)e_b(k-1,i) \tag{7.98}$$

$$\kappa_f(k,i) = \frac{\delta(k,i)}{\xi^d_{b_{\min}}(k-1,i)}$$

$$\kappa_b(k,i) = \frac{\delta(k,i)}{\xi^d_{f_{\min}}(k,i)}$$

$$\xi^d_{f_{\min}}(k,i+1) = \xi^d_{f_{\min}}(k,i) - \delta(k,i)\kappa_f(k,i) \tag{7.31}$$

$$\xi^d_{b_{\min}}(k,i+1) = \xi^d_{b_{\min}}(k-1,i) - \delta(k,i)\kappa_b(k,i) \tag{7.27}$$

Feedforward Filtering

$$\delta_D(k,i) = \lambda\delta_D(k-1,i) + \boxed{\gamma(k,i)e_b(k,i)}\, e(k,i) \tag{7.101}$$

$$e(k,i+1) = e(k,i) - v_i(k-1)e_b(k,i) \tag{7.102}$$

$$v_i(k) = \frac{\delta_D(k,i)}{\xi^d_{b_{\min}}(k,i)} \tag{7.103}$$

End
End

shown that the bias error in the reflection coefficients was more significant than the variance of the estimate error. The bias in the estimated reflection coefficients is mainly caused by the quantization error associated with the calculation of the square roots of $[1 - \overline{\varepsilon}_b^2(k-1,i)]$ and $[1 - \overline{\varepsilon}_f^2(k,i)]$, assuming they are calculated separately. An upper bound for this quantization error is given by

$$m_{sq} = 2^{-b} \tag{7.104}$$

assuming that b is the number of bits after the sign bit and that quantization is performed through rounding. In the analysis, the basic assumption that $1 - \lambda \gg 2^{-b+1}$ was used. The upper bound of the bias error in the reflection coefficients is given by [7]

$$\Delta\overline{\delta}(k,i) = \frac{2^{-b+1}\overline{\delta}(k,i)}{1-\lambda} \tag{7.105}$$

Obviously, the accuracy of this result depends on the validity of the assumptions used in the analysis [7]. However it is a good indication of how the bias is generated in the reflection coefficients. It should also be noted that the above result is valid as long as the updating of the related reflection coefficient does not stop. An analysis for the case in which the updating stops is also included in [7].

The bias error of a given stage of the lattice realization propagates to the succeeding stages and its accumulation in the prediction errors can be expressed as

$$\Delta\overline{\varepsilon}_b^2(k,i+1) = \Delta\overline{\varepsilon}_f^2(k,i+1) \approx 2^{-b+2}\sum_{l=0}^{i}\frac{\overline{\delta}^2(k,l)}{1-\overline{\delta}^2(k,l)} \tag{7.106}$$

for $i = 0, 1, \ldots, N$. This equation indicates that whenever the value of the parameter $\overline{\delta}^2(k,l)$ is small, the corresponding term in the summation is also small. On the other hand, if the value of this parameter tends to one, the corresponding term of the summation is large. Also note that the accumulated error tends to grow as the number of sections of the lattice is increased. In a finite-precision implementation, it is possible to determine the maximum order that the lattice can have such that the error signals at the end of the realization still represent actual signals and not only accumulated quantization noise.

The lattice algorithms remain stable even when using quite short wordlength in fixed- and floating-point implementations. In terms of accuracy the error-feedback algorithms are usually better than the conventional LRLS algorithms [3]. The reduction in the quantization effects of the error-feedback LRLS algorithms is verified in [3], where a number of examples show satisfactory performance for implementation with less than 10 bits in fixed-point arithmetic.

Another investigation examines the finite-wordlength implementation employing floating-point arithmetic of the unnormalized lattice with and without error feedback [8]. As expected, the variance of the accumulated error in the reflection coefficients of the error-feedback algorithms are smaller than that for the conventional LRLS algorithm. Another important issue relates to the so-called self-generated noise that originates in the internal stages of the lattice realization when the order of adaptive filter is greater than necessary. In the cases where the signal-to-noise ratio is high in the desired signal, the internal signals of the last stages of the lattice realization can reach the quantization level and start self-generated noise, leading to an excess mean-square error and possibly to instability. The stability problem can be avoided by turning off the stages after the one in which the weighted forward and backward squared errors are smaller than a given threshold.

Example 7.1

The system identification problem described in Chapter 3 (subsection 3.6.2) is solved using the lattice algorithms presented in this chapter. The main objective is to compare the performance of the algorithms when implemented in finite precision.

Solution:

We present here the results of using the unnormalized, the normalized and error-feedback *a posteriori* lattice RLS algorithms in the system identification example. All results presented are obtained by running 200 independent experiments and calculating the average of the quantities of interest. We consider the case of eigenvalue spread 20, and $\lambda = 0.99$. Parameter ϵ is 0.1, 0.01, and 0.1 for the unnormalized, the normalized, and the error-feedback lattice filters, respectively. The measured misadjustments of the lattice algorithms are given in Table 7.1. As expected, the results are close to those obtained by the conventional RLS algorithm, where in the latter the misadjustment is 0.0421. Not included is the result for the normalized lattice because the *a posteriori* error is not available, in this case the measured normalized MSE is 0.00974.

Table 7.2 summarizes the results obtained by the implementation of the lattice algorithms with finite precision. Parameter ϵ in the finite-precision implementation is 0.1, 0.04 and 0.5 for the unnormalized, normalized and error-feedback lattices, respectively. These values assure a good convergence behavior of the algorithms in this experiment. In short-wordlength implementation of the lattice algorithms, it is advisable to test if the denominator expressions of the algorithm steps involving division are not rounded to zero. In the case of the detection of a zero denominator, replace its value by the value of the least significant bit. Table 7.2 shows that for the unnormalized and error-feedback lattices, the mean-squared errors are comparable to the case of the conventional RLS previously shown in Table 5.2. The normalized lattice is more sensitive to quantization errors due to its higher computational complexity. The errors introduced by the calculations to obtain $\mathbf{w}(k)_Q$, starting with the lattice coefficients, is the main reason for the increased values of $E[||\Delta\mathbf{w}(k)_Q||^2]$ shown in Table 7.2. Therefore, this result should not be considered as an indication of poor performance of the normalized lattice implemented with finite precision.

□

Table 7.1 Evaluation of the Lattice RLS Algorithms

Algorithm	**Misadjustment**
Unnorm.	0.0416
Error Feed.	0.0407

Table 7.2 Results of the Finite-Precision Implementation of the Lattice RLS Algorithms

	$\xi(k)_Q$			$E[\|\|\Delta\mathbf{w}(k)_Q\|\|^2]$		
No. of bits	**Unnorm.**	**Norm.**	**Error Feed.**	**Unnorm.**	**Norm.**	**Error Feed.**
16	$1.563\ 10^{-3}$	$8.081\ 10^{-3}$	$1.555\ 10^{-3}$	$9.236\ 10^{-4}$	$2.043\ 10^{-3}$	$9.539\ 10^{-4}$
12	$1.545\ 10^{-3}$	$8.096\ 10^{-3}$	$1.567\ 10^{-3}$	$9.317\ 10^{-4}$	$2.201\ 10^{-3}$	$9.271\ 10^{-4}$
10	$1.587\ 10^{-3}$	$10.095\ 10^{-3}$	$1.603\ 10^{-3}$	$9.347\ 10^{-4}$	$4.550\ 10^{-3}$	$9.872\ 10^{-4}$

Example 7.2

The channel equalization example first described in subsection 3.6.3 is used in simulations using the lattice RLS algorithm with error feedback. The present example uses a 25th-order equalizer.

Solution:

Applying the error-feedback lattice RLS algorithm, using $\lambda = 0.99$ with a 25th-order equalizer, we obtain after 100 iterations the equalizer whose impulse response is shown in Fig. 7.4. The appropriate value of L for this case is 18. The algorithm is initialized with $\epsilon = 0.1$.

The convolution of this response with the channel impulse response is depicted in Fig. 7.5, which clearly approximates an impulse. In this case, the measured MSE was 0.3056, a value comparable with that obtained with the LMS algorithm in the example of subsection 3.6.3. Note that in the LMS case a 50th-order equalizer was used.

□

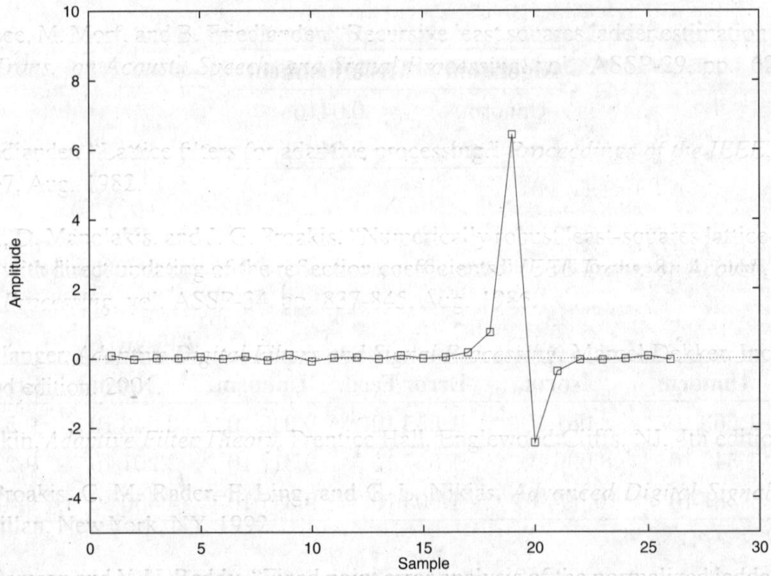

Figure 7.4 Equalizer impulse response, lattice RLS algorithm with error feedback.

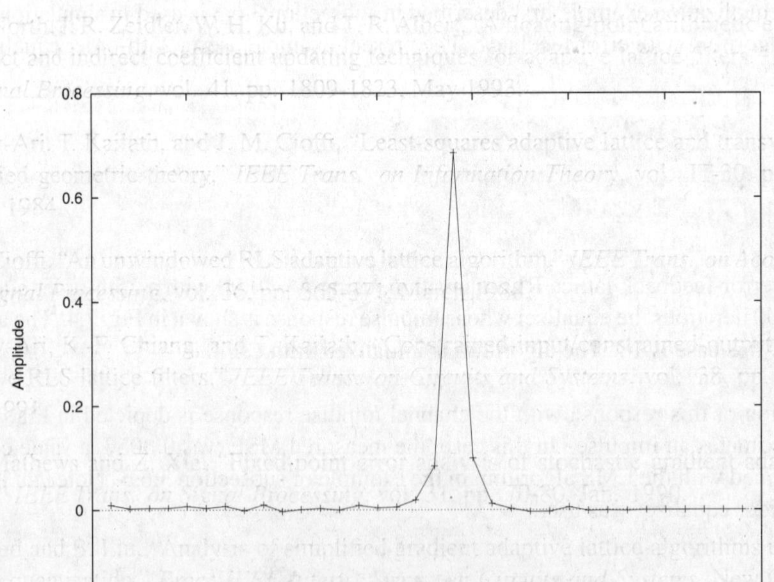

Figure 7.5 Convolution result, lattice RLS algorithm with error feedback.

7.11 CONCLUDING REMARKS

A number of alternative RLS algorithms based on the lattice realization were introduced. These algorithms consist of stages where growing-order forward and backward predictors of the input signal are built from stage to stage. This feature makes the lattice-based algorithms attractive in a number of applications where information about the statistics of the input signal, such as the order of the input signal model, is useful. Another important feature of the lattice-based algorithms is their robust performance when implemented in finite-precision arithmetic.

Also, their computational complexity of at least $16N$ multiplications per output sample is acceptable in a number of practical situations. However, by starting from the lattice formulation without making extensive use of order updating, it is possible to derive the fast transversal RLS algorithms, which can reduce the computational complexity to orders of $7N$ multiplications per output sample. The derivation of these algorithms is the subject of the Chapter 8.

Several interesting topics related to the lattice formulation of adaptive filters have been addressed in the open literature [9]-[13]. The geometric formulation of the least-squares estimation problem can be used to derive the lattice-based algorithms [9] in an elegant manner. Also, an important situation that we usually find in practice is the case where the input data cannot be considered zero before the first iteration of the adaptive algorithm. The derivation of the lattice algorithms that account for nonzero initial conditions for the input data is found in [10]. Another important problem is the characterization of the conditions under which the stability of the lattice algorithm is maintained when perturbations to the normal operation occur [11]. There is also a family of lattice-based algorithms employing gradient-type updating equations. These algorithms present reduced computational complexity and good behavior when implemented with finite-precision arithmetic [12]-[13].

A number of simulation examples involving the lattice algorithms were presented. In these examples the performance of the lattice algorithm was evaluated in different applications as well as in finite-precision implementations.

7.12 REFERENCES

1. D. L. Lee, M. Morf, and B. Friedlander, "Recursive least squares ladder estimation algorithms," *IEEE Trans. on Acoust., Speech, and Signal Processing*, vol. ASSP-29, pp. 627-641, June 1981.

2. B. Friedlander, "Lattice filters for adaptive processing," *Proceedings of the IEEE*, vol. 70, pp. 829-867, Aug. 1982.

3. F. Ling, D. Manolakis, and J. G. Proakis, "Numerically robust least-squares lattice-ladder algorithms with direct updating of the reflection coefficients," *IEEE Trans. on Acoust., Speech, and Signal Processing*, vol. ASSP-34, pp. 837-845, Aug. 1986.

4. M. Bellanger, *Adaptive Digital Filters and Signal Processing*, Marcel Dekker, Inc., New York, NY, 2nd edition, 2001.

5. S. Haykin, *Adaptive Filter Theory*, Prentice Hall, Englewood Cliffs, NJ, 4th edition, 2002.

6. J. G. Proakis, C. M. Rader, F. Ling, and C. L. Nikias, *Advanced Digital Signal Processing*, MacMillan, New York, NY, 1992.

7. C. G. Samson and V. U. Reddy, "Fixed point error analysis of the normalized ladder algorithm," *IEEE Trans. on Acoust., Speech, and Signal Processing*, vol. ASSP-31, pp. 1177-1191, Oct. 1983.

8. R. C. North, J. R. Zeidler, W. H. Ku, and T. R. Albert, "A floating-point arithmetic error analysis of direct and indirect coefficient updating techniques for adaptive lattice filters," *IEEE Trans. on Signal Processing*, vol. 41, pp. 1809-1823, May 1993.

9. H. Lev-Ari, T. Kailath, and J. M. Cioffi, "Least-squares adaptive lattice and transversal filters: A unified geometric theory," *IEEE Trans. on Information Theory*, vol. IT-30, pp. 222-236, March 1984.

10. J. M. Cioffi, "An unwindowed RLS adaptive lattice algorithm," *IEEE Trans. on Acoust., Speech, and Signal Processing*, vol. 36, pp. 365-371, March 1988.

11. H. Lev-Ari, K.-F. Chiang, and T. Kailath, "Constrained-input/constrained-output stability for adaptive RLS lattice filters," *IEEE Trans. on Circuits and Systems*, vol. 38, pp. 1478-1483, Dec. 1991.

12. V. J. Mathews and Z. Xie, "Fixed-point error analysis of stochastic gradient adaptive lattice filters," *IEEE Trans. on Signal Processing*, vol. 31, pp. 70-80, Jan. 1990.

13. M. Reed and B. Liu, "Analysis of simplified gradient adaptive lattice algorithms using power-of-two quantization," *Proc. IEEE Intern. Symp. on Circuits and Systems*, New Orleans, LA, pp. 792-795, May 1990.

7.13 PROBLEMS

1. Deduce the time-updating formula for the backward predictor coefficients.

2. Given a square matrix

$$\mathbf{P} = \begin{bmatrix} \mathbf{A} & \mathbf{B} \\ \mathbf{C} & \mathbf{D} \end{bmatrix}$$

where **A** and **D** are also square matrices, the inverse of **P** can be expressed as

$$\mathbf{P}^{-1} = \begin{bmatrix} \mathbf{A}^{-1}[\mathbf{I}+\mathbf{B}(\mathbf{D}-\mathbf{C}\mathbf{A}^{-1}\mathbf{B})^{-1}\mathbf{C}\mathbf{A}^{-1}] & -\mathbf{A}^{-1}\mathbf{B}(\mathbf{D}-\mathbf{C}\mathbf{A}^{-1}\mathbf{B})^{-1} \\ -(\mathbf{D}-\mathbf{C}\mathbf{A}^{-1}\mathbf{B})^{-1}\mathbf{C}\mathbf{A}^{-1} & (\mathbf{D}-\mathbf{C}\mathbf{A}^{-1}\mathbf{B})^{-1} \end{bmatrix}$$
$$= \begin{bmatrix} (\mathbf{A}-\mathbf{B}\mathbf{D}^{-1}\mathbf{C})^{-1} & -(\mathbf{A}-\mathbf{B}\mathbf{D}^{-1}\mathbf{C})^{-1}\mathbf{B}\mathbf{D}^{-1} \\ -\mathbf{D}^{-1}\mathbf{C}(\mathbf{A}-\mathbf{B}\mathbf{D}^{-1}\mathbf{C})^{-1} & \mathbf{D}^{-1}[\mathbf{I}+\mathbf{C}(\mathbf{A}-\mathbf{B}\mathbf{D}^{-1}\mathbf{C})^{-1}\mathbf{B}\mathbf{D}^{-1}] \end{bmatrix}$$

(a) Show the validity of this result.

(b) Use the appropriate partitioned forms of $\mathbf{R}_D(k-1,i)$ to derive the partitioned forms of $\mathbf{S}_D(k-1,i)$ of equations (7.56) and (7.59).

3. Derive the time-updating formula of $\delta_D(k,i)$.

4. Demonstrate that the backward *a posteriori* prediction errors $\varepsilon_b(k,i)$ and $\varepsilon_b(k,j)$ for $i \neq j$ are uncorrelated when the average is calculated over time.

5. Justify the initialization of $\xi^d_{b_{\min}}(0)$ and $\xi^d_{f_{\min}}(0)$ in the lattice RLS algorithm.

6. Derive the *a posteriori* lattice RLS algorithm for complex input signals.

7. Derive equation (7.71).

8. Derive the order-updating equation of the normalized forward and backward errors.

9. Demonstrate the validity of the order-updating formula of the weighted least-squares error of the joint-process estimation described in equation (7.88).

10. Derive equation (7.89).

11. Derive the error-feedback LRLS algorithm based on *a priori* errors.

12. Derive the normalized LRLS algorithm based on *a priori* errors.

13. The lattice RLS algorithm based on *a posteriori* errors is used to predict the signal $x(k) = \sin\frac{\pi k}{4}$. Given $\lambda = 0.99$, calculate the error and the tap coefficients for the first 10 iterations.

14. The normalized lattice RLS algorithm based on *a posteriori* errors is used to predict the signal $x(k) = \sin\frac{\pi k}{4}$. Given $\lambda = 0.99$, calculate the error and the multiplier coefficients for the first 10 iterations.

15. The error-feedback LRLS algorithm is applied to identify a 7th-order time-varying unknown system whose coefficients are first-order Markov processes with $\lambda_{\mathbf{w}} = 0.999$ and $\sigma_{\mathbf{w}}^2 = 0.033$. The initial time-varying system multiplier coefficients are

$$\mathbf{w}_o^T = [0.03490 \ -0.01100 \ -0.06864 \ 0.22391 \ 0.55686 \ 0.35798 \ -0.02390 \ -0.07594]$$

The input signal is Gaussian white noise with variance $\sigma_x^2 = 1$ and the measurement noise is also Gaussian white noise independent of the input signal and of the elements of $\mathbf{n_w}(k)$ with variance $\sigma_n^2 = 0.01$.

Simulate the experiment above described and measure the excess MSE for $\lambda = 0.97$ and $\lambda = 0.99$.

16. Repeat the experiment described in problem 15 using the normalized lattice algorithm.

17. Suppose that a 15th-order FIR digital filter with the multiplier coefficients given below is identified through an adaptive FIR filter of the same order using the unnormalized LRLS algorithm. Considering that fixed-point arithmetic is used, simulate the identification problem described using the following specifications:

Additional noise : white noise with variance	$\sigma_n^2 = 0.0015$
Coefficients wordlength:	$b_c = 16$ bits
Signal wordlength:	$b_d = 16$ bits
Input signal: Gaussian white noise with variance	$\sigma_x^2 = 0.7$
	$\lambda = 0.98$

$$\mathbf{w}_o^T = [0.0219360 \ 0.0015786 \ -0.0602449 \ -0.0118907 \ 0.1375379 \ 0.0574545 \ -0.3216703 \ -0.5287203 \ -0.2957797 \ 0.0002043 \ 0.290670 \ -0.0353349 \ -0.0068210 \ 0.0026067 \ 0.0010333 \ -0.0143593]$$

Plot the learning curves for the finite- and infinite-precision implementations. Also plot $E[||\Delta\kappa_f(k,0)||^2]$ and $E[||\Delta\kappa_b(k,0)||^2]$ versus k in both cases.

18. Repeat the above problem for the following cases:
(a) $\sigma_n^2 = 0.01, b_c = 9$ bits, $b_d = 9$ bits, $\sigma_x^2 = 0.7, \lambda = 0.98$.
(b) $\sigma_n^2 = 0.1, b_c = 10$ bits, $b_d = 10$ bits, $\sigma_x^2 = 0.8, \lambda = 0.98$.
(c) $\sigma_n^2 = 0.05, b_c = 8$ bits, $b_d = 16$ bits, $\sigma_x^2 = 0.8, \lambda = 0.98$.

19. In problem 17, rerun the simulations for $\lambda = 1, \lambda = 0.940$. Comment on the results.

20. Repeat problem 18, using the normalized and error-feedback LRLS algorithms. Compare the results for the different algorithms.

21. Repeat problem 17 for the case where the input signal is a first-order Markov process with $\lambda_{\mathbf{x}} = 0.98$.

22. Given a channel with impulse response

$$h(k) = 0.9^k + 0.4^k$$

for $k = 0, 1, 2, \ldots, 25$, design an adaptive equalizer. The input signal is white noise with unit variance and the adaptive-filter input signal-to-noise ratio is 30 dB. Use the unnormalized lattice algorithm of order 35.

23. The unnormalized lattice algorithm is used to perform the forward prediction of a signal $x(k)$ generated by applying zero-mean Gaussian white noise signal with unit variance to the input of a linear filter with transfer function given by

$$H(z) = \frac{0.5}{(1 - 1.512z^{-1} + 0.827z^{-2})(1 - 1.8z^{-1} + 0.87z^{-2})}$$

Calculate the zeros of the resulting predictor error transfer function and compare with the poles of the linear filter.

24. Determine the computational complexity of the Algorithms 7.1, 7.2, 7.3, and 7.4.

8

FAST TRANSVERSAL RLS ALGORITHMS

8.1 INTRODUCTION

Among the large number of algorithms that solve the least-squares problem in a recursive form, the fast transversal recursive least-squares (FTRLS) algorithms are very attractive due to their reduced computational complexity [1]-[7].

The FTRLS algorithms can be derived by solving simultaneously the forward and backward linear prediction problems, along with two other transversal filters: the joint-process estimator and an auxiliary filter whose desired signal vector has one as its first and unique nonzero element (i.e., $d(0) = 1$). Unlike the lattice-based algorithms, the FTRLS algorithms require only time-recursive equations. However, a number of relations required to derive some of the FTRLS algorithms can be taken from the previous chapter on LRLS algorithms. The FTRLS algorithm can also be considered a fast version of an algorithm to update the transversal filter for the solution of the RLS problem, since a fixed-order update for the transversal adaptive filter coefficient vector is computed at each iteration.

The relations derived for the backward and forward prediction in the lattice-based algorithms can be used to derive the FTRLS algorithms. The resulting algorithms have computational complexity of order N making them especially attractive for practical implementation. When compared to the lattice-based algorithms, the computational complexity of the FTRLS algorithms is lower due to the absence of order-updating equations. In particular, FTRLS algorithms typically require $7N$ to $11N$ multiplications and divisions per output sample, as compared to $14N$ to $29N$ for the LRLS algorithms. Therefore, FTRLS algorithms are considered the fastest implementation solutions of the RLS problem [1]-[7].

Several alternative FTRLS algorithms have been proposed in the literature. The so-called fast Kalman algorithm [1], which is certainly one of the earlier fast transversal RLS algorithms, has computational complexity of $11N$ multiplications and divisions per output sample. In a later stage of research development in the area of fast transversal algorithms, the fast *a posteriori* error sequential technique (FAEST) [2], and the fast transversal filter (FTF) [3] algorithms were proposed, both requiring an order of $7N$ multiplications and divisions per output sample. The FAEST and FTF algorithms have

P.S.R. Diniz, *Adaptive Filtering*, DOI: 10.1007/978-0-387-68606-6_8,

the lowest complexity known for RLS algorithms, and are useful for problems where the input vector elements consist of delayed versions of a single input signal. Unfortunately, these algorithms are very sensitive to quantization effects and become unstable if certain actions are not taken [5]-[7], [9].

In this chapter, a particular form of the FTRLS algorithm is presented, where most of the derivations are based on those presented for the lattice algorithms. It is well known that the quantization errors in the FTRLS algorithms present exponential divergence [1]-[7]. Since the FTRLS algorithms have unstable behavior when implemented with finite-precision arithmetic, we discuss the implementation of numerically stable FTRLS algorithms, and provide the description of a particular algorithm [8]-[10].

8.2 RECURSIVE LEAST-SQUARES PREDICTION

All fast algorithms explore some structural property of the information data in order to achieve low computational complexity. In the particular case of the fast RLS algorithms discussed in this text, the reduction in the computational complexity is achieved for the cases where the input signal consists of consecutively delayed samples of the same signal. In this case, the patterns of the fast algorithms are similar in the sense that the forward and backward prediction filters are essential parts of these algorithms. The predictors perform the task of modeling the input signal, which as a result allows the replacement of matrix equations by vector and scalar relations.

In the derivation of the FTRLS algorithms, the solutions of the RLS forward and backward prediction problems are required in the time-update equations. In this section, these solutions are reviewed with emphasis on the results that are relevant to the FTRLS algorithms. As previously mentioned, we will borrow a number of derivations from the previous chapter on lattice algorithms. It is worth mentioning that the FTRLS could be introduced through an independent derivation, however the derivation based on the lattice is probably more insightful and certainly more straightforward at this point.

8.2.1 Forward Prediction Relations

The instantaneous *a posteriori* forward prediction error for an Nth-order predictor is given by

$$\begin{aligned}\varepsilon_f(k,N) &= x(k) - \mathbf{w}_f^T(k,N)\mathbf{x}(k-1,N)\\ &= \mathbf{x}^T(k,N+1)\begin{bmatrix}1\\ -\mathbf{w}_f(k,N)\end{bmatrix}\end{aligned} \tag{8.1}$$

The relationship between *a posteriori* and *a priori* forward prediction error, first presented in equation (7.49) and repeated here for convenience, is given by

$$e_f(k,N) = \frac{\varepsilon_f(k,N)}{\gamma(k-1,N)} \tag{8.2}$$

A simple manipulation of equation (7.73), leads to the following relation for the time updating of the minimum weighted least-squares error, which will be used in the FTRLS algorithm:

$$\xi^d_{f_{\min}}(k, N) = \lambda \xi^d_{f_{\min}}(k-1, N) + e_f(k, N)\varepsilon_f(k, N) \tag{8.3}$$

From the same equation (7.73), we can obtain the following equality that will also be required in the FTRLS algorithm:

$$\gamma(k, N+1) = \frac{\lambda \xi^d_{f_{\min}}(k-1, N)}{\xi^d_{f_{\min}}(k, N)} \gamma(k-1, N) \tag{8.4}$$

The updating equation of the forward prediction tap-coefficient vector can be performed through equation (7.40) of the previous chapter, i.e.,

$$\mathbf{w}_f(k, N) = \mathbf{w}_f(k-1, N) + \boldsymbol{\phi}(k-1, N) e_f(k, N) \tag{8.5}$$

where $\boldsymbol{\phi}(k-1, N) = \mathbf{S}_D(k-1, N)\mathbf{x}(k-1, N)$.

As will be seen, the updating of vector $\boldsymbol{\phi}(k-1, N)$ to $\boldsymbol{\phi}(k, N+1)$ is needed to update the backward predictor coefficient vector. Also, the last element of $\boldsymbol{\phi}(k, N+1)$ is used to update the backward prediction *a priori* error and to obtain $\gamma(k, N)$. Vector $\boldsymbol{\phi}(k, N+1)$ can be obtained by post-multiplying both sides of equation (7.56), at instant k and for order N, by $\mathbf{x}(k, N+1) = [x(k)\mathbf{x}^T(k-1, N)]^T$. The result can be expressed as

$$\boldsymbol{\phi}(k, N+1) = \begin{bmatrix} 0 \\ \boldsymbol{\phi}(k-1, N) \end{bmatrix} + \frac{1}{\xi^d_{f_{\min}}(k, N)} \begin{bmatrix} 1 \\ -\mathbf{w}_f(k, N) \end{bmatrix} \varepsilon_f(k, N) \tag{8.6}$$

However, it is not convenient to use the above equation in the FTRLS algorithm because when deriving the backward prediction part, it would lead to extra computation. The solution is to use an alternative recursion involving $\hat{\boldsymbol{\phi}}(k, N+1) = \frac{\boldsymbol{\phi}(k,N+1)}{\gamma(k,N+1)}$ instead of $\boldsymbol{\phi}(k, N+1)$ (see problem 7 for further details). The resulting recursion can be derived after some algebraic manipulations of equations (8.6) and (8.3) to (8.5), leading to

$$\hat{\boldsymbol{\phi}}(k, N+1) = \begin{bmatrix} 0 \\ \hat{\boldsymbol{\phi}}(k-1, N) \end{bmatrix} + \frac{1}{\lambda \xi^d_{f_{\min}}(k-1, N)} \begin{bmatrix} 1 \\ -\mathbf{w}_f(k-1, N) \end{bmatrix} e_f(k, N) \tag{8.7}$$

The forward prediction tap-coefficient vector should then be updated using $\hat{\boldsymbol{\phi}}(k-1, N)$ as

$$\mathbf{w}_f(k, N) = \mathbf{w}_f(k-1, N) + \hat{\boldsymbol{\phi}}(k-1, N)\varepsilon_f(k, N) \tag{8.8}$$

8.2.2 Backward Prediction Relations

In this subsection, the relations involving the backward prediction problem that are used in the FTRLS algorithm are derived.

The relationship between *a posteriori* and *a priori* backward prediction errors can be expressed as

$$\varepsilon_b(k, N) = e_b(k, N)\gamma(k, N) \tag{8.9}$$

It is also known that the ratio of conversion factors for different orders is given by

$$\frac{\gamma(k, N+1)}{\gamma(k, N)} = \frac{\lambda \xi^d_{b_{\min}}(k-1, N)}{\xi^d_{b_{\min}}(k, N)} \tag{8.10}$$

see equation (7.79) of the previous chapter.

We rewrite for convenience the last equality of equation (7.70), i.e.,

$$\xi^d_{b_{\min}}(k, N) = \lambda \xi^d_{b_{\min}}(k-1, N) + \frac{\varepsilon_b^2(k, N)}{\gamma(k, N)} \tag{8.11}$$

This equation can be rewritten as

$$1 + \frac{\varepsilon_b^2(k, N)}{\lambda \gamma(k, N) \xi^d_{b_{\min}}(k-1, N)} = \frac{\xi^d_{b_{\min}}(k, N)}{\lambda \xi^d_{b_{\min}}(k-1, N)} \tag{8.12}$$

Now we recall that the time updating for the backward predictor filter is given by

$$\begin{aligned}\mathbf{w}_b(k, N) &= \mathbf{w}_b(k-1, N) + \boldsymbol{\phi}(k, N) e_b(k, N)\\ &= \mathbf{w}_b(k-1, N) + \hat{\boldsymbol{\phi}}(k, N) \varepsilon_b(k, N)\end{aligned} \tag{8.13}$$

Following a similar approach to that used to derive equation (8.7), by first post-multiplying both sides of equation (7.59), at instant k and for order N, by $\mathbf{x}(k, N+1) = [\mathbf{x}^T(k, N)\, x(k-N)]^T$, and using relations (8.10), (8.11), and (8.13), we have

$$\begin{bmatrix} \hat{\boldsymbol{\phi}}(k, N) \\ 0 \end{bmatrix} = \hat{\boldsymbol{\phi}}(k, N+1) - \frac{1}{\lambda \xi^d_{b_{\min}}(k-1, N)} \begin{bmatrix} -\mathbf{w}_b(k-1, N) \\ 1 \end{bmatrix} e_b(k, N) \tag{8.14}$$

Note that in this equation the last element of $\hat{\boldsymbol{\phi}}(k, N+1)$ was already calculated in equation (8.7). In any case, it is worth mentioning that the last element of $\hat{\boldsymbol{\phi}}(k, N+1)$ can alternatively be expressed as

$$\hat{\phi}_{N+1}(k, N+1) = \frac{e_b(k, N)}{\lambda \xi^d_{b_{\min}}(k-1, N)} \tag{8.15}$$

By applying equations (8.9), (8.15), and (8.10) in equation (8.12), we can show that

$$1 + \hat{\phi}_{N+1}(k, N+1) \varepsilon_b(k, N) = \frac{\gamma(k, N)}{\gamma(k, N+1)} \tag{8.16}$$

By substituting equation (8.9) into the above equation, we can now derive an updating equation that can be used in the FTRLS algorithm as

$$\gamma^{-1}(k,N) = \gamma^{-1}(k,N+1) - \hat{\phi}_{N+1}(k,N+1)e_b(k,N) \tag{8.17}$$

The updating equations related to the forward and backward prediction problems and for the conversion factor $\gamma(k,N)$ are now available. We can thus proceed with the derivations to solve the more general problem of estimating a related process represented by the desired signal $d(k)$, known as joint-process estimation.

8.3 JOINT-PROCESS ESTIMATION

As for all previously presented adaptive-filter algorithms, it is useful to derive a FTRLS algorithm that can match a desired signal $d(k)$ through the minimization of the weighted squared error. Starting with the *a priori* error

$$e(k,N) = d(k) - \mathbf{w}^T(k-1,N)\mathbf{x}(k,N) \tag{8.18}$$

we can calculate the *a posteriori* error as

$$\varepsilon(k,N) = e(k,N)\gamma(k,N) \tag{8.19}$$

As in the conventional RLS algorithm, the time updating for the output tap coefficients of the joint-process estimator can be performed as

$$\begin{aligned}\mathbf{w}(k,N) &= \mathbf{w}(k-1,N) + \boldsymbol{\phi}(k,N)e(k,N)\\ &= \mathbf{w}(k-1,N) + \hat{\boldsymbol{\phi}}(k,N)\varepsilon(k,N)\end{aligned} \tag{8.20}$$

All the updating equations are now available to describe the fast transversal RLS algorithm. The FRLS algorithm consists of equations (8.1)-(8.3), (8.7)-(8.8), and (8.4) related to the forward predictor; equations (8.15), (8.17), (8.9), (8.11), (8.14), and (8.13) related to the backward predictor and the conversion factor; and (8.18)-(8.20) related to the joint-process estimator. The FTRLS algorithm is in step-by-step form as Algorithm 8.1. The computational complexity of the FTRLS algorithm is $7(N)+14$ multiplications per output sample. The key feature of the FTRLS algorithm is that it does not require matrix multiplications. Because of this, the implementation of the FTRLS algorithm has complexity of order N multiplications per output sample.

The initialization procedure consists of setting the tap coefficients of the backward prediction, forward prediction, and joint-process estimation filters to zero, namely

$$\mathbf{w}_f(-1,N) = \mathbf{w}_b(-1,N) = \mathbf{w}(-1,N) = \mathbf{0} \tag{8.21}$$

Vector $\hat{\boldsymbol{\phi}}(-1,N)$ is set to $\mathbf{0}$ assuming that the input and desired signals are zero for $k<0$, i.e., prewindowed data. The conversion factor should be initialized as

$$\gamma(-1,N) = 1 \tag{8.22}$$

Algorithm 8.1

Fast Transversal RLS Algorithm

Initialization

$\mathbf{w}_f(-1,N) = \mathbf{w}_b(-1,N) = \mathbf{w}(-1,N) = \mathbf{0}$
$\hat{\boldsymbol{\phi}}(-1,N) = \mathbf{0},\ \gamma(-1,N) = 1$
$\xi^d_{b_{\min}}(-1,N) = \xi^d_{f_{\min}}(-1,N) = \epsilon$ (a small positive constant)

Prediction Part

Do for each $k \geq 0$,

$$e_f(k,N) = \mathbf{x}^T(k,N+1)\begin{bmatrix} 1 \\ -\mathbf{w}_f(k-1,N) \end{bmatrix}$$
$$\varepsilon_f(k,N) = e_f(k,N)\gamma(k-1,N) \quad (8.2)$$
$$\xi^d_{f_{\min}}(k,N) = \lambda\xi^d_{f_{\min}}(k-1,N) + e_f(k,N)\varepsilon_f(k,N) \quad (8.3)$$
$$\mathbf{w}_f(k,N) = \mathbf{w}_f(k-1,N) + \hat{\boldsymbol{\phi}}(k-1,N)\varepsilon_f(k,N) \quad (8.8)$$
$$\hat{\boldsymbol{\phi}}(k,N+1) = \begin{bmatrix} 0 \\ \hat{\boldsymbol{\phi}}(k-1,N) \end{bmatrix} + \frac{1}{\lambda\xi^d_{f_{\min}}(k-1,N)}\begin{bmatrix} 1 \\ -\mathbf{w}_f(k-1,N) \end{bmatrix} e_f(k,N) \quad (8.7)$$
$$\gamma(k,N+1) = \frac{\lambda\xi^d_{f_{\min}}(k-1,N)}{\xi^d_{f_{\min}}(k,N)}\gamma(k-1,N) \quad (8.4)$$
$$e_b(k,N) = \lambda\xi^d_{b_{\min}}(k-1,N)\hat{\phi}_{N+1}(k,N+1) \quad (8.15)$$
$$\gamma^{-1}(k,N) = \gamma^{-1}(k,N+1) - \hat{\phi}_{N+1}(k,N+1)e_b(k,N) \quad (8.17)$$
$$\varepsilon_b(k,N) = e_b(k,N)\gamma(k,N) \quad (8.9)$$
$$\xi^d_{b_{\min}}(k,N) = \lambda\xi^d_{b_{\min}}(k-1,N) + \varepsilon_b(k,N)e_b(k,N) \quad (8.11)$$
$$\begin{bmatrix} \hat{\boldsymbol{\phi}}(k,N) \\ 0 \end{bmatrix} = \hat{\boldsymbol{\phi}}(k,N+1) - \hat{\phi}_{N+1}(k,N+1)\begin{bmatrix} -\mathbf{w}_b(k-1,N) \\ 1 \end{bmatrix} \quad (8.14)$$
$$\mathbf{w}_b(k,N) = \mathbf{w}_b(k-1,N) + \hat{\boldsymbol{\phi}}(k,N)\varepsilon_b(k,N) \quad (8.13)$$

Joint-Process Estimation

$$e(k,N) = d(k) - \mathbf{w}^T(k-1,N)\mathbf{x}(k,N) \quad (8.18)$$
$$\varepsilon(k,N) = e(k,N)\gamma(k,N) \quad (8.19)$$
$$\mathbf{w}(k,N) = \mathbf{w}(k-1,N) + \hat{\boldsymbol{\phi}}(k,N)\varepsilon(k,N) \quad (8.20)$$

End

since no difference between *a priori* and *a posteriori* errors exists during the initialization period. The weighted least-square errors should be initialized with a positive constant ϵ

$$\epsilon = \xi^d_{f_{\min}}(-1,N) = \xi^d_{b_{\min}}(-1,N) \quad (8.23)$$

in order to avoid division by zero in the first iteration. The reason for introducing this initialization parameter suggests that it should be a small value. However, for stability reasons, the value of ϵ should not be small (see the examples at the end of this chapter).

It should be mentioned that there are exact initialization procedures for the fast transversal RLS filters with the aim of minimizing the objective function at all instants during the initialization period [3].

These procedures explore the fact that during the initialization period the number of data samples in both $d(k)$ and $x(k)$ is less than $N + 1$. Therefore the objective function can be made zero since there are more parameters than needed. The exact initialization procedure of [3] replaces the computationally intensive backsubstitution algorithm and is rather simple when the adaptive-filter coefficients are initialized with zero. The procedure can also be generalized to the case where some nonzero initial values for the tap coefficients are available.

As previously mentioned, several fast RLS algorithms based on the transversal realization exist; the one presented here corresponds to the so-called FTF proposed in [3]. A number of alternative algorithms are introduced in the problems.

8.4 STABILIZED FAST TRANSVERSAL RLS ALGORITHM

Although the fast transversal algorithms proposed in the literature provide a nice solution to the computational complexity burden inherent to the conventional RLS algorithm, these algorithms are unstable when implemented with finite-precision arithmetic. Increasing the wordlength does not solve the instability problem. The only effect of employing a longer wordlength is that the algorithm will take longer to diverge. Earlier solutions to this problem consisted of restarting the algorithm when the accumulated errors in chosen variables reached prescribed thresholds [3]. Although the restart procedure would use past information, the resulting performance is suboptimal due to the discontinuity of information in the corresponding deterministic correlation matrix.

The cause for the unstable behavior of the fast transversal algorithms is the inherent positive feedback mechanism. This explanation led to the idea that if some specific measurements of the numerical errors were available, they could conveniently be fed back in order to make the negative feedback dominant in the error propagation dynamics. Fortunately, some measurements of the numerical errors can be obtained by introducing computational redundancy into the fast algorithm. Such a computational redundancy would involve calculating a given quantity using two different formulas. In finite-precision implementation, the resulting values for the quantity calculated by these formulas are not equal and their difference is a good measurement of the accumulated errors in that quantity. This error can then be fed back in an attempt to stabilize the algorithm. The key problem is to determine the quantities where the computational redundancy should be introduced such that the error propagation dynamics can be stabilized. In the early proposed solutions [6]-[7], only a single quantity was chosen to introduce the redundancy. Later, it was shown that at least two quantities are required in order to guarantee the stability of the FTRLS algorithm [9]. Another relevant question is where should the error be fed back inside the algorithm. Note that any point could be chosen without affecting the behavior of the algorithm when implemented with infinite precision, since the feedback error is zero in this case. A natural choice is to feed the error back into the expressions of the quantities that are related to it. That means for each quantity in which redundancy is introduced, its final value is a combination of the two forms of computing it.

The FTRLS algorithm can be seen as a discrete-time nonlinear dynamic system [9]: when finite precision is used in the implementation, quantization errors will rise. In this case, the internal quantities will be perturbed when compared with the infinite-precision quantities. When modeling the error propagation, a nonlinear system can be described that, if properly linearized, allows the study of the error propagation mechanism. Using an averaging analysis, which is meaningful for stationary input signals, it is possible to obtain a system characterized by its set of eigenvalues whose dynamic behavior is similar to that of the error propagation behavior when $k \to \infty$ and $(1-\lambda) \to 0$. Through these eigenvalues, it is possible to determine the feedback parameters as well as the quantities to choose for the introduction of redundancy. The objective here is to modify the unstable modes through the error feedback in order to make them stable [9]. Fortunately, it was found in [9] that the unstable modes can be modified and stabilized by the introduced error feedback. The unstable modes can be modified by introducing redundancy in $\gamma(k,N)$ and $e_b(k,N)$. These quantities can be calculated using different relations and in order to distinguish them an extra index is included in their description.

The *a priori* backward error can be described in a number of alternative forms such as

$$e_b(k,N,1) = \lambda\xi^d_{b_{\min}}(k-1,N)\hat{\phi}_{N+1}(k,N+1) \tag{8.24}$$

$$e_b(k,N,2) = \left[-\mathbf{w}_b^T(k-1,N)\ 1\right]\mathbf{x}(k,N+1) \tag{8.25}$$

and

$$\begin{aligned} e_{b,i}(k,N,3) &= e_b(k,N,2)\kappa_i + e_b(k,N,1)[1-\kappa_i] \\ &= e_b(k,N,1) + \kappa_i[e_b(k,N,2) - e_b(k,N,1)] \end{aligned} \tag{8.26}$$

where the first form was employed in the FTRLS algorithm and the second form corresponds to the inner product implementation of the *a priori* backward error. The third form corresponds to a linear combination of the first two forms where the numerical difference between these forms is fed back to determine the final value of $e_{b,i}(k,N,3)$ which will be used at different places in the stabilized algorithm. For each κ_i, $i = 1,2,3$, we choose a different value in order to guarantee that the related eigenvalues are less than one.

The conversion factor $\gamma(k,N)$ is probably the first parameter to show signs that the algorithm is becoming unstable. This parameter can also be calculated through different relations. These alternative relations are required to guarantee that all modes of the error propagation system become stable. The first equation is given by

$$\begin{aligned} \gamma^{-1}(k,N+1,1) &= \gamma^{-1}(k-1,N,3)\frac{\xi^d_{f_{\min}}(k,N)}{\lambda\xi^d_{f_{\min}}(k-1,N)} \\ &= \gamma^{-1}(k-1,N,3)\left[1+\frac{e_f(k,N)\varepsilon_f(k,N)}{\lambda\xi^d_{f_{\min}}(k-1,N)}\right] \\ &= \gamma^{-1}(k-1,N,3) + \frac{e_f^2(k,N)}{\lambda\xi^d_{f_{\min}}(k-1,N)} \\ &= \gamma^{-1}(k-1,N,3) + \hat{\phi}_0(k,N+1)e_f(k,N) \end{aligned} \tag{8.27}$$

where $\hat{\phi}_0(k,N+1)$ is the first element of $\hat{\phi}(k,N+1)$. The above equalities are derived from equations (8.4), (8.3), (8.2) and (8.7), respectively. The second expression for the conversion factor is derived from equation (8.14) and given by

$$\gamma^{-1}(k,N,2) = \gamma^{-1}(k,N+1,1) - \hat{\phi}_{N+1}(k,N+1)e_{b,3}(k,N,3) \tag{8.28}$$

The third expression is

$$\gamma^{-1}(k,N,3) = 1 + \hat{\boldsymbol{\phi}}^T(k,N)\mathbf{x}(k,N) \tag{8.29}$$

In equation (8.27), the conversion factor was expressed in different ways, one of which was first presented in the FTRLS algorithm of [9]. The second form already uses an *a priori* backward error with redundancy. The third form can be derived from equation (7.48) for the lattice RLS algorithms (see problem 10).

An alternative relation utilized in the stabilized fast transversal algorithm involves the minimum forward least-squares error. From equations (8.3) and (8.7), we can write

$$\begin{aligned}[\xi^d_{f_{\min}}(k,N)]^{-1} &= \lambda^{-1}[\xi^d_{f_{\min}}(k-1,N)]^{-1} - \frac{e_f(k,N)\varepsilon_f(k,N)}{\lambda\xi^d_{f_{\min}}(k-1,N)\xi^d_{f_{\min}}(k,N)} \\ &= \lambda^{-1}[\xi^d_{f_{\min}}(k-1,N)]^{-1} - \frac{\hat{\phi}_0(k,N)\varepsilon_f(k,N)}{\xi^d_{f_{\min}}(k,N)}\end{aligned}$$

From (8.6), we can deduce that

$$\frac{\varepsilon_f(k,N)}{\xi^d_{f_{\min}}(k,N)} = \phi_0(k,N) = \hat{\phi}_0(k,N)\gamma(k,N+1,1)$$

With this relation, we can obtain the desired equation as

$$[\xi^d_{f_{\min}}(k,N)]^{-1} = \lambda^{-1}[\xi^d_{f_{\min}}(k-1,N)]^{-1} - \gamma(k,N+1,1)\hat{\phi}_0^2(k,N+1) \tag{8.30}$$

where the choice of $\gamma(k,N+1,1)$ is used to keep the error-system modes stable [9].

Using the equations for the conversion factor and for the *a priori* backward error with redundancy, we can obtain the stabilized fast transversal RLS algorithm (SFTRLS) whose step-by-step implementation is given as Algorithm 8.2. The parameters κ_i for $i = 1,2,3$ were determined through computer simulation search [9] where the optimal values found were $\kappa_1 = 1.5$, $\kappa_2 = 2.5$, and $\kappa_3 = 1$. It was also found in [9] that the numerical behavior is quite insensitive to values of κ_i around the optimal and that optimal values chosen for a given situation work well for a wide range of environments and algorithm setup situations (for example, for different choices of the forgetting factor).

Another issue related to the SFTRLS algorithm concerns the range of values for λ such that stability is guaranteed. Results of extensive simulation experiments [9] indicate that the range is

$$1 - \frac{1}{2(N+1)} \leq \lambda < 1 \tag{8.31}$$

where N is the order of the adaptive filter. It was also verified that the optimal numerical behavior is achieved when the value of λ is chosen as

$$\lambda = 1 - \frac{0.4}{N+1} \tag{8.32}$$

Algorithm 8.2

Stabilized Fast Transversal RLS Algorithm

Initialization

$$\mathbf{w}_f(-1,N) = \mathbf{w}_b(-1,N) = \mathbf{w}(-1,N) = \mathbf{0}$$
$$\hat{\boldsymbol{\phi}}(-1,N) = \mathbf{0},\ \gamma(-1,N,3) = 1$$
$$\xi^d_{b_{\min}}(-1,N) = \xi^d_{f_{\min}}(-1,N) = \epsilon \text{ (a small positive constant)}$$
$$\kappa_1 = 1.5, \kappa_2 = 2.5, \kappa_3 = 1$$

Prediction Part

Do for each $k \geq 0$,

$$e_f(k,N) = \mathbf{x}^T(k,N+1)\begin{bmatrix} 1 \\ -\mathbf{w}_f(k-1,N)\end{bmatrix}$$
$$\varepsilon_f(k,N) = e_f(k,N)\gamma(k-1,N,3) \quad (8.2)$$
$$\hat{\boldsymbol{\phi}}(k,N+1) = \begin{bmatrix} 0 \\ \hat{\boldsymbol{\phi}}(k-1,N)\end{bmatrix} + \frac{1}{\lambda\xi^d_{f_{\min}}(k-1,N)}\begin{bmatrix} 1 \\ -\mathbf{w}_f(k-1,N)\end{bmatrix} e_f(k,N) \quad (8.7)$$
$$\gamma^{-1}(k,N+1,1) = \gamma^{-1}(k-1,N,3) + \hat{\phi}_0(k,N+1)e_f(k,N) \quad (8.27)$$
$$[\xi^d_{f_{\min}}(k,N)]^{-1} = \lambda^{-1}[\xi^d_{f_{\min}}(k-1,N)]^{-1} - \gamma(k,N+1,1)\hat{\phi}^2_0(k,N+1) \quad (8.30)$$
$$\mathbf{w}_f(k,N) = \mathbf{w}_f(k-1,N) + \hat{\boldsymbol{\phi}}(k-1,N)\varepsilon_f(k,N) \quad (8.8)$$
$$e_b(k,N,1) = \lambda\xi^d_{b_{\min}}(k-1,N)\hat{\phi}_{N+1}(k,N+1) \quad (8.15)$$
$$e_b(k,N,2) = \left[-\mathbf{w}^T_b(k-1,N)\ \ 1\right]\mathbf{x}(k,N+1) \quad (8.25)$$
$$e_{b,i}(k,N,3) = e_b(k,N,2)\kappa_i + e_b(k,N,1)[1-\kappa_i] \ \ for\ i = 1,2,3 \quad (8.25)$$
$$\gamma^{-1}(k,N,2) = \gamma^{-1}(k,N+1,1) - \hat{\phi}_{N+1}(k,N+1)e_{b,3}(k,N,3) \quad (8.28)$$
$$\varepsilon_{b,j}(k,N,3) = e_{b,j}(k,N,3)\gamma(k,N,2)\ \ j = 1,2$$
$$\xi^d_{b_{\min}}(k,N) = \lambda\xi^d_{b_{\min}}(k-1,N) + \varepsilon_{b,2}(k,N,3)e_{b,2}(k,N,3) \quad (8.11)$$
$$\begin{bmatrix} \hat{\boldsymbol{\phi}}(k,N) \\ 0\end{bmatrix} = \hat{\boldsymbol{\phi}}(k,N+1) - \hat{\phi}_{N+1}(k,N+1)\begin{bmatrix} -\mathbf{w}_b(k-1,N) \\ 1\end{bmatrix} \quad (8.14)$$
$$\mathbf{w}_b(k,N) = \mathbf{w}_b(k-1,N) + \hat{\boldsymbol{\phi}}(k,N)\varepsilon_{b,1}(k,N,3) \quad (8.13)$$
$$\gamma^{-1}(k,N,3) = 1 + \hat{\boldsymbol{\phi}}^T(k,N)\mathbf{x}(k,N) \quad (8.29)$$

Joint-Process Estimation

$$e(k,N) = d(k) - \mathbf{w}^T(k-1,N)\mathbf{x}(k,N) \quad (8.18)$$
$$\varepsilon(k,N) = e(k,N)\gamma(k,N,3) \quad (8.19)$$
$$\mathbf{w}(k,N) = \mathbf{w}(k-1,N) + \hat{\boldsymbol{\phi}}(k,N)\varepsilon(k,N) \quad (8.20)$$

End

The range of values for λ as well as its optimal value can be very close to one for high-order filters. This can be a potential limitation for the use of the SFTRLS algorithm, especially in nonstationary environments where smaller values for λ are required.

The computational complexity of the SFTRLS algorithm is of order $9N$ multiplications per output sample. There is an alternative algorithm with computational complexity of order $8N$ (see problem 9).

Before leaving this section, it is worth mentioning a nice interpretation for the fast transversal RLS algorithm. The FTRLS algorithm can be viewed as four transversal filters working in parallel and exchanging quantities with each other, as depicted in Fig. 8.1. The first filter is the forward prediction filter that utilizes $\mathbf{x}(k-1,N)$ as the input signal vector, $\mathbf{w}_f(k,N)$ as the coefficient vector, and provides quantities $\varepsilon_f(k,N), e_f(k,N)$, and $\xi^d_{f_{\min}}(k,N)$ as outputs. The second filter is the backward prediction filter that utilizes $\mathbf{x}(k,N)$ as the input signal vector, $\mathbf{w}_b(k,N)$ as the coefficient vector, and provides quantities $\varepsilon_b(k,N), e_b(k,N)$, and $\xi^d_{b_{\min}}(k,N)$ as outputs. The third filter is an auxiliary filter whose coefficients are given by $-\hat{\phi}(k,N)$, whose input signal vector is $\mathbf{x}(k,N)$, and whose output parameter is $\gamma^{-1}(k,N)$. For this filter, the desired signal vector is constant and equal to $[1\,0\,0\ldots0]^T$. The fourth and last filter is the joint-process estimator whose input signal vector is $\mathbf{x}(k,N)$, whose coefficient vector is $\mathbf{w}(k,N)$, and which provides the quantities $\varepsilon(k,N)$ and $e(k,N)$ as outputs.

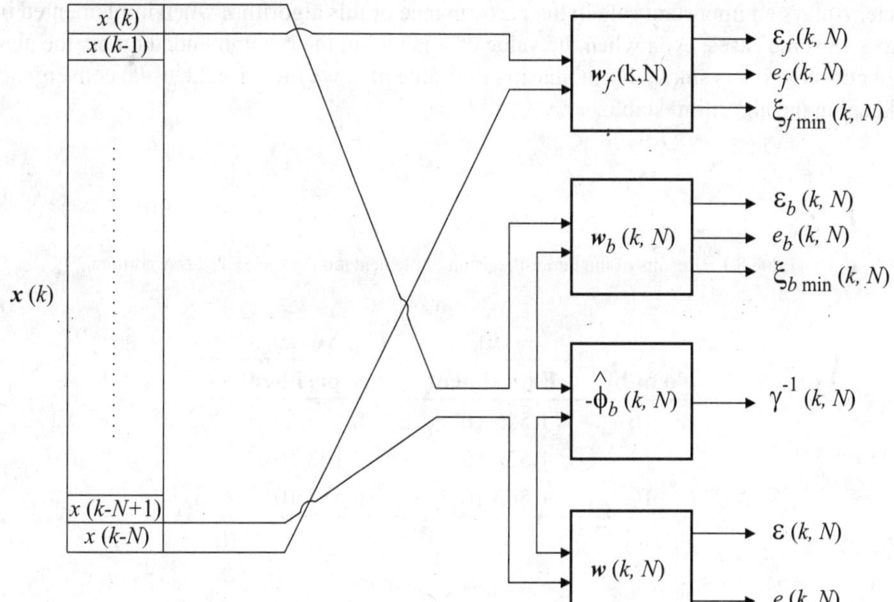

Figure 8.1 Fast transversal RLS algorithm: block diagram.

Example 8.1

The system identification problem described in subsection 3.6.2 is solved using the stabilized fast transversal algorithm presented in this chapter. The main objective is to check the stability of the algorithm when implemented in finite precision.

Solution:

According to equation (8.31), the lower bound for λ in this case is 0.9375. A value $\lambda = 0.99$ is chosen. The stabilized fast transversal algorithm is applied to solve the identification problem and the measured MSE is 0.0432.

Using $\epsilon = 2$, we ran the algorithm with finite precision and the results are summarized in Table 8.1. No sign of instability is found for $\lambda = 0.99$. These results are generated by ensemble averaging 200 experiments. A comparison of the results of Table 8.1 with those of Tables 5.2 and 7.2 shows that the SFTRLS algorithm has similar performance compared to the conventional and lattice-based RLS algorithms, in terms of quantization error accumulation. The question is which algorithm remains stable in most situations. Regarding the SFTRLS, for large-order filters we are left with a limited range of values to choose λ. Also, it was found in our experiments that the choice of the initialization parameter ϵ plays an important role in the performance of this algorithm when implemented in finite precision. In some cases, even when the value of λ is within the recommended range, the algorithm does not converge if ϵ is small. By increasing the value of ϵ, we increase the usual convergence time while keeping the algorithm stable.

□

Table 8.1 Results of the Finite-Precision Implementation of the SFTRLS Algorithm

	$\xi(k)_Q$	$E[\|\|\Delta\mathbf{w}(k)_Q\|\|^2]$
No of bits	**Experiment**	**Experiment**
16	$1.545\ 10^{-3}$	$6.089\ 10^{-5}$
12	$1.521\ 10^{-3}$	$3.163\ 10^{-5}$
10	$1.562\ 10^{-3}$	$6.582\ 10^{-5}$

Example 8.2

The channel equalization example described in subsection (3.6.3) is also used in simulations to test the SFTRLS algorithm. We use a 25th-order equalizer and a forgetting factor $\lambda = 0.99$.

Solution:

In order to solve the equalization problem the stabilized fast transversal RLS algorithm is initialized with $\epsilon = 0.5$. The results presented here were generated by ensemble averaging 200 experiments. The resulting learning curve of the MSE is shown in Fig. 8.2, and the measured MSE is 0.2973. The overall performance of the SFTRLS algorithm for this particular example is as good as any other RLS algorithm, such as lattice-based algorithms.

□

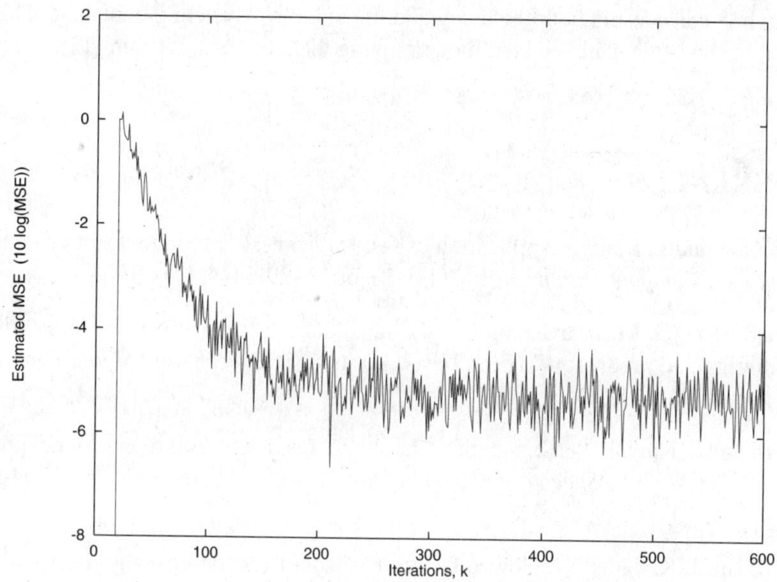

Figure 8.2 Learning curves for the stabilized fast transversal RLS algorithm.

8.5 CONCLUDING REMARKS

In this chapter we have presented some fast transversal RLS algorithms. This class of algorithms is computationally more efficient than conventional and lattice-based RLS algorithms. Some simulation examples were included where the SFTRLS algorithm was employed. The finite-wordlength simulations are of special interest for the reader.

A number of alternative FTRLS algorithms as well as theoretical results can be found in [3]. The derivation of normalized versions of the FTRLS algorithm is also possible and was not addressed in the present chapter, for this result refer to [4]. The most computationally efficient FTRLS algorithms are known to be unstable. The error-feedback approach was briefly introduced that allowed the stabilization of the FTRLS algorithm. The complete derivation and justification for the error-feedback approach is given in [9].

In nonstationary environments, it might be useful to employ a time-varying forgetting factor. Therefore it is desirable to obtain FTRLS algorithms allowing the use of variable λ. This problem was first addressed in [11]. However a computationally more efficient solution was proposed in [8] where the concept of data weighting was introduced to replace the concept of error weighting.

The FTRLS algorithm has potential for a number of applications. In particular, the problem in which the signals available from the environment are noisy version of a transmitted signal and noisy filtered versions of the same transmitted signal is an interesting application. In this problem, both the delay and unknown filter coefficients have to be estimated. The weighted squared errors have to be minimized while considering both the delay and the unknown system parameters. This problem of joint estimation can be elegantly solved by employing the FTRLS algorithm [12].

8.6 REFERENCES

1. D. D. Falconer and L. Ljung, "Application of fast Kalman estimation to adaptive equalization," *IEEE Trans. on Communications*, vol. COM-26, pp. 1439-1446, Oct. 1978.

2. G. Carayannis, D. G. Manolakis, and N. Kalouptsidis, "A fast sequential algorithm for least-squares filtering and prediction," *IEEE Trans. on Acoust., Speech, and Signal Processing*, vol. ASSP-31, pp. 1394-1402, Dec. 1983.

3. J. M. Cioffi and T. Kailath, "Fast, recursive-least-squares transversal filters for adaptive filters," *IEEE Trans. on Acoust., Speech, and Signal Processing*, vol. ASSP-32, pp. 304-337, April 1984.

4. J. M. Cioffi and T. Kailath, "Windowed fast transversal filters adaptive algorithms with normalization," *IEEE Trans. on Acoust., Speech, and Signal Processing*, vol. ASSP-33, pp. 607-627, June 1985.

5. S. Ljung and L. Ljung, "Error propagation properties of recursive least-squares adaptation algorithms," *Automatica*, vol. 21, pp. 157-167, 1985.

6. J.-L. Botto and G. V. Moustakides, "Stabilizing the fast Kalman algorithms," *IEEE Trans. on Acoust., Speech, and Signal Processing*, vol. 37, pp. 1342-1348, Sept. 1989.

7. M. Bellanger, "Engineering aspects of fast least squares algorithms in transversal adaptive filters," *Proc. IEEE Intern. Conf. on Acoust., Speech, Signal Processing*, pp. 49.14.1-49.14.4, 1987.

8. D. T. M. Slock and T. Kailath, "Fast transversal filters with data sequence weighting," *IEEE Trans. on Acoust., Speech, and Signal Processing*, vol. 37, pp. 346-359, March 1989.

9. D. T. M. Slock and T. Kailath, "Numerically stable fast transversal filters for recursive least squares adaptive filtering," *IEEE Trans. on Signal Processing*, vol. 39, pp. 92-113, Jan. 1991.

10. J. G. Proakis, C. M. Rader, F. Ling, and C. L. Nikias, *Advanced Digital Signal Processing*, MacMillan, New York, NY, 1992.

11. B. Toplis and S. Pasupathy, "Tracking improvements in fast RLS algorithms using a variable forgetting factor," *IEEE Trans. on Acoust., Speech, and Signal Processing*, vol. 36, pp. 206-227, Feb. 1988.

12. D. Boudreau and P. Kabal, "Joint-time delay estimation and adaptive recursive least squares filtering," *IEEE Trans. on Signal Processing*, vol. 41, pp. 592-601, Feb. 1993.

8.7 PROBLEMS

1. Show that

$$\phi(k,N) = \mathbf{S}_D(k,N)\mathbf{x}(k,N) = \frac{\mathbf{S}_D(k-1,N)\mathbf{x}(k,N)}{\lambda + \mathbf{x}^T(k,N)\mathbf{S}_D(k-1,N)\mathbf{x}(k,N)}$$

   Hint: Use the matrix inversion lemma for $\mathbf{S}_D(k,N)$.

2. Show that

$$\phi_N(k-1,N) - \frac{\mathbf{w}_{f,N}(k)\varepsilon_f(k,N)}{\xi^d_{f_{\min}}(k,N)} = \frac{-\varepsilon_b(k,N)}{\xi^d_{b_{\min}}(k,N)} = \phi_{N+1}(k,N+1)$$

   where $\mathbf{w}_{f,N}(k)$ represents the last element of $\mathbf{w}_f(k,N)$.

3. Using a proper mixture of relations of the lattice RLS algorithm based on *a posteriori* and the FTRLS algorithm, derive a fast exact initialization procedure for the transversal filter coefficients.

4. Show that the following relations are valid, assuming the input signals are prewindowed:

$$\frac{\det[\mathbf{S}_D(k,N+1)]}{\det[\mathbf{S}_D(k-1,N)]} = \frac{1}{\xi^d_{f_{\min}}(k,N)}$$

$$\frac{\det[\mathbf{S}_D(k,N+1)]}{\det[\mathbf{S}_D(k,N)]} = \frac{1}{\xi^d_{b_{\min}}(k,N)}$$

5. Show that

$$\gamma^{-1}(k,N) = \frac{\det[\mathbf{R}_D(k,N)]}{\lambda^N \det[\mathbf{R}_D(k-1,N)]}.$$

   Hint: $\det[\mathbf{I}+\mathbf{AB}] = \det[\mathbf{I}+\mathbf{BA}]$.

6. Using the results of problems 4 and 5, prove that

$$\gamma^{-1}(k,N) = \frac{\xi^d_{f_{\min}}(k,N)}{\lambda^N \xi^d_{b_{\min}}(k,N)}$$

7. Derive equations (8.7) and (8.14). Also show that the use of $\phi(k,N)$ would increase the computational complexity of the FTRLS algorithm.

8. If one avoids the use of the conversion factor $\gamma(k,N)$, it is necessary to use inner products to derive the *a posteriori* errors in the fast algorithm. Derive a fast algorithm without the conversion factor.

9. By replacing the relation for $\gamma(k, N, 3)$ in the SFTRLS algorithm by the relation

$$\gamma(k, N) = \frac{\lambda^N \xi^d_{b_{\min}}(k, N)}{\xi^d_{f_{\min}}(k, N)}$$

derived in problem 6, describe the resulting algorithm and show that it requires order $8N$ multiplications per output sample.

10. Derive the equation (8.29).

11. The FTRLS algorithm is applied to predict the signal $x(k) = \sin(\frac{\pi k}{4} + \frac{\pi}{3})$. Given $\lambda = 0.98$, calculate the error and the tap coefficients for the first 10 iterations.

12. The SFTRLS algorithm is applied to predict the signal $x(k) = \sin(\frac{\pi k}{4} + \frac{\pi}{3})$. Given $\lambda = 0.98$, calculate the error and the tap coefficients for the first 10 iterations.

13. The FTRLS algorithm is applied to identify a 7th-order unknown system whose coefficients are $\mathbf{w}^T = [0.0272 \quad 0.0221 \quad -0.0621 \quad 0.1191 \quad 0.6116 \quad -0.3332 \quad -0.0190 \quad -0.0572]$ The input signal is Gaussian white noise with variance $\sigma_x^2 = 1$ and the measurement noise is also Gaussian white noise independent of the input signal with variance $\sigma_n^2 = 0.01$.

    Simulate the experiment above described and measure the excess MSE for $\lambda = 0.97$ and $\lambda = 0.98$.

14. Repeat problem 13 for the case where the input signal is a first-order Markov process with $\lambda_{\mathbf{x}} = 0.98$.

15. Redo problem 13 using a fixed-point implementation with the FTRLS and SFTRLS algorithms. Use 12 bits in the fractional part of the signal and parameter representations.

16. Suppose a 15th-order FIR digital filter with the multiplier coefficients given below is identified through an adaptive FIR filter of the same order using the FTRLS algorithm. Assuming fixed-point arithmetic, simulate the identification problem described in terms of the following specifications:

Additional noise : white noise with variance	$\sigma_n^2 = 0.0015$
Coefficients wordlength:	$b_c = 16$ bits
Signal wordlength:	$b_d = 16$ bits
Input signal: Gaussian white noise with variance	$\sigma_x^2 = 0.7$
	$\lambda = 0.98$

$\mathbf{w}_o^T = [0.0219360 \quad 0.0015786 \quad -0.0602449 \quad -0.0118907 \quad 0.1375379$
$0.0574545 \quad -0.3216703 \quad -0.5287203 \quad -0.2957797 \quad 0.0002043 \quad 0.290670$
$-0.0353349 \quad -0.0068210 \quad 0.0026067 \quad 0.0010333 \quad -0.0143593]$

    Plot the learning curves for the finite- and infinite-precision implementations.

17. Repeat the above problem for the SFTRLS algorithm. Also reduce the wordlength used until a noticeable (10 percent increase) excess MSE is observed at the output.

18. Repeat problem 16 for the SFTRLS algorithm, using $\lambda = 0.999$ and $\lambda = 0.960$. Comment on the results.

19. The SFTRLS algorithm is used to perform the forward prediction of a signal $x(k)$ generated by applying zero-mean Gaussian white noise with unit variance to the input of a linear filter with transfer function given by

$$H(z) = \frac{0.5}{(1 - 1.512z^{-1} + 0.827z^{-2})(1 - 1.8z^{-1} + 0.87z^{-2})}$$

Calculate the zeros of the resulting predictor error transfer function and compare with the poles of the linear filter.

20. Perform the equalization of a channel with impulse response given by

$$h(k) = 0.96^k + (-0.9)^k$$

for $k = 0, 1, 2, \ldots, 15$. The transmitted signal is zero-mean Gaussian white noise with unit variance and the adaptive filter input signal-to-noise ratio is 30 dB. Use the SFTRLS algorithm of order 100.

9

QR-DECOMPOSITION-BASED RLS FILTERS

9.1 INTRODUCTION

The application of QR decomposition [1] to triangularize the input data matrix results in an alternative method for the implementation of the recursive least-squares (RLS) method previously discussed. The main advantages brought about by the recursive least-squares algorithm based on QR decomposition are its possible implementation in systolic arrays [2]-[4] and its improved numerical behavior when quantization effects are taken into account [5].

The earlier proposed RLS algorithms based on the QR decomposition [2]-[3] focused on the triangularization of the information matrix in order to avoid the use of matrix inversion. However, their computational requirement was of $O[N^2]$ multiplications per output sample. Later, fast versions of the QR-RLS algorithms were proposed with a reduced computational complexity of $O[N]$ [4]-[11].

In this chapter, the QR-RLS algorithms based on Givens rotations are presented together with some stability considerations. Two families of fast algorithms are also discussed [4]-[11], and one fast algorithm is presented in detail. These fast algorithms are related to the tapped delay line FIR filter realization of the adaptive filter.

9.2 TRIANGULARIZATION USING QR-DECOMPOSITION

The RLS algorithm provides in a recursive way the coefficients of the adaptive filter which lead to the minimization of the following cost function

$$\xi^d(k) = \sum_{i=0}^{k} \lambda^{k-i} \varepsilon^2(i) = \sum_{i=0}^{k} \lambda^{k-i} [d(i) - \mathbf{x}^T(i)\mathbf{w}(k)]^2 \tag{9.1}$$

where

$$\mathbf{x}(k) = [x(k)\, x(k-1) \ldots x(k-N)]^T$$

P.S.R. Diniz, *Adaptive Filtering*, DOI: 10.1007/978-0-387-68606-6_9,

is the input signal vector,

$$\mathbf{w}(k) = [w_0(k)\, w_1(k) \ldots w_N(k)]^T$$

is the coefficient vector at instant k, $\varepsilon(i)$ is the *a posteriori* error at instant i, and λ is the forgetting factor.

The same problem can be rewritten as a function of increasing dimension matrices and vectors which contain all the weighted signal information available so far to the adaptive filter. These matrices are redefined here for convenience:

$$\begin{aligned}\underline{\mathbf{X}}^T(k) &= \mathbf{X}(k) \\ &= \begin{bmatrix} x(k) & \lambda^{1/2}x(k-1) & \cdots & \lambda^{(k-1)/2}x(1) & \lambda^{k/2}x(0) \\ x(k-1) & \lambda^{1/2}x(k-2) & \cdots & \lambda^{(k-1)/2}x(0) & 0 \\ \vdots & \vdots & \ddots & \vdots & \vdots \\ x(k-N) & \lambda^{1/2}x(k-N-1) & \cdots & 0 & 0 \end{bmatrix} \\ &= [\mathbf{x}(k)\, \lambda^{1/2}\mathbf{x}(k-1) \ldots \lambda^{k/2}\mathbf{x}(0)]\end{aligned} \tag{9.2}$$

$$\mathbf{y}(k) = \underline{\mathbf{X}}(k)\mathbf{w}(k) = \begin{bmatrix} y(k) \\ \lambda^{1/2}y(k-1) \\ \vdots \\ \lambda^{k/2}y(0) \end{bmatrix} \tag{9.3}$$

$$\mathbf{d}(k) = \begin{bmatrix} d(k) \\ \lambda^{1/2}d(k-1) \\ \vdots \\ \lambda^{k/2}d(0) \end{bmatrix} \tag{9.4}$$

$$\boldsymbol{\varepsilon}(k) = \begin{bmatrix} \varepsilon(k) \\ \lambda^{1/2}\varepsilon(k-1) \\ \vdots \\ \lambda^{k/2}\varepsilon(0) \end{bmatrix} = \mathbf{d}(k) - \mathbf{y}(k) \tag{9.5}$$

The objective function of equation (9.1) can now be rewritten as

$$\xi^d(k) = \boldsymbol{\varepsilon}^T(k)\boldsymbol{\varepsilon}(k) \tag{9.6}$$

As shown in Chapter 5, equation (5.15), the optimal solution to the least-squares problem at a given instant of time k can be found by solving the following equation

$$\underline{\mathbf{X}}^T(k)\underline{\mathbf{X}}(k)\mathbf{w}(k) = \underline{\mathbf{X}}^T(k)\mathbf{d}(k) \tag{9.7}$$

However, solving this equation by using the conventional RLS algorithm can be a problem when the matrix $\mathbf{R}_D(k) = \underline{\mathbf{X}}^T(k)\underline{\mathbf{X}}(k)$ and its correspondent inverse estimate become ill-conditioned due to loss of persistence of excitation of the input signal or to quantization effects.

The QR decomposition approach avoids inaccurate solutions to the RLS problem, and allows easy monitoring of the positive definiteness of a transformed information matrix in ill-conditioned situations.

9.2.1 Initialization Process

During the initialization period, i.e., from $k = 0$ to $k = N$, the solution of equation (9.7) can be found exactly without using any matrix inversion. From equation (9.7), it can be found that for $k = 0$ and $x(0) \neq 0$

$$w_0(0) = \frac{d(0)}{x(0)} \tag{9.8}$$

for $k = 1$

$$\begin{aligned} w_0(1) &= \frac{d(0)}{x(0)} \\ w_1(1) &= \frac{-x(1)w_0(1) + d(1)}{x(0)} \end{aligned} \tag{9.9}$$

for $k = 2$

$$\begin{aligned} w_0(2) &= \frac{d(0)}{x(0)} \\ w_1(2) &= \frac{-x(1)w_0(2) + d(1)}{x(0)} \\ w_2(2) &= \frac{-x(2)w_0(2) - x(1)w_1(2) + d(2)}{x(0)} \end{aligned} \tag{9.10}$$

at the instant k, we can show by induction that

$$w_i(k) = \frac{-\sum_{j=1}^{i} x(j)w_{i-j}(k) + d(i)}{x(0)} \tag{9.11}$$

The above equation represents the so-called back-substitution algorithm.

9.2.2 Input Data Matrix Triangularization

After the instant $k = N$, the above equation (9.11) is no longer valid and the inversion of $\mathbf{R}_D(k)$ or the calculation of $\mathbf{S}_D(k)$ is required to find the optimal solution for the coefficients $\mathbf{w}(k)$. This is exactly what makes the conventional RLS algorithm more sensitive to quantization effects and input signal conditioning. The matrix $\underline{\mathbf{X}}(k)$ at instant $k = N+1$ is given by

$$\underline{\mathbf{X}}(N+1) = \begin{bmatrix} x(N+1) & x(N) & \cdots & x(1) \\ \lambda^{1/2}x(N) & \lambda^{1/2}x(N-1) & \cdots & \lambda^{1/2}x(0) \\ \lambda x(N-1) & \lambda x(N-2) & \cdots & 0 \\ \vdots & \vdots & & \vdots \\ \lambda^{\frac{N+1}{2}}x(0) & 0 & \cdots & 0 \end{bmatrix}$$
$$= \begin{bmatrix} x(N+1)x(N)\cdots x(1) \\ \lambda^{1/2}\underline{\mathbf{X}}(N) \end{bmatrix} = \begin{bmatrix} \mathbf{x}^T(N+1) \\ \lambda^{1/2}\underline{\mathbf{X}}(N) \end{bmatrix} \quad (9.12)$$

As it is noted, the matrix $\underline{\mathbf{X}}(k)$ is no longer upper triangular, and, therefore, the back-substitution algorithm cannot be employed to find the tap-weight coefficients.

The matrix $\underline{\mathbf{X}}(N+1)$ can be triangularized through an orthogonal triangularization approach such as Givens rotations, Householder transformation, or Gram-Schmidt orthogonalization [1]. Since here the interest is to iteratively apply the triangularization procedure to each new data vector added to $\underline{\mathbf{X}}(k)$, the Givens rotation seems to be the most appropriate approach.

In the Givens rotation approach, each element of the first line of equation (9.12) can be eliminated by premultiplying the matrix $\underline{\mathbf{X}}(N+1)$ by a series of Givens rotation matrices given by

$$\tilde{\mathbf{Q}}(N+1) = \mathbf{Q}'_N(N+1)\cdot\mathbf{Q}'_{N-1}(N+1)\cdots\mathbf{Q}'_0(N+1)$$
$$= \begin{bmatrix} \cos\theta_N(N+1) & \cdots & 0 & \cdots & -\sin\theta_N(N+1) \\ \vdots & & & & \vdots \\ 0 & & \mathbf{I}_N & & 0 \\ \vdots & & & & \vdots \\ \sin\theta_N(N+1) & \cdots & 0 & \cdots & \cos\theta_N(N+1) \end{bmatrix}$$
$$\cdot \begin{bmatrix} \cos\theta_{N-1}(N+1) & \cdots & 0 & \cdots & -\sin\theta_{N-1}(N+1) & 0 \\ \vdots & & & & \vdots & \vdots \\ 0 & & \mathbf{I}_{N-1} & & 0 & 0 \\ \vdots & & & & \vdots & \vdots \\ \sin\theta_{N-1}(N+1) & \cdots & 0 & \cdots & \cos\theta_{N-1}(N+1) & 0 \\ 0 & \cdots & 0 & \cdots & 0 & 1 \end{bmatrix}$$
$$\cdots \begin{bmatrix} \cos\theta_0(N+1) & -\sin\theta_0(N+1) & \cdots & 0 & \cdots & 0 \\ \sin\theta_0(N+1) & \cos\theta_0(N+1) & \cdots & 0 & \cdots & 0 \\ \vdots & \vdots & & & & \\ 0 & 0 & & \mathbf{I}_N & & \\ \vdots & \vdots & & & & \\ 0 & 0 & & & & \end{bmatrix} \quad (9.13)$$

where $\mathbf{I}_i$ is an i by i identity matrix. The rotation angles θ_i are chosen such that each entry of the first row of the resulting matrix is zero. Consider first the matrix product $\mathbf{Q}'_0(N+1)\underline{\mathbf{X}}(N+1)$. If:

$$\cos\theta_0(N+1)x(1) - \sin\theta_0(N+1)\lambda^{1/2}x(0) = 0 \quad (9.14)$$

the element in the position $(1, N+1)$ of the resulting matrix product will be zero. If it is further considered that $\cos^2\theta_0(N+1) + \sin^2\theta_0(N+1) = 1$, it can be easily deduced that

$$\cos\theta_0(N+1) = \frac{\lambda^{1/2}x(0)}{\sqrt{\lambda x^2(0) + x^2(1)}} \tag{9.15}$$

$$\sin\theta_0(N+1) = \frac{x(1)}{\sqrt{\lambda x^2(0) + x^2(1)}} \tag{9.16}$$

Next, $\mathbf{Q}'_1(N+1)$ premultiplies $\mathbf{Q}'_0(N+1)\underline{\mathbf{X}}(N+1)$ with the objective of generating a zero element at the position $(1, N)$ in the resulting product matrix. Note that the present matrix product does not remove the zero of the element $(1, N+1)$. The required rotation angle can be calculated by first noting that the elements $(1, N)$ and $(3, N)$ of $\mathbf{Q}'_0(N+1)\underline{\mathbf{X}}(N+1)$ are respectively

$$a = \cos\theta_0(N+1)x(2) - \lambda^{1/2}x(1)\sin\theta_0(N+1) \tag{9.17}$$

$$b = \lambda x(0) \tag{9.18}$$

From these expressions we can compute the elements required in the following rotation, which are given by

$$\cos\theta_1(N+1) = \frac{b}{\sqrt{a^2+b^2}} \tag{9.19}$$

$$\sin\theta_1(N+1) = \frac{a}{\sqrt{a^2+b^2}} \tag{9.20}$$

In this manner, after the last Givens rotation the input signal information matrix will be transformed in a matrix with null first row

$$\tilde{\mathbf{Q}}(N+1)\underline{\mathbf{X}}(N+1) = \begin{bmatrix} 0 \quad 0 \quad \cdots \quad 0 \\ \mathbf{U}(N+1) \end{bmatrix} \tag{9.21}$$

where $\mathbf{U}(N+1)$ is an upper triangular matrix.

In the next iteration, the input signal matrix $\underline{\mathbf{X}}(N+2)$ receives a new row that should be replaced by a zero vector through a QR decomposition. In this step, the matrices involved are the following

$$\underline{\mathbf{X}}(N+2) = \begin{bmatrix} x(N+2)\; x(N+1) \cdots x(2) \\ \lambda^{1/2}\underline{\mathbf{X}}(N+1) \end{bmatrix} \tag{9.22}$$

and

$$\begin{bmatrix} 1 & 0 \quad \cdots \quad \cdots \\ 0 & \\ \vdots & \tilde{\mathbf{Q}}(N+1) \\ \vdots & \end{bmatrix} \underline{\mathbf{X}}(N+2) = \begin{bmatrix} x(N+2) & x(N+1) & \cdots & x(2) \\ 0 & 0 & \cdots & 0 \\ & \lambda^{1/2}\mathbf{U}(N+1) & & \end{bmatrix} \tag{9.23}$$

In order to eliminate the new input vector through rotations with the corresponding rows of the triangular matrix $\lambda^{1/2}\mathbf{U}(N+1)$, we apply the QR decomposition to equation (9.23) as follows:

$$\tilde{\mathbf{Q}}(N+2)\begin{bmatrix} 1 & \mathbf{0} \\ \mathbf{0} & \tilde{\mathbf{Q}}(N+1) \end{bmatrix}\underline{\mathbf{X}}(N+2) = \begin{bmatrix} 0 \quad 0 \quad \cdots \quad 0 \\ 0 \quad 0 \quad \cdots \quad 0 \\ \mathbf{U}(N+2) \end{bmatrix} \tag{9.24}$$

where again $\mathbf{U}(N+2)$ is an upper triangular matrix and $\tilde{\mathbf{Q}}(N+2)$ is given by

$$\tilde{\mathbf{Q}}(N+2) = \mathbf{Q}'_N(N+2)\mathbf{Q}'_{N-1}(N+2)\cdots\mathbf{Q}'_0(N+2)$$

$$= \begin{bmatrix} \cos\theta_N(N+2) & \cdots & 0 & \cdots & -\sin\theta_N(N+2) \\ \vdots & & & & \vdots \\ 0 & & \mathbf{I}_{N+1} & & 0 \\ \vdots & & & & \vdots \\ \sin\theta_N(N+2) & \cdots & 0 & \cdots & \cos\theta_N(N+2) \end{bmatrix}$$

$$\cdot \begin{bmatrix} \cos\theta_{N-1}(N+2) & \cdots & 0 & \cdots & -\sin\theta_{N-1}(N+2) & 0 \\ \vdots & & & & & \vdots \\ 0 & & \mathbf{I}_N & & & 0 \\ \vdots & & & & & \vdots \\ \sin\theta_{N-1}(N+2) & & & & \cos\theta_{N-1}(N+2) & 0 \\ 0 & \cdots & 0 & \cdots & 0 & 1 \end{bmatrix}$$

$$\cdots \begin{bmatrix} \cos\theta_0(N+2) & 0 & -\sin\theta_0(N+2) & \cdots & 0 \\ 0 & 1 & 0 & \cdots & 0 \\ \sin\theta_0(N+2) & 0 & \cos\theta_0(N+2) & \cdots & 0 \\ \vdots & \vdots & \vdots & & \\ \vdots & \vdots & \vdots & \mathbf{I}_N & \\ 0 & 0 & 0 & & \end{bmatrix} \tag{9.25}$$

The above procedure should be repeated for each new incoming input signal vector as follows:

$$\mathbf{Q}(k)\underline{\mathbf{X}}(k) = \tilde{\mathbf{Q}}(k)\begin{bmatrix} 1 & \mathbf{0} \\ \mathbf{0} & \tilde{\mathbf{Q}}(k-1) \end{bmatrix}\begin{bmatrix} \mathbf{I}_2 & \mathbf{0} \\ \mathbf{0} & \tilde{\mathbf{Q}}(k-2) \end{bmatrix}$$

$$\cdots \begin{bmatrix} \mathbf{I}_{k-N} & \mathbf{0} \\ \mathbf{0} & \tilde{\mathbf{Q}}(k-N) \end{bmatrix} \underline{\mathbf{X}}(k) = \begin{bmatrix} \mathbf{0} \\ \mathbf{U}(k) \end{bmatrix} \begin{matrix} \}\, k-N \\ \}\, N+1 \end{matrix} \tag{9.26}$$

$$\underbrace{\qquad}_{N+1}$$

where $\mathbf{Q}(k)$ is a $(k+1)$ by $(k+1)$ matrix which represents the overall triangularization matrix via elementary Givens rotations matrices $\mathbf{Q}'_i(m)$ for all $m \leq k$ and $0 \leq i \leq N$.

Since each Givens rotation matrix is orthogonal, then it can easily be proved that $\mathbf{Q}(k)$ is also orthogonal (actually orthonormal), i.e.,

$$\mathbf{Q}(k)\mathbf{Q}^T(k) = \mathbf{I}_{k+1} \tag{9.27}$$

Also, from equation (9.26), it is straightforward to note that

$$\mathbf{Q}(k) = \tilde{\mathbf{Q}}(k) \begin{bmatrix} 1 & \mathbf{0} \\ \mathbf{0} & \mathbf{Q}(k-1) \end{bmatrix} \tag{9.28}$$

where $\tilde{\mathbf{Q}}(k)$ is responsible for zeroing the latest input vector $\mathbf{x}^T(k)$ in the first row of $\underline{\mathbf{X}}(k)$. The matrix $\tilde{\mathbf{Q}}(k)$ is given by

$$\tilde{\mathbf{Q}}(k) = \begin{bmatrix} \cos\theta_N(k) & \cdots & 0 & \cdots & -\sin\theta_N(k) \\ \vdots & & & & \vdots \\ 0 & & \mathbf{I}_{k-1} & & 0 \\ \vdots & & & & \vdots \\ \sin\theta_N(k) & \cdots & 0 & \cdots & \cos\theta_N(k) \end{bmatrix}$$

$$\cdot \begin{bmatrix} \cos\theta_{N-1}(k) & \cdots & 0 & \cdots & -\sin\theta_{N-1}(k) & 0 \\ \vdots & & & & \vdots & \vdots \\ 0 & & \mathbf{I}_{k-2} & & 0 & 0 \\ \vdots & & & & \vdots & \vdots \\ \sin\theta_{N-1}(k) & \cdots & 0 & \cdots & \cos\theta_{N-1}(k) & 0 \\ 0 & \cdots & 0 & \cdots & 0 & 1 \end{bmatrix}$$

$$\cdots \begin{bmatrix} \cos\theta_0(k) & \cdots & 0 & \cdots & -\sin\theta_0(k) & 0 \\ \vdots & & & & \vdots & \vdots \\ 0 & & \mathbf{I}_{k-N-1} & & 0 & 0 \\ \vdots & & & & \vdots & \vdots \\ \sin\theta_0(k) & \cdots & 0 & \cdots & \cos\theta_0(k) & 0 \\ & & \mathbf{0} & & & \mathbf{I}_N \end{bmatrix}$$

$$
= \left[\begin{array}{ccccc}
\prod_{i=0}^{N}\cos\theta_i(k) & \cdots & 0 & \cdots & -\prod_{i=1}^{N}\cos\theta_i(k)\sin\theta_0(k) \\
\vdots & & & & \vdots \\
0 & & \mathbf{I}_{k-N-1} & & 0 \\
\vdots & & & & \vdots \\
\sin\theta_0(k) & & & & \cos\theta_0(k) \\
\vdots & & \vdots & & \vdots \\
\prod_{i=0}^{j-1}\cos\theta_i(k)\sin\theta_j(k) & \cdots & 0 & \cdots & \vdots \\
\vdots & & \vdots & & -\sin\theta_N(k)\prod_{i=1}^{N-1}\cos\theta_i(k)\sin\theta_0(k)
\end{array} \right.
$$

$$
\left. \begin{array}{ccccc}
\cdots & -\prod_{i=j+1}^{N}\cos\theta_i(k)\sin\theta_j(k) & & \cdots & -\sin\theta_N(k) \\
 & & & 0 & \\
\ddots & & & & \\
 & \ddots & & & \mathbf{0} \\
 & & & \cos\theta_{N-1}(k) & \\
 & & & & \cos\theta_N(k)
\end{array} \right] \tag{9.29}
$$

Note that the matrix $\tilde{\mathbf{Q}}(k)$ has the following general form

$$
\tilde{\mathbf{Q}}(k) = \begin{bmatrix}
* & 0 & \cdots & 0 & \cdots & 0 & \overbrace{* \quad \cdots \quad *}^{N+1} \\
0 & & & & & & \\
\vdots & & & \mathbf{I}_{k-N-1} & & & \mathbf{0} \\
* & & & & & & * \\
\vdots & & & \mathbf{0} & & & \ddots \\
* & & & & & & * \quad\quad *
\end{bmatrix}
\left.\begin{matrix} \\ \\ \\ \end{matrix}\right\} N+1 \tag{9.30}
$$

where $*$ represents a nonzero element. This structure of $\tilde{\mathbf{Q}}(k)$ is useful for developing some fast QR-RLS algorithms.

Returning to equation (9.26), we can conclude that

$$\mathbf{Q}(k)\underline{\mathbf{X}}(k) = \tilde{\mathbf{Q}}(k)\begin{bmatrix} x(k) & x(k-1) & \cdots & x(k-N) \\ 0 & 0 & \cdots & 0 \\ \vdots & \vdots & \ddots & \vdots \\ 0 & 0 & \cdots & 0 \\ & \lambda^{1/2}\mathbf{U}(k-1) & & \end{bmatrix} \tag{9.31}$$

The first Givens rotation angle required to replace $x(k-N)$ by a zero is $\theta_0(k)$ such that

$$\cos\theta_0(k)x(k-N) - \sin\theta_0(k)\lambda^{1/2}u_{1,N+1}(k-1) = 0 \tag{9.32}$$

where $u_{1,N+1}(k-1)$ is the element $(1, N+1)$ of $\mathbf{U}(k-1)$. Then, it follows that

$$\cos\theta_0(k) = \frac{\lambda^{1/2}u_{1,N+1}(k-1)}{u_{1,N+1}(k)} \tag{9.33}$$

$$\sin\theta_0(k) = \frac{x(k-N)}{u_{1,N+1}(k)} \tag{9.34}$$

where

$$u_{1,N+1}^2(k) = x^2(k-N) + \lambda u_{1,N+1}^2(k-1) \tag{9.35}$$

From equation (9.35), it is worth noting that the $(1, N+1)$ element of $\mathbf{U}(k)$ is the square root of the exponentially weighted input signal energy, i.e.,

$$u_{1,N+1}^2(k) = \sum_{i=0}^{k-N} \lambda^i x^2(k-N-i) \tag{9.36}$$

In the triangularization process, all the submatrices multiplying each column of $\underline{\mathbf{X}}(k)$ are orthogonal matrices and as a consequence the norm of each column in $\underline{\mathbf{X}}(k)$ and $\mathbf{Q}(k)\underline{\mathbf{X}}(k)$ should be the same. This confirms that equation (9.36) is valid. Also, it can be shown that

$$\sum_{i=1}^{k+1} \underline{x}_{i,j}^2(k) = \sum_{i=1}^{N+2-j} u_{i,j}^2(k) = \sum_{i=1}^{k+1} \lambda^{i-1}x^2(k+2-i-j) \tag{9.37}$$

for $j = 1, 2, \ldots, N+1$.

Now consider that the intermediate calculations of equation (9.31) are performed as follows:

$$\tilde{\mathbf{Q}}(k)\begin{bmatrix} \mathbf{x}^T(k) \\ \mathbf{0} \\ \lambda^{1/2}\mathbf{U}(k-1) \end{bmatrix} = \mathbf{Q}'_N(k)\mathbf{Q}'_{N-1}(k)\cdots\mathbf{Q}'_i(k)\begin{bmatrix} \mathbf{x}'_i(k) \\ \mathbf{0} \\ \mathbf{U}'_i(k) \end{bmatrix} \tag{9.38}$$

where $\mathbf{x}'_i(k) = [x'_i(k)x'_i(k-1)\ldots x'_i(k-N-i)0\ldots 0]$ and $\mathbf{U}'_i(k)$ is an intermediate upper triangular matrix. Note that $\mathbf{x}'_0(k) = \mathbf{x}^T(k)$, $\mathbf{U}'_0(k) = \lambda^{1/2}\mathbf{U}(k-1)$, and $\mathbf{U}'_{N+1}(k) = \mathbf{U}(k)$. In practice, the multiplication by the zero elements in equation (9.38) should be avoided. We start by removing the increasing $\mathbf{I}_{k-N-1}$ section of $\tilde{\mathbf{Q}}(k)$ (see equation (9.30)), thereby generating a matrix with reduced dimension denoted by $\mathbf{Q}_\theta(k)$. The resulting equation is

$$\mathbf{Q}_\theta(k)\begin{bmatrix}\mathbf{x}^T(k)\\ \lambda^{1/2}\mathbf{U}(k-1)\end{bmatrix} = \mathbf{Q}'_{\theta_N}(k)\mathbf{Q}'_{\theta_{N-1}}(k)\cdots\mathbf{Q}'_{\theta_i}(k)\begin{bmatrix}\mathbf{x}'_i(k)\\ \mathbf{U}'_i(k)\end{bmatrix}$$
$$= \begin{bmatrix}\mathbf{0}\\ \mathbf{U}(k)\end{bmatrix} \tag{9.39}$$

where $\mathbf{Q}'_{\theta_i}(k)$ is derived from $\mathbf{Q}'_i(k)$ by removing the $\mathbf{I}_{k-N-1}$ section of $\mathbf{Q}'_i(k)$ along with the corresponding rows and columns, resulting in the following form

$$\mathbf{Q}'_{\theta_i}(k) = \begin{bmatrix}\cos\theta_i(k) & \cdots & 0 & \cdots & -\sin\theta_i(k) & \cdots & 0\\ \vdots & & & & \vdots & & \vdots\\ 0 & & \mathbf{I}_i & & 0 & \cdots & 0\\ \vdots & & & & \vdots & & \vdots\\ \sin\theta_i(k) & \cdots & 0 & \cdots & \cos\theta_i(k) & \cdots & 0\\ \vdots & & \vdots & & \vdots & \mathbf{I}_{N-i} & \\ 0 & \cdots & 0 & \cdots & 0 & & \end{bmatrix} \tag{9.40}$$

The Givens rotation elements are calculated by

$$\cos\theta_i(k) = \frac{[\mathbf{U}'_i(k)]_{i+1,N+1-i}}{c_i} \tag{9.41}$$

$$\sin\theta_i(k) = \frac{x'_i(k-N-i)}{c_i} \tag{9.42}$$

where $c_i = \sqrt{[\mathbf{U}'_i(k)]^2_{i+1,N+1-i} + x'^2_i(k-N-i)}$ and $[\cdot]_{i,j}$ is the (i,j) element of the matrix.

9.2.3 QR-Decomposition RLS Algorithm

The triangularization procedure above discussed can be applied to generate the QR-RLS algorithm that avoids the calculation of the $\mathbf{S}_D(k)$ matrix of the conventional RLS algorithm. The weighted *a posteriori* error vector can be written as a function of the input data matrix, that is

$$\boldsymbol{\varepsilon}(k) = \begin{bmatrix}\varepsilon(k)\\ \lambda^{1/2}\varepsilon(k-1)\\ \vdots\\ \lambda^{k/2}\varepsilon(0)\end{bmatrix} = \begin{bmatrix}d(k)\\ \lambda^{1/2}d(k-1)\\ \vdots\\ \lambda^{k/2}d(0)\end{bmatrix} - \underline{\mathbf{X}}(k)\mathbf{w}(k) \tag{9.43}$$

By premultiplying the above equation by $\mathbf{Q}(k)$, it follows that

$$\begin{aligned}\varepsilon_q(k) = \mathbf{Q}(k)\varepsilon(k) &= \mathbf{Q}(k)\mathbf{d}(k) - \mathbf{Q}(k)\underline{\mathbf{X}}(k)\mathbf{w}(k) \\ &= \mathbf{d}_q(k) - \begin{bmatrix} \mathbf{0} \\ \mathbf{U}(k) \end{bmatrix} \mathbf{w}(k)\end{aligned} \tag{9.44}$$

where

$$\varepsilon_q(k) = \begin{bmatrix} \varepsilon_{q_1}(k) \\ \varepsilon_{q_2}(k) \\ \vdots \\ \varepsilon_{q_{k+1}}(k) \end{bmatrix}$$

and

$$\mathbf{d}_q(k) = \begin{bmatrix} d_{q_1}(k) \\ d_{q_2}(k) \\ \vdots \\ d_{q_{k+1}}(k) \end{bmatrix}$$

Since $\mathbf{Q}(k)$ is an orthogonal matrix, equation (9.6) is equivalent to

$$\xi^d(k) = \varepsilon_q^T(k)\varepsilon_q(k) \tag{9.45}$$

because

$$\varepsilon_q^T(k)\varepsilon_q(k) = \varepsilon^T(k)\mathbf{Q}^T(k)\mathbf{Q}(k)\varepsilon(k) = \varepsilon^T(k)\varepsilon(k)$$

The weighted-square error can be minimized in equation (9.45) by calculating $\mathbf{w}(k)$ such that $\varepsilon_{q_{k-N+1}}(k)$ to $\varepsilon_{q_{k+1}}(k)$ are made zero using a back-substitution algorithm such as

$$w_i(k) = \frac{-\sum_{j=1}^{i} u_{N+1-i,i-j+1}(k)w_{i-j}(k) + d_{q\,k+1-i}(k)}{u_{N+1-i,i+1}(k)} \tag{9.46}$$

for $i = 0, 1, \ldots, N$, where $\sum_{j=i}^{i-1}[\cdot] = 0$. With this choice for $\mathbf{w}(k)$, the minimum weighted-square error at instant k is given by

$$\xi_{\min}^d(k) = \sum_{i=1}^{k-N} \varepsilon_{q_i}^2(k) \tag{9.47}$$

An important relation can be deduced by rewriting equation (9.44) as

$$\mathbf{d}_q(k) = \begin{bmatrix} \mathbf{d}_{q1}(k) \\ --- \\ \mathbf{d}_{q2}(k) \end{bmatrix} = \begin{bmatrix} d_{q_1}(k) \\ \vdots \\ d_{q_{k-N}}(k) \\ ----- \\ d_{q_{k-N+1}}(k) \\ \vdots \\ d_{q_{k+1}}(k) \end{bmatrix}$$

$$= \begin{bmatrix} \varepsilon_{q_1}(k) \\ \vdots \\ \varepsilon_{q_{k-N}}(k) \\ 0 \\ \vdots \\ 0 \end{bmatrix} + \begin{bmatrix} \mathbf{0} \\ \mathbf{U}(k) \end{bmatrix} \mathbf{w}(k) \tag{9.48}$$

where $\mathbf{w}(k)$ is the optimum coefficient vector at instant k. By examining the equations (9.31) and (9.44), the right-most side of equation (9.48) can then be expressed as

$$\begin{bmatrix} \boldsymbol{\varepsilon}_{q1}(k) \\ \mathbf{d}_{q2}(k) \end{bmatrix} = \begin{bmatrix} \varepsilon_{q_1}(k) \\ \vdots \\ \varepsilon_{q_{k-N}}(k) \\ \mathbf{d}_{q2}(k) \end{bmatrix} = \tilde{\mathbf{Q}}(k) \begin{bmatrix} d(k) \\ \lambda^{1/2} \begin{bmatrix} \varepsilon_{q_1}(k-1) \\ \vdots \\ \varepsilon_{q_{k-N-1}}(k-1) \\ \mathbf{d}_{q2}(k-1) \end{bmatrix} \end{bmatrix} \tag{9.49}$$

Using similar arguments around equations (9.38) to (9.40), and starting from equation (9.49), the transformed weighted-error vector can be updated as described below:

$$\tilde{\mathbf{Q}}(k) \begin{bmatrix} d(k) \\ \lambda^{1/2} \begin{bmatrix} \boldsymbol{\varepsilon}_{q1}(k-1) \\ \mathbf{d}_{q2}(k-1) \end{bmatrix} \end{bmatrix} = \mathbf{Q}'_N(k)\mathbf{Q}'_{N-1}(k)\cdots\mathbf{Q}'_i(k) \begin{bmatrix} d'_i(k) \\ \boldsymbol{\varepsilon}'_{q_i}(k) \\ \mathbf{d}'_{q2_i}(k) \end{bmatrix} \tag{9.50}$$

where $d'_i(k)$, $\boldsymbol{\varepsilon}'_{q_i}(k)$, and $\mathbf{d}'_{q2_i}(k)$ are intermediate quantities generated during the rotations. Note that $\boldsymbol{\varepsilon}'_{q_{N+1}}(k) = [\varepsilon_{q_2}(k)\ \varepsilon_{q_3}(k) \ldots \varepsilon_{q_{k-N}}(k)]^T$, $d'_{N+1}(k) = \varepsilon_{q_1}(k)$, and $\mathbf{d}'_{q2_{N+1}} = \mathbf{d}_{q_2}(k)$.

If we delete all the columns and rows of $\tilde{\mathbf{Q}}(k)$ whose elements are zeros and ones, i.e., the $\mathbf{I}_{k-N-1}$ section of $\tilde{\mathbf{Q}}(k)$ with the respective bands of zeros below, above, and on each side of it in equation (9.30), one would obtain matrix $\mathbf{Q}_\theta(k)$. In this case, the resulting equation corresponding to equation (9.49) is given by

$$\underline{\mathbf{d}}(k) = \begin{bmatrix} \varepsilon_{q_1}(k) \\ \mathbf{d}_{q_2}(k) \end{bmatrix} = \mathbf{Q}_\theta(k) \begin{bmatrix} d(k) \\ \lambda^{1/2}\mathbf{d}_{q_2}(k-1) \end{bmatrix} \tag{9.51}$$

Therefore, we eliminate the vector $\varepsilon'_{q_{N+1}}(k)$ which is always increasing, such that in real-time implementation the updating is performed through

$$\begin{aligned}\underline{\mathbf{d}}(k) &= \mathbf{Q}_\theta(k)\begin{bmatrix} d(k) \\ \lambda^{1/2}\mathbf{d}_{q_2}(k-1)\end{bmatrix} \\ &= \mathbf{Q}'_{\theta_N}(k)\mathbf{Q}'_{\theta_{N-1}}(k)\cdots\mathbf{Q}'_{\theta_i}(k)\begin{bmatrix} d'_i(k) \\ \mathbf{d}'_{q_{2i}}(k)\end{bmatrix}\end{aligned} \tag{9.52}$$

Another important relation can be derived from equation (9.44) by premultiplying both sides by $\mathbf{Q}^T(k)$, transposing the result, and postmultiplying the result by the pinning vector

$$\varepsilon_q^T(k)\mathbf{Q}(k)\begin{bmatrix}1\\0\\\vdots\\0\end{bmatrix} = \varepsilon^T(k)\begin{bmatrix}1\\0\\\vdots\\0\end{bmatrix} = \varepsilon(k) \tag{9.53}$$

Then, from the definition of $\mathbf{Q}(k)$ in equations (9.28) and (9.29), the following relation is obtained

$$\begin{aligned}\varepsilon(k) &= \varepsilon_{q_1}(k)\prod_{i=0}^{N}\cos\theta_i(k) \\ &= \varepsilon_{q_1}(k)\gamma(k)\end{aligned} \tag{9.54}$$

This relation shows that the *a posteriori* output error can be computed without the explicit calculation of $\mathbf{w}(k)$. The only information needed is the Givens rotation cosines. In applications where only the *a posteriori* output error is of interest, the computationally intensive back-substitution algorithm of equation (9.46) to obtain $\mathbf{w}_i(k)$ can be avoided.

Now, all the mathematical background to develop the QR-RLS algorithm has been derived. After initialization, the Givens rotation elements are computed using equations (9.41) and (9.42). These rotations are then applied to the information matrix and the desired signal vector respectively as indicated in equations (9.39) and (9.52). The next step is to compute the error signal using equation (9.54). Finally, if the tap-weight coefficients are required we should calculate them using equation (9.46). Algorithm 9.1 summarizes the algorithm with all essential computations.

Example 9.1

In this example, we solve the system identification problem described in subsection 3.6.2 by using the QR-RLS algorithm described in this section.

Solution:

In the present example, we are mainly concerned in testing the algorithm implemented in finite precision, since the remaining characteristics (such as: misadjustment, convergence speed etc.) should follow the same pattern of the conventional RLS algorithm. We considered the case where

Algorithm 9.1
QR-RLS Algorithm

$\mathbf{w}(-1) = [0\,0 \ldots 0]^T, \quad w_0(0) = \frac{d(0)}{x(0)}$
For $k = 1$ to N (Initialization)
Do for $i = 1$ to k

$$w_i(k) = \frac{-\sum_{j=1}^{i} x(j) w_{i-j}(k) + d(i)}{x(0)} \quad (9.11)$$

End
End

$\mathbf{U}'_0(N+1) = \lambda^{1/2} \underline{\mathbf{X}}(N)$ (9.12)
$\mathbf{d}'_{q2\,0}(N+1) = [\,\lambda^{1/2} d(N)\; \lambda d(N-1) \ldots \lambda^{(N+1)/2} d(0)]^T$
For $k \geq N+1$
Do for each k
$\gamma'_{-1} = 1$
$d'_0(k) = d(k)$
$\mathbf{x}'_0(k) = \mathbf{x}^T(k)$
Do for $i = 0$ to N

$c_i = \sqrt{[\mathbf{U}'_i(k)]^2_{i+1,N+1-i} + x'^2_i(k-N-i)}$

$\cos\theta_i = \frac{[\mathbf{U}'_i(k)]_{i+1,N+1-i}}{c_i}$ (9.41)

$\sin\theta_i = \frac{x'_i(k-N-i)}{c_i}$ (9.42)

$$\begin{bmatrix} \mathbf{x}'_{i+1}(k) \\ \mathbf{U}'_{i+1}(k) \end{bmatrix} = \mathbf{Q}'_{\theta_i}(k) \begin{bmatrix} \mathbf{x}'_i(k) \\ \mathbf{U}'_i(k) \end{bmatrix} \quad (9.39)$$

$\gamma'_i = \gamma'_{i-1} \cos\theta_i$ (9.54)

$$\begin{bmatrix} d'_{i+1}(k) \\ \mathbf{d}'_{q2i+1}(k) \end{bmatrix} = \mathbf{Q}'_{\theta_i}(k) \begin{bmatrix} d'_i(k) \\ \mathbf{d}'_{q2i}(k) \end{bmatrix} \quad (9.51)$$

End
$\mathbf{d}'_{q20}(k+1) = \lambda^{1/2} \mathbf{d}'_{q2N+1}(k)$
$\mathbf{U}'_0(k+1) = \lambda^{1/2} \mathbf{U}'_{N+1}(k)$
$\gamma(k) = \gamma'_N$
$\varepsilon(k) = d'_{N+1}(k)\gamma(k)$ (9.51)
If required compute

$$\underline{\mathbf{d}}(k) = \begin{bmatrix} d'_{N+1}(k) \\ \mathbf{d}'_{q2N+1}(k) \end{bmatrix} \quad (9.51)$$

$w_0(k) = \frac{\underline{d}_{N+2}(k)}{u_{N+1,1}(k)}$
Do for $i = 1$ to N

$$w_i(k) = \frac{-\sum_{j=1}^{i} u_{N+1-i,i-j+1}(k) w_{i-j}(k) + \underline{d}_{N+2-i}(k)}{u_{N+1-i,i+1}(k)} \quad (9.46)$$

End
End

eigenvalue spread of the input signal correlation matrix is 20, with $\lambda = 0.99$. The presented results were obtained by averaging the outcomes of 200 independent runs. Table 9.1 summarizes the results, where it can be noticed that the MSE is comparable to the case of the conventional RLS algorithm (consult Table 5.2). On the other hand, the quantization error introduced by the calculations to obtain $\mathbf{w}(k)_Q$ is considerable. After leaving the algorithm running for a large number of iterations, we found no sign of divergence.

In the infinite-precision implementation, the misadjustment measured was 0.0429. As expected (consult Table 5.1) this result is close to the misadjustment obtained by the conventional RLS algorithm.

□

Table 9.1 Results of the Finite-Precision Implementation of the QR-RLS Algorithm

	$\xi(k)_Q$	$E[\|\|\Delta\mathbf{w}(k)_Q\|\|^2]$
No. of bits	**Experiment**	**Experiment**
16	$1.544\ 10^{-3}$	0.03473
12	$1.563\ 10^{-3}$	0.03254
10	$1.568\ 10^{-3}$	0.03254

9.3 SYSTOLIC ARRAY IMPLEMENTATION

The systolic array implementation of a given algorithm consists of mapping the algorithm in a pipelined sequence of basic computation cells. These basic cells perform their task in parallel, such that in each clock period all the cells are activated. An in depth treatment of systolic array implementation and parallelization of algorithms is beyond the scope of this text. Our objective in this section is to demonstrate in a summarized form that the QR-RLS algorithm can be mapped in a systolic array. Further details regarding this subject can be found in references [2]-[4], [13]-[14].

A Givens rotation requires two basic steps. The first step is the calculation of the sine and cosine which are the elements of the rotation matrix. The second step is the application of the rotation matrix to given data. Therefore, the basic computational elements required to perform the systolic array implementation of the QR-RLS algorithm introduced in the last section are the angle and the rotation processors shown in Fig. 9.1. The angle processor computes the cosine and sine, transferring the results to outputs 1 and 2 respectively, whereas in output 3 the cell delivers a partial product of cosines meant to generate the error signal as in equation (9.54). The rotation processor performs the rotation between the data coming from input 1 with the internal element of the matrix $\mathbf{U}(l)$ and transfers the result to output 3. This processor also updates the elements of $\mathbf{U}(l)$ and transfers the cosine and sine values to the neighboring cell on the left.

Now, imagine that we have the upper triangular matrix $\mathbf{U}(k)$ arranged below the row consisting of the new information data vector as in equation (9.31), or equivalently as in equation (9.39). Following the same pattern, we can arrange the basic cells in order to compute the rotations of the QR-RLS algorithm as shown in Fig. 9.2, with the input signal $x(k)$ entering the array serially. In this figure, do not consider for this moment the time indexes and the left-hand side column. The input data weighting is performed by the processors of the systolic array.

Basically, the computations corresponding to the triangularization of equation (9.31) are performed through the systolic array shown in Fig. 9.2, where at each instant of time an element of the matrix $\mathbf{U}(k)$ is stored in the basic processor as shown inside the building blocks. Note that these stored elements are skewed in time, and are initialized with zero. The left-hand cells store the elements of the vector $\underline{\mathbf{d}}(k)$ defined in equation (9.51), which are also initialized with zero and updated in each clock cycle. The column on the left-hand side of the array performs the rotation and stores the rotated values of the desired signal vector which are essential to compute the error signal.

In order to allow the pipelining, the outputs of each cell are computed at the present clock period and made available to the neighboring cells in the following clock period. Note that the neighboring cells on the left and below a given cell are performing computations related to a previous iteration, whereas the cells on the right and above are performing the computations of one iteration in advance. This is the pipelining scheme of Fig. 9.2.

Each row of cells in the array implements a basic Givens rotation between one row of $\lambda\mathbf{U}(k-1)$ and a vector related to the new incoming data $\mathbf{x}(k)$. The top row of the systolic array performs the zeroing of the last element of the most recent incoming $\mathbf{x}(k)$ vector. The result of the rotation is then passed to the second row of the array. This second row performs the zeroing of the second-to-last element in the rotated input signal. The zeroing processing continues in the following rows by eliminating the remaining elements of the intermediate vectors $\mathbf{x}'_i(k)$, defined in equation (9.38), through Givens rotations. The angle processors compute the rotation angles that are passed to each row to perform the rotations.

More specifically, returning to equation (9.31), at the instant k, the element $x(k-N)$ of $\mathbf{x}(k)$ is eliminated by calculating the angle $\theta_0(k)$ in the upper angle processor. The same processor also performs the computation of $u_{1,N+1}(k)$ that will be stored and saved for later elimination of $x(k-N+1)$, which occurs during the triangularization of $\underline{\mathbf{X}}(k+1)$. In the same period of time, the neighboring rotation processor performs the computation of $u_{1,N}(k-1)$ using the angle $\theta_0(k-1)$ that was received from the angle processor in the beginning of the present clock period k. The modifications to the first row of the $\mathbf{U}(k)$ matrix and to the vector $\underline{\mathbf{d}}(k)$ related to the desired signal are performed in the first row of the array, due to the rotation responsible for the elimination of $x(k-N)$. Note that the effect of the angle $\theta_0(k)$ in the remaining elements of the first row of $\mathbf{U}(k)$ will be felt only in the following iterations, one element each time, starting from the right- to the left-hand side.

The second row of the systolic array is responsible for the rotation corresponding to $\theta_1(l)$ that eliminates the element $x'_1(l-N+1)$ of $\mathbf{x}'_1(l)$ defined in equation (9.38). The rotation $\theta_1(l)$ of course modifies the remaining nonzero elements of $\mathbf{x}'_1(l)$ generating $\mathbf{x}'_2(l)$, whose elements are calculated by the rotation processor and forwarded to the next row through output 3.

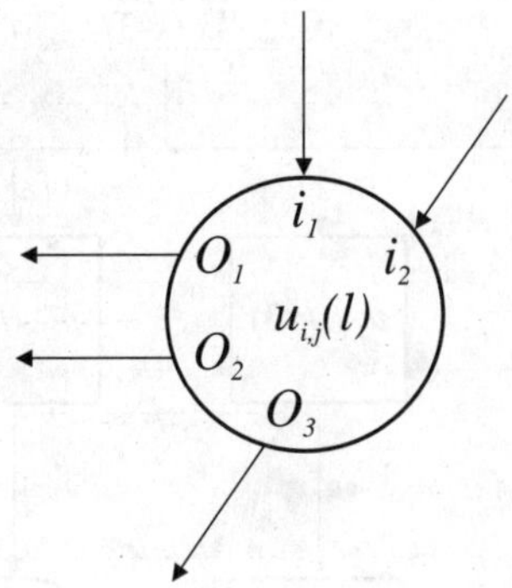

If i_1=0 then
$O_1 \leftarrow 1, O_2 \leftarrow 0, O_3 \leftarrow i_2$
$u_{i,j} = \lambda u_{i,j}$
Otherwise
$c \leftarrow \sqrt{\lambda^2 u_{i,j}^2(l) + i_1^2}$
$O_1 \leftarrow \cos\theta_{i-1} = \frac{\lambda u_{i,j}(l)}{c}$
$O_2 \leftarrow \sin\theta_{i-1} = \frac{i_1}{c}$
$O_3 \leftarrow i_2 O_1$
$u_{i,j}(l+1) \leftarrow c$
End

(a)

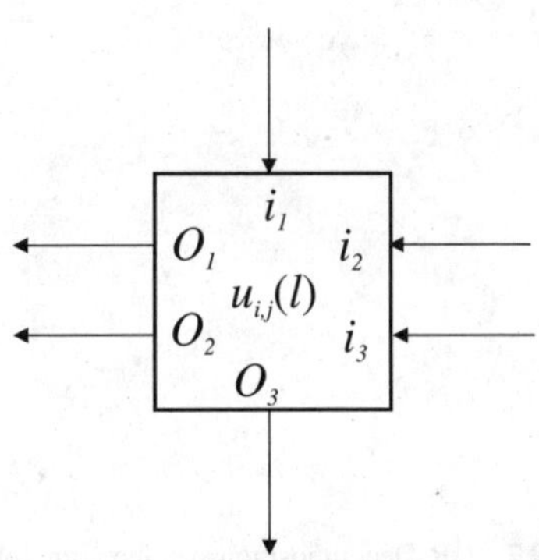

$O_1 \leftarrow i_2$
$O_2 \leftarrow i_3$
$O_3 \leftarrow i_1 i_2 - i_3 \lambda u_{i,j}(l)$
$u_{i,j}(l+1) \leftarrow i_1 i_3 + i_2 \lambda u_{i,j}(l)$

(b)

Figure 9.1 Basic cells: (a) Angle processor, (b) Rotation processor.

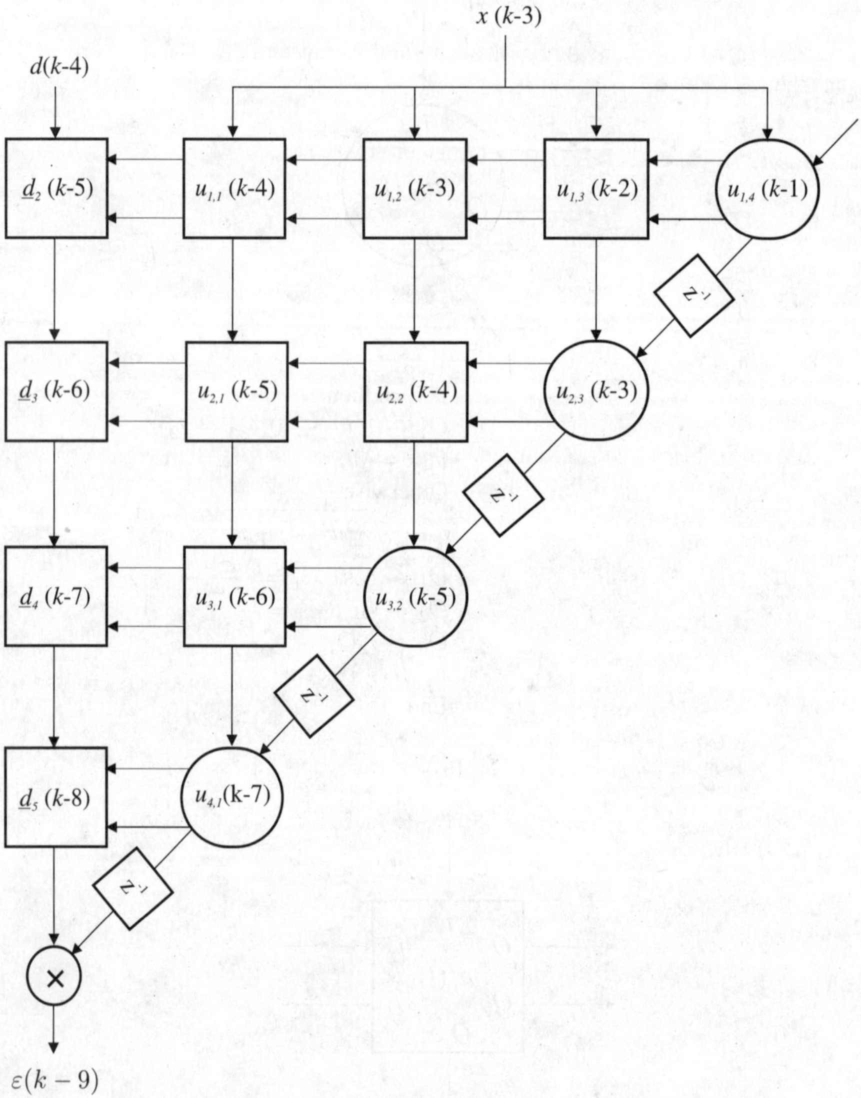

Figure 9.2 QR-Decomposition systolic array for N=3.

Likewise, the $(i+1)$th row performs the rotation $\theta_i(l)$ that eliminates $x'_i(l-N+i)$ and also the rotation in the vector $\underline{\mathbf{d}}(l)$.

In the bottom part of the systolic array, the product of $\varepsilon_{q_1}(l)$ and $\gamma(l)$ is calculated at each clock instant, in order to generate *a posteriori* output error given by $\varepsilon(l)$. The output error obtained in a given sample period k corresponds to the error related to the input data vector of $2(N+1)$ clock periods before.

The systolic array of Fig. 9.2 exhibits several desirable features such as local interconnection, regularity, and simple control circuitry, that yields a simple implementation. A possible problem, as pointed out in [13], is the need to distribute a single clock throughout a large array, without incurring any clock skew.

The presented systolic array does not allow the computation of the tap-weight coefficients. A solution pointed out in [13] employs the array of Fig. 9.2 by freezing the array and applying an appropriate input signal sequence such that the tap-weight coefficients are made available at the array output $\varepsilon(l)$. An alternative way is add a systolic array to solve the back-substitution problem [13]. The array is shown in Fig. 9.3 with the corresponding algorithm. The complete computation of the coefficient vector $\mathbf{w}(k)$ requires 2^{N+1} clock samples. In this array, the square cells produce the partial products involved in equation (9.11). The round cell performs the subtraction of the sum of the product result with an element of the vector $\underline{\mathbf{d}}(k-8)$, namely $\underline{d}_{5-i}(k-8)$. This cell also performs the division of the subtraction result by the element $u_{N+1-i,i+1}(k-8)$ of the matrix $\mathbf{U}(k-8)$. Starting with $i=0$, the sum of products has no elements and as a consequence the round cell just performs the division $\frac{\underline{d}_{5-i}(k-8)}{u_{N+1-i,i+1}(k-8)}$. On the other hand, for $i=N$ all the square cells are actually taking part in the computation of the sum of products. Note that in this case, in order to obtain $w_N(k-8)$, the results of all the cells starting from left to right must be ready, i.e., there is no pipelining involved.

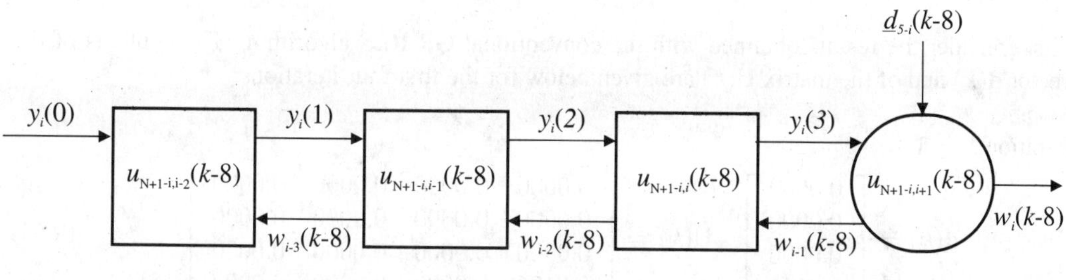

$w_i = 0$ for $i < 0$
Do for $i = 0, 1, \ldots, N$
$y_i(N-i) = 0$
Do for l= $N-i+1, \ldots, N$
$y_i(l) = y_i(l-1) + u_{N+1-i,i-N+l}(k-8)w_{i-N+l-1}(k-8)$
End

$$w_i(k-8) = \frac{\underline{d}_{5-i}(k-8) - y_i(3)}{u_{N+1-i,i+1}(k-8)}$$

End

Figure 9.3 Systolic array and algorithm for the computation of $\mathbf{w}(k)$.

Example 9.2

Let us choose a simple example, in order to illustrate how the systolic array implementation works, and compare the results with those belonging to the standard implementation of the QR-RLS algorithm. The chosen order is $N = 3$ and the forgetting factor is $\lambda = 0.99$.

Suppose that in an adaptive-filtering environment, the input signal consists of

$$x(k) = \sin(\omega_0 k)$$

where $\omega_0 = \frac{\pi}{250}$.

The desired signal is generated by applying the same sinusoid to an FIR filter whose coefficients are given by

$$\mathbf{w}_o = [1.0\; 0.9\; 0.1\; 0.2]^T$$

Solution:

First consider the results obtained with the conventional QR-RLS algorithm. The contents of the vector $\underline{\mathbf{d}}(k)$ and of the matrix $\mathbf{U}(k)$ are given below for the first four iterations.

Iteration $k = 1$

$$\underline{\mathbf{d}}(k) = \begin{bmatrix} 0.0000 \\ 0.0000 \\ 0.0000 \\ 0.0126 \end{bmatrix} \quad \mathbf{U}(k) = \begin{bmatrix} 0.0000 & 0.0000 & 0.0000 & 0.0000 \\ 0.0000 & 0.0000 & 0.0000 & 0.0000 \\ 0.0000 & 0.0000 & 0.0000 & 0.0000 \\ 0.0126 & 0.0000 & 0.0000 & 0.0000 \end{bmatrix} \tag{9.55}$$

Iteration $k = 2$

$$\underline{\mathbf{d}}(k) = \begin{bmatrix} 0.0000 \\ 0.0000 \\ 0.0364 \\ 0.0125 \end{bmatrix} \quad \mathbf{U}(k) = \begin{bmatrix} 0.0000 & 0.0000 & 0.0000 & 0.0000 \\ 0.0000 & 0.0000 & 0.0000 & 0.0000 \\ 0.0251 & 0.0126 & 0.0000 & 0.0000 \\ 0.0125 & 0.0000 & 0.0000 & 0.0000 \end{bmatrix} \tag{9.56}$$

Iteration $k = 3$

$$\underline{\mathbf{d}}(k) = \begin{bmatrix} 0.0000 \\ 0.0616 \\ 0.0363 \\ 0.0124 \end{bmatrix} \quad \mathbf{U}(k) = \begin{bmatrix} 0.0000 & 0.0000 & 0.0000 & 0.0000 \\ 0.0377 & 0.0251 & 0.0126 & 0.0000 \\ \boxed{0.0250} & \boxed{0.0125} & 0.0000 & 0.0000 \\ 0.0124 & 0.0000 & 0.0000 & 0.0000 \end{bmatrix} \tag{9.57}$$

Iteration $k = 4$

$$\underline{\mathbf{d}}(k) = \begin{bmatrix} 0.0892 \\ 0.0613 \\ 0.0361 \\ 0.0124 \end{bmatrix} \quad \mathbf{U}(k) = \begin{bmatrix} 0.0502 & 0.0377 & 0.0251 & 0.0126 \\ 0.0375 & 0.0250 & 0.0125 & 0.0000 \\ 0.0249 & 0.0124 & 0.0000 & 0.0000 \\ \boxed{0.0124} & 0.0000 & 0.0000 & 0.0000 \end{bmatrix} \tag{9.58}$$

Iteration $k = 5$

$$\underline{\mathbf{d}}(k) = \begin{bmatrix} 0.1441 \\ 0.0668 \\ 0.0359 \\ 0.0123 \end{bmatrix} \quad \mathbf{U}(k) = \begin{bmatrix} 0.0785 & 0.0617 & 0.0449 & 0.0281 \\ 0.0409 & 0.0273 & 0.0136 & 0.0000 \\ 0.0248 & 0.0124 & 0.0000 & 0.0000 \\ 0.0123 & 0.0000 & 0.0000 & 0.0000 \end{bmatrix} \tag{9.59}$$

The data stored in the systolic array implementation represent the elements of the vector $\underline{\mathbf{d}}(k)$ and of the matrix $\mathbf{U}(k)$ skewed in time. This data is shown below starting from the the fourth iteration, since before that no data is available to the systolic array.

Observe when the elements of the $\mathbf{U}(k)$ appear stored at the systolic array. For example, consider the highlighted elements. In particular, the element $(4, 1)$ at instant $k = 4$ appears stored in the systolic array at instant $k = 10$, whereas the elements $(3, 1)$ and $(3, 2)$ at instant $k = 3$ appear stored in the systolic array at instants $k = 8$ and $k = 7$, respectively. Following the same line of thought, it is straightforward to understand how the remaining elements of the systolic array are calculated.

Iteration $k = 4$

$$\begin{bmatrix} 0. \\ 0. \\ 0. \\ 0. \end{bmatrix} \begin{bmatrix} 0. & 0. & 0. & 0.0126 \\ 0. & 0. & 0. & \\ 0. & 0. & & \\ 0. & & & \end{bmatrix} \tag{9.60}$$

Iteration $k = 5$

$$\begin{bmatrix} 0. \\ 0. \\ 0. \\ 0. \end{bmatrix} \begin{bmatrix} 0. & 0. & 0.0251 & 0.0281 \\ 0. & 0. & 0.0126 & \\ 0. & 0. & & \\ 0. & & & \end{bmatrix} \tag{9.61}$$

Iteration $k = 6$

$$\begin{bmatrix} 0. \\ 0. \\ 0. \\ 0. \end{bmatrix} \begin{bmatrix} 0. & 0.0377 & 0.0449 & 0.0469 \\ 0. & 0.0251 & 0.0125 & \\ 0. & 0.0126 & & \\ 0. & & & \end{bmatrix} \tag{9.62}$$

Iteration $k = 7$

$$\begin{bmatrix} 0. \\ 0. \\ 0. \\ 0. \end{bmatrix} \begin{bmatrix} 0.0502 & 0.0617 & 0.0670 & 0.0686 \\ 0.0377 & 0.0250 & 0.0136 & \\ 0.0251 & \boxed{0.0125} & & \\ 0.0126 & & & \end{bmatrix} \tag{9.63}$$

Iteration $k = 8$

$$\begin{bmatrix} 0.0892 \\ 0.0616 \\ 0.0364 \\ 0.0126 \end{bmatrix} \begin{bmatrix} 0.0785 & 0.0870 & 0.0913 & 0.0927 \\ 0.0375 & 0.0273 & 0.0148 & \\ \boxed{0.0250} & 0.0124 & & \\ 0.0125 & & & \end{bmatrix} \tag{9.64}$$

Iteration $k = 9$

$$\begin{bmatrix} 0.1441 \\ 0.0613 \\ 0.0363 \\ 0.0125 \end{bmatrix} \begin{bmatrix} 0.1070 & 0.1141 & 0.1179 & 0.1191 \\ 0.0409 & 0.0297 & 0.0160 & \\ 0.0249 & 0.0124 & & \\ 0.0124 & & & \end{bmatrix} \tag{9.65}$$

Iteration $k = 10$

$$\begin{bmatrix} 0.2014 \\ 0.0668 \\ 0.0361 \\ 0.0124 \end{bmatrix} \begin{bmatrix} 0.1368 & 0.1430 & 0.1464 & 0.1475 \\ 0.0445 & 0.0319 & 0.0170 & \\ 0.0248 & 0.0123 & & \\ \boxed{0.0124} & & & \end{bmatrix} \tag{9.66}$$

Iteration $k = 11$

$$\begin{bmatrix} 0.2624 \\ 0.0726 \\ 0.0359 \\ 0.0124 \end{bmatrix} \begin{bmatrix} 0.1681 & 0.1737 & 0.1768 & 0.1778 \\ 0.0479 & 0.0340 & 0.0180 & \\ 0.0246 & 0.0123 & & \\ 0.0123 & & & \end{bmatrix} \tag{9.67}$$

It is a good exercise for the reader to examine the elements of the vectors and matrices in equations (9.60)-(9.67) and detect when these elements appear in the corresponding vectors $\underline{\mathbf{d}}(k)$ and matrices $\mathbf{U}(k)$ of equations (9.55)-(9.59).

□

9.4 SOME IMPLEMENTATION ISSUES

Several articles related to implementation issues of the QR-RLS algorithm such as the elimination of square root computation [16], stability and quantization error analyses [17]-[20] are available in the open literature. In this section, some of these results are briefly reviewed.

The stability of the QR-RLS algorithm is the first issue to be concerned when considering a real implementation. Fortunately, the QR-RLS algorithm implemented in finite precision was proved stable in the bounded input/bounded output sense in [18]. The proof was based on the analysis of the bounds for the internal recursions of the algorithm [18]-[19]. From another study based on the quantization-error propagation in the finite-precision implementation of the QR-RLS algorithm, it was possible to derive the error recursions for the main quantities of the algorithm, leading to the stability conditions of the QR-RLS algorithm [20]. The convergence on average of the QR-RLS algorithm can be guaranteed if the following inequality is satisfied [20]:

$$\lambda^{1/2} \parallel \tilde{\mathbf{Q}}_Q(k) \parallel_2 \leq 1 \tag{9.68}$$

where the two norm $\parallel \cdot \parallel_2$ of a matrix used here is the square root of the largest eigenvalue and the notation $[\cdot]_Q$ denotes the finite-precision version of $[\cdot]$. Therefore,

$$\parallel \tilde{\mathbf{Q}}_Q(k) \parallel_2 = \text{MAX}_i \sqrt{\cos_Q^2 \theta_i(k) + \sin_Q^2 \theta_i(k)} \tag{9.69}$$

where $\text{MAX}_i[\cdot]$ is the maximum value of $[\cdot]$. The stability condition can be rewritten as follows:

$$\lambda \leq \frac{1}{\text{MAX}_i \left[\cos_Q^2 \theta_i(k) + \sin_Q^2 \theta_i(k)\right]} \tag{9.70}$$

It can then be concluded that keeping the product of the forgetting factor and the maximum eigenvalue of the Givens rotations smaller than unity is a sufficient condition to guarantee the stability.

For the implementation of any adaptive algorithm, it is necessary to estimate quantitatively the dynamic range of all internal variables of the algorithm in order to determine the length of all the registers required in the actual implementation. Although this issue should be considered in the implementation of any adaptive-filtering algorithm, it is particularly relevanι in the QR-RLS algorithms due to their large number of internal variables. The first attempt to address this problem was reported in [19], where expressions for the steady-state values of the cosines and sines of the Givens rotations were determined, as well as the bounds for the dynamic range of the information stored in the processing cells. The full quantitative analysis of the dynamic range of all internal quantities of the QR-RLS algorithm was presented in [20] for the conventional and systolic-array forms. For fixed-point implementation, it is important to determine the internal signal with the largest energy such that frequent overflow in the internal variables of the QR-RLS algorithm can be avoided. The first entry of the triangularized information matrix can be shown to have the largest energy [20] and its steady-state value is approximately

$$u_{0,0}(k) \approx \frac{\sigma_x}{\sqrt{1-\lambda}} \tag{9.71}$$

where σ_x^2 is the variance of the input signal.

The procedure to derive the results above discussed consists of first analyzing the QR-RLS algorithm for ideal infinite-precision implementation. The second step is modeling the quantization errors and deriving the recursive equations that include the overall error in each quantity of the QR-RLS algorithm [20]. Then conditions to guarantee the stability of the algorithm could be derived. A further step is to derive closed-form solutions to the mean-squared values of the deviations in the internal variables of the algorithm due to finite-precision operations. The main objective in this step is to obtain the excess mean-square error and the variance of the deviation in the tap-weight coefficients. Analytical expressions for these quantities are not very simple unless a number of assumptions about the input and reference signals are assumed [20]. However, they are useful to the designer.

9.5 FAST QR-RLS ALGORITHM

For the derivation of the fast QR-RLS algorithms, it is first necessary to study the solutions of the forward and backward prediction problems. As seen in Chapters 7 and 8, the predictor solutions were also required in the derivation of the lattice-based and the fast transversal RLS algorithms.

A family of fast QR-RLS algorithms can be generated depending on the following aspects of their derivation:

- The type of triangularization applied to the input signal matrix, taking into consideration the notation adopted in this book where the first element of the data vectors corresponds to the most recent data. The upper triangularization is related to the updating of forward prediction errors, whereas the lower triangularization involves the updating of backward prediction errors.
- The type of error utilized in the updating process, namely, if it is *a priori* or *a posteriori* error.

Table 9.2 shows the classification of the fast QR-RLS algorithms indicating the references where the specific algorithms can be found. Although these algorithms are comparable in terms of computational complexity, those based on backward prediction errors (which utilize lower triangularization of the information matrix) are numerically stable when implemented in finite precision. This good numerical behavior is related to backward consistency and minimal properties inherent to these algorithms [21].

Table 9.2 Classification of the Fast QR-RLS Algorithms

Error Type	Prediction Forward	Prediction Backward
A Priori	[9]	[10], [11]
A Posteriori	[4]	[8], [12]

In this section, we start with the application of the QR decomposition to the lower triangularization of the input signal information matrix. Then, the decomposition is applied to the backward and forward prediction problems. This type of triangularization is related to the updating of backward prediction errors.

A fast QR-RLS algorithm is derived by performing the triangularization of the information matrix in this alternative form, namely by generating a lower triangular matrix, and by first applying the triangularization to the backward linear prediction problem. Originally, the algorithm to be presented here was proposed in [5] and later detailed in [7] and [8]. The derivations are quite similar to those presented for the standard QR-RLS algorithm. Therefore, we will use the previous results in order to avoid unnecessary repetition. In order to accomplish this objective while avoiding confusion, the following notations are respectively used for the triangularization matrix and the lower triangular matrices $\mathcal{Q}$ and $\mathcal{U}$. These matrices have the following forms

$$\mathcal{U}(k) = \begin{bmatrix} 0 & 0 & \cdots & 0 & u_{1,N+1} \\ 0 & 0 & \cdots & u_{2,N} & u_{2,N+1} \\ \vdots & & & \vdots & \vdots \\ u_{N+1,1} & u_{N+1,2} & \cdots & u_{N+1,N} & u_{N+1,N+1} \end{bmatrix} \tag{9.72}$$

$$\tilde{\mathcal{Q}}(k) = \begin{bmatrix} \cos\theta_N(k) & \cdots & 0 & \cdots & -\sin\theta_N(k) & \mathbf{0} \\ \vdots & & & & \vdots & \vdots \\ 0 & & \mathbf{I}_{k-N-1} & & 0 & \vdots \\ \vdots & & & & \vdots & \vdots \\ \sin\theta_N(k) & \cdots & 0 & \cdots & \cos\theta_N(k) & \mathbf{0} \\ & & \mathbf{0} & & & \mathbf{I}_N \end{bmatrix}$$

$$\cdot \begin{bmatrix} \cos\theta_{N-1}(k) & \cdots & 0 & \cdots & -\sin\theta_{N-1}(k) & 0 \\ \vdots & & & & \vdots & \vdots \\ 0 & & \mathbf{I}_{k-N} & & 0 & \vdots \\ \vdots & & & & \vdots & \vdots \\ \sin\theta_{N-1}(k) & \cdots & 0 & \cdots & \cos\theta_{N-1}(k) & 0 \\ 0 & \cdots & 0 & \cdots & 0 & \mathbf{I}_{N-1} \end{bmatrix}$$

$$\cdots \begin{bmatrix} \cos\theta_0(k) & \cdots & 0 & \cdots & -\sin\theta_0(k) \\ \vdots & & & & \vdots \\ 0 & & \mathbf{I}_{k-1} & & 0 \\ \vdots & & & & \vdots \\ \sin\theta_0(k) & \cdots & 0 & \cdots & \cos\theta_0(k) \end{bmatrix} \tag{9.73}$$

The triangularization procedure has the following general form

$$\begin{aligned} \mathcal{Q}(k)\underline{\mathbf{X}}(k) &= \tilde{\mathcal{Q}}(k) \begin{bmatrix} 1 & \mathbf{0} \\ \mathbf{0} & \tilde{\mathcal{Q}}(k-1) \end{bmatrix} \begin{bmatrix} \mathbf{I}_2 & \mathbf{0} \\ \mathbf{0} & \tilde{\mathcal{Q}}(k-2) \end{bmatrix} \\ &\quad \cdots \begin{bmatrix} \mathbf{I}_{k-N} & \mathbf{0} \\ \mathbf{0} & \tilde{\mathcal{Q}}(k-N) \end{bmatrix} \underline{\mathbf{X}}(k) \\ &= \underbrace{\begin{bmatrix} \mathbf{0} \\ \mathcal{U}(k) \end{bmatrix}}_{N+1} \begin{matrix} \}\, k-N \\ \}\, N+1 \end{matrix} \end{aligned} \tag{9.74}$$

where $\mathcal{Q}(k)$ is a $(k+1)$ by $(k+1)$ matrix which represents the overall triangularization matrix.

As usual the multiplication by zero elements can be avoided by replacing $\tilde{\mathcal{Q}}(k)$ by $\mathcal{Q}_\theta(k)$, where the increasing $\mathbf{I}_{k-N-1}$ section of $\tilde{\mathcal{Q}}(k)$ is removed very much like in equations (9.38) and (9.39). The resulting equation is

$$\mathcal{Q}_\theta(k) \begin{bmatrix} \mathbf{x}^T(k) \\ \lambda^{1/2}\mathcal{U}(k-1) \end{bmatrix} = \mathcal{Q}'_{\theta_N}(k)\mathcal{Q}'_{\theta_{N-1}}(k)\cdots\mathcal{Q}'_{\theta_i}(k) \begin{bmatrix} \mathbf{x}'_i(k) \\ \mathcal{U}'_i(k) \end{bmatrix} \tag{9.75}$$

where $\mathcal{Q}'_{\theta_i}(k)$ is derived from $\mathcal{Q}'_i(k)$ by removing the $\mathbf{I}_{k-N-1}$ section of $\mathcal{Q}'_i(k)$ along with the corresponding rows and columns, resulting in the following form

$$\mathcal{Q}'_{\theta_i}(k) = \begin{bmatrix} \cos\theta_i(k) & \cdots & 0 & \cdots & -\sin\theta_i(k) & \cdots & 0 \\ \vdots & & & & \vdots & & \vdots \\ 0 & & \mathbf{I}_{N-i} & & 0 & \cdots & 0 \\ \vdots & & & & \vdots & & \vdots \\ \sin\theta_i(k) & \cdots & 0 & \cdots & \cos\theta_i(k) & \cdots & 0 \\ \vdots & & \vdots & & \vdots & & \mathbf{I}_i \\ 0 & \cdots & 0 & \cdots & 0 & & \end{bmatrix} \tag{9.76}$$

The Givens rotation elements are calculated by

$$\cos\theta_i(k) = \frac{[\mathcal{U}'_i(k)]_{N+1-i,i+1}}{c_i} \tag{9.77}$$

$$\sin\theta_i(k) = \frac{x'_i(k-i)}{c_i} \tag{9.78}$$

where $c_i = \sqrt{[\mathcal{U}'_i(k)]^2_{N+1-i,i+1} + x'^2_i(k-i)}$, and $[\cdot]_{i,j}$ denotes the (i,j) element of the matrix.

9.5.1 Backward Prediction Problem

In the backward prediction problem, the desired signal and vector are respectively

$$d_b(k+1) = x(k-N) \tag{9.79}$$

$$\mathbf{d}_b(k+1) = \begin{bmatrix} x(k-N) \\ \lambda^{1/2}x(k-N-1) \\ \vdots \\ \lambda^{\frac{k-N}{2}}x(0) \\ 0 \\ \vdots \\ 0 \end{bmatrix} \tag{9.80}$$

The reader should note that in the present case an extra row was added to the vector $\mathbf{d}_b(k+1)$. For example, the dimension of $\mathbf{d}_b(k+1)$ is now $(k+2)$ by 1. The backward-prediction-error vector is given by

$$\begin{aligned} \boldsymbol{\varepsilon}_b(k+1) &= \mathbf{d}_b(k+1) - \underline{\mathbf{X}}(k+1)\mathbf{w}_b(k+1) \\ &= [\underline{\mathbf{X}}(k+1)\ \mathbf{d}_b(k+1)] \begin{bmatrix} -\mathbf{w}_b(k+1) \\ 1 \end{bmatrix} \end{aligned} \tag{9.81}$$

The triangularization matrix $\mathcal{Q}(k+1)$ of the input data matrix can be applied to the backward prediction error resulting in

$$\mathcal{Q}(k+1)\varepsilon_b(k+1) = \mathcal{Q}(k+1)\mathbf{d}_b(k+1) - \begin{bmatrix} \mathbf{0} \\ \mathcal{U}(k+1) \end{bmatrix} \mathbf{w}_b(k+1) \tag{9.82}$$

or equivalently

$$\varepsilon_{bq}(k+1) = \mathbf{d}_{bq}(k+1) - \begin{bmatrix} \mathbf{0} \\ \mathcal{U}(k+1) \end{bmatrix} \mathbf{w}_b(k+1) \tag{9.83}$$

From equations and (9.81) and (9.83), it follows that

$$\begin{aligned}\varepsilon_{bq}(k+1) &= \mathcal{Q}(k+1)[\underline{\mathbf{X}}(k+1)\ \mathbf{d}_b(k+1)] \begin{bmatrix} -\mathbf{w}_b(k+1) \\ 1 \end{bmatrix} \\ &= \begin{bmatrix} & \varepsilon_{bq_1}(k+1) \\ \mathbf{0} & \varepsilon_{bq_2}(k+1) \\ & \vdots \\ & \varepsilon_{bq_{k-N+1}}(k+1) \\ \mathcal{U}(k+1) & \mathbf{x}_{q_3}(k+1) \end{bmatrix} \begin{bmatrix} -\mathbf{w}_b(k+1) \\ 1 \end{bmatrix}\end{aligned} \tag{9.84}$$

Also note that

$$[\underline{\mathbf{X}}(k+1)\ \mathbf{d}_b(k+1)] = \underline{\mathbf{X}}^{(N+2)}(k+1) \tag{9.85}$$

where $\underline{\mathbf{X}}^{(N+2)}(k+1)$ is an extended version of $\underline{\mathbf{X}}(k+1)$, with one input signal information vector added. In other words, $\underline{\mathbf{X}}^{(N+2)}(k+1)$ is the information matrix that would be obtained if one additional delay was added at the end of the delay line.

In order to avoid increasing vectors in the algorithm, $\varepsilon_{bq_1}(k+1), \varepsilon_{bq_2}(k+1), \ldots, \varepsilon_{bq_{k-N}}(k+1)$ can be eliminated in equation (9.84) through Givens rotations, as follows:

$$\begin{aligned}\mathbf{Q}_b(k+1)\varepsilon_{bq}(k+1) &= \mathbf{Q}_b(k+1) \begin{bmatrix} & \varepsilon_{bq_1}(k+1) \\ \mathbf{0} & \varepsilon_{bq_2}(k+1) \\ & \vdots \\ & \varepsilon_{bq_{k-N+1}}(k+1) \\ \mathcal{U}(k+1) & \mathbf{x}_{q_3}(k+1) \end{bmatrix} \begin{bmatrix} -\mathbf{w}_b(k+1) \\ 1 \end{bmatrix} \\ &= \begin{bmatrix} \mathbf{0} & \mathbf{0} \\ & ||\varepsilon_b(k+1)|| \\ \mathcal{U}(k+1) & \mathbf{x}_{q_3}(k+1) \end{bmatrix} \begin{bmatrix} -\mathbf{w}_b(k+1) \\ 1 \end{bmatrix}\end{aligned} \tag{9.86}$$

Note that by induction $[\mathcal{U}]_{N+1-i,i+1}(k+1) = ||\varepsilon_{b,i}(k+1)||$, where $||\varepsilon_{b,i}(k+1)||^2$ corresponds to the least-square backward prediction error of an $(i-1)$th-order predictor.

9.5.2 Forward Prediction Problem

In the forward prediction problem, the following relations are valid:[1]

$$d_f(k) = x(k+1) \tag{9.87}$$

$$\mathbf{d}_f(k) = \begin{bmatrix} x(k+1) \\ \lambda^{1/2}x(k) \\ \vdots \\ \lambda^{\frac{k+1}{2}}x(0) \end{bmatrix} \tag{9.88}$$

$$\varepsilon_f(k) = \mathbf{d}_f(k) - \begin{bmatrix} \underline{\mathbf{X}}(k) \\ \mathbf{0} \end{bmatrix} \mathbf{w}_f(k) \tag{9.89}$$

where $d_f(k)$ is the desired signal, $\mathbf{d}_f(k)$ is the desired signal vector, and $\varepsilon_f(k)$ is the error signal vector.

Now, we can consider applying the QR decomposition, as was previously done in equation (9.74) to the forward prediction error above defined. It should be noted that in the present case an extra row was added to the vectors $\varepsilon_f(k)$ and $\mathbf{d}_f(k)$, as can be verified in the following relations:

$$\varepsilon_f(k) = \left[\begin{array}{c|c} \mathbf{d}_f(k) & \begin{array}{c} \underline{\mathbf{X}}(k) \\ \mathbf{0} \end{array} \end{array} \right] \begin{bmatrix} 1 \\ -\mathbf{w}_f(k) \end{bmatrix} \tag{9.90}$$

and

$$\begin{aligned} \varepsilon_{fq}(k) &= \begin{bmatrix} \mathcal{Q}(k) & \mathbf{0} \\ \mathbf{0} & 1 \end{bmatrix} \left[\begin{array}{c|c} \mathbf{d}_f(k) & \begin{array}{c} \underline{\mathbf{X}}(k) \\ \mathbf{0} \end{array} \end{array} \right] \begin{bmatrix} 1 \\ -\mathbf{w}_f(k) \end{bmatrix} \\ &= \begin{bmatrix} \varepsilon_{fq_1}(k) & \\ \vdots & \mathbf{0} \\ \varepsilon_{fq_{k-N}}(k) & \\ \mathbf{x}_{q2}(k) & \mathcal{U}(k) \\ \lambda^{\frac{k+1}{2}}x(0) & \mathbf{0} \end{bmatrix} \begin{bmatrix} 1 \\ -\mathbf{w}_f(k) \end{bmatrix} \end{aligned} \tag{9.91}$$

Note that:

$$\left[\begin{array}{c|c} \mathbf{d}_f(k) & \begin{array}{c} \underline{\mathbf{X}}(k) \\ \mathbf{0} \end{array} \end{array} \right] = \underline{\mathbf{X}}^{(N+2)}(k+1) \tag{9.92}$$

which is an order extended version of $\underline{\mathbf{X}}(k+1)$ and has dimension $(k+2)$ by $(N+2)$.

In order to recursively solve equation (9.91) without dealing with ever increasing matrices, a set of Givens rotations are applied in order to eliminate $\varepsilon_{fq_1}(k)$, $\varepsilon_{fq_2}(k)$, $\ldots$, $\varepsilon_{fq_{k-N}}(k)$, such that

[1] The reader should note that here the definition of forward prediction error is slightly different from that used in Chapters 7 and 8, where in the present case we are using the input and desired signals one step ahead. This allows us to use the same information matrix as the conventional QR-Decomposition algorithm of subsection 9.2.3.

the information matrix that premultiplies the vector $[1 \quad -\mathbf{w}_f(k)]^T$ is triangularized. The Givens rotations can recursively be obtained by

$$\begin{aligned}\mathbf{Q}_f(k) &= \tilde{\mathbf{Q}}_f(k)\begin{bmatrix}1 & \mathbf{0}\\ \mathbf{0} & \mathbf{Q}_f(k-1)\end{bmatrix}\\ &= \tilde{\mathbf{Q}}_f(k)\begin{bmatrix}1 & \mathbf{0}\\ \mathbf{0} & \tilde{\mathbf{Q}}_f(k-1)\end{bmatrix}\cdots\begin{bmatrix}\mathbf{I}_{k-N-1} & \mathbf{0}\\ \mathbf{0} & \tilde{\mathbf{Q}}_f(N+1)\end{bmatrix}\end{aligned} \tag{9.93}$$

where $\tilde{\mathbf{Q}}_f(k)$ is defined as

$$\tilde{\mathbf{Q}}_f(k) = \begin{bmatrix}\cos\theta_f(k) & \cdots & 0 & \cdots & -\sin\theta_f(k)\\ \vdots & & & & \vdots\\ 0 & & \mathbf{I}_k & & 0\\ \vdots & & & & \vdots\\ \sin\theta_f(k) & \cdots & 0 & \cdots & \cos\theta_f(k)\end{bmatrix} \tag{9.94}$$

If in each iteration, the above rotation is applied to equation (9.91), we have

$$\begin{aligned}\varepsilon'_{fq}(k) &= \tilde{\mathbf{Q}}_f(k)\begin{bmatrix}1 & \mathbf{0}\\ \mathbf{0} & \mathbf{Q}_f(k-1)\end{bmatrix}\begin{bmatrix}\varepsilon_{fq_1}(k) & \\ \vdots & \mathbf{0}\\ \varepsilon_{fq_{k-N}}(k) & \\ \mathbf{x}_{q2}(k) & \mathcal{U}(k)\\ \lambda^{\frac{k+1}{2}}x(0) & \mathbf{0}\end{bmatrix}\begin{bmatrix}1\\ -\mathbf{w}_f(k)\end{bmatrix}\\ &= \tilde{\mathbf{Q}}_f(k)\begin{bmatrix}\varepsilon_{fq_1}(k) & \\ 0 & \mathbf{0}\\ \vdots & \\ 0 & \\ \mathbf{x}_{q_2}(k) & \mathcal{U}(k)\\ \lambda^{1/2}||\varepsilon_f(k-1)|| & \mathbf{0}\end{bmatrix}\begin{bmatrix}1\\ -\mathbf{w}_f(k)\end{bmatrix}\\ &= \begin{bmatrix}0 & \\ \vdots & \mathbf{0}\\ 0 & \\ \mathbf{x}_{q_2}(k) & \mathcal{U}(k)\\ ||\varepsilon_f(k)|| & \mathbf{0}\end{bmatrix}\begin{bmatrix}1\\ -\mathbf{w}_f(k)\end{bmatrix}\end{aligned} \tag{9.95}$$

where

$$\cos\theta_f(k) = \frac{\lambda^{1/2}||\varepsilon_f(k-1)||}{\sqrt{\lambda||\varepsilon_f(k-1)||^2 + \varepsilon^2_{fq_1}(k)}} \tag{9.96}$$

$$\sin\theta_f(k) = \frac{\varepsilon_{fq_1}(k)}{\sqrt{\lambda||\varepsilon_f(k-1)||^2 + \varepsilon^2_{fq_1}(k)}} \tag{9.97}$$

and $||\varepsilon_f(k)||$ is the norm of the forward prediction error vector shown in equation (9.91). This result can be shown by evoking the fact that the last element of $\varepsilon'_{fq}(k)$ is equal to $||\varepsilon_f(k)||$, since $||\varepsilon'_{fq}(k)|| = ||\varepsilon_{fq}(k)|| = ||\varepsilon_f(k)||$, because these error vectors are related through unitary transformations.

Also, it is worthwhile to recall that in equation (9.95) the relation $[\mathcal{U}]_{N+1-i,i+1}(k) = ||\varepsilon_{b,i}(k)||$ is still valid (see equation (9.86)). Also, by induction, it can easily be shown from equation (9.91) that:

For $k = 0, 1, \ldots, N$

$$||\varepsilon_f(k)|| = \lambda^{\frac{k+1}{2}} x(0)$$

for $k = N + 1$

$$||\varepsilon'_{fq}(k)|| = ||\varepsilon_f(k)|| = \sqrt{\lambda^{k+1}x^2(0) + \varepsilon^2_{fq_1}(k)}$$

for $k = N + 2$

$$\begin{aligned} ||\varepsilon_f(k)|| &= \sqrt{\lambda^{k+1}x^2(0) + \lambda\varepsilon^2_{fq_1}(k-1) + \varepsilon^2_{fq_1}(k)} \\ &= \sqrt{\lambda||\varepsilon_f(k-1)||^2 + \varepsilon^2_{fq_1}(k)} \end{aligned}$$

for $k > N + 2$

$$||\varepsilon_f(k)||^2 = \lambda||\varepsilon_f(k-1)||^2 + \varepsilon^2_{fq_1}(k) \tag{9.98}$$

In the present case, it can be assumed that the partial triangularization can be performed at each iteration as follows:

$$\begin{bmatrix} 0 & \\ 0 & \mathbf{0} \\ \vdots & \\ 0 & \\ \mathbf{x}_{q2}(k) & \mathcal{U}(k) \\ ||\varepsilon_f(k)|| & \mathbf{0} \end{bmatrix} = \tilde{\mathbf{Q}}_f(k) \begin{bmatrix} \tilde{\mathcal{Q}}(k) & \mathbf{0} \\ \mathbf{0} & 1 \end{bmatrix} \begin{bmatrix} x(k+1) & \mathbf{x}^T(k) \\ & \\ \mathbf{0} & \mathbf{0} \\ & \\ \lambda^{1/2}\mathbf{x}_{q2}(k-1) & \lambda^{1/2}\mathcal{U}(k-1) \\ \lambda^{1/2}||\varepsilon_f(k-1)|| & \mathbf{0} \end{bmatrix} \tag{9.99}$$

Now we can eliminate $\mathbf{x}_{q_2}(k)$ through a set of rotations $\mathbf{Q}'_f(k+1)$ such that

$$\mathcal{U}^{(N+2)}(k+1) = \mathbf{Q}'_f(k+1) \begin{bmatrix} \mathbf{x}_{q2}(k) & \mathcal{U}(k) \\ ||\varepsilon_f(k)|| & \mathbf{0} \end{bmatrix} \tag{9.100}$$

where the superscript $(N + 2)$ in the above matrices denotes rotation matrices applied to data with $(N + 2)$ elements.

From the above equation, we can realize that $\mathbf{Q}'_f(k+1)$ consists of a series of rotations in the following order

$$\mathbf{Q}'_f(k+1) = \begin{bmatrix} \mathbf{I}_N & \mathbf{0} & \\ \mathbf{0} & \cos\theta'_{f_1}(k+1) & -\sin\theta'_{f_1}(k+1) \\ & \sin\theta'_{f_1}(k+1) & \cos\theta'_{f_1}(k+1) \end{bmatrix}$$

$$\cdots \begin{bmatrix} 1 & 0 & \cdots & \cdots & \cdots & \cdots & \cdots & 0 \\ 0 & \cos\theta'_{f_N}(k+1) & 0 & \cdots & 0 & \cdots & 0 & -\sin\theta'_{f_N}(k+1) \\ \vdots & 0 & & & & & & 0 \\ \vdots & & & \mathbf{I}_{N-1} & & & \vdots & \vdots \\ 0 & \sin\theta'_{f_N}(k+1) & 0 & \cdots & 0 & \cdots & 0 & \cos\theta'_{f_N}(k+1) \end{bmatrix}$$

$$\cdot \begin{bmatrix} \cos\theta'_{f_{N+1}}(k+1) & 0 & \cdots & 0 & \cdots & 0 & -\sin\theta'_{f_{N+1}}(k+1) \\ 0 & & & & & & 0 \\ \vdots & & & \mathbf{I}_N & & & \vdots \\ \vdots & & & & & & \vdots \\ \sin\theta'_{f_{N+1}}(k+1) & 0 & \cdots & 0 & \cdots & 0 & \cos\theta'_{f_{N+1}}(k+1) \end{bmatrix} \quad (9.101)$$

where the rotation entries of $\mathbf{Q}'_f(k+1)$ are calculated as follows:

$$\mu_i = \sqrt{\mu_{i-1}^2 + x_{q2i}^2(k)}$$
$$\cos\theta'_{f_{N+2-i}}(k+1) = \frac{\mu_{i-1}}{\mu_i}$$
$$\sin\theta'_{f_{N+2-i}}(k+1) = \frac{x_{q2i}(k)}{\mu_i} \quad (9.102)$$

for $i = 1, \ldots, N+1$, where $\mu_0 = ||\varepsilon_f(k)||$. Note that μ_{N+1} is the norm of the weighted backward prediction error $||\varepsilon_{b,0}(k+1)||$, for a zero-order predictor (see equation (9.86)). The quantity $x_{q2i}(k)$ denotes the ith element of the vector $\mathbf{x}_{q2}(k)$.

Since the above rotations, at instant k, are actually completing the triangularization of $\underline{\mathbf{X}}^{(N+2)}(k+1)$, it follows that

$$\tilde{\mathcal{Q}}^{(N+2)}(k+1) = \begin{bmatrix} \mathbf{I}_{k-N} & \mathbf{0} \\ \mathbf{0} & \mathbf{Q}'_f(k+1) \end{bmatrix} \tilde{\mathbf{Q}}_f(k) \begin{bmatrix} \tilde{\mathcal{Q}}(k) & \mathbf{0} \\ \mathbf{0} & 1 \end{bmatrix} \quad (9.103)$$

If the pinning vector, $[1\,0\ldots 0]^T$, is postmultiplied on both sides of the above equation, we obtain the following relation

$$\tilde{\mathcal{Q}}^{(N+2)}(k+1)\begin{bmatrix}1\\0\\\vdots\\0\end{bmatrix} = \begin{bmatrix}\mathbf{I}_{k-N} & \mathbf{0}\\ \mathbf{0} & \mathbf{Q}'_f(k+1)\end{bmatrix}\tilde{\mathbf{Q}}_f(k)\begin{bmatrix}\tilde{\mathcal{Q}}(k) & \mathbf{0}\\ \mathbf{0} & 1\end{bmatrix}\begin{bmatrix}1\\0\\\vdots\\0\end{bmatrix}$$

$$= \begin{bmatrix}\gamma^{(N+2)}(k+1)\\0\\\vdots\\\mathbf{r}^{(N+2)}(k+1)\end{bmatrix}\}N+2$$

$$= \begin{bmatrix}\mathbf{I}_{k-N} & \mathbf{0}\\ \mathbf{0} & \mathbf{Q}'_f(k+1)\end{bmatrix}\tilde{\mathbf{Q}}_f(k)\begin{bmatrix}\gamma(k)\\0\\\vdots\\\mathbf{r}(k)\\0\end{bmatrix}\}N+1 \tag{9.104}$$

where $\mathbf{r}^{(N+2)}(k)$ and $\mathbf{r}(k)$ are vectors representing the last nonzero elements in the first column of $\tilde{\mathcal{Q}}^{(N+2)}(k)$ and $\tilde{\mathcal{Q}}(k)$, respectively, as can be seen in equation (9.73). Now, we can proceed by taking the product involving the matrix $\tilde{\mathbf{Q}}_f(k)$ resulting in the following relation

$$\begin{matrix}1\{\\ k-N\{\\ N+1\{\end{matrix}\begin{bmatrix}\gamma(k)\cos\theta_f(k)\\0\\\vdots\\\mathbf{r}(k)\\\gamma(k)\sin\theta_f(k)\end{bmatrix} = \begin{bmatrix}\mathbf{I}_{k-N} & \mathbf{0}\\ \mathbf{0} & \mathbf{Q}'^T_f(k+1)\end{bmatrix}\begin{bmatrix}\gamma^{(N+2)}(k+1)\\0\\\vdots\\\mathbf{r}^{(N+2)}(k+1)\end{bmatrix}\begin{matrix}\}1\\ \}k-N-1\\ \}N+2\end{matrix} \tag{9.105}$$

Since our interest is to calculate $\mathbf{r}(k+1)$, the above equation can be reduced to

$$\mathbf{Q}'_f(k+1)\begin{bmatrix}\mathbf{r}(k)\\\gamma(k)\sin\theta_f(k)\end{bmatrix} = \mathbf{r}^{(N+2)}(k+1) \tag{9.106}$$

where the unused $k-N$ rows and columns were deleted and $\mathbf{r}(k+1)$ is the last $N+1$ rows of $\mathbf{r}^{(N+2)}(k+1)$. Now, since we have $\mathbf{r}(k+1)$ available as a function of known quantities, it is possible to calculate the angles of the reduced rotation matrix $\mathcal{Q}_\theta(k+1)$ using the following relation.

$$\begin{bmatrix}\gamma(k+1)\\\mathbf{r}(k+1)\end{bmatrix} = \mathcal{Q}_\theta(k+1)\begin{bmatrix}1\\0\\\vdots\\0\end{bmatrix} \tag{9.107}$$

By examining the definition of $\mathcal{Q}_\theta(k+1)$ in equations (9.75) and (9.76), it is possible to conclude that it has the following general form (see equations (9.29) and (9.30) for similar derivation)

$$\mathcal{Q}_\theta(k+1) = \overbrace{\begin{bmatrix} * & * & \cdots & * \\ * & * & & \\ \vdots & & \ddots & \\ * & * & \cdots & * \end{bmatrix}}^{N+1} \Big\} N+1 \tag{9.108}$$

where $*$ represents a nonzero element, with the first column given by

$$\begin{bmatrix} \prod_{i=0}^{N} \cos\theta_i(k+1) \\ \prod_{i=0}^{N-1} \cos\theta_i(k+1)\sin\theta_N(k+1) \\ \vdots \\ \prod_{i=0}^{j-1} \cos\theta_i(k+1)\sin\theta_j(k+1) \\ \vdots \\ \sin\theta_0(k+1) \end{bmatrix} \tag{9.109}$$

Although $\gamma(k+1)$ is not known, referring back to equation (9.107) and considering that each angle θ_i is individually responsible for an element in the vector $\mathbf{r}(k+1)$, it is possible to show that equation (9.107) can be solved by the following algorithm:

Initialize $\gamma'_0 = 1$

For $i = 1$ to $N+1$ calculate

$$\sin\theta_{i-1}(k+1) = \frac{r_{N+2-i}(k+1)}{\gamma'_0} \tag{9.110}$$

$$\begin{aligned} \gamma'^2_1 &= \gamma'^2_0[1 - \sin^2\theta_{i-1}(k+1)] \\ &= \gamma'^2_0 - r^2_{N+2-i}(k+1) \end{aligned} \tag{9.111}$$

$$\cos\theta_{i-1}(k+1) = \frac{\gamma'_1}{\gamma'_0} \tag{9.112}$$

$$\gamma'_0 = \gamma'_1 \tag{9.113}$$

After computation is finished make $\gamma(k+1) = \gamma'_1$.

In the fast QR-RLS algorithm, we first calculate the rotated forward prediction error as in equation (9.99), followed by the calculation of the energy of the forward prediction error using equation

(9.98) and the elements of $\tilde{\mathbf{Q}}_f(k)$ given in equations (9.96) and (9.97), respectively. The rotation entries of $\mathbf{Q}'_f(k+1)$ are calculated using the relations of (9.102), which in turn allow us to calculate $\mathbf{r}^{(N+2)}(k+1)$ through equation (9.106). Given $\mathbf{r}^{(N+2)}(k+1)$, the rotation angles θ_i can be calculated via equations (9.110)-(9.112). The remaining equations of the algorithm are the joint-processor section and the computation of the forward prediction error given by equations (9.51) and (9.54), respectively.

The resulting Algorithm 9.2 is almost the same as the hybrid QR-lattice algorithm of [8]. The main difference is the order the of computation of the angles θ_i. In [8] the computation starts from θ_N by employing the relation

$$\gamma(k+1) = \sqrt{1 - ||\mathbf{r}(k+1)||^2} \tag{9.114}$$

This algorithm is closely related to the normalized lattice algorithm (see [8]). Some key results are needed to establish the relation between these algorithms. For example it can be shown that the parameter $\gamma(k, N+1)$ of the lattice algorithms corresponds to $\gamma^2(k)$ in the fast QR algorithm.

In problem 17, it is proved that the elements of $\mathbf{r}(k+1)$ in equation (9.106) correspond to normalized backward prediction *a posteriori* errors of distinct orders [8]. This is the explanation for the classification of Algorithm 9.2 in Table 9.2 as one which updates the *a posteriori* backward prediction errors.

Example 9.3

In this example, the system identification problem described in subsection 3.6.2 is solved using the QR-RLS algorithm described in this section. We implemented the fast QR-RLS algorithm with finite precision.

Solution:

The main objective of this example is to test the stability of the fast QR-RLS algorithm. For that we run the algorithm implemented with fixed-point arithmetic. The wordlengths used are 16, 12, and 10 bits respectively. We force the rotations to be kept passive. In other words, for each rotation the sum of the squares of the quantized sine and cosine are kept less or equal to one. Also, we test γ'_1 to prevent it from becoming less than zero. With these measures, we did not notice any sign of divergence in our experiments. Table 9.3 shows the measured MSE in the finite-precision implementation, where the expected MSE for the infinite-precision implementation is 0.0015. The analysis of these results shows that the fast QR-RLS has low sensitivity to quantization effects and is comparable to the other stable RLS algorithms presented in this text.

□

9.6 CONCLUSIONS AND FURTHER READING

Motivated by the numerically well conditioned Givens rotations, two types of rotation-based algorithms were presented in this chapter. In both cases the QR decomposition implemented with orthogonal Givens rotations were employed. The first algorithm is computationally intensive (order

Algorithm 9.2

Fast QR-RLS Algorithm
Based on *a Posteriori* Backward Prediction Error

Initialization

$||\boldsymbol{\varepsilon}_f(-1)|| = \delta \quad \delta$ small

All cosines with 1 (use for $k \leq N+1$)

All other variables with zero.

Do for each $k \geq 0$

$$\begin{bmatrix} \varepsilon_{fq_1}(k) \\ \mathbf{x}_{q_2}(k) \end{bmatrix} = \mathcal{Q}_\theta(k) \begin{bmatrix} x(k+1) \\ \lambda^{1/2}\mathbf{x}_{q_2}(k-1) \end{bmatrix} \quad (9.99)$$

$$||\boldsymbol{\varepsilon}_f(k)||^2 = \lambda||\boldsymbol{\varepsilon}_f(k-1)||^2 + \varepsilon_{fq_1}^2(k) \quad (9.98)$$

$$\sin\theta_f(k) = \frac{\varepsilon_{fq_1}(k)}{||\boldsymbol{\varepsilon}_f(k)||} \quad (9.97)$$

$\mu_0 = ||\boldsymbol{\varepsilon}_f(k)||$

Do for $i = 1$ to $N+1$

$$\mu_i = \sqrt{\mu_{i-1}^2 + x_{q_2 i}^2(k)} \quad (9.102)$$

$$\cos\theta'_{f_{N+2-i}}(k+1) = \frac{\mu_{i-1}}{\mu_i} \quad (9.102)$$

$$\sin\theta'_{f_{N+2-i}}(k+1) = \frac{x_{q_2 i}(k)}{\mu_i} \quad (9.102)$$

End

$$\mathbf{r}^{(N+2)}(k+1) = \mathbf{Q}'_f(k+1) \begin{bmatrix} \mathbf{r}(k) \\ \gamma(k)\sin\theta_f(k) \end{bmatrix} \quad (9.106)$$

$\mathbf{r}(k+1) =$ last $N+1$ elements of $\mathbf{r}^{(N+2)}(k+1)$

$\gamma'_0 = 1$

Do for $i = 1$ to $N+1$

$$\sin\theta_{i-1}(k+1) = \frac{r_{N+2-i}(k+1)}{\gamma'_0} \quad (9.110)$$

$$\gamma'^2_1 = \gamma'^2_0 - r^2_{N+2-i}(k+1) \quad (9.111)$$

$$\cos\theta_{i-1}(k+1) = \frac{\gamma'_1}{\gamma'_0} \quad (9.112)$$

$\gamma'_0 = \gamma'_1$

End

$\gamma(k+1) = \gamma'_1$

Filter evolution

$$\begin{bmatrix} \varepsilon_{q_1}(k+1) \\ \mathbf{d}_{q_2}(k+1) \end{bmatrix} = \mathcal{Q}_\theta(k+1) \begin{bmatrix} d(k+1) \\ \lambda^{1/2}\mathbf{d}_{q_2}(k) \end{bmatrix} \quad (9.51)$$

$$\varepsilon(k+1) = \varepsilon_{q_1}(k+1)\gamma(k+1) \quad (9.54)$$

End

Table 9.3 Results of the Finite-Precision Implementation of the Fast QR-RLS Algorithm

	$\xi(k)_Q$
No. of bits	Experiment
16	$1.7\ 10^{-3}$
12	$2.0\ 10^{-3}$
10	$2.1\ 10^{-3}$

N^2) and is mainly useful in applications where the input signal vector does not consist of time delayed elements. The advantages of this algorithm are its numerical stability and its systolic array implementation. The second class of algorithms explores the time-shift property of the input signal vector which is inherent to a number of applications, yielding the fast QR-RLS algorithms with order N numerical operations per output sample.

It should be noticed that the subject of QR-decomposition-based algorithms is not fully covered here. A complete approach to generating fast QR-RLS algorithm using lattice formulation is known [23]-[26]. In [23], the author applied QR decomposition to avoid inversion of covariance matrices in the multichannel problem employing lattice RLS formulation. A full orthogonalization of the resulting algorithm was later proposed in [25]. By using different formulations, the works of [24], [25], and [26], propose virtually identical QR-decomposition-based lattice RLS algorithms. In terms of computational complexity, the fast QR-RLS algorithm presented in this chapter is more efficient. Although not discussed here, a solution to compute the adaptive-filter weights from the internal quantities of the fast QR-RLS algorithm is currently available [27].

Another family of algorithms employing QR decomposition are those that replace the Givens rotation by the Householder transformation [1]. The Householder transformation can be considered an efficient method to compute the QR decomposition and is known to yield more accurate results than the Givens rotations in finite-precision implementations. In [28], the fast Householder RLS adaptive-filtering algorithm was proposed and shown to require computational complexity on the order of $7N$. However, no stability proof for this algorithm exists so far. In another work, the Householder transformation is employed to derive a block-type RLS algorithm that can be mapped on a systolic-block Householder transformation [29]. In [30], by employing the Householder transformation, a QR-based LMS algorithm was proposed as a numerically stable and fast converging algorithm with $O[N]$ computational complexity.

A major drawback of the conventional QR-RLS algorithm is the backsubstitution algorithm which is required for computing the weight vector. In a systolic array, it can be implemented as shown in this chapter, through a bidirectional array that requires extra clock cycles. Alternatively, a two-dimensional array can be employed despite being more computationally expensive [13]. An approach called inverse QR method can be used to derive a QR-based RLS algorithm such that the weight vector can be calculated without backsubstitution [31]-[32], however, no formal proof of stability for this algorithm is known.

The QR decomposition has also been shown to be useful for the implementation of numerically stable nonlinear adaptive-filtering algorithms. In [33], a QR-based RLS algorithm for adaptive nonlinear filtering has been proposed.

Some performance evaluations of the QR-RLS and fast QR-RLS algorithms are found in this chapter where these algorithms were employed in some simulation examples.

9.7 REFERENCES

1. G. H. Golub and C. F. Van Loan, *Matrix Computations,* John Hopkins University Press, Baltimore, MD, 2nd edition, 1989.

2. W. H. Gentleman and H. T. Kung, "Matrix triangularization by systolic arrays," *Proc. of SPIE, Real Time Signal Processing* IV, vol. 298, pp. 19-26, 1981.

3. J. G. McWhirter, "Recursive least-squares minimization using a systolic array," *Proc. of SPIE, Real Time Signal Processing* VI, vol. 431, pp. 105-112, 1983.

4. J. M. Cioffi, "The fast adaptive ROTOR's RLS algorithm," *IEEE Trans. on Acoust., Speech, and Signal Processing*, vol. 38, pp. 631-653, April 1990.

5. I. K. Proudler, J. G. McWhirter, and Y. J. Shepherd, "Fast QRD-based algorithms for least squares linear prediction," *Proc. IMA Conference on Mathematics in Signal Processing*, Warwick, England, pp. 465-488, Dec. 1988.

6. M. G. Bellanger, "The FLS-QR algorithm for adaptive filtering," *Signal Processing*, vol. 17, pp. 291-304, Aug. 1984.

7. M. G. Bellanger and P. A. Regalia, "The FLS-QR algorithm for adaptive filtering: The case of multichannel signals," *Signal Processing*, vol. 22, pp. 115-126, March 1991.

8. P. A. Regalia and M. G. Bellanger, "On the duality between fast QR methods and lattice methods in least squares adaptive filtering," *IEEE Trans. on Signal Processing*, vol. 39, pp. 879-891, April 1991.

9. J. A. Apolinário, Jr., and P. S. R. Diniz, "A new fast QR algorithm based on *a priori* errors," *IEEE Signal Processing Letters*, vol. 4, pp. 307-309, Nov. 1997.

10. M. D. Miranda and M. Gerken, "A hybrid QR-lattice least squares algorithm using *a priori* errors," *IEEE Trans. on Signal Processing*, vol. 45, pp. 2900-2911, Dec. 1997.

11. A. A. Rontogiannis and S. Theodoridis, "New fast QR decomposition least squares adaptive algorithms," *IEEE Trans. on Signal Processing*, vol. 46, pp. 2113-2121, Aug. 1998.

12. J. A. Apolinário, Jr., M. G. Siqueira, and P. S. R. Diniz, "On fast QR algorithm based on backward prediction errors: new result and comparisons," *Proc. first IEEE Balkan Conf. on Signal Processing, Communications, Circuits, and Systems*, Istanbul, Turkey, pp. 1-4, CD-ROM, June 2000.

13. C. R. Ward, P. J. Hargrave, and J. G. McWhirter, "A novel algorithm and architecture for adaptive digital beamforming," *IEEE Trans. on Antennas and Propagation*, vol. 34, pp. 338-346, March 1986.

14. Z. Chi, J. Ma, and K. Parhi, "Hybrid annihilation transformation (HAT) for pipelining QRD-based least-square adaptive filters," *IEEE Trans. on Circuits and Systems-II: Analog and Digital Signal Processing*, vol. 48, pp. 661-674, July 2001.

15. C. A. Mead and L. A. Conway, *Introduction to VLSI Systems*, Addison-Wesley, Reading, MA, 1980.

16. W. H. Gentleman, "Least squares computations by Givens transformations without square roots," *Inst. Maths. Applics.*, vol. 12, pp. 329-336, 1973.

17. W. H. Gentleman, "Error analysis of QR decompositions by Givens transformations," *Linear Algebra and its Applications* , vol. 10, pp. 189-197, 1975.

18. H. Leung and S. Haykin, "Stability of recursive QRD-LS algorithms using finite-precision systolic array implementation," *IEEE Trans. on Acoust., Speech, and Signal Processing*, vol. 37, pp. 760-763, May 1989.

19. K. J. R. Liu, S.-F. Hsieh, K. Yao, and C.-T. Chiu, "Dynamic range, stability, and fault-tolerant capability of finite-precision RLS systolic array based on Givens rotations," *IEEE Trans. on Circuits and Systems*, vol. 38, pp. 625-636, June 1991.

20. P. S. R. Diniz and M. G. Siqueira, "Fixed-point error analysis of the QR-recursive least squares algorithm," *IEEE Trans. on Circuits and Systems II: Analog and Digital Signal Processing*, vol. 43, pp. 334-348, May 1995.

21. P. A. Regalia, "Numerical stability properties of a QR-based fast least squares algorithm," *IEEE Trans. on Signal Processing*, vol. 41, pp. 2096-2109, June 1993.

22. M. G. Siqueira, P. S. R. Diniz, and A. Alwan,"Infinite precision analysis of the fast QR-recursive least squares algorithm," *Proc. IEEE Intern. Symposium on Circuits and Systems*, London, England, pp. 2.293-2.296 , May 1994.

23. P. S. Lewis, "QR-based algorithms for multichannel adaptive least squares lattice filters," *IEEE Trans. on Acoust., Speech, and Signal Processing*, vol. 38, pp. 421-432, May 1990.

24. I. K. Proudler, J. G. McWhirter, and T. J. Shepherd, "Computationally efficient QR decomposition approach to least squares adaptive filtering," *IEE Proceedings-Part F*, vol. 148, pp. 341-353, Aug. 1991.

25. B. Yang and J. F. Böhme, "Rotation-based RLS algorithms: Unified derivations, numerical properties, and parallel implementations," *IEEE Trans. on Signal Processing*, vol. 40, pp. 1151-1166, May 1992.

26. F. Ling, "Givens rotation based least squares lattice and related algorithms," *IEEE Trans. on Signal Processing*, vol. 39, pp. 1541-1551, July 1991.

27. M. Shoaib, S. Werner, J. A. Apolinário, Jr., and T. I. Laakso, "Solution to the weight extraction problem in fast QR-decomposition RLS algorithms," *Proc. IEEE Intern. Conf. on Acoust., Speech, Signal Processing*, Toulouse, France, pp. III-572-III-575, 2006.

28. J. M. Cioffi, "The fast Householder filters RLS adaptive filter," *Proc. IEEE Intern. Conf. on Acoust., Speech, Signal Processing*, Albuquerque, NM, pp. 1619-1622, 1990.

29. K. J. R. Liu, S.-F. Hsieh, and K. Yao, "Systolic block Householder transformation for RLS algorithm with two-level pipelined implementation," *IEEE Trans. on Signal Processing*, vol. 40, pp. 946-957, April 1992.

30. Z.-S. Liu and J. Li, "A QR-based least mean squares algorithm for adaptive parameter estimation," *IEEE Trans. on Circuits and Systems-II: Analog and Digital Signal Processing*, vol. 45, pp. 321- 329, March 1998.

31. A. Ghirnikar and S. T. Alexander,"Stable recursive least squares filtering using an inverse QR decomposition," *Proc. IEEE Intern. Conf. on Acoust., Speech, Signal Processing*, Albuquerque, NM, pp. 1623-1626, 1990.

32. S. T. Alexander and A. Ghirnikar, "A method for recursive least squares filtering based upon an inverse QR decomposition," *IEEE Trans. on Signal Processing*, vol. 41, pp. 20-30, Jan. 1993.

33. M. Syed and V. J. Mathews, "QR-Decomposition based algorithms for adaptive Volterra filtering," *IEEE Trans. on Circuits and Systems I: Fundamental Theory and Applications*, vol. 40, pp. 372-382, June 1993.

9.8 PROBLEMS

1. If we consider each anti-diagonal element of $\lambda^{\frac{1}{2}}\mathbf{U}(k)$ as a scaling constant d_i, and we divide the input signal vector initially by a constant δ, we can derive a QR-decomposition algorithm without square roots as described below:

   The first two rows to be rotated are

   $$\begin{array}{llll} \delta\tilde{x}(k) & \delta\tilde{x}(k-1) & \cdots & \delta\tilde{x}(k-N) \\ d_1\lambda^{1/2}\tilde{u}_{1,1}(k-1) & d_1\lambda^{1/2}\tilde{u}_{1,2}(k-1) & \cdots & d_1 \end{array}$$

   where $d_1 = \lambda^{1/2}u_{1,N+1}(k-1)$. The parameter δ can be initialized with 1.
   Applying the Givens rotation to the rows above results in

   $$\begin{array}{llll} \delta' x_1'(k) & \delta' x_1'(k-1)\cdots & \delta' x_1'(k-N+1) & 0 \\ d_1'\tilde{u}_{1,1}(k) & d_1'\tilde{u}_{1,2}(k)\cdots & d_1'\tilde{u}_{1,N}(k) & d_1' \end{array}$$

   where

   $$d_1'^2 = d_1^2 + \delta^2\tilde{x}^2(k-N)$$
   $$c = \frac{d_1^2}{d_1^2+\delta^2\tilde{x}^2(k-N)}$$
   $$\delta'^2 = \frac{d_1^2\delta^2}{d_1^2+\delta^2\tilde{x}^2(k-N)}$$
   $$s = \frac{\delta^2\tilde{x}(k-N)}{d_1^2+\delta^2\tilde{x}^2(k-N)}$$
   $$x_1'(k-N+i) = \tilde{x}(k-N+i) - \tilde{x}(k-N)\lambda^{1/2}\tilde{u}_{1,N-i+1}(k-1)$$

$\tilde{u}'_{1,N-i+1}(k) = c\lambda^{1/2}\tilde{u}_{1,N+1-i}(k-1) + s\tilde{x}(k-N+i)$
The same procedure can be used to triangularize completely the input signal matrix.

(a) Using the above procedure derive a QR-RLS algorithm without square roots.

(b) Compare the computational complexity of the QR-RLS algorithms with and without square roots.

(c) Show that the triangularized matrix $\tilde{\mathbf{U}}(k)$ is related with $\mathbf{U}(k)$ through

$$\mathbf{U}(k) = \mathbf{D}'\tilde{\mathbf{U}}(k)$$

where $\mathbf{D}'$ is a diagonal matrix with the diagonal elements given by d'_i for $i = 1, 2, \ldots, N+1$.

2. Since $\mathbf{Q}^T(k)\mathbf{Q}(k) = \mathbf{I}_{k+1}$, the following identity is valid for any matrix $\mathbf{A}$ and $\mathbf{B}$:
$\mathbf{C}^T\mathbf{D} = \mathbf{A}^T\mathbf{B}$ for $\mathbf{Q}(k)\,[\mathbf{A} \mid \mathbf{B}] = [\mathbf{C} \mid \mathbf{D}]$
where $\mathbf{Q}(k), \mathbf{A}, \mathbf{B}, \mathbf{C}$,and $\mathbf{D}$ have the appropriate dimensions. By choosing $\mathbf{A}, \mathbf{B}, \mathbf{C}$, and $\mathbf{D}$ appropriately, derive the following relations.
(a)$\mathbf{U}^T(k)\mathbf{U}(k) = \lambda\mathbf{U}^T(k-1)\mathbf{U}(k-1) + \mathbf{x}(k)\mathbf{x}^T(k)$
(b)$\mathbf{p}_D(k) = \lambda\mathbf{p}_D(k-1) + \mathbf{x}(k)d(k)$
where $\mathbf{p}_D(k) = \Sigma_{i=0}^{k}\lambda^k\mathbf{x}(i)d(i)$
(c)$\mathbf{U}^T(k)\mathbf{U}^{-T}(k)\mathbf{x}(k) = \mathbf{x}(k)$
where $\mathbf{U}^{-T}(k) = \left[\mathbf{U}^{-1}(k)\right]^T$
(d)$\mathbf{p}_D^T(k)\mathbf{U}^{-1}(k)\mathbf{U}^{-T}(k)\mathbf{x}(k) + \varepsilon_{q1}(k)\gamma(k) = d(k)$

3. Partitioning $\mathbf{Q}_\theta(k)$ as follows:

$$\mathbf{Q}_\theta(k) = \begin{bmatrix} \gamma(k) & \mathbf{q}_\theta^T(k) \\ \mathbf{q}'_\theta(k) & \mathbf{Q}_{\theta r}(k) \end{bmatrix}$$

show from equations (9.51) and (9.39) that
$\mathbf{q}_\theta^T(k)\lambda^{1/2}\mathbf{U}(k-1) + \gamma(k)\mathbf{x}^T(k) = \mathbf{0}^T$

$\mathbf{q}_\theta^T(k)\lambda^{1/2}\mathbf{d}_{q2}(k-1) + \gamma(k)d(k) = \varepsilon_{q1}(k)$

4. Using the relations of the previous two problems and the fact that $\mathbf{U}(k)\mathbf{w}(k) = \mathbf{d}_{q2}(k)$, show that
(a)$e(k) = \frac{\varepsilon_{q1}(k)}{\gamma(k)}$
(b)$\varepsilon(k) = e(k)\gamma^2(k)$
(c)$\varepsilon_{q1}(k) = \sqrt{\varepsilon(k)e(k)}$

5. Show that $\mathbf{U}^T(k)\mathbf{d}_{q2}(k) = \mathbf{p}_D(k)$.

6. Using some of the formulas of the conventional RLS algorithm show that
$\gamma^2(k) = 1 - \mathbf{x}^T(k)\mathbf{R}_D^{-1}(k)\mathbf{x}(k)$.

7. The QR-RLS algorithm is used to predict the signal $x(k) = \cos(\pi k/3)$ using a second-order FIR filter with the first tap fixed at 1. Note that we are interested in minimizing the MSE of the FIR output error. Given $\lambda = 0.985$, calculate $y(k)$ and the filter coefficients for the first 10 iterations.

8. Use the QR-RLS algorithm to identify a system with the transfer function given below. The input signal is uniformly distributed white noise with variance $\sigma_x^2 = 1$ and the measurement noise is Gaussian white noise uncorrelated with the input with variance $\sigma_n^2 = 10^{-3}$. The adaptive filter has 12 coefficients.

$$H(z) = \frac{1 - z^{-12}}{1 - z^{-1}}$$

(a) Run the algorithm for $\lambda = 1$, $\lambda = 0.99$, and $\lambda = 0.97$. Comment on the convergence behavior in each case.
(b) Plot the obtained FIR filter frequency response at any iteration after convergence is achieved and compare with the unknown system.

9. Perform the equalization of a channel with the following impulse response

$$h(k) = \sum_{l=k}^{10} (l - 10)[u(k) - u(k - 10)]$$

where $u(k)$ is a step sequence.
Use a known training signal that consists of a binary (-1,1) random signal. An additional Gaussian white noise with variance 10^{-2} is present at the channel output.
(a) Apply the QR-RLS with an appropriate λ and find the impulse response of an equalizer with 50 coefficients.
(b) Convolve the equalizer impulse response at a given iteration after convergence, with the channel impulse response and comment on the result.

10. In a system identification problem the input signal is generated by an autoregressive process given by

$$x(k) = -1.2x(k - 1) - 0.81x(k - 2) + n_x(k)$$

where $n_x(k)$ is zero-mean Gaussian white noise with variance such that $\sigma_x^2 = 1$. The unknown system is described by

$$H(z) = 1 + 0.9z^{-1} + 0.1z^{-2} + 0.2z^{-3}$$

The adaptive filter is also a third-order FIR filter. Using the QR-RLS algorithm:
Choose an appropriate λ, run an ensemble of 20 experiments, and plot the average learning curve.

11. The QR-RLS algorithm is applied to identify a 7th-order time-varying unknown system whose coefficients are first-order Markov processes with $\lambda_{\mathbf{w}} = 0.999$ and $\sigma_{\mathbf{w}}^2 = 0.001$. The initial time-varying system multiplier coefficients are

$$\mathbf{w}_o^T = [0.03490 \; -0.01100 \; -0.06864 \; 0.22391 \; 0.55686 \; 0.35798 \; -0.02390 \; -0.07594]$$

The input signal is Gaussian white noise with variance $\sigma_x^2 = 0.7$, and the measurement noise is also Gaussian white noise independent of the input signal and of the elements of $\mathbf{n_w}(k)$, with variance $\sigma_n^2 = 0.01$.

(a) For $\lambda = 0.97$ measure the excess MSE.

(b) Repeat (a) for $\lambda = \lambda_{\text{opt}}$.

12. Suppose a 15th-order FIR digital filter with multiplier coefficients given below is identified through an adaptive FIR filter of the same order using the QR-RLS algorithm. Considering that fixed-point arithmetic is used and for 10 independent runs, calculate an estimate of the expected value of $||\Delta\mathbf{w}(k)_Q||^2$ and $\xi(k)_Q$ for the following case.

Additional noise : white noise with variance	$\sigma_n^2 = 0.0015$
Coefficients wordlength:	$b_c = 16$ bits
Signal wordlength:	$b_d = 16$ bits
Input signal: Gaussian white noise with variance	$\sigma_x^2 = 0.7$
	$\lambda = 0.99$

$\mathbf{w}_o^T = [0.0219360\ \ 0.0015786\ \ -0.0602449\ \ -0.0118907\ \ 0.1375379$
$0.0574545\ \ -0.3216703\ \ -0.5287203\ \ -0.2957797\ \ 0.0002043\ \ 0.290670$
$-0.0353349\ \ -0.0068210\ \ 0.0026067\ \ 0.0010333\ \ -0.0143593]$

Plot the learning curves for the finite- and infinite-precision implementations.

13. Repeat the above problem for the following cases

(a) $\sigma_n^2 = 0.01, b_c = 9$ bits, $b_d = 9$ bits, $\sigma_x^2 = 0.7, \lambda = 0.98$.

(b) $\sigma_n^2 = 0.1, b_c = 10$ bits, $b_d = 10$ bits, $\sigma_x^2 = 0.8, \lambda = 0.98$.

(c) $\sigma_n^2 = 0.05, b_c = 8$ bits, $b_d = 16$ bits, $\sigma_x^2 = 0.8, \lambda = 0.98$.

14. Repeat problem 12 for the case where the input signal is a first-order Markov process with $\lambda_{\mathbf{x}} = 0.95$.

15. Repeat problem 9 using the fast QR-RLS algorithm.

16. From equation (9.74) it is straightforward to show that

$$\underline{\mathbf{X}}(k) = \mathcal{Q}^T(k) \begin{bmatrix} \mathbf{0} \\ \mathcal{U}(k) \end{bmatrix}$$

$$= [\mathcal{Q}_u(k)\ \mathcal{Q}_d(k)] \begin{bmatrix} \mathbf{0} \\ \mathcal{U}(k) \end{bmatrix}$$

where $\mathcal{Q}(k) = [\mathcal{Q}_u(k)\mathcal{Q}_d(k)]^T$.

(a) Using the above relation show that the elements of $\mathbf{x}_{q_2}(k)$ in equation (9.95) are given by

$$x_{q_2 i}(k) = [\mathbf{q}_{di}^T(k)\ 0]\mathbf{d}_f(k)$$

where $\mathbf{q}_{di}(k)$ is the ith column of $\mathcal{Q}_d(k)$.

(b) Show that the *a posteriori* error vector for an Nth-order forward predictor can be given by

$$\varepsilon_f(k, N+1) = \mathbf{d}_f(k) - \sum_{i=1}^{N+1} x_{q_2 i}(k) \begin{bmatrix} \mathbf{q}_{di}(k) \\ 0 \end{bmatrix}$$

(c) Can the above expression be generalized to represent the *a posteriori* error vector for an $(N-j)$th-order forward predictor? See the expression below

$$\varepsilon_f(k, N+1-j) = \mathbf{d}_f(k) - \sum_{i=j}^{N+1} x_{q_2 i}(k) \begin{bmatrix} \mathbf{q}_{di}(k) \\ 0 \end{bmatrix}$$

17. For the fast QR-RLS algorithm, show that the elements of $\mathbf{r}(k+1)$ correspond to a normalized backward prediction *a posteriori* error defined as

$$r_{N+1-i}(k) = \bar{\varepsilon}_b(k, i) = \frac{\varepsilon_b(k, i)}{||\varepsilon_{b,i}(k)||} = \frac{\varepsilon_{bq_i}(k, i)}{||\varepsilon_{b,i}(k)||} \prod_{j=0}^{i-1} \cos\theta_j(k)$$

where $\prod_{j=0}^{-1} = 1$, and $\varepsilon_b(k, i+1)$ is the *a posteriori* backward prediction error for a predictor of order i, with $i = 0, 1, \ldots$. Note that $||\varepsilon_{b,i}(k)||^2$ corresponds to $\xi^d_{b_{\min}}(k, i+1)$ used in the lattice derivations of Chapter 7.

10

ADAPTIVE IIR FILTERS

10.1 INTRODUCTION

Adaptive infinite impulse response (IIR) filters are those in which the zeros and poles of the filter can be adapted. For that benefit the adaptive IIR filters usually[1] have adaptive coefficients on the transfer function numerator and denominator. Adaptive IIR filters present some advantages as compared with the adaptive FIR filters, including reduced computational complexity. If both have the same number of coefficients, the frequency response of the IIR filter can approximate much better a desired characteristic. Therefore, an IIR filter in most cases requires fewer coefficients, mainly when the desired model has poles and zeros, or sharp resonances [2]-[1]. There are applications requiring hundreds and sometimes thousands of taps in an FIR filter where the use of an adaptive IIR filter is highly desirable. Among these applications are satellite-channel and mobile-radio equalizers, acoustic echo cancellation, etc.

The advantages of the adaptive IIR filters come with a number of difficulties, some of them not encountered in the adaptive FIR counterparts. The main drawbacks are: possible instability of the adaptive filter, slow convergence, and error surface with local minima or biased global minimum depending on the objective function [3].

In this chapter, several strategies to implement adaptive IIR filters will be discussed. First, adaptive IIR filters having as objective function the minimization of the mean-square output error are introduced. Several alternative structures are presented and some properties of the error surface are addressed. In addition, some algorithms based on the minimization of alternative objective functions are discussed. The algorithms are devised to avoid the multimodality inherent to the methods based on the output error.

[1] There are adaptive filtering algorithms with fixed poles.

P.S.R. Diniz, *Adaptive Filtering*, DOI: 10.1007/978-0-387-68606-6_10,

10.2 OUTPUT-ERROR IIR FILTERS

In the present section, we examine strategies to reduce a function of the output error given by

$$\xi(k) = F[e(k)] \tag{10.1}$$

using an adaptive filter with IIR structure. The output error is defined by

$$e(k) = d(k) - y(k) \tag{10.2}$$

as illustrated in Fig. 10.1.a. As usual, an adaptation algorithm determines how the coefficients of the adaptive IIR filter should change in order to get the objective function reduced.

Let us consider that the adaptive IIR filter is realized using the direct-form structure of Fig. 10.1.b. The signal information vector in this case is defined by

$$\boldsymbol{\phi}(k) = [y(k-1)\; y(k-2) \ldots y(k-N)\; x(k)\; x(k-1) \ldots x(k-M)]^T \tag{10.3}$$

where N and M are the adaptive filter denominator and numerator orders, respectively.

The direct-form adaptive filter can be characterized in time domain by the following difference equation

$$y(k) = \sum_{j=0}^{M} b_j(k)x(k-j) - \sum_{j=1}^{N} a_j(k)y(k-j) \tag{10.4}$$

In the system identification field [8], the above difference equation is in general described through polynomial operator as follows:

$$y(k) = \frac{B(k, q^{-1})}{A(k, q^{-1})} x(k) \tag{10.5}$$

where

$$B(k, q^{-1}) = b_0(k) + b_1(k)q^{-1} + \cdots + b_M(k)q^{-M}$$
$$A(k, q^{-1}) = 1 + a_1(k)q^{-1} + \cdots + a_N(k)q^{-N}$$

and q^{-j} denotes a delay operation in a time domain signal of j samples, i.e., $q^{-j}x(k) = x(k-j)$. The difference equation (10.4) can also be rewritten in a vector form, which is more convenient for the algorithm description and implementation, as described below

$$y(k) = \boldsymbol{\theta}^T(k)\boldsymbol{\phi}(k) \tag{10.6}$$

where $\boldsymbol{\theta}(k)$ is the adaptive-filter coefficient vector given by

$$\boldsymbol{\theta}(k) = [-a_1(k)\; -a_2(k) \ldots -a_N(k)\; b_0(k)\; b_1(k) \ldots b_M(k)]^T \tag{10.7}$$

In a given iteration k, the adaptive-filter transfer function can be expressed as follows:

$$\begin{aligned} H_k(z) &= z^{N-M} \frac{b_0(k)z^M + b_1(k)z^{M-1} + \cdots + b_{M-1}(k)z + b_M(k)}{z^N + a_1(k)z^{N-1} + \cdots + a_{N-1}(k)z + a_N(k)} \\ &= z^{N-M} \frac{N_k(z)}{D_k(z)} \end{aligned} \tag{10.8}$$

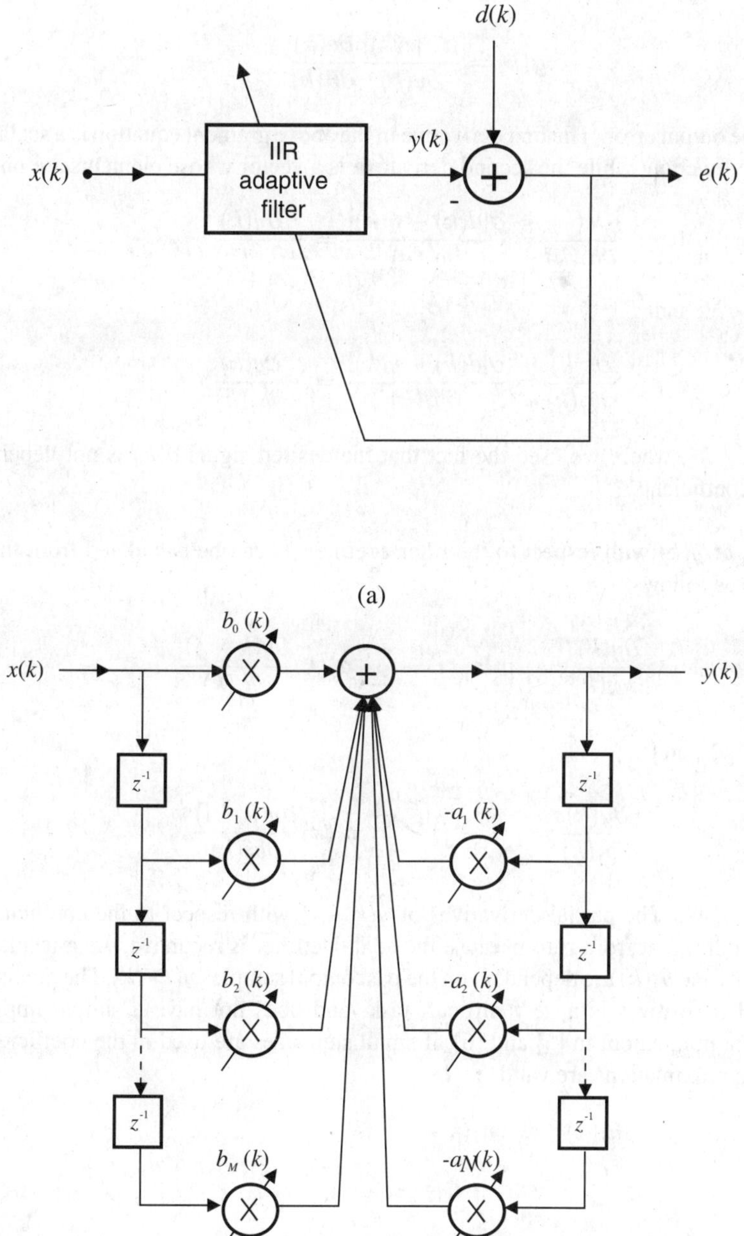

Figure 10.1 Adaptive IIR Filtering: (a) General configuration, (b) Adaptive IIR direct-form realization.

Given the objective function $F[e(k)]$, the gradient vector required to be employed in the adaptive algorithm is given by

$$\mathbf{g}(k) = \frac{\partial F[e(k)]}{\partial e(k)} \frac{\partial e(k)}{\partial \boldsymbol{\theta}(k)} \tag{10.9}$$

where $e(k)$ is the output error. The first derivative in the above gradient equation is a scalar dependent on the objective function, while the second derivative is a vector whose elements are obtained by

$$\frac{\partial e(k)}{\partial a_i(k)} = \frac{\partial[d(k) - y(k)]}{\partial a_i(k)} = -\frac{\partial y(k)}{\partial a_i(k)}$$

for $i = 1, 2, \ldots, N$, and

$$\frac{\partial e(k)}{\partial b_j(k)} = \frac{\partial[d(k) - y(k)]}{\partial b_j(k)} = -\frac{\partial y(k)}{\partial b_j(k)} \tag{10.10}$$

for $j = 0, 1, \ldots, M$, where we used the fact that the desired signal $d(k)$ is not dependent on the adaptive-filter coefficients.

The derivatives of $y(k)$ with respect to the filter coefficients can be calculated from the difference equation (10.4) as follows:

$$\frac{\partial y(k)}{\partial a_i(k)} = -y(k-i) - \sum_{j=1}^{N} a_j(k) \frac{\partial y(k-j)}{\partial a_i(k)}$$

for $i = 1, 2, \ldots, N$, and

$$\frac{\partial y(k)}{\partial b_j(k)} = x(k-j) - \sum_{i=1}^{N} a_i(k) \frac{\partial y(k-i)}{\partial b_j(k)} \tag{10.11}$$

for $j = 0, 1, \ldots, M$. The partial derivatives of $y(k-i)$ with respect to the coefficients, for $i = 1, 2, \ldots, N$, are different from zero because the adaptive filter is recursive. As a result, the present coefficients $a_i(k)$ and $b_j(k)$ are dependent on the past output samples $y(k-i)$. The precise evaluation of these partial derivatives is a very difficult task, and does not have a simple implementation. However, as first pointed out in [5] and [6], if small step sizes are used in the coefficient updating, the following approximations are valid

$$a_i(k) \approx a_i(k-j) \text{ for } i, j = 1, 2, \ldots, N$$

and

$$b_j(k) \approx b_j(k-i) \text{ for } j = 0, 1, \ldots, M \text{ and } i = 1, 2, \ldots, N \tag{10.12}$$

As a consequence, equations (10.11) can be rewritten as

$$-\frac{\partial y(k)}{\partial a_i(k)} \approx +y(k-i) - \sum_{j=1}^{N} a_j(k) \left[\frac{-\partial y(k-j)}{\partial a_i(k-j)}\right]$$

for $i = 1, 2, \ldots, N$, and

$$\frac{\partial y(k)}{\partial b_j(k)} \approx x(k-j) - \sum_{i=1}^{N} a_i(k) \frac{\partial y(k-i)}{\partial b_j(k-i)} \tag{10.13}$$

for $j = 0, 1, \ldots, M$. Note that these equations are standard difference equations.

The above equations can be implemented by all-pole filters having as input signals $-y(k-i)$ and $x(k-j)$ for the first and second set of equations, respectively. The implementation of the derivative signals of equations (10.13) is depicted in Fig. 10.2. The all-pole sections realization can be performed through IIR direct-form structure, with transfer function given by

$$S^{a_i}(z) = \mathcal{Z}\left[\frac{\partial y(k)}{\partial a_i(k)}\right] = \frac{-z^{N-i}}{D_k(z)} Y(z)$$

for $i = 1, 2, \ldots, N$, and

$$S^{b_j}(z) = \mathcal{Z}\left[\frac{\partial y(k)}{\partial b_i(k)}\right] = \frac{z^{N-j}}{D_k(z)} X(z) \tag{10.14}$$

for $j = 0, 1, \ldots, M$, respectively, where $\mathcal{Z}[\cdot]$ denotes the $\mathcal{Z}$-transform of $[\cdot]$.

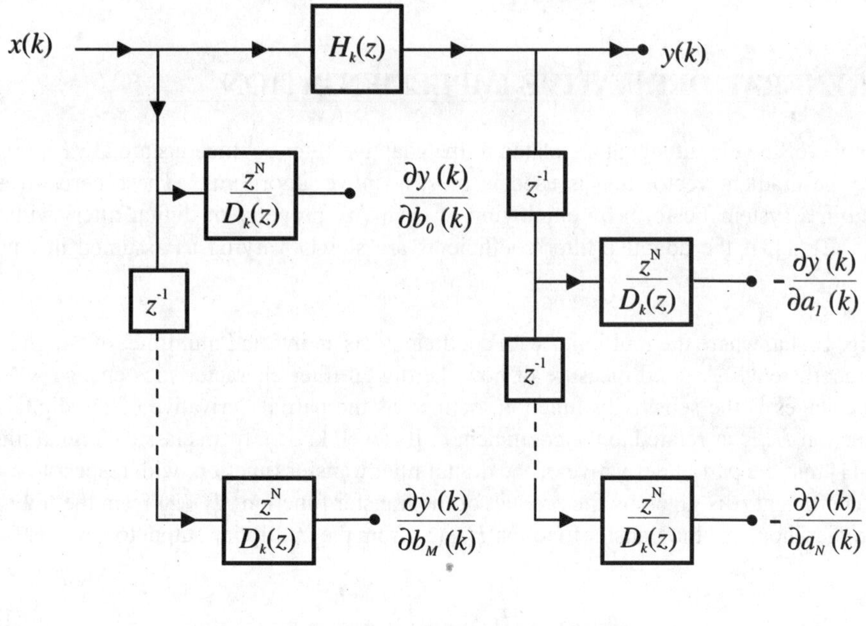

Figure 10.2 Derivative implementation.

The amount of computation spent to obtain the derivatives is relatively high, as compared with the adaptive-filter computation itself. A considerable reduction in the amount of computation can be achieved, if it is considered that the coefficients of the adaptive-filter denominator polynomial are slowly varying, such that

$$D_k(z) \approx D_{k-i}(z) \qquad \text{for } i = 1, 2, \ldots, \max(N, M) \tag{10.15}$$

where $\max(a, b)$ denotes maximum between a and b. The interpretation is that the denominator polynomial is kept almost constant for a number of iterations. With this approximation, it is possible to eliminate the duplicating all-pole filters of Fig. 10.2, and replace them by a single all-pole in front of the two sets of delays as depicted in Fig. 10.3.a. In addition, if the recursive part of the adaptive filter is implemented before the numerator part, one more all-pole section can be saved as illustrated in Fig. 10.3.b [7].

Note that in the time domain, the approximations of equation (10.15) imply the following relations

$$\frac{\partial y(k)}{\partial a_i(k)} \approx q^{-i+1} \frac{\partial y(k)}{\partial a_1(k)}$$

for $i = 1, 2, \ldots, N$, and

$$\frac{\partial y(k)}{\partial b_j(k)} \approx q^{-j} \frac{\partial y(k)}{\partial b_0(k)} \tag{10.16}$$

for $j = 0, 1, \ldots, M$, where $\frac{\partial y(k)}{\partial a_1(k)}$ represents the partial derivative of $y(k)$ with respect to the first non unit coefficient of the denominator polynomial, whereas $\frac{\partial y(k)}{\partial b_0(k)}$ is the partial derivative of $y(k)$ with respect to the first coefficient of the numerator polynomial.

10.3 GENERAL DERIVATIVE IMPLEMENTATION

The derivatives of the output signal as related to the adaptive-filter coefficients are always required to generate the gradient vector that is used in most adaptive algorithms. These derivatives can be obtained in a systematic form by employing a sensitivity property of digital filters with fixed coefficients [2]-[1], if the adaptive-filter coefficients are slowly varying as assumed in equation (10.12).

Refer to Fig. 10.4.a, where the multiplier with coefficient c is an internal multiplier of a digital filter with fixed coefficients. A good measure of how the digital filter characteristics change when the value of c changes is the sensitivity function, defined as the partial derivative of the digital filter transfer function $H(z)$ as related to the coefficient c. It is well known from classical digital filtering theory [2]-[1] that the partial derivative of the digital filter transfer function, with respect to a given multiplier coefficient c, is given by the product of the transfer function $H_{13}(z)$ from the filter input to the multiplier input and the transfer function $H_{42}(z)$ from the multiplier output to the filter output, that is

$$S^c(z) = H_{13}(z) \cdot H_{42}(z) \tag{10.17}$$

Fig. 10.4.b illustrates the derivative implementation. It can be noted that the implementation of the derivatives for the direct-form structure shown in Fig. 10.2 can be obtained by employing equation (10.17). In the time domain, the filtering operation performed in the implementation of Fig. 10.4.b is given by

$$\frac{\partial y(k)}{\partial c} = h_{13}(k) * h_{42}(k) * x(k) \tag{10.18}$$

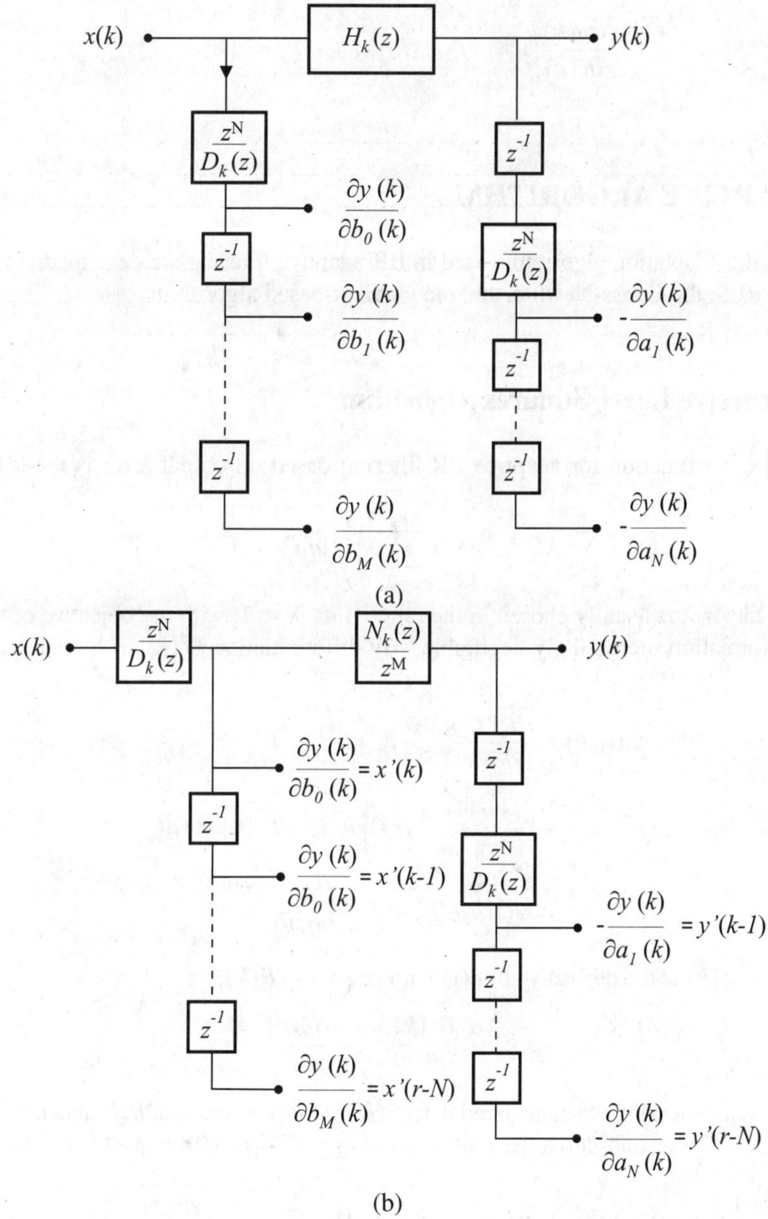

Figure 10.3 Simplified derivative implementation: (a) Simplification I, (b) Simplification II.

where $*$ denotes convolution and $h_{ij}(k)$ is the impulse response related to $H_{ij}(z)$. When the digital filter coefficients are slowly varying, the desired derivatives can be derived as in Fig. 10.4 for each adaptive coefficient. In this case, only an approximated derivative is obtained

$$\frac{\partial y(k)}{\partial c(k)} \approx h_{13k}(k) * h_{42k}(k) * x(k) \tag{10.19}$$

10.4 ADAPTIVE ALGORITHMS

In this section, the adaptation algorithms used in IIR adaptive filtering are described. In particular, we present the RLS, the Gauss-Newton, and the gradient-based algorithms.

10.4.1 Recursive Least-Squares Algorithm

A possible objective function for adaptive IIR filtering based on output error is the least-squares function[2]

$$\xi^d(k) = \sum_{i=0}^{k} \lambda^{k-i} e^2(i) = \sum_{i=0}^{k} \lambda^{k-i}[d(i) - \boldsymbol{\theta}^T(k)\boldsymbol{\phi}(i)]^2 \tag{10.20}$$

The forgetting factor λ is usually chosen in the range $0 \ll \lambda < 1$, with the objective of turning the distant past information increasingly negligible. By differentiating $\xi^d(k)$ with respect to $\boldsymbol{\theta}(k)$, it follows that

$$\begin{aligned} 2\mathbf{g}_D(k) &= \frac{\partial \xi^d(k)}{\partial \boldsymbol{\theta}(k)} \\ &= 2\sum_{i=0}^{k} \lambda^{k-i} \boldsymbol{\varphi}(i)[d(i) - \boldsymbol{\theta}^T(k)\boldsymbol{\phi}(i)] \\ &= 2\boldsymbol{\varphi}(k)e(k) + \lambda \frac{\partial \xi^d(k-1)}{\partial \boldsymbol{\theta}(k)} \end{aligned} \tag{10.21}$$

where the vector $\boldsymbol{\varphi}(k)$ is the derivative of $e(i)$ with respect to $\boldsymbol{\theta}(k)$, i.e.,

$$\boldsymbol{\varphi}(k) = \frac{\partial e(k)}{\partial \boldsymbol{\theta}(k)} = -\frac{\partial y(k)}{\partial \boldsymbol{\theta}(k)} \tag{10.22}$$

and without loss of generality we considered that $\xi^d(k-1)$ is a function of $\boldsymbol{\theta}(k)$ and not of $\boldsymbol{\theta}(k-1)$ as in the FIR case. The second-derivative matrix $2\mathbf{R}_D(k)$ of $\xi^d(k)$[3] with respect to $\boldsymbol{\theta}(k)$ is then given by

$$\frac{\partial^2 \xi^d(k)}{\partial \boldsymbol{\theta}^2(k)} = 2\mathbf{R}_D(k) = 2\lambda \mathbf{R}_D(k-1) + 2\boldsymbol{\varphi}(k)\boldsymbol{\varphi}^T(k) - 2\frac{\partial^2 y(k)}{\partial \boldsymbol{\theta}^2(k)} e(k) \tag{10.23}$$

[2]The reader should note that this definition of the deterministic weighted least squares utilizes the *a priori* error with respect to the latest data pair $d(k)$ and $x(k)$, unlike the FIR RLS case.

[3]By differentiating $2\mathbf{g}_D(k)$ in equation (10.21) with respect to $\boldsymbol{\theta}(k)$.

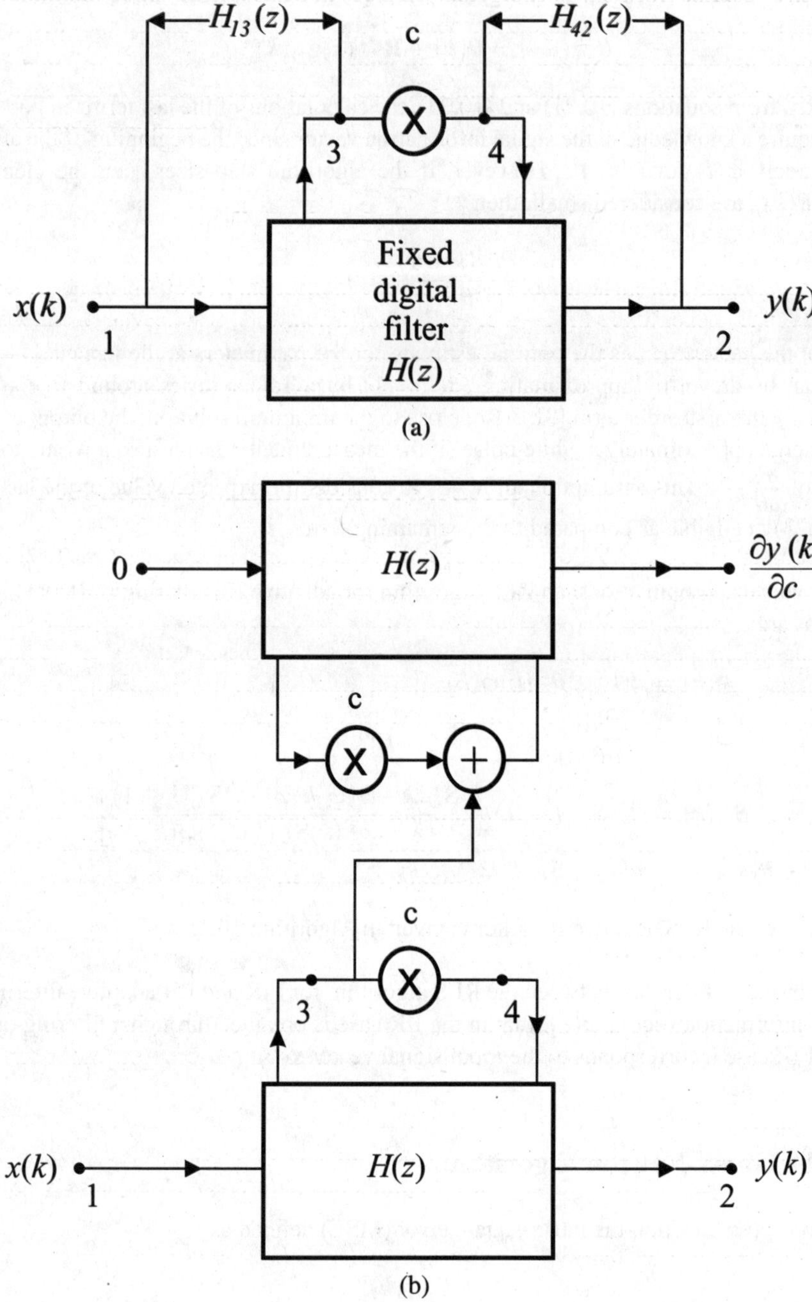

Figure 10.4 General derivative implementation: (a) General structure, (b) Derivative implementation.

Now, several assumptions are made to generate a recursive algorithm. The adaptive-filter parameters are considered to be updated by

$$\boldsymbol{\theta}(k+1) = \boldsymbol{\theta}(k) - \mathbf{R}_D^{-1}(k)\mathbf{g}_D(k) \tag{10.24}$$

As can be noted from equations (10.21) and (10.23), the calculations of the last terms in both $\mathbf{R}_D(k)$ and $\mathbf{g}_D(k)$ require a knowledge of the signal information vector since the beginning of the algorithm operation, namely $\boldsymbol{\varphi}(i)$ for $i < k$. However, if the algorithm step sizes, i.e., the elements of $|\boldsymbol{\theta}(k+1) - \boldsymbol{\theta}(k)|$, are considered small, then

$$\frac{\partial \xi^d(k-1)}{\partial \boldsymbol{\theta}(k)} \approx 0 \tag{10.25}$$

assuming that the vector $\boldsymbol{\theta}(k)$ is the optimal estimate for the parameters at the instant $k-1$. This conclusion can be drawn by approximating $\xi^d(k-1)$ by a Taylor series around $\boldsymbol{\theta}(k-1)$ and considering only the first-order term [8]. Also, close to the minimum solution, the output error $e(k)$ can be considered approximately a white noise (if the measurement noise is also a white noise) and independent of $\frac{\partial^2 y(k)}{\partial \boldsymbol{\theta}^2(k)}$. This assumption allows us to consider the expected value of the last term in equation (10.23) negligible as compared to the remaining terms.

Applying the above approximations, an RLS algorithm for adaptive IIR filtering is derived in which the basic steps are:

$$e(k) = d(k) - \boldsymbol{\theta}^T(k)\boldsymbol{\phi}(k) \tag{10.26}$$

$$\boldsymbol{\varphi}(k) = -\frac{\partial y(k)}{\partial \boldsymbol{\theta}(k)} \tag{10.27}$$

$$\mathbf{S}_D(k) = \frac{1}{\lambda}\left[\mathbf{S}_D(k-1) - \frac{\mathbf{S}_D(k-1)\boldsymbol{\varphi}(k)\boldsymbol{\varphi}^T(k)\mathbf{S}_D(k-1)}{\lambda + \boldsymbol{\varphi}^T(k)\mathbf{S}_D(k-1)\boldsymbol{\varphi}(k)}\right] \tag{10.28}$$

$$\boldsymbol{\theta}(k+1) = \boldsymbol{\theta}(k) - \mathbf{S}_D(k)\boldsymbol{\varphi}(k)e(k) \tag{10.29}$$

The description of the RLS adaptive IIR filter is given in Algorithm 10.1.

Note that the primary difference between the RLS algorithm for FIR and IIR adaptive filtering relies on the signal information vector, $\boldsymbol{\varphi}(k)$, that in the IIR case is obtained through a filtering operation while in the FIR case it corresponds to the input signal vector $\mathbf{x}(k)$.

10.4.2 The Gauss-Newton Algorithm

Consider as objective function the mean-square error (MSE) defined as

$$\xi = E[e^2(k)] \tag{10.30}$$

In the Gauss-Newton algorithm, the minimization of the objective function is obtained by performing searches in the Newton direction, using estimates of the inverse Hessian matrix and the gradient vector.

Algorithm 10.1

Output Error Algorithm, RLS Version

Initialization
$a_i(k) = b_i(k) = e(k) = 0$
$y(k) = x(k) = 0\,,\ k\ < 0$
$\mathbf{S}_D(0) = \delta^{-1}\mathbf{I}$
Definition
$\boldsymbol{\varphi}^T(k) = [-y'(k-1) \ldots -y'(k-N)\ \ -x'(k)\ \ -x'(k-1) \ldots -x'(k-M)]$
For each $x(k)$, $d(k)$, $k \geq 0$, do
$y(k) = \boldsymbol{\phi}^T(k)\boldsymbol{\theta}(k)$
$y'(k) = -y(k) - \sum_{i=1}^{N} a_i(k)y'(k-i)$
$x'(k) = x(k) - \sum_{i=1}^{N} a_i(k)x'(k-i)$
$e(k) = d(k) - y(k)$
$\mathbf{S}_D(k) = \frac{1}{\lambda}\left[\mathbf{S}_D(k-1) - \frac{\mathbf{S}_D(k-1)\boldsymbol{\varphi}(k)\boldsymbol{\varphi}^T(k)\mathbf{S}_D(k-1)}{\lambda+\boldsymbol{\varphi}^T(k)\mathbf{S}_D(k-1)\boldsymbol{\varphi}(k)}\right]$
$\boldsymbol{\theta}(k+1) = \boldsymbol{\theta}(k) - \mathbf{S}_D(k)\boldsymbol{\varphi}(k)e(k)$
Stability test

The gradient vector is calculated as follows:

$$\frac{\partial \xi}{\partial \boldsymbol{\theta}(k)} = E[2e(k)\boldsymbol{\varphi}(k)] \tag{10.31}$$

where $\boldsymbol{\varphi}(k) = \frac{\partial e(k)}{\partial \boldsymbol{\theta}(k)}$ as defined in equation (10.22).

The Hessian matrix is then given by

$$\frac{\partial^2 \xi}{\partial \boldsymbol{\theta}^2(k)} = 2E\left[\boldsymbol{\varphi}(k)\boldsymbol{\varphi}^T(k) + \frac{\partial^2 e(k)}{\partial \boldsymbol{\theta}^2(k)}e(k)\right] \tag{10.32}$$

where the expected value of the second term in the above equation is approximately zero, since close to a solution the output error $e(k)$ is "almost" a white noise independent of the following term

$$\frac{\partial^2 e(k)}{\partial \boldsymbol{\theta}^2(k)} = -\frac{\partial^2 y(k)}{\partial \boldsymbol{\theta}^2(k)}$$

The determination of the gradient vector and the Hessian matrix requires statistical expectation calculations. In order to derive a recursive algorithm, estimates of the gradient vector and Hessian matrix have to be used. For the gradient vector, the most commonly used estimation is the stochastic gradient given by

$$\frac{\partial \hat{\xi}}{\partial \boldsymbol{\theta}(k)} = 2e(k)\boldsymbol{\varphi}(k) \tag{10.33}$$

where $\hat{\xi}$ is an estimate of ξ. Such approximation was also used in the derivation of the LMS algorithm. The name stochastic gradient originates from the fact that the estimates point to random directions around the true gradient direction.

The Hessian estimate can be generated by employing a weighted summation as follows:

$$\begin{aligned}\hat{\mathbf{R}}(k+1) &= \alpha\boldsymbol{\varphi}(k)\boldsymbol{\varphi}^T(k) + \alpha\sum_{i=0}^{k-1}(1-\alpha)^{k-i}\boldsymbol{\varphi}(i)\boldsymbol{\varphi}^T(i) \\ &= \alpha\boldsymbol{\varphi}(k)\boldsymbol{\varphi}^T(k) + (1-\alpha)\hat{\mathbf{R}}(k)\end{aligned} \tag{10.34}$$

where α is a small factor chosen in the range $0 < \alpha < 0.1$. By taking the expected value on both sides of the above equation and assuming that $k \to \infty$, it follows that

$$\begin{aligned}E[\hat{\mathbf{R}}(k+1)] &= \alpha\sum_{i=0}^{k}(1-\alpha)^{k-i}E[\boldsymbol{\varphi}(i)\boldsymbol{\varphi}^T(i)] \\ &\approx E[\boldsymbol{\varphi}(k)\boldsymbol{\varphi}^T(k)]\end{aligned} \tag{10.35}$$

Applying the approximation discussed and the matrix inversion lemma to calculate the inverse of $\hat{\mathbf{R}}(k+1)$, i.e., $\hat{\mathbf{S}}(k+1)$, the Gauss-Newton algorithm for IIR adaptive filtering is derived, consisting of the following basic steps

$$e(k) = d(k) - \boldsymbol{\theta}^T(k)\boldsymbol{\phi}(k) \tag{10.36}$$

$$\boldsymbol{\varphi}(k) = \frac{\partial e(k)}{\partial \boldsymbol{\theta}(k)} \tag{10.37}$$

$$\hat{\mathbf{S}}(k+1) = \frac{1}{1-\alpha}\left[\hat{\mathbf{S}}(k) - \frac{\hat{\mathbf{S}}(k)\boldsymbol{\varphi}(k)\boldsymbol{\varphi}^T(k)\hat{\mathbf{S}}(k)}{\frac{1-\alpha}{\alpha} + \boldsymbol{\varphi}^T(k)\hat{\mathbf{S}}(k)\boldsymbol{\varphi}(k)}\right] \tag{10.38}$$

$$\boldsymbol{\theta}(k+1) = \boldsymbol{\theta}(k) - \mu\hat{\mathbf{S}}(k+1)\boldsymbol{\varphi}(k)e(k) \tag{10.39}$$

where μ is the convergence factor. In most cases, μ is chosen approximately equal to α.

In the updating of the $\hat{\mathbf{R}}(k)$ matrix, the factor $(1-\alpha)$ plays the role of a forgetting factor that determines the effective memory of the algorithm when computing the present estimate. The closer α is to zero the more important is the past information, in other words, the longer is the memory of the algorithm.

10.4.3 Gradient-Based Algorithm

If in the Gauss-Newton algorithm, the estimate of the Hessian matrix is replaced by the identity matrix, the resulting basic algorithm is given by

$$e(k) = d(k) - \boldsymbol{\theta}^T(k)\boldsymbol{\phi}(k) \tag{10.40}$$

$$\boldsymbol{\varphi}(k) = \frac{\partial e(k)}{\partial \boldsymbol{\theta}(k)} \tag{10.41}$$

$$\boldsymbol{\theta}(k+1) = \boldsymbol{\theta}(k) - \mu\boldsymbol{\varphi}(k)e(k) \tag{10.42}$$

These are the steps of a gradient-based algorithm for IIR filtering. The computational complexity is much lower in gradient-based algorithm than in the Gauss-Newton algorithm. With the latter, however, faster convergence is in general achieved.

10.5 ALTERNATIVE ADAPTIVE FILTER STRUCTURES

The direct-form structure is historically the most widely used realization for the IIR adaptive filter. The main advantages of the direct form are the minimum number of multiplier coefficients required to realize a desired transfer function and the computationally efficient implementation for the gradient (which is possible under the assumption that the denominator coefficients are slowly varying, as illustrated in Fig. 10.3). On the other hand, the stability monitoring of the direct form is difficult because it requires either the factorization of a high-order denominator polynomial in each algorithm step or the use of a sophisticated stability test. In addition, the coefficient sensitivities and output quantization noise are known to be high in the direct form [2].

Alternate solutions are the cascade and parallel realizations using first- or second-order sections as building blocks [9]-[10]. Also, the lattice structures are popular in the implementation of adaptive filters [13]-[19]. All these structures allow easy stability monitoring while the parallel form appears to be most efficient in the gradient computation. The standard parallel realization, however, may converge slowly if two poles approach each other, as will be discussed later and, when a Newton-based algorithm is employed, the estimated Hessian matrix becomes ill-conditioned bringing convergence problems. This problem can be alleviated by applying a preprocessing to the input signal [10]-[11].

10.5.1 Cascade Form

Any Nth-order transfer function can be realized by connecting several first- or second-order sections in series, generating the so-called cascade form. Here we consider that all subfilters are second-order sections without loss of generality, and if an odd-order adaptive filter is required we add a single first-order section. Also, only filters with real multiplier coefficients are discussed. The cascade realization transfer function is given by

$$H_k(z) = \prod_{i=1}^{m} \frac{b_{0i}z^2 + b_{1i}(k)z + b_{2i}(k)}{z^2 + a_{1i}(k)z + a_{2i}(k)} = \prod_{i=1}^{m} H_{ki}(z) \tag{10.43}$$

where m denotes the number of sections.

The parameter vector in the cascade form is

$$\theta(k) = [-a_{11}(k) \; -a_{21}(k) \; b_{01}(k) \; b_{11}(k) \; b_{21}(k) \ldots -a_{1m}(k) \; -a_{2m}(k) \; b_{0m}(k) \; b_{1m}(k) \; b_{2m}(k)]^T$$

The transfer function derivatives as related to the multiplier coefficients can be generated by employing the general result of Fig. 10.4. Fig. 10.5 depicts the cascade realization along with the generation of the derivative signals of interest, where the sections were realized through the direct form of Fig. 10.1.

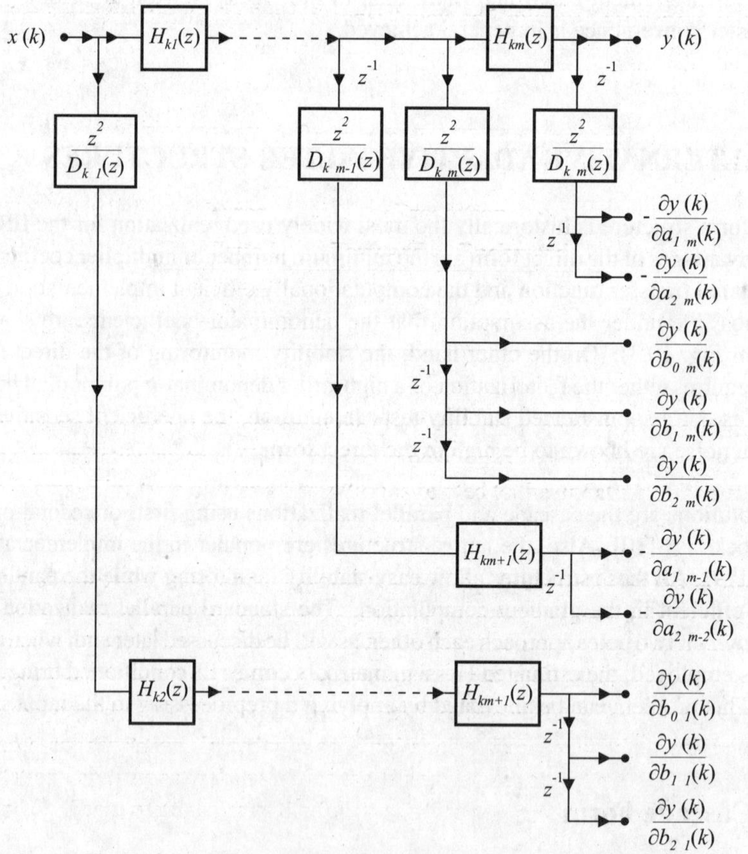

Figure 10.5 Cascade form.

Any alternative second-order section can be used in the cascade form and the appropriate choice depends on a trade-off between quantization effects, hardware resources, computation time, and other factors. The main drawbacks of the cascade form are the amount of extra computations required to generate the gradients, and the manifolds (see sections 10.6 and 10.7) generated on the error surface which may result in slow convergence of the gradient-based algorithms.

10.5.2 Lattice Structure

In this subsection we discuss the lattice algorithm starting from its realization. Although this might appear to be a recipe approach, the development presented here allows us to access the nice properties of the lattice realization. The book by Regalia [12] provides a detailed presentation of the various forms of lattice realization.

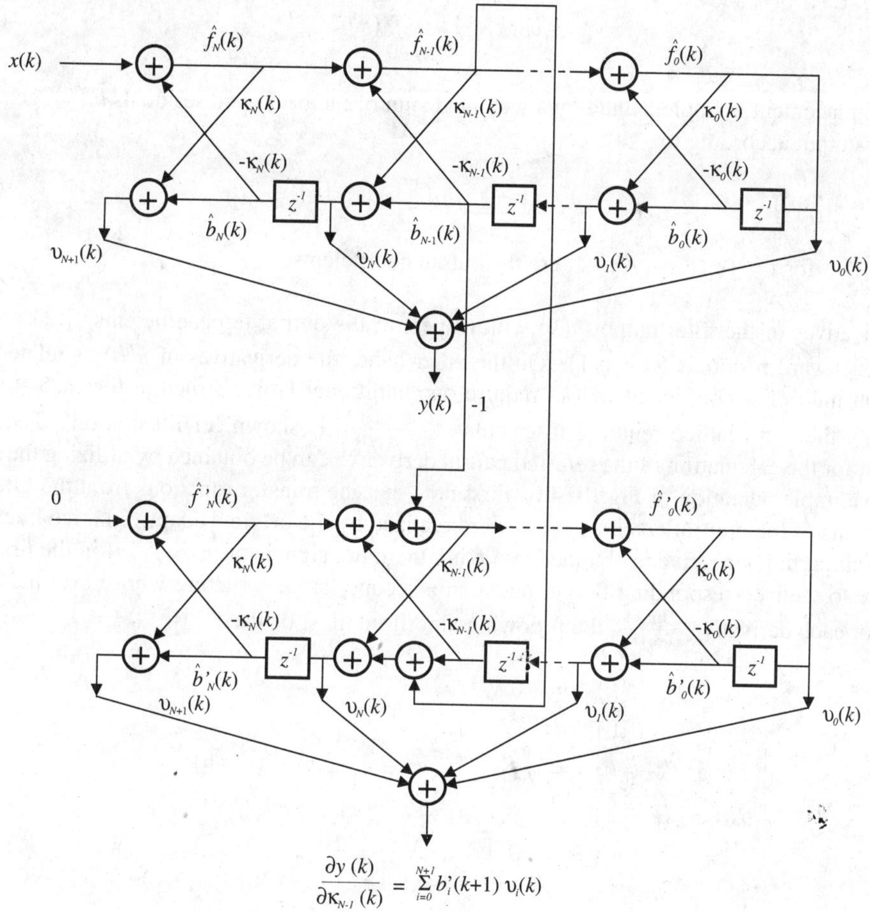

Figure 10.6 Lattice structure including a sample of gradient computation.

The two-multiplier lattice structure [13]-[18] for IIR filters is depicted in Fig. 10.6 with a sample of gradient computation. The coefficients $\kappa_i(k)$ in the recursive part of the structure are called reflection coefficients. The internal signals $\hat{f}_i(k)$ and $\hat{b}_i(k)$ are the forward and backward residuals, respectively. These internal signals are calculated as follows:

$$\hat{f}_{N+1}(k) = x(k)$$
$$\hat{f}_{N-i}(k) = \hat{f}_{N-i+1}(k) - \kappa_{N-i}(k)\hat{b}_{N-i}(k)$$
$$\hat{b}_{N-i+1}(k+1) = \kappa_{N-i}(k)\hat{f}_{N-i}(k) + \hat{b}_{N-i}(k)$$

for $i = 0, 1, \ldots, N$, and

$$\hat{b}_0(k+1) = \hat{f}_0(k) \tag{10.44}$$

The zero placement is implemented by a weighted sum of the backward residuals $\hat{b}_i(k)$, generating the filter output according to

$$y(k) = \sum_{i=0}^{N+1} \hat{b}_i(k+1)v_i(k) \tag{10.45}$$

where $v_i(k)$, for $i = 0, 1, \ldots, N+1$, are the output coefficients.

The derivatives of the filter output $y(k)$ with respect to the output tap coefficients $v_i(k)$ are given by the backward residuals $\hat{b}_i(k+1)$. On the other hand, the derivatives of $y(k)$ as related to the reflection multiplier coefficients $\kappa_i(k)$ require one additional lattice structure for each $\kappa_i(k)$. In Fig. 10.6, the extra lattice required to calculate $\frac{\partial y(k)}{\partial \kappa_{N-1}(k)}$ is shown for illustration. The overall structure for the calculation of the referred partial derivative can be obtained by utilizing the general derivative implementation of Fig. 10.4.b. First note that the transfer functions from the filter input to the inputs of the multipliers $\pm\kappa_{N-1}(k)$ were realized by the original adaptive lattice filter. Next, the overall partial derivative is obtained by taking the input signals of $\pm\kappa_{N-1}(k)$ in the first lattice structure to their corresponding output nodes in a second lattice structure whose external input is zero. For each derivative $\frac{\partial y(k)}{\partial \kappa_j(k)}$, the following algorithm must be used

$$\begin{aligned}
\hat{f}'_{N+1}(k) &= 0 \\
\text{If } i &\neq N-j \\
\hat{f}'_{N-i}(k) &= \hat{f}'_{N-i+1}(k) - \kappa_{N-i}(k)\hat{b}'_{N-i}(k) \\
\hat{b}'_{N-i+1}(k+1) &= \kappa_{N-i}(k)\hat{f}'_{N-i}(k) + \hat{b}'_{N-i}(k) \\
\text{for } i &= 0, 1, \ldots, N-j-1, N-j+1, \ldots, N \\
\text{If } i &= N-j \\
\hat{f}'_j(k) &= \hat{f}'_{j+1}(k) - \kappa_j(k)\hat{b}'_j(k) - \hat{b}_j(k) \\
\hat{b}'_{j+1}(k+1) &= \kappa_j(k)\hat{f}'_j(k) + \hat{b}'_j(k) + \hat{f}_j(k) \\
\hat{b}'_o(k+1) &= \hat{f}_o(k) \\
\text{Then} & \\
\frac{\partial y(k)}{\partial \kappa_j(k)} &= \sum_{i=0}^{N+1} \hat{b}'_i(k+1)v_i(k)
\end{aligned} \tag{10.46}$$

The main desirable feature brought about by the lattice IIR realization is the simple stability test. The stability requires only that reflection coefficients $\kappa_i(k)$ be maintained with modulus less than one [17]. However, the gradient computations are extremely complex, and of order N^2 in terms of multiplication count. An approach for the gradient computations with order N multiplications and divisions was proposed [16], which is still more complex than for the direct-form realization. It should be noticed that in the direct form, all the signals at the multiplier's input are delayed versions of each other, and the transfer function from the multiplier's output to the filter output are the same. These properties make the gradient computational complexity in the direct form low. The lattice IIR realization does not have these features.

When the two-multiplier lattice structure is realizing a transfer function with poles close to the unit circle, the internal signals may present a large dynamic range, resulting in poor performance due to quantization effects. In this case, the normalized lattice [19] is a better choice despite its higher computational complexity. There are alternative lattice structures, such as the two-multiplier with distinct reflection coefficients and the one-multiplier structures [15], that can also be employed in adaptive filtering. For all these options the stability test is trivial, retaining the main feature of the two-multiplier lattice structure.

An application where adaptive IIR filtering is the natural choice is sinusoid detection using notch filters. A notch transfer function using direct-form structure is given by

$$H_{\mathrm{N}}(z) = \frac{1 - 2\cos\omega_0 z^{-1} + z^{-2}}{1 - 2r\cos\omega_0 z^{-1} + r^2 z^{-2}} \tag{10.47}$$

where ω_0 is the notch frequency and r is the pole radius [20]. The closer the pole radius is to the unit circle the narrower is the notch transfer function, leading to better estimate of the sinusoid frequency in a noisy environment. However, in the direct form the noise gain, caused by the notch transfer function, varies with the sinusoid frequency, causing a bias in the frequency estimate [12].

An alternative is to construct a notch filter by using a lattice structure. A second-order notch filter can be generated by

$$H_{\mathrm{N}}(z) = \frac{1}{2}\left[1 + H_{\mathrm{AP}}(z)\right] \tag{10.48}$$

where $H_{\mathrm{AP}}(z)$ is an all-pass transfer function which can be realized by a lattice structure by setting $v_2 = 1$ and $v_1 = v_0 = 0$ in Fig. 10.6. In this case,

$$H_{\mathrm{AP}}(z) = \frac{\kappa_1 + \kappa_0(1+\kappa_1)z^{-1} + z^{-2}}{1 + \kappa_0(1+\kappa_1)z^{-1} + \kappa_1 z^{-2}} \tag{10.49}$$

The notch frequency ω_0 and the relation between -3 dB attenuation bandwidth $\Delta\omega_{3\mathrm{dB}}$ and κ_1 are given by

$$\omega_0 = \cos^{-1}(-\kappa_0) \tag{10.50}$$

and

$$\kappa_1 = \frac{1 - \tan\frac{\Delta\omega_{3\mathrm{dB}}}{2}}{1 + \tan\frac{\Delta\omega_{3\mathrm{dB}}}{2}} \tag{10.51}$$

respectively. The main feature of the notch filter based on the lattice structure is the independent control of the notch frequency and the -3 dB attenuation bandwidth.

It is worth mentioning that an enhanced version of the sinusoid signal can be obtained by applying the noisy input signal to the bandpass filter whose transfer function is given by

$$H_{\mathrm{BP}}(z) = \frac{1}{2}\left[1 - H_{\mathrm{AP}}(z)\right] \tag{10.52}$$

For identification of multiple sinusoids the most widely used structure is the cascade of second-order sections, where each section identifies one of the sinusoids removing the corresponding sinusoid from the input to the following sections.

Sinusoid detection in noise utilizing adaptive notch filter has rather simple implementation as compared with other methods, and finds application in synchronization, tone detection and tracking for music signals among others.

Example 10.1

Apply an IIR notch adaptive filter using the second-order lattice structure to detect a sinusoid buried in noise.

The input signal noise is a Gaussian white noise with variance $\sigma_x^2 = 1$, whereas the sampling frequency is 10000Hz and the sinusoid to be detected is at 1000Hz. Use a gradient-based algorithm.
(a) Choose the appropriate value of μ.
(b) Run the algorithm using for signal to noise ratios of 0 and -5dB, respectively.
(c) Show the learning curves for the detected frequency, the input and the bandpass filtered output signal.

Solution:

A rather small convergence factor $\mu = 0.000001$ is used in this example. Higher values can be used for lower ratio between the sampling frequency and the sinusoid frequency. The starting search frequency is 1100Hz. A quality factor of 10 is used, where this factor measures ratio between the notch frequency and the frequencies with -3dB of attenuation with respect to the gain in the pass band of filter. The stopband width is then 100 Hz. Figs. 10.7.a and 10.7.b depict the input signals for the cases where the signal to noise ratios are 0 and -5 dB's, respectively. Figs. 10.8.a and 10.8.b show the learning curves for the sinusoid frequencies where in both cases the correct frequencies are detected in less than one second which is equivalent to 1000 iterations. As can be observed, the noisier input leads to noisier output. Figs. 10.9.a and 10.9.b depict the bandpass output signal where the sinusoidal components are clearly seen, and again the higher signal to noise ratio results in cleaner sinusoids. In these plots we froze the value of κ_0 at a given iteration after convergence in order to generate the band-passed signals.

□

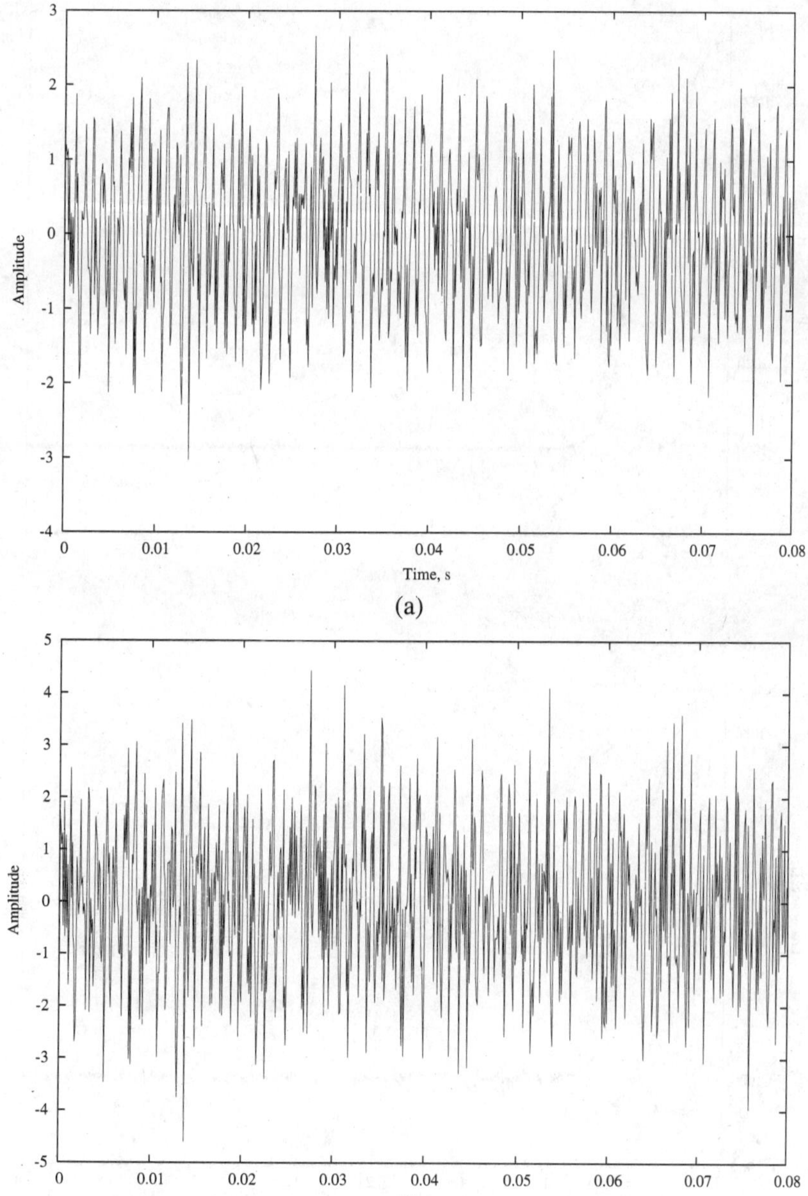

Figure 10.7 Sinusoid buried in noise for signal to noise ratio (a) 0dB, (b) −5dB.

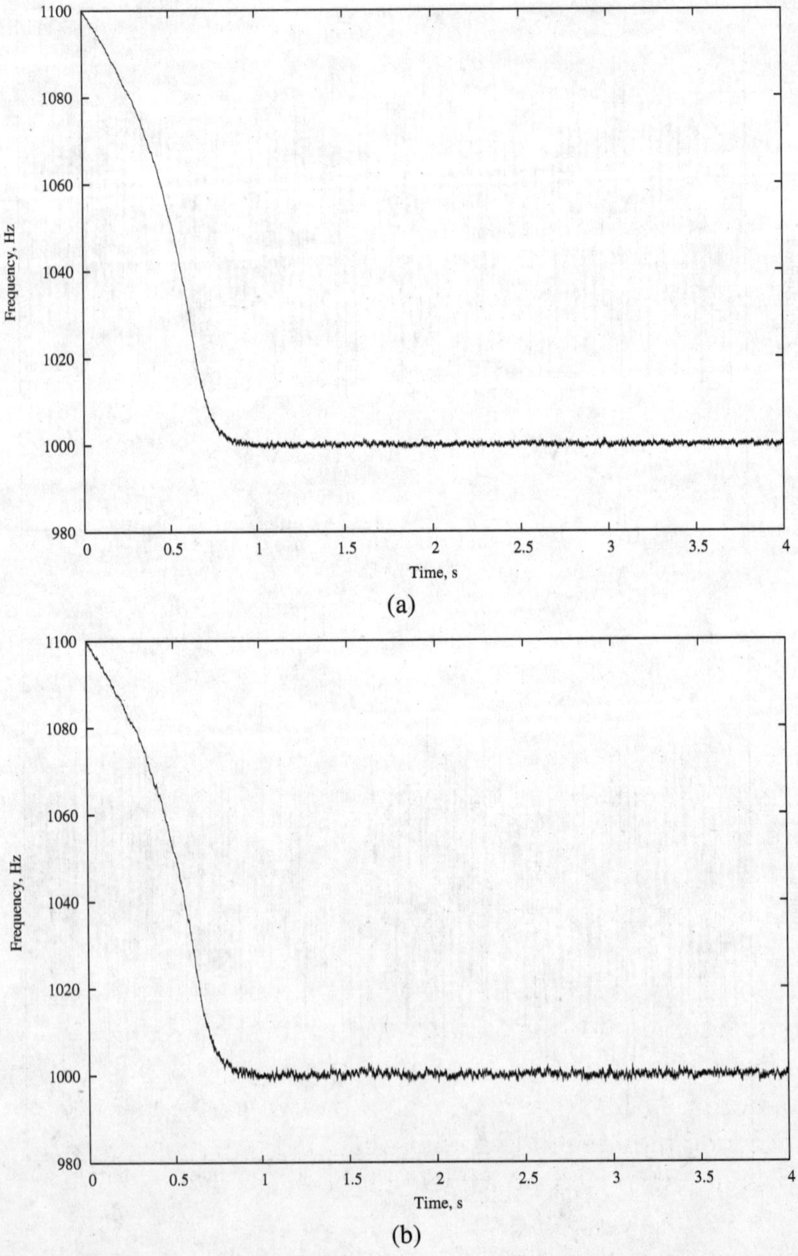

Figure 10.8 Learning curves of the sinusoid frequency (a) 0dB, (b) −5dB.

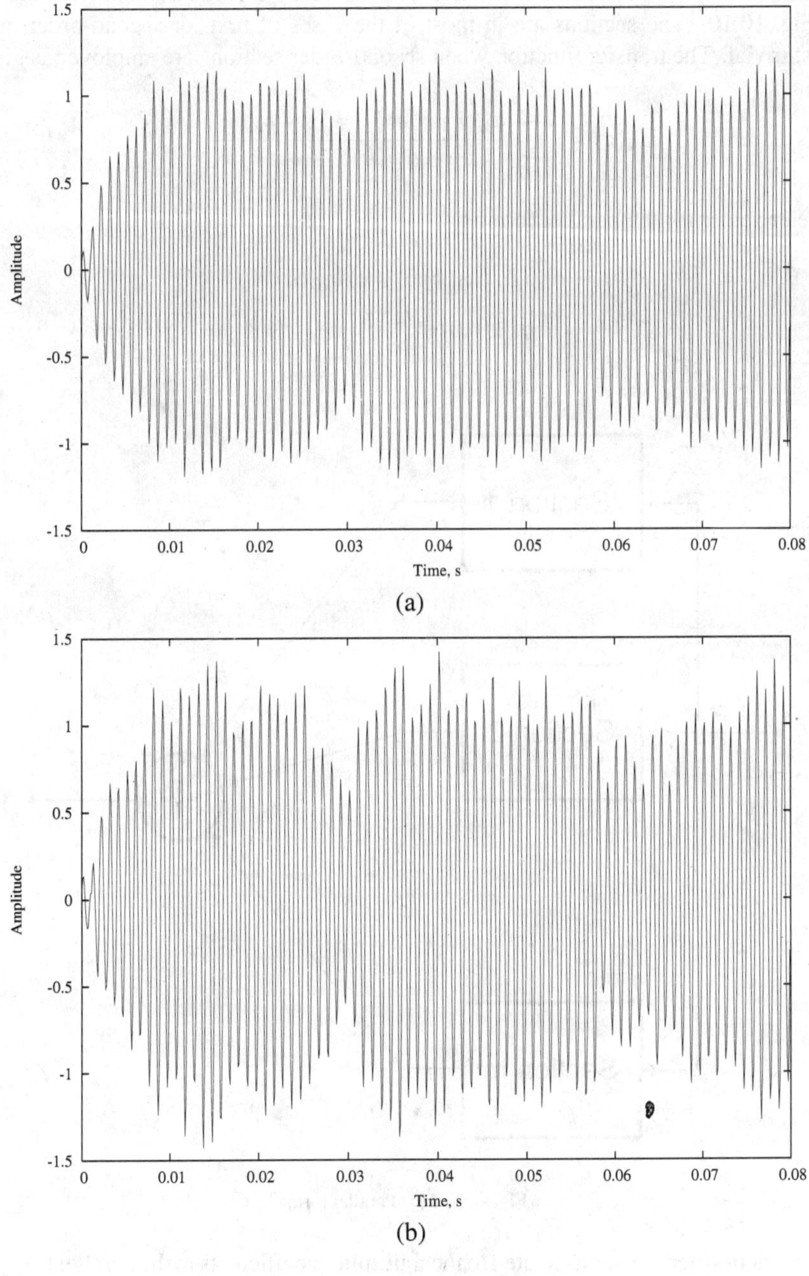

Figure 10.9 Band-passed output signals (a) 0dB, (b) −5dB.

10.5.3 Parallel Form

In the parallel realization, the transfer function is realized by a parallel connection of sections as shown in Fig. 10.10. The sections are in most of the cases of first- or second-order, making the stability test trivial. The transfer function when second-order sections are employed is given by

$$H_k(z) = \sum_{i=0}^{m-1} \frac{b_{0i}(k)z^2 + b_{1i}(k)z + b_{2i}(k)}{z^2 + a_{1i}(k)z + a_{2i}(k)} \tag{10.53}$$

The parameter vector for the parallel form is

$$\begin{aligned}\boldsymbol{\theta}(k) &= [-a_{10}(k) \ -a_{20}(k) \ b_{00}(k) \ b_{10}(k) \ b_{20}(k) \\ &\quad \ldots - a_{1\,m-1}(k) \ -a_{2\,m-1}(k) \ b_{0\,m-1}(k) \ b_{1\,m-1}(k) \ b_{2\,m-1}(k)]^T\end{aligned} \tag{10.54}$$

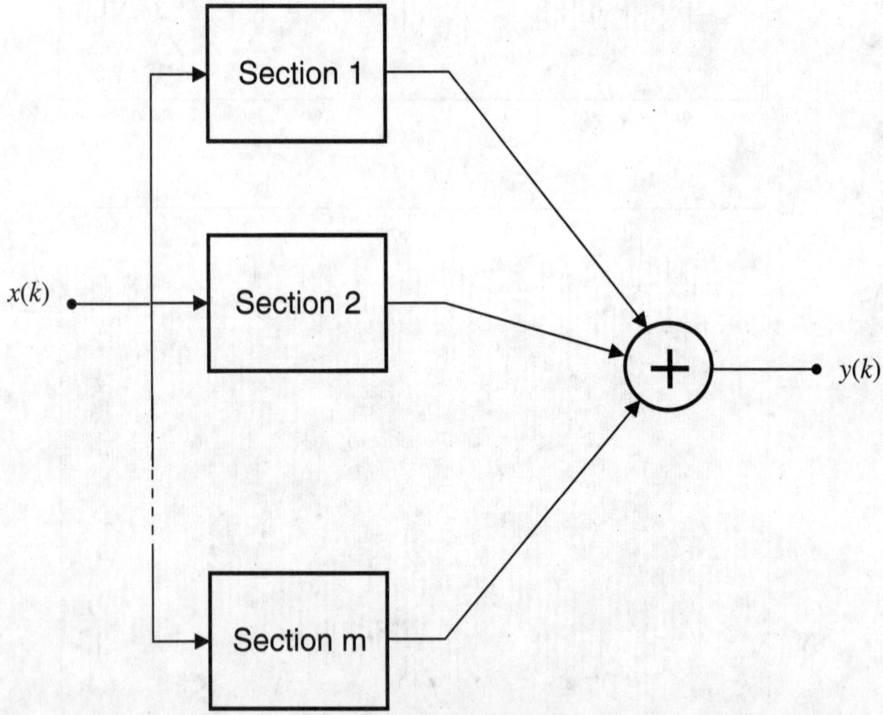

Figure 10.10 Parallel form.

The transfer function derivatives as related to the multiplier coefficients in the parallel form are simple to calculate, because they depend on the derivative of the individual section transfer function with respect to the multiplier coefficients belonging to that section. Basically, the technique of Fig. 10.4 can be applied to each section individually.

Since the interchange of sections in the parallel form does not alter the transfer function, there are $m!$ global minimum points each located in separate subregions of the MSE surface. These subregions are separated by boundaries that are reduced-order manifolds as will be discussed in section 10.7. These boundaries contain saddle points and if the filter parameters are initialized on a boundary, the convergence rate is most probably slow. Consider that the internal signals cross-correlation matrix is approximately estimated by

$$\hat{\mathbf{R}}(k+1) = \alpha \sum_{i=0}^{k} (1-\alpha)^{k-i} \boldsymbol{\varphi}(i) \boldsymbol{\varphi}^T(i) \tag{10.55}$$

when k is large. In this case, if the sections coefficients are identical the information vector consists of a set of identical subvectors $\boldsymbol{\varphi}(i)$, which in turn makes $\hat{\mathbf{R}}(k+1)$ ill-conditioned. The above discussion suggests that the sections in the parallel realization should be initialized differently, although there is no guarantee that this will avoid the ill-conditioning problems.

10.5.4 Frequency-Domain Parallel Structure

A possible alternative parallel realization first proposed in [10] incorporates a preprocessing of the input signal using a discrete-time Fourier transform, generating m signals that are individually applied as input to first-order complex-coefficients sections. With this strategy, the matrix $\hat{\mathbf{R}}(k)$ is more unlikely to become ill-conditioned. Also, it is more difficult for a gradient-based algorithm to get stuck on a reduced-order manifold, resulting in faster convergence. The parallel realization can also be implemented using a real-coefficient transform for the preprocessing and second-order sections.

The frequency-domain parallel structure is illustrated in Fig. 10.11, where $d(k)$ is the reference signal, $x(k)$ is the input signal, $n(k)$ is an additive noise source, and $y(k)$ is the output. The ith parallel section is represented by the transfer function

$$H_i(z) = \frac{b_{0i}(k)z^2 + b_{1i}(k)z + b_{2i}(k)}{z^2 + a_{1i}(k)z + a_{2i}(k)} \qquad k = 0, 1, \ldots, m-1 \tag{10.56}$$

where $a_{1i}(k), a_{2i}(k), b_{0i}(k), b_{1i}(k)$, and $b_{2i}(k)$ are adjustable real coefficients. The inputs of the filter sections are preprocessed as shown in Fig. 10.11.

The purpose of preprocessing in Fig. 10.11 is to generate a set of uncorrelated signals $x_1(k), x_2(k), \ldots, x_m(k)$ in order to reduce the probability that two or more sections converge to the same solution, to simplify the adaptation algorithm, and to improve the rate of convergence.

On employing the discrete-time cosine transform (DCT), the input signals to the subfilters in Fig. 10.11 are given by

$$x_0(k) = \frac{\sqrt{2}}{m} \sum_{l=0}^{m-1} x(k-l)$$

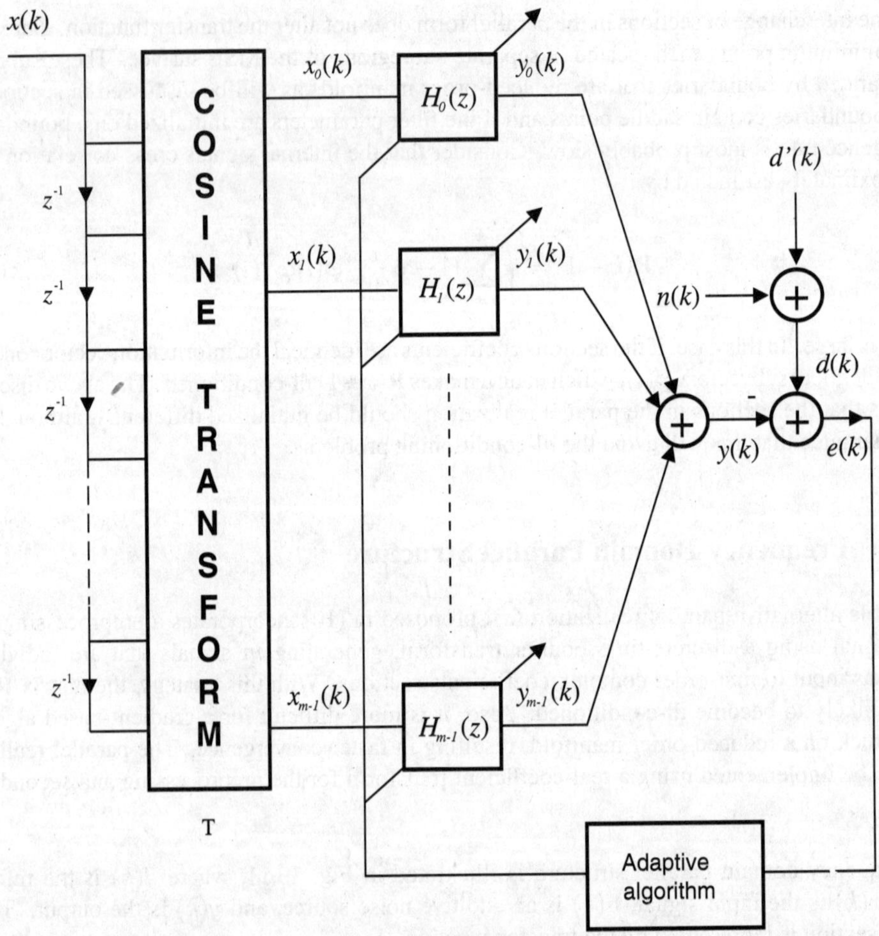

Figure 10.11 Real coefficient frequency-domain adaptive filter.

and

$$x_i(k) = \sqrt{\frac{2}{m}} \sum_{l=0}^{m-1} x(k-l) \cos[\pi i(2l+1)/(2m)] \tag{10.57}$$

The transfer function from the input to the outputs of the DCT preprocessing filter (or prefilter) can be described through the recursive frequency-domain description given by

$$T_i(z) = \frac{k_0}{m} \cos \tau_i \frac{[z^m - (-1)^i](z-1)}{z^{m-1}[z^2 - (2\cos 2\tau_i)z + 1]} \tag{10.58}$$

where

$$k_0 = \begin{cases} \sqrt{2} & \text{if } i = 0 \\ \sqrt{2m} & \text{if } i = 1, 2, \ldots, m-1 \end{cases}$$

and $\tau_i = \pi i/(2m)$. The DCT can be efficiently implemented through some fast algorithms, or by employing equation (10.58). In the latter case, special consideration must be given to the poles on the unit circle.

Alternatively, the transfer functions of the prefilter can be expressed as

$$T_i(z) = \frac{1}{m}\sum_{j=0}^{m-1} t_{ij} z^{-j} = \frac{1}{m}\prod_{r=0}^{m-2}\frac{(z-\tau_{ir})}{z} = \frac{1}{z^{m-1}}\frac{(z-1)[z^m-(-1)^i]}{[z^2-(2\cos\frac{\pi i}{m})z+1]} \tag{10.59}$$

where the t_{ij} are the coefficients of the transform matrix $\mathbf{T}$, and the τ_{ir} are the zeros of $T_i(z)$. The gain constants k_0 and $\cos\tau$ were dropped in equation (10.59) and will not be considered from now on, since they can be absorbed by the numerator coefficients $b_{0i}(k)$, $b_{1i}(k)$, and $b_{2i}(k)$ of $H_i(z)$.

The overall transfer function of the frequency-domain adaptive filter of Fig. 10.11 is given by

$$\begin{aligned} H(z) &= \sum_{i=0}^{m-1} T_i(z)H_i(z) \\ &= \frac{1}{m}(\frac{1}{z^{m-1}})\left[\sum_{i=0}^{m-1}\left(\frac{b_{0i}z^2+b_{1i}z+b_{2i}}{z^2+a_{1i}z+a_{2i}}\right)\prod_{r=0}^{m-2}(z-\tau_{ir})\right] \\ &= \frac{1}{m}\frac{1}{z^{3m+1}}\left[\sum_{i=0}^{m-1}(b_{0i}z^2+b_{1i}z+b_{2i})\frac{\prod_{j=0,\neq i}^{m-1}(z^2+a_{1j}z+a_{2j})\prod_{r=0}^{m-2}(z-\tau_{ir})}{\prod_{l=0}^{m-1}(z^2+a_{1l}z+a_{2l})}\right] \end{aligned} \tag{10.60}$$

Now assume that the realization discussed is used to identify a system of order $2N_p$ described by

$$H_D(z) = Kz^{2N_p-P}\frac{\prod_{r=0}^{P-1}(z-\gamma_r)}{\prod_{i=0}^{N_p-1}(z^2+\alpha_{1i}z+\alpha_{2i})} \tag{10.61}$$

where K is a gain constant, p_{0i} and p_{1i} are the poles of section i, and γ_r are the zeros of $H_D(z)$ such that

$$\gamma_r \neq p_{0i}, p_{1i} \text{ for } r = 0, \ldots, P-1 \text{ and for } i = 0, \ldots, N_p - 1$$

It can be shown that if the conditions outlined below are satisfied, the filter of Fig. 10.11 can identify exactly systems with $N_p \leq m$ and $P \leq 3m+1$. The sufficient conditions are:

i) The transformation matrix **T** of the prefilter is square and has linearly independent rows.

ii) $a_{1i} \neq a_{1j}$, and $a_{2i} \neq a_{2j}$ for $i \neq j$; a_{1i} and a_{2i} are not simultaneously zero for all i.

iii) The zeros of the prefilter do not coincide with the system's poles, i.e., $\tau_{ij} \neq p_{0l}$, $\tau_{ij} \neq p_{1l}$, for all i, j, and l.

Adaptation Algorithm

The adaptation algorithm entails the manipulation of a number of vectors, namely, the coefficient vector

$$\boldsymbol{\theta}(k) = \left[\boldsymbol{\theta}_0^T(k) \ldots \boldsymbol{\theta}_{m-1}^T(k)\right]^T$$

where

$$\boldsymbol{\theta}_i(k) = [-a_{1i}(k) \; -a_{2i}(k) \; b_{0i}(k) \; b_{1i}(k) \; b_{2i}(k)]^T$$

the internal data vector

$$\boldsymbol{\phi}(k) = \left[\boldsymbol{\phi}_0^T(k) \ldots \boldsymbol{\phi}_{m-1}^T(k)\right]^T$$

where

$$\boldsymbol{\phi}_i(k) = [y_i(k-1) \; y_i(k-2) \; x_i(k) \; x_i(k-1) \; x_i(k-2)]^T$$

the gradient vector

$$\tilde{\boldsymbol{\varphi}}(k) = [\boldsymbol{\varphi}_0^T(k) \ldots \boldsymbol{\varphi}_{m-1}^T(k)]^T$$

where

$$\boldsymbol{\varphi}_i(k) = [-y_i'(k-1) \; -y_i'(k-2) \; -x_i'(k) \; -x_i'(k-1) \; -x_i'(k-2)]^T$$

and the matrix $\hat{\mathbf{S}}(k)$ which is an estimate of the inverse Hessian $\hat{\mathbf{R}}^{-1}(k)$.

The elements of the gradient vector can be calculated by using the relations

$$x_i'(k) = x_i(k) - a_{1i}(k)x_i'(k-1) - a_{2i}(k)x_i'(k-2)$$

and

$$y_i'(k) = -y_i(k) - a_{1i}(k)y_i'(k-1) - a_{2i}(k)y_i'(k-2)$$

An adaptation algorithm for updating the filter coefficients based on the Gauss-Newton algorithm is summarized in Algorithm 10.2. The algorithm includes the updating of matrix $\hat{\mathbf{S}}(k)$, which is obtained through the matrix inversion lemma.

The stability monitoring consists of verifying whether each set of coefficients $a_{1i}(k)$ and $a_{2i}(k)$ defines a point outside the stability triangle [2], i.e., by testing whether

$$1 - a_{1i}(k) + a_{2i}(k) < 0 \quad \text{or} \quad 1 + a_{1i}(k) + a_{2i}(k) < 0 \quad \text{or} \quad |a_{2i}(k)| \geq 1 \tag{10.62}$$

If instability is detected in a particular section, the poles must be projected back inside the unit circle. A possible strategy is to project each pole by keeping its angle and inverting its modulus. In this case, a_{2i} and a_{1i} should be replaced by $1/a_{2i}(k)$ and $-a_{1i}(k)/a_{2i}(k)$, respectively.

Algorithm 10.2

Frequency-Domain Parallel Algorithm, RLS Version

Initialization

$\hat{\mathbf{S}}(0) = \delta \mathbf{I}(\delta > 0)$

$\boldsymbol{\theta}_i(k), 0 \leq i \leq m-1$

For each $x(k)$ and $d(k)$ given for $k \geq 0$, compute:

$X_{\mathrm{DCT}}(k) = \mathrm{DCT}[x(k) \ldots x(k-m+1)]$

Do for $i = 0, 1, \ldots, m-1$:

$x_i'(k) = x_i(k) - a_{1i}(k)x_i'(k-1) - a_{2i}(k)x_i'(k-2)$

$y_i(k) = \boldsymbol{\theta}_i^T(k)\boldsymbol{\phi}_i(k)$

$y_i'(k) = -y_i(k) - a_{1i}(k)y_i'(k-1) - a_{2i}(k)y_i'(k-2)$

End

$e(k) = d(k) - \sum_{i=0}^{m-1} y_i(k)$

$\boldsymbol{h}(k) = \hat{\mathbf{S}}(k)\tilde{\boldsymbol{\varphi}}(k)$

$\hat{\mathbf{S}}(k+1) = \left[\hat{\mathbf{S}}(k) - \frac{\boldsymbol{h}(k)\boldsymbol{h}^T(k)}{(\frac{1}{\alpha}-1)+\boldsymbol{h}^T(k)\tilde{\boldsymbol{\varphi}}(k)}\right](\frac{1}{1-\alpha})$

$\boldsymbol{\theta}(k+1) = \boldsymbol{\theta}(k) - \mu\hat{\mathbf{S}}(k+1)\tilde{\boldsymbol{\varphi}}(k)e(k)$

Carry out stability test.
End

If the outputs of the DCT prefilter $x_i(k)$ are sufficiently uncorrelated, the Hessian matrix is approximately block-diagonal consisting of 5×5 submatrices $\hat{\mathbf{R}}_i(k)$. In this case, instead of computing a $5m \times 5m$ inverse Hessian estimate $\hat{\mathbf{S}}(k)$, several 5×5 submatrices are computed and applied in the above algorithm as follows:

For $i = 0, 1, \ldots, m-1$

$$\boldsymbol{h}_i(k) = \hat{\mathbf{S}}_i(k)\boldsymbol{\varphi}_i(k)$$

$$\hat{\mathbf{S}}_i(k+1) = \left[\hat{\mathbf{S}}_i(k) - \frac{\boldsymbol{h}_i(k)\boldsymbol{h}_i^T(k)}{(\frac{1}{\alpha}-1)+\boldsymbol{h}_i^T(k)\boldsymbol{\varphi}_i(k)}\right](\frac{1}{1-\alpha})$$

$$\boldsymbol{\theta}_i(k+1) = \boldsymbol{\theta}_i(k) - \mu\hat{\mathbf{S}}_i(k+1)\boldsymbol{\varphi}_i(k)e(k)$$

□

The choice of the adaptive-filter realization has implications on the computational complexity as well as on the convergence speed. Some studies exploring this aspect related to the frequency-domain realization can be found in [21]. The exploration of realization related properties of the IIR adaptive MSE surface led to a fast parallel realization where no transform preprocessing is required [22]. In this approach, the reduced-order manifolds are avoided by properly configuring the parallel sections which are implemented with general purpose second-order sections [23]. An analysis of the asymptotic convergence speed of some adaptive IIR filtering algorithms from the realization point of view can be found in [24]. Another approach proposes a cascade/parallel orthogonal realization, with simplified gradient computation, by utilizing some of the ideas behind the derivation of improved parallel realizations [25].

Example 10.2

An IIR adaptive filter of sufficient order is used to identify a system with the transfer function given below.

$$H(z) = \frac{0.8(z^2 - 1.804z + 1)^2}{(z^2 - 1.512z + 0.827)(z^2 - 1.567z + 0.736)}$$

The input signal is a uniformly distributed white noise with variance $\sigma_x^2 = 1$, and the measurement noise is Gaussian white noise uncorrelated with the input with variance $\sigma_n^2 = 10^{-1.5}$. Use a gradient-based algorithm.
(a) Choose the appropriate values of μ.
(b) Run the algorithm using the direct-form structure, the lattice structure, the parallel realization with preprocessing, and the cascade realization with direct-form sections. Compare their convergence speed.
(c) Measure the MSE.
(d) Plot the obtained IIR filter frequency response at any iteration after convergence is achieved and compare with the unknown system. Consider for this item only the direct-form realization.

Solution:

A convergence factor $\mu = 0.004$ is used in all examples, except for the lattice realization where $\mu = 0.0002$ is employed for the internal coefficients and a larger $\mu = 0.002$ is employed for the updating of the feedforward coefficients, for stability reasons. Although the chosen value of μ is not an optimal value in any sense, it led to the convergence of all algorithms. Fig. 10.12 depicts the magnitude response of the adaptive filter using the direct form at a given iteration after convergence. For comparison the magnitude response of the system being modeled is also plotted. As can be seen, the responses are close outside the frequency range where the unknown system has a notch. Fig. 10.13 shows the learning curves of the algorithms obtained by averaging the results of 200 independent runs. As can be seen the faster algorithms led to higher MSE. The cascade realization presented faster convergence, followed by the parallel and lattice realizations. The measured MSEs are given in Table 10.1.

There are very few results published in the literature addressing the finite-precision implementation of IIR adaptive filters. For this particular example, all algorithms are also implemented with fixed point arithmetic, with 12 and 16 bits. No sign of divergence is detected during the early 2000 iterations. However, the reader should not take this result as conclusive.

□

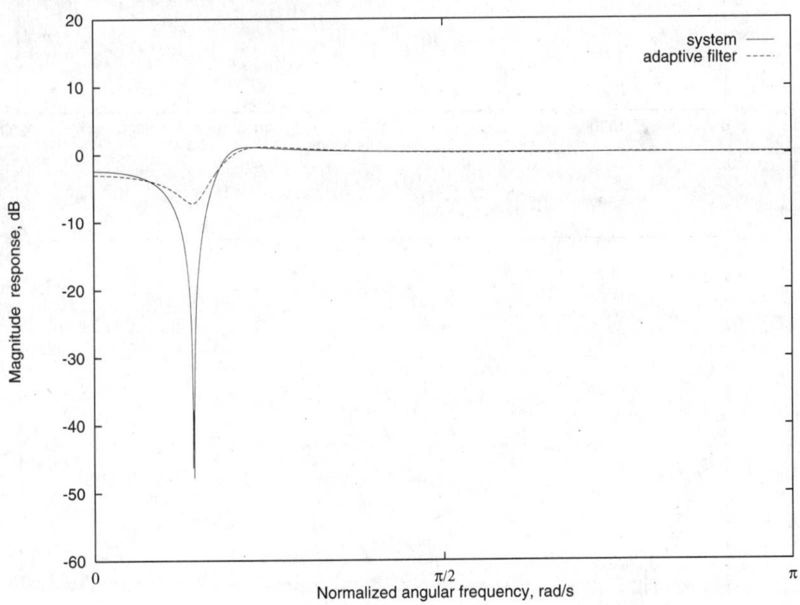

Figure 10.12 Magnitude response of the IIR adaptive filter with direct form at a given iteration after convergence.

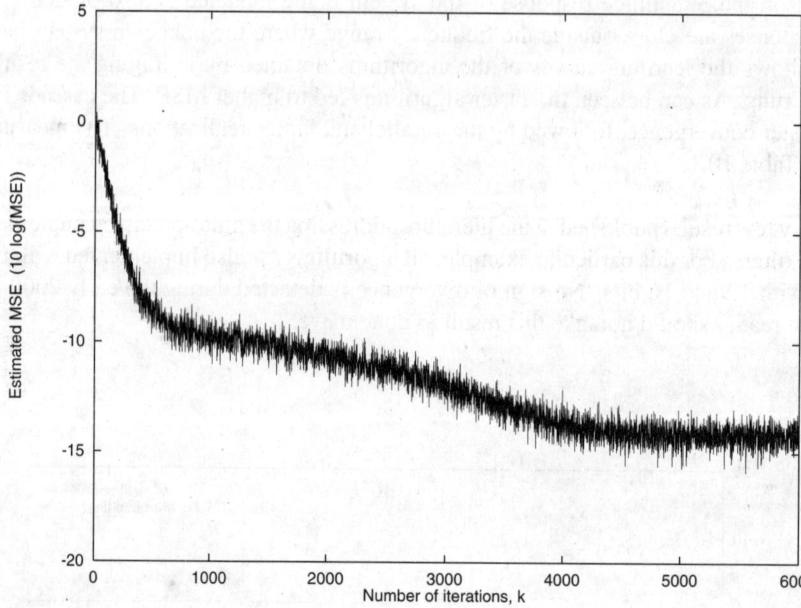
Estimated MSE (10 log(MSE))
Number of iterations, k
5
0
-5
-10
-15
-20
0
1000
2000
3000
4000
5000
6000

(a)

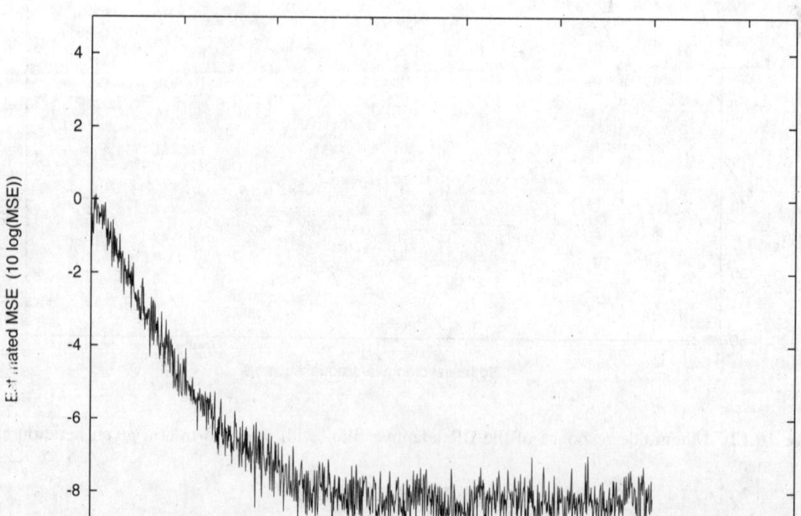
Estimated MSE (10 log(MSE))
Number of iterations, k
4
2
0
-2
-4
-6
-8
-10
0
200
400
600
800
1000
1200
1400

(b)

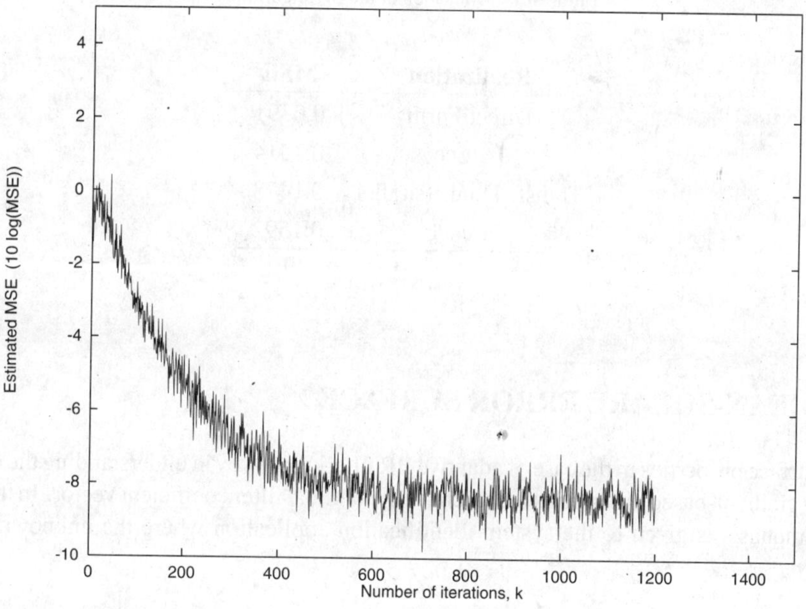

(c)

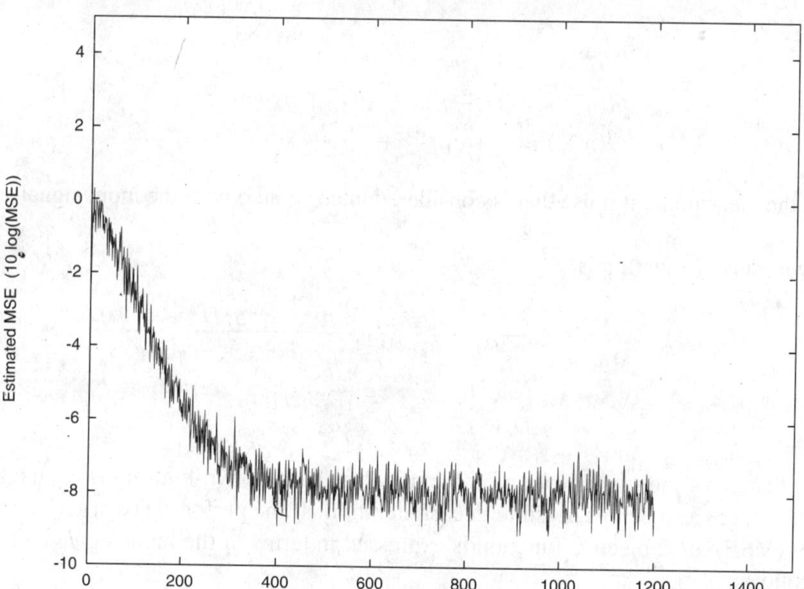

(d)

Figure 10.13 Learning curves for IIR adaptive filters with (a) Direct form, (b) Parallel form with preprocessing, (c) Lattice, and (d) Cascade realizations.

Table 10.1 Evaluation of the IIR Algorithms

Realization	MSE
Direct Form	0.0391
Lattice	0.1514
Transf. Dom. Parallel	0.1478
Cascade	0.1592

10.6 MEAN-SQUARE ERROR SURFACE

The error surface properties in the case of adaptive IIR filtering are key in understanding the difficulties in applying gradient-based algorithms to search for the optimal filter coefficient vector. In this section, the main emphasis is given to the system identification application where the unknown system is modeled by

$$d(k) = \frac{G(q^{-1})}{C(q^{-1})}x(k) + n(k) \tag{10.63}$$

where

$$\begin{aligned} G(q^{-1}) &= g_0 + g_1 q^{-1} + \cdots + g_{M_d} q^{-M_d} \\ C(q^{-1}) &= 1 + c_1 q^{-1} + \cdots + c_{N_d} q^{-N_d} \end{aligned}$$

and $n(k)$ is the measurement noise that is considered uncorrelated with the input signal $x(k)$.

The unknown transfer function is

$$\begin{aligned} H_o(z) &= z^{N_d - M_d} \frac{g_0 z^{M_d} + g_1 z^{M_d - 1} + \cdots + g_{M_d - 1} z + g_{M_d}}{z^{N_d} + c_1 z^{N_d - 1} + \cdots + c_{N_d - 1} z + c_{N_d}} \\ &= z^{N_d - M_d} \frac{N_o(z)}{D_o(z)} \end{aligned} \tag{10.64}$$

The desired feature of the identification problem is that the adaptive-filter transfer function $H_k(z)$ approximates $H_o(z)$ as much as possible in each iteration. If the performance criterion is the mean-square error (MSE), the objective function is expressed in terms of the input signal and the desired signals as follows:

$$\begin{aligned} \xi &= E[e^2(k)] = E\{[d(k) - y(k)]^2\} \\ &= E[d^2(k) - 2d(k)y(k) + y^2(k)] \\ &= E\left\{\left[\left(\frac{G(q^{-1})}{C(q^{-1})}x(k) + n(k)\right) - \frac{B(k, q^{-1})}{A(k, q^{-1})}x(k)\right]^2\right\} \end{aligned} \tag{10.65}$$

Since $n(k)$ is not correlated to $x(k)$ and $E[n(k)] = 0$, equation (10.65) can be rewritten as

$$\xi = E\left\{\left[\left(\frac{G(q^{-1})}{C(q^{-1})} - \frac{B(k,q^{-1})}{A(k,q^{-1})}\right) x(k)\right]^2\right\} + E[n^2(k)] \tag{10.66}$$

The interest here is to study the relation between the objective function ξ and the model filter coefficients, independently if these coefficients are adaptive or not. The polynomials operators $B(k, q^{-1})$ and $A(k, q^{-1})$ will be considered fixed, denoted respectively by $B(q^{-1})$ and $A(q^{-1})$.

The power spectra of the signals involved in the identification process are given by

$$\begin{aligned} R_{xx}(z) &= \mathcal{Z}[r_{xx}(l)] \\ R_{nn}(z) &= \mathcal{Z}[r_{nn}(l)] \\ R_{dd}(z) &= H_o(z)\, H_o(z^{-1})\, R_{xx}(z) + R_{nn}(z) \\ R_{yy}(z) &= H_k(z)\, H_k(z^{-1})\, R_{xx}(z) \\ R_{dy}(z) &= H_o(z)\, H_k(z^{-1})\, R_{xx}(z) \end{aligned} \tag{10.67}$$

By noting that for any processes $x_1(k)$ and $x_2(k)$

$$E[x_1(k)x_2(k)] = \frac{1}{2\pi \jmath} \oint R_{x_1x_2}(z)\frac{dz}{z} \tag{10.68}$$

where the integration path is the counterclockwise unit circle, the objective function, as in equation (10.65), can be rewritten as

$$\begin{aligned} \xi &= \frac{1}{2\pi \jmath} \oint \left[|H_o(z) - H_k(z)|^2 R_{xx}(z) + R_{nn}(z)\right] \frac{dz}{z} \\ &= \frac{1}{2\pi \jmath} \left[\oint H_o(z)H_o(z^{-1})R_{xx}(z)\frac{dz}{z} - 2\oint H_o(z)H_k(z^{-1})R_{xx}(z)\frac{dz}{z} \right. \\ &\quad \left. + \oint H_k(z)H_k(z^{-1})R_{xx}(z)\frac{dz}{z} + \oint R_{nn}(z)\frac{dz}{z}\right] \end{aligned} \tag{10.69}$$

For the case the input and additional noise signals are white with variances respectively given by σ_x^2 and σ_n^2, the equation (10.69) can be simplified to

$$\xi = \frac{\sigma_x^2}{2\pi \jmath} \oint \left[H_o(z)H_o(z^{-1}) - 2H_o(z)H_k(z^{-1}) + H_k(z)H_k(z^{-1})\right] \frac{dz}{z} + \sigma_n^2 \tag{10.70}$$

This expression provides the relation between the MSE surface represented by ξ and the coefficients of the adaptive filter. The following example illustrates the use of the above equation.

Example 10.3

An all-pole adaptive filter of second-order is used to identify a system with transfer function

$$H_o(z) = \frac{1}{z^2 + 0.9z + 0.81}$$

The input signal and the measurement (additional) noise are white with $\sigma_x^2 = 1$ and $\sigma_n^2 = 0.1$, respectively. Compute the MSE as a function of the adaptive-filter multiplier coefficients.

Solution

The adaptive-filter transfer function is given by

$$H_k(z) = \frac{b_2}{z^2 + a_1 z + a_2}$$

Equation (10.70) can be solved by employing the residue theorem [1] which results in

$$\begin{aligned}\xi = & \frac{b_2^2(1+a_2)}{(1-a_2)(1+a_2-a_1)(1+a_2+a_1)} \\ & - \frac{2b_2(1-0.81a_2)}{1-0.9a_1-0.81a_2-0.729a_1a_2+0.81a_1^2+0.6561a_2^2} \\ & +3.86907339+0.1 \end{aligned} \tag{10.71}$$

If the adaptive-filter coefficients are set to their optimal values, i.e., $b_2 = 1$, $a_1 = 0.9$ and $a_2 = 0.81$, indicating a perfect identification of the unknown system, the resulting MSE is

$$\begin{aligned}\xi &= 3.86907339 - 7.73814678 + 3.86907339 + 0.1 \\ &= 0.1\end{aligned}$$

Note that the minimum MSE is equal to the measurement noise variance.

□

Equations (10.69) and (10.70), and more specifically equation (10.71), indicate clearly that the MSE surface is a nonquadratic function of the multiplier coefficients of the adaptive filter. This is particularly true for the multiplier coefficients pertaining to the denominator of the adaptive filter. As a consequence, the MSE surface may have several local minima, some of those corresponding to the desired global minimum. The multiplicity of minimum points depends upon the order of the adaptive IIR filter as compared to the unknown system that shapes the desired signal, and also upon the input signal properties when it is a colored noise.

Note that when the adaptive filter is FIR there is only a minimum point because the MSE surface is quadratic, independently of the unknown system and input signal characteristics. If the input or the desired signal are not stationary, the minimum point of the MSE surface moves in time but it is still unique.

The main problem brought about by the multimodality of the MSE surface is that gradient and Newton direction search algorithms will converge to a local minimum. Therefore, the adaptive filter may converge to a very bad point where the MSE assumes a large and unacceptable value. For example, in the system identification application, the generated transfer function may differ significantly from the unknown system transfer function.

Example 10.4
An unknown system with transfer function

$$H_o(z) = \frac{z - 0.85}{z + 0.99}$$

is supposed to be identified by a first-order adaptive filter described by

$$H_k(z) = \frac{bz}{z - a}$$

Plot the error surface, considering the input signal variance $\sigma_x^2 = 1$.

Solution
The expression for the MSE is given by

$$\xi = 171.13064 - \frac{(2 - 1.7a)b}{1 + 0.99a} + \frac{b^2}{1 - a^2}$$

The MSE surface is depicted in Fig. 10.14, where the MSE is clipped at 1 for a better view.

□

Several results regarding the uniqueness of the minimum point in the MSE surface are available in the literature [26]-[31]. Here, some of these results are summarized without proof, in order to give the designer some tools to support the appropriate choice of the adaptive IIR filter order.

First consider the case of inverse filtering or equalization, where the adaptive filter is placed in cascade with an unknown system and the desired signal is a delayed version of the overall cascade input signal. This case had been originally explored by Åström and Söderström [26], and they proved that if the adaptive filter is of sufficient order to find the inverse filter of the unknown system all the local minima will correspond to global minima if the input signal is a white noise. The sufficient order means that

$$N \geq M_d$$

and

$$M \geq N_d \tag{10.72}$$

where N and M are the numerator and denominator orders of the adaptive filter as indicated in equation (10.5), N_d and M_d are the corresponding orders for the unknown system as indicated in equation (10.64).

When $N > M_d$ and $M > N_d$, there are infinitely many solutions given by

$$N(z) = L(z)D_o(z)$$

and

$$D(z) = L(z)N_o(z) \tag{10.73}$$

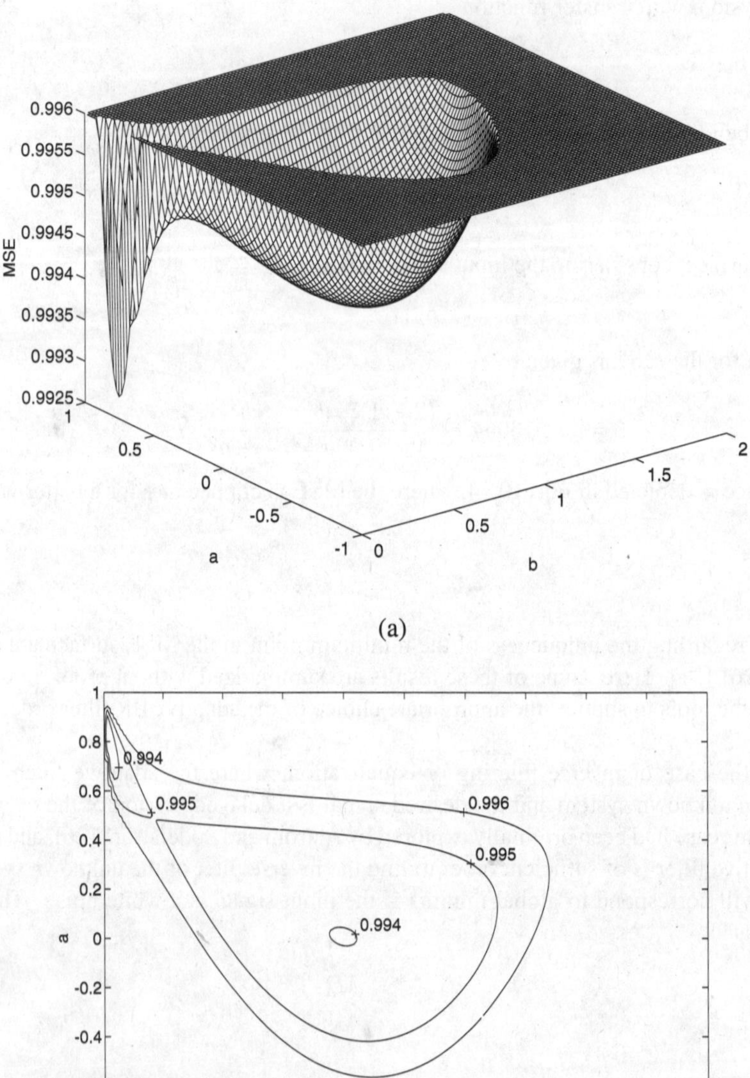

Figure 10.14 (a) MSE error surface, (b) MSE contours.

where $L(z) = z^{-N_l}(z^{N_l} + l_1 z^{N_l-1} + \cdots + l_{N_l})$, $N_l = \min(N - M_d, M - N_d)$, and l_i, for $i = 1, 2, \ldots, N_l$, are arbitrary.

The input signal can be colored noise generated for example by applying an IIR filter to a white noise. In this case, the adaptive filter must have order sufficient to generate the inverse of the unknown system and the input signal must be persistently exciting of order $\max(N + M_d, M + N_d)$, see for example [26]-[27], in order to guarantee that all local minima correspond to global minima.

For insufficient-order equalization, several local minima that do not correspond to a global minimum may occur. In this case, the MSE may not attain its minimum value after the algorithm convergence.

The situation is not the same in system identification application, as thought in the early investigations [28]. For this application, the sufficient order means

$$N \geq N_d$$

and

$$M \geq M_d \tag{10.74}$$

since the desired feature is to reproduce the unknown system frequency response, and not its inverse as in the equalization case. For $N > N_d$ and $M > M_d$, the local minima corresponding to global minima must satisfy the following conditions

$$N(z) = L(z)N_o(z)$$

and

$$D(z) = L(z)D_o(z) \tag{10.75}$$

where $L(z) = z^{-N_l}(z^{N_l} + l_i z^{N_l-1} + \cdots + l_{N_l})$, $N_l = \min(N - M_d, M - N_d)$, and l_i, for $i = 1, 2, \ldots, N_l$, are arbitrary.

The strongest result derived so far regarding the error surface property in system identification was derived by Söderström and Stoica [29]. The result states: For white noise input, all the stationary points correspond to global minima if

$$M \geq N_d - 1$$

and

$$\min(N - N_d, M - M_d) \geq 0 \tag{10.76}$$

Suppose that the input signal is an ARMA process generated by filtering a white noise with an IIR filter of orders M_n by N_n, and that there are no common zeros between the unknown system denominator and the input coloring IIR filter. In this case, all stationary points correspond to global minima if

$$M - N_d + 1 \geq N_n$$

and

$$\min(N - N_d, M - M_d) \quad \geq \quad M_n \tag{10.77}$$

The conditions summarized by equations (10.76) and (10.77) are sufficient but not necessary to guarantee that all stationary solutions correspond to the minimum MSE.

For $N = N_d = 1,\ M \geq M_d \geq 0$ and the input signal persistently exciting of order M_d there is a unique solution given by [29]

$$D(z) = D_o(z)$$

and

$$N(z) = N_o(z) \tag{10.78}$$

Also, when the adaptive filter and unknown system are all-pole second-order sections the unique solution is given by equation (10.78) [30].

Another particular result of some interest presented in [31], states that if

$$N - N_d = M - M_d = 0$$

and

$$M \geq N_d - 2 \tag{10.79}$$

the MSE surface has a unique stationary point corresponding to a global minimum.

For the case of insufficient-order identification [32], i.e., $\min(N - N_d, M - M_d) < 0$, or of sufficient order not satisfying the condition related to equations (10.77)-(10.79), the MSE surface may have local minima not attaining the minimum MSE, i.e., that are not global minima.

To satisfy any of the conditions of equations (10.77)-(10.79) a knowledge of the unknown system numerator and denominator orders is required. This information is not in general available or easy to obtain. This is one of the reasons adaptive IIR filters are not as popular as their FIR counterparts. However, there are situations where either a local minimum is acceptable or some information about the unknown system is available.

It should be noted that a vast literature is available for system identification [8],[33]-[34]. Here, the objective was to summarize some properties of the MSE surface, when the unknown system is modeled as an IIR filter with additive, white, and uncorrelated measurement noise. The assumptions regarding the measurement noise are quite reasonable for most applications of adaptive filtering.

10.7 INFLUENCE OF THE FILTER STRUCTURE ON THE MSE SURFACE

Some characteristics of the MSE surface differ when alternative structures are used in the realization of the adaptive filter. Each realization has a different relation between the filter transfer function and the multiplier coefficients, originating modifications in the MSE surface [35].

The MSE surfaces related to two alternative realizations for the adaptive filter can be described as functions of the filter multiplier coefficients by $F_1(\boldsymbol{\theta}_1)$ and $F_2(\boldsymbol{\theta}_2)$, respectively. Note that no index was used to indicate the varying characteristics of the adaptive-filter parameters, since this simplifies the notation while keeping the relevant MSE surface properties. It is assumed that the desired signal and the input signal are the same in the alternative experiments. Also, it is considered that for any set of parameters $\boldsymbol{\theta}_1$ leading to a stable filter, there is a continuous mapping given by $\boldsymbol{f}_3(\boldsymbol{\theta}_1) = \boldsymbol{\theta}_2$, where $\boldsymbol{\theta}_2$ also leads to a stable filter. Both $\boldsymbol{\theta}_1$ and $\boldsymbol{\theta}_2$ are N' by 1 vectors.

The two alternative structures are equivalent if the objective functions are equal, i.e.,

$$F_1(\boldsymbol{\theta}_1) = F_2(\boldsymbol{\theta}_2) = F_2[\boldsymbol{f}_3(\boldsymbol{\theta}_1)] \tag{10.80}$$

First consider the case where $\boldsymbol{f}_3$ is differentiable, and then from the above equation it follows that

$$\frac{\partial F_1(\boldsymbol{\theta}_1)}{\partial \boldsymbol{\theta}_1} = \frac{\partial F_2[\boldsymbol{f}_3(\boldsymbol{\theta}_1)]}{\partial \boldsymbol{\theta}_1} = \frac{\partial F_2[\boldsymbol{f}_3(\boldsymbol{\theta}_1)]}{\partial \boldsymbol{f}_3(\boldsymbol{\theta}_1)} \frac{\partial \boldsymbol{f}_3(\boldsymbol{\theta}_1)}{\partial \boldsymbol{\theta}_1} \tag{10.81}$$

where the first partial derivative on the rightmost side of the above equation is an 1 by N' vector while the second partial derivative is a matrix with dimensions N' by N', where N' is the number of parameters in $\boldsymbol{\theta}_1$. Suppose that $\boldsymbol{\theta}_2'$ is a stationary point of $F_2(\boldsymbol{\theta}_2)$, it then follows that

$$\frac{\partial F_2(\boldsymbol{\theta}_2)}{\partial \boldsymbol{\theta}_2}\Big|_{\boldsymbol{\theta}_2=\boldsymbol{\theta}_2'} = \mathbf{0} = \frac{\partial F_1(\boldsymbol{\theta}_1)}{\partial \boldsymbol{\theta}_1}\Big|_{\boldsymbol{\theta}_1=\boldsymbol{\theta}_1'} \tag{10.82}$$

where $\boldsymbol{\theta}_2' = \boldsymbol{f}_3(\boldsymbol{\theta}_1')$. Note that the type of the stationary points of $F_1(\boldsymbol{\theta}_1)$ and $F_2(\boldsymbol{\theta}_2)$ are the same, since their second derivatives have the same properties at these stationary points (see problem 1).

Now consider the case where

$$\frac{\partial F_2[\boldsymbol{f}_3(\boldsymbol{\theta}_1)]}{\partial \boldsymbol{f}_3(\boldsymbol{\theta}_1)}\Big|_{\boldsymbol{\theta}_1=\boldsymbol{\theta}_1''} = \mathbf{0} \tag{10.83}$$

but

$$\frac{\partial F_1(\boldsymbol{\theta}_1)}{\partial \boldsymbol{\theta}_1}\Big|_{\boldsymbol{\theta}_1=\boldsymbol{\theta}_1''} \neq \mathbf{0} \tag{10.84}$$

that can happen only when $\boldsymbol{f}_3(\boldsymbol{\theta}_1)$ is not differentiable at $\boldsymbol{\theta}_1 = \boldsymbol{\theta}_1''$. In this case, the chain rule of equation (10.81) does not apply. The new generated stationary points in $F_2(\boldsymbol{\theta}_2)$ can be shown to be saddle points (see problem 2).

Example 10.5

An unknown second-order system described by

$$H_o(z) = \frac{2z + c_1}{z^2 + c_1 z + c_2}$$

is to be identified by using two different structures for the adaptive filter, namely the direct form and the parallel form described respectively by

$$H_d(z) = \frac{2z + a_1}{z^2 + a_1 z + a_2}$$

and

$$H_p(z) = \frac{1}{z + p_1} + \frac{1}{z + p_2} = \frac{2z + p_1 + p_2}{z^2 + (p_1 + p_2)z + p_1 p_2}$$

verify the existence of new saddle points in the parallel realization.

Solution

The function relating the parameters of the two realizations can be given by

$$\boldsymbol{\theta}_2 = \begin{bmatrix} \frac{a_1 + \sqrt{a_1^2 - 4a_2}}{2} \\ \frac{a_1 - \sqrt{a_1^2 - 4a_2}}{2} \end{bmatrix} = \boldsymbol{f}_3(\boldsymbol{\theta}_1)$$

where function $\boldsymbol{f}_3(\boldsymbol{\theta}_1)$ is not differentiable when $a_2 = \frac{a_1^2}{4}$.

The inverse of the matrix $\frac{\partial \boldsymbol{f}_3(\boldsymbol{\theta}_1)}{\partial \boldsymbol{\theta}_1}$ is given by

$$\left[\frac{\partial \boldsymbol{f}_3(\boldsymbol{\theta}_1)}{\partial \boldsymbol{\theta}_1}\right]^{-1} = \begin{bmatrix} 1 & 1 \\ p_2 & p_1 \end{bmatrix}$$

and, if $p_1 = p_2$, the above matrix is singular, which makes it possible that $\frac{\partial F_1(\boldsymbol{\theta}_1)}{\partial \boldsymbol{\theta}_1} \neq 0$ when $\frac{\partial F_2(\boldsymbol{\theta}_2)}{\partial \boldsymbol{\theta}_2} = 0$, as previously mentioned in equations (10.81) and (10.82).

Note that, as expected, $p_1 = p_2$ only when $a_2 = \frac{a_1^2}{4}$. On this parabola, the objective function $F_1(\boldsymbol{\theta}_1)$ has a minimum that corresponds to a saddle point of the function $F_2(\boldsymbol{\theta}_2)$. Also, this is the situation where the parallel realization is of reduced order, i.e., first order.

□

Basically, the manifold generated by the parallel realization is due to the fact that a given section can identify any pole of the unknown system, leaving the other poles to the remaining sections in parallel. This means that in a sufficient-order identification problem, if for the direct-form realization there is a unique global minimum point, in the case of parallel realization with first-order sections there will be $N!$ global minima, where N is the number of poles in the unknown system. When using a parallel realization it is assumed that no multiple poles exist in the unknown system.

In the initialization of the algorithm, the adaptive-filter parameters should not be in a reduced-order manifold, because by employing a gradient-based algorithm the parameters may be kept in the manifold and eventually reach a saddle point. The measurement noise, that is in general present in the adaptive-filtering process, will help the parameters to skip the manifolds, but despite that the convergence will be slowed. A similar phenomenon occurs with the cascade realization of the adaptive filter.

10.8 ALTERNATIVE ERROR FORMULATIONS

The error signal (in some cases the regressor) can be chosen in alternative ways in order to avoid some of the drawbacks related to the output error formulation, as for example the multiple local minima. Several formulations have been investigated in the literature [36]-[37], [39], [40]-[42], [45]-[46], [51]-[52], where each of them has its own advantages and disadvantages. The choice of the best error formulation depends on the application and on the information available about the adaptive-filtering environment. In this section, we present two alternative error formulations, namely the equation error and Steiglitz-McBride methods, and discuss some of their known properties. Throughout the section other error formulations are briefly mentioned.

10.8.1 Equation Error Formulation

In the equation error (EE) formulation, the information vector instead of having past samples of the adaptive-filter output, uses delayed samples of the desired signal as follows:

$$\boldsymbol{\phi}_e(k) = [d(k-1)\, d(k-2) \ldots d(k-N)\, x(k)\, x(k-1) \ldots x(k-M)]^T \tag{10.85}$$

The equation error is defined by

$$e_e(k) = d(k) - \boldsymbol{\theta}^T(k)\boldsymbol{\phi}_e(k) \tag{10.86}$$

as illustrated in Fig. 10.15. The parameter vector $\boldsymbol{\theta}(k)$ is given by

$$\boldsymbol{\theta}(k) = [-a_1(k) \; - a_2(k) \ldots - a_N(k)\, b_0(k) \ldots b_M(k)]^T \tag{10.87}$$

The equation error can be described in a polynomial form as follows:

$$e_e(k) = A(k, q^{-1})d(k) - B(k, q^{-1})x(k) \tag{10.88}$$

where, once again

$$B(k, q^{-1}) = b_0(k) + b_1(k)q^{-1} + \cdots + b_M(k)q^{-M}$$
$$A(k, q^{-1}) = 1 + a_1(k)q^{-1} + \cdots + a_N(k)q^{-N}$$

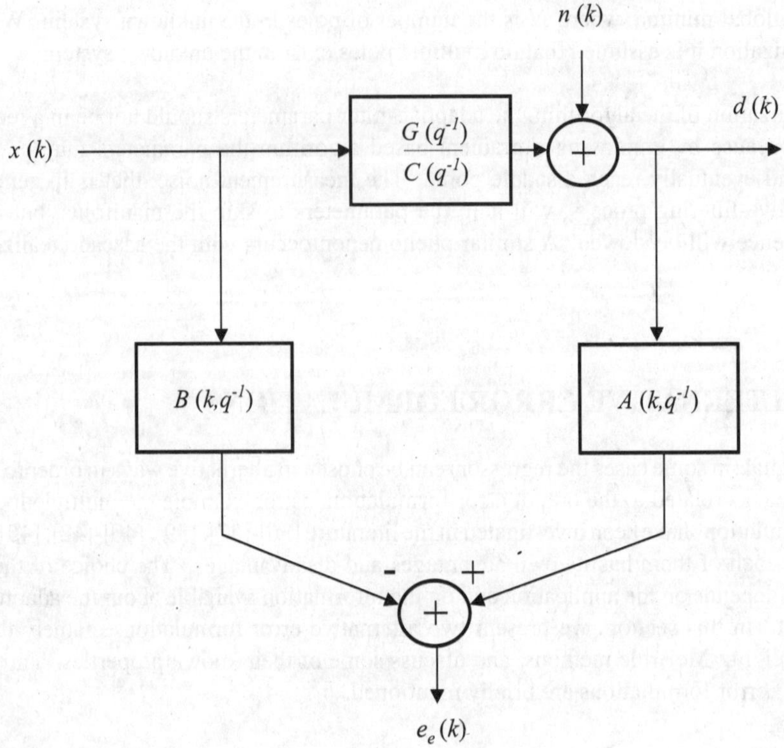

Figure 10.15 Equation error configuration.

The output signal related to the EE formulation is obtained through the following linear difference equation

$$
\begin{aligned}
y_e(k) &= \sum_{j=0}^{M} b_j(k)x(k-j) - \sum_{j=1}^{N} a_j(k)d(k-j) \\
&= \boldsymbol{\theta}^T(k)\boldsymbol{\phi}_e(k)
\end{aligned}
\tag{10.89}
$$

As can be noted, the adaptive filter does not have feedback and $y_e(k)$ is a linear function of the parameters.

In the EE formulation, the adaptation algorithm determines how the coefficients of the adaptive IIR filter should change in order to minimize an objective function which involves $e_e(k)$ defined as

$$\xi_e = F[e_e(k)] \tag{10.90}$$

Usually, the objective function to be minimized is the mean-squared value of the EE (MSEE), i.e.,

$$\xi_e(k) = E[e_e^2(k)] \tag{10.91}$$

Since the input and desired signals are not functions of the adaptive-filter parameters, it can be expected that the sole approximation in the gradient computation is due to the estimate of the expected value required in practical implementations. The key point is to note that since the MSEE is a quadratic function of the parameters, only a global minimum exists provided the signals involved are persistently exciting. When the estimate of the MSEE is the instantaneous squared equation error, the gradient vector is proportional to minus the information vector. In this case, the resulting algorithm is called LMSEE algorithm whose coefficient updating equation is given by

$$\boldsymbol{\theta}(k+1) = \boldsymbol{\theta}(k) + 2\mu\boldsymbol{\phi}_e(k)e_e(k) \tag{10.92}$$

A number of approaches with different points of view are available to analyze the convergence properties of this method. A particularly interesting result is that if the convergence factor is chosen in the range

$$0 < \mu < \frac{1}{\lambda_{\max}} \tag{10.93}$$

the convergence in the mean of the LMSEE algorithm can be guaranteed [37], where $\lambda_{\max}$ is the maximum eigenvalue of $E[\boldsymbol{\phi}_e(k)\boldsymbol{\phi}_e^T(k)]$. This result can be easily proved by exploring the similarity between the LMSEE algorithm and the standard FIR LMS algorithm. Some stability results of the LMSEE algorithm can be found in [38].

An alternative objective function for adaptive IIR filtering based on equation error is the least-squares function

$$\xi_e(k) = \sum_{i=0}^{k} \lambda^{k-i} e_e^2(i) = \sum_{i=0}^{k} \lambda^{k-i}[d(i) - \boldsymbol{\theta}^T(k)\boldsymbol{\phi}_e(i)]^2 \tag{10.94}$$

The forgetting factor λ, as usual is chosen in the range $0 << \lambda < 1$, allowing the distant past information to be increasingly negligible. In this case, the corresponding RLS algorithm consists of the following basic steps

$$e(k) = d(k) - \boldsymbol{\theta}^T(k)\boldsymbol{\phi}_e(k) \tag{10.95}$$

$$\mathbf{S}_{De}(k+1) = \frac{1}{\lambda}\left[\mathbf{S}_{De}(k) - \frac{\mathbf{S}_{De}(k)\boldsymbol{\phi}_e(k)\boldsymbol{\phi}_e^T(k)\mathbf{S}_{De}(k)}{\lambda + \boldsymbol{\phi}_e^T(k)\mathbf{S}_{De}(k)\boldsymbol{\phi}_e(k)}\right] \tag{10.96}$$

$$\boldsymbol{\theta}(k+1) = \boldsymbol{\theta}(k) + \mathbf{S}_{De}(k+1)\boldsymbol{\phi}_e(k)e_e(k) \tag{10.97}$$

In a given iteration k, the adaptive IIR filter transfer function related to the EE formulation can be expressed as follows:

$$H_k(z) = z^{N-M}\frac{b_0(k)z^M + b_1(k)z^{M-1} + \cdots + b_{M-1}(k)z + b_M(k)}{z^N + a_1(k)z^{N-1} + \cdots + a_{N-1}(k)z + a_N(k)} \tag{10.98}$$

In Fig. 10.16 an alternative structure for the EE approach where the IIR adaptive filter appears explicitly is depicted. Note that the structure shows clearly that the polynomial $A(k, q^{-1})$ is meant to model the denominator polynomial of the unknown system, in system identification applications. During the adaptation process, it is necessary to monitor the stability of the poles, as described for the output error method. The full description of the RLS equation error algorithm is given in Algorithm 10.3.

The basic problem related to this method is the parameter bias induced by the measurement noise [37]-[38], even for sufficient-order case. The bias is caused by the fact that the additional noise $n(k)$

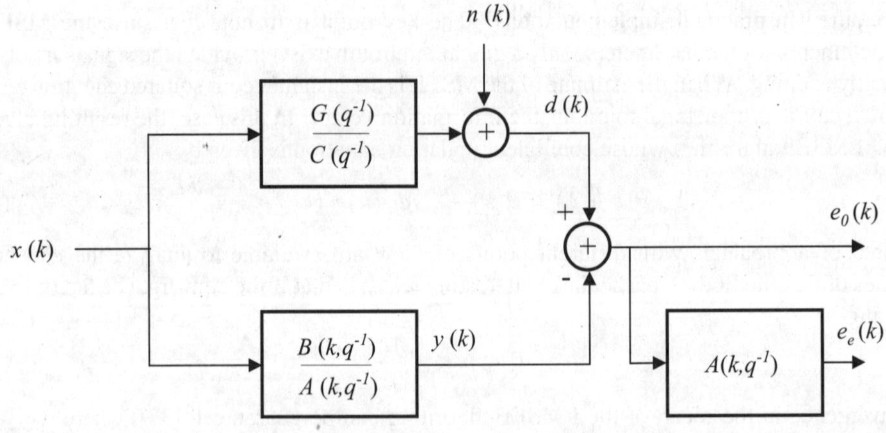

Figure 10.16 Basic configuration for system identification using equation error.

Algorithm 10.3

EE Algorithm, RLS Version

Initialization

$a_i(k) = b_i(k) = e(k) = 0$

$y(k) = x(k) = 0\,,\ k\ < 0$

$\mathbf{S}_{De}(0) = \delta^{-1}\mathbf{I}$

For each $x(k)$, $d(k)$, $k \geq 0$, do

$e_e(k) = d(k) - \boldsymbol{\phi}_e^T(k)\boldsymbol{\theta}(k)$

$\mathbf{S}_{De}(k+1) = \frac{1}{\lambda}\left[\mathbf{S}_{De}(k) - \frac{\mathbf{S}_{De}(k)\boldsymbol{\phi}_e(k)\boldsymbol{\phi}_e^T(k)\mathbf{S}_{De}(k)}{\lambda+\boldsymbol{\phi}_e^T(k)\mathbf{S}_{De}(k)\boldsymbol{\phi}_e(k)}\right]$

$\boldsymbol{\theta}(k+1) = \boldsymbol{\theta}(k) + \mathbf{S}_{De}(k+1)\boldsymbol{\phi}_e(k)e_e(k)$

Stability test

is filtered by the FIR filter represented by the polynomial $A(k, q^{-1})$. Since the coefficients of this polynomial are updated with the objective of minimizing the EE signal, they also attempt to minimize the contribution of $n(k)$ to the EE power. The bias is induced by the fact that the additional noise does not belong to the unknown system model. An increase in the power of $n(k)$ leads to higher bias in the parameter estimate.

The Instrumental Variable methods [39] were proposed to solve the bias problem. In these methods the stability cannot be guaranteed under the same general conditions as for the LMSEE method.

Another approach was proposed in [40], and extended in [41] and [42], where a family of asymptotically stable algorithms was introduced. The resulting algorithms are based on a modification of the basic LMSEE updating equations, that within sufficiently general conditions lead to consistent parameter estimates. These algorithms employ a type of output error feedback to the information vector. There are also algorithms that combine different algorithms to define the objective function [43]-[44].

10.8.2 The Steiglitz-McBride Method

The Steiglitz-McBride (SM) error formulation [45], by employing some extra all-pole filtering, leads to algorithms whose behavior resembles the EE approach in the initial iterations and the output error approach after convergence. The main motivation of the SM method is the global convergence behavior for some cases of insufficient-order system identification. Such interest sparked investigations which resulted in a number of on-line algorithms based on the SM method that are suitable for adaptive IIR filtering [46]. The main problem associated with the SM method is the inconsistent behavior when the measurement noise is colored [47]. Since the on-line method converges asymptotically to the off-line solution, the bias error also affects the on-line algorithms proposed in [46].

In order to introduce the SM method, consider the identification of a system whose model is described by

$$d(k) = \frac{G(q^{-1})}{C(q^{-1})} x(k) + n(k) = y_d(k) + n(k) \tag{10.99}$$

where $d(k)$ is the reference signal, $x(k)$ is the input signal, $n(k)$ is the measurement noise, and $y_d(k)$ is the output signal of the plant, with $C(q^{-1}) = 1 - \sum_{i=1}^{N_d} c_i q^{-i}$ and $G(q^{-1}) = \sum_{i=0}^{M_d} g_i q^{-i}$ coprime. The polynomial $C(q^{-1})$ has zeros inside the unit circle, and the input signal $x(k)$ and the measurement noise $n(k)$ are assumed independent. The estimation of the parameters associated with the polynomials $C(q^{-1})$ and $G(q^{-1})$ through the SM method is based on the minimization of the following criterion [45]

$$\xi_s(\boldsymbol{\theta}(k+1)) = E\left\{\left[A(k+1, q^{-1})\frac{d(k)}{A(k, q^{-1})} - B(k+1, q^{-1})\frac{x(k)}{A(k, q^{-1})}\right]^2\right\} \tag{10.100}$$

where $A(k, q^{-1}) = 1 + \sum_{i=1}^{N} a_i(k) q^{-i}$ and $B(k, q^{-1}) = \sum_{i=0}^{M} b_i(k) q^{-i}$ are the denominator and numerator estimator polynomials, respectively, and

$$\boldsymbol{\theta}(k) = [-a_1(k)\ -a_2(k)\ \ldots\ -a_N(k)\ b_0(k)\ \ldots\ b_M(k)]^T \tag{10.101}$$

is the adaptive-filter parameter vector.

The estimate $\boldsymbol{\theta}(k+1)$ is obtained by minimizing equation (10.100) assuming $\boldsymbol{\theta}(k)$ known. The solution of this MSE minimization problem at iteration $(k+1)$ is

$$\begin{aligned}\boldsymbol{\theta}(k+1) &= \left[E\left\{\boldsymbol{\phi}_s(k)\boldsymbol{\phi}_s^T(k)\right\}\right]^{-1} E\left[\boldsymbol{\phi}_s(k)\frac{d(k)}{A(k, q^{-1})}\right] \\ &= \left[E\left\{\boldsymbol{\phi}_s(k)\boldsymbol{\phi}_s^T(k)\right\}\right]^{-1} E\left[\boldsymbol{\phi}_s(k) d_f(k)\right]\end{aligned} \tag{10.102}$$

where

$$\phi_s(k) = \left[\frac{d(k-1)}{A(k,q^{-1})} \cdots \frac{d(k-N)}{A(k,q^{-1})} \frac{x(k)}{A(k,q^{-1})} \cdots \frac{x(k-M)}{A(k,q^{-1})}\right]^T$$
$$= [d_f(k-1) \ldots d_f(k-N)\; x_f(k) \ldots x_f(k-M)]^T \qquad (10.103)$$

is the regressor related to the SM method.

If the input signal is persistently exciting of sufficient order and the adaptive filter has strictly sufficient order, some properties of the estimate resulting from equation (10.102) are known [47]: a) The estimate that minimizes equation (10.100) is unique; b) If the measurement noise is not white, the estimate resulting from equation (10.102) is biased.

In real-time signal processing applications, it is important to consider an on-line version of the SM method. In this case, some approximations are necessary. First note that the error criterion whose variance is to be minimized in equation (10.102) is

$$e_s(k) = \frac{d(k)}{A(k,q^{-1})} - \theta^T(k+1)\phi_s(k) \qquad (10.104)$$

The SM error is computed as illustrated in Fig. 10.17. Assuming a sufficiently slow parameter variation, we can consider that $\theta(k+1) \approx \theta(k)$. Therefore, equation (10.104) can be rewritten as follows:

$$e_s(k) \approx \frac{d(k)}{A(k,q^{-1})} - \theta^T(k)\phi_s(k) \qquad (10.105)$$

The exact implementation of the regressor $\phi_s(k)$ requires an independent filtering of each component by an all-pole filter with denominator polynomial $A(k, q^{-1})$. A useful approximation that reduces considerably the computational complexity is possible by assuming slow parameter variation [46] in such a way that

$$\theta(k-1) \approx \theta(k-2) \ldots \approx \theta(k-N) \qquad (10.106)$$

With these simplifications only one all-pole filtering is required. Note that a hypothesis similar to equation (10.106) was utilized in the output error method in order to simplify the implementation. However, in the case of the output error method, the measurement noise does not affect the regressor, since the regressor vector is composed of delayed samples of the adaptive-filter input and output. For the SM method, except for white measurement noise, the simplification in equation (10.106) is not easily justified.

The updating equation of the on-line SM algorithm for system identification employing a stochastic gradient search is given by

$$\theta(k+1) = \theta(k) + 2\mu\phi_s(k)\left[\frac{d(k)}{A(k,q^{-1})} - \phi_s^T(k)\theta(k)\right]$$
$$= \theta(k) + 2\mu\phi_s(k)e_s(k) \qquad (10.107)$$

The description of a gradient SM algorithm is given in the Algorithm 10.4.

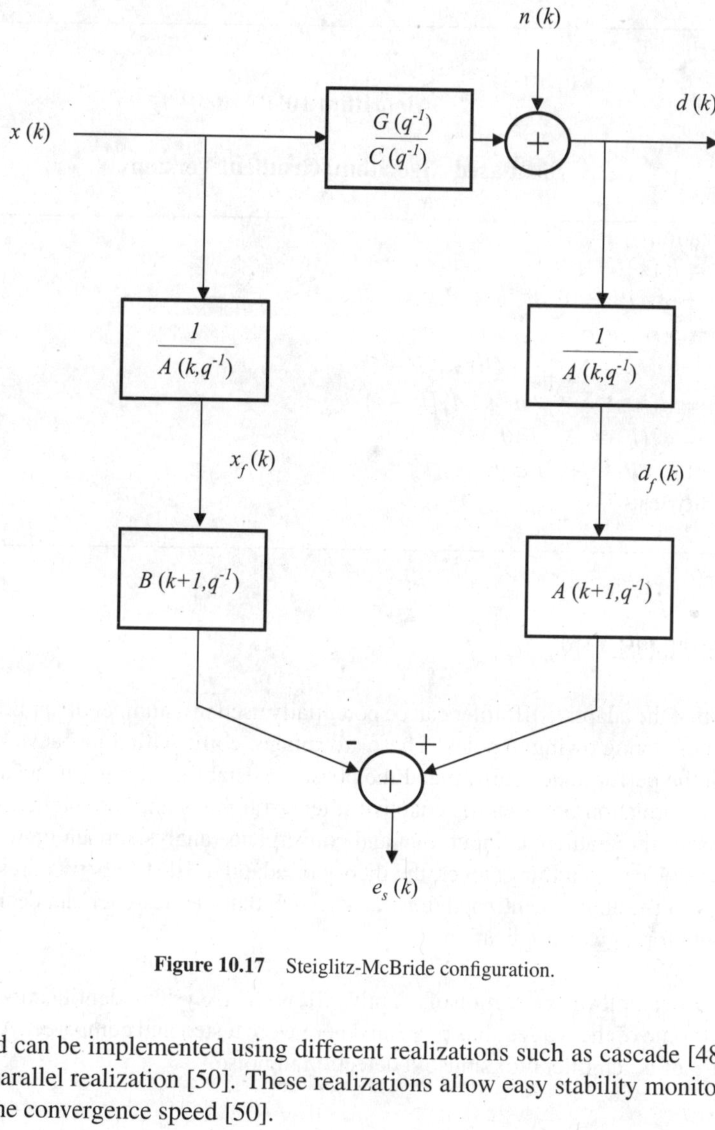

Figure 10.17 Steiglitz-McBride configuration.

The SM method can be implemented using different realizations such as cascade [48], lattice [49], and the series-parallel realization [50]. These realizations allow easy stability monitoring, and their choice affects the convergence speed [50].

It should be mentioned that a family of algorithms based on the SM method that solves the problem of inconsistency of the parameter estimates was proposed in [51]-[52]. These algorithms are very attractive for adaptive IIR filtering due to their behavior in terms of consistency (i.e., definition of stationary points) and convergence properties. In [55], simulation results as well as an alternative implementation for the consistent SM method was presented.

The interested reader can also find some interesting results about the convergence behavior of the SM-based algorithms in [53]-[54] and in the references therein. Also, applications of the SM algorithm to equalization can be found in [56].

Algorithm 10.4

SM Based Algorithm, Gradient Version

Initialization

$a_i(k) = b_i(k) = 0$

$d_f(k) = x_f(k) \doteq 0\,,\; k\; < 0$

For each $x(k),\ d(k),\ k \geq 0$ do

$x_f(k) = x(k) - \sum_{i=1}^{N} a_i(k)\, x_f(k-i)$

$d_f(k) = d(k) - \sum_{i=1}^{N} a_i(k)\, d_f(k-i)$

$e_s(k) = d_f(k) - \boldsymbol{\phi}_s^T(k)\boldsymbol{\theta}(k)$

$\boldsymbol{\theta}(k+1) = \boldsymbol{\theta}(k) + 2\mu\boldsymbol{\phi}_s(k)e_s(k)$

Stability test

10.9 CONCLUSION

It is recognized that the adaptive IIR filter can be potentially used in a number of applications due to its superior system modeling owing to poles. These advantages come with drawbacks such as possible local minima in the performance surface and the possible instability during the adaptation process. Also, the nonlinear relation between the adaptive-filter parameters and the internal signals in some formulations makes the gradient computation and convergence analysis much more complicated as compared to the FIR case. In this chapter, the theory of adaptive IIR filters was presented exposing several solutions to the above mentioned drawbacks, such that the designer can decide which is the best configuration for a given application.

In this chapter, an example of application of adaptive IIR filters in system identification was presented. In this example, some of the realizations presented here were tested and compared. Another example exploited the use of notch filters for sinusoid detection in noise.

10.10 REFERENCES

1. P. S. R. Diniz, E. A. B. da Silva, and S. L. Netto, *Digital Signal Processing: System Analysis and Design*, Cambridge University Press, Cambridge, UK, 2002.

2. A. Antoniou, *Digital Signal Processing: Signals, Systems, and Filters*, McGraw Hill, New York, NY, 2005.

3. C. R. Johnson, Jr., "Adaptive IIR filtering: Current results and open issues," *IEEE Trans. on Information Theory*, vol. IT-30, pp. 237-250, Mar. 1984.

4. J. J. Shynk, "Adaptive IIR filtering," *IEEE ASSP Magazine*, vol. 6, pp. 4-21, April 1989.

5. S. A. White, "An adaptive recursive digital filter," *Proc. 9th Asilomar Conf. on Circuits, Systems, and Computers*, Pacific Grove, CA, pp. 21-25, Nov. 1975.

6. S. Hovarth, Jr., "A new adaptive recursive LMS filter," *Proc. Digital Signal Processing Conference*, Florence, Italy, pp. 21-26, 1980.

7. T. C. Hsia, "A simplified adaptive recursive filter design," *Proceedings of the IEEE*, vol. 69, pp. 1153-1155, Sept. 1981.

8. L. Ljung and T. Söderström, *Theory and Practice of Recursive Identification*, MIT Press, Cambridge, MA, 1983.

9. R. A. David, "A modified cascade structure for IIR adaptive algorithms," *Proc. 15th Asilomar Conf. on Circuits, Systems, and Computers*, Pacific Grove, CA, pp. 175-179, Nov. 1981.

10. J. J. Shynk, "Adaptive IIR filtering using parallel-form realization," *IEEE Trans. on Acoust., Speech, and Signal Processing*, vol. 37, pp. 519-533, April 1989.

11. P. S. R. Diniz, J. E. Cousseau, and A. Antoniou, "Improved parallel realization of IIR adaptive filters," *Proceedings of the IEE Part G: Circuits, Devices, and Systems*, vol. 140, pp. 322-328, Oct. 1993.

12. P. A. Regalia, *Adaptive IIR Filtering for Signal Processing and Control*, Marcel Dekker, New York, NY, 1995.

13. D. Parikh, N. Ahmed, and S. D. Stearns, "An adaptive lattice algorithm for recursive filters," *IEEE Trans. on Acoust., Speech, and Signal Processing*, vol. ASSP-28, pp. 110-112, Feb. 1980.

14. I. L. Ayala, "On a new adaptive lattice algorithm for recursive filters," *IEEE Trans. on Acoust., Speech, and Signal Processing*, vol. ASSP-30, pp. 316-319, April 1982.

15. J. J. Shynk, "On lattice-form algorithms for adaptive IIR filtering," *Proc. IEEE Intern. Conf. on Acoust., Speech, Signal Processing*, New York, NY, pp. 1554-1557, April 1988.

16. J. A. Rodríguez-Fonollosa and E. Masgrau, "Simplified gradient calculation in adaptive IIR lattice filters," *IEEE Trans. on Signal Processing*, vol. 39, pp. 1702-1705, July 1991.

17. A. H. Gray, Jr., and J. D. Markel, "Digital lattice and ladder filter synthesis," *IEEE Trans. on Audio Electroacoust.*, vol. AU-21, pp. 492-500, Dec. 1973.

18. F. Itakura and S. Saito, "Digital filtering techniques for speech analysis and synthesis," *Proc. 7th Intern. Congr. Acoustics*, Paper 25C-1, Budapest, Hungary, pp. 261-264, 1971.

19. M. Tummala, "New adaptive normalised lattice algorithm for recursive filters," *Electronics Letters*, vol. 24, pp. 659-661, May 1988.

20. J. M. Romano and M. G. Bellanger, "Fast least-squares adaptive notch filtering," *IEEE Trans. on Acoust. Speech and Signal Processing*, vol. 36, pp. 1536-1540, Sept. 1988.

21. H. Fan and Y. Yang, "Analysis of a frequency-domain adaptive IIR filter," *IEEE Trans. on Acoust. Speech, and Signal Processing*, vol. 38, pp. 864-870, May 1990.

22. P. S. R. Diniz, J. E. Cousseau, and A. Antoniou, "Fast parallel realization for IIR adaptive filters," *IEEE Trans. on Circuits and Systems-II: Analog and Digital Signal Processing*, vol. 41, pp. 561-567, Aug. 1994.

23. P. S. R. Diniz and A. Antoniou, "Digital-filter structures based on the concept of the voltage-conversion generalized immittance converter," *Canadian J. of Electrical and Computer Engineering*, vol. 13, pp. 90-98, April 1988.

24. H. Fan, "A structural view of asymptotic convergence speed of adaptive IIR filtering algorithms: Part I-Infinite precision implementation," *IEEE Trans. on Signal Processing*, vol. 41, pp. 1493-1517, April 1993.

25. J. E. Cousseau, P. S. R. Diniz, G. Sentoni, and O. Agamennoni, "On orthogonal realizations for adaptive IIR filters," *International Journal Circuit Theory and Applications*, John Wiley & Sons, NY, vol. 28, pp. 481-500, Sept. 2000.

26. K. J. Åström and T. Söderström, "Uniqueness of the maximum likelihood estimates of the parameters of an ARMA model," *IEEE Trans. on Automatic Control*, vol. AC-19, pp. 769-773, Dec. 1974.

27. T. Söderström, "On the uniqueness of maximum likelihood identification," *Automatica*, vol. 11, pp. 193-197, Mar. 1975.

28. S. D. Stearns, "Error surfaces of recursive adaptive filters," *IEEE Trans. on Acoust., Speech, and Signal Processing*, vol. ASSP-29, pp. 763-766, June 1981.

29. T. Söderström and P. Stoica, "Some properties of the output error model," *Automatica*, vol. 18, pp. 93-99, Jan. 1982.

30. H. Fan and M. Nayeri, "On error surfaces of sufficient order adaptive IIR filters: Proofs and counterexamples to a unimodality conjecture," *IEEE Trans. on Acoust., Speech, and Signal Processing*, vol. 37, pp. 1436-1442, Sept. 1989.

31. M. Nayeri, "Uniqueness of MSOE estimates in IIR adaptive filtering; a search for necessary conditions," *Proc. IEEE Intern. Conf. on Acoust., Speech, Signal Processing*, Glasgow, Scotland, pp. 1047-1050, May 1989.

32. M. Nayeri, H. Fan, and W. K. Jenkins, "Some characteristics of error surfaces for insufficient order adaptive IIR filters," *IEEE Trans. on Acoust., Speech, and Signal Processing*, vol. 38, pp. 1222-1227, July 1990.

33. L. Ljung, *System Identification: Theory for the User*, Prentice Hall Inc., Englewood Cliffs, NJ, 1987.

34. T. Söderström and P. Stoica, *System Identification*, Prentice Hall International, Hempstead, Hertfordshire, 1989.

35. M. Nayeri and W. K. Jenkins, "Alternate realizations to adaptive IIR filters and properties of their performance surfaces," *IEEE Trans. on Circuits and Systems*, vol. 36, pp. 485-496, April 1989.

36. S. L. Netto, P. S. R. Diniz, and P. Agathoklis, "Adaptive IIR filtering algorithms for system identification: A general framework," *IEEE Trans. on Education*, vol. 26, pp. 54-66, Feb. 1995.

37. J. M. Mendel, *Discrete Techniques of Parameter Estimation: The Equation Error Formulation*, Marcel Dekker, New York, NY, 1973.

38. T. Söderström and P. Stoica, "On the stability of dynamic models obtained by least squares identification," *IEEE Trans. on Automatic Control*, vol. AC-26, pp. 575-577, April 1981.

39. T. Söderström and P. Stoica, *Instrumental Variable Methods for System Identification*, Springer Verlag, New York, NY, 1983.

40. J. -N. Lin and R. Unbehauen, "Bias-remedy least mean square equation error algorithm for IIR parameter recursive estimation," *IEEE Trans. on Signal Processing*, vol. 40, pp. 62-69, Jan. 1992.

41. P. S. R. Diniz and J. E. Cousseau, "A family of equation-error based IIR adaptive algorithms," *IEEE Proc. Midwest Symposium of Circuits and Systems*, Lafayette, LA, pp. 1083-1086, 1994.

42. J. E. Cousseau and P. S. R. Diniz, "A general consistent equation-error algorithm for adaptive IIR filtering," *Signal Processing*, vol. 56, pp. 121-134, 1997.

43. J. B. Kenney and C. E. Rohrs, "The composite regressor algorithm for IIR adaptive systems," *IEEE Trans. on Signal Processing*, vol. 41, pp. 617-628, Feb. 1993.

44. S. L. Netto and P. S. R. Diniz, "Composite algorithms for adaptive IIR filtering," *IEE Electronics Letters*, vol. 28, pp. 886-888, April 1992.

45. K. Steiglitz and L. E. McBride, "A technique for the identification of linear systems," *IEEE Trans. on Automatic Control*, vol. AC-10, pp. 461-464, Oct. 1965.

46. H. Fan and W. K. Jenkins, "A new adaptive IIR filter," *IEEE Trans. on Circuits and Systems*, vol. CAS-33, pp. 939-947, Oct. 1986.

47. P. Stoica and T. Söderström, "The Steiglitz-McBride identification algorithm revisited–convergence analysis and accuracy aspects," *IEEE Trans. on Automatic Control*, vol. AC-26, pp. 712-717, June 1981.

48. B. E. Usevitch and W. K. Jenkins, "A cascade implementation of a new IIR adaptive digital filter with global convergence and improved convergence rates," *Proc. IEEE Intern. Symposium of Circuits and Systems*, Portland, OR, pp. 2140-2142, 1989.

49. P. A. Regalia, "Stable and efficient lattice algorithms for adaptive IIR filtering," *IEEE Trans. on Signal Processing*, vol. 40, pp. 375-388, Feb. 1992.

50. J. E. Cousseau and P. S. R. Diniz, "A new realization of IIR echo cancellers using the Steiglitz-McBride method," *Proc. IEEE Intern. Telecommunication Symposium*, Rio de Janeiro, Brazil, pp. 11-14, 1994.

51. J. E. Cousseau and P. S. R. Diniz, "A consistent Steiglitz-McBride algorithm," *Proc. IEEE Intern. Symposium of Circuits and Systems*, Chicago, IL, pp. 52-55, 1993.

52. J. E. Cousseau and P. S. R. Diniz, "New adaptive IIR filtering algorithms based on Steiglitz-McBride method," *IEEE Trans. on Signal Processing*, vol. 45, pp. 1367-1371, May 1997.

53. H. Fan and M. Doroslovački, "On 'global convergence' of Steiglitz-McBride adaptive algorithm," *IEEE Trans. on Circuits and Systems-II: Analog and Digital Signal Processing*, vol. 40, pp. 73-87, Feb. 1993.

54. M. H. Cheng and V. L. Stonick, "Convergence, convergence point and convergence rate for Steiglitz-McBride method: A unified approach," *Proc. IEEE Intern. Conf. on Acoust., Speech and Signal Processing*, Adelaide, Australia, 1994.

55. V. L. Stonick and M. H. Cheng, "Adaptive IIR filtering: composite regressor method," *Proc. IEEE Intern. Conf. on Acoust., Speech and Signal Processing*, Adelaide, Australia, 1994.

56. P. M. Crespo and M. L. Honig, "Pole-zero decision feedback equalization with a rapidly converging adaptive IIR algorithm," *IEEE J. on Selected Areas in Communications*, vol. 9, pp. 817-828, Aug. 1991.

10.11 PROBLEMS

1. Show that the stationary points related to two equivalent adaptive realizations of the type in equation (10.82) have the same nature, i.e., are minimum, maximum or saddle point.

2. Show that the new stationary points generated by the discontinuity in $f_3(\theta_1)$ as discussed after equation (10.84) are saddle points.

3. Describe how the manifolds are formed in the MSE surface when a cascade realization is used for the adaptive-filter implementation. Give a generic example.

4. Derive a general expression for the transfer function of the two-multiplier lattice structure.

5. Derive an adaptive-filtering algorithm which employs the canonic direct-form structure shown in Fig. 10.18. Consider that the adaptive-filter parameters are slowly varying in order to derive an efficient implementation for the gradient vector.

6. A second-order all-pole adaptive filter is used to find the inverse model of the signal $x(k) = 1.7n(k-1)+0.81n(k-2)+n(k)$, where $n(k)$ is Gaussian white noise with variance 0.1. Using the gradient algorithm, calculate the error and the filter coefficients for the first 10 iterations. Start with $a_1(0) = 0, a_2(0) = 0$.

7. Repeat the problem 6 using the Gauss-Newton algorithm.

8. Use an IIR adaptive filter of sufficient order to identify a system with the transfer function given below. The input signal is a uniformly distributed white noise with variance $\sigma_x^2 = 1$, and the measurement noise is Gaussian white noise uncorrelated with the input with variance $\sigma_n^2 = 10^{-2}$. Use a Gauss-Newton based algorithm and the direct-form structure.

$$H(z) = \frac{0.000058(z^2 - 2z + 1)^3}{(z^2 + 1.645z + 0.701)(z^2 + 1.575z + 0.781)(z^2 + 1.547z + 0.917)}$$

(a) Run the algorithm for three values of μ. Comment on the convergence behavior in each case.
(b) Measure the MSE in each example.
(c) Plot the obtained IIR filter frequency response at any iteration after convergence is achieved and compare with the unknown system.

9. Repeat the previous problem using a second-order adaptive filter and interpret the results.

10. A sinusoid of normalized frequency equal to $\frac{\pi}{4}$ with unit amplitude is buried in noise. The signal to noise ratio is 0 dB. Detect the sinusoid with notch filters using the lattice and the direct-form structures.

(a) After convergence compute an estimate of the frequency by averaging the result of ten samples for each structure and comment on the result.

(b) Depict the input signal and the output signal for the bandpass filter based on the lattice structure.

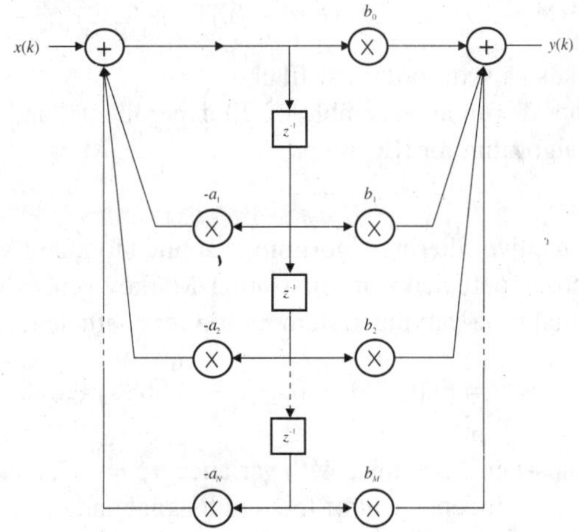

Figure 10.18 Direct form of Problem 5.

11. Replace the direct-form structure in problem 8 by the parallel realization with preprocessing.

12. Replace the direct-form structure in problem 8 by the cascade realization.

13. Repeat problem 8 in case the input signal is a uniformly distributed white noise with variance $\sigma^2_{n_x} = 0.1$, filtered by all-pole filter given by

$$H(z) = \frac{z}{z - 0.95}$$

14. In problem 8 consider that the additional noise has the following variances (a) $\sigma^2_n = 0$, (b) $\sigma^2_n = 1$. Comment on the results obtained in each case.

15. Perform the equalization of a channel with the following transfer function

$$H(z) = \frac{z^2 - 1.359z + 0.81}{z^2 - 1.919z + 0.923}$$

using a known training signal that consists of a binary (-1,1) random signal. An additional Gaussian white noise with variance 10^{-2} is present at the channel output.
(a) Apply a Newton-based algorithm with direct-form structure.
(b) Plot the magnitude response of the cascade of the channel and the adaptive-filter transfer functions. Comment on the result.

16. In a system identification problem the input signal is generated by an autoregressive process given by

$$x(k) = -1.2x(k-1) - 0.81x(k-2) + n_x(k)$$

where $n_x(k)$ is zero-mean Gaussian white noise with variance such that $\sigma^2_x = 1$. The unknown system is described by

$$H(z) = \frac{80z^3(z^2 + 0.81)(z - 0.9)}{(z^2 - 0.71z + 0.25)(z^2 + 0.75z + 0.56)(z^2 - 0.2z + 0.81)}$$

The adaptive filter is also a sixth-order IIR filter.
Choose an appropriate λ, run an ensemble of 20 experiments, and plot the average learning curve. Use the RLS algorithm for IIR filters.

17. A second-order IIR adaptive-filtering algorithm is applied to identify a 3rd-order time-varying unknown system whose coefficients are first-order Markov processes with $\lambda_{\mathbf{w}} = 0.999$ and $\sigma^2_{\mathbf{w}} = 0.001$. The initial time-varying system multiplier coefficients are

$$\mathbf{w}_o^T = [0.03490 \ -0.011 \ -0.06864 \ 0.22391]$$

The input signal is Gaussian white noise with variance $\sigma^2_x = 0.7$, and the measurement noise is also Gaussian white noise independent of the input signal and of the elements of $\mathbf{n}_{\mathbf{w}}(k)$, with variance $\sigma^2_n = 0.01$.
Simulate the experiment described and plot the learning curve, by using the direct-form structure with a gradient-type algorithm.

18. Suppose a second-order IIR digital filter, with multiplier coefficients given below, is identified by an adaptive IIR filter of the same order using the gradient algorithm. Considering that fixed-point arithmetic is used, measure the values of $E[||\Delta\boldsymbol{\theta}(k)_Q||^2]$ and $\xi(k)_Q$ for the case described below. Plot the learning curves for the finite- and infinite-precision implementations. Also plot an estimate of the expected value of $||\Delta\boldsymbol{\theta}(k)||^2$ versus k in both cases.

    Additional noise: white noise with variance $\sigma_n^2 = 0.0015$
    Coefficient wordlength: $b_c = 16$ bits
    Signal wordlength: $b_d = 16$ bits
    Input signal: Gaussian white noise with variance $\sigma_x^2 = 0.7$

$$H(z) = \frac{z^2 - 1.804z + 1}{z^2 - 1.793z + 0.896}$$

19. Repeat the above problem for the following cases
    (a) $\sigma_n^2 = 0.01$, $b_c = 9$ bits, $b_d = 9$ bits, $\sigma_x^2 = 0.7$.
    (b) $\sigma_n^2 = 0.1$, $b_c = 10$ bits, $b_d = 10$ bits, $\sigma_x^2 = 0.8$.
    (c) $\sigma_n^2 = 0.05$, $b_c = 8$ bits, $b_d = 16$ bits, $\sigma_x^2 = 0.8$.

20. Replace the direct-form structure in problem 18 by the lattice structure, and comment on the results.

21. Repeat problem 8 using the LMSEE algorithm.

22. Show the inequality in equation (10.93).

23. Repeat problem 15 using the LMSEE algorithm.

24. Repeat problem 8 using a gradient-type algorithm based on the SM method.

25. Repeat problem 15 using a gradient-type algorithm based on the SM method.

26. Derive the RLS-type algorithm based on the SM method.

11

NONLINEAR ADAPTIVE FILTERING

11.1 INTRODUCTION

The classic adaptive-filtering algorithms, such as those discussed in the remaining chapters of this book, consist of adapting the coefficients of linear filters in real time. These algorithms have applications in a number of situations where the signals measured in the environment can be well modeled as Gaussian noises applied to linear systems, and their combinations are of additive type. In digital communication systems, most of the classical approaches model the major impairment affecting the transmission with a linear model. For example, channel noise is considered additive Gaussian noise, intersymbol and co-channel interferences are also considered of additive type, and channel models are assumed to be linear frequency selective filters. While these models are accurate, there is nothing wrong with the use of linear adaptive filters[1] to remedy these impairments. However, the current demand for higher-speed communications leads to the exploration of the channel resources beyond the range their models can be considered linear. For example, when the channel is the pair of wires of the telephone system, it is widely accepted that linear models are not valid for data transmission above 4.8 Kb/s. Signal companding, amplifier saturation, multiplicative interaction between Gaussian signals, and nonlinear filtering of Gaussian signals are typical phenomena occurring in communication systems that cannot be well modeled with linear adaptive systems. In addition, if the channel transfer function does not have minimum phase and/or the signal to noise ratio is not high enough, the use of linear adaptive-filtering equalizer yields poor performance measured in terms of bit error rate. A major drawback of dealing with nonlinear models is the lack of mathematical tools that, on the other hand, are widely available for linear models. The lack of analytical tools originates in the high degrees and dimensionality of the nonlinearities. The improved performance of the nonlinear equalizer is mainly justified by extensive simulation results available in the literature, where the bit error rate is used as a performance measure.

In this chapter, we will describe some of the techniques available to model nonlinear systems using nonlinear adaptive systems using the general structure depicted in Fig. 11.1. In particular, the following approaches for nonlinear adaptive filtering will be discussed here:

[1]The reader should bear in mind that adaptive filters are nonlinear filters, even if we are adapting the coefficients of a linear filter structure, therefore the term linear adaptive filter means that we are adapting the coefficients of a linear filter structure.

P.S.R. Diniz, *Adaptive Filtering*, DOI: 10.1007/978-0-387-68606-6_11,

1. The nonrecursive polynomial model based on the Volterra series expansion.
2. The recursive polynomial model based on nonlinear difference equations.
3. The multilayer perceptron (MLP) neural network.
4. The radial basis function (RBF) neural network.

In the following sections, we will introduce the methods above mentioned for modeling nonlinear systems and for each approach adaptive algorithms for updating the corresponding nonlinear filter coefficients will be described. The chapter includes examples aimed at comparing the different structures and algorithms.

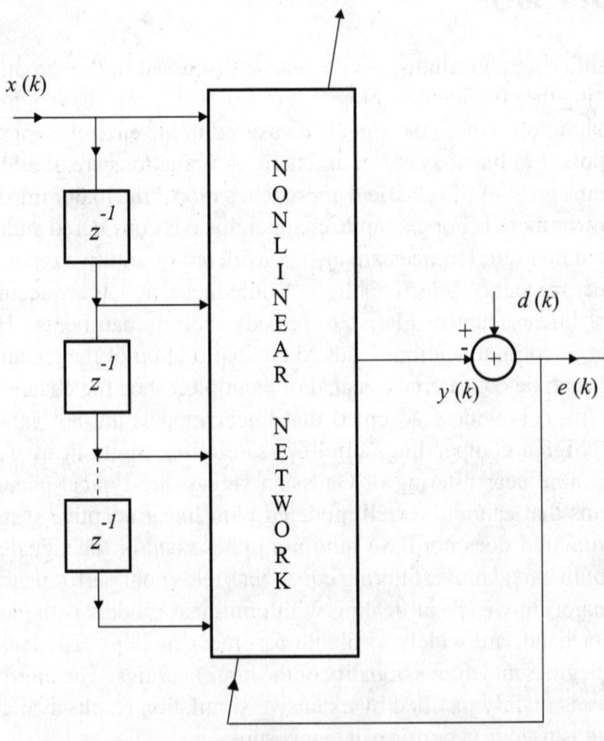

Figure 11.1 Adaptive nonlinear filter.

11.2 THE VOLTERRA SERIES ALGORITHM

The Volterra series model is the most widely used model for nonlinear systems for several reasons. In particular, this model is useful for nonlinear adaptive filtering because the classical formulation of linear adaptive filters can be easily extended to fit this model. The Volterra series expansion of a nonlinear system consists of a nonrecursive series in which the output signal is related to the input signal as follows[2]

$$
\begin{aligned}
d'(k) &= \sum_{l_1=0}^{\infty} w_{o1}(l_1)x(k-l_1) \\
&+ \sum_{l_1=0}^{\infty}\sum_{l_2=0}^{\infty} w_{o2}(l_1,l_2)x(k-l_1)x(k-l_2) \\
&+ \sum_{l_1=0}^{\infty}\sum_{l_2=0}^{\infty}\sum_{l_3=0}^{\infty} w_{o3}(l_1,l_2,l_3)x(k-l_1)x(k-l_2)x(k-l_3) \\
&+ \sum_{l_1=0}^{\infty}\sum_{l_2=0}^{\infty}\cdots \\
& \sum_{l_i=0}^{\infty} w_{oi}(l_1,l_2,\ldots,l_i)x(k-l_1)x(k-l_2)\cdots x(k-l_i) \\
&+\cdots
\end{aligned} \tag{11.1}
$$

where $w_{oi}(l_1, l_2, \ldots, l_i)$, for $i = 0, 1, \ldots, \infty$, are the coefficients of the nonlinear filter model based on the Volterra series, and $d'(k)$ represents, in the context of system identification application, the unknown system output when no measurement noise exists. The term $w_{oi}(l_1, l_2, \ldots, l_i)$ is also known as the Volterra kernel of the system. Note that the input signals in this case are assumed to consist of a tapped-delay line. For the general case, where the signals of the input signal vector come from different origins, such as in an antenna array, the Volterra series representation is given by

$$
\begin{aligned}
d'(k) &= \sum_{l_1=0}^{\infty} w_{o1}(l_1)x_{l_1}(k) \\
&+ \sum_{l_1=0}^{\infty}\sum_{l_2=0}^{\infty} w_{o2}(l_1,l_2)x_{l_1}(k)x_{l_2}(k) \\
&+ \sum_{l_1=0}^{\infty}\sum_{l_2=0}^{\infty}\sum_{l_3=0}^{\infty} w_{o3}(l_1,l_2,l_3)x_{l_1}(k)x_{l_2}(k)x_{l_3}(k) \\
&+ \sum_{l_1=0}^{\infty}\sum_{l_2=0}^{\infty}\cdots \\
& \sum_{l_i=0}^{\infty} w_{oi}(l_1,l_2,\ldots,l_i)x_{l_1}(k)x_{l_2}(k)\cdots x_{l_i}(k) \\
&+\cdots
\end{aligned} \tag{11.2}
$$

[2]The reader should note that the Volterra series expansion includes a constant term w_{o0} which is irrelevant for our discussions here, and will not be further included in the expansion.

where $w_{oi}(l_1, l_2, \ldots, l_i)$, for $i = 0, 1, \ldots, \infty$, are the coefficients of the nonlinear combiner model based on the Volterra series.

As discussed by Mathews [1], the Volterra series expansion can be interpreted as a Taylor series expansion with memory. As such, the Volterra series representation is not suitable to model systems containing discontinuities on their models, as occurs with the Taylor series representation of functions with discontinuities. Another clear drawback of the Volterra series representation is the computational complexity, if the complete series is employed. By truncating the series one can reduce the computational complexity by sacrificing the accuracy of the series expansion. With reduced order, the Volterra series representation is quite complex even when the orders of the series and the filter are moderate. The interested reader can also refer to [2] for a deeper treatment of fixed and adaptive polynomial signal processing.

11.2.1 LMS Volterra Filter

In this subsection, the Volterra LMS algorithm is presented for a second-order series and Nth-order filter. This choice reduces the computational complexity to an acceptable level for some applications and also simplifies the derivations. The extension for higher-order cases is straightforward. The adaptive filter that estimates the signal $d'(k)$ using a truncated Volterra series expansion of second order, can be described by

$$y(k) = \sum_{l_1=0}^{N} w_{l_1}(k)x(k-l_1) + \sum_{l_1=0}^{N}\sum_{l_2=0}^{N} w_{l_1,l_2}(k)x(k-l_1)x(k-l_2) \tag{11.3}$$

where $w_{l_1}(k)$ and $w_{l_1,l_2}(k)$, for $l_1, l_2 = 0, 1, \ldots, N$, are the coefficients of the nonlinear filter model based on the second-order Volterra series expansion, and $y(k)$ represents the adaptive-filter output signal.

The standard approach to derive the LMS algorithm is to use as estimate of the mean-square error (MSE) defined as

$$F[e(k)] = \xi(k) = E[e^2(k)] = E[d^2(k) - 2d(k)y(k) + y^2(k)] \tag{11.4}$$

the instantaneous square error given by

$$e^2(k) = d^2(k) - 2d(k)y(k) + y^2(k) \tag{11.5}$$

Most of the analyses and algorithms presented for the linear LMS apply equally to the nonlinear LMS filter case, if we interpret the information and coefficient vectors as follows

$$\mathbf{x}(k) = \begin{bmatrix} x(k) \\ x(k-1) \\ \vdots \\ x(k-N) \\ x^2(k) \\ x(k)x(k-1) \\ \vdots \\ x(k)x(k-N) \\ \vdots \\ x(k-N)x(k-N+1) \\ x^2(k-N) \end{bmatrix} \tag{11.6}$$

$$\mathbf{w}(k) = \begin{bmatrix} w_0(k) \\ w_1(k) \\ \vdots \\ w_N(k) \\ w_{0,0}(k) \\ w_{0,1}(k) \\ \vdots \\ w_{0,N}(k) \\ \vdots \\ w_{N,N-1}(k) \\ w_{N,N}(k) \end{bmatrix} \tag{11.7}$$

As illustrated in Fig. 11.2, the adaptive-filter output is given by

$$y(k) = \mathbf{w}^T(k)\mathbf{x}(k) \tag{11.8}$$

The estimate of the MSE objective function can now be rewritten as

$$e^2(k) = d^2(k) - 2d(k)\mathbf{w}^T(k)\mathbf{x}(k) + \mathbf{w}^T(k)\mathbf{x}(k)\mathbf{x}^T(k)\mathbf{w}(k) \tag{11.9}$$

An LMS-based algorithm can be used to minimize the objective function as follows:

$$\begin{aligned} \mathbf{w}(k+1) &= \mathbf{w}(k) - \mu\hat{\mathbf{g}}_{\mathbf{w}}(k) \\ &= \mathbf{w}(k) - 2\mu e(k)\frac{\partial e(k)}{\partial \mathbf{w}(k)} \end{aligned} \tag{11.10}$$

for $k = 0, 1, 2, \ldots$, where $\hat{\mathbf{g}}_{\mathbf{w}}(k)$ represents an estimate of the gradient vector of the objective function with respect to the filter coefficients. However, it is wise to have different convergence factors for the first- and second-order terms of the LMS Volterra filter. In this case, the updating equations are

Algorithm 11.1

Volterra LMS Algorithm

Initialization

$\mathbf{x}(0) = \mathbf{w}(0) = [0\,0 \ldots 0]^T$

Do for $k \geq 0$

$e(k) = d(k) - \mathbf{x}^T(k)\mathbf{w}(k)$

$$\mathbf{w}(k+1) = \mathbf{w}(k) + 2\begin{bmatrix} \mu_1 & \cdots & 0 & 0 & \cdots & 0 \\ 0 & \ddots & 0 & 0 & \ddots & 0 \\ 0 & \cdots & \mu_1 & 0 & \cdots & 0 \\ 0 & \cdots & 0 & \mu_2 & \cdots & 0 \\ 0 & \ddots & 0 & 0 & \ddots & 0 \\ 0 & \cdots & 0 & 0 & \cdots & \mu_2 \end{bmatrix} e(k)\mathbf{x}(k)$$

given by

$$w_{l_1}(k+1) = w_{l_1}(k) + 2\mu_1 e(k)x(k-l_1) \tag{11.11}$$

$$w_{l_1,l_2}(k+1) = w_{l_2}(k) + 2\mu_2 e(k)x(k-l_1)x(k-l_2) \tag{11.12}$$

where $l_1 = 0, 1, \ldots, N$ and $l_2 = 0, 1, \ldots, N$. As can be observed in Algorithm 11.1, the Volterra LMS algorithm has the same form as the conventional LMS algorithm except for the form of the input vector $\mathbf{x}(k)$. In order to guarantee convergence of the coefficients in the mean, the convergence factor of the Volterra LMS algorithm must be chosen in the range

$$0 < \mu_1 < \frac{1}{tr(\mathbf{R})} < \frac{1}{\lambda_{\max}} \tag{11.13}$$

$$0 < \mu_2 < \frac{1}{tr(\mathbf{R})} < \frac{1}{\lambda_{\max}} \tag{11.14}$$

where $\lambda_{\max}$ is the largest eigenvalue of the input signal vector autocorrelation matrix $\mathbf{R} = E[\mathbf{x}(k)\mathbf{x}^T(k)]$. It should be noted that this matrix involves high-order statistics of the input signal, leading to high eigenvalue spread of the matrix $\mathbf{R}$ even if the input signal is a white noise. As a consequence, the Volterra LMS algorithm has in general slow convergence. As an alternative, we can consider implementing a Volterra adaptive filter using an RLS algorithm.

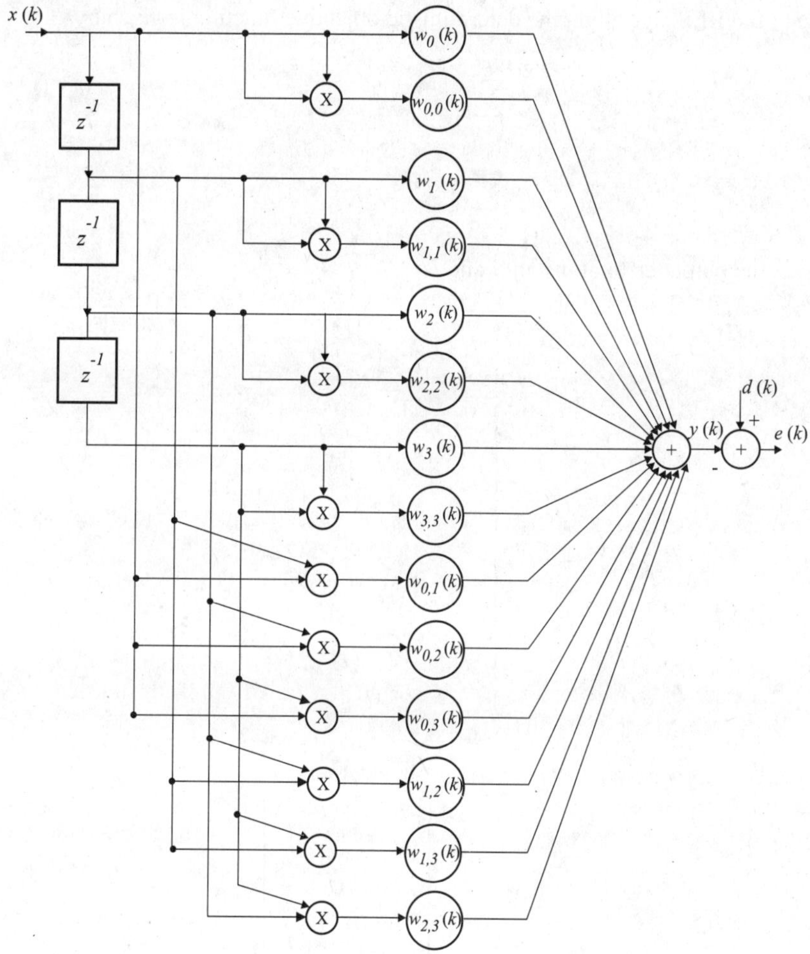

Figure 11.2 Adaptive Volterra filter.

11.2.2 RLS Volterra Filter

The RLS algorithms are known to achieve fast convergence even when the eigenvalue spread of the input vector correlation matrix is large. The objective of the RLS algorithm is to choose the coefficients of the adaptive filter such that the output signal $y(k)$, during the period of observation, will match the desired signal as closely as possible in the least-squares sense. This minimization process can be easily adapted to the nonlinear adaptive filtering case by reinterpreting the entries of the input signal vector and the coefficient vector, as done in the LMS case.

In the case of the RLS algorithm, the deterministic objective function is given by

$$\begin{aligned}\xi^d(k) &= \sum_{i=0}^{k} \lambda^{k-i}\varepsilon^2(i) \\ &= \sum_{i=0}^{k} \lambda^{k-i}\left[d(i) - \mathbf{x}^T(i)\mathbf{w}(k)\right]^2\end{aligned} \tag{11.15}$$

where $\varepsilon(i)$ is the output error at instant i and

$$\mathbf{x}(i) = \begin{bmatrix} x(i) \\ x(i-1) \\ \vdots \\ x(i-N) \\ x^2(i) \\ x(i)x(i-1) \\ \vdots \\ x(i)x(i-N) \\ \vdots \\ x(i-N)x(i-N+1) \\ x^2(i-N) \end{bmatrix} \tag{11.16}$$

$$\mathbf{w}(k) = \begin{bmatrix} w_0(k) \\ w_1(k) \\ \vdots \\ w_N(k) \\ w_{0,0}(k) \\ w_{0,1}(k) \\ \vdots \\ w_{0,N}(k) \\ \vdots \\ w_{N,N-1}(k) \\ w_{N,N}(k) \end{bmatrix} \tag{11.17}$$

are the input and the adaptive-filter coefficient vectors, respectively. The parameter λ is an exponential weighting factor that should be chosen in the range $0 \ll \lambda \leq 1$.

By differentiating $\xi^d(k)$ with respect to $\mathbf{w}(k)$ and equating the result to zero, the optimal vector $\mathbf{w}(k)$ that minimizes the least-squares error can be shown to be given by

$$\begin{aligned}\mathbf{w}(k) &= \left[\sum_{i=0}^{k} \lambda^{k-i}\mathbf{x}(i)\mathbf{x}^T(i)\right]^{-1} \sum_{i=0}^{k} \lambda^{k-i}\mathbf{x}(i)d(i) \\ &= \mathbf{R}_D^{-1}(k)\mathbf{p}_D(k)\end{aligned} \tag{11.18}$$

where $\mathbf{R}_D(k)$ and $\mathbf{p}_D(k)$ are called the deterministic correlation matrix of the input vector and the deterministic cross-correlation vector between the input vector and the desired signal, respectively.

The Volterra RLS algorithm has the same form as the conventional RLS algorithm as shown in Algorithm 11.2, where the only difference is the form of the input vector $\mathbf{x}(k)$.

Algorithm 11.2

Volterra RLS Algorithm

Initialization

$\mathbf{S}_D(-1) = \delta\mathbf{I}$

where δ can be the inverse of an estimate of the input signal power

$\mathbf{x}(-1) = \mathbf{w}(-1) = [0\,0\ldots 0]^T$

Do for $k \geq 0$

$e(k) = d(k) - \mathbf{x}^T(k)\mathbf{w}(k-1)$

$\boldsymbol{\psi}(k) = \mathbf{S}_D(k-1)\mathbf{x}(k)$

$\mathbf{S}_D(k) = \frac{1}{\lambda}[\mathbf{S}_D(k-1) - \frac{\boldsymbol{\psi}(k)\boldsymbol{\psi}^T(k)}{\lambda+\boldsymbol{\psi}^T(k)\mathbf{x}(k)}]$

$\mathbf{w}(k) = \mathbf{w}(k-1) + e(k)\mathbf{S}_D(k)\mathbf{x}(k)$

If necessary compute

$y(k) = \mathbf{w}^T(k)\mathbf{x}(k)$

$\varepsilon(k) = d(k) - y(k)$

A clear disadvantage of the Volterra RLS algorithm is the high computational complexity which requires an order of N^4 multiplications per output sample. However, by examining closely the form of the input data vector it is possible to conclude that the nonlinear filtering problem can be recast into a linear multichannel adaptive-filtering problem for which fast RLS algorithms exist. Using this strategy, several fast RLS algorithms for Volterra filters have been proposed, namely the fast transversal [3], the lattice and QR-based lattice algorithms [4], and the QR-decomposition-based algorithm [5]. Other strategies to reduce computation while trying to retain fast convergence, include the orthogonal lattice-based structures tailored for Gaussian input signals [6].

Example 11.1

A digital channel model can be represented by the following system of equations

$$v(k) = x(k) + 0.5x(k-1)$$
$$y(k) = v(k) + 0.2v^2(k) + 0.1v^3(k) + n(k)$$

The channel is corrupted by Gaussian white noise with variance σ_n^2, varying from -10dB to -25dB. The training signal and the actual input signal, consist of independent binary samples (-1,1). The training period depends on the algorithm but our first attempt is 200 iterations, and after that one can start normal operation.

(a) Design an equalizer for this problem. Use a filter of appropriate order and plot the learning curves.

(b) Using the same number of adaptive-filter coefficients, implement a DFE equalizer as shown in Fig. 11.3 and compare the results with those obtained with the FIR equalizer.

We start with the normalized LMS and after making it work, we compare it with the:

1. DFE normalized LMS algorithm
2. Volterra normalized LMS algorithm
3. DFE Volterra normalized LMS algorithm

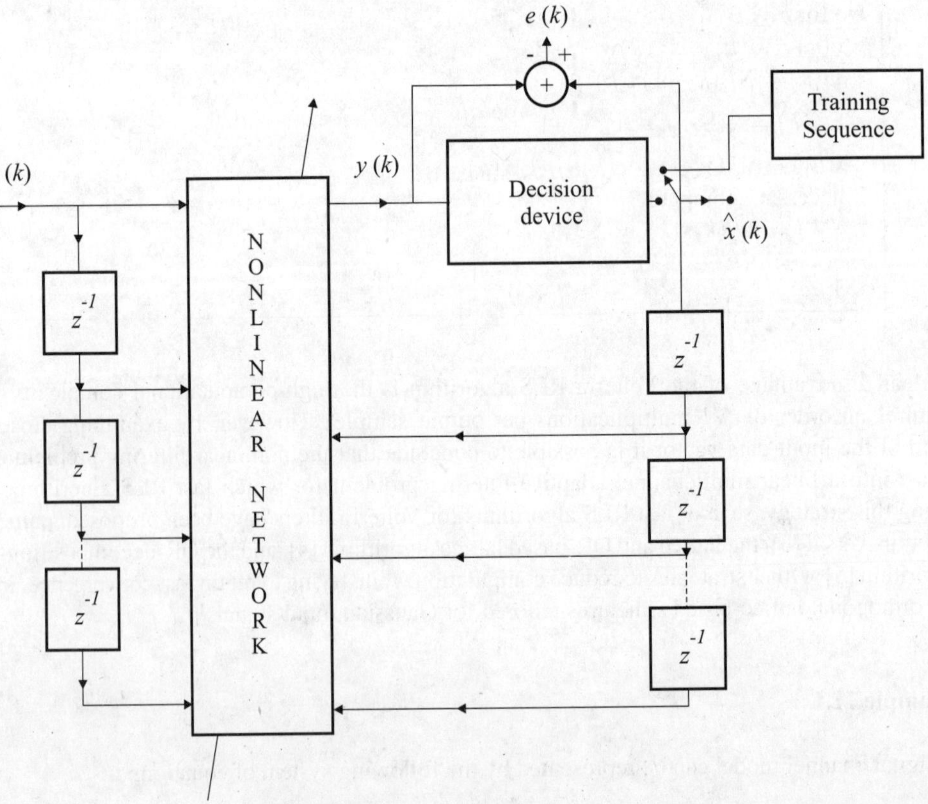

Figure 11.3 Decision feedback equalizer.

Solution:

In the DFE of Fig. 11.3, we initially utilize a training sequence which consists of a properly delayed version of the transmitted signal which is known to the receiver. Obviously, this is an overhead to the communication system since in the beginning no information is actually being transmitted. After the training period no actual reference signal is available, and the equalizer replaces the training sequence by the output of the decision device by moving the switch to its output. The average of square error to be presented corresponds to average of a hundred experiments, whereas the number of errors are measured in single run experiments.

For the normalized LMS algorithm the number of coefficients is 10 with convergence factor $\mu = 0.2$. The square errors for the different levels of channel noise are depicted in Fig. 11.4. As can be observed, the normalized LMS algorithm converges fast for this example where only few training samples are required to train the filter, when the signal to noise ratio is high. However, since the channel is nonlinear the square error after convergence does not reach low levels.

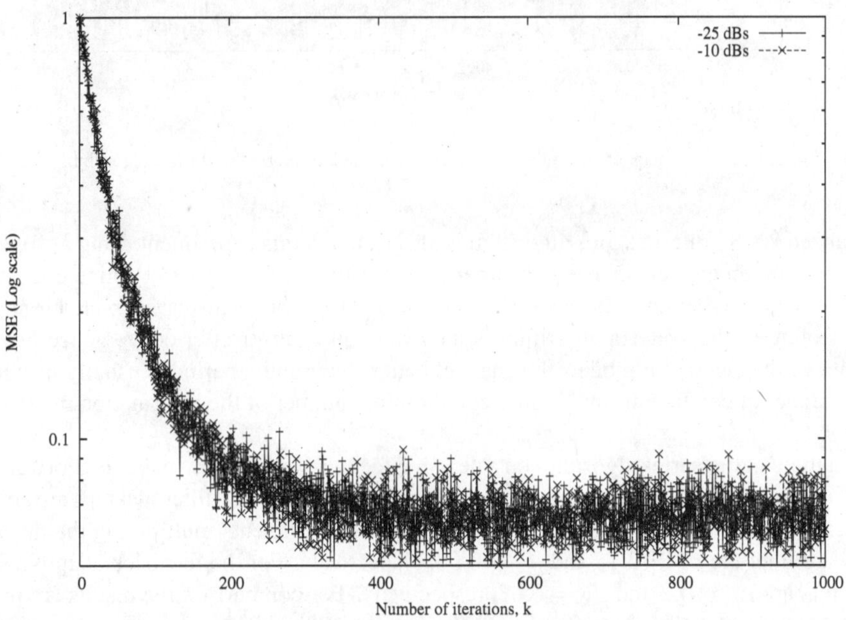

Figure 11.4 Square error, normalized LMS algorithm.

In the next experiment, the decision feedback equalizer is tested using the normalized LMS algorithm with convergence factors $\mu = 0.2$ for the forward and feedback adaptive filters. The forward filter has eight coefficients whereas the feedback filter has two coefficients. For comparison, the results presented are the same as in the previous case for the same levels of channel noise. The resulting square errors are depicted in Fig. 11.5. In this case, the algorithm requires a somewhat comparable training period and also leads to similar square error after convergence. When the signal to noise ratio is poor the standard and the DFE algorithms perform poorly.

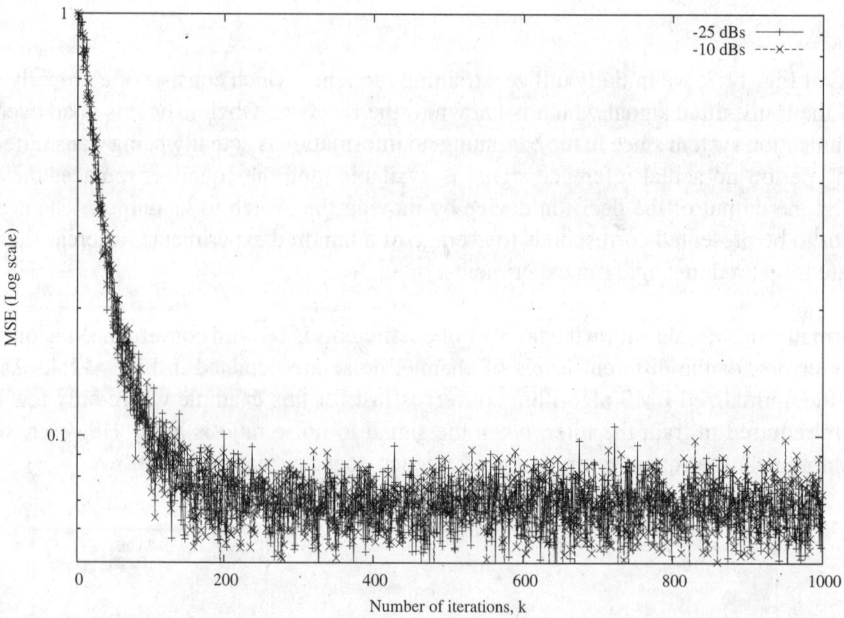

Figure 11.5 Square error for the experiments with the DFE normalized LMS algorithm.

The normalized LMS Volterra series algorithm is also tested in this experiment using a tapped delay line as input with ten elements. The convergence factor for the first-order adaptive coefficients is $\mu_1 = 0.51$ and for the second-order coefficients is $\mu_2 = 0.08$. The results are depicted in Fig. 11.6. A distinct feature of the Volterra algorithm is its lower square error after convergence, which is a consequence of the fact that it models the channel better. Its training period is usually longer due to the larger number of coefficients and higher conditioning number of the information matrix.

We also test the Volterra series algorithm on a decision feedback equalizer. In the feedforward filter a tapped-delay line with eight coefficients is used whereas in the feedback filter two taps are employed. For these experiments the convergence factors used in the coefficients multiplying the linear terms of the forward filter are $\mu_1 = 0.15$ and $\mu_2 = 0.08$, respectively. For the feedback adaptive filter the chosen factors are $\mu_1 = 0.2$ and $\mu_2 = 0.08$, respectively. For comparison the results are presented for the same levels of channel noise as the previous examples. These square errors are seen in Fig. 11.7. The comparison between the DFE and non DFE Volterra filter implementation shows that the DFE requires comparable training period while achieving lower square error and requiring less computational effort. As expected, in all examples the lower additional noise leads to lower MSE after convergence.

Table 11.1 illustrates the number of decision errors made in a single run of the algorithms analyzed in this example. The table also contains the iteration number after which no decision errors are noticed. As can be observed the DFE algorithms usually take longer to converge. Also, the Volterra algorithms have longer learning periods.

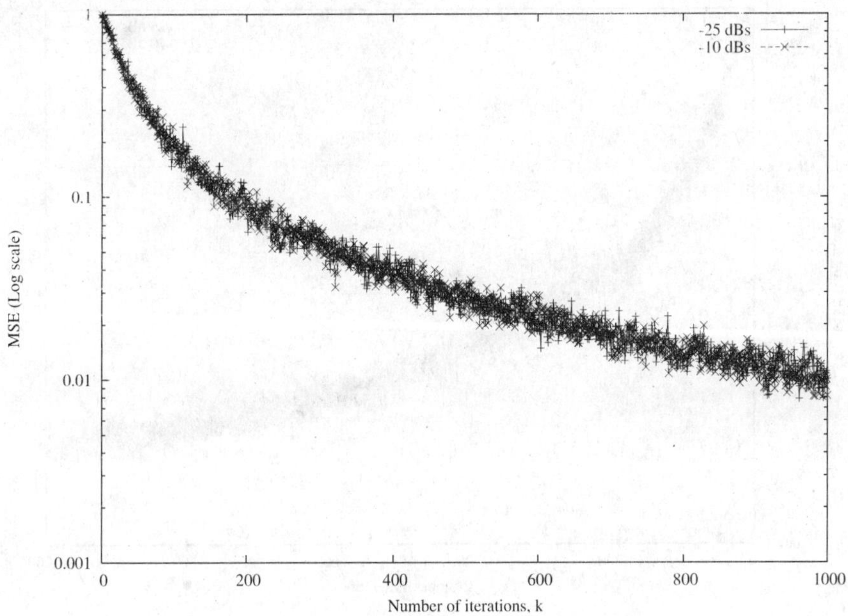

Figure 11.6 Square error for the experiments with the Volterra normalized LMS algorithm.

□

Table 11.1 Evaluation of the Volterra LMS Algorithms

	Noise Level	NLMS	DFE NLMS	Volterra	DFE Volterra
No. of Errors	-25 dBs	2	8	7	9
No. of Errors	-10 dBs	9	11	12	17
Last Error Iter.	-25 dBs	4	30	26	50
Last Error Iter.	-10 dBs	23	25	102	168

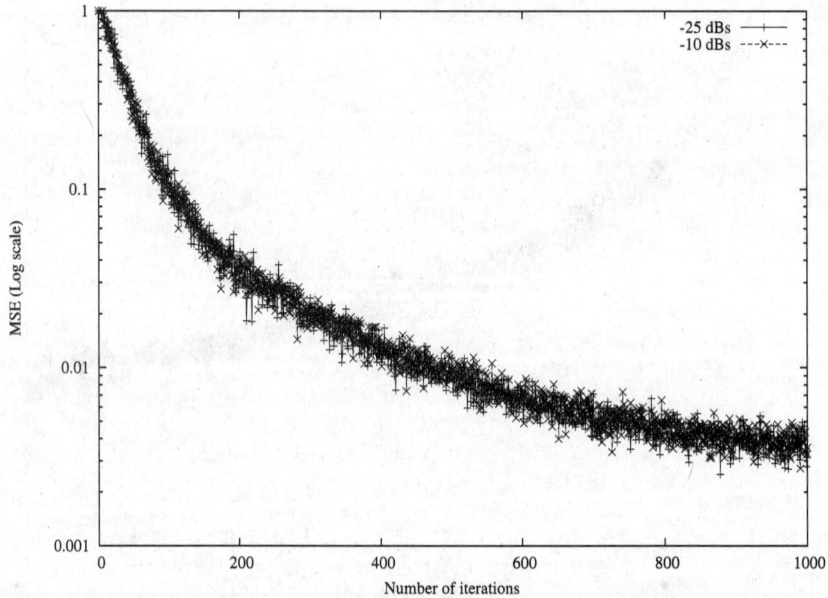

Figure 11.7 Square error for the experiments with DFE Volterra series algorithm.

11.3 ADAPTIVE BILINEAR FILTERS

As it is widely known, the reduction in the computational complexity is the main advantage the adaptive IIR filters present when compared with the adaptive FIR filters. Motivated by this observation, we can consider implementing nonlinear adaptive filters via a nonlinear difference equation, in order to reduce the computational burden related to the Volterra series expansion. The most widely accepted nonlinear difference equation model used for adaptive filtering is the so-called bilinear equation given by

$$y(k) = \sum_{m=0}^{M} b_m(k)x(k-m) - \sum_{j=1}^{N} a_j(k)y(k-j) + \sum_{i=0}^{I}\sum_{l=1}^{L} c_{i,l}x(k-i)y(k-l) \quad (11.19)$$

where $y(k)$ is the adaptive-filter output.

A bilinear adaptive filter in most cases requires fewer coefficients than the Volterra series adaptive filter in order to achieve a given performance. The advantages of the adaptive bilinear filters come with a number of difficulties, some of them not encountered in the Volterra series adaptive filters.

In the present case, the signal information vector is defined by

$$\phi(k) = \begin{bmatrix} x(k) \\ x(k-1) \\ \vdots \\ x(k-M) \\ y(k-1) \\ y(k-2) \\ \vdots \\ y(k-N) \\ x(k)y(k-1) \\ \vdots \\ x(k-I)y(k-L+1) \\ x(k-I)y(k-L) \end{bmatrix} \tag{11.20}$$

where N, M, I and L are the orders of the adaptive-filter difference equations. The coefficient vector can then be described as

$$\boldsymbol{\theta}(k) = \begin{bmatrix} b_0(k) \\ b_1(k) \\ \vdots \\ b_M(k) \\ -a_1(k) \\ -a_2(k) \\ \vdots \\ -a_N(k) \\ c_{0,1}(k) \\ \vdots \\ c_{I,L-1}(k) \\ c_{I,L}(k) \end{bmatrix} \tag{11.21}$$

A possible objective function for adaptive bilinear filtering based on output error is the least-squares function[3]

$$\begin{aligned} \xi^d(k) &= \sum_{i=0}^{k} \lambda^{k-i} e^2(i) \\ &= \sum_{i=0}^{k} \lambda^{k-i} [d(i) - \boldsymbol{\theta}^T(k)\boldsymbol{\phi}(i)]^2 \end{aligned} \tag{11.22}$$

[3] Like in Chapter 10, the reader should note that this definition of the deterministic weighted least squares utilizes the *a priori* error with respect to the latest data pair $d(k)$ and $x(k)$, unlike the FIR RLS case.

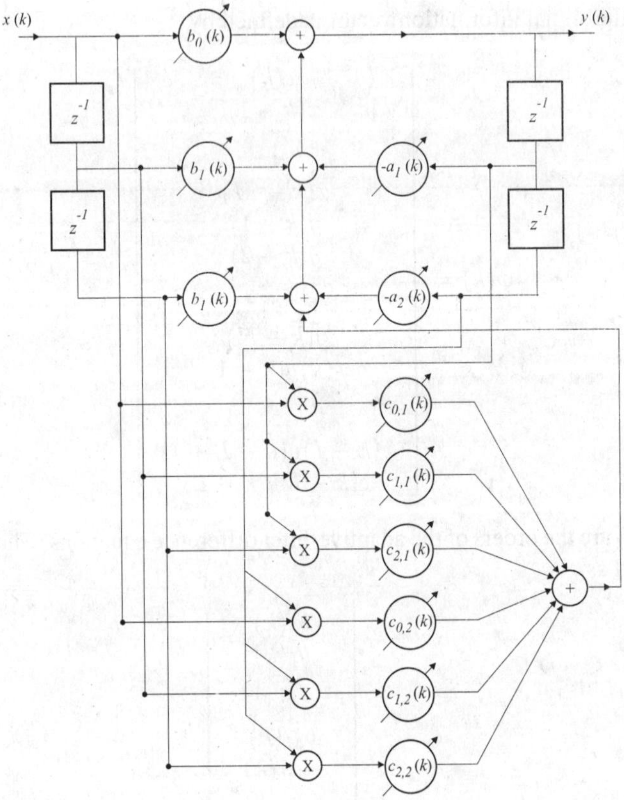

Figure 11.8 Adaptive bilinear filter.

The forgetting factor λ as usual is chosen in the range $0 << \lambda < 1$. By differentiating $\xi^d(k)$ with respect to $\boldsymbol{\theta}(k)$, and by using the same arguments used to deduce the output error RLS algorithm for linear IIR adaptive filters, we conclude that the RLS algorithm for adaptive bilinear filtering consists of the following basic steps:

$$e(k) = d(k) - \boldsymbol{\theta}^T(k)\boldsymbol{\phi}(k) \tag{11.23}$$

$$\boldsymbol{\varphi}(k) = -\frac{\partial y(k)}{\partial \boldsymbol{\theta}(k)} \approx -\boldsymbol{\phi}(k) \tag{11.24}$$

$$\mathbf{S}_D(k+1) = \frac{1}{\lambda}\left[\mathbf{S}_D(k) - \frac{\mathbf{S}_D(k)\boldsymbol{\varphi}(k)\boldsymbol{\varphi}^T(k)\mathbf{S}_D(k)}{\lambda + \boldsymbol{\varphi}^T(k)\mathbf{S}_D(k)\boldsymbol{\varphi}(k)}\right] \tag{11.25}$$

$$\boldsymbol{\theta}(k+1) = \boldsymbol{\theta}(k) - \mathbf{S}_D(k+1)\boldsymbol{\varphi}(k)e(k) \tag{11.26}$$

The approximation of equation (11.24) is not accurate, however it is computationally simple and simulation results confirm that it works. The reader should notice that the partial derivatives used in this algorithm are only approximations, leading to a suboptimal RLS solution. More accurate approximations can be derived by following the same reasonings in which the partial derivatives were calculated for the output error RLS algorithm for linear IIR adaptive filters. The description of the bilinear RLS algorithm is given in Algorithm 11.3.

Algorithm 11.3

Bilinear RLS Algorithm

Initialization

$a_i(k) = b_i(k) = c_{i,l}(k) = e(k) = 0$

$y(k) = x(k) = 0\,,\ k\ < 0$

$\mathbf{S}_D(0) = \delta^{-1}\mathbf{I}$

For each $x(k),\ d(k),\ k \geq 0,$ do

$y(k) = \boldsymbol{\phi}^T(k)\boldsymbol{\theta}(k)$

$e(k) = d(k) - y(k)$

$\mathbf{S}_D(k+1) = \frac{1}{\lambda}\left[\mathbf{S}_D(k) - \frac{\mathbf{S}_D(k)\boldsymbol{\varphi}(k)\boldsymbol{\varphi}^T(k)\mathbf{S}_D(k)}{\lambda+\boldsymbol{\varphi}^T(k)\mathbf{S}_D(k)\boldsymbol{\varphi}(k)}\right]$

$\boldsymbol{\theta}(k+1) = \boldsymbol{\theta}(k) - \mathbf{S}_D(k+1)\boldsymbol{\varphi}(k)e(k)$

Stability test

If we consider as objective function the mean-square error (MSE) defined as

$$\xi = E[e^2(k)] \tag{11.27}$$

we can derive a gradient-based algorithm, by using $e^2(k)$ as an estimate for ξ, leading to an updating equation given by

$$\boldsymbol{\theta}(k+1) = \boldsymbol{\theta}(k) - 2\begin{bmatrix} \mu_1 & \cdots & 0 & 0 & \cdots & 0 & 0 & \cdots & 0 \\ 0 & \ddots & 0 & 0 & \ddots & 0 & 0 & \ddots & 0 \\ 0 & \cdots & \mu_1 & 0 & \cdots & 0 & 0 & \cdots & 0 \\ 0 & \cdots & 0 & \mu_2 & \cdots & 0 & 0 & \cdots & 0 \\ 0 & \ddots & 0 & 0 & \ddots & 0 & 0 & \ddots & 0 \\ 0 & \cdots & 0 & 0 & \cdots & \mu_2 & 0 & \cdots & 0 \\ 0 & \cdots & 0 & 0 & \cdots & 0 & \mu_3 & \cdots & 0 \\ 0 & \ddots & 0 & 0 & \ddots & 0 & 0 & \ddots & 0 \\ 0 & \cdots & 0 & 0 & \cdots & 0 & 0 & \cdots & \mu_3 \end{bmatrix}\boldsymbol{\varphi}(k)e(k) \tag{11.28}$$

where

$$e(k) = d(k) - \boldsymbol{\theta}^T(k)\boldsymbol{\phi}(k) \tag{11.29}$$

and

$$\boldsymbol{\varphi}(k) = \frac{\partial e(k)}{\partial \boldsymbol{\theta}(k)} \tag{11.30}$$

Again, the calculation of an accurate gradient vector can be quite cumbersome.

The main drawbacks of the adaptive bilinear filters based on the output error are: possible instability of the adaptive filter [25], slow convergence, and convergence to local minima of the error surface. It is also possible in the case of adaptive bilinear filter to apply an equation error formulation. In the presence of additional noise, the equation error algorithm may also lead to instability or to a biased global minimum solution.

Example 11.2

Identify an unknown system with the following model

$$d(k) = -0.3d(k-1) + x(k) + 0.04x^2(k) + 0.1x^3(k) + n(k)$$

using the bilinear algorithm, and compare the results with those obtained with the Volterra normalized LMS algorithm. The additional noise is Gaussian white noise with variance $\sigma_n^2 = -10$dB. Use Gaussian white noise with unit variance as input.

Solution:

Three coefficients are sufficient for the bilinear algorithm to perform well. The chosen convergence factor is $\mu = 0.005$. For the Volterra normalized LMS algorithm we use six coefficients and $\mu = 0.1$. As can be observed in Fig. 11.9, the bilinear algorithm converges faster and leads to a lower square error after convergence than the Volterra normalized LMS algorithm, since the unknown system has a bilinear model.

□

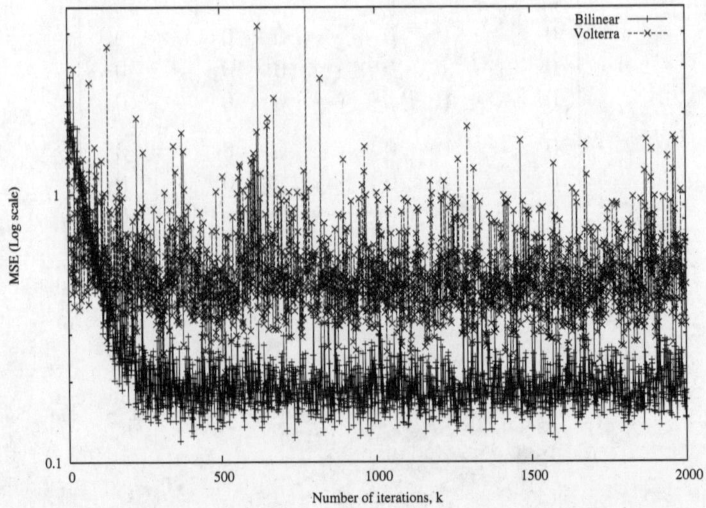

Figure 11.9 Square error for the experiment with the bilinear and Volterra normalized LMS algorithms.

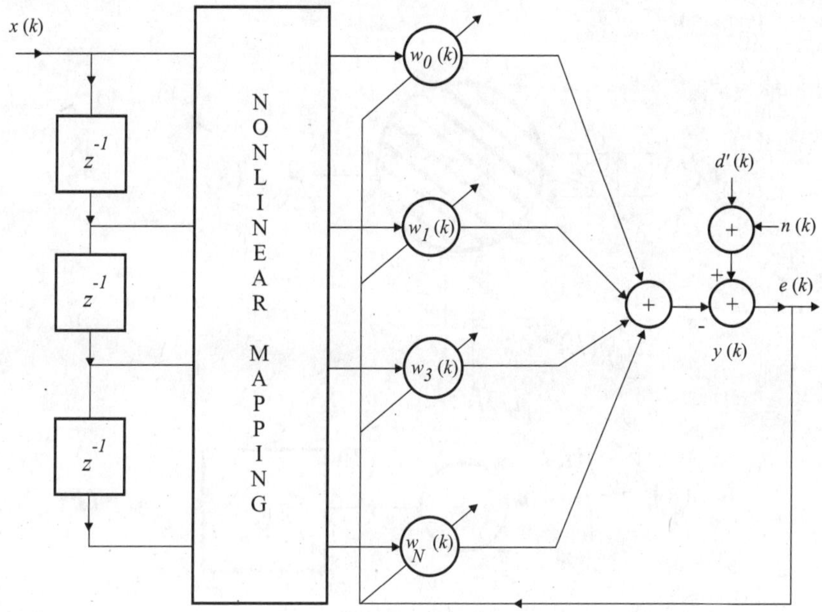

Figure 11.10 Neural network based adaptive filter.

11.4 MULTILAYER PERCEPTRON ALGORITHM

In this section, the multilayer perceptron algorithm is briefly presented [24]. This algorithm belongs to a class of nonlinear adaptive filters where the input signal vector is mapped into another signal vector through a multiport network containing several local nonlinearities, as depicted in Fig. 11.10. Usually, the nonlinear multiport network consists of feedforward neural networks with several layers, where the nonlinearities (neurons) are placed inside the network in a structurally modular form.

The multilayer perceptron structure consists of several layers including an input layer, an output layer and several internal layers usually called hidden layers. Fig. 11.11 illustrates a multilayer perceptron-based adaptive filter with three layers. In communication applications the output layer usually has a single neuron, with $y(k)$ representing the nonlinear adaptive-filter output signal. The mathematical description for each neuron is

$$y_{l,i}(k) = f_{l,i}\left\{\sum_{j=0}^{N_{l-1}-1} w_{l,i,j}(k)y_{l-1,j}(k) - bs_{l,i}(k)\right\} \tag{11.31}$$

where $w_{l,i,j}(k)$ are the weight coefficients connecting the output signal $y_{l-1,j}(k)$ of the jth neuron from layer $l-1$ to input of neuron i of layer l, for $l = 0, 1, \ldots, L-1$; $i = 0, \ldots, N_l - 1$. Note that N_l is the number of neurons in the lth layer and the index L is the number of layers. Each constant $bs_{l,i}(k)$ is the bias term of the ith neuron at layer l, that is also known as the threshold. It is a well known result that the multilayer perceptron network is able to implement any desired nonlinear mapping by properly choosing the weights, the thresholds and the nonlinear activation function $f\{\cdot\}$

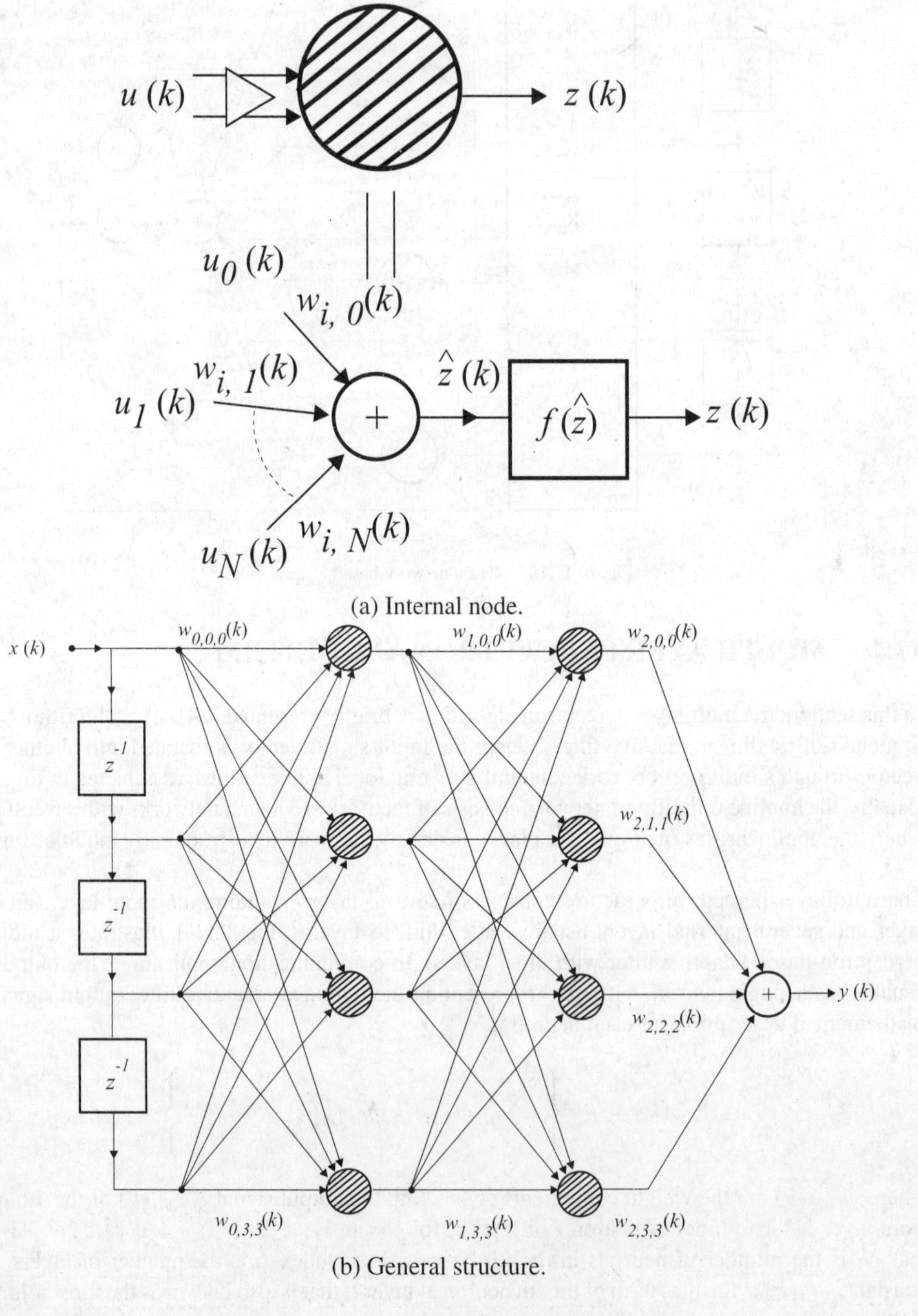

(a) Internal node.

(b) General structure.

Figure 11.11 Multilayer perceptron adaptive filter.

[26]. Although, the activation function and the threshold could be chosen to be different for each layer, we will not consider this general case here. Also, it is possible to show that three layers is always enough for practical purposes. However, the use of more than three layers is desirable in many applications, where in the three layers case the hidden layer requires a large number of neurons in order to achieve an acceptable nonlinear mapping.

The most widely used activation function is the sigmoid function, defined as

$$\mathrm{sgm}(x) = \frac{2c_1}{1 + \mathrm{e}^{-c_2 x}} - c_1 \tag{11.32}$$

where c_1 and c_2 are suitably chosen constants. The derivative of the sigmoid function is given by

$$\mathrm{sgd}(x) = \frac{c_2}{2c_1}[c_1^2 - \mathrm{sgm}^2(x)] \tag{11.33}$$

A popular updating algorithm for the multilayer perceptron is the so-called backpropagation algorithm. The objective function is to minimize the instantaneous output square error, that is

$$e^2(k) = [d(k) - y(k)]^2 \tag{11.34}$$

In order to minimize the above objective function, the backpropagation algorithm uses a steepest-decent updating, with the gradient calculated from the output layer to the input layer as following presented. The derivation of the backpropagation algorithm falls beyond the scope of this book, the interested reader should consult [26] or [27]. In the output layer the error signal is given by $e(k)$ itself, as a result the coefficient updating for the coefficients of the output layer is given by

$$w_{L-1,i,j}(k+1) = w_{L-1,i,j}(k) + 2\mu_{L-1}e(k)y_{L-1,j}(k) \tag{11.35}$$

where $i = 0, 1, \ldots, N_{L-2} - 1$ and $j = 0, 1, \ldots, N_{L-1} - 1$. Notice that in our case we are considering a single output multilayer perceptron, therefore $N_{L-1} = 1$. The parameter μ_{L-1} is the convergence factor for the output layer. Also the simplified updating equation above resulted from not using an activation function at the output node. If the activation function is included at the output node the updating equation is given by

$$w_{L-1,i,j}(k+1) = w_{L-1,i,j}(k) + 2\mu_{L-1}e(k)\mathrm{sgd}\left\{\mathrm{sgm}^{-1}[y_{L-1,j}(k)]\right\}\mathrm{sgm}[y_{L-2,j}(k)] \tag{11.36}$$

Since we know the error in the output layer, we can propagate this error backwards, and calculate the corresponding errors in the output of the internal neurons. By examining Fig. 11.11 closely, after applying the chain rule for derivative and performing a number of manipulations (see [26] and [27] for details) it is possible to show that the error signal at the jth neuron from layer l is given by

$$\begin{aligned} e_{l,j}(k) &= \mathrm{sgd}\left\{\mathrm{sgm}^{-1}[y_{l,j}(k)]\right\}\sum_{i=0}^{N_l-1} w_{l+1,i,j}(k)e_{l+1,i}(k) \\ &= \mathrm{sgd}\left[\sum_{j=0}^{N_{l-1}-1} w_{l,i,j}(k)y_{l-1,j}(k)\right]\sum_{i=0}^{N_l-1} w_{l+1,i,j}(k)e_{l+1,i}(k) \end{aligned} \tag{11.37}$$

The updating equations for the coefficients of the internal layers and the bias terms are given by

$$\begin{aligned} w_{l,i,j}(k+1) &= w_{l,i,j}(k) + 2\mu_l e_{l,j}(k) y_{l-1,j}(k) \\ bs_{l,i}(k+1) &= bs_{l,i}(k) - 2\mu_l e_{l,j}(k) \end{aligned} \tag{11.38}$$

for $i = 0, 1, \ldots, N_{l-1} - 1$ and $j = 0, 1, \ldots, N_l - 1$.

The description of the multilayer perceptron algorithm for nonlinear adaptive filtering is given in Algorithm 11.4.

Algorithm 11.4

Multilayer Perceptron Algorithm

Initialization
  Choose each $w_{l,i,j}(0)$ randomly
Do for $k \geq 0$
  Choose $y_{-1,j}(k) = x_j(k)$
  Do for $l = 0, \ldots, L-1$
    Do for $i = 0, \ldots, N_l - 1$
      Do for $j = 0, \ldots, N_{l-1} - 1$
        $y_{l,j}(k) = f_{l,j}\{\sum_{i=0}^{N_{l-1}-1} w_{l,j,i}(k) y_{l-1,i}(k) - bs_{l,j}(k)\}$
      End
    End
  End
  $e(k) = d(k) - y_{L-1,0}(k)$
  Do for $l = L-1, \ldots, 0$
    Do for $i = 0, \ldots, N_l - 1$
      Do for $j = 0, \ldots, N_{l-1} - 1$
      If $l = L - 1$
        $w_{L-1,i,j}(k+1) = w_{L-1,i,j}(k) + 2\mu_{L-1} e(k) \text{sgd}\left\{\text{sgm}^{-1}[y_{L-1,j}(k)]\right\} \text{sgm}[y_{L-2,j}(k)]$
      Else
        $e_{l,j}(k) = \text{sgd}\left[\sum_{j=0}^{N_{l-1}-1} w_{l,i,j}(k) y_{l-1,j}(k)\right] \sum_{i=0}^{N_l-1} w_{l+1,i,j}(k) e_{l+1,i}(k)$
        $w_{l,i,j}(k+1) = w_{l,i,j}(k) + 2\mu_l e_{l,j}(k) y_{l-1,j}(k)$
        $bs_{l,i}(k+1) = bs_{l,i}(k) - 2\mu_l e_{l,j}(k)$
      End if
    End
  End
End

This algorithm has an increased computational complexity as compared with the linear adaptive filters, for a given number of adaptive coefficients. In addition, the convergence speed is likely to be slow, because we are employing a gradient-based algorithm to search an objective function with a nonquadratic surface.

Some attempts to improve the convergence speed have been proposed, see for example [20]. Despite that, nonlinear adaptive filters based on multilayer perceptron require long training periods, and have no methodology to appropriately define the number of layers and the number of neurons, rendering these algorithms difficult to apply in practical problems. However, it is worth it to search for improved nonlinear solutions for the adaptive-filtering problem, because in many communication applications the linear adaptive filter does not yield good enough performance.

Example 11.3

Identify the same system described in example 11.2 using the multilayer perceptron method, and compare the results with those obtained with the Volterra normalized LMS algorithm.

Solution:

In order to identify the same system of example 11.2 with the multilayer perceptron method, we use a network with 3 inputs and 8 neurons in each of the two hidden layers. The chosen convergence factor is μ=0.1. As can be observed in Figs. 11.9 and 11.12, the multilayer perceptron algorithm has worse performance than the bilinear algorithm, but converges slightly faster and reaches a lower square error after convergence than the Volterra normalized LMS algorithm.

□

11.5 RADIAL BASIS FUNCTION ALGORITHM

The radial basis function network is an attractive alternative to the multilayer perceptron for nonlinear adaptive filtering for a number of reasons. As mentioned in [27], the learning process of the radial basis function neural network is the same as finding a surface in the multidimensional space which is a best fit to the training data. In particular, in the case of communication applications this technique is attractive because its learning allows the division of a multidimensional space in appropriate subregions where each received data fits in.

For equalization problems [23], [21], it is well known that the maximum likelihood equalizer using the Viterbi algorithm provides the best solution, with high computational cost. As a compromise, the radial basis function has been proposed as an attractive alternative because of its lower computational complexity and due to its close relationship with Bayesian methods [22]. The Bayesian methods are effective in interference cancellation and channel equalization [9], [11]-[15]. In fact, the Bayesian design leads to the optimal nonlinear adaptive equalizer [8]. In the Bayesian approach, the decision in favor of a symbol is made only if the probability that the referred symbol had caused the current input signal vector exceeds the probability that any other symbol had caused the same input. The optimal decision boundaries are determined by the values of the input signal vector where these

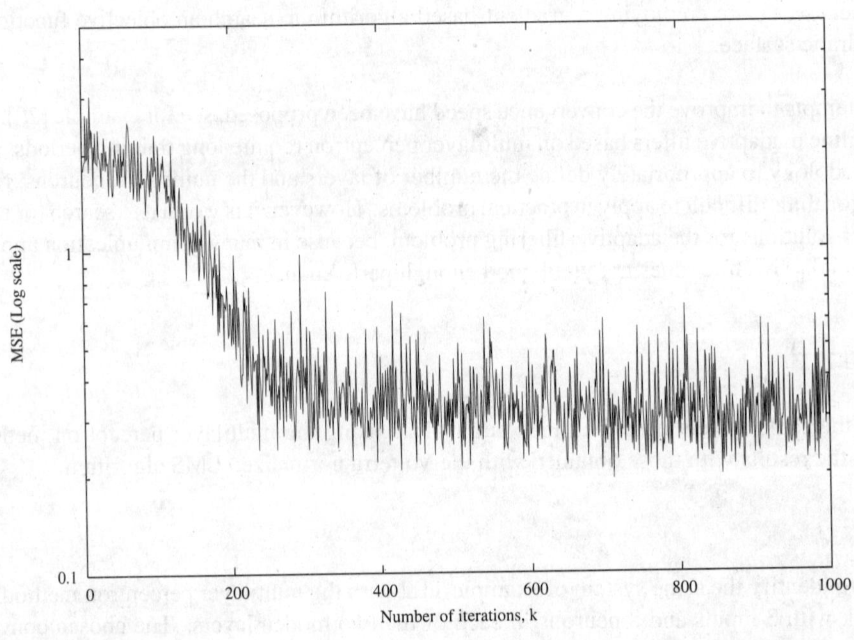

Figure 11.12 Square error for the experiment with the multilayer perceptron algorithm.

probabilities are the same. The Bayesian theory shows that in a number of situations the optimal decision boundaries are not given by hyperplanes (the only ones realizable with linear equalizers), but by nonplanar boundaries. This is exactly what happens when the channel model in communication systems cannot be well modeled with linear adaptive systems, or if the channel transfer function does not have minimum phase. Also, the linear adaptive equalizer does not explore the fact that the input signal originates from transmitted signals consisting of a finite set of symbols.

Since the radial basis function can approximate the Bayesian solution within a reasonable training time, it is a potential candidate to be employed in a number of communication applications where nonlinear adaptive filters are required.

The radial basis function network consists of three layers where the first feeds the second layer directly without any weighting (weights are equal to one), and the output layer is just a linear combiner as depicted in Fig. 11.13.b. The hidden layer implements a nonlinear mapping on the input vector, as represented in Fig. 11.13.a, and consists of two steps. In the first step, the input signal vector is compared with a set of reference vectors $\mathbf{r}_i(k)$, for $i = 0, 1, \ldots, N_N - 1$, where N_N is the number of (hidden) neurons. These vectors are called centers. The comparison between the input signal vector and the centers are usually measured through the Euclidean norm as follows

$$d_i(k) = ||\mathbf{x}(k) - \mathbf{r}_i(k)|| \tag{11.39}$$

These distances are then applied to a nonlinear activation function, which is scalar and radially symmetric. Typical choices are the Gaussian and thin-plate-spline functions, respectively given by

$$
\begin{aligned}
f(d_i(k)) &= \mathrm{e}^{\frac{-d_i^2(k)}{\sigma_i^2(k)}} \\
f(d_i(k)) &= \frac{d_i^2(k)}{\sigma_i^2(k)} \log[\frac{d_i(k)}{\sigma_i(k)}]
\end{aligned}
\tag{11.40}
$$

The parameter $\sigma_i(k)$ controls the spread of the function, related to the radius of influence of radial basis function $f[d_i(k)]$. The output signal is computed by

$$
F[\mathbf{x}(k)] = f_2 \left\{ \sum_{i=0}^{N_N-1} w_i(k) f[d_i(k)] \right\}
\tag{11.41}
$$

where $f_2\{\cdot\}$ is the activation function of the output signal. This function is usually of the following form

$$
f_2(x) = \frac{1 - \mathrm{e}^{-cx}}{1 + \mathrm{e}^{-cx}}
\tag{11.42}
$$

where c is a suitably chosen constant. In most cases, no activation function is used at the output in order to simplify the algorithm, that is $f_2(x) = x$. As a result we will not consider it further.

Usually the training for the radial basis function adaptive filter is done in three steps, where the radius parameters, the centers and the weights are trained separately and in sequence. By using a stochastic gradient algorithm and Gaussian activation function, the radial basis function updating equations are given by

$$
\begin{aligned}
w_i(k+1) &= w_i(k) + 2\mu_w e(k) f[d_i(k)] \\
\sigma_i(k+1) &= \sigma_i(k) + 2\mu_\sigma e(k) f[d_i(k)] w_i(k) \frac{d_i^2(k)}{\sigma_i^3(k)} \\
\mathbf{r}_i(k+1) &= \mathbf{r}_i(k) + 2\mu_r e(k) f[d_i(k)] w_i(k) \frac{\mathbf{x}(k) - \mathbf{r}_i(k)}{\sigma_i^2(k)}
\end{aligned}
\tag{11.43}
$$

for $i = 0, 1, \ldots, N_N - 1$. In Algorithm 11.5, the adaptive nonlinear filter based on the radial basis function is detailed. In many cases the parameters $\sigma_i(k)$, that control the spread of the function in each neuron, are kept constant, where in this case they are chosen as the expected channel noise power.

In a number of communication applications the signals involved are originally complex. In those cases, we need to use a complex radial basis function algorithm whose configuration is depicted in Fig. 11.14. The complex algorithm is described in Algorithm 11.6, where the derivations are omitted for the sake of brevity, for details consult [16]-[17], [18]-[19].

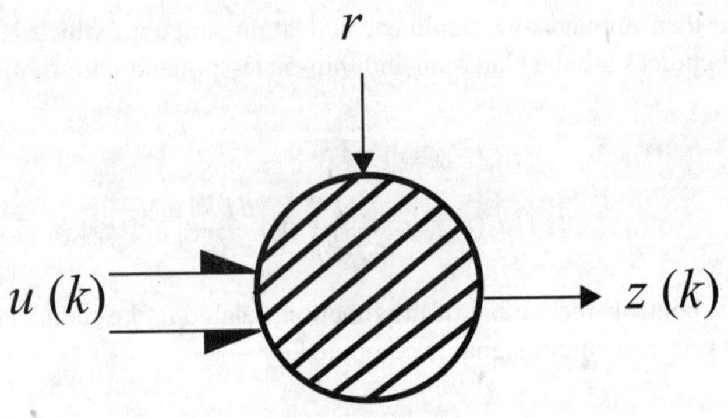

(a) Internal node.

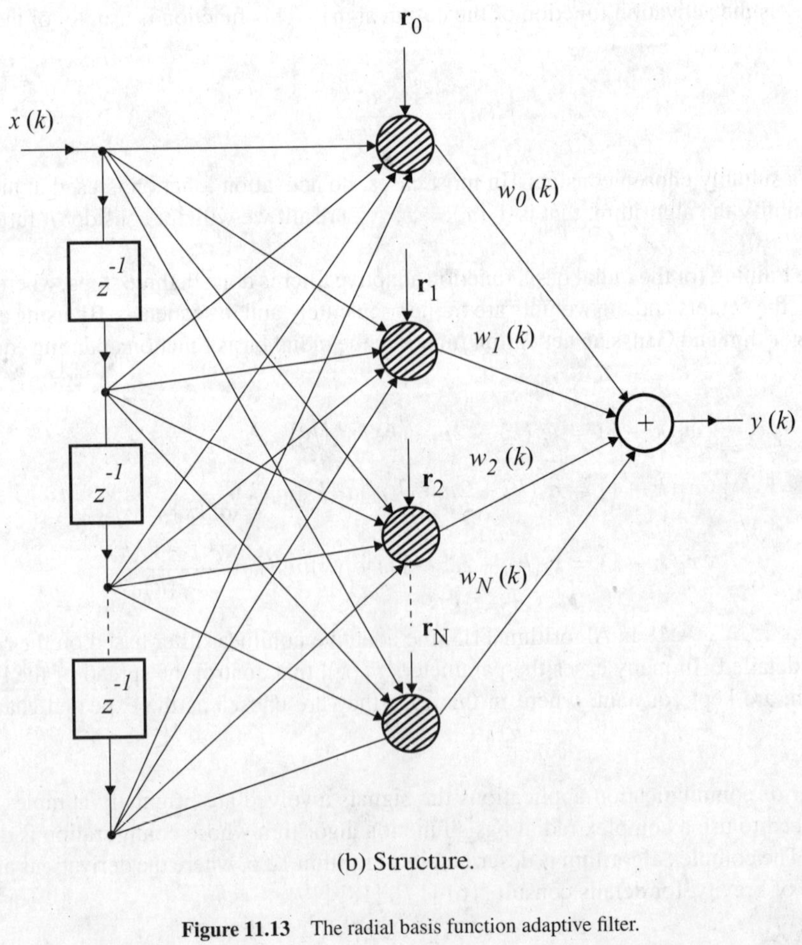

(b) Structure.

Figure 11.13 The radial basis function adaptive filter.

Algorithm 11.5

Radial Basis Function Algorithm

Initialization
  Choose each $w_i(0)$ randomly
Do for $k \geq 0$
  $y(k) = F(\mathbf{x}(k)) = \sum_{i=0}^{N_N-1} w_i(k) f[d_i(k)]$
  $e(k) = d(k) - y(k)$
  Do for $i = 0, 1, \ldots, N_N - 1$
    $w_i(k+1) = w_i(k) + 2\mu_w e(k) f[d_i(k)]$
    $\sigma_i(k+1) = \sigma_i(k) + 2\mu_\sigma f[d_i(k)] e(k) w_i(k) \frac{d_i^2(k)}{\sigma_i^3(k)}$
    $\mathbf{r}_i(k+1) = \mathbf{r}_i(k) + 2\mu_r f[d_i(k)] e(k) w_i(k) \frac{\mathbf{x}(k) - \mathbf{r}_i(k)}{\sigma_i^2(k)}$
  End
End

Algorithm 11.6

Complex Radial Basis Function Algorithm

Initialization
  Choose each $w_i(0)$ randomly
  $\mathbf{r}_i(k)$, $\mathbf{x}_i(k)$ are complex vectors
  $e(k)$, is a complex scalar
Do for $k \geq 0$
  $y(k) = F(\mathbf{x}(k)) = \sum_{i=0}^{N_N-1} w_i^*(k) f(d_i(k))$
  $e(k) = d(k) - y(k)$
  Do for $i = 0, 1, \ldots, N_N - 1$
    $w_i(k+1) = w_i(k) + 2\mu_w e(k) f[d_i(k)]$
    $\sigma_i(k+1) = \sigma_i(k) + 2\mu_\sigma f[d_i(k)] \{\mathrm{re}[e(k)] w_{R_i}(k) + \mathrm{im}[e(k)] w_{I_i}(k)\} \frac{d_i^2(k)}{\sigma_i^3(k)}$
    $\mathbf{r}_i(k+1) = \mathbf{r}_i(k) + 2\mu_r f[d_i(k)] \frac{\mathrm{re}[e(k)] w_{R_i}(k) \mathrm{re}[\mathbf{x}(k) - \mathbf{r}_i(k)] + \jmath \mathrm{im}[e(k)] w_{I_i}(k) \mathrm{im}[\mathbf{x}(k) - \mathbf{r}_i(k)]}{\sigma_i^2(k)}$
  End
End

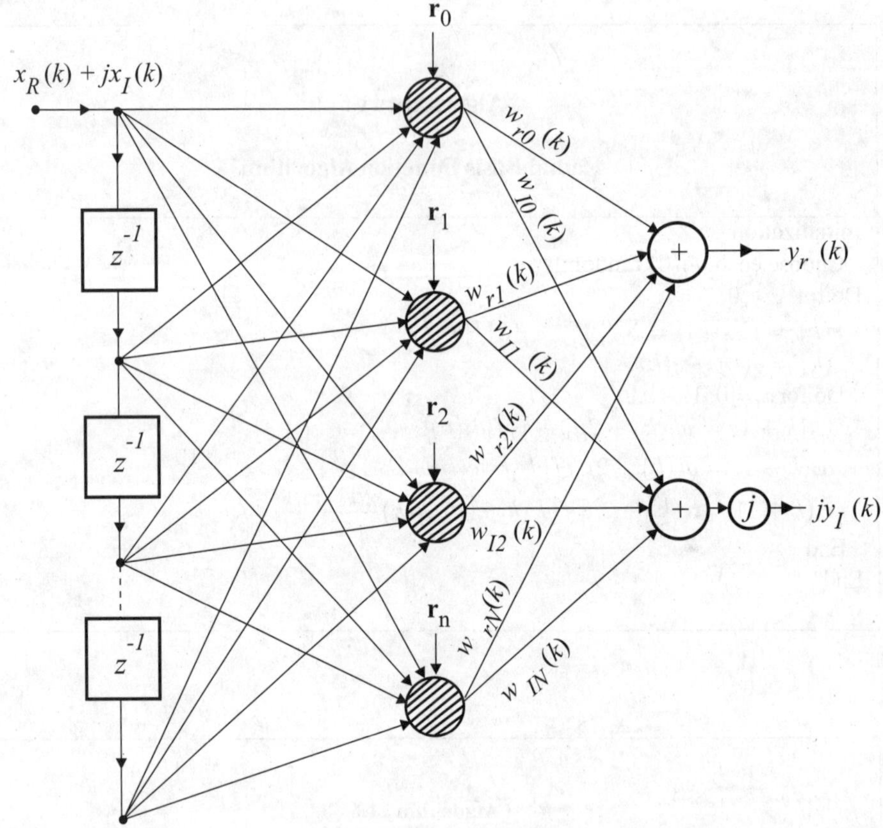

Figure 11.14 The radial basis function adaptive filter for complex signals.

Example 11.4

Solve the problem described in Example 11.1 using:

1. Radial basis function algorithm
2. DFE radial basis function algorithm

Solution:

In order to solve the problem, the following two experiments use neural network equalizers of the radial basis function type with ten delays in the input tap-delay line and ten hidden neurons. In the first experiments the standard radial basis approach is applied using a convergence factor for the linear combiner of $\mu_w = 0.1$, a convergence factor for the radius of $\mu_r = 0.9$, and a spread factor of $\sigma = 0.8$. Fig. 11.15 shows the learning curves for the square errors. As can be observed, the radial basis algorithm requires longer training period than the previous algorithms. This is the price paid by its generality in approximating nonlinear functions.

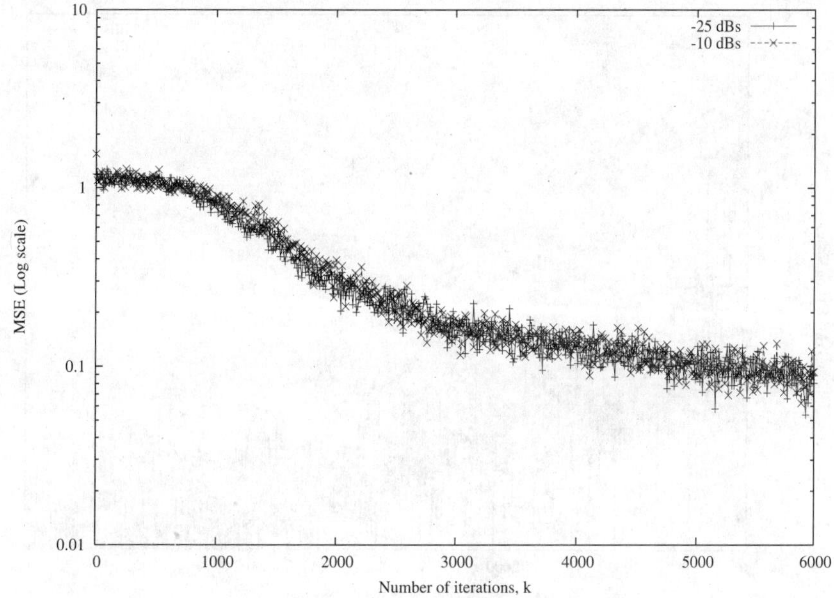

Figure 11.15 Square errors for the experiments with the radial basis algorithm.

The final experiment uses a neural network DFE of the radial basis function type with eight taps and hidden neurons in the forward filter and two in the feedback filter. The convergence factor for the forward filter is $\mu_w = 0.5$, the convergence factor for the radius is $\mu_r = 0.9$, and the spread factor is $\sigma = 0.8$. For the backward filter, these parameters are $\mu_w = 0.04$, $\mu_r = 0.9$, and $\sigma = 1.2$, respectively. These results are depicted in Fig. 11.16 for an ensemble of a hundred experiments. The results with DFE are better than in the case without DFE.

Table 11.2 illustrates the number of decision errors made in a single run of the radial basis function algorithms for this example, including the iteration number after which no decision errors are noticed. As can be observed, the radial basis function algorithms take longer to converge than the Volterra algorithms for this example.

Table 11.2 Evaluation of the Radial Basis Function Algorithms

	Radial Basis Algorithm		**DFE Radial Basis Algorithm**	
Noise level	**-25 dBs**	**-10 dBs**	**-25 dBs**	**-10 dBs**
No. of Errors	74	113	79	92
Iter. of Last Error	318	387	287	370

Fig. 11.17 depicts the results of an experiment with the radial basis function algorithm with DFE where the training is done for a long period. The graphs show that after the learning is complete the

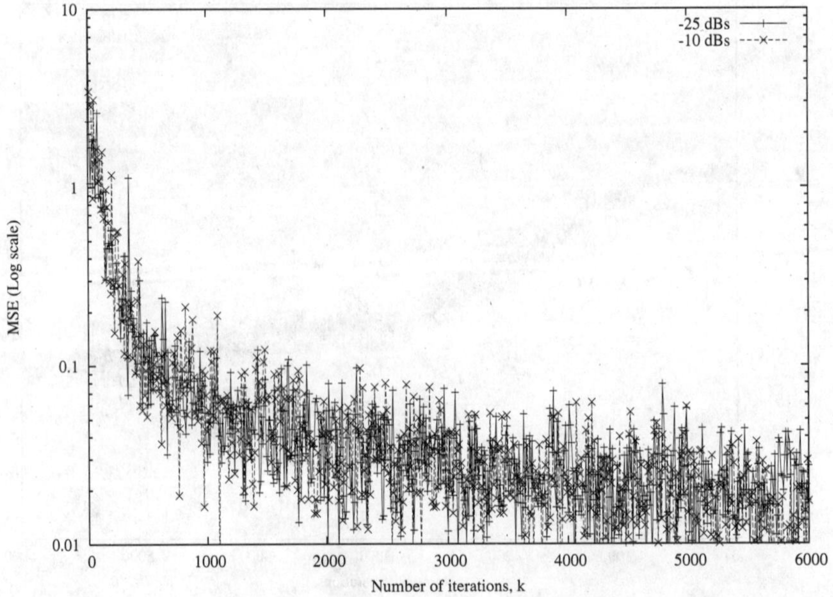

Figure 11.16 Square errors for the experiments with DFE radial basis function algorithm.

algorithm enables perfect bit detection, reaching a lower square error level than the algorithms not based on neural networks.

□

11.6 CONCLUSION

In this chapter, we introduced some nonlinear adaptive-filtering methods which can be applied in communication systems, as well as in many other fields. The methods discussed here are far from consisting of a complete set, many other methods have been investigated using different points of view, see for example [28] and [29]. The emphasis was to describe methods allowing a training period and suitable for channel equalization and co-channel interference. No attempt was made to discuss blind equalization methods that are nonlinear adaptive filters which usually utilize high-order statistics, see Chapter 13.

The wide use of these algorithms in modern communication systems, while required, remains to be seen. However, with a deep knowledge of the type of nonlinearities affecting the given communication environment, one can come up with a nonlinear adaptive-filtering algorithm tailored for that particular application, where a good compromise concerning computational complexity, training period and performance in terms of bit error rate can be reached.

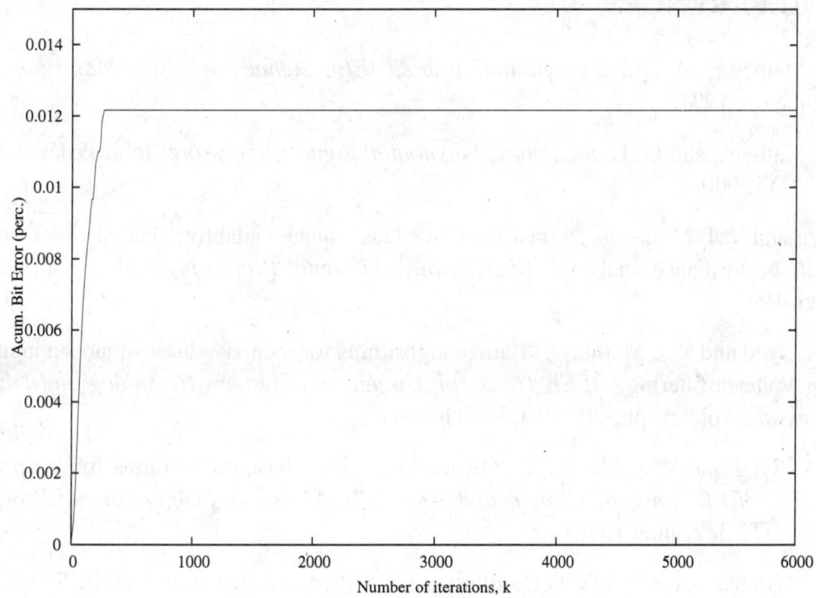

(a) Accumulated bit error

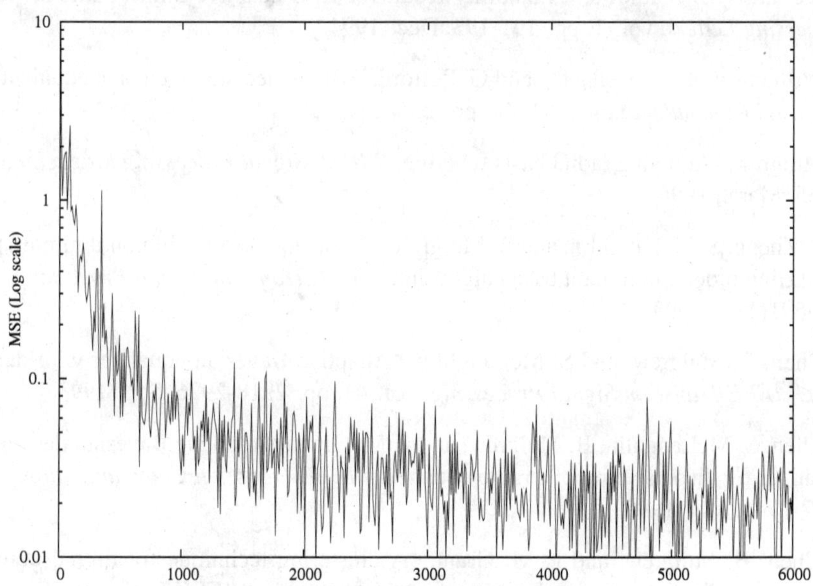

(b) Square error

Figure 11.17 Experiments with DFE radial basis function algorithm, noise level −25dBs.

11.7 REFERENCES

1. V. J. Mathews, "Adaptive polynomial filters," *IEEE Signal Processing Magazine*, vol. 8, pp. 10-26, Nov. 1991.

2. V. J. Mathews and G. L. Sicuranza, *Polynomial Signal Processing*, John Wiley & Sons, New York, NY, 2000.

3. J. Lee and V. J. Mathews, "A fast recursive least squares adaptive second-order Volterra filter and its performance analysis," *IEEE Trans. on Signal Processing*, vol. 41, pp. 1087-1101, March 1993.

4. M. A. Syed and V. J. Mathews, "Lattice algorithms for recursive least squares adaptive second-order Volterra filtering," *IEEE Trans. on Circuits and Systems II: Analog and Digital Signal Processing*, vol. 41, pp. 202-214, March 1994.

5. M. A. Syed and V. J. Mathews, "QR-decomposition based algorithms for adaptive Volterra filtering," *IEEE Trans. on Circuits and Systems II: Analog and Digital Signal Processing*, vol. 40, pp. 372-382, June 1993.

6. V. J. Mathews, "Adaptive Volterra filters using orthogonal structures," *IEEE Signal Processing Letters*, vol. 3, pp. 307-309, Dec. 1996.

7. J. Lee and V. J. Mathews, "A stability result for RLS adaptive bilinear filters," *IEEE Signal Processing Letters*, vol. 1, pp. 191-193, Dec. 1994.

8. D. Williamson, R. A. Kennedy, and G. Pulford, " Block decision feedback equalization," *IEEE Trans. on Communications*, vol. 40, pp. 255-264, Feb. 1992.

9. B. Mulgrew, "Applying radial basis functions," *IEEE Signal Processing Magazine*, vol. 13, pp. 50-65, March 1996.

10. F.-C. Zheng, S. McLaughlin, and B. Mulgrew, "Blind equalization of nonminimum phase channels: High order cummulant based algorithm," *IEEE Trans. on Signal Processing*, vol. 41, pp. 681-691, Feb. 1993.

11. S. Chen, B. Mulgrew, and S. McLaughlin, "Adaptive Bayesian equalizer with decision feedback," *IEEE Trans. on Signal Processing*, vol. 41, pp. 2918-2926, Sep. 1993.

12. S. Chen, S. McLaughlin, B. Mulgrew, and P. M. Grant, "Adaptive Bayesian decision feedback equalizer for dispersive mobile radio channels," *IEEE Trans. on Communications*, vol. 43, pp. 1937-1946, May 1995.

13. S. Chen, B. Mulgrew, and P. M. Grant, "A clustering technique for digital communication channel equalization using radial basis function networks," *IEEE Trans. on Neural Networks*, vol. 4, pp. 570-579, July 1993.

14. S. Chen and B. Mulgrew, "Reconstruction of binary signals using an adaptive-radial-basis function equaliser," *Signal Processing*, vol. 22, pp. 77-93, Jan. 1991.

15. S. Chen and B. Mulgrew, "Overcoming co-channel interference using an adaptive radial basis function equaliser," *Signal Processing*, vol. 28, pp. 91-107, July 1992.

16. S. Chen, S. McLaughlin, and B. Mulgrew, "Complex valued radial basis function network, Part I: Network architecture and learning algorithms," *Signal Processing*, vol. 35, pp. 19-31, Jan. 1994.

17. S. Chen, S. McLaughlin, and B. Mulgrew, "Complex valued radial basis function network, Part II: Application to digital communications channel equalisation," *Signal Processing*, vol. 36, pp. 175-188, March 1994.

18. I. Cha and S. Kassam, "Channel equalization using adaptive complex radial basis function networks," *IEEE Trans. on Selected Areas in Communications*, vol. 13, pp. 122-131, Jan. 1995.

19. I. Cha and S. Kassam, "Interference cancellation using radial basis function networks," *Signal Processing*, vol. 47, pp. 247-268, Dec. 1995.

20. D. Gonzaga, M. L. R. de Campos, and S. L. Netto, "Composite squared-error algorithm for training feedforward neural networks," *Proc. of the 1998 IEEE Digital Filtering and Signal Processing Conference*, Victoria, B.C., June 1998.

21. B. Widrow and S. D. Stearns, *Adaptive Signal Processing*, Prentice Hall, Englewood Cliffs, NJ, 1985.

22. A. Papoulis, *Probability, Random Variables, and Stochastic Processes*, McGraw Hill, New York, NY, 3rd edition, 1991.

23. S. U. Qureshi, "Adaptive equalization," *Proceedings of the IEEE*, vol-73, pp. 1349-1387, Sept. 1985.

24. B. Widrow and E. Walach, *Adaptive Inverse Control*, Prentice Hall, Englewood Cliffs, NJ, 1996.

25. K. K. Johnson and I. W. Sandberg, "Notes on the stability of bilinear filters," *IEEE Trans. on Signal Processing*, vol. 46, pp. 2056-2058, July 1998.

26. F.-L. Luo and R. Unbehauen, *Applied Neural Networks for Signal Processing*, Cambridge University Press, Cambridge, U.K, 1996.

27. S. Haykin, *Neural Networks: A Comprehensive Foundation*, Prentice Hall, Englewood Cliffs, NJ, 2nd edition, 1999.

28. L.-X. Wang, *Adaptive Fuzzy Systems and Control: Design and Stability Analysis*, Prentice Hall, Englewood Cliffs, NJ, 1994.

29. C. Nikias and A. P. Petropulu, *Higher-order Spectra Analysis: A Nonlinear Signal Processing Framework*, Prentice Hall, Englewood Cliffs, NJ, 1993.

11.8 PROBLEMS

1. Perform the equalization of a nonlinear channel described by the following relation

$$r(k) = 0.9x(k) + 0.1x^2(k) - 0.3x^3(k) + n(k)$$

using a known training signal that consists of a binary (-1,1) random signal. An additional Gaussian white noise with variance 10^{-2} is present at the channel output.
Apply the LMS and RLS Volterra series algorithms.

2. Repeat problem 1 using the adaptive bilinear structure.

3. Repeat problem 1 using the multilayer perceptron algorithm.

4. Repeat problem 1 using the adaptive radial basis function structure.

5. Utilize a DFE equalizer to problem 1, also using the LMS and RLS Volterra series algorithms, and comment on the results.

6. Compare the performances of Volterra LMS and RLS algorithms in the identification of the following system.

$$\begin{aligned} d(k) &= -0.76x(k) - 1.0x(k-1) + 1.0x(k-2) + 0.5x^2(k) \\ &\quad +2.0x(k)x(k-2) - 1.6x^2(k-1) + 1.2x^2(k-2) \\ &\quad +0.8x(k-1)x(k-2) + n(k) \end{aligned}$$

The input signal is a uniformly distributed white noise with variance $\sigma_{n_x}^2 = 0.1$, filtered by all-pole filter given by

$$H(z) = \frac{z}{z - 0.95}$$

An additional Gaussian white noise with variance 10^{-2} is present at unknown system output.

7. Identify an unknown system with the following model

$$d(k) = -0.6d(k-1) + x(k) + 0.01x(k)d(k-1) + 0.02x(k-1)d(k-1) + n(k)$$

using the bilinear algorithm. The additional noise is Gaussian white noise with variance $\sigma_n^2 = -20$dB. Use Gaussian white noise with unit variance as input.

8. Repeat problem 7 using the multilayer perceptron algorithm.

9. Identify a system with the following nonlinear input to output relation

$$\begin{aligned} d(k) &= -0.08x(k) - 0.15x(k-1) + 0.14x(k-2) + 0.055x^2(k) \\ &\quad +0.30x(k)x(k-2) - 0.16x^2(k-1) + 0.14x^2(k-2) + n(k) \end{aligned}$$

The input signal is Gaussian white noise with variance $\sigma_x^2 = 0.7$, and the measurement noise is also Gaussian white noise independent of the input signal with variance $\sigma_n^2 = 0.01$.
Apply the radial basis function algorithm.

10. Repeat problem 9 using the multilayer perceptron algorithm.

12

SUBBAND ADAPTIVE FILTERS

12.1 INTRODUCTION

There are many applications where the required adaptive-filter order is high, as for example, in acoustic echo cancellation where the unknown system (echo) model has a long impulse response, on the order of a few thousand samples [1]-[5]. In such applications, the adaptive-filtering algorithm entails a large number of computations. In addition, the high order of the adaptive filter affects the convergence speed.

A solution to problems where long-impulse-response filters are needed is to employ adaptive filtering in subbands. In subband adaptive filtering, both the input signal and the desired signal are split into frequency subbands via an analysis filter bank. Assuming that the signal decomposition in subchannels is effective, we can decimate (subsample) these subband signals and apply adaptive filtering to the resulting signals. Each subband adaptive filter usually has shorter impulse response than its fullband counterpart. If a gradient type algorithm is used to update the adaptive filters, we can adjust the step size in the adaptation algorithm individually for each subband, which leads to higher convergence speed than in the case of fullband adaptive filter.

Decimation allows the reduction in computational complexity. Mainly if critical subsampling (i.e., decimation by a factor equal to the number of subbands) is employed, aliasing effects may impair the obtained filter estimates. This issue will be discussed during this chapter. Therefore, by judicious use of adaptive filtering in subbands we can reduce the computational complexity, as well as increase the algorithm convergence speed [1]-[7].

This chapter starts with a brief introduction to multirate systems, where the concepts of decimation, interpolation, and filter banks are presented. Then, the basic structures for adaptive filtering in subbands are presented along with a discussion regarding their main features. The concept of delayless subband adaptive filtering is also addressed, where the adaptive-filter coefficients are updated in subbands and mapped to an equivalent fullband filter. Finally, we point out the relation between subband and block adaptive-filtering (also known as frequency-domain adaptive filters) algorithms.

P.S.R. Diniz, *Adaptive Filtering*, DOI: 10.1007/978-0-387-68606-6_12,

12.2 MULTIRATE SYSTEMS

In this section, we briefly review the fundamentals of multirate systems which are essential to implement adaptive filters in subbands. For further details related to multirate systems and filter banks the reader can refer to the review article [8] or the comprehensive textbook [9].

12.2.1 Decimation and Interpolation

Decimation (also known as down-sampling or compression) of a digital signal $x(k)$ by a factor of L means reducing its sampling rate L times. Decimation is achieved by retaining only every Lth sample of the signal. The decimator symbol is depicted in Fig. 12.1.a.

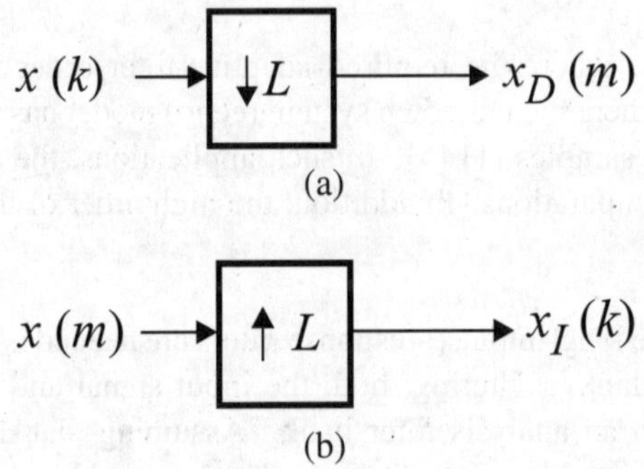

Figure 12.1 (a) Decimation by a factor L, (b) Interpolation by a factor L.

The decimated signal is then $x_D(m) = x(mL)$. In the frequency domain, if the spectrum of $x(k)$ is $X(e^{\jmath\omega})$, the spectrum of the sub-sampled signal, $X_D(e^{\jmath\omega})$ is given by [9]

$$X_D(e^{\jmath\omega}) = \frac{1}{L}\sum_{k=0}^{L-1} X(e^{\jmath\frac{\omega-2\pi k}{L}}) \tag{12.1}$$

The above equation indicates that the spectrum of $x_D(m)$ is composed of copies of the spectrum of $x(k)$ expanded by L and repeated with period 2π. Figs. 12.2.a and 12.2.b depict the effect of sub-sampling on the spectrum of $x(k)$, for $L = 2$. This implies that, in order to avoid aliasing after sub-sampling, the bandwidth of the signal $x(k)$ must be limited to the interval $[-\frac{\pi}{L}, \frac{\pi}{L}]$. In fact, the sub-sampling operation is generally preceded by a lowpass filter that approximates the following frequency response

$$H_D(e^{\jmath\omega}) = \begin{cases} 1, & \omega \in [-\frac{\pi}{L}, \frac{\pi}{L}] \\ 0, & \text{otherwise} \end{cases} \tag{12.2}$$

It should be noted that the decimation operation is shift varying, i.e., if the input signal $x(k)$ is shifted, the output signal will not in general be a shifted version of the previous output. More precisely, the decimation is a periodically shift-invariant operation.

The interpolation (or up-sampling) of a digital signal $x(m)$ by a factor of L entails including $L - 1$ zeros in between samples. The interpolator symbol is depicted in Fig. 12.1.b.

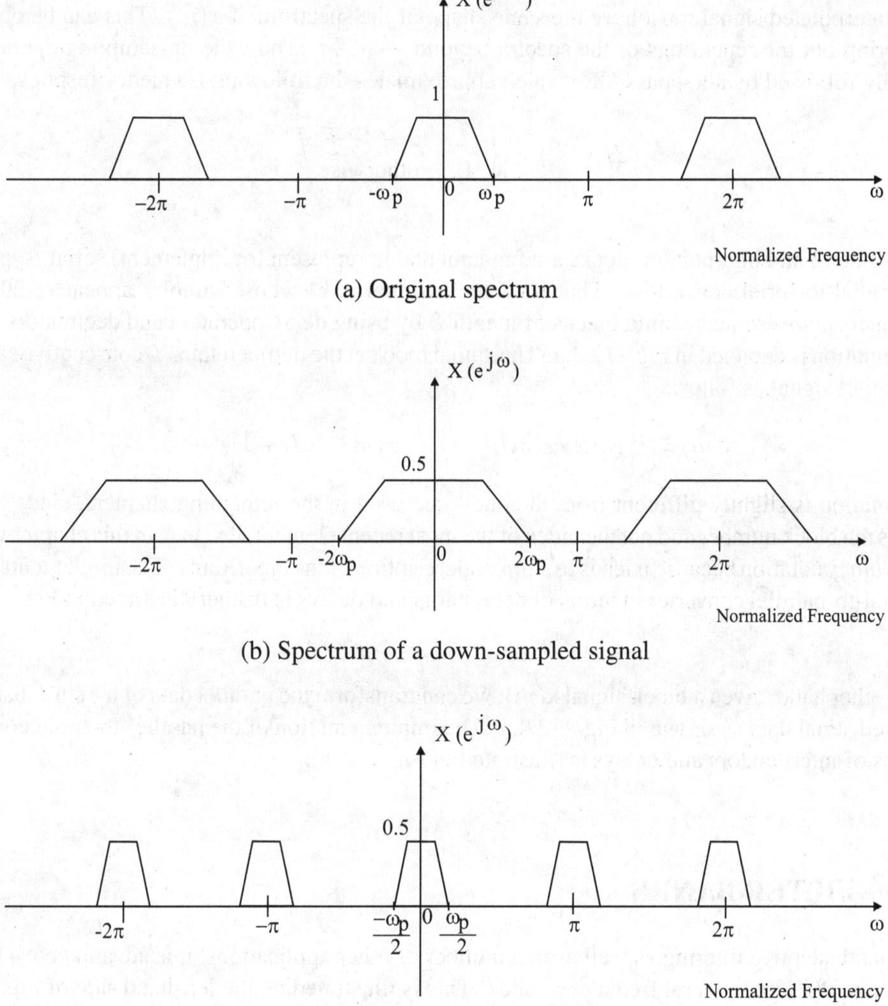

(a) Original spectrum

(b) Spectrum of a down-sampled signal

(c) Spectrum of an up-sampled signal

Figure 12.2 Spectra of up- and down-sampled signals.

The interpolated signal is then

$$x_I(k) = \begin{cases} x(\frac{k}{L}), & k = mL, m \in \mathcal{Z} \\ 0, & \text{otherwise} \end{cases} \tag{12.3}$$

If the spectrum of $x(m)$ is $X(e^{\jmath\omega})$, it is straightforward to show that the spectrum of the up-sampled signal, $X_I(e^{\jmath\omega})$, is given by

$$X_I(e^{\jmath\omega}) = X(e^{\jmath\omega L}) \tag{12.4}$$

Since the spectrum of the input signal is periodic with period 2π, the spectrum of the interpolated signal will have period $\frac{2\pi}{L}$. Fig. 12.2.c illustrates how the signal spectrum is modified after the up-sampling operation. If we wish to obtain a smooth interpolated version of $x(m)$, the spectrum of the interpolated signal must have the same shape of the spectrum of $x(m)$. This can be obtained by filtering out the repetitions of the spectra beyond $[-\frac{\pi}{L}, \frac{\pi}{L}]$. Thus, the up-sampling operation is generally followed by a lowpass filter which approximates the following frequency response

$$H_I(e^{\jmath\omega}) = \begin{cases} L, & \omega \in [-\frac{\pi}{L}, \frac{\pi}{L}] \\ 0, & \text{otherwise} \end{cases} \tag{12.5}$$

The decimator and interpolator blocks are fundamental to represent (or implement) serial-to-parallel and parallel-to-serial converters. That is, given a signal $x(k)$ whose samples appear serially, we can transform this sequence into blocks of length L by using delay operators and decimators whose representation is depicted in Fig. 12.3.a. The signal block at the output retains L consecutive samples of the input signal as follows

$$\mathbf{x}(m) = [x(mL)\ x(mL-1) \ldots x(mL-L+1)]^T \tag{12.6}$$

This notation is slightly different from the one to be used in the remaining chapters, since m here denotes the block number and not the index of the most recent element of $\mathbf{x}(m)$. In this chapter we will use the *block* notation because it leads to simpler description of the algorithms. The implementation of the serial-to-parallel converter in terms of decimators and delays is further illustrated in Fig. 12.3.b.

On the other hand, given a block signal $\mathbf{x}(m)$, we can transform the parallel data of length L back into a delayed serial data as shown in Fig. 12.4.a. The implementation of the parallel-to-serial converter in terms of interpolators and delays is illustrated in Fig. 12.4.b.

12.3 FILTER BANKS

In subband adaptive filtering as well as in a number of other applications, it is advantageous to split a sequence $x(k)$ into several frequency bands. This is illustrated on the left-hand side of Fig. 12.5.

The analysis filters, represented by the transfer functions $F_i(z)$ for $i = 0, 1, \ldots, M-1$, comprise of a lowpass filter $F_0(z)$, bandpass filters $F_i(z)$ for $i = 1, 2, \ldots, M-2$, and a highpass filter $F_{M-1}(z)$. Ideally these filters have nonoverlapping passbands, while they together cover the entire spectrum of the input signal. Since each of the analysis filter outputs $x_i(k)$, $i = 0, 1, \ldots, M-1$ has the same number of samples as the original signal $x(k)$, after the M-band decomposition, all signals $x_i(k)$ together have M times more samples than the original one. This expansion on the number of samples is undesirable because of the resulting computational burden.

In most cases, the input signal is uniformly split into subbands, where each of the frequency bands has the same bandwidth. Since the bandwidth of each analysis filter output band is M times smaller than in the original signal, we can decimate each $x_i(k)$ by a factor of L smaller or equal to M without destroying the original information. For $L = M$, the amount of data after the decimators in Fig. 12.5 is maintained when compared to the number of samples of the input signal. This case is called

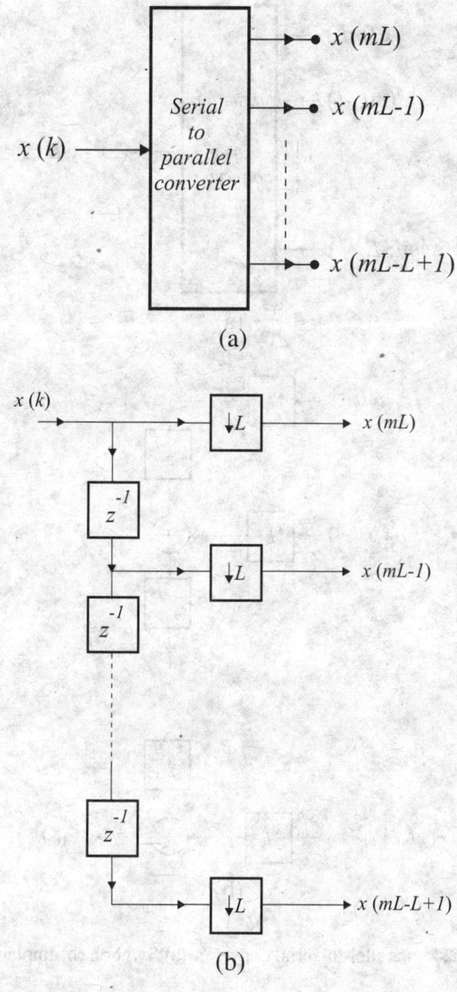

Figure 12.3 Serial-to-parallel converter (a) Symbol, (b) Implementation.

maximally (or critically) decimated analysis filter bank. If $L > M$, there is a loss of information due to aliasing which does not allow the recovery of the original signal. For $L \leq M$, it is possible to retain all information contained in the input signal by properly designing the analysis filters in conjunction with the synthesis filters $G_i(z)$, for $i = 0, 1, \ldots, M-1$. If no signal processing task is performed in the subbands (see Fig. 12.5), the filter bank output $y(k)$ can be made to be a delayed version of the input signal $x(k)$, where the delay is due to the causality of the subband filters. In this case, we have a perfect reconstruction filter bank. In fact, there are several methods for designing the analysis filters $F_i(z)$ and the synthesis filters $G_i(z)$ such that perfect reconstruction is achieved or arbitrarily approximated. These filters can be finite-length (FIR) filters with overlapping frequency responses, which are designed to cancel out the aliasing effects and results in the perfect reconstruction.

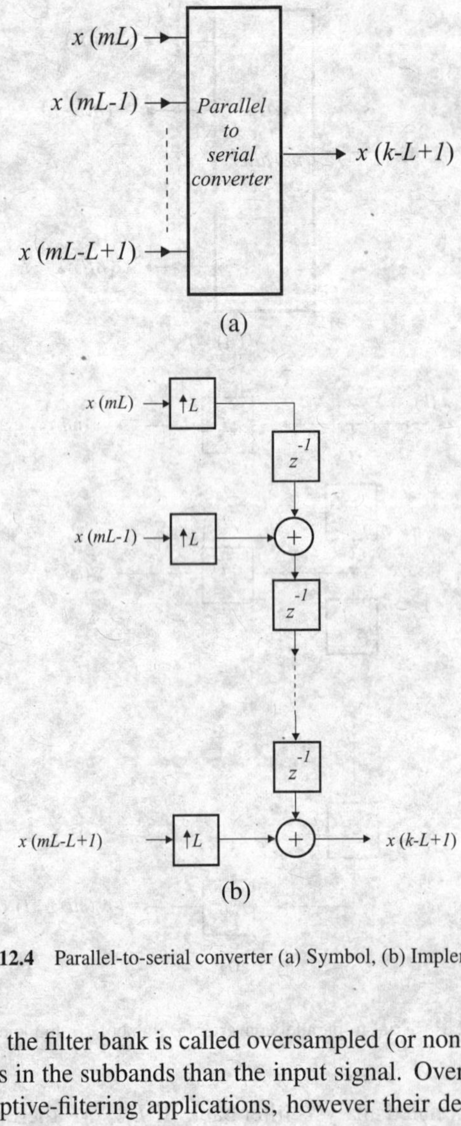

Figure 12.4 Parallel-to-serial converter (a) Symbol, (b) Implementation.

In the case where $L < M$, the filter bank is called oversampled (or noncritically sampled) since we are retaining more samples in the subbands than the input signal. Oversampled filter banks appear frequently in subband adaptive-filtering applications, however their design is beyond the scope of this book.

We will now discuss the polyphase representation of a transfer function which is quite useful in describing filter banks. Defining $E_{ij}(z) = \sum_{l=0}^{N_p-1} f_i(Ll + j)z^{-l}$ as the polyphase components of the analysis filter $F_i(z)$, and N_p as the length of the polyphase components of the analysis filters, we can express the transfer function of the filter $F_i(z)$ as follows

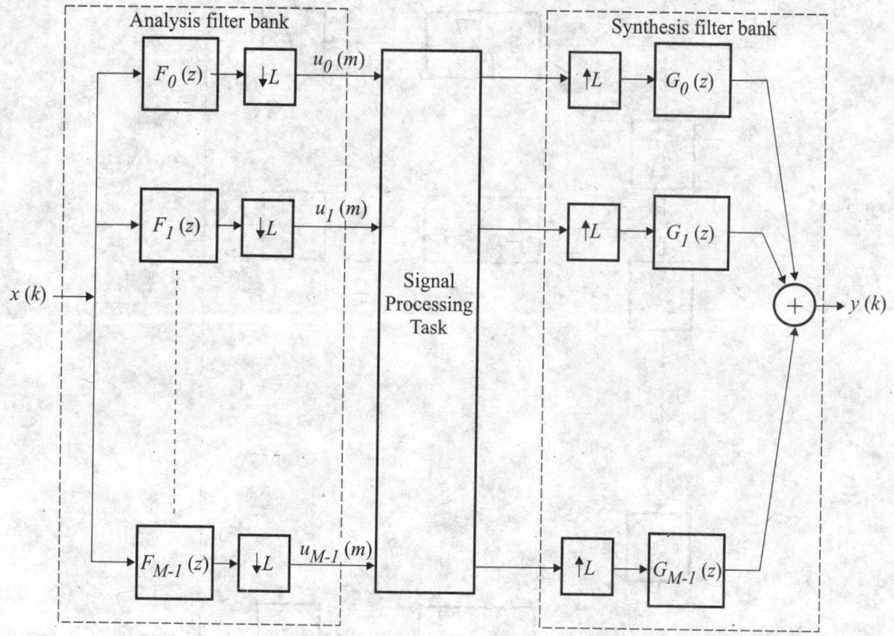

Figure 12.5 Signal processing in subbands.

$$\begin{aligned} F_i(z) &= \sum_{k=0}^{N_pL-2L+1} f_i(k)z^{-k} \\ &= \sum_{l=0}^{N_p-1} f_i(Ll)z^{-Ll} + z^{-1}\sum_{l=0}^{N_p-1} f_i(Ll+1)z^{-Ll} + \cdots + z^{-L+1}\sum_{l=0}^{N_p-1} f_i(Ll+L-1)z^{-Ll} \\ &= \sum_{j=0}^{L-1} z^{-j}E_{ij}(z^L) \end{aligned} \tag{12.7}$$

In the polyphase decomposition we decompose each analysis filter $F_i(z)$ into L filters, the first one has an impulse response consisting of every sample of $f_i(k)$ whose indexes are multiples of L, the second one has every sample of $f_i(k)$ whose indexes are one plus a multiple of L, and so on. The resulting representation for an analysis subfilter, along with decimation, is depicted in Fig. 12.6. By means of a noble identity [9], the cascade connection of $E_{ij}(z^L)$ and the decimators can be replaced by decimators followed by the polynomials $E_{ij}(z)$.

For the synthesis filter bank we can employ an alternative polyphase decomposition which matches the interpolation operation. That is, each synthesis filter can be described in the following polyphase

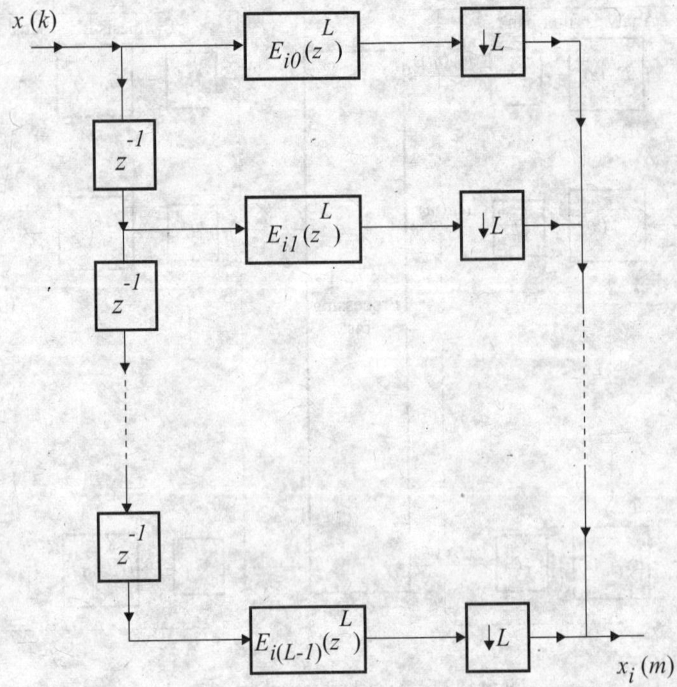

Figure 12.6 Polyphase representation.

form

$$G_i(z) = \sum_{j=0}^{M-1} z^{-(L-1-j)} R_{ji}(z^L) \tag{12.8}$$

Again by means of a noble identity [9], the polynomials $R_{ji}(z^L)$ preceded by interpolators can be replaced by interpolators preceded by the polynomials $R_{ji}(z)$.

By replacing each of the filters $F_i(z)$ and $G_i(z)$ by their polyphase components, the M-band filter bank of Fig. 12.5 can be transformed in the structure of Fig. 12.7. The matrices $\mathbf{E}(z)$ and $\mathbf{R}(z)$ are formed from the polyphase components of $F_i(z)$ and $G_i(z)$. $E_{ij}(z)$ is the jth polyphase component of $F_i(z)$ and $R_{ji}(z)$ is the jth polyphase component of $G_i(z)$. From Fig. 12.7 we conclude that if $\mathbf{R}(z)\mathbf{E}(z) = z^{-\Delta}\mathbf{I}$, where Δ is an arbitrary delay and $\mathbf{I}$ is the identity matrix, the M-band filter bank holds the perfect reconstruction property.

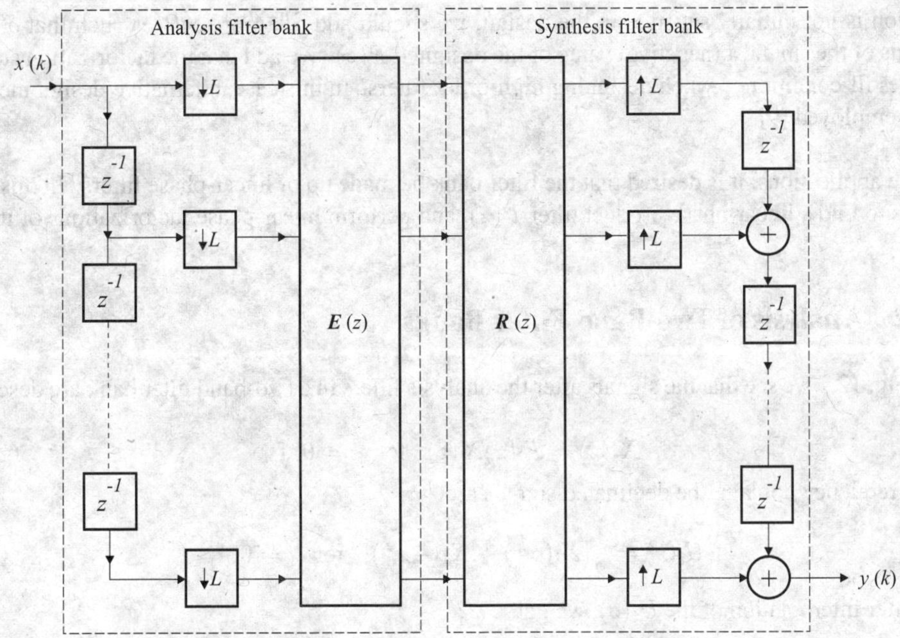

Figure 12.7 M-band filter bank with polyphase representation.

12.3.1 Two-Band Perfect Reconstruction Filter Banks

For a two-band perfect reconstruction filter bank with FIR analysis and synthesis filters, the following conditions must be satisfied

$$F_0(-z)F_1(z) - F_0(z)F_1(-z) = 2cz^{-2l-1} \tag{12.9}$$

$$G_0(z) = -\frac{z^{2(l-\Delta)}}{c}F_1(-z) \tag{12.10}$$

$$G_1(z) = \frac{z^{2(l-\Delta)}}{c}F_0(-z) \tag{12.11}$$

where equation (12.9) guarantees that the synthesis filters are FIR, while equations (12.10) and (12.11) guarantee perfect reconstruction. The delay Δ is included in equations (12.10) and (12.11) in order to guarantee that the subfilters in the filter bank are causal.

Equations (12.9) to (12.11) lead to the following design procedure for the two-band perfect reconstruction filter bank [10]: (1) Find a polynomial $P(z)$ such that $P(-z) - P(z) = 2z^{-2l-1}$; (2) Factorize $P(z)$ into two factors, $F_0(z)$ and $F_1(-z)$, such that $F_0(z)$ and $F_1(-z)$ are lowpass filters; (3) Design $G_0(z)$ and $G_1(z)$ using equations (12.10) and (12.11). In step (1) $P(z)$ is an approximation to a half-band filter[1], whose amplitude response should be positive everywhere. In case this

[1] The amplitude response of a half-band filter is symmetric with respect to $\frac{\pi}{2}$, with $\omega_p + \omega_s = \pi$, where ω_p is the passband edge and ω_s is the stopband edge.

condition is not initially satisfied in the design, we should add δz^{-2l-1} to $P(z)$ such that δ is the modulus of the smallest (negative) value of the designed $P(z)$. We add that the factorization step (2) becomes ill-conditioned when designing high-order filters. In this case, alternative design methods can be employed [9].

In some applications, it is desired that the filter bank be made up of linear-phase filters. In this case, one has to find a linear-phase product filter $P(z)$, and perform linear-phase factorizations of it.

12.3.2 Analysis of Two-Band Filter Banks

From Fig. 12.5 we see that the signals after the analysis filters in a two-band filter bank are described by

$$X_i(z) = F_i(z)X(z) \quad \text{for} \quad i = 0, 1 \tag{12.12}$$

In the frequency domain, the decimated signals are

$$U_i(z) = \frac{1}{2}[X_i(z^{\frac{1}{2}}) + X_i(-z^{\frac{1}{2}})] \quad \text{for} \quad i = 0, 1 \tag{12.13}$$

Thus after interpolation of the $U_i(z)$, we get

$$\begin{aligned} U_i(z^2) &= \frac{1}{2}[X_i(z) + X_i(-z)] \\ &= \frac{1}{2}[F_i(z)X(z) + F_i(-z)X(-z)] \end{aligned} \tag{12.14}$$

The reconstructed signal is then expressed as

$$\begin{aligned} Y(z) &= G_0(z)U_0(z^2) + G_1(z)U_1(z^2) \\ &= \frac{1}{2}[F_0(z)G_0(z) + F_1(z)G_1(z)]X(z) + \frac{1}{2}[F_0(-z)G_0(z) + F_1(-z)G_1(z)]X(-z) \\ &= \frac{1}{2}\begin{bmatrix} X(z) & X(-z) \end{bmatrix} \begin{bmatrix} F_0(z) & F_1(z) \\ F_0(-z) & F_1(-z) \end{bmatrix} \begin{bmatrix} G_0(z) \\ G_1(z) \end{bmatrix} \end{aligned} \tag{12.15}$$

The last equality is called modulation-matrix representation of a two-band filter bank. In this case, the aliasing effect caused by the decimation operation is represented by the terms containing $X(-z)$.

Note that it is possible to avoid aliasing at the output by properly choosing the synthesis filter, as for example in the perfect reconstruction case.

12.3.3 Analysis of M-Band Filter Banks

The expression for two-band case can be easily generalized to M bands by noting that, after decimation by L, the signals will have $L-1$ aliased components. That is

$$X_d(z) = \frac{1}{L}\sum_{k=0}^{L-1} X(z^{\frac{1}{L}} e^{-\frac{\jmath 2\pi k}{L}}) \tag{12.16}$$

The kth aliased component of $X(z)$ is $X(z^{\frac{1}{L}}e^{-\frac{\jmath 2\pi k}{L}})$.

Therefore, the modulation matrix for the M-band filter bank is given by

$$Y(z) = \frac{1}{2}\left[\begin{array}{cccc} X(z) & X(zW) & \dots & X(zW^{L-1}) \end{array}\right] \begin{bmatrix} F_0(z) & F_1(z) & \dots & F_{M-1}(z) \\ F_0(zW) & F_1(zW) & \dots & F_{M-1}(zW) \\ \vdots & \vdots & \ddots & \vdots \\ F_0(zW^{L-1}) & F_1(zW^{L-1}) & \dots & F_{M-1}(zW^{L-1}) \end{bmatrix} \begin{bmatrix} G_0(z) \\ G_1(z) \\ \vdots \\ G_{M-1}(z) \end{bmatrix} \quad (12.17)$$

where $W = e^{-\frac{\jmath 2\pi}{L}}$.

12.3.4 Hierarchical M-Band Filter Banks

By connecting two-band filter banks in series, we can produce many different kinds of maximally decimated decompositions. For example, we can design a 2^n-band uniform decomposition filter bank as illustrated in Fig. 12.8 for $n = 3$. It is also possible to implement nonuniform filter banks by using two-band filter banks in series, but using a different type of hierarchical decomposition [9]. A commonly used one is the octave-band decomposition.

12.3.5 Cosine-Modulated Filter Banks

Cosine-modulated filter banks are a class of filters efficient for the design and implementation of filter banks with large number of subbands. A cosine-modulated filter bank is easy to design because it is based on a single lowpass prototype filter whose impulse response satisfies some constraints required to achieve perfect reconstruction. It also leads to low computational complexity because the analysis and synthesis filter banks make use of the so-called discrete-time cosine transform (DCT), for which there are many fast implementations available for its computation.

The design of the maximally decimated cosine-modulated filter bank starts with a linear-phase prototype lowpass filter $F(z)$ whose passband edge is $\frac{\pi}{2L} - \delta$ and the stop-band edge is $\frac{\pi}{2L} + \delta$, where 2δ is the transition band. The length of the prototype filter is usually chosen to be an even multiple of the number of subbands: $N_{pr} = 2KL$, for K, an integer. Then, we generate cosine modulated versions of the prototype filter in order to obtain the analysis and synthesis filter banks. The impulse responses of the subfilters are given by

$$f_l(n) = 2f(n)\cos\left[(2l+1)\frac{\pi}{2L}\left(n - \frac{N_{pr}-1}{2}\right) + (-1)^l\frac{\pi}{4}\right] \quad (12.18)$$

$$g_l(n) = 2f(n)\cos\left[(2l+1)\frac{\pi}{2L}\left(n - \frac{N_{pr}-1}{2}\right) - (-1)^l\frac{\pi}{4}\right] \quad (12.19)$$

for $1 \le n \le N_{pr}$ and $0 \le l \le L-1$, where $f(n)$, for $n = 1, 2, \dots, N_{pr}$, denotes the elements of the prototype impulse response. The constraints required to achieve perfect reconstruction are given

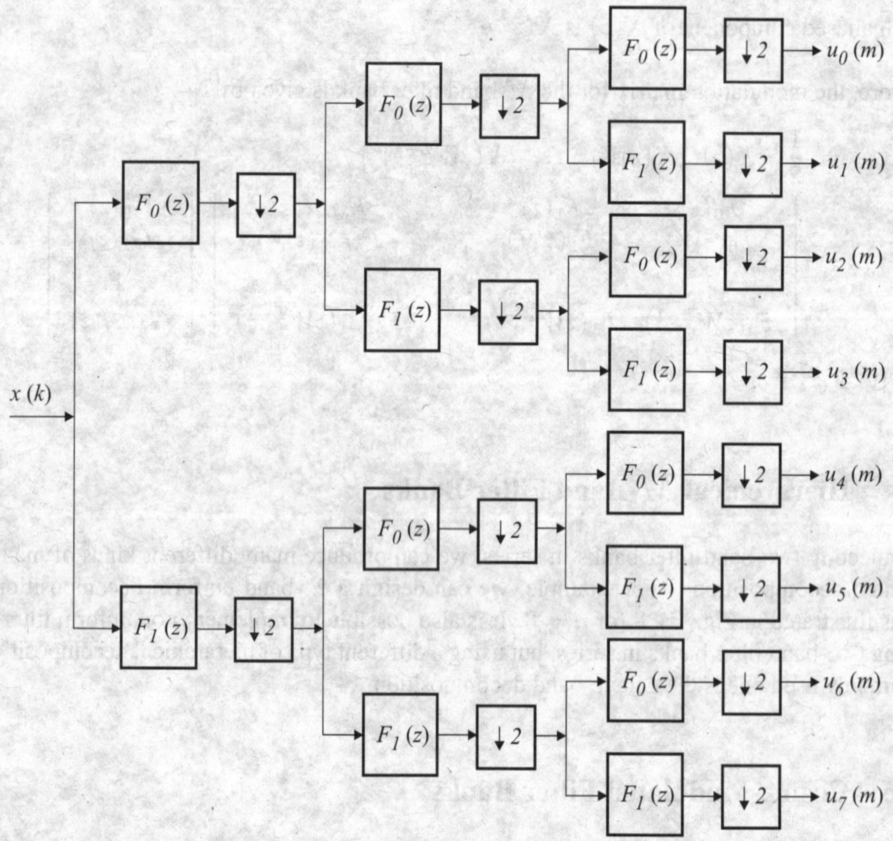

Figure 12.8 Hierarchical uniform filter bank.

by

$$E_j(z^{-1})E_j(z) + E_{j+L}(z^{-1})E_{j+L}(z) = \frac{1}{2L} \tag{12.20}$$

where $E_j(z)$ for any $j = 0, 1, \ldots, L-1$ is the jth polyphase component of the prototype filter $F(z)$.

There are computationally efficient implementations for the cosine-modulated filter bank which make use of the polyphase decomposition of the prototype filter. For further details refer to [9], [8]. Also, it is possible to design oversampled cosine-modulated filter banks with perfect reconstruction [11], which can be used in nonmaximally decimated subband adaptive filtering.

12.3.6 Block Representation

By using the polyphase concept we can show that any scalar linear time-invariant transfer function $H(z)$ can be implemented through a pseudocirculant matrix $\mathbf{H}(z)$, where the particular case of a 3×3 matrix $\mathbf{H}(z)$ is given by

$$\mathbf{H}(z) = \begin{bmatrix} H_0(z) & H_1(z) & H_2(z) \\ z^{-1}H_2(z) & H_0(z) & H_1(z) \\ z^{-1}H_1(z) & z^{-1}H_2(z) & H_0(z) \end{bmatrix} \tag{12.21}$$

where the $H_i(z)$, $i = 0, 1, 2$, are the polyphase components of $H(z)$.

The overall realization of $H(z)$ is equivalent to a cascade connection of the serial-to-parallel converter of Fig. 12.3.b, the transfer matrix $\mathbf{H}(z)$, and the parallel-to-serial converter of Fig. 12.4.b, except for a delay of z^{-L+1} since the converter of Fig. 12.4.b is causal (i.e., it utilizes negative powers of z). See the implementation of Fig. 12.7 with $\mathbf{H}(z)$ replacing the cascade of $\mathbf{E}(z)$ and $\mathbf{R}(z)$. This realization is known as blocked implementation of a scalar transfer function [32].

We note that the cascade of the unblock/block mechanisms of Figs. 12.4.b (noncausal case) and 12.4.a result in an identity matrix (see section 12.5). The reader is encouraged to verify this result.

12.4 SUBBAND ADAPTIVE FILTERS

A number of adaptive-filtering structures based on multirate techniques have been proposed in the literature [1]-[7], [14]-[28]. In most of these structures, the input signal is decomposed into subbands via an analysis filter bank, and the resulting signals are downsampled and filtered by adaptive filters. Each of these adaptive filters has order smaller than the equivalent full-band adaptive filter (by a factor approximately equal to the decimation rate). The subsampling operations create aliased versions of the decimated signal which will affect the performance of the adaptive filter. The aliasing effect is more severe when critically sampled filter banks are employed. An obvious solution is to allow frequency gaps between adjacent subbands, which for sure degrades the original signal quality. Some other structures apply subband decomposition only to the error signal in order to improve tracking ability in nonstationary environments [29]-[30].

Several adaptive subband structures have been suggested. One early approach uses pseudo-QMF[2] banks with overlapping subfilters and critical subsampling [1], i.e., with $L = M$. This results in undesirable aliased components at the output, which causes severe degradation. A second approach uses QMF banks with critical subsampling [2]. In order to avoid aliasing problems, it is shown that additional adaptive cross terms among the subbands are necessary. These cross terms, however, increase the computational complexity and reduce the convergence rate of the adaptive algorithm.

An alternative solution is to employ oversampling, that is, to use a decimation factor in the filtered signals smaller than the critical subsampling factor (or number of bands), i.e., with $L < M$. In the

[2] Quadrature-mirror filter.

oversampling case, the computational complexity is higher than needed because after decimation the number of samples retained in the subbands is larger than that of the filter bank input. Despite this problem, oversampled adaptive filters are often used in practice [3]-[7], [15]-[16]. In this chapter, we focus on the critically decimated case, although some analysis is also carried out for the general oversampled case.

In all subband structures above described, the convergence rate can be improved for colored input signals by using a normalized gradient algorithm in the update of the coefficients of each subband filter. This improvement is justified in Fig. 12.9, where considering that the filter bank consists of ideal subfilters, the spectrum of each signal in the subbands after critical decimation will be closer to that of white noise than that of the original fullband signal. If the spectral separation is perfect, the subband structure allows the transformation of the fullband adaptive-filtering problem into several independent narrowband adaptive-filtering subproblems. In general, the subband separation will be effective when the order of each subband adaptive filter is much smaller than the order of the fullband filter. The justification is that the speed of convergence becomes faster for all subbands, and the overall computational complexity is further reduced due to decimation.

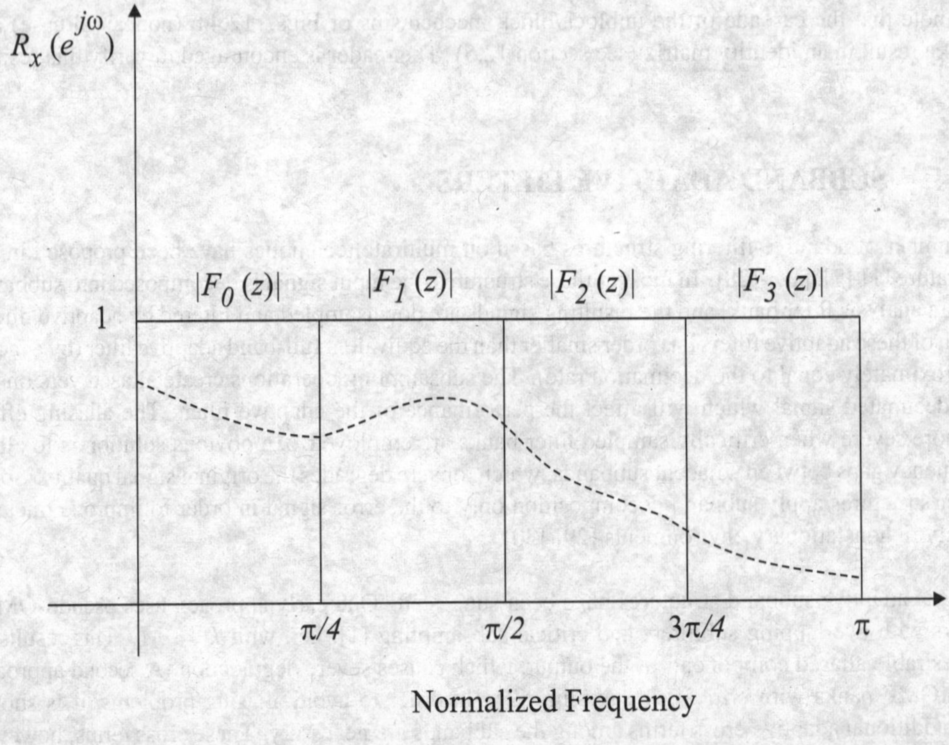

Figure 12.9 Spectrum split in subbands.

In the conventional subband adaptive filters, error signals are locally evaluated in each subband and an objective function taking into account all these local errors is minimized during adaptation. Fig. 12.10 illustrates the open-loop structure, where we can see that both input and reference signals are first split into subbands by an analysis filter bank. Then, the subband signals are filtered by an adaptive-filter matrix in order to generate the output signals to be compared with the desired signals in the subbands. In the open-loop scheme, we aim to minimize the subband error energy.

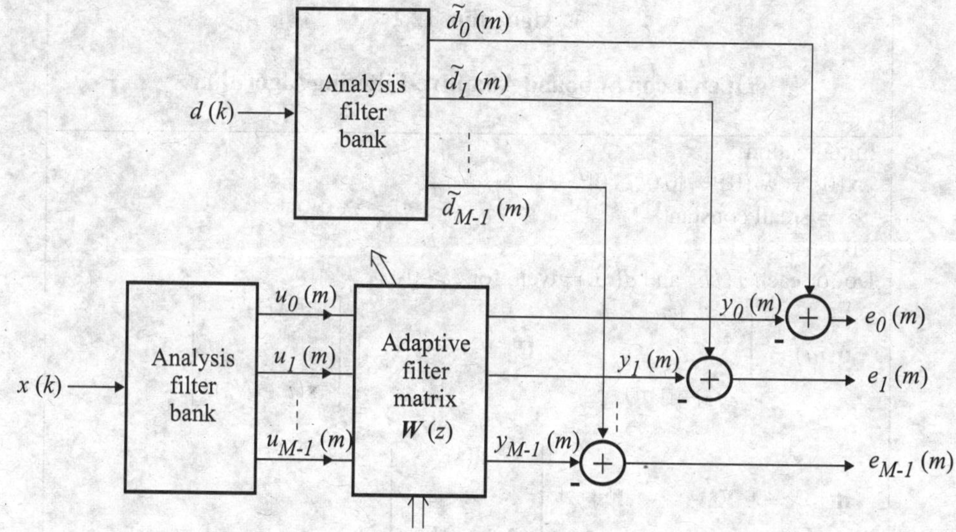

Figure 12.10 Open-loop subband structure.

For the open-loop structure, the objective function can be a linear combination of the magnitude square of the local errors as follows

$$\xi = \sum_{i=0}^{M-1} E[|e_i(m)|^2] \tag{12.22}$$

If we assume that the adaptive-filter matrix is diagonal, and that the subband signals are complex, the updating equation for the subband adaptive filters based on the normalized LMS algorithm is given by

$$e_i(m) = \tilde{d}_i(m) - \mathbf{w}_i^T(m)\mathbf{u}_i(m) \tag{12.23}$$

$$\mathbf{w}_i(m+1) = \mathbf{w}_i(m) + \frac{2\mu}{\gamma + N_s\sigma_i^2(m)} e_i(m)\mathbf{u}_i^*(m) \tag{12.24}$$

where N_s is the length of the adaptive filter in the ith subband (which we consider the same for all subbands in order to simplify the notation). In addition, $\sigma_i^2(m) = (1-\alpha)\sigma_i^2(m-1) + \alpha\,|u_i(m)|^2$, with α being a small factor chosen in the range $0 < \alpha \leq 0.1$, and γ is a small constant to prevent the updating factor from getting too large. The signal $e_i(m)$ is the subband error signal at the ith subband, and $\mathbf{u}_i(m)$ is the input signal vector to the ith adaptive filter.

Based on our knowledge of the normalized LMS algorithms, we can conjecture that the range of values for the convergence factor is typically[3]

$$0 < \mu < 1 \tag{12.25}$$

[3]The upper bound can be tighter depending on the input signal statistics.

Algorithm 12.1

Open-Loop Subband Adaptive-Filtering Algorithm

Initialization

$\mathbf{x}(0) = \mathbf{w}_l(0) = [0\,0 \ldots 0]^T$

$\gamma =$ small constant

$0 < \alpha \leq 0.1$

Do for each $x(iL)$ and $d(iL)$ given, for $i \geq 0$

$$\mathbf{u}(m) = \begin{bmatrix} u_0(m) \\ \vdots \\ u_{M-1}(m) \end{bmatrix} = \begin{bmatrix} \mathbf{E}_0 & \cdots & \mathbf{E}_{N_p} \end{bmatrix} \begin{bmatrix} \mathbf{x}(i) \\ \vdots \\ \mathbf{x}(i-N_p) \end{bmatrix}$$

$$\tilde{\mathbf{d}}(m) = \begin{bmatrix} \mathbf{E}_0 & \cdots & \mathbf{E}_{N_p} \end{bmatrix} \begin{bmatrix} \mathbf{d}(i) \\ \vdots \\ \mathbf{d}(i-N_p) \end{bmatrix}$$

$$\mathbf{y}(m) = \begin{bmatrix} y_0(m) \\ \vdots \\ y_{M-1}(m) \end{bmatrix} = \begin{bmatrix} \mathbf{W}_0 & \cdots & \mathbf{W}_{N_s} \end{bmatrix} \begin{bmatrix} \mathbf{u}(m) \\ \vdots \\ \mathbf{u}(m-N_s) \end{bmatrix}$$

$\mathbf{e}(m) = \tilde{\mathbf{d}}(m) - \mathbf{y}(m)$

Do for each for $0 \leq l \leq M-1$

$\sigma_l^2(m) = (1-\alpha)\sigma_l^2(m-1) + \alpha\,|u_l(m)|^2$

$\mathbf{w}_l(m+1) = \mathbf{w}_l(m) + \frac{2\mu}{\gamma + N_s\sigma_l^2(m)} e_l(m)\mathbf{u}_l^*(m)$

The steps of the open-loop algorithm are described in Algorithm 12.1, where $\mathbf{x}(iL-l)$ and $\mathbf{d}(iL-l)$ represent a block of the input and desired signals respectively, $\mathbf{E}_l$, for $l = 0, 1, \ldots, N_p$, are the matrices containing the coefficients of the polyphase representation of the analysis filter bank, that is,

$$\mathbf{E}(z) = \sum_{l=0}^{N_p} \mathbf{E}_l z^{-l}$$

The coefficient matrices $\mathbf{W}_l$, for $l = 0, 1, \ldots, N_s$, are the entries of the adaptive filter matrices.

Since the frequency responses of the subfilters that compose the filter bank are not ideal, the minimization of an objective function based on local errors will not necessarily reduce the fullband error energy to a minimum MSE. In this case, the unknown system might not be identified accurately.

12.4.1 Subband Identification

Define the $\mathcal{Z}$-transforms of the blocked versions of input and desired signals $x(k)$ and $d(k)$ as

$$\begin{aligned}\mathbf{X}(z) &= \sum_m \mathbf{x}(m)z^{-m} \\ \mathbf{D}(z) &= \sum_m \mathbf{d}(m)z^{-m}\end{aligned} \tag{12.26}$$

where $\mathbf{x}(m)$ is given in equation (12.6), and

$$\mathbf{d}(m) = [d(mL) \;\; d(mL-1) \ldots d(mL-L+1)]^T \tag{12.27}$$

If we describe the analysis filter transfer functions $F_i(z)$, for $i = 0, 1, \ldots, L-1$, in terms of their polyphase components, the subband input and desired signals, described in the $\mathcal{Z}$-domain for the critically decimated case (i.e. $L = M$), can be written in vector form as

$$\begin{aligned}\mathcal{Y}(z) &= \mathbf{W}(z)\mathbf{E}(z)\mathbf{X}(z) \\ \mathcal{D}(z) &= \mathbf{E}(z)\mathbf{D}(z)\end{aligned} \tag{12.28}$$

where $\mathcal{D}(z)$ is the desired signal split into subbands, and $\mathcal{Y}(z)$ is the adaptive system output (refer to Figs. 12.7 and 12.10).

By describing the unknown system model in the block form, as explained in subsection 12.3.6, the blocked desired signal is given by

$$\mathbf{D}(z) = \mathbf{H}(z)\mathbf{X}(z) \tag{12.29}$$

By substituting the above expression into equation (12.28), we obtain

$$\mathcal{D}(z) = \mathbf{E}(z)\mathbf{H}(z)\mathbf{X}(z) \tag{12.30}$$

By defining the channel error vector as $\mathcal{E}(z) = \mathcal{D}(z) - \mathcal{Y}(z)$ and setting it to zero, for $\mathbf{X}(z) \neq 0$, we generate the optimal solution for the adaptive-filter coefficient matrix

$$\mathbf{E}(z)\mathbf{H}(z) = \mathbf{W}_o(z)\mathbf{E}(z) \tag{12.31}$$

whose expression is given by

$$\mathbf{W}_o(z) = \mathbf{E}(z)\mathbf{H}(z)\mathbf{E}^{-1}(z) \tag{12.32}$$

Note that since $\mathbf{W}_o(z)$ is nondiagonal, it requires cross filters among channels in order to model the unknown system perfectly.

12.4.2 Two-Band Identification

The two-band case is easier to analyze in closed form, leading to interesting insights into the problem of cross filters. Using the relations described in equations (12.12) and (12.13), and considering the error signals equal to zero in Fig. 12.10, we can show that for the identification of an unknown transfer function $H(z)$, the optimal coefficients for the two-band adaptive filter are given by

$$\begin{aligned}\mathbf{W}_o(z) &= \begin{bmatrix} F_0(z^{\frac{1}{2}}) & F_0(-z^{\frac{1}{2}}) \\ F_1(z^{\frac{1}{2}}) & F_1(-z^{\frac{1}{2}}) \end{bmatrix} \begin{bmatrix} H(z^{\frac{1}{2}}) & 0 \\ 0 & H(-z^{\frac{1}{2}}) \end{bmatrix} \frac{1}{\Xi(z)} \begin{bmatrix} F_1(-z^{\frac{1}{2}}) & -F_0(-z^{\frac{1}{2}}) \\ -F_1(z^{\frac{1}{2}}) & F_0(z^{\frac{1}{2}}) \end{bmatrix} \\ &= \frac{1}{\Xi(z)} \begin{bmatrix} A(z) & -F_0(z^{\frac{1}{2}})F_0(-z^{\frac{1}{2}})[H(z^{\frac{1}{2}}) - H(-z^{\frac{1}{2}})] \\ F_1(z^{\frac{1}{2}})F_1(-z^{\frac{1}{2}})[H(z^{\frac{1}{2}}) - H(-z^{\frac{1}{2}})] & B(z) \end{bmatrix}\end{aligned} \tag{12.33}$$

where

$$\begin{aligned}\Xi(z) &= F_0(z^{\frac{1}{2}})F_1(-z^{\frac{1}{2}}) - F_0(-z^{\frac{1}{2}})F_1(z^{\frac{1}{2}}) \\ A(z) &= F_0(z^{\frac{1}{2}})F_0(-z^{\frac{1}{2}})H(z^{\frac{1}{2}}) - F_0(-z^{\frac{1}{2}})F_1(z^{\frac{1}{2}})H(-z^{\frac{1}{2}}) \\ B(z) &= -F_0(-z^{\frac{1}{2}})F_1(z^{\frac{1}{2}})H(z^{\frac{1}{2}}) + F_0(z^{\frac{1}{2}})F_1(-z^{\frac{1}{2}})H(-z^{\frac{1}{2}})\end{aligned}$$

The right-hand side of equation (12.33) shows that nonzero off-diagonal elements are required in order to model the unknown system. Note that the products of $F_0(z^{\frac{1}{2}})F_0(-z^{\frac{1}{2}})$ and $F_1(z^{\frac{1}{2}})F_1(-z^{\frac{1}{2}})$ would be null if the analysis filter bank were ideal. In the case of a nonideal filter bank, ill-conditioned signals appear in the adaptive part of the cross filters (which model the term $[H(z^{\frac{1}{2}}) - H(-z^{\frac{1}{2}})]$), leading to slow convergence of the adaptive cross filters.

12.4.3 Closed-Loop Structure

An alternative subband adaptive-filtering realization is the closed-loop structure depicted in Fig. 12.11. In the closed-loop structure, the fullband output signal of the adaptive filter is reconstructed through a synthesis filter bank, and the overall error signal is computed and utilized in the objective function. The overall error is split into subbands, which are then used in the adaptation algorithm. In the closed-loop scheme we aim to minimize the fullband error energy. In this case, the NLMS updating equation is given by

$$\mathbf{w}_i(m+1) = \mathbf{w}_i(m) + \frac{2\mu}{\gamma + N_s\sigma_i^2(m)}\mathbf{u}_i^*(m-\Delta)e_i'(m) \tag{12.34}$$

where the fullband error is evaluated as $e(k) = d(k - \Delta L) - y(k)$, and $e_i'(m)$ corresponds to the ith component of the fullband error signal split into subbands. The delay Δ is key to compensate for the extra delay the input signal faces, due to the analysis and synthesis filter bank, with respect to the desired signal. The delay value is given by

$$\Delta = \lfloor \frac{2KM - 1}{L} \rfloor \tag{12.35}$$

where $\lfloor(\cdot)\rfloor$ denotes the integer part of $(\cdot)$, $2KM$ is the length of the subfilters of the analysis and synthesis filter banks, and K is a positive integer number. The closed-loop scheme allows for the

Algorithm 12.2

Closed-Loop Subband Adaptive-Filtering Algorithm

Initialization

$\mathbf{x}(0) = \mathbf{w}_l(0) = [0\,0 \ldots 0]^T$

$\gamma =$ small constant

$0 < \alpha \leq 0.1$

Do for each $x(iL)$ and $d(iL)$ given, for $i \geq 0$

$$\mathbf{u}(m) = \begin{bmatrix} \mathbf{E}_0 & \cdots & \mathbf{E}_{N_p} \end{bmatrix} \begin{bmatrix} \mathbf{x}(i) \\ \vdots \\ \mathbf{x}(i - N_p) \end{bmatrix}$$

$$\mathbf{y}(m) = \begin{bmatrix} \mathbf{W}_0 & \cdots & \mathbf{W}_{N_s} \end{bmatrix} \begin{bmatrix} \mathbf{u}(m) \\ \vdots \\ \mathbf{u}(m - N_s) \end{bmatrix}$$

$$y(k) = \begin{bmatrix} 1 & \cdots & 1 \end{bmatrix} \begin{bmatrix} \mathbf{R}_0 & \cdots & \mathbf{R}_{N_p} \end{bmatrix} \begin{bmatrix} \mathbf{y}(m) \\ \vdots \\ \mathbf{y}(m - N_p) \end{bmatrix}$$

$$e(k) = d(k - \Delta L) - y(k)$$

$$\mathbf{e}'(m) = \begin{bmatrix} \mathbf{E}_0 & \cdots & \mathbf{E}_{N_p} \end{bmatrix} \begin{bmatrix} \mathbf{e}(i) \\ \vdots \\ \mathbf{e}(i - N_p) \end{bmatrix}$$

Do for each for $0 \leq l \leq M - 1$

$$\sigma_l^2(m) = (1 - \alpha)\sigma_l^2(m - 1) + \alpha\,|u_l(m)|^2$$

$$\mathbf{w}_l(m + 1) = \mathbf{w}_l(m) + \frac{2\mu}{\gamma + N_s\sigma_l^2(m)}\mathbf{u}_l^*(m - \Delta)e_l'(m)$$

minimization of a cost function based on the fullband error signal, and guarantees that the algorithm converges to a minimum MSE.

The closed-loop algorithm is described in detail in Algorithm 12.2. Note that the matrix coefficient $\mathbf{R}_l$, for $l = 0, 1, \ldots, N_p$, represents the element of order l of the synthesis filter polyphase matrix, and $\mathbf{y}(m - l)$ is the subband adaptive-filter output vector at time instant $(m - l)$. A comparison between the two schemes shows that the open-loop scheme generates an excess MSE because it actually minimizes the subband error energy, whereas the closed-loop scheme minimizes the fullband error. On the other hand, since in the closed-loop scheme a delay is introduced by the synthesis filter bank, and by the analysis filter bank applied to the error signal $e(k)$, the adaptation algorithm uses past information about the error signal, which can be shown to slow down the convergence. In fact, this delay reduces the upper bound of μ that can be employed in the closed-loop algorithm. The recursive equations governing the convergence of the adaptive-filter coefficients of the closed-loop algorithm have the following general characteristics polynomial (see problem 8)

$$p(\Delta) = z^{\Delta+1} - z^{\Delta} + 2\mu\lambda_i = 0 \tag{12.36}$$

where Δ is the delay introduced by the filter banks and λ_i is related to the maximum eigenvalue of the autocorrelation matrix of the input signal in the ith subband. Considering the critical case of maximum eigenvalue $\lambda_{\max}$, the critical value of μ such that the zeros of equation (12.36) meet at the real axis is

$$\mu_{\text{crit}} = \frac{(\Delta-1)^{\Delta-1}}{2\lambda_{\max}\Delta^{\Delta}} \tag{12.37}$$

For higher values of μ the zeros move away from the real axis and eventually reach the unit circle at $\mu \approx 4.5\mu_{\text{crit}}$, see [17] and [16] for further details. Higher delays lead to lower values of μ. As a consequence, the closed-loop structures are more susceptible to convergence problems and less used in practice.

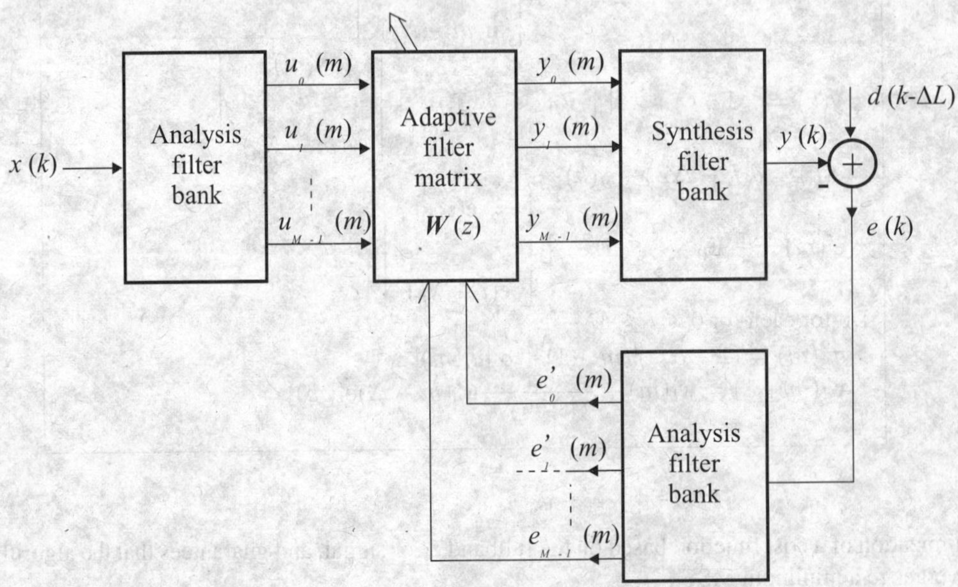

Figure 12.11 Closed-loop subband structure.

For the closed-loop structure, the excess MSE due to gradient noise (which tends to zero as $\mu \to 0$) is not related to the additional error resulting from the use of non-ideal filter banks. By making some simplifying assumptions, we can easily estimate the excess MSE in the closed-loop structure (the open-loop scheme follows similar analysis). The final result will closely follow the one for the standard LMS algorithm. If we consider that the input signal in each subband and the adaptive-filter coefficients are uncorrelated, and that the subfilters in the filter bank are frequency selective, we can calculate the excess MSE individually in each subband, and combine them to derive the overall excess MSE. The result is given by

$$\xi_{exc} \approx \sum_{i=0}^{M-1} \frac{\mu_i \sigma_{n_i}^2 \mathrm{tr}[\mathbf{U}_i]}{1 - \mu_i \mathrm{tr}[\mathbf{U}_i]} \tag{12.38}$$

where $\mathbf{U}_i = E[\mathbf{u}_i(k)\mathbf{u}_i^H(k)]$, $\sigma_{n_i}^2 \approx \sigma_n^2/M$, and $\mu_i = \frac{\mu}{\gamma + N_s \sigma_i^2}$. This equation provides a good estimate to the excess MSE when the assumptions are closely met. A more accurate estimate is not straightforward to obtain.

Example 12.1

Identify an unknown system with the following transfer function

$$H(z) = \frac{0.1z}{(z+0.9)} + \frac{0.08z}{(z^2+0.92)} + \frac{0.1z}{(z-0.9)}$$

The input signal is a uniformly distributed white noise with variance $\sigma_x^2 = 1$, and the measurement noise is Gaussian white noise uncorrelated with the input with variance $\sigma_n^2 = 10^{-3}$. The filter bank is a cosine-modulated type of length 32.

(a) Start with a fullband filter using the normalized LMS algorithm.

(b) Compare the results obtained with those using an open-loop subband adaptive filter with three bands. Plot the MSE for an average of five independent runs, including the local errors and the overall error.

Solution:

Fig. 12.12 shows the MSE for the fullband normalized LMS algorithm.

The impulse response of the unknown system has infinite length. However, since the samples after 90 are rather small, we use three subband filters of length 30 each. No cross filters are employed. The convergence factor in all subbands is $\mu = 0.1$, and the parameters of the normalized updating equation are given by: $\alpha = 0.1$ and $\gamma = 0.001$. The prototype filter coefficients of the cosine-modulated filter bank are given in Table 12.1. Figs. 12.13 and 12.14 depict the MSE measured in the subbands and the global error computed after reconstruction of the adaptive-filter output through the synthesis filter bank. As can be observed, the convergence speed of global and local errors are not reduced due to the aliasing effects caused by the analysis filter banks. The aliasing errors appear at the global error and cannot be cancelled by the synthesis filter bank. As we can observe in Fig. 12.12, the fullband normalized LMS algorithm achieves a larger reduction in the excess of MSE since in this case there are no aliasing effects. In both examples, some excess MSE is expected since the unknown system has infinite length.

□

Table 12.1 Coefficients of the prototype filter of the cosine modulated filter bank

n	$f(n)$	n	$f(n)$	n	$f(n)$	n	$f(n)$
0	0.000689	8	-0.023394	16	0.188567	24	-0.015614
1	-0.000316	9	-0.023179	17	0.163319	25	-0.005030
2	0.001608	10	-0.008268	18	0.119646	26	0.001726
3	0.003180	11	0.023394	19	0.069041	27	0.004631
4	0.004631	12	0.069041	20	0.023394	28	0.003180
5	0.001726	13	0.119646	21	-0.008268	29	0.001608
6	-0.005030	14	0.163319	22	-0.023179	30	-0.000316
7	-0.015614	15	0.188567	23	-0.023394	31	0.000689

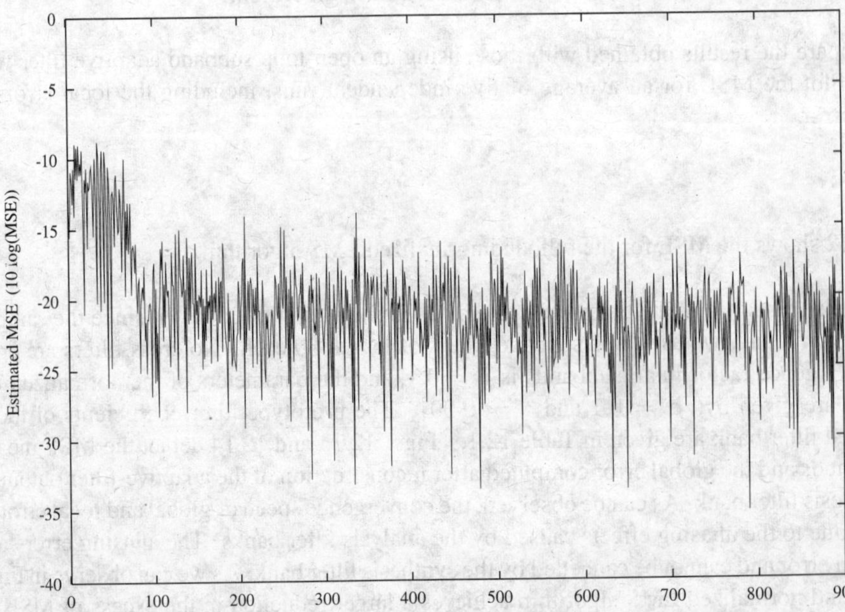

Figure 12.12 MSE in the fullband normalized LMS algorithm.

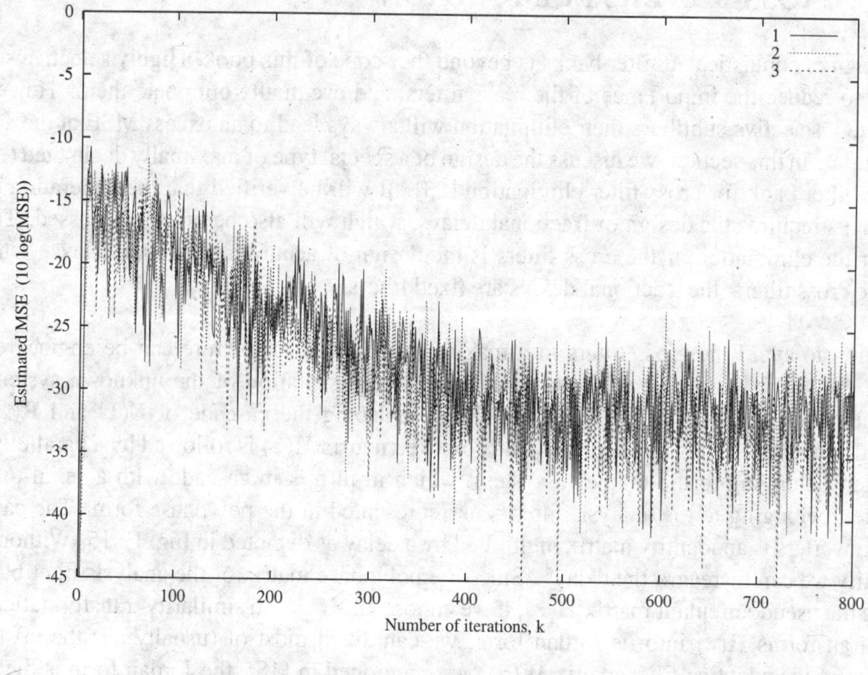

Figure 12.13 Subband errors in the open-loop structure.

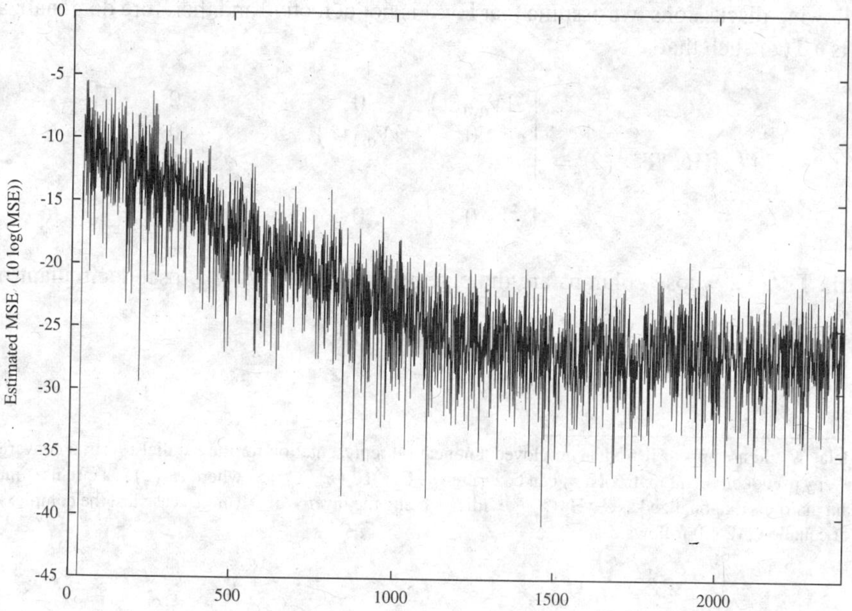

Figure 12.14 Global errors in the open-loop structure.

12.5 CROSS-FILTERS ELIMINATION

The design of sophisticated filter banks is beyond the scope of this book. Highly selective subfilters are key to reduce the importance of the cross filters, and eventually eliminate them. However, for moderately selective subfilters their elimination will always lead to an excess MSE at the adaptive-filter output. In this section, we discuss the design of a special type of maximally decimated ($M = L$) analysis filter bank for cross-filter elimination [23]. It will be verified that the generation of these filter banks requires the design of fractional delays, which will also be briefly discussed. The price paid for the elimination of the cross filters is the design of accurate fractional delays. Unlike the adaptive cross filters, the fractional delays are fixed filters.

A solution to avoid the cross filters in a maximally decimated structure can be engineered if we explore the special structure of the blocked matrix representation of the unknown system. This implementation is given in Fig. 12.7 with $\mathbf{H}(z)$ replacing the cascade of $\mathbf{E}(z)$ and $\mathbf{R}(z)$. In a subband adaptive-filtering configuration, this blocked matrix $\mathbf{H}(z)$ is followed by a parallel-to-serial converter, belonging to the unknown system, which in turn is in cascade with a serial-to-parallel converter, belonging to the analysis filter bank represented in the polyphase form. The cascade of these converters is an identity matrix multiplied by a delay as depicted in Fig. 12.15. Without loss of generality we can disregard the delay[4]. Since the polyphase matrix of the analysis filter bank $\mathbf{E}(z)$ follows the pseudocirculant matrix $\mathbf{H}(z)$, if we choose an $\mathbf{E}(z)$ as a similarity transformation matrix which transforms $\mathbf{H}(z)$ into its Jordan form, we can avoid most of (usually all) the off-diagonal elements of the adaptive-filter matrix $\mathbf{W}(z)$. As mentioned in [18], the Jordan form is the extreme effort in diagonalizing a matrix. The full diagonalization is impossible only for defective matrices.

In the following discussions, we assume that $\mathbf{H}(z)$ is not defective and therefore diagonalizable, that is, there is a $\mathbf{T}(z)$ such that

$$\mathbf{T}(z)\mathbf{H}(z)\mathbf{T}^{-1}(z) = \begin{bmatrix} \mathcal{W}_{o,0}(z) & 0 & \cdots & 0 \\ 0 & \mathcal{W}_{o,1}(z) & \cdots & 0 \\ \vdots & \vdots & \ddots & \vdots \\ 0 & 0 & \cdots & \mathcal{W}_{o,L-1}(z) \end{bmatrix} \tag{12.39}$$

The matrix $\mathbf{T}^{-1}(z)$, whose columns are the eigenvectors of any $L \times L$ pseudocirculant matrix, is given by[5]

$$\mathbf{T}^{-1}(z) = \mathbf{\Gamma}(z)\mathcal{F} \tag{12.40}$$

[4]This delay would not appear if we had employed a noncausal representation for the parallel-to-serial converter.

[5]In fact, any pseudocirculant matrix $\mathbf{H}(z)$ can be written as $\mathbf{\Gamma}(z)\mathbf{H}_c(z)\mathbf{\Gamma}^{-1}(z)$ where $\mathbf{H}_c(z)$ is a circulant matrix. Since any circulant matrix is diagonalized as $\mathcal{F}^*\mathbf{H}_c(z)\mathcal{F}$, with $\mathcal{F}^*$ being the inverse of $\mathcal{F}$ (in this case just the complex conjugate), the result of equation (12.40) follows.

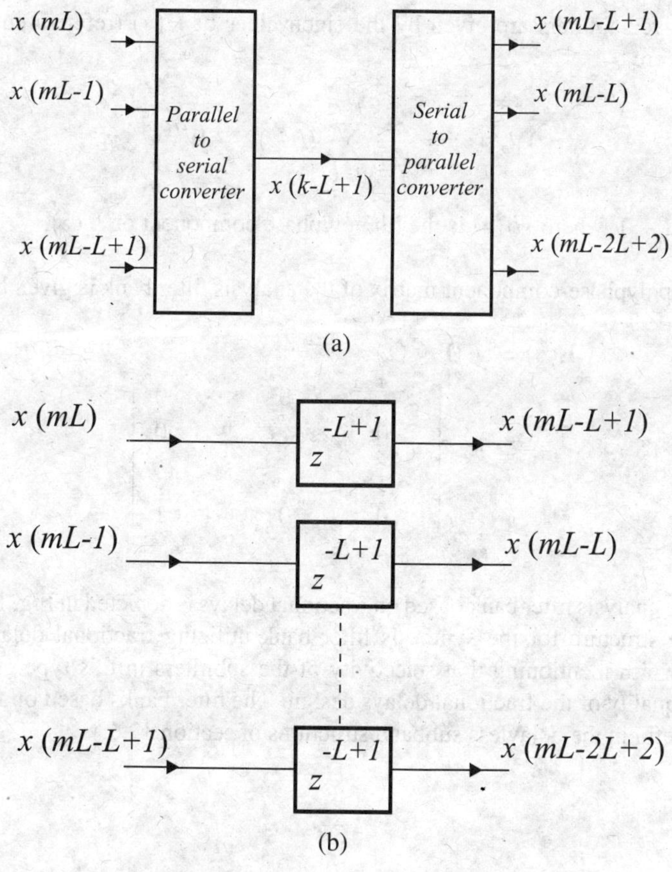

Figure 12.15 (a) Cascade connection of block converters, (b) Equivalent circuit.

where $\mathcal{F}$ is the $L \times L$ DFT matrix whose element (i, j), for $i, j = 0, 1, \ldots, L-1$, is given by $\frac{W^{ij}}{\sqrt{L}}$, where $W = \mathrm{e}^{\frac{-\jmath 2\pi}{L}}$, and

$$\mathbf{\Gamma}(z) = \begin{bmatrix} 1 & 0 & \cdots & 0 \\ 0 & z^{-\frac{1}{L}} & \cdots & 0 \\ \vdots & \vdots & \ddots & \vdots \\ 0 & 0 & \cdots & z^{-\frac{L-1}{L}} \end{bmatrix} \tag{12.41}$$

Now if we examine equations (12.32) and (12.39) more closely, we conclude that by choosing the polyphase matrix as $\mathbf{E}(z) = \mathbf{T}(z) z^{-\frac{L-1}{L}}$, where the delay was included in order to guarantee causality of the analysis filter bank, the cross filters are eliminated.

The optimal adaptive subfilters are given by the eigenvalues of $\mathbf{H}(z)$ (refer to equation (12.39)), whose expressions are

$$\mathcal{W}_{o,i}(z) = \frac{1}{\sqrt{L}} \sum_{l=0}^{L-1} H_l(z) z^{-\frac{l}{L}} W^{li} \tag{12.42}$$

for $i = 0, 1, \ldots, M-1$, where $H_l(z)$ is the lth polyphase component of $H(z)$.

In conclusion, the polyphase-component matrix of the analysis filter bank is given by

$$\begin{aligned} \mathbf{E}(z) &= \mathcal{F}^* \mathbf{\Gamma}^{-1}(z) z^{-\frac{L-1}{L}} \\ &= \mathcal{F}^* \begin{bmatrix} z^{-\frac{L-1}{L}} & 0 & \cdots & 0 \\ 0 & z^{-\frac{L-2}{L}} & \cdots & 0 \\ \vdots & \vdots & \ddots & \vdots \\ 0 & 0 & \cdots & 1 \end{bmatrix} \end{aligned} \tag{12.43}$$

The structure of the analysis filter bank based on fractional delays is depicted in Fig. 12.16. Similarly, we can derive the structure for the synthesis filter bank utilizing fractional delays illustrated in Fig. 12.17. It is worth mentioning that selectivity of the subfilters in this type of bank is highly dependent on the quality of the fractional delays design. The filter banks based on fractional delays are particularly useful in the delayless subband structures of section 12.6.

12.5.1 Fractional Delays

The review article about fractional delays [31] proposes several techniques for the approximation of a fractional delay. One of them consists of designing a symmetric Lth band filter (also known as a Nyquist filter), and keeping its lth polyphase component to represent the fractional delay $\tilde{\Delta} + l/L$. The delay $\tilde{\Delta}$ is the integer part of the group delay inherent to the FIR filter approximating a fractional delay. The Lth band filter has an impulse response that satisfies

$$h(kL) = \begin{cases} K, & k = 0 \\ 0, & \text{otherwise} \end{cases} \tag{12.44}$$

where K is a constant value. In the $\mathcal{Z}$-domain, the representation of $h(k)$ is

$$H(z) = K + z^{-1} E_1(z^L) + \cdots + z^{-(L-1)} E_{L-1}(z^L) \tag{12.45}$$

If $H(z)$ satisfies the above condition, it can be shown that [9]

$$\sum_{l=0}^{L-1} H(zW^l) = LK \tag{12.46}$$

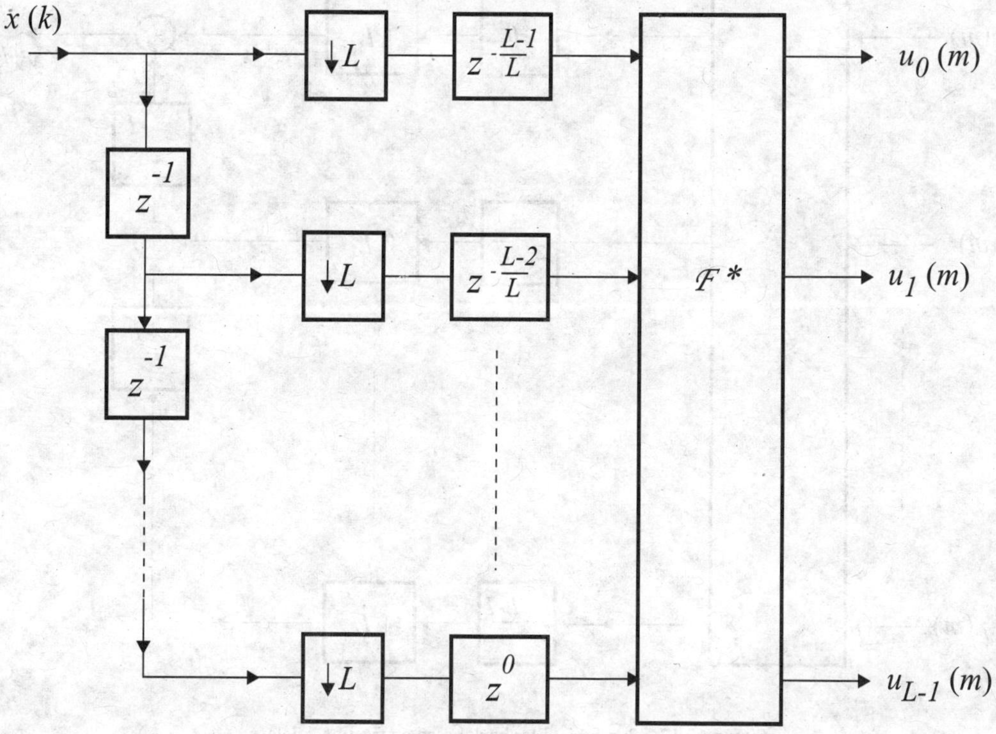

Figure 12.16 Analysis filter bank based on fractional delays.

where $W = e^{-\frac{j2\pi}{L}}$. The proof for the above relation is straightforward, if we just replace z by zW^l in equation (12.45), and compute the summation in equation (12.46).

Therefore, a natural proposition to eliminate adaptive cross filters is to design a DFT filter bank with a lowpass prototype filter given by an Lth band filter whose polyphase components approximate the fractional delays. The Lth band filter can be easily designed by using the so called eigenfilter approach for FIR filter approximation [9]. This approach allows the incorporation of the constraints inherent in the Nyquist filters. The Lth band filter is usually designed as a lowpass filter whose passband (ω_p) and stopband (ω_s) edges are symmetric with respect to the normalized frequency $\frac{\pi}{L}$, that is $\omega_p + \omega_s = 2\frac{\pi}{L}$. Although the fractional delays designed using Lth band filters are not very accurate, they can be considered acceptable for the delayless structures discussed in section 12.6.

Another simple FIR filter design to approximate the fractional delay is through the classical Lagrange interpolation formula. The interested reader should refer to [31].

Example 12.2

Repeat example 12.1 with a filter bank using fractional delays.

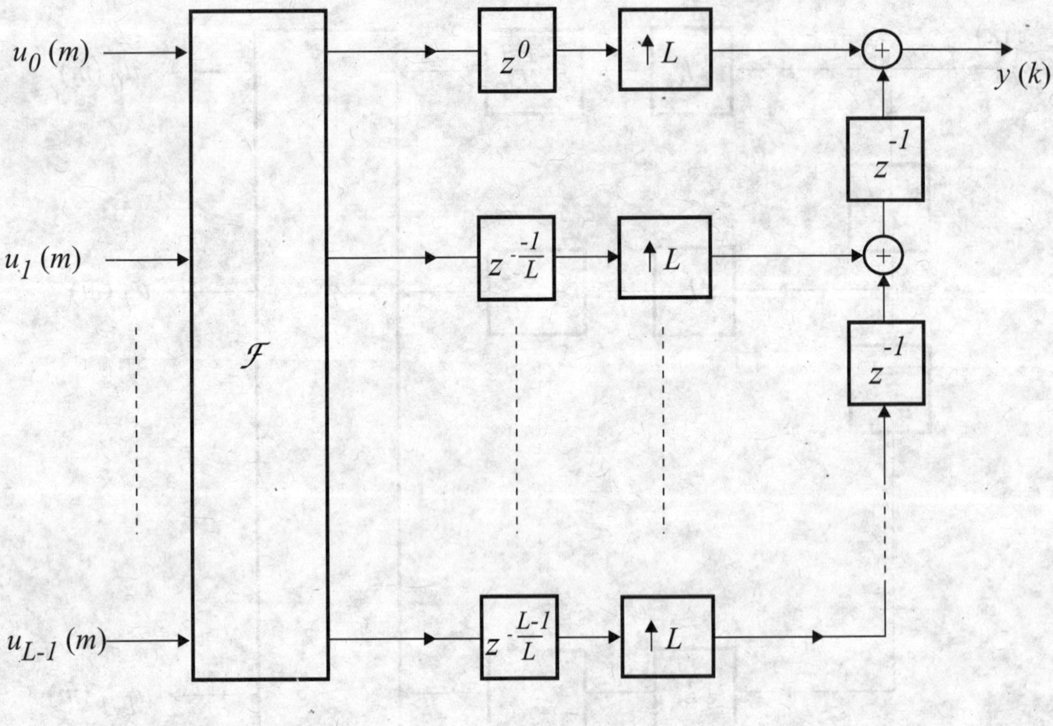

Figure 12.17 Synthesis filter bank based on fractional delays.

Solution:

For this example we design the fractional delays via a three-band filter. The length of the polyphase components is 9, with values given in Table 12.5.1. The length of the adaptive filters in the subbands is $N = 30$, the convergence factor in all subbands is $\mu = 0.1$, and parameters of the normalized updating equation are given by $\alpha = 0.1$ and $\gamma = 0.001$.

As can be observed in Figs. 12.18 and 12.19, the errors measured in the subbands and the global error are rather high due to the aliasing effects. Due to these effects, we can see in Fig. 12.20 that the magnitude response obtained after convergence resembles the unknown system response although the approximation is not very close.

□

Table 12.2 Coefficients of the fractional delays of the analysis filter bank

n	E_0	E_1	E_2
0	0.0000	0.0000	0.0000
1	-0.0072	-0.0117	0.0000
2	0.0320	0.0497	0.0000
3	-0.1090	-0.1592	0.0000
4	0.3880	0.8140	1.0000
5	0.8140	0.3880	0.0000
6	-0.1592	-0.1090	0.0000
7	0.0497	0.0320	0.0000
8	-0.0117	-0.0072	0.0000

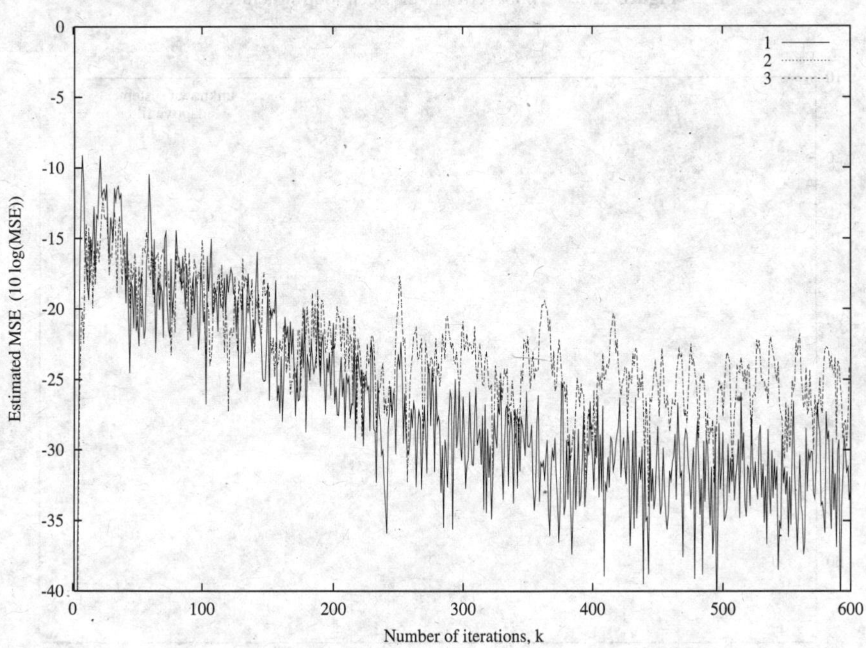

Figure 12.18 Subband errors in the open-loop structure.

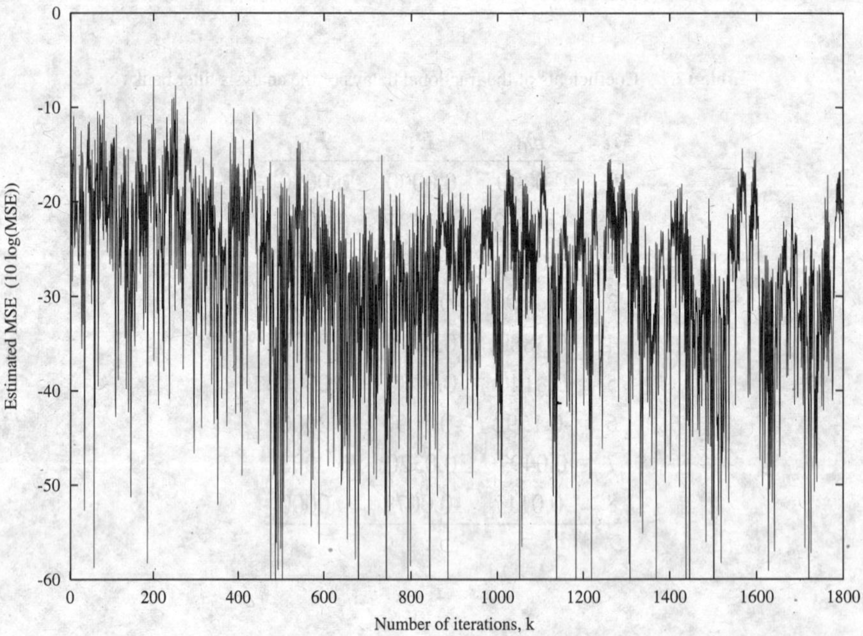

Figure 12.19 Global error in the open-loop structure.

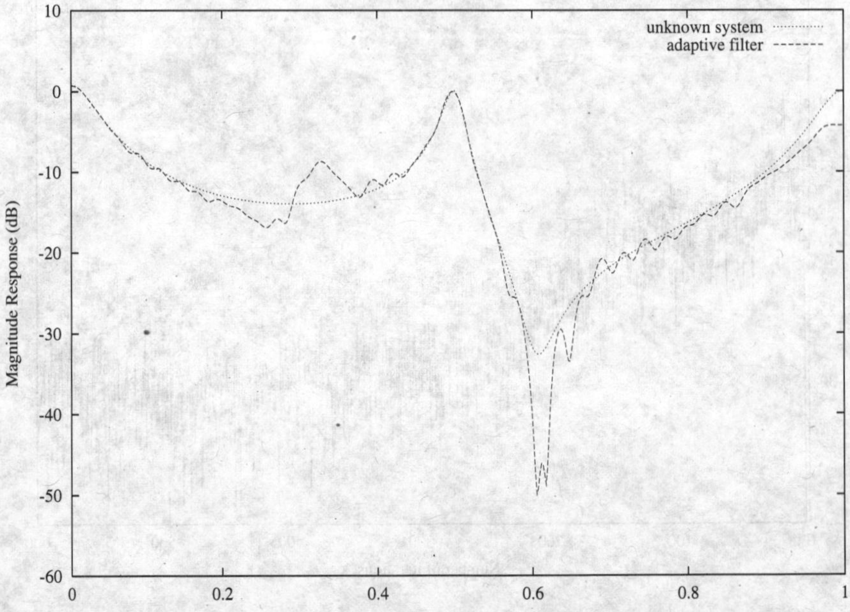

Figure 12.20 Magnitude responses of the unknown system and the obtained model at a given iteration.

12.6 DELAYLESS SUBBAND ADAPTIVE FILTERING

In the subband adaptive-filtering schemes presented so far, a delay is always introduced in the signal path due to the filter bank analysis and synthesis. In applications such as acoustic echo cancellation and active noise control, the delay is highly undesirable. In acoustic echo cancellation, the echo is not fully canceled and can be perceptually unacceptable. In active noise control the delay reduces the cancellation bandwidth [21].

In order to avoid the effect of signal path delay in these applications, we can avoid the synthesis filter bank and map the subband adaptive filters into a wideband filter, leading to the so called *delayless subband adaptive filters*. Several techniques to perform this mapping have been proposed [22]-[27], where the distinctive feature among them is the construction of each analysis filter bank and its corresponding subband to fullband mapping. In this section, we describe the delayless subband adaptive filter proposed in [23] which utilizes DFT-based filter banks with fractional delays discussed in this chapter. Fig. 12.21 depicts the general configuration of a delayless adaptive filter in subbands, employing a maximally decimated filter bank.

Equation (12.42) gives the coefficients of the optimal subband adaptive filters in each subband, for the open-loop scheme. The transfer functions of these subfilters represent the eigenvalues of a pseudocirculant matrix. Therefore, if we apply the inverse DFT to a vector whose elements are the transfer functions of the adaptive subfilters, we can recover the polyphase components estimates of the unknown system multiplied by fractional delays as described in the equation below

$$\frac{1}{\sqrt{L}}\begin{bmatrix} \hat{H}_0(z) \\ \hat{H}_1(z)z^{-\frac{1}{L}} \\ \vdots \\ \hat{H}_{L-1}(z)z^{-\frac{(L-1)}{L}} \end{bmatrix} = \mathcal{F}^*\left(\begin{bmatrix} \mathcal{W}_0(z) \\ \mathcal{W}_1(z) \\ \vdots \\ \mathcal{W}_{L-1}(z) \end{bmatrix}\right) \tag{12.47}$$

It should be noticed that in most cases the length of the adaptive subfilters is chosen as $\frac{N}{L}$, where N is the unknown system length. However, from the above equation some extra coefficient should be allotted to the subband adaptive filters in order to account for the fractional delays.

Since in our case any subfilter of the bank $F_i(z)$ has an inherent fractional delay, it is reasonable to conjecture that the product $F_i(z)z^{-\frac{i}{L}}$ represents a filter with one more sample than $F_i(z)$. Through a number of simulations, we concluded that a single coefficient is enough to perform this task in closed-loop schemes. As a consequence, the adaptive subfilters have length $N_s = \frac{N}{L} + 1$.

By denoting each element of the time-domain representation of $\mathcal{W}_i(z)$ as $w_{i,l}$, we can compute the previous equation in parts as follows

$$\begin{bmatrix} w'_{0,l} \\ w'_{1,l} \\ \vdots \\ w'_{L-1,l} \end{bmatrix} = \mathcal{F}^*\left(\begin{bmatrix} w_{0,l} \\ w_{1,l} \\ \vdots \\ w_{L-1,l} \end{bmatrix}\right) \tag{12.48}$$

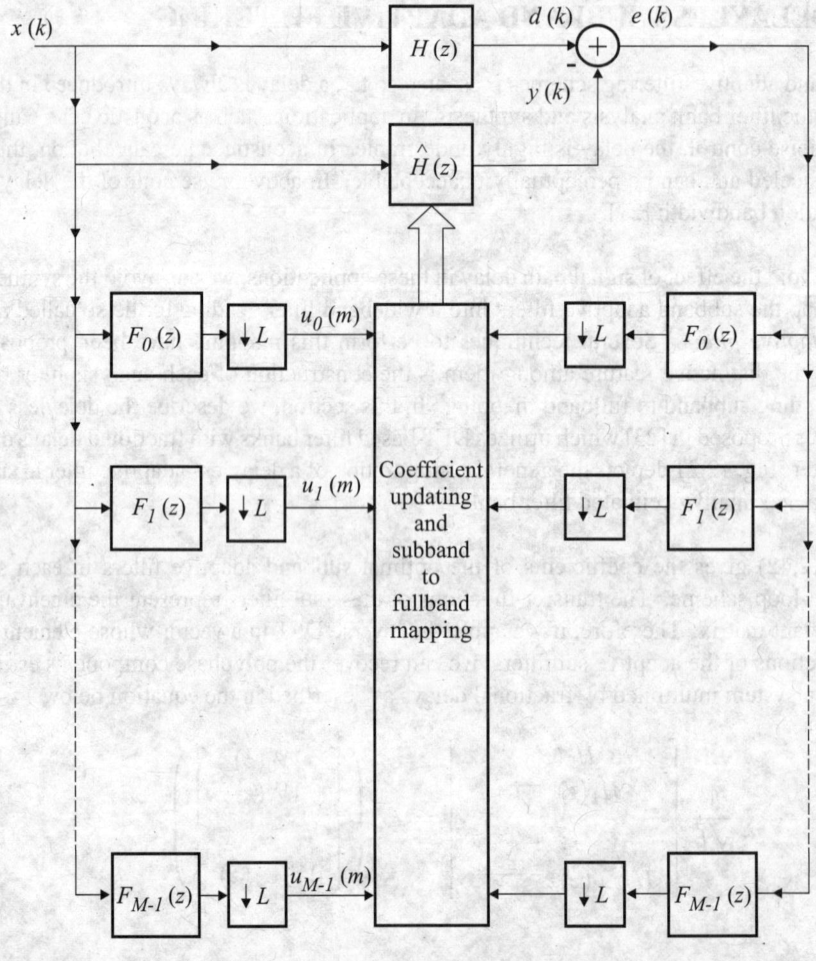

Figure 12.21 Delayless closed-loop subband structure.

for $l = 0, 1, \ldots, N_s - 1$. The polyphase component of the corresponding fullband adaptive filter is then given by $\frac{1}{\sqrt{L}}\hat{H}_i(z)z^{-\frac{i}{L}} = W_i'(z)$, where $W_i'(z)$ represents the $\mathcal{Z}$-transform of $w'_{i,l}$, and $\hat{H}_i(z)$ represents an estimate of the ith polyphase component of the unknown system. We can obtain the polyphase components $\hat{H}_i(z)$ from $W_i'(z)$, if we note that

$$\hat{H}_i(z)z^{-\frac{i}{L}}z^{-\frac{L-i}{L}} = \hat{H}_i(z)z^{-1} \tag{12.49}$$

for $i = 0, 1, \ldots, L-1$. The above discussion indicates that the cascade of $W_i'(z)$ with the fractional delay $E_{i-1}(z)$, $i = 1, \ldots, L-1$, leads to the polyphase component $\hat{H}_i(z)$ delayed by $\tilde{\Delta}+1$ samples and scaled by $\frac{1}{\sqrt{L}}$. Recall that $\tilde{\Delta}$ is the integer part of the group delay introduced by the design of the fractional delays.

Note that the impulse response of $\hat{H}_0(z)$ is represented by $w'_{0,l}$. Similarly, we can infer that

$$\mathcal{W}'_0(z) \approx \frac{1}{\sqrt{L}}\hat{H}_0(z)$$
$$\mathcal{W}'_i(z)E_{i-1}(z) \approx \frac{1}{\sqrt{L}}\hat{H}_i(z)z^{-(\tilde{\Delta}+1)} \tag{12.50}$$

for $i = 1, \ldots, M-1$. In conclusion, to obtain the first polyphase filter $\hat{H}_0(z)$ we simply discard the last sample of $w'_{0,l}$. For $\hat{H}_i(z)$, with $i = 1, \ldots, M-1$, we discard the first $\tilde{\Delta}+1$ samples and retain the next $N_s - 1$ samples (here the reader should recall that we used an extra coefficient for the adaptive subfilters to compensate for the fractional delay in the subfilter of the analysis bank). The fullband filter is then formed by

$$\hat{H}(z) = \sum_{i=0}^{L-1} \hat{H}_i(z^L)z^{-i} \tag{12.51}$$

The delayless closed-loop algorithm is described in detail in Algorithm 12.3, where $\mathbf{e}(mL)$ represents a length L block of the error signal at instant mL. The detailed structure is shown in Fig. 12.22. It is worth mentioning that the delayless closed-loop structure does not suffer as much from the stability problems inherent in the standard closed-loop subband structure. This is because we do not have to reconstruct the adaptive-filter output through a synthesis filter bank in order to generate the global error. This reconstruction originates part of the convergence problems of the standard closed-loop structure.

Example 12.3

Repeat example 12.1 using the closed-loop delayless structure whose filter banks employ fractional delays.

Solution:

For this example, we use the same parameters as Example 12.2. As can be observed in Figs. 12.23 and 12.24, the errors measured in the subbands and the global error are reduced despite the fact that the subfilters of the filter bank are not very selective. In this case, the delayless closed-loop structure is able to compensate for the limitations of the filter bank. Fig. 12.25 shows that the magnitude response obtained after convergence is very close to the unknown system response.

□

12.6.1 Computational Complexity

An interesting issue to illustrate the results of this chapter is to assess the overall computational complexity of the subband structure. The computational complexity is counted in multiplications per input sample, and considering that the product of complex values is implemented through four real multiplications. In the delayless subband structure, the overall computation consists of the components described below.

- The subband decomposition: It consists of one convolution of an N_{pr}-length prototype filters, which is the total number of coefficients required to realize all the fractional delays, and one L-point FFT for each block of L input samples. Assuming that the number of complex

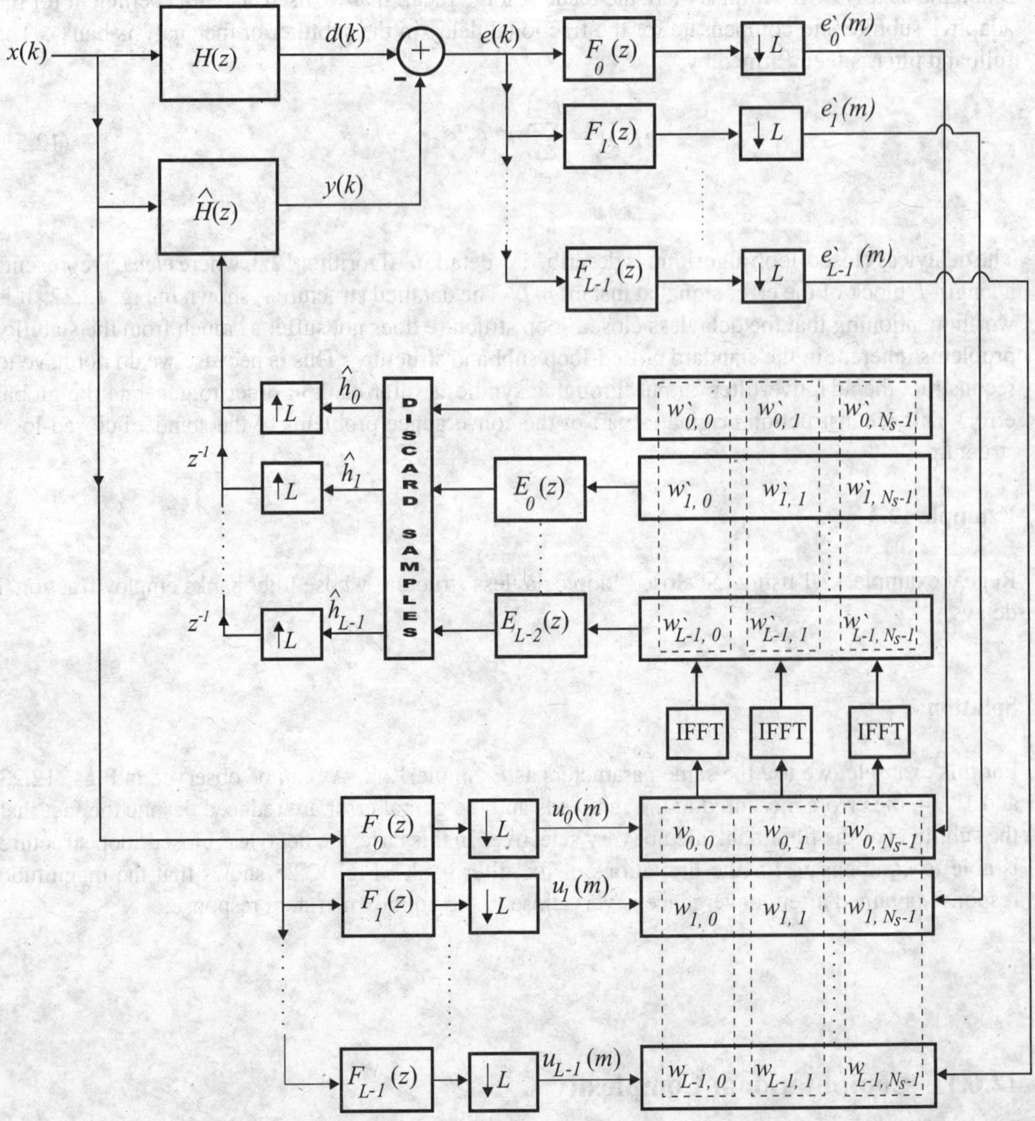

Figure 12.22 Detailed delayless closed-loop subband structure.

Algorithm 12.3

Delayless Closed-Loop Subband Adaptive-Filtering Algorithm

Initialization

$\mathbf{x}(0) = \mathbf{w}_l(0) = [0\,0 \ldots 0]^T$

$\gamma =$ small constant

$0 < \alpha \leq 0.1$

Do for each $x(iL)$ and $d(iL)$ given, for $i \geq 0$

$\mathbf{u}(m) = \mathcal{F}^* \left[\begin{array}{ccc} \mathbf{E}_0 & \cdots & \mathbf{E}_{N_p} \end{array}\right] \left[\begin{array}{ccc} \mathbf{x}(i) & \cdots & \mathbf{x}(i - N_p) \end{array}\right]^T$

where $\mathbf{E}_l$, for $l = 0, 1, \ldots, N_p$ are diagonal matrices whose elements are the lth element of the impulse response of the filter implementing the fractional delays, and N_p is the order of fractional delays implementation.

$$\begin{bmatrix} w'_{0,l} \\ w'_{1,l} \\ \vdots \\ w'_{L-1,l} \end{bmatrix} = \mathcal{F}^* \left(\begin{bmatrix} w_{0,l} \\ w_{1,l} \\ \vdots \\ w_{L-1,l} \end{bmatrix} \right)$$

Get $\frac{1}{\sqrt{L}} \hat{H}_0(z)$ by discarding the last sample of $w'_{0,l}$.

For $\frac{1}{\sqrt{L}} \hat{H}_i(z)$, with $i = 1, \ldots, L-1$, we discard the first $\tilde{\Delta} + 1$ samples of the impulse response corresponding to equation (12.50) and retain the following $N_s - 1$ samples.

$\hat{\mathbf{h}}(k)$ is the impulse response of $\hat{H}(z) = \sum_{i=0}^{L-1} \hat{H}_i(z^L) z^{-i}$.

$e(k) = d(k) - \hat{\mathbf{h}}^H(k)\mathbf{x}(k)$

$\mathbf{e}'(m) = \mathcal{F}^* \left[\begin{array}{ccc} \mathbf{E}_0 & \cdots & \mathbf{E}_{N_p} \end{array}\right] \left[\begin{array}{ccc} \mathbf{e}(i) & \cdots & \mathbf{e}(i - N_p) \end{array}\right]^T$

Do for each for $0 \leq l \leq L-1$

$\sigma_l^2(m) = (1-\alpha)\sigma_l^2(m-1) + \alpha\,|u_l(m)|^2$

$\mathbf{w}_l(m+1) = \mathbf{w}_l(m) + \frac{2\mu}{\gamma + N_s \sigma_l^2(m)} \mathbf{u}_l^*(m) e'_l(m)$

multiplications required to compute a L-point FFT is $\frac{L}{2} \log_2 L$, see [33], we obtain

$$\frac{2N_{pr}}{L} + 2\log_2 L \tag{12.52}$$

real multiplications per input sample for the two analysis filter banks. The symmetry of the IDFT for real signals allows us to process only half of the L channel complex signals. Therefore, we have to update $\frac{L}{2}$ adaptive filters.

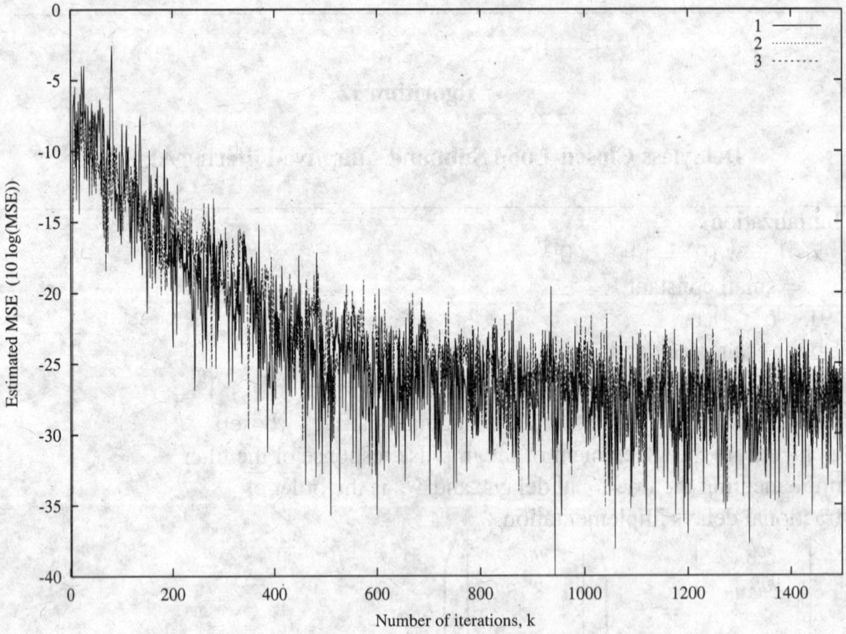

Figure 12.23 Subband errors in the closed-loop structure.

- The subband NLMS algorithm: Considering that we have to update $\frac{L}{2}$ adaptive filters of length $N_s = \frac{N}{L} + 1$ for every L input samples, the computational complexity entails

$$2\frac{N+L}{L} \tag{12.53}$$

real multiplications per input sample.

For the open-loop scheme an additional of $2\frac{N+L}{L}$ is required to evaluate the adaptive-filters outputs $y(m)$.

- The wideband filter convolution: There are some approaches to reduce the computational complexity of the wideband convolution as discussed in [22]. Here we consider only the direct implementation which entails N multiplications per output sample.

- The transformation from the subband adaptive filters to the wideband adaptive filter: It consists of N_s IFFTs and $L-1$ convolutions with the polyphase filters as indicated in equation (12.50). However, there is no need to perform the transformation for every L input samples, since in most applications the fullband adaptive-filter output cannot vary much faster then the length of filter impulse response. The computational cost is then given by

$$\frac{1}{r}\left[(\frac{N}{L}+1)\log_2 L + \frac{NN_{pr}(L-1)}{L^3}\right] \tag{12.54}$$

real multiplications per input sample, where rL represents how often the transformation is performed in terms of the number of input samples.

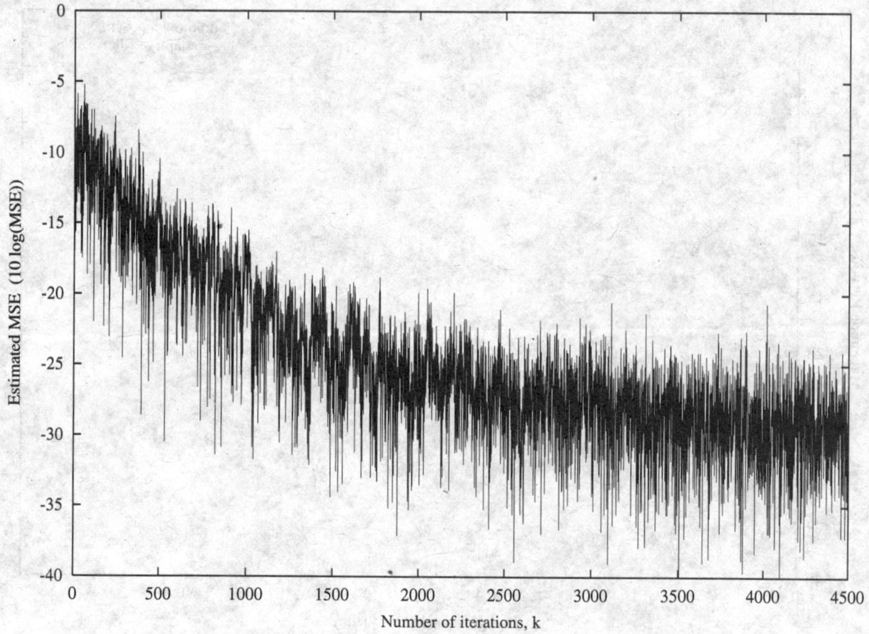

Figure 12.24 Global error in the closed-loop structure.

The overall computational complexity for the closed-loop scheme is

$$P_c = \frac{2N_{pr}}{L} + 2\log_2 L + \frac{2(N+L)}{L} + \frac{1}{r}\left[(\frac{N}{L}+1)\log_2 L + \frac{NN_{pr}(L-1)}{L^3}\right] + N \quad (12.55)$$

while for the open-loop scheme we have

$$P_o = \frac{2N_{pr}}{L} + 2\log_2 L + \frac{4(N+L)}{L} + \frac{1}{r}\left[(\frac{N}{L}+1)\log_2 L + \frac{NN_{pr}(L-1)}{L^3}\right] + N \quad (12.56)$$

12.7 FREQUENCY-DOMAIN ADAPTIVE FILTERING

Frequency-domain adaptive algorithms, which employ block processing in order to reduce the computational complexity associated with high-order adaptive filters, have been suggested in [34]. Such algorithms utilize FFTs to implement convolutions (for filtering) and correlations (for coefficient updating). More general block algorithms, in which the block size can be smaller than the order of the adaptive filter, have also been investigated [35]. Such approach, called *multidelay adaptive filter* (MDF) [37]-[40], utilizes adaptive filters in the bins (equivalent to the subbands), unlike the original frequency-domain adaptive-filtering algorithms that use a single adaptive coefficient in each bin. Like the subband adaptive filters discussed so far, frequency-domain adaptive filters can increase the convergence speed by decreasing the eigenvalue spread of the autocorrelation matrices of the signals

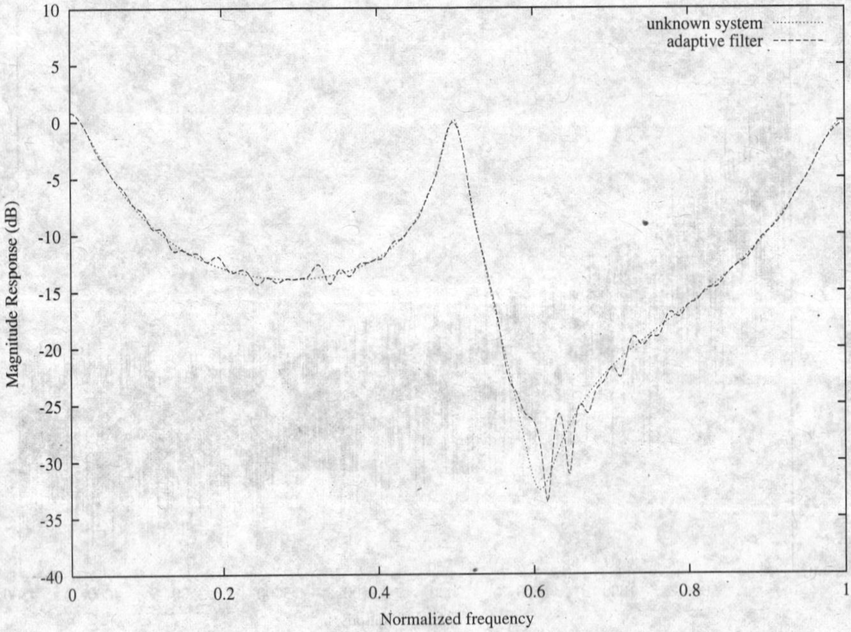

Figure 12.25 Magnitude responses of the unknown system and the obtained model at a given iteration.

at the inputs of the adaptive filters. In fact, the subband and the frequency-domain adaptive filters are closely related as will become clear in the sequel.

Let us consider the case where both the input and desired signals are presented in their corresponding blocked versions as described in subsection 12.3.6. The adaptive filter transfer function is represented by a blocked matrix denoted by $\hat{\mathbf{H}}(z)$. In this case, the adaptive-filter output is also represented in block form $\mathbf{y}(m)$, which in turn is compared with the desired signal block $\mathbf{d}(m)$. These vectors are defined as

$$\begin{aligned}\mathbf{y}(m) &= [y(mL)\, y(mL-1) \ldots y(mL-L+1)]^T \\ \mathbf{x}(m) &= [x(mL)\, x(mL-1) \ldots x(mL-L+1)]^T \\ \mathbf{d}(m) &= [d(mL)\, d(mL-1) \ldots d(mL-L+1)]^T\end{aligned} \tag{12.57}$$

In the particular case where the matrix $\hat{\mathbf{H}}(z)$ is 3×3, we have

$$\begin{aligned}\hat{\mathbf{H}}(z) &= \begin{bmatrix} \hat{H}_0(z) & \hat{H}_1(z) & \hat{H}_2(z) \\ z^{-1}\hat{H}_2(z) & \hat{H}_0(z) & \hat{H}_1(z) \\ z^{-1}\hat{H}_1(z) & z^{-1}\hat{H}_2(z) & \hat{H}_0(z) \end{bmatrix} \\ &= \hat{\mathbf{H}}_0(z)\hat{\mathbf{H}}_1(z)\end{aligned} \tag{12.58}$$

where $\hat{H}_i(z)$, $i = 0, 1, 2$, are the polyphase components of $W(z)$, and

$$\hat{\mathbf{H}}_0(z) = \begin{bmatrix} \hat{H}_0(z) & \hat{H}_1(z) & \hat{H}_2(z) & 0 & 0 & 0 \\ 0 & \hat{H}_0(z) & \hat{H}_1(z) & \hat{H}_2(z) & 0 & 0 \\ 0 & 0 & \hat{H}_0(z) & \hat{H}_1(z) & \hat{H}_2(z) & 0 \end{bmatrix}$$

$$\hat{\mathbf{H}}_1(z) = \begin{bmatrix} 1 & 0 & 0 \\ 0 & 1 & 0 \\ 0 & 0 & 1 \\ z^{-1} & 0 & 0 \\ 0 & z^{-1} & 0 \\ 0 & 0 & z^{-1} \end{bmatrix} \tag{12.59}$$

The last column of $\hat{\mathbf{H}}_0(z)$ and the last row of $\hat{\mathbf{H}}_1(z)$ were artificially added to generate a square circulant matrix in the sequel whose dimension can be designed to be a power of two, allowing the use of FFTs. The overall factorization of $\hat{\mathbf{H}}(z)$ as above described is crucial to derive the frequency-domain algorithm and the MDF in the sequel. It is worth noting that our presentation follows the embedding approach which was generalized in [41], and was indirectly employed in [42]. The embedding approach leads to a simpler derivation than those presented in early references [35]-[40].

The embedding approach starts by defining a circulant matrix $\hat{\mathbf{H}}_2(z)$ as follows

$$\hat{\mathbf{H}}_2(z) = \begin{bmatrix} \hat{H}_0(z) & \hat{H}_1(z) & \hat{H}_2(z) & 0 & 0 & 0 \\ 0 & \hat{H}_0(z) & \hat{H}_1(z) & \hat{H}_2(z) & 0 & 0 \\ 0 & 0 & \hat{H}_0(z) & \hat{H}_1(z) & \hat{H}_2(z) & 0 \\ 0 & 0 & 0 & \hat{H}_0(z) & \hat{H}_1(z) & \hat{H}_2(z) \\ \hat{H}_2(z) & 0 & 0 & 0 & \hat{H}_0(z) & \hat{H}_1(z) \\ \hat{H}_1(z) & \hat{H}_2(z) & 0 & 0 & 0 & \hat{H}_0(z) \end{bmatrix} \tag{12.60}$$

The matrix $\hat{\mathbf{H}}_0(z)$ is embedded into $\hat{\mathbf{H}}_2(z)$, that is:

$$\hat{\mathbf{H}}_0(z) = \begin{bmatrix} \mathbf{I}_L & \mathbf{0} \end{bmatrix} \hat{\mathbf{H}}_2(z) \tag{12.61}$$

where in the above equation we treat the general case, i.e., for block length equal to L instead of 3. Since the matrix $\hat{\mathbf{H}}_2(z)$ is circulant, it can be diagonalized by a DFT matrix as follows

$$\hat{\mathbf{H}}_2(z) = \mathcal{F}^*\mathbf{W}(z)\mathcal{F} \tag{12.62}$$

where $\mathbf{W}(z)$ is a diagonal matrix. If these diagonal elements are given by single complex coefficients, the resulting algorithm is the so-called frequency-domain algorithm, whereas for higher order filters the resulting algorithm is called MDF.

From equations (12.58), (12.61) and (12.62), we can relate the blocked matrix of the overall adaptive filter to the adaptive filter in the bins as follows

$$\hat{\mathbf{H}}(z) = \begin{bmatrix} \mathbf{I}_L & \mathbf{0} \end{bmatrix} \mathcal{F}^*\mathbf{W}(z)\mathcal{F}\hat{\mathbf{H}}_1(z) \tag{12.63}$$

In the frequency domain, the block output is given by

$$\mathcal{Z}[\mathbf{y}(m)] = \hat{\mathbf{H}}(z)\mathcal{Z}[\mathbf{x}(m)] \tag{12.64}$$

whereas the error signal vector is given by

$$\mathcal{Z}[\mathbf{e}(m)] = \mathcal{Z}[\mathbf{d}(m)] - \mathcal{Z}[\mathbf{y}(m)] \tag{12.65}$$

We use as an objective function the squared values of the error vector elements, that is

$$\xi = \sum_{i=0}^{L-1} |e_i(m)|^2 \tag{12.66}$$

In problem 16, the resulting gradient estimate for the set of coefficients placed at each bin is shown to be given by

$$\begin{aligned}\hat{\mathbf{g}}_{\mathbf{w},i}(m) &= -2\mathbf{u}_i^*(m)\left(\mathcal{F}\begin{bmatrix}\mathbf{I}_L\\ \mathbf{0}\end{bmatrix}\mathbf{e}(m)\right)_i\\ &= -2\mathbf{u}_i^*(m)\left(\tilde{\mathbf{e}}(m)\right)_i\\ &= -2\mathbf{u}_i^*(m)\tilde{e}_i(m)\end{aligned} \tag{12.67}$$

where $\mathbf{u}_i^*(m)$ represents the complex conjugate of the data vector stored in ith bin, at instant m, and $(\tilde{\mathbf{e}}(m))_i$ denotes the ith element of vector $\tilde{\mathbf{e}}(m)$ with

$$\tilde{\mathbf{e}}(m) = \mathcal{F}\begin{bmatrix}\mathbf{I}_L\\ \mathbf{0}\end{bmatrix}\mathbf{e}(m)$$

It is worth mentioning that the data vectors are calculated as follows

$$\mathcal{Z}\begin{bmatrix}\mathbf{u}_0^T(m)\\ \vdots\\ \mathbf{u}_{2L-1}^T(m)\end{bmatrix} = \mathcal{F}\hat{\mathbf{H}}_1(z)\mathcal{Z}\left[\mathbf{x}(m)\cdots\mathbf{x}(m-N_s+1)\right] \tag{12.68}$$

In this case, the NLMS updating equation is given by

$$\mathbf{w}_i(m+1) = \mathbf{w}_i(m) + \frac{2\mu}{\gamma + \sigma_i^2(m)}\mathbf{u}_i^*(m)\tilde{e}_i(m) \tag{12.69}$$

for $i = 0, 1, \ldots, N_s$, where N_s is the length of the adaptive filter at the output of bin i, and $\sigma_i^2(m) = (1-\alpha)\sigma_i^2(m-1) + \alpha\left|u_i(m)\right|^2$, with $0 < \alpha \leq 0.1$ and γ is a small constant as established before.

If we examine the first row of the matrices in equation (12.62) and use the fact that $\mathcal{F}$ is a symmetric matrix, it is straightforward to infer that

$$\begin{bmatrix} \hat{H}_0(z) \\ \hat{H}_1(z) \\ \vdots \\ \hat{H}_{L-1}(z) \\ 0 \\ \vdots \\ 0 \end{bmatrix} = \frac{1}{\sqrt{2L}} \mathcal{F} \begin{bmatrix} W_0(z) \\ W_1(z) \\ \vdots \\ W_{2L-1}(z) \end{bmatrix} \tag{12.70}$$

where $W_i(z)$, for $i = 0, 1, \ldots, 2L-1$, are the transfer functions of the subfilters of $\mathbf{W}(z)$. The above equation shows that the adaptive filters in the bins must be constrained such that $\hat{\mathbf{H}}_2(z)$ contains the estimates of the polyphase components of the unknown system. Note that, in the update equation (12.69), it is not guaranteed that this constraint is satisfied.

As a solution, we can enforce the constraint in the adaptive-filter updating with the help of equation (12.70), as follows. First define the matrices that include all the coefficients and data of all subfilters:

$$\mathcal{W}(m) = \begin{bmatrix} \mathbf{w}_0^T(m) \\ \mathbf{w}_1^T(m) \\ \vdots \\ \mathbf{w}_{2L-1}^T(m) \end{bmatrix}$$

$$\mathcal{U}(m) = \begin{bmatrix} \mathbf{u}_0^H(m) \\ \mathbf{u}_1^H(m) \\ \vdots \\ \mathbf{u}_{2L-1}^H(m) \end{bmatrix} \tag{12.71}$$

and a diagonal matrix $\mathcal{E}(m)$ whose nonzero elements are the entries of vector $\tilde{\mathbf{e}}(m)$.

In matrix form the updating equation (12.69) can be rewritten as

$$\mathcal{W}(m+1) = \mathcal{W}(m) + 2\mu\mathbf{\Sigma}^{-2}(m)\mathcal{E}(m)\mathcal{U}(m) \tag{12.72}$$

where $\mathbf{\Sigma}^{-2}(m)$ is a diagonal matrix whose elements are $\frac{1}{\gamma+\sigma_i^2(m)}$, with $\sigma_i^2(m) = (1-\alpha)\sigma_i^2(m-1) + \alpha\,|u_i(m)|^2$.

A constrained version of the above equation can be derived by observing equation (12.70). The resulting algorithm consists of enforcing the constraint in the update equation as follows (see problem 17)

$$\mathcal{W}_c(m+1) = \mathcal{F}^* \begin{bmatrix} \mathbf{I}_L \\ \mathbf{0} \end{bmatrix} \begin{bmatrix} \mathbf{I}_L & \mathbf{0} \end{bmatrix} \mathcal{F}\mathcal{W}(m+1) \tag{12.73}$$

The above algorithm is widely known as the constrained frequency-domain algorithm. The original constrained algorithm was derived for a single coefficient per bin, not for the more general MDF.

Also, the particular version of the algorithm presented here corresponds to the overlap-save version, in which the constraints are included in order to guarantee that the internal DFTs perform linear convolutions on the signals involved. By examining equation (12.63), the reader should note that the transform applied to the input signal after it is filtered by $\hat{\mathbf{H}}_1(z)$ has length $2L$, whereas in the calculation of the adaptive-filter output block, L signals are discarded due to the product by $[\mathbf{I}_L \ \mathbf{0}]$. This reflects the overlap-save characteristic of the algorithm. The block diagram related to this algorithm is depicted in Fig. 12.26.

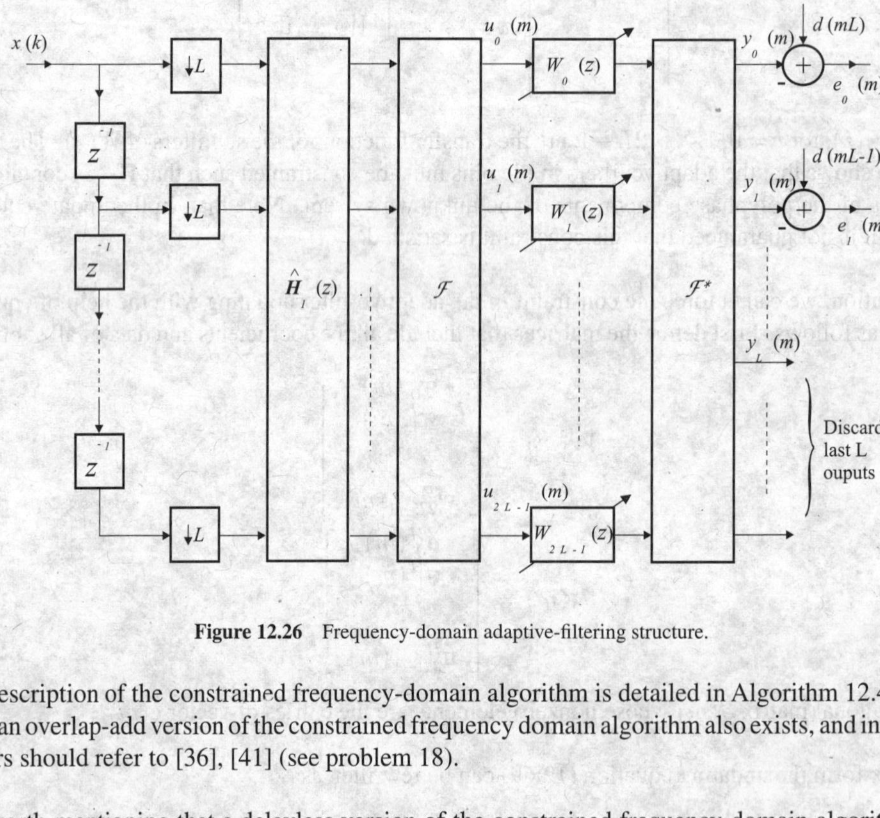

Figure 12.26 Frequency-domain adaptive-filtering structure.

The description of the constrained frequency-domain algorithm is detailed in Algorithm 12.4. Likewise, an overlap-add version of the constrained frequency domain algorithm also exists, and interested readers should refer to [36], [41] (see problem 18).

It is worth mentioning that a delayless version of the constrained frequency-domain algorithm follows directly from equation (12.70) which implements the mapping from the subband filter to the polyphase components of the fullband estimate. It is also important to note that although the embedding approach presented here was based on the DFT, it can also be employed using other class of transforms such as DCT, DST and Hartley transform. Though these alternative transforms require more cumbersome embedding formulations, they do not require complex arithmetic when environment signals are not represented by complex numbers [41].

Example 12.4

Repeat example 12.1 using the multidelay structure with $L = 64$ and the frequency-domain structure. Choose the appropriate order for the subfilters in the multidelay case.

Algorithm 12.4

Constrained Frequency-Domain Algorithm

Initialization

$\gamma =$ small constant

$0 < \alpha \leq 0.1$

Do for each $x(iL)$ and $d(iL)$ given, for $i \geq 0$

$$\begin{bmatrix} \mathbf{u}_0^T(m) \\ \vdots \\ \mathbf{u}_{2L-1}^T(m) \end{bmatrix} = \mathcal{F}\hat{\mathbf{H}}_1(z)\left[\mathbf{x}(m)\cdots\mathbf{x}(m-N_s+1)\right]$$

the dimension of $\mathcal{F}$ is $2L$.

$$\mathbf{e}(m) = \mathbf{d}(m) - \begin{bmatrix} \mathbf{I}_L & \mathbf{0} \end{bmatrix}\mathcal{F}^* \begin{bmatrix} \mathbf{w}_{c,0}^T(m)\mathbf{u}_0^*(m) \\ \mathbf{w}_{c,1}^T(m)\mathbf{u}_1^*(m) \\ \vdots \\ \mathbf{w}_{c,2L-1}^T(m)\mathbf{u}_{2L-1}^*(m) \end{bmatrix}$$

where $\mathbf{w}_{c,l}$ are the constrained adaptive-filter coefficients of the $(l-1)$th subband, that is the $(l-1)$th row of $\mathcal{W}_c(m)$.

$$\tilde{\mathbf{e}}(m) = \mathcal{F}\begin{bmatrix} \mathbf{I}_L \\ \mathbf{0} \end{bmatrix}\mathbf{e}(m)$$

$$\sigma_i^2(m) = (1-\alpha)\sigma_i^2(m-1) + \alpha\left|u_i(m)\right|^2$$

$$\mathcal{W}(m+1) = \mathcal{W}(m) + 2\mu\mathbf{\Sigma}^{-2}(m)\mathcal{E}(m)\mathcal{U}(m)$$

$$\mathcal{W}_c(m+1) = \mathcal{F}^*\begin{bmatrix} \mathbf{I}_L \\ \mathbf{0} \end{bmatrix}\begin{bmatrix} \mathbf{I}_L & \mathbf{0} \end{bmatrix}\mathcal{F}\mathcal{W}(m+1)$$

Solution:

For the frequency-domain algorithm we use a block size of 90 and the following parameters: $\alpha = 0.5$, $\gamma = 0.001$, and $\mu = 0.2$. The average MSE obtained from five runs is -29.2 dB.

Fig. 12.27 depicts the global MSE where the algorithm converges rather fast to the minimum MSE. Fig. 12.28 shows that the magnitude response obtained after convergence approaches the unknown system response.

For the multidelay filter, we use a block size of 18 with five coefficients in each bin and the following parameters: $\alpha = 0.1$, $\gamma = 0.001$, and $\mu = 0.4$. The average MSE obtained from five runs is -29.0 dB. Fig. 12.29 depicts the global MSE where we observe that the MDF algorithm also converges fast to the minimum MSE. Fig. 12.30 shows that the magnitude response obtained after convergence does not approach so closely the unknown system response.

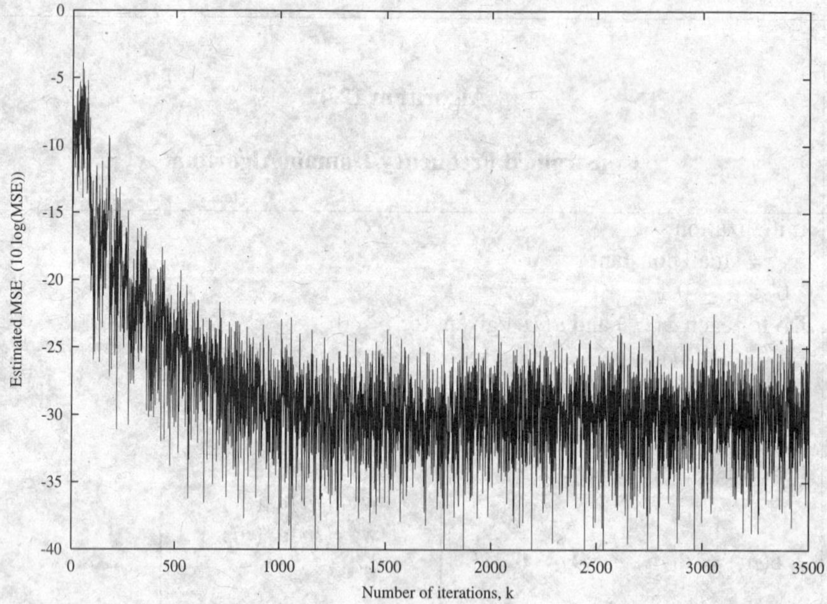

Figure 12.27 Global error of the frequency-domain structure.

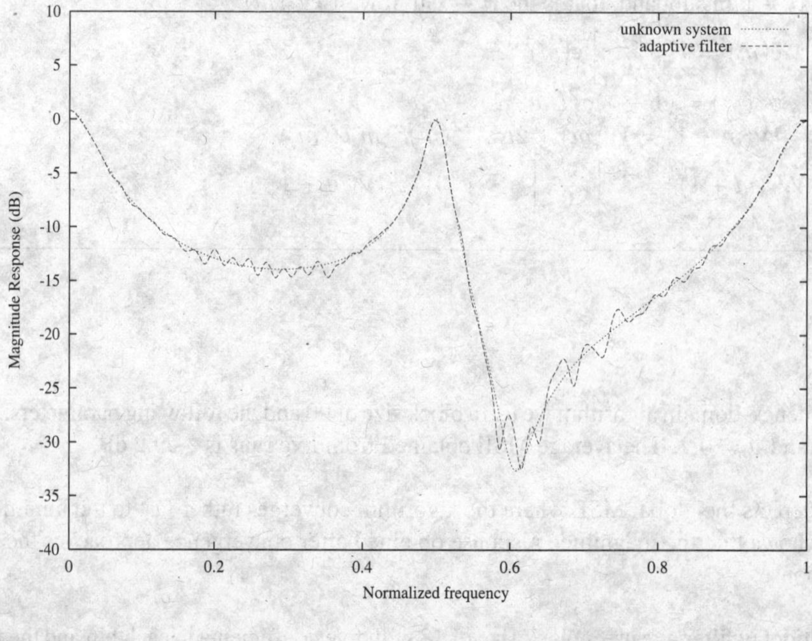

Figure 12.28 Magnitude responses of the unknown system and the obtained model at a given iteration.

□

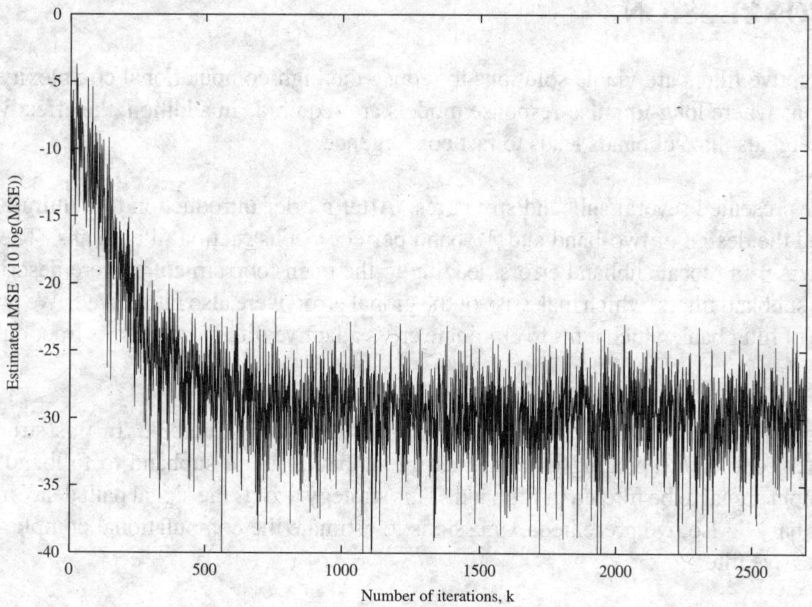

Figure 12.29 Global error of the MDF structure.

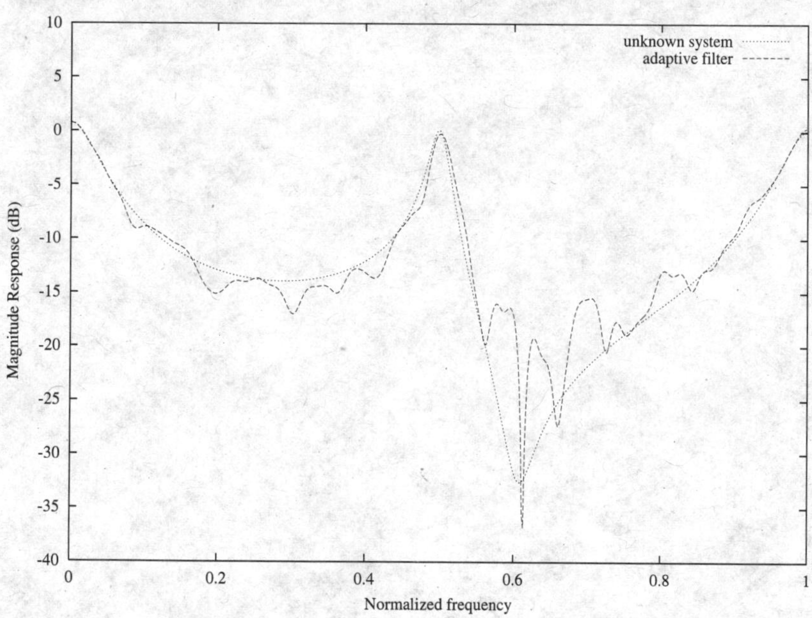

Figure 12.30 Magnitude responses of the unknown system and the obtained model at a given iteration, MDF case.

12.8 CONCLUSION

Subband adaptive filters are viable solutions to reduce the high-computational complexity inherent in applications where long-impulse-response models are required. In addition, the effective split of the internal signals into subbands leads to fast convergence.

This chapter presented several subband structures. After a brief introduction to multirate systems, we discussed the design of two-band and M-band perfect reconstruction filter banks. The subband adaptive filters using local subband errors, leading to the open-loop structure, were described. The closed-loop subband filters, which make use of the global error, were also introduced. We presented a special type of filter bank which aims to eliminate cross adaptive filters and utilizes fractional delays.

Another type of subband adaptive filter is based on a delayless structure. In this structure, the adaptive-filter coefficient updating is performed in subbands and a subband to fullband mapping allows the input signal to be filtered in fullband. This strategy avoids the signal path delay introduced by the filter bank. Also, we presented expressions to estimate the computational complexity of the subband adaptive filters.

Finally, we presented the frequency-domain and multidelay structures, which employ block processing and are closely related to subband adaptive filters. These structures further lead to reduced computational complexity.

12.9 REFERENCES

1. A. Gilloire, "Experiments with sub-band acoustic echo cancellers for teleconferencing," *Proc. IEEE Intern. Conf. Acoust., Speech, Signal Processing*, pp. 2141-2144, Dallas, TX, April 1987.

2. A. Gilloire and M. Vetterli, "Adaptive filtering in subbands with critical sampling: analysis, experiments, and application to acoustic echo cancellation," *IEEE Trans. on Signal Processing*, vol. 40, pp. 1862-1875, Aug. 1992.

3. W. Kellermann, "Analysis and design of multirate systems for cancellation of acoustical echoes," *Proc. IEEE Intern. Conf. Acoust., Speech, Signal Processing*, pp. 2570-2573, New York, NY, April 1988.

4. Y. Lu and J. M. Morris, "Gabor expansion for adaptive echo cancellation," *IEEE Signal Processing Magazine*, vol.16, pp. 68-80, March 1999.

5. E. Hänsler and G. U. Schmidt, "Hands-free telephones - joint control of echo cancellation and post filtering," *Signal Processing*, vol. 80, pp. 2295-2305, Nov. 2000.

6. M. R. Petraglia and S. K. Mitra, "Performance analysis of adaptive filter structures based on subband decomposition," *Proc. IEEE Intern. Symp. on Circuits and Systems*, pp. 60-63, Chicago, IL, May 1993.

7. P. L. De León, II and D. M. Etter, "Experimental results with increased bandwidth analysis filters in oversampled, subband acoustic echo cancellers," *IEEE Signal Processing Letters*, vol. 2, pp. 1-3, Jan. 1995.

8. E. A. B. da Silva and P. S. R. Diniz, "Time-Varying Filters," *Encyclopedia of Electrical and Electronics Engineering*, Editor: John G. Webster, John Wiley & Sons, New York, NY, vol. 22, pp. 249-274, 1999.

9. P. P. Vaidyanathan, *Multirate Systems and Filter Banks*, Prentice-Hall, Englewood Cliffs, NJ, 1993.

10. M. Vetterli and J. Kovačević, *Wavelets and Subband Coding*, Prentice-Hall, Englewood Cliffs, NJ, 1995.

11. H. Bölcskei and F. Hlawatsch, "Oversampled cosine modulated filter banks with perfect reconstruction," *IEEE Trans. on Signal Processing*, vol. 45, pp. 1057-1071, Aug. 1998.

12. V. P. Sathe and P. P. Vaidyanathan, "Effects of multirate systems on the statistical properties of random signals," *IEEE Trans. on Signal Processing*, vol. 41, pp. 131-146, Jan. 1993.

13. Y. G. Yang, N. I. Cho, and S. U. Lee, "On the performance analysis and applications of subband adaptive digital filters," *Signal Processing*, vol. 41, pp. 295-307, 1995.

14. M. R. Petraglia, R. G. Alves, and P. S. R. Diniz, "New structures for adaptive filtering in subbands with critical sampling," *IEEE Trans. on Signal Processing*, vol.48 , pp. 3316-3327, Dec. 2000.

15. M. R. Petraglia, R. G. Alves, and P. S. R. Diniz, "Convergence analysis of an oversampled subband adaptive filtering structure with local errors," *Proc. IEEE Intern. Symp. on Circuits and Systems*, pp. I-563-I-566, Geneve, Switzerland, May 2000.

16. M. R. Petraglia, R. G. Alves, and P. S. R. Diniz, "Convergence analysis of an oversampled subband adaptive filtering structure with global error," *Proc. IEEE Intern. Conf. Acoust., Speech, Signal Processing*, pp. 468-471, Istanbul, Turkey, June 2000.

17. J. R. Treichler, S. L. Wood, and M. G. Larimore, "Convergence rate limitations in certain frequency-domain adaptive filters," *Proc. IEEE Intern. Conf. Acoust., Speech, Signal Processing*, pp. 960-963, Scotland, May 1989.

18. G. Strang, *Linear Algebra and Its Applications*, Academic Press, New York, NY, 1980.

19. S. S. Pradhan and V. U. Reddy, "A new approach to subband adaptive filtering," *IEEE Trans. on Signal Processing*, vol. 47, pp. 655-664, March 1999.

20. Y. Higa, H. Ochi, and S. Kinjo,"A subband adaptive filter with the statistically optimum analysis filter bank," *IEEE Trans. on Circuits and Systems II: Analog and Digital Signal Processing*, vol. 45, pp. 1150-1154, Aug. 1998.

21. S. M. Kuo and D. R. Morgan, *Active Noise Control Systems*, John Wiley & Sons, New York, NY, 1996.

22. D. R. Morgan and M. J. C. Thi, "A delayless subband adaptive filter architecture," *IEEE Trans. on Signal Processing*, vol. 43, pp. 1819-1830, Aug. 1995.

23. R. Merched, P. S. R. Diniz, and M. R. Petraglia, "A delayless alias-free subband adaptive filter structure," *IEEE Trans. on Signal Processing*, vol. 47, pp. 1580-1591, June 1999.

24. R. Merched, P. S. R. Diniz, and M. R. Petraglia, "A delayless alias-free subband adaptive filter structure," *Proc. 1997 IEEE Intern. Symposium on Circuits and Systems*, Hong-Kong, pp. 2329-2332, June 1997.

25. N. Hirayama, H. Sakai, and S. Miyagi, "Delayless subband adaptive filtering using the Hadamard transform," *IEEE Trans. on Signal Processing*, vol. 47, pp. 1731-1734, June 1999.

26. S. Ohno and H. Sakai, "On Delayless subband adaptive filtering by subband/fullband transforms," *IEEE Signal Processing Letters*, vol. 6, pp. 236-239, Sept. 1999.

27. K. Nishikawa and H. Kiya, "Conditions for convergence of a delayless subband adaptive filter and its efficient implementation," *IEEE Trans. on Signal Processing*, vol. 46, pp. 1158-1167, April 1998.

28. U. Iyer, M. Nayeri, and H. Ochi,"Polyphase based adaptive structure for adaptive filtering and tracking," *IEEE Trans. on Circuits and Systems II: Analog and Digital Signal Processing*, vol. 43, pp. 220-232, March 1996.

29. F. G. V. Resende, Jr., P. S. R. Diniz, K. Tokuda, M. Kaneko, and A. Nishihara, "LMS-based algorithms with multi-band decomposition of the estimation error applied to system identification," *IEICE Trans. Fundamentals*, Special Issue on Digital Signal Processing, Japan, vol. E00-A, pp. 1376-1383, Aug. 1997.

30. F. G. V. Resende, Jr., P. S. R. Diniz, K. Tokuda, M. Kaneko, and A. Nishihara, "New adaptive algorithms based on multi-band decomposition of the error signal," *IEEE Trans. on Circuits and Systems II: Analog and Digital Signal Processing*, vol. 45, pp. 592-599, May 1998.

31. T. I. Laakso, V. Välimäki, M. Karjalainen, and U. K. Laine, "Splitting the unit delay," *IEEE Signal Processing Magazine*, vol.13, pp. 30-60, Jan. 1996.

32. I.-S. Lin and S. K. Mitra,"Overlapped block digital filtering," *IEEE Trans. on Circuits and Systems II: Analog and Digital Signal Processing*, vol. 43, pp. 586-596, Aug. 1996.

33. P. S. R. Diniz, E. A. B. da Silva, and S. L. Netto, *Digital Signal Processing: System Analysis and Design*, Cambridge University Press, New York, NY, 2002.

34. G. A. Clark, S. R. Parker, and S. K. Mitra, "A unified approach to time- and frequency-domain realization of FIR adaptive digital filters," *IEEE Trans. on Acoust., Speech, Signal Processing*, vol. ASSP-31, pp. 1073-1083, Oct. 1983.

35. P. C. Sommen, "On the convergence properties of a partitioned block frequency domain adaptive filter (PBFDAF)," *Proc. European Signal Processing Conf.*, pp. 201-203, Barcelona, Spain, Sept. 1990.

36. J. J. Shynk, "Frequency-domain and multirate adaptive filtering," *IEEE Signal Processing Magazine*, vol. 9, pp. 15-37, Jan. 1992.

37. J.-S. Soo and K. Pang, "Multidelay block frequency domain adaptive filter," *IEEE Trans. on Acoust., Speech, Signal Processing*, vol. 38, pp. 373-376, Feb. 1990.

38. B. Fahang-Boroujeny, "Analysis and efficient implementation of partitioned block LMS filters," *IEEE Trans. on Signal Processing*, vol. 44, pp. 2865-2868, Nov. 1996.

39. E. Moulines, O. A. Amrane, and Y. Grenier, "The generalized multidelay adaptive filter: structure and convergence analysis," *IEEE Trans. on Signal Processing*, vol. 43, pp. 14-28, Jan. 1995.

40. M. de Couville and P. Duhamel, "Adaptive filtering in subbands using a weighted criterion," *IEEE Trans. on Signal Processing*, vol. 46, pp. 2359-2371, Sep. 1998.

41. R. Merched and A. H. Sayed, "An embedding approach to frequency-domain and subband adaptive filtering," *IEEE Trans. on Signal Processing*, vol. 48, pp. 2607-2619, Sept. 2000.

42. K. Eneman and M. Moonen, "Hybrid subband/frequency-domain adaptive filters," *Signal Processing*, vol. 81, pp. 117-136, 2001.

12.10 PROBLEMS

1. Show the validity of equation (12.1).
2. Design a linear-phase two-band filter bank of order 42 using the approach described in subsection 12.3.1.

3. Design a uniform linear-phase 8-band filter bank having at least 40dB of stopband attenuation, using a hierarchical filter bank.

4. Design a uniform 8-band filter bank having at least 40dB of stopband attenuation, using the cosine-modulated method.

5. Design a fractional delay via a Nyquist filter having at least 60dB of stopband attenuation.

6. Use an open-loop subband adaptive filter with four bands to identify a system with the transfer function given below. The input signal is a uniformly distributed white noise with variance $\sigma_x^2 = 1$, and the measurement noise is Gaussian white noise uncorrelated with the input with variance $\sigma_n^2 = 10^{-3}$. The filter bank is a cosine-modulated type of length 64.

$$H(z) = \frac{0.1z}{(z+0.9)} + \frac{0.1z}{(z-0.9)}$$

Choose the order of the equivalent FIR filter as the one for which the envelop of the unknown system impulse response falls below $\frac{1}{1000}$ of its leading term. Plot the MSE for an average of five independent runs, including the local errors and the overall error.

7. Repeat the previous problem using a closed-loop algorithm and interpret the results.

8. Show that the recursive equation governing the convergence of the adaptive coefficients in the closed-loop structure has the characteristic polynomial of equation (12.36).

9. For a prototype filter of length 256, and 32 subbands, calculate and plot the ratio between the computational complexities of the subband and fullband implementations, for $N = 256, 512, 1024, 2048$ and 4096. Consider the maximally decimated case as well as the cases where $L = M - 1$, $L = \frac{3M}{4}$, and $L = \frac{M}{2}$. Assume we are using a simple DFT filter bank (which is similar to the filter bank using fractional delays where these delays are replaced by a transfer function equal to one) and consider the cases of open-loop and closed-loop structures.

10. Replace the structure in problem 6 by the closed-loop and open-loop delayless structures with the fractional delays designed via a Nyquist filter of order 64.

11. In a system identification problem, the input signal is a uniformly distributed white noise with variance $\sigma_{n_x}^2 = 0.1$, filtered by an all-pole filter given by

$$H_{n_x}(z) = \frac{z}{z-0.95}$$

The unknown system is a 300th-order FIR filter whose impulse response is identical to the first 301 impulse response samples of the transfer function described by

$$H(z) = \frac{0.00756z^2}{(z^2 - 1.960636z + 0.9849357)}$$

Choose the appropriate parameters, run an ensemble of 5 experiments, and plot the average learning curve. Use the delayless subband filter using fractional delays in the open-loop scheme, with 8-bands.

12. Repeat problem 11 using the closed-loop structure.

13. Prove that the expressions for the computational complexity of the subband adaptive filters in equations (12.52) to (12.56) are valid.

14. Solve problem 6 using the frequency-domain structure with $L = 64$.

15. Solve problem 6 using the multidelay structure with $L = 16$. Choose the appropriate order for the subfilters.

16. Prove the validity of equations (12.67), (12.68), and (12.70). Hint: Create a block-diagonal matrix of subband input signals consisting of

$$\text{diag}\{\mathbf{u}_0^T(m)\ \mathbf{u}_1^T(m)\ \cdots\ \mathbf{u}_{2L-1}^T(m)\}$$

and a vector containing all the elements of the subband adaptive filters

$$\begin{bmatrix} \mathbf{w}_0(m) \\ \mathbf{w}_1(m) \\ \vdots \\ \mathbf{w}_{2L-1}(m) \end{bmatrix}$$

17. Demonstrate how the relation below enforces the constraint of equation (12.70).

$$\mathcal{F}^* \begin{bmatrix} \mathbf{I}_L \\ \mathbf{0} \end{bmatrix} \begin{bmatrix} \mathbf{I}_L & \mathbf{0} \end{bmatrix} \mathcal{F}$$

18. Derive an overlap-add version of the frequency-domain algorithm using the embedding strategy in which a 3×3 matrix $\hat{\mathbf{H}}(z)$ can be written as

$$\hat{\mathbf{H}}(z) = \hat{\mathbf{H}}_3(z)\hat{\mathbf{H}}_4(z)$$

where

$$\hat{\mathbf{H}}_3(z) = \begin{bmatrix} 0 & 0 & 1 & 0 & 0 \\ z^{-1} & 0 & 0 & 1 & 0 \\ 0 & z^{-1} & 0 & 0 & 1 \end{bmatrix}$$

$$\hat{\mathbf{H}}_4(z) = \begin{bmatrix} \hat{H}_2(z) & 0 & 0 \\ \hat{H}_1(z) & \hat{H}_2(z) & 0 \\ \hat{H}_0(z) & \hat{H}_1(z) & \hat{H}_2(z) \\ 0 & \hat{H}_0(z) & \hat{H}_1(z) \\ 0 & 0 & \hat{H}_0(z) \end{bmatrix}$$

13

BLIND ADAPTIVE FILTERING

13.1 INTRODUCTION

There are a number of applications where the reference signal is either not available or consists of a training signal that in communication systems implies in reduction of useful data transmission. In those cases, we should utilize some alternative objective functions applied to the available data as well as some knowledge related to the nature (properties) of the signals involved.

In this chapter, some adaptive-filtering algorithms are presented which do not utilize reference signal that are collectively known as blind adaptive-filtering algorithms. The algorithms are also called training-less or unsupervised algorithms since their learning do not include any reference or training signal. This chapter makes no attempt to cover this subject in breadth and in depth, but the interested reader can consult some books [1]-[4] for further details.

There are two main types of blind signal processing procedures widely discussed in the literature, namely blind source separation and blind deconvolution. In the former case several signal sources are mixed by an unknown environment and the objective of the blind signal processor is to separate these signal sources [2]. On the other hand, the blind deconvolution aims at removing the effect of a linear time-invariant system on a signal source where the only assumptions are the observation of the signal before the deconvolution process and the probability density of the input signal source.

Blind deconvolution is obviously closely related to blind equalization, and the distinction lies on the fact that in the equalization case it is usually assumed that the input signal belongs to a prescribed finite set (constellation) and the channel is a continuous-time channel. These features of the equalization setup are assets that can be exploited by allowing nonlinear channel equalization solutions, whereas blind deconvolution employs linear solutions because its input signal cannot be considered to belong to a finite set constellation. However, it is fact that several solutions for both problems are closely related and here we emphasize the blind equalization case.

P.S.R. Diniz, *Adaptive Filtering*, DOI: 10.1007/978-0-387-68606-6_13,

In blind equalization the channel model is either identified explicitly or implicitly. The algorithms utilizing as objective function the minimization of the MSE or generating a zero-forcing (ZF) solution[1] in general do not estimate the channel model explicitly. On the other hand, nonlinear solutions for channel equalization such as maximum likelihood sequence detector (MLSD) [8] and the DFE require explicit estimation of the channel model.

As a rule, the blind signal processing algorithms utilize second and higher order statistics indirectly or explicitly. The high-order statistics are directly employed in algorithms based on cummulants, see [9] for details, and they usually have slow convergence and high complexity. There is yet another class of algorithms based on models originated from information theory [3].

This chapter deals with blind algorithms utilizing high-order statistics implicitly for the single-input single-output (SISO) equalization case, e.g. constant modulus algorithm (CMA), and algorithms employing second-order statistics for the single-input multi-output (SIMO) equalization case. Unfortunately the SISO blind solutions have some drawbacks related to the multiple minima solutions, slow convergence, and difficulties in equalizing channels with nonminimum phase[2]. In the SIMO case we are usually dealing with oversampled received signal, that is, the received signal is sampled at rate multiple of the symbol rate (at least twice). Another SIMO situation is whenever we use multiple receive antennas that can be proved to be equivalent to oversampling. Such sampling higher than baud rate results in received signals which are cyclostationary allowing the extraction of phase information of the channel. In the case of baud rate sampling and WSS inputs, the received signal is also WSS and only minimum-phase channels can be identified from second-order statistics since the channel phase information is lost. Under certain assumptions the SIMO configuration allows the identification of the channel model as well as blind channel equalization utilizing only second-order statistics. In particular, this chapter presents the Godard, CM, and Sato algorithms for the SISO case. We also discuss some properties related to the error surface of the CMA. Then we derive the blind CM affine projection algorithm which is then applied to the SISO and SIMO setups.

13.2 CONSTANT-MODULUS ALGORITHM

In this section we present a family of blind adaptive-filtering algorithms that minimizes the distance between the modulus of the equalizer output and some prescribed constant values, without utilizing a reference signal. These constant values are related to the modulus of constellation symbols, denoted by $\mathcal{C}$, of typical modulations utilized in many digital communication systems. The earlier blind equalization proposals addressed the case of Pulse Amplitude Modulation (PAM) for the case the channel model is considered a linear time-invariant Single-Input Single-Output (SISO) system [5]-[6], operating at symbol rate. This approach was latter generalized in [7] by modifying the objective function to consider higher order statistics of the adaptive-filter output signal that accommodates the case of Quadrature Amplitude Modulation (QAM).

[1] In the ZF solution the equalized signal is forced to be equal to the transmitted signal, a solution not recommended whenever the environment noise is not negligible, due to noise enhancement. The ZF equalizer aims at estimating a channel inverse in order to eliminate intersymbol interference.

[2] Channels whose discrete-time models have poles and zeros outside the unit circle.

Let's assume here that symbols denoted by $s(k)$ are transmitted through a communication channel. The channel impulse response described by $h(k)$ convolves with the sequence $s(k)$ generating the received signal given by

$$x(k+J) = s(k)h(J) + \left(\sum_{l=-\infty,\, l\neq k}^{k+J} s(l)h(k+J-l) \right) + n(k+J) \tag{13.1}$$

where J denotes the channel time delay which will be considered zero without loss of generality. The transmitted signals $s(k)$ belong to a set of possible symbols, that is $s(k) \in \mathcal{C}$, with $\mathcal{C}$ representing the constellation set, defined by the chosen constellation such as PAM[3] and the complex QAM. The symbol occurrence is uniformly distributed over the defined elements of the constellation. In the following we present the Godard algorithm which relies on a high-order statistics property of the chosen constellation to define its updating mechanism.

13.2.1 Godard Algorithm

The general objective of the Godard algorithm utilizing the criterion proposed in [7] is to minimize

$$\begin{aligned} \xi_{\text{Godard}} &= E\left[(|\mathbf{w}^H(k)\mathbf{x}(k)|^q - r_q)^p\right] \\ &= E\left[(|y(k)|^q - r_q)^p\right] \\ &= E\left[e^p_{\text{Godard}}(k)\right] \end{aligned} \tag{13.2}$$

with

$$r_q = \frac{E[|s(k)|^{2q}]}{E[|s(k)|^q]} \tag{13.3}$$

where q and p are positive integers. The value of r_q defines the level which $|y(k)|^q$ should approach, with a penalization error powered by p.

The simple stochastic gradient version of this algorithm can be obtained by differentiating the objective function of equation (13.2) with respect to $\mathbf{w}^*(k)$. The resulting updating equation is given by

$$\begin{aligned} \mathbf{w}(k+1) &= \mathbf{w}(k) - \frac{1}{2}\,\mu\, p\, q\, (|y(k)|^q - r_q)^{p-1}\, |y(k)|^{q-2}\, y^*(k)\, \mathbf{x}(k) \\ &= \mathbf{w}(k) - \frac{1}{2}\,\mu\, p\, q\, e^{p-1}_{\text{Godard}}(k)\, |y(k)|^{q-2}\, y^*(k)\, \mathbf{x}(k) \end{aligned} \tag{13.4}$$

The detailed description of the Godard algorithm is provided by Algorithm 13.1.

[3] The M-ary PAM constellation points are represented by $s_i = \tilde{a}_i$, with $\tilde{a}_i = \pm\tilde{d}, \pm 3\tilde{d}, \ldots, \pm(\sqrt{M}-1)\tilde{d}$. The parameter $\tilde{d}$ represents half of the distance between two points in the constellation.

Algorithm 13.1 **Godard Algorithm**
Initialization Choose p and q $\mathbf{x}(0) = \mathbf{w}(0) = \text{random vectors}$ $r_q = \frac{E[\lvert s(k)\rvert^{2q}]}{E[\lvert s(k)\rvert^{q}]}$ Do for $k > 0$ $y(k) = \mathbf{w}^H(k)\mathbf{x}(k)$ $e_{\text{Godard}}(k) = \lvert y(k)\rvert^q - r_q$ $\mathbf{w}(k+1) = \mathbf{w}(k) - \frac{1}{2}\,\mu\, p\, q\, e_{\text{Godard}}^{p-1}(k)\,\lvert y(k)\rvert^{q-2}\, y^*(k)\,\mathbf{x}(k)$

13.2.2 Constant-Modulus Algorithm

For $q = p = 2$ in the Godard framework, the objective function of equation (13.2) corresponds to the constant-modulus algorithm (CMA) whose objective function is described by

$$\begin{aligned} E\left[e_{\text{CMA}}^2(k)\right] &= E\left[(|\mathbf{w}^H(k)\mathbf{x}(k)|^2 - r_2)^2\right] \\ &= E\left[(|y(k)|^2 - r_2)^2\right] \end{aligned} \tag{13.5}$$

In this case,

$$r_2 = \frac{E[|s(k)|^4]}{E[|s(k)|^2]} \tag{13.6}$$

meaning that whenever the input symbols have constant modulus, the CM error minimization aims at keeping the modulus $|y(k)|^2$ as close as possible to the constant value of r_2. For the CMA, the stochastic gradient update equation is given by

$$\begin{aligned} \mathbf{w}(k+1) &= \mathbf{w}(k) - 2\mu\,(|y(k)|^2 - r_2)\, y^*(k)\,\mathbf{x}(k) \\ &= \mathbf{w}(k) - 2\mu\, e_{\text{CMA}}(k)\, y^*(k)\,\mathbf{x}(k) \end{aligned} \tag{13.7}$$

Algorithm 13.2 describes in detail the CM algorithm.

13.2.3 Sato Algorithm

A historically important objective function somewhat related to the case of the Godard algorithm above is the so-called Sato algorithm whose objective function is defined as

$$e_{\text{Sato}}(k) = y(k) - \text{sgn}[y(k)]r_1 \tag{13.8}$$

Algorithm 13.2

Constant-Modulus Algorithm

Initialization
$\mathbf{x}(0) = \mathbf{w}(0) = \text{random vectors}$
$r_2 = \frac{E[|s(k)|^4]}{E[|s(k)|^2]}$
Do for $k \geq 0$
$y(k) = \mathbf{w}^H(k)\mathbf{x}(k)$
$e_{\text{CMA}}(k) = |y(k)|^2 - r_2$
$\mathbf{w}(k+1) = \mathbf{w}(k) - 2\mu\, e_{\text{CMA}}(k)\, y^*(k)\, \mathbf{x}(k)$

Algorithm 13.3

Sato Algorithm

Initialization
$\mathbf{x}(0) = \mathbf{w}(0) = \text{random vectors}$
$r_1 = \frac{E[|s(k)|^2]}{E[|s(k)|]}$
Do for $k \geq 0$
$y(k) = \mathbf{w}^H(k)\mathbf{x}(k)$
$e_{\text{Sato}}(k) = y(k) - \text{sgn}[y(k)]r_1$
$\mathbf{w}(k+1) = \mathbf{w}(k) - \mu\, e_{\text{Sato}}(k)\, \mathbf{x}(k)$

where $\text{sgn}[y] = \frac{y}{|y|}$ such that for $y = 0$, $\text{sgn}[y] = 1$. Its update equation is described by

$$\begin{aligned}\mathbf{w}(k+1) &= \mathbf{w}(k) - \mu\,(y(k) - \text{sgn}[y(k)]r_1)\,\mathbf{x}(k) \\ &= \mathbf{w}(k) - \mu\, e_{\text{Sato}}(k)\, \mathbf{x}(k)\end{aligned} \tag{13.9}$$

In this case, the target is that the equalized signal $y(k)$ follows the sign of the transmitted symbol, that is, this algorithm follows the decision direction whenever the input signal is a binary PAM signal. The Sato algorithm was the first blind adaptive equalizer taking into consideration PAM transmission signals with multilevel. Algorithm 13.3 describes step by step the Sato algorithm.

13.2.4 Error Surface of CMA

In this subsection we derive an expression for the CMA error surface for a simple and yet illustrative case, where both the symbol constellation as well as the adaptive-filter coefficients are real valued. Let's assume the simplest equalization problem where the unknown channel is modeled as

$$H(z) = \frac{\kappa z}{z + a} \tag{13.10}$$

In a noiseless environment this channel has an ideal equalizer (zero forcing) given by

$$\begin{aligned} W(z) &= \pm z^{-i}\left(w_0 + w_1 z^{-1}\right) \\ &= \pm\frac{z^{-i}}{\kappa}[1 + az^{-1}] \end{aligned} \tag{13.11}$$

where i is a nonnegative integer. For $i = 0$ it leads to an equalized signal with zero delay. For the CMA case, the objective function in this particular example can be written as

$$\begin{aligned} \xi_{\text{CMA}} &= E\left\{[|y(k)|^2 - r_2]^2\right\} \\ &= E[|y(k)|^4] - 2E[|y(k)|^2]r_2 + r_2^2 \end{aligned} \tag{13.12}$$

The required expected values for the above equation are given by

$$E[|y(k)|^2] = (w_0^2 + w_1^2)\frac{\kappa^2 E[|s(k)|^2]}{1-a^2} - 2w_0 w_1 \frac{a\kappa^2 E[|s(k)|^2]}{1-a^2} \tag{13.13}$$

$$\begin{aligned} E[|y(k)|^4] &= (w_0^4 + w_1^4)\left[\frac{\kappa^4 E[|s(k)|^4]}{1-a^4} + \frac{6a^2\kappa^4\{E[|s(k)|^2]\}^2}{(1-a^4)(1-a^2)}\right] \\ &\quad +6w_0^2 w_1^2\left\{a^2\left[\frac{\kappa^4 E[|s(k)|^4]}{1-a^4} + \frac{6a^2\kappa^4\{E[|s(k)|^2]\}^2}{(1-a^4)(1-a^2)}\right] + \frac{\kappa^2\{E[|s(k)|^2]\}^2}{1-a^2}\right\} \\ &\quad -4w_0 w_1^3 a\left\{\left[\frac{\kappa^4 E[|s(k)|^4]}{1-a^4}\right] + \frac{6a^2\kappa^4\{E[|s(k)|^2]\}^2}{(1-a^4)(1-a^2)}\right\} \\ &\quad -4w_0^3 w_1\left\{a^3\left[\frac{\kappa^4 E[|s(k)|^4]}{1-a^4} + \frac{6a^2\kappa^4\{E[|s(k)|^2]\}^2}{(1-a^4)(1-a^2)}\right] + \frac{3a\kappa^4\{E[|s(k)|^2]\}^2}{1-a^2}\right\} \end{aligned} \tag{13.14}$$

where the detailed derivations pertaining to the above equations can be found in problem 2.

Example 13.1

Assume a QAM signal with four symbols is transmitted through an AR channel whose transfer function is

$$H(z) = \frac{0.36z}{z + a}$$

for the cases where $a = 0.4$ and $a = 0.8$, respectively. The equalizer is a first-order FIR adaptive filter as described in equation (13.11). For a signal to noise ratio of 10dB, plot the CMA error surface and its corresponding contours.

Solution:

Fig. 13.1 depicts the error surface and its contours for the CM objective function, with $a = 0.4$, where the surface is flattened for certain ranges of w_0 and w_1 in order to allow a better view of valleys and local minima and maxima. As can be verified the surface presents multiple minima, the ones at $w_0 = 0$ do not correspond to global minima. The surface shape indicates that if a good initial point is not given to a CM-based algorithm, the parameters will converge to an undesirable local minima where the equalization performance might be very poor. In addition, if the algorithm traverses a region in the neighborhood of a saddle point the convergence of stochastic gradient algorithms can be particularly slow. Fig. 13.2 shows the error surface and its contours for $a = 0.8$, where in this case the local minima are not so visible but they do exist.

□

Example 13.2

In this example we consider an equalization problem. Perform the equalization of a channel with the following impulse response

$$\mathbf{h} = [1.1 + \jmath 0.5 \;\; 0.1 - \jmath 0.3 \;\; -0.2 - \jmath 0.1]^T$$

The transmitted signals are uniformly distributed four QAM samples with unitary power. An additional Gaussian white noise with variance $10^{-2.5}$ is present at the channel output. Utilize the CMA.

(a) Find the Wiener solution for an equalizer with five coefficients and convolve with the channel impulse response.

(b) Perform a blind equalization also with five coefficients and depict the detected symbols before and after the equalization.

Solution:

(a) In the first step, we compute the Wiener solution and perform the convolution with the channel impulse response in order to verify the effectiveness of the equalizer order in the present example. For a delay of 1, the convolution samples are given by

$$\mathbf{y} = \begin{bmatrix} 0.0052 + \jmath 0.0104 \\ 0.9675 + \jmath 0.0000 \\ 0.0074 + \jmath 0.0028 \\ -0.0548 - \jmath 0.0014 \\ 0.0129 + \jmath 0.0222 \\ -0.0939 - \jmath 0.0075 \\ 0.0328 - \jmath 0.0098 \end{bmatrix}^T$$

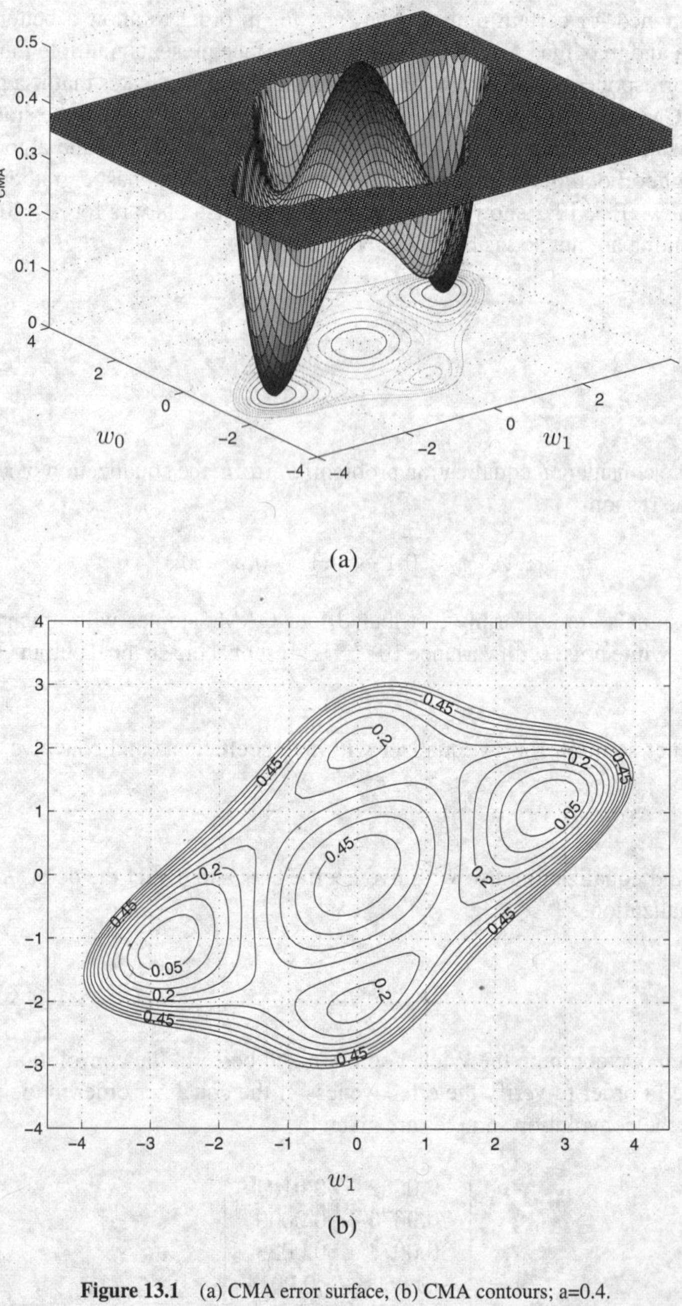

Figure 13.1 (a) CMA error surface, (b) CMA contours; a=0.4.

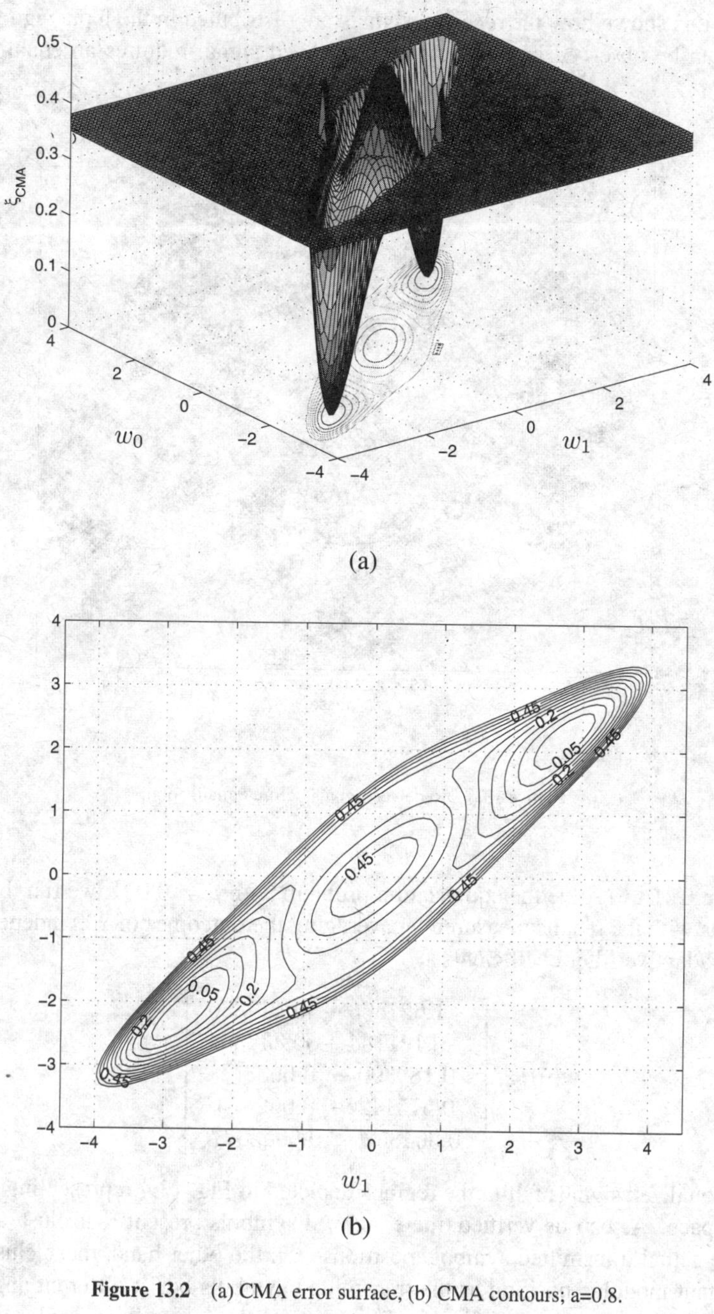

Figure 13.2 (a) CMA error surface, (b) CMA contours; a=0.8.

where as can be observed the real part of the second sample is much higher than the remaining samples, showing that the equalization is successful.

(b) In Fig. 13.3 it is shown how the received signals are distributed in the input signal constellation space, and as can be observed and expected the received signal requires an equalizer for proper detection.

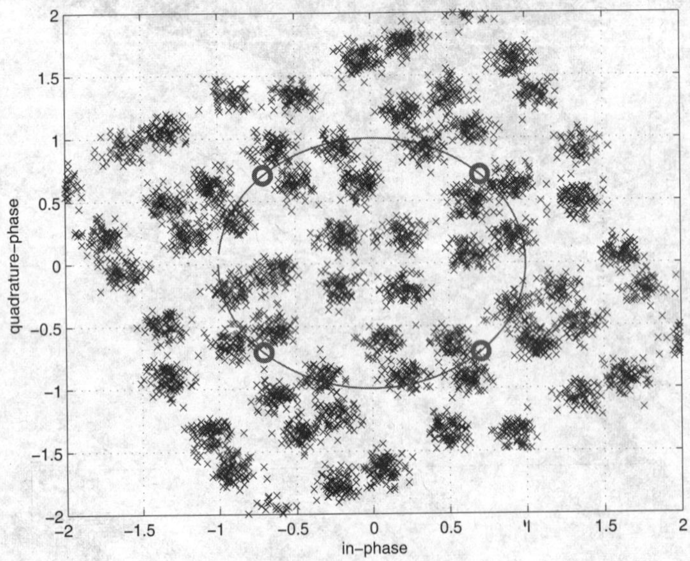

Figure 13.3 Receiver signals before equalization.

By applying the CMA to solve the equalization problem with $\mu = 0.001$, we run the algorithm for 10000 iterations with the results measured by averaging the outcomes of 200 independent runs. By initializing the adaptive-filter coefficients at

$$\mathbf{w}(0) = \begin{bmatrix} -1.627563 - \jmath 0.443856 \\ -0.121194 + \jmath 0.338364 \\ 0.189390 + \jmath 0.063311 \\ 0.575142 - \jmath 0.062878 \\ 0.364852 - \jmath 0.6053977 \end{bmatrix}$$

the last 1000 equalized signals fall in the regions depicted in Fig. 13.4 representing the input signal constellation space. As can be verified, the equalized symbols present four clusters which are not centered at the actual transmitted symbols positions. On the other hand, these clusters are around the same constant modulus position as the transmitted symbols but at different angles, that is, the transmitted constellation is received after equalization rotated by an arbitrary angle. For differentially encoded symbols the mentioned phase shift can be eliminated, allowing proper decoding of the received symbols.

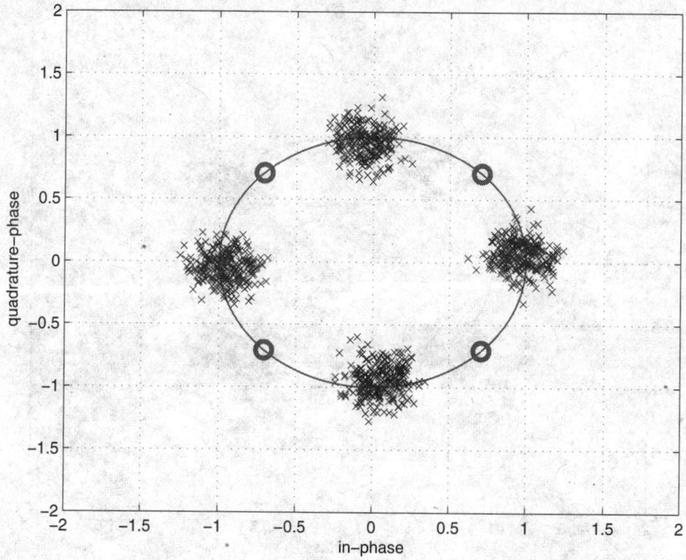

Figure 13.4 Equalized signals for the CM algorithm using the first coefficient initialization.

If the CMA filter coefficients are initialized at

$$\mathbf{w}(0) = \begin{bmatrix} 2.011934 + \jmath 0.157299 \\ 0.281061 + \jmath 0.324327 \\ -0.017917 + \jmath 0.836021 \\ -0.391982 + \jmath 1.144051 \\ -0.185579 - \jmath 0.898060 \end{bmatrix}$$

the resulting clusters are shown in Fig. 13.5, where it is possible to verify that in this case the clusters occur at the right positions with respect to the transmitted symbols.

For illustration, Fig. 13.6 shows the equalization results when using the Wiener solution, where it can be observed by comparing it with Fig. 13.5 that the CMA can lead to Wiener like solutions when properly initialized.

The typical learning curve for the CM algorithm in the present example is illustrated in 13.7 where in this case we utilized random initial coefficients for the adaptive filter.

□

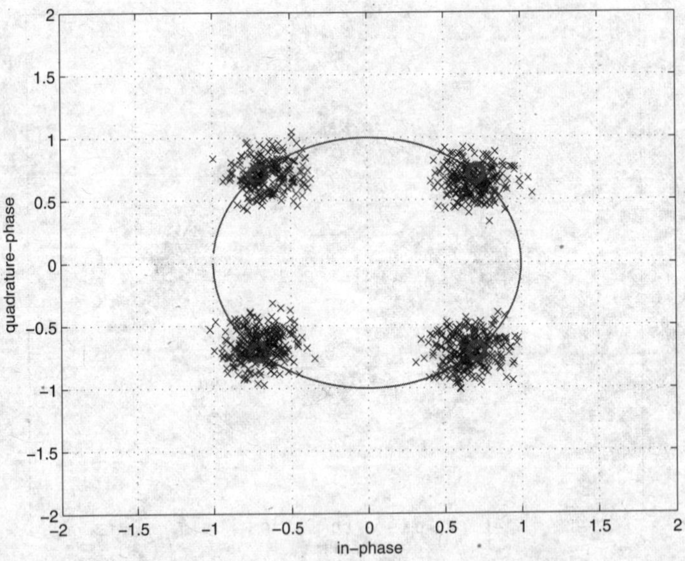

Figure 13.5 Equalized signals for the CM algorithm using the second coefficient initialization.

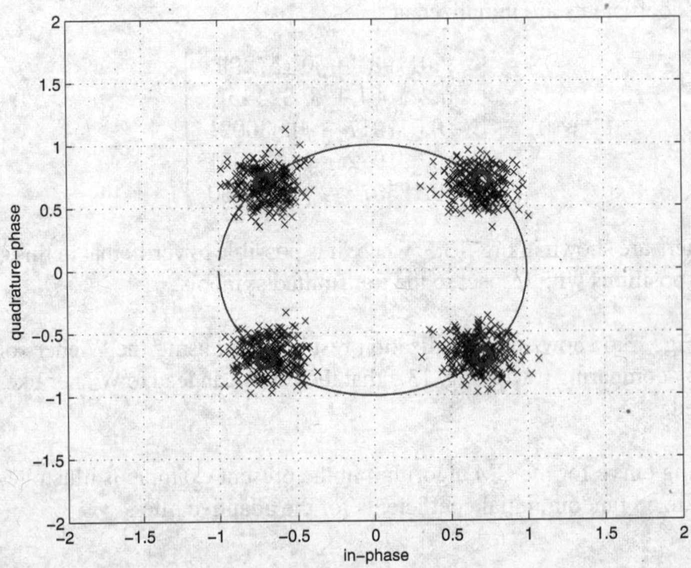

Figure 13.6 Equalized signals for the Wiener filter.

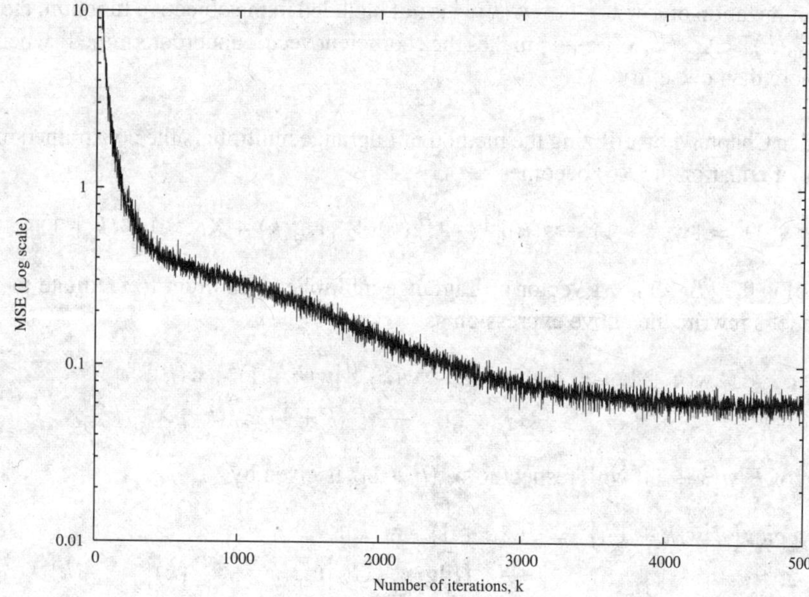

Figure 13.7 Learning curve of the CM algorithm.

13.3 AFFINE PROJECTION CM ALGORITHM

In general the CMA like algorithms present slow convergence when the update equation has a stochastic gradient form. A possible alternative solution when the convergence speed is not acceptable is to utilize the affine projection form. Let's consider the cases where either the desired vector is a CMA like function at each entry of a vector $\mathbf{r}_{\text{ap}}(k)$ or represents a nonlinear function $G_1[\cdot]$ applied to the adaptive-filter output, that is,

$$\mathbf{r}_{\text{ap}}(k) = G_1\left[\mathbf{y}_{\text{ap}}(k)\right] = G_1\left[\mathbf{X}_{\text{ap}}^T(k)\mathbf{w}^*(k)\right] \tag{13.15}$$

where the definitions of the data matrix and vectors of the affine projection algorithm are defined in equations (4.74) and (4.77).

The objective function that the affine projection algorithm minimizes in this case is

$$\begin{aligned}&\|\mathbf{w}(k+1) - \mathbf{w}(k)\|^2\\&\text{subject to :}\\&G_2\left\{\mathbf{r}_{\text{ap}}(k) - \mathbf{X}_{\text{ap}}^T(k)\mathbf{w}^*(k+1)\right\} = \mathbf{0}\end{aligned} \tag{13.16}$$

where $\mathbf{r}_{\text{ap}}(k)$ is a vector replacing $\mathbf{d}_{\text{ap}}(k)$ in the blind formulation whose elements are determined by the type of blind objective function at hand. $G_2[\cdot]$ represents another nonlinear operation applied elementwise on $[\cdot]$, usually given by $(\cdot)^2$ as in the CM algorithm. In any situation, $G_2(\mathbf{0}) = \mathbf{0}$. Also in this case the affine projection algorithm keeps the next coefficient vector $\mathbf{w}(k+1)$ as close as possible to the current one and aims at making the *a posteriori* error to be zero. It is worth mentioning that if the minimization of $\|\mathbf{w}(k+1)-\mathbf{w}(k)\|^2$ is not included in the objective function, the problem of keeping $\mathbf{r}_{\mathrm{ap}}(k) = \mathbf{X}_{\mathrm{ap}}^T(k)\mathbf{w}^*(k+1)$ makes the coefficient vector underdetermined[4] whenever this vector has more than one entry.

As described in Chapter 4 by utilizing the method of Lagrange multipliers the constrained minimization problem of equation (13.16) becomes

$$F[\mathbf{w}(k+1)] = \|\mathbf{w}(k+1)-\mathbf{w}(k)\|^2 + \boldsymbol{\lambda}_{\mathrm{ap}}^H(k)G_2\left\{\mathbf{r}_{\mathrm{ap}}(k) - \mathbf{X}_{\mathrm{ap}}^T(k)\mathbf{w}^*(k+1)\right\} \tag{13.17}$$

where $\boldsymbol{\lambda}_{\mathrm{ap}}(k)$ is the $(L+1)\times 1$ vector of Lagrange multipliers. In order to facilitate the gradient computation let's rewrite the above expression as

$$\begin{aligned} F[\mathbf{w}(k+1)] &= [\mathbf{w}(k+1)-\mathbf{w}(k)]^H[\mathbf{w}(k+1)-\mathbf{w}(k)] \\ &\quad + G_2\left\{\mathbf{r}_{\mathrm{ap}}^T(k) - \mathbf{w}^H(k+1)\mathbf{X}_{\mathrm{ap}}(k)\right\}\boldsymbol{\lambda}_{\mathrm{ap}}^*(k) \end{aligned} \tag{13.18}$$

The gradient of $F[\mathbf{w}(k+1)]$ with respect to $\mathbf{w}^*(k+1)$[5] is given by

$$\begin{aligned} \mathbf{g}_{\mathbf{w}^*}\{F[\mathbf{w}(k+1)]\} &= [\mathbf{w}(k+1)-\mathbf{w}(k)] \\ &\quad + \mathbf{X}_{\mathrm{ap}}(k)\mathbf{g}_{\bar{\mathbf{y}}_{\mathrm{ap}}}\left\{G_2\left[\mathbf{r}_{\mathrm{ap}}^T(k) - \bar{\mathbf{y}}_{\mathrm{ap}}^T(k)\right]\right\}\boldsymbol{\lambda}_{\mathrm{ap}}^*(k) \end{aligned} \tag{13.19}$$

where $\bar{\mathbf{y}}_{\mathrm{ap}}(k)$ represents the *a posteriori* adaptive-filter output signal. After setting the gradient of $F[\mathbf{w}(k+1)]$ with respect to $\mathbf{w}^*(k+1)$ equal to zero, we get

$$\mathbf{w}(k+1) = \mathbf{w}(k) - \mathbf{X}_{\mathrm{ap}}(k)\mathbf{g}_{\bar{\mathbf{y}}_{\mathrm{ap}}}\left\{G_2\left[\mathbf{r}_{\mathrm{ap}}^T(k) - \bar{\mathbf{y}}_{\mathrm{ap}}^T(k)\right]\right\}\boldsymbol{\lambda}_{\mathrm{ap}}^*(k) \tag{13.20}$$

By premultiplying equation (13.20) by $\mathbf{X}_{\mathrm{ap}}^H(k)$, using the constraint relation of equation (13.16), and considering the fact that $G_2(\mathbf{0}) = \mathbf{0}$ so that $\mathbf{X}_{\mathrm{ap}}^H(k)\mathbf{w}(k+1) = \mathbf{r}_{\mathrm{ap}}^*(k)$, we obtain

$$-\mathbf{X}_{\mathrm{ap}}^H(k)\mathbf{X}_{\mathrm{ap}}(k)\mathbf{g}_{\bar{\mathbf{y}}_{\mathrm{ap}}}\left\{G_2\left[\mathbf{r}_{\mathrm{ap}}^T(k) - \bar{\mathbf{y}}_{\mathrm{ap}}^T(k)\right]\right\}\boldsymbol{\lambda}_{\mathrm{ap}}^*(k) + \mathbf{X}_{\mathrm{ap}}^H(k)\mathbf{w}(k) = \mathbf{r}_{\mathrm{ap}}^*(k) \tag{13.21}$$

This expression leads to

$$\mathbf{g}_{\bar{\mathbf{y}}_{\mathrm{ap}}}\left\{G_2\left[\mathbf{r}_{\mathrm{ap}}^T(k) - \bar{\mathbf{y}}_{\mathrm{ap}}^T(k)\right]\right\}\boldsymbol{\lambda}_{\mathrm{ap}}^*(k) = \left[\mathbf{X}_{\mathrm{ap}}^H(k)\mathbf{X}_{\mathrm{ap}}(k)\right]^{-1}\left\{-\mathbf{r}_{\mathrm{ap}}^*(k) + \mathbf{X}_{\mathrm{ap}}^H(k)\mathbf{w}(k)\right\} \tag{13.22}$$

By substituting equation (13.22) in equation (13.20), the update equation can be rewritten as

$$\begin{aligned} \mathbf{w}(k+1) &= \mathbf{w}(k) + \mathbf{X}_{\mathrm{ap}}(k)\left(\mathbf{X}_{\mathrm{ap}}^H(k)\mathbf{X}_{\mathrm{ap}}(k)\right)^{-1}\left\{\mathbf{r}_{\mathrm{ap}}^*(k) - \mathbf{X}_{\mathrm{ap}}^H(k)\mathbf{w}(k)\right\} \\ &= \mathbf{w}(k) + \mathbf{X}_{\mathrm{ap}}(k)\left(\mathbf{X}_{\mathrm{ap}}^H(k)\mathbf{X}_{\mathrm{ap}}(k)\right)^{-1}\mathbf{e}_{\mathrm{ap}}^*(k) \end{aligned} \tag{13.23}$$

From the above equation it follows that

$$\|\mathbf{w}(k+1)-\mathbf{w}(k)\|^2 = \mathbf{e}_{\mathrm{ap}}^T(k)\left(\mathbf{X}_{\mathrm{ap}}^H(k)\mathbf{X}_{\mathrm{ap}}(k)\right)^{-1}\mathbf{e}_{\mathrm{ap}}^*(k) \tag{13.24}$$

---

[4] A solution exists but it is not unique.

[5] We could also formulate this solution employing the gradient with respect to $\mathbf{w}(k+1)$, leading to the same results.

such that the minimization of the terms on the left- and right-hand sides are equivalent. However, the minimization of the right-hand side term does not mean minimizing $\|\mathbf{e}_{\text{ap}}^*(k)\|$ unless the matrix $\left(\mathbf{X}_{\text{ap}}^H(k)\mathbf{X}_{\text{ap}}(k)\right)^{-1}$ is a diagonal matrix with equal nonzero values in the main diagonal. Despite of that, in order to generate a tractable solution we minimize $\|\mathbf{e}_{\text{ap}}^*(k)\|$ and interpret the objective function that is actually minimized.

If we assume $\mathbf{r}_{\text{ap}}^*(k)$ has constant modulus elementwise, the minimization of

$$\|\mathbf{e}_{\text{ap}}^*(k)\|^2 = \|\mathbf{r}_{\text{ap}}^*(k) - \mathbf{X}_{\text{ap}}^H(k)\mathbf{w}(k)\|^2$$

occurs when $\mathbf{r}_{\text{ap}}^*(k)$ is in the same direction as (is colinear with) $\mathbf{X}_{\text{ap}}^H(k)\mathbf{w}(k)$. In this case the following choice should be made

$$\mathbf{r}_{\text{ap}}^*(k) = \text{sgn}[\mathbf{X}_{\text{ap}}^H(k)\mathbf{w}(k)] \tag{13.25}$$

where for a complex number y, $\text{sgn}[y] = \frac{y}{|y|}$, and whenever $y = 0$, $\text{sgn}[y] = 1$.

In the update equation (13.24) the convergence factor is unity, and as previously discussed a trade-off between final misadjustment and convergence speed is achieved by including convergence factor as follows

$$\mathbf{w}(k+1) = \mathbf{w}(k) + \mu\mathbf{X}_{\text{ap}}(k)\left(\mathbf{X}_{\text{ap}}^H(k)\mathbf{X}_{\text{ap}}(k)\right)^{-1}\{\mathbf{r}_{\text{ap}}^*(k) - \mathbf{X}_{\text{ap}}^H(k)\mathbf{w}(k)\} \tag{13.26}$$

As before, with a convergence factor different from one (smaller than one) *a posteriori* error is no longer zero. The reader might question why $G_2[\cdot]$ did not appear in the final update expression of equation (13.22), the reason is the assumption that the constraint in equation (13.16) is satisfied exactly leading to a zero *a posteriori* error.

The objective function that equation (13.26) actually minimizes is given by

$$\begin{aligned}&\left(\frac{1}{\mu} - 1\right)\|\mathbf{w}(k+1) - \mathbf{w}(k)\|^2 + \|\mathbf{r}_{\text{ap}}(k) - \mathbf{X}_{\text{ap}}^T(k)\mathbf{w}^*(k+1)\|_{\mathbf{P}}^2 = \\ &\left(\frac{1}{\mu} - 1\right)\|\mathbf{w}(k+1) - \mathbf{w}(k)\|^2 + \|\text{sgn}[\mathbf{X}_{\text{ap}}^H(k)\mathbf{w}(k)] - \mathbf{X}_{\text{ap}}^T(k)\mathbf{w}^*(k+1)\|_{\mathbf{P}}^2\end{aligned} \tag{13.27}$$

where $\mathbf{P} = \left(\mathbf{X}_{\text{ap}}^H(k)\mathbf{X}_{\text{ap}}(k)\right)^{-1}$ and $\|\mathbf{a}\|_{\mathbf{P}}^2 = \mathbf{a}^H\mathbf{P}\mathbf{a}$.

Proof:

In order to simplify the derivations let's define

$$\alpha = \left(\frac{1}{\mu} - 1\right)$$

The objective function to be minimized with respect to the coefficients $\mathbf{w}^*(k+1)$ is given by

$$\xi(k) = \alpha\|\mathbf{w}(k+1) - \mathbf{w}(k)\|^2 + \|\mathbf{r}_{\text{ap}}(k) - \mathbf{X}_{\text{ap}}^T(k)\mathbf{w}^*(k+1)\|_{\mathbf{P}}^2$$

The derivative of the objective function is then given by

$$\frac{\partial \xi(k)}{\partial \mathbf{w}^*(k+1)} = \alpha[\mathbf{w}(k+1) - \mathbf{w}(k)] - \mathbf{X}_{\text{ap}}(k)\mathbf{P}\left[\mathbf{r}_{\text{ap}}^*(k) - \mathbf{X}_{\text{ap}}^H(k)\mathbf{w}(k+1)\right]$$

By setting this result to zero it follows that

$$\left[\alpha\mathbf{I} + \mathbf{X}_{\text{ap}}(k)\mathbf{P}\mathbf{X}_{\text{ap}}^H(k)\right]\mathbf{w}(k+1) = \alpha\mathbf{w}(k) + \mathbf{X}_{\text{ap}}(k)\mathbf{P}\mathbf{r}_{\text{ap}}^*(k) \tag{13.28}$$

By applying the matrix inversion lemma we obtain

$$\begin{aligned}
\left[\alpha\mathbf{I} + \mathbf{X}_{\text{ap}}(k)\mathbf{P}\mathbf{X}_{\text{ap}}^H(k)\right]^{-1} &= \frac{1}{\alpha}\mathbf{I} - \frac{1}{\alpha}\mathbf{I}\mathbf{X}_{\text{ap}}(k)\left[\mathbf{X}_{\text{ap}}^H(k)\frac{1}{\alpha}\mathbf{I}\mathbf{X}_{\text{ap}}(k) + \mathbf{P}^{-1}\right]^{-1}\mathbf{X}_{\text{ap}}^H(k)\frac{1}{\alpha}\mathbf{I} \\
&= \frac{1}{\alpha}\mathbf{I} - \frac{1}{\alpha}\mathbf{I}\mathbf{X}_{\text{ap}}(k)\left[\frac{\mathbf{P}^{-1}}{\alpha} + \mathbf{P}^{-1}\right]^{-1}\mathbf{X}_{\text{ap}}^H(k)\frac{1}{\alpha}\mathbf{I} \\
&= \frac{1}{\alpha}\left[\mathbf{I} - \mathbf{X}_{\text{ap}}(k)\frac{\alpha}{1+\alpha}\mathbf{P}\mathbf{X}_{\text{ap}}^H(k)\frac{1}{\alpha}\mathbf{I}\right] \\
&= \frac{1}{\alpha}\left[\mathbf{I} - \frac{\mathbf{X}_{\text{ap}}(k)\mathbf{P}\mathbf{X}_{\text{ap}}^H(k)}{1+\alpha}\right]
\end{aligned}$$

By replacing the last expression in the updating equation (13.28), we obtain

$$\begin{aligned}
\mathbf{w}(k+1) &= \left[\mathbf{I} - \frac{\mathbf{X}_{\text{ap}}(k)\mathbf{P}\mathbf{X}_{\text{ap}}^H(k)}{1+\alpha}\right]\mathbf{w}(k) + \frac{1}{\alpha}\left[\mathbf{I} - \frac{\mathbf{X}_{\text{ap}}(k)\mathbf{P}\mathbf{X}_{\text{ap}}^H(k)}{1+\alpha}\right]\mathbf{X}_{\text{ap}}(k)\mathbf{P}\mathbf{r}_{\text{ap}}^*(k) \\
&= \mathbf{w}(k) - \frac{\mathbf{X}_{\text{ap}}(k)\mathbf{P}\mathbf{y}_{\text{ap}}^*(k)}{1+\alpha} + \frac{1}{\alpha}\mathbf{X}_{\text{ap}}(k)\mathbf{P}\mathbf{r}_{\text{ap}}^*(k) - \frac{1}{\alpha}\frac{\mathbf{X}_{\text{ap}}(k)\mathbf{P}\mathbf{r}_{\text{ap}}^*(k)}{1+\alpha} \\
&= \mathbf{w}(k) - \mu\mathbf{X}_{\text{ap}}(k)\mathbf{P}\mathbf{y}_{\text{ap}}^*(k) + \mu\mathbf{X}_{\text{ap}}(k)\mathbf{P}\mathbf{r}_{\text{ap}}^*(k) \\
&= \mathbf{w}(k) + \mu\mathbf{X}_{\text{ap}}(k)\left(\mathbf{X}_{\text{ap}}^H(k)\mathbf{X}_{\text{ap}}(k)\right)^{-1}\mathbf{e}_{\text{ap}}^*(k)
\end{aligned}$$

□

The description of the affine projection CM algorithm is provided in Algorithm 13.4, where as standard an identity matrix multiplied by a small constant was added to the matrix $\mathbf{X}_{\text{ap}}^H(k)\mathbf{X}_{\text{ap}}(k)$ in order to avoid numerical problems in the matrix inversion.

It is worth mentioning that the update equation (13.22) represents other important application such as the case where $\mathbf{r}_{\text{ap}}^*(k) = \text{dec}[\mathbf{X}_{\text{ap}}^H(k)\mathbf{w}(k)]$, which corresponds to a decision directed blind algorithm, where $\text{dec}[\cdot]$ represents a hard limiter where each entry of its argument is mapped into the closest symbol of the constellation used in the transmission [10].

Now let's consider the special scalar case where the nonlinear operations to be applied to the output error of the normalized LMS algorithm are as following described. The objective function to be

Algorithm 13.4

The Affine Projection CM Algorithm

Initialization
$\mathbf{x}(0) = \mathbf{w}(0) = \text{random vectors}$
choose μ in the range $0 < \mu \leq 2$
$\gamma = $ small constant
Do for $k > 0$
$\mathbf{y}^*_{\text{ap}}(k) = \mathbf{X}^H_{\text{ap}}(k)\mathbf{w}(k)$
$\mathbf{r}^*_{\text{ap}}(k) = \text{sgn}[\mathbf{X}^H_{\text{ap}}(k)\mathbf{w}(k)]$
$\mathbf{e}^*_{\text{ap}}(k) = \mathbf{r}^*_{\text{ap}}(k) - \mathbf{y}^*_{\text{ap}}(k)$
$\mathbf{w}(k+1) = \mathbf{w}(k) + \mu\mathbf{X}_{\text{ap}}(k)\left(\mathbf{X}^H_{\text{ap}}(k)\mathbf{X}_{\text{ap}}(k) + \gamma\mathbf{I}\right)^{-1}\mathbf{e}^*_{\text{ap}}(k)$

minimized is

$$\begin{aligned} &\|\mathbf{w}(k+1) - \mathbf{w}(k)\|^2 \\ &\text{subject to :} \\ &|1 - |\mathbf{x}^H(k)\mathbf{w}(k+1)|^q|^p = 0 \end{aligned} \tag{13.29}$$

The resulting update equation is

$$\mathbf{w}(k+1) = \mathbf{w}(k) + \mu\mathbf{x}(k)\left(\mathbf{x}^H(k)\mathbf{x}(k)\right)^{-1}\left\{\text{sgn}\left[\mathbf{x}^H(k)\mathbf{w}(k)\right] - \mathbf{x}^H(k)\mathbf{w}(k)\right\} \tag{13.30}$$

corresponding to a scalar normalized LMS CM algorithm.

Example 13.3

Repeat Example 13.2 for the case of the affine projection CM algorithm, for $L = 1$ and $L = 3$ and compare the result with the CM algorithm with $q = 2$.

Solution:

Using $\mu = 0.001$ and the CM algorithm, the equalizer took well over 1000 iterations to converge as depicted in Fig. 13.8. The same figure shows that the affine projection CM algorithm with $L = 3$ has the fastest convergence, around 100 iterations, while leading to higher MSE after convergence when compared with the cases of $L = 1$ and the CMA. For the affine projection cases the convergence factor is $\mu = 0.1$. Fig. 13.9 depicts the equalized signals after convergence for the case where $L = 3$. All these figures were generated by averaging the outcomes of 50 independent runs.

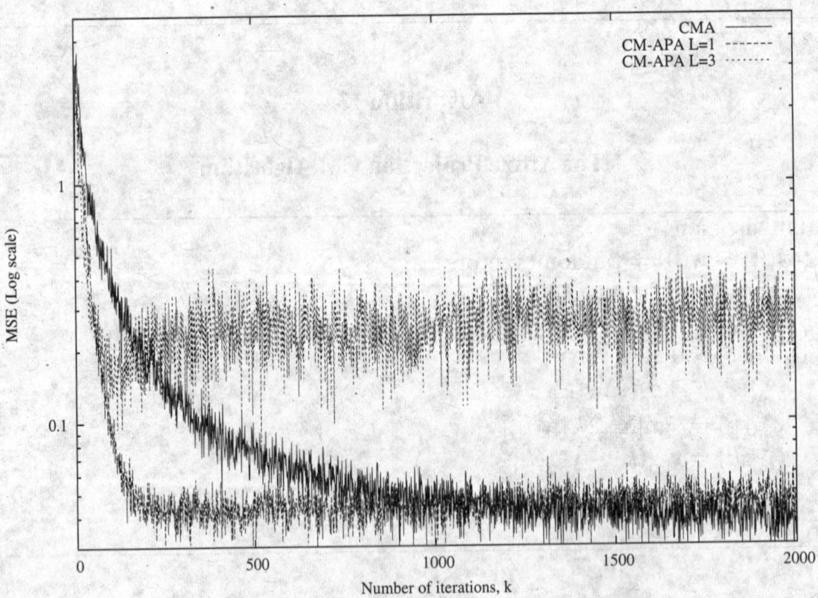

Figure 13.8 Learning curves for the CM and affine projection CM algorithms, with $L = 1$ and $L = 3$.

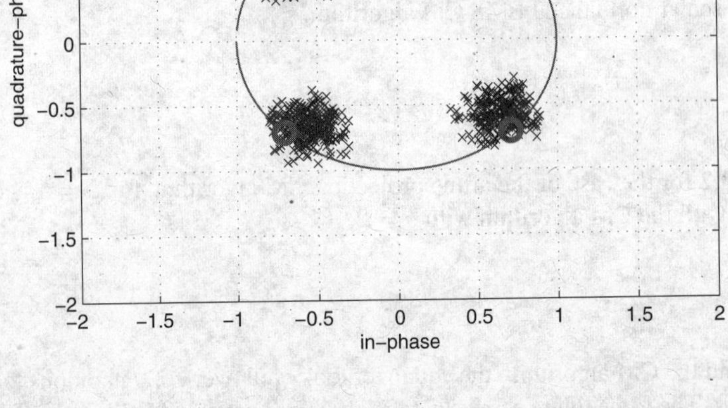

Figure 13.9 Equalized signals for the affine projection CM algorithm, with $L = 3$.

□

13.4 BLIND SIMO EQUALIZERS

The symbol spaced blind CMA equalizer methods described in previous section may converge to unacceptable local minima induced by the finite-length of the FIR equalizers, despite these minima being correct whenever the equalizer is a double sided filter with infinite order [1]. This situation changes favorably in the case a fractionally spaced equalizer is employed as following discussed. Many of the early blind equalizer methods utilized SISO channel model and relied on high-order (greater than second-order) statistics which lead to multiple minima and slow convergence. These equalizers are more sensitive to noise than those using second-order statistics. On the other hand, the availability of multiple measures of the received signal gives rise to SIMO configuration that in turn allows for blind channel equalization using second-order statistics. For example, oversampling the channel output signal by an integer factor l leads to a cyclostationary process with period l, such that the received discrete signal has cyclic correlation function allowing, under certain conditions, the identification of the channel modulus and phase [1] blindly. The SIMO configuration can be obtained by exploring diversity of antennas or by oversampling (also known as fractionally sampling) the received signal.

It is worth mentioning that the SIMO methods are not only useful to estimate a SIMO channel inverse filter but can be also used to perform channel identification. Many identification and equalization approaches can be constructed from the observed data such as subspace methods [11] and prediction methods [12]-[14] among others. The subspace methods are in general computationally complex. Furthermore they are sensitive to the channel order uncertainty causing dimension errors in the constructed signal and noise subspaces. Prediction error methods (PEM) are robust to overmodeling [15] and lend themselves to adaptive implementations.

These SIMO approaches can be extended in a rather straightforward way to device CDMA receivers [21] where blind multiuser detections are required [22]-[28], and in the cases semi-blind solutions are possible [29]. In addition, in multiple transmitter and receiver antennas systems several types of blind MIMO receivers can be derived [30]-[33]. In this section we briefly introduce the formulation for SIMO blind equalization [1], [16], and point out how this formulation brings useful solutions to blind equalization problems.

Let's consider the single-input I-output linear system model depicted in Fig. 13.10, representing an oversampling and/or the presence of multiple antenna at the receiver. In this case, the received signal can be described by

$$\mathbf{r}(k) = \sum_{i=0}^{M} x(k-i)\mathbf{h}(i) + \mathbf{n}(k) \tag{13.31}$$

where

$$\begin{aligned}
\mathbf{r}(k) &= [r_1(k)\ r_2(k) \cdots r_I(k)]^T \\
\mathbf{n}(k) &= [n_1(k)\ n_2(k) \cdots n_I(k)]^T \\
\mathbf{h}(m) &= [h_1(m)\ h_2(m) \cdots h_I(m)]^T
\end{aligned}$$

The elements of vector $\mathbf{r}(k)$ represent the I received signals at instant k, $\mathbf{n}(k)$ collects the noise samples from each subchannel at the same instant. The elements of vector $\mathbf{h}(m)$, that is $h_i(m)$, represent the mth sample of the ith subchannel model, for $m = 0, 1, \ldots, M$ and $i = 1, 2, \ldots, I$.

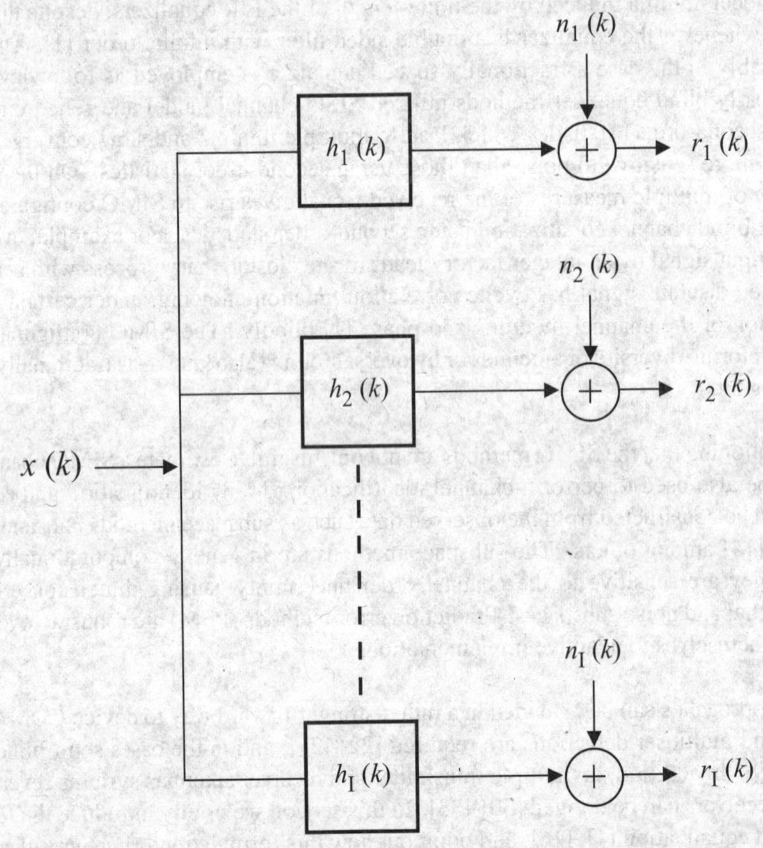

Figure 13.10 Single-input multiple-output model.

Now let's collect N samples of information vectors and pile them up in long vectors such that the received signal vector is function of the input signal block as follows

$$\bar{\mathbf{r}}(k) = \mathbf{H}\mathbf{x}(k) + \bar{\mathbf{n}}(k) \tag{13.32}$$

where

$$\bar{\mathbf{r}}(k) = \left[\mathbf{r}^T(k)\ \mathbf{r}^T(k-1)\cdots\mathbf{r}^T(k-N+1)\right]^T$$

$$\bar{\mathbf{n}}(k) = \left[\mathbf{n}^T(k)\ \mathbf{n}^T(k-1)\cdots\mathbf{n}^T(k-N+1)\right]^T$$

$$\bar{\mathbf{x}}(k) = [x(k)\ x(k-1)\cdots x(k-M-N+1)]^T$$

$$\mathbf{H} = \begin{bmatrix} \mathbf{h}(0) & \cdots & \mathbf{h}(M) & \mathbf{0} & \cdots & \mathbf{0} \\ \mathbf{0} & \mathbf{h}(0) & \cdots & \mathbf{h}(M) & \mathbf{0} & \mathbf{0} \\ \vdots & \ddots & \ddots & \ddots & \ddots & \vdots \\ \mathbf{0} & \cdots & \mathbf{0} & \mathbf{h}(0) & \cdots & \mathbf{h}(M) \end{bmatrix}$$

Vectors $\bar{\mathbf{r}}(k)$ and $\bar{\mathbf{n}}(k)$ have dimension NI, the input signal vector $\bar{\mathbf{x}}(k)$ has dimension $N + M$ whereas the channel model matrix $\mathbf{H}$ has dimension $NI \times M + N$ and is a block Toeplitz matrix.

Applying a linear combiner equalizer to the system of equation (13.32) the following relation results

$$y(k) = \bar{\mathbf{w}}^H(k)\bar{\mathbf{r}}(k) = \bar{\mathbf{w}}^H(k)\mathbf{H}\bar{\mathbf{x}}(k) + \bar{\mathbf{w}}^H(k)\bar{\mathbf{n}}(k) \tag{13.33}$$

The coefficient vector $\bar{\mathbf{w}}(k)$ is the equalizer vector of length NI described as

$$\bar{\mathbf{w}}(k) = \left[\tilde{\mathbf{w}}_0^T(k)\ \tilde{\mathbf{w}}_1^T(k) \cdots \tilde{\mathbf{w}}_{N-1}^T(k)\right]^T \tag{13.34}$$

where the vector $\tilde{\mathbf{w}}_n(k)$ represents the weights applied to $\mathbf{r}(k-n)$, for $n = 0, 1, \ldots, N-1$. The ith element of $\tilde{\mathbf{w}}_n(k)$, for $i = 1, 2, \ldots, I$, represents the ith weight applied to the corresponding element of $\mathbf{r}(k-n)$.

In a noiseless environment the zero-forcing equalizer is the desired solution such that

$$\bar{\mathbf{w}}^H(k)\mathbf{H} = [0 \ldots 0\ 1\ 0 \ldots 0]^T \tag{13.35}$$

However, the possible noise enhancement originated by $\bar{\mathbf{w}}^T(k)\bar{\mathbf{n}}(k)$ makes the zero-forcing solution not practical in many situations.

13.4.1 Identification Conditions

An FIR channel is identifiable utilizing second-order statistics whenever the block Toeplitz matrix $\mathbf{H}$ in equation (13.32) has full column rank, such that there is a left inverse. Alternatively, we can say that the system of equation (13.32) can be equalized according to some objective function, if for a set of subchannels each with order M the following conditions are met

1. $\text{rank}[\mathbf{H}] = M + N$.

   This means that matrix $\mathbf{H}$ has full column rank.

2. $NI \geq N + M$, i.e., $\mathbf{H}$ is a tall matrix in the case $NI > N + M$.

   In the latter case, this means that matrix $\mathbf{H}$ has more rows than columns.

For the case $N \geq M$, condition 1 is equivalent to say that the transfer functions

$$H_i(z) = \sum_{m=0}^{M} h_i(m) z^{-m} \tag{13.36}$$

for $i = 1, 2, \ldots, I$, have no common zeros [1], that is, the polynomials $H_i(z)$ are coprime. In the case $\frac{M}{I-1} \leq N < M$, we cannot infer that whenever $H_i(z)$, for $i = 1, 2, \ldots, I$, have no common zeros, the matrix $\mathbf{H}$ will have full column rank. In case the $H_i(z)$ have common zeros there is no left inverse matrix for $\mathbf{H}$. In addition, it can also be shown that even if the subchannels are coprime, the matrix $\mathbf{H}$ has its rank reduced if $N < M$. Condition 2 is equivalent to say that the channel matrix $\mathbf{H}$ has full column rank, making possible the channel equalization as well as identification using second-order statistics. Several alternative proofs related to the identifiability of a SIMO system are available in the literature such as in [17]-[19], no proof is included here.

Once satisfied the conditions for identifiability in the SIMO system, the finite-length input signal included in $\bar{\mathbf{x}}(k)$ should contain a large number of modes meaning it should have rich spectral content. This way, in a noiseless environment the SIMO channel can be perfectly identified, except for a gain ambiguity[6], through several methods available in the literature [1], [11]-[14]. The requirements on the channel input signal statistics vary from method to method, with some requiring that it is uncorrelated while others not.

The same type of results applies for the SIMO blind equalizers, that is, a single-input I-output channel can be equalized whenever:

- At least one of the subchannels has length $M + 1$, i.e., $h_i(0) \neq 0$ and $h_i(M) \neq 0$, for any $i = 1, 2, \ldots, I$.
- $H_i(z)$ for $i = 1, 2, \ldots, I$, have no common zeros.
- $N \geq M$.

These conditions are necessary and sufficient for the SIMO channel identifiability or equalization utilizing second-order statistics of the I outputs.

Many of the available solutions for blind channel identification and equalization based on second-order statistics are very sensitive to channel order or rank estimation. Some of them rely on singular value decomposition(s) (SVD) which are very computationally complex and are usually meant for batch form of implementation. The emphasis here is to present a recursive solution which is more robust to order estimation errors and is computationally attractive such that it can be applied to track time-varying channels. An online blind SIMO equalizer is introduced in the following section.

13.5 SIMO-CMA EQUALIZER

This section discusses an important result that suggests that by combining the techniques implicitly utilizing high-order statistics such as the CMA, with SIMO systems using second-order statistics can be very beneficial. Let's start by stating the following result whose proof can be found in [1], [20]:

In a noiseless channel, if the Multiple-Input Single-Output (MISO) FIR equalizer has length $N \geq M$, then the SIMO CMA equalizer is globally convergent if the subchannels $H_i(z)$ for $i = 1, 2, \ldots, I$, have no common zeros.

[6] A constant value multiplying the channel model.

The reader should notice that a SIMO setup utilizing a CM objective function can be interpreted as fractionally spaced constant-modulus equalizer.

The expression for the SIMO equalizer output signal as described in equation (13.33) can rewritten as

$$y(k) = \sum_{i=1}^{I} \mathbf{w}_i^H(k)\mathbf{r}_i(k) \tag{13.37}$$

where the nth element of vector $\mathbf{w}_i(k)$ corresponds to the $(i+n-1)$th element of $\bar{\mathbf{w}}(k)$, and the nth element of vector $\mathbf{r}_i(k)$ corresponds to $r_i(k-n)$, for $i = 1, 2, \ldots, I$, and $n = 0, 1, \ldots, N-1$. The equivalent SIMO system is depicted in Fig. 13.11, where it can be observed that the overall equalization consists of using a separated sub-equalizer for each sub-channel with a global output signal used in the blind adaptation algorithm.

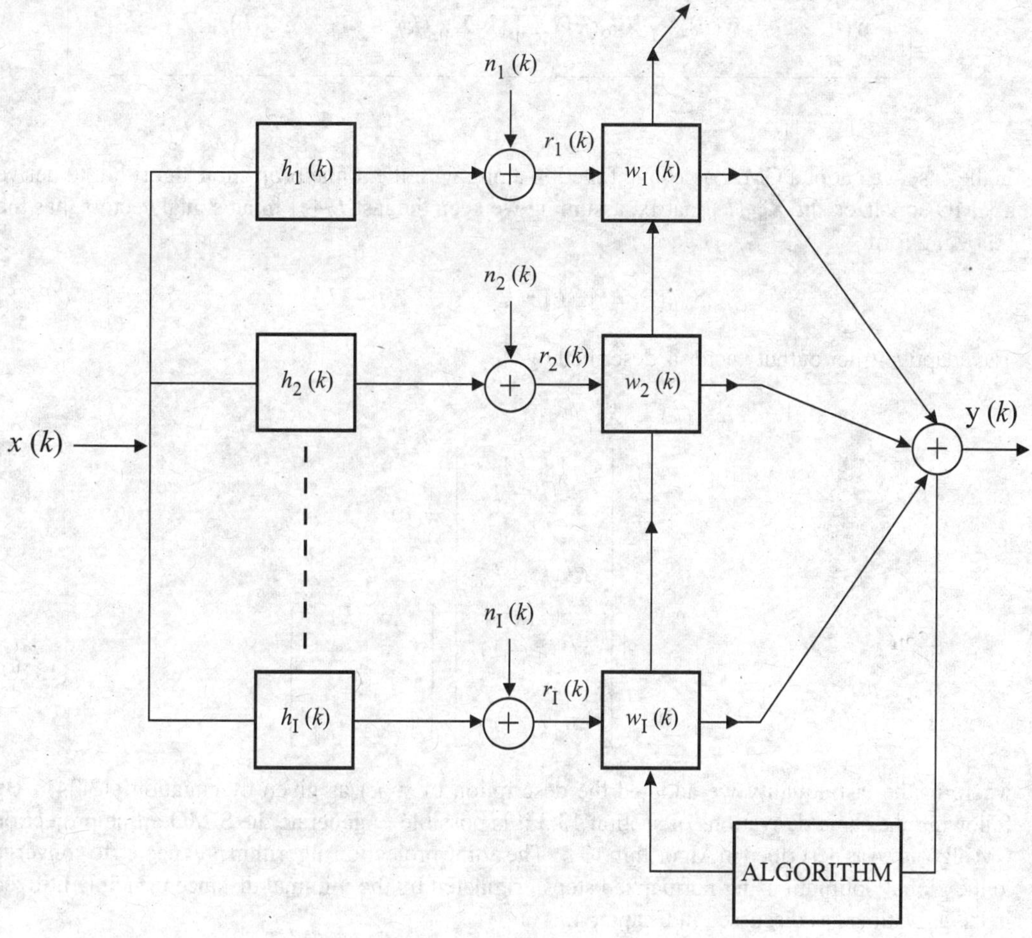

Figure 13.11 SIMO equalizer.

Algorithm 13.5

SIMO Affine Projection CM Algorithm

Initialization
$\bar{\mathbf{r}}(0) = \bar{\mathbf{w}}(0) = \text{random vectors}$
choose μ in the range $0 < \mu \leq 2$
$\gamma =$ small constant
Do for $k > 0$
$\mathbf{y}_{\text{ap}}^{*}(k) = \mathbf{X}_{\text{ap}}^{H}(k)\bar{\mathbf{w}}(k)$
$\mathbf{r}_{\text{ap}}^{*}(k) = \text{sgn}[\mathbf{X}_{\text{ap}}^{H}(k)\bar{\mathbf{w}}(k)]$
$\mathbf{e}_{\text{ap}}^{*}(k) = \mathbf{r}_{\text{ap}}^{*}(k) - \mathbf{y}_{\text{ap}}^{*}(k)$
$\bar{\mathbf{w}}(k+1) = \bar{\mathbf{w}}(k) + \mu \mathbf{X}_{\text{ap}}(k)\left(\mathbf{X}_{\text{ap}}^{H}(k)\mathbf{X}_{\text{ap}}(k) + \gamma \mathbf{I}\right)^{-1} \mathbf{e}_{\text{ap}}^{*}(k)$

In the case we adopt a CMA objective function along with the affine projection algorithm to derive a SIMO equalizer, the $\mathbf{X}_{\text{ap}}(k)$ matrix, assuming we keep the last $L+1$ input signal vectors, has the following form

$$\mathbf{X}_{\text{ap}}(k) = [\bar{\mathbf{r}}(k)\, \bar{\mathbf{r}}(k-1) \ldots \bar{\mathbf{r}}(k-L)] \tag{13.38}$$

The adaptive-filter output vector is described by

$$\begin{aligned} \mathbf{y}_{\text{ap}}^{*}(k) &= \mathbf{X}_{\text{ap}}^{H}(k)\bar{\mathbf{w}}(k) \\ &= \begin{bmatrix} \bar{\mathbf{r}}^{H}(k) \\ \bar{\mathbf{r}}^{H}(k-1) \\ \vdots \\ \bar{\mathbf{r}}^{H}(k-L) \end{bmatrix} \bar{\mathbf{w}}(k) \\ &= \begin{bmatrix} \bar{\mathbf{r}}^{H}(k) \\ \bar{\mathbf{r}}^{H}(k-1) \\ \vdots \\ \bar{\mathbf{r}}^{H}(k-L) \end{bmatrix} \begin{bmatrix} \tilde{\mathbf{w}}_{0}(k) \\ \tilde{\mathbf{w}}_{1}(k) \\ \vdots \\ \tilde{\mathbf{w}}_{N-1}(k) \end{bmatrix} \end{aligned} \tag{13.39}$$

where in the last equality we adopted the description of $\bar{\mathbf{w}}(k)$ as given by equation (13.34). By following the same derivations of section 13.3 it is possible to generate the SIMO affine projection CM algorithm as described in Algorithm 13.5. The affine projection algorithm is expected to converge to the global optimum using normalized steps originated by the minimal distance principle utilized in its derivations, as discussed in Chapter 4.

Example 13.4

Given the one-input two-output channel whose model is described below. Assume a QAM signal with four symbols is transmitted through these channels and simulate a blind equalization using the SIMO affine projection CM algorithm of order 12, for a signal to noise ratio of 20dB measured at the receiver input.

$$\begin{bmatrix} \mathbf{h}_1^T \\ \mathbf{h}_2^T \end{bmatrix} = \begin{bmatrix} 0.1823 & -0.7494 & -0.4479 & 0.2423 & 0.0047 & -0.41 \\ 0.3761 & -0.1612 & -0.1466 & 0.6437 & 0.5952 & -0.2060 \end{bmatrix}$$

Solution:

We utilize the affine projection CM algorithm to solve the SIMO equalization problem with $\mu = 0.1$, $L = 2$ and $\gamma = 10^{-6}$. The symbol error rate is measured by averaging the outcoming results of 50 independent runs, and the initial conditions utilized correspond to the Wiener solution randomly disturbed. Fig. 13.12 shows the evolution of the errors in the symbols, and as can be observed minimum symbol error rate occurs after 500 iterations. This result is expected since the conditions for the correct channel equalization is met in this case, see subsection 13.4.1, and there is some channel noise. Fig. 13.13 depicts the MSE between the equalized signal and the transmitted symbols where the convergence of the affine projection CM algorithm takes places in around 1000 iterations. Fig. 13.14 illustrates the effectiveness of the equalizer through the appropriate combination of signals measured in each antenna.

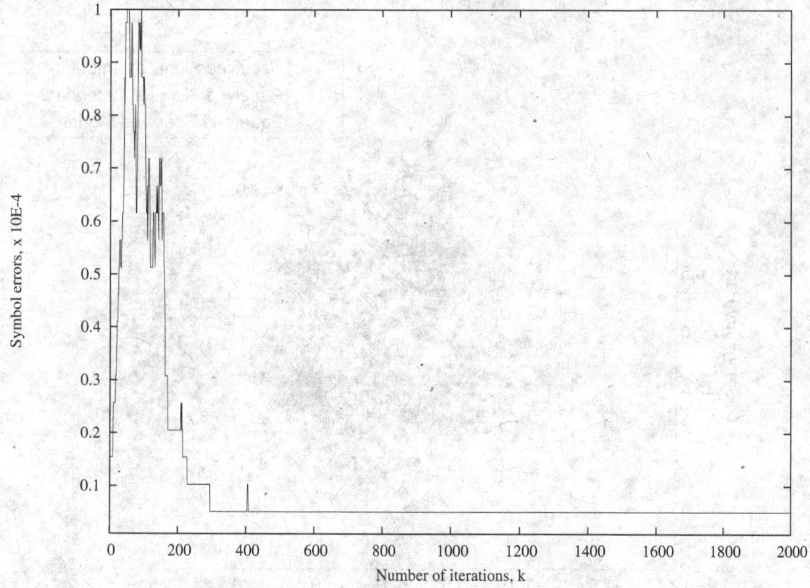

Figure 13.12 Symbol errors; affine projection CM algorithm.

□

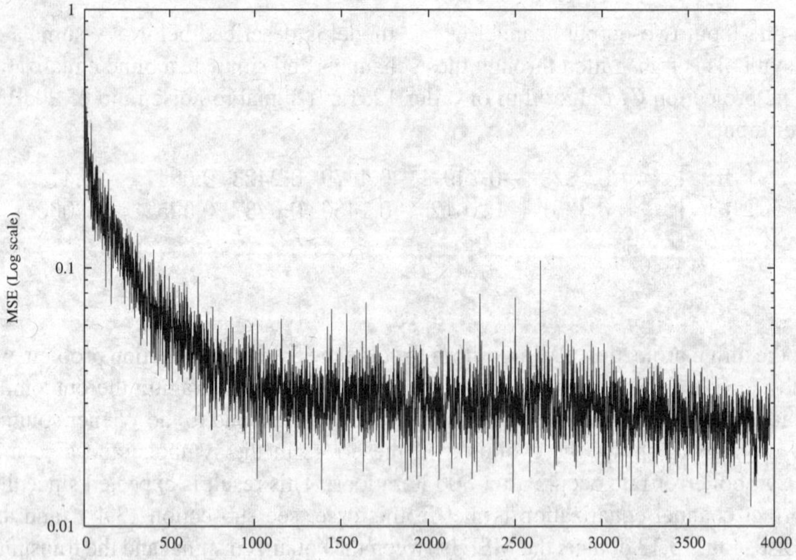

Figure 13.13 Learning curve (MSE), $\mu = 0.1$, SNR = 20 dB, order=12.

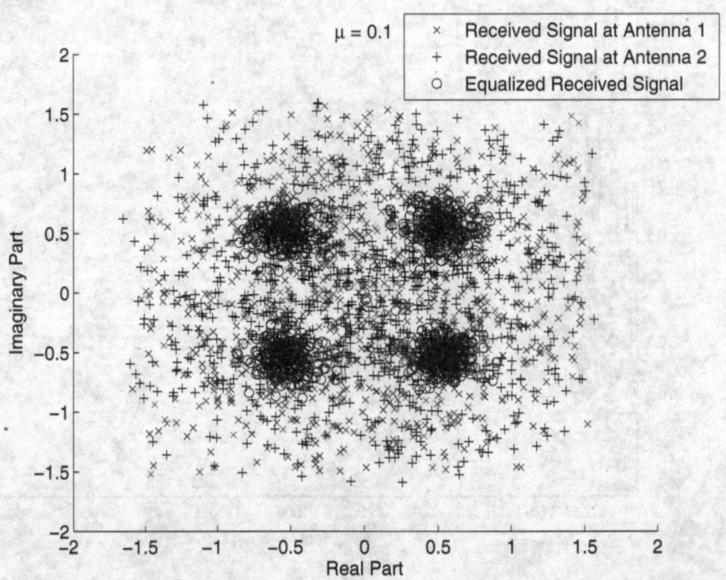

Figure 13.14 Equalized signals for the SIMO affine projection CM algorithm, with $L = 2$.

Example 13.5

Repeat the Example 6.3 by measuring through simulations the MSE performance of an equalizer implemented with the SIMO affine projection CM algorithm, when it is available two received signals obtained through different antennas. Choose the appropriate parameters and comment on the results.

Solution:

The channels available for the detection of the transmitted symbols correspond to the transfer function from the transmitter to each antenna. The blind affine projection CM algorithm is employed to update the sub equalizers of the SIMO system. The parameters chosen after some simulation trials are $\mu = 0.3$, $L = 1$, and $\gamma = 10^{-6}$. The measures of MSE reflect an average taken from the outcomes of 50 independent runs, where in the initialization one of the receiver filters is set to the Wiener solution during the first 350 iterations. Each sub equalizer has order 30. Fig. 13.15 illustrates the MSE evolution and as can be observed only after a few thousand iterations the curve shows a non decreasing behavior. In comparison with the results from Example 6.3, the learning process takes a lot more iterations when compared to the algorithms employing some sort of training. However, despite of slower convergence the equalization is feasible since the conditions for the correct channel equalization are met.

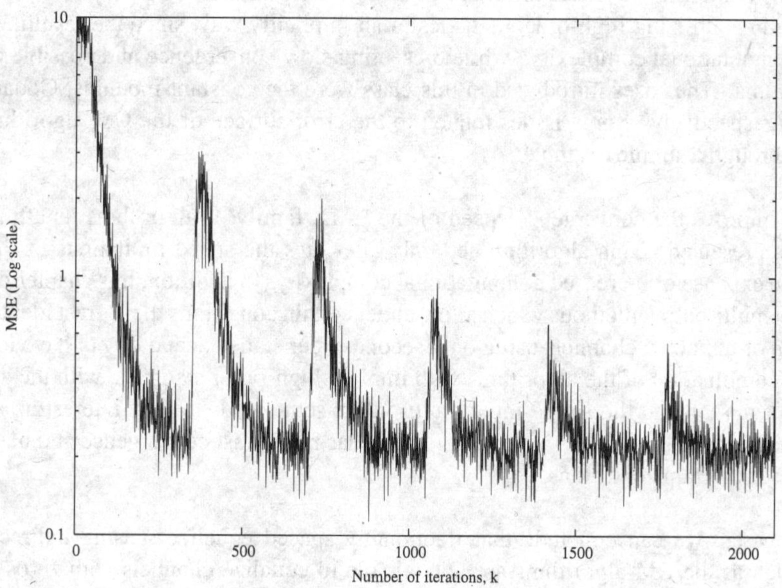

Figure 13.15 Learning curve of the SIMO affine projection CM algorithm, $L = 1$.

□

The SIMO formulation presented in this chapter can be extended to the Multi-Input Multi-Output (MIMO) case in rather straightforward way, under some assumptions such as independence of the sources. There are several communication system setups that can be modeled as MIMO systems by properly stacking the transmitted and received information. In some applications the setup is originally MIMO such as in multiuser communication systems [21]-[28], and in case we use antenna array at transmitter and receiver in order to increase the communication capacity [30]-[33]. In many MIMO applications adaptive-filtering algorithms are often utilized with training or in a blind form.

The affine projection CM algorithm presented in this chapter can be extended to include selective updating using the set-membership approach presented in Chapter 6. In addition, for multiuser environments such as CDMA systems, it is possible to incorporate some blind measurements related to the multi-access and additional noise interferences in order to improve the overall performance of blind receivers based on the set-membership affine projection CM algorithm, as discussed in [34]. The set-membership affine projection algorithm can be very efficient in SIMO as well as in MIMO setups.

13.6 CONCLUDING REMARKS

This chapter presented some blind adaptive-filtering algorithms mostly aimed at direct blind channel equalization. The subject of blind signal processing is quite extensive, as a result our emphasis was to present the related issues and to introduce some useful algorithms. In particular it was introduced some algorithms utilizing high-order statistics in an implicitly way, since the resulting algorithms have low computational complexity[7] while presenting slow convergence and possible convergence to local minima. The cases introduced in this class were the constant-modulus, Godard, and Sato algorithms, respectively. Some issues related to the error surface of the CM algorithm were also illustrated through a simple example.

In order to improve the convergence speed of the CMA family of algorithms its affine projection version was presented. This algorithm certainly alleviates the speed limitations of the CM algorithms at the expense of increased computational complexity. In addition, this chapter discussed the single-input multi-output methods which allow under certain conditions the correct identification and equalization of unknown channels using only second-order statistics and do not have local minima. In fact, the combination of the algorithms with implicit high-order statistics, with the affine projection update equation and the single-input multi-output setup leads to very interesting solutions for blind channel equalization. The resulting algorithm has rather fast convergence and has only global solutions under certain conditions.

In specific cases, we can conclude that fractionally spaced equalizers using indirect high-order statistics such as the CM algorithms are not suitable to equalize channels with zeros in common. In case this happens an additional equalizer after the SIMO equalizer might help in combating the remaining intersymbol interference. On the other hand, the SIMO equalizers are suitable to equalize channels with zeros on the unit circle, a rough situation for symbol spaced equalizers. In this case, the SIMO equalizer can be used with an implicit high-order statistics objective function or with training signal, as long as the subchannels do not have common zeros. For situations with common zeros on the unit circle, or close to it, the standard way out is to employ DFE.

[7] In comparison with the algorithms using high-order statistics explicitly.

13.7 REFERENCES

1. Z. Ding and Y. Li, *Blind Equalization and Identification*, Marcel Dekker, New York, N.Y., 2001.
2. S. Haykin, Editor, *Unsupervised Adaptive Filtering, Vol. I: Blind Source Separation*, John Wiley & Sons, New York, NY, 2000.
3. S. Haykin, Editor, *Unsupervised Adaptive Filtering, Vol. II: Blind Deconvolution*, John Wiley & Sons, New York, NY, 2000.
4. C.-Y. Chi, C.-C. Feng, C.-H. Chen, and C.-Y. Chen, *Blind Equalization and System Identification*, Springer, London, UK, 2006.
5. Y. Sato, "A method of self-recovering equalization for multi-level amplitude modulation," *IEEE Trans. on Communications*, vol. COM-23, pp. 679-682, June 1975.
6. A. Benveniste, M. Gousat, and R. Ruget, "Robust identification nonminimum phase system: Blind adjustment of a linear equalizer in data communications," *IEEE Trans. on Automatic Control*, vol. AC-25, pp. 385-399, June 1980.
7. D. N. Godard, "Self-recovering equalization and carrier tracking in two-dimensional data communication system," *IEEE Trans. on Communications*, vol. COM-28, pp. 1867-1875, Nov. 1980.
8. J. R. Barry, E. A. Lee, and D. G. Messerschmitt, *Digital Communication*, Kluwer Academic Publishers, Boston, MA, 3rd edition, 2004.
9. C. L. Nikias and A. P. Petropulu, *Higher-order spectra analysis: A nonlinear signal processing framework*, Prentice Hall, Englewood Cliffs, NJ, 1993.
10. C. B. Papadias and D. T. M. Slock, "Normalized sliding window constant-modulus and decision-directed algorithms: A link between blind equalization and classical adaptive filtering," *IEEE Trans. on Signal Processing*, vol. 45, pp. 231-235, Jan. 1997.
11. E. Moulines, P. Duhamel, J.-F. Cardoso, and S. Mayarargue, "Subspace methods for the blind identification of multichannel FIR filters," *IEEE Trans. on Signal Processing*, vol. 43, pp. 516-525, Feb. 1995.
12. D. Gesbert and P. Duhamel, "Unbiased blind adaptive channel identification," *IEEE Trans. on Signal Processing*, vol. 48, pp. 148-158, Jan. 2000.
13. X. Li and H. Fan, "Direct estimation of blind zero-forcing equalizers based on second-order statistics," *IEEE Trans. on Signal Processing*, vol. 48, pp. 2211-2218, Aug. 2000.
14. L. Tong, and Q. Zhao, "Jointly order detection and blind channel estimation by least squares smoothing," *IEEE Trans. on Signal Processing*, vol. 47, pp. 2345-2355, Sept. 1999.
15. A. P. Liavas, P. A. Regalia, and J.-P. Delmas, "Blind channel approximation: effective channel order determination," *IEEE Trans. on Signal Processing*, vol. 47, pp. 3336-3344, Dec. 1999.
16. L. Tong and Z. Ding, "Single-user channel estimation and equalization," *IEEE Signal Processing Magazine*, vol. 17, pp. 17-28, May 2000.

17. L. Tong, G. Xu, and T. Kailath, "Blind identification and equalization based on second-order statistics: A time domain approach," *IEEE Trans. on Information Theory*, vol. 40, pp. 340-349, March 1994.

18. L. Tong, G. Xu, B. Hassibi, and T. Kailath, "Blind identification and equalization based on second-order statistics: A frequency-domain approach," *IEEE Trans. on Information Theory*, vol. 41, pp. 329-334, March 1994.

19. E. Serpedin and G. B. Giannakis, "A simple proof of a known blind channel identifiability result," *IEEE Trans. on Signal Processing*, vol. 47, pp. 591-593, Feb. 1999.

20. Y. Li and Z. Ding, "Global convergence of fractionally spaced Godard (CMA) adaptive equalizers," *IEEE Trans. on Signal Processing*, vol. 44, pp. 818-826, April 1996.

21. A. J. Viterbi, *Principles of Spread Spectrum Communication*, Addison Wesley, Reading, MA, 1995.

22. M. Honig and M. K. Tsatsanis, "Adaptive techniques for multiuser CDMA receivers," *IEEE Signal Processing Magazine*, vol. 17, pp. 49-61, May 2000.

23. S. Verdu, *Multiuser Detection*, Cambridge University Press, Cambridge, UK, 1998.

24. M. L. Honig, U. Madhow, and S. Verdú, "Blind adaptive multiuser detection," *IEEE Trans. on Information Theory*, vol. 41, pp. 944-960, July 1995.

25. M. K. Tsatsanis, "Inverse filtering criteria for CDMA systems," *IEEE Trans. on Signal Processing*, vol. 45, pp. 102-112, Jan. 1997.

26. Z. Xu and M. K. Tsatsanis, "Blind adaptive algorithms for minimum variance CDMA receivers," *IEEE Trans. on Signal Processing*, vol. 49, pp. 180-194, Jan. 2001.

27. Z. Xu, P. Liu, and X. Wang, "Blind multiuser detection: From MOE to subspace methods," *IEEE Trans. on Signal Processing*, vol. 52, pp. 510-524, Feb. 2004.

28. X. Wang and H. V. Poor, *Wireless Communication Systems: Advanced Techniques for Signal Reception*, Prentice Hall, Upper Saddle River, NJ, 2003.

29. E. de Carvalho and D. T. M. Slock, "Blind and semi-blind FIR multichannel estimation: (Global) Identifiability conditions," *IEEE Trans. on Signal Processing*, vol. 52, pp. 1053-1064, April 2004.

30. A. Paulraj, R. Nabar, and D. Gore, *Introduction to Space-Time Wireless Communications*, Cambridge University Press, Cambridge, UK, 2003.

31. E. G. Larsson and P. Stoica, *Space-Time Block Coding for Wireless Communications*, Cambridge University Press, Cambridge, UK, 2003.

32. A. Hottinen, O. Tirkkonen, and R. Wichman, *Multi-Antenna Transceiver Techniques for 3G and Beyond*, John Wiley, New York, NY, 2003.

33. H. L. Van Trees, *Optimum Array Processing: Detection, Estimation, and Modulation Theory, Part IV*, John Wiley Interscience, New York, NY, 2002.

34. R. C. de Lamare and P. S. R. Diniz, "Blind constrained set-membership algorithms with time-varying bounds for CDMA interference suppression," *Proc. IEEE Intern. Conf. on Acoust. Speech and Signal Processing*, Toulouse, France, pp. IV-617 - IV-620, May 2006.

13.8 PROBLEMS

1. Derive the Godard algorithm for real input signal constellations.

2. Derive equations (13.13) and (13.14).

   Hint: Utilize the difference equation that describes $x(k)$.

3. Perform the equalization of a channel with the following impulse response

$$h(k) = ku(k) - (2k - 9)u(k - 5) + (k - 9)u(k - 10)$$

   using as transmitted signal a binary (-1,1) random signal. An additional Gaussian white noise with variance 10^{-2} is present at the channel output.
   (a) Apply the Godard algorithm for $p = q = 4$ with an appropriate μ and find the impulse response of an equalizer with 15 coefficients.
   (b) Plot the detected equalized signal before the decision after the algorithm has converged for a number of iterations (over 50 samples) and comment on the result.

4. Repeat the problem 3 for the Sato algorithm.

5. Repeat the problem 3 for the CMA.

6. Assume a PAM signal with four symbols is transmitted through an AR channel whose transfer function is

$$H(z) = \frac{0.25z}{z + 0.5}$$

   The equalizer is a first-order FIR adaptive filter. For a signal to noise ratio of 5dB, plot the error surface and contours for Godard with $p = q = 4$.

7. Assume a QAM signal with four symbols is transmitted through an AR channel whose transfer function is

$$H(z) = \frac{0.25z}{z + 0.5}$$

   Simulate a blind equalization using a first-order FIR adaptive filter, for a signal to noise ratio of 10dB, using the CMA.

8. Given the channel model below whose input is a binary PAM signal.

$$\begin{aligned} H(z) &= 0.2816 + 0.5622z^{-1} + 0.2677z^{-2} - 0.3260z^{-3} - 0.4451z^{-4} \\ &\quad +0.3102z^{-5} - 0.2992z^{-6} - 0.2004z^{-7} \end{aligned}$$

Our objective is to equalize this channel with a blind affine projection CM algorithm. The equalizer has order 10 and its objective is to shorten the effective impulse response of the equalized signal. That means the channel-equalizer impulse response has most of its energy concentrated in a few samples. Simulate this experiment for a signal to noise ratio of 15dB, and comment on the channel shortening process.

9. Derive the set-membership affine projection CM algorithm.

10. (a) Show that recursion of equation (13.30) minimizes the objective function of equation (13.29).

(b) Show that recursion of equation (13.30) also minimizes the objective function

$$\|\mathbf{w}(k+1) - \mathbf{w}(k)\|^2$$
$$\text{subject to :}$$
$$|\text{sgn}\left[\mathbf{x}^H(k)\mathbf{w}(k+1)\right] - |\mathbf{x}^H(k)\mathbf{w}(k+1)|^q|^p = 0$$

11. Derive a constrained minimum variance (CMV) affine projection algorithm for equalization, whose objective function is to minimize

$$\frac{1}{2}\|\mathbf{w}(k+1) - \mathbf{w}(k)\|^2$$
$$\text{and}$$
$$\frac{1}{2}\mathbf{w}^T(k+1)\mathbf{r}(k)\mathbf{r}^T(k)\mathbf{w}(k+1)$$
$$\text{subject to :}$$
$$\mathbf{w}^T(k+1)\mathbf{c} = c$$

where $\mathbf{r}(k)$ is a vector that in the present case represents the received signal vector, c is an arbitrary constant, and $\mathbf{c}$ is a constraint vector.

12. Assume a PAM signal with two symbols is transmitted through a noiseless AR channel whose transfer function is

$$H(z) = \frac{0.25z}{z + 0.5}$$

Simulate a blind equalization using a first-order FIR adaptive filter, using affine projection CM algorithm as well as the stochastic gradient version CMA. Plot the convergence trajectories of $w_0(k)$ and $w_1(k)$ for 20 distinct initialization points (on the same figure) for $w_0(0)$ and $w_1(0)$ corresponding to zeros in the interior of unit circle. Interpret the results.

13. Equalize the one-input two-output channel described below using the SIMO affine projection CM algorithm. The input signal is a two PAM signal representing a randomly generated bit stream with the signal to noise ratios $\frac{\sigma_{r_i}^2}{\sigma_n^2} = 20$, for $i = 1, 2$, at the receiver end, that is, $r_i(k)$ is the received signal without taking into consideration the additional channel noise. Choose the appropriate equalizer order and the number of reuses such that the bit error rate falls below 0.01.

$$\begin{bmatrix} \mathbf{h}_1 & \mathbf{h}_2 \end{bmatrix} = \begin{bmatrix} 0.345 & -0.715 \\ -0.016 & 0.690 \\ -0.324 & 0.625 \\ 0.209 & 0.120 \\ 0.253 & 0.388 \\ -0.213 & 0.132 \\ 0.254 & -0.120 \\ 0.118 & -0.388 \\ 0.483 & 0.451 \\ -0.034 & -0.204 \\ 0.462 & 0.560 \\ -0.111 & -0.675 \\ -0.285 & 0.147 \end{bmatrix}$$

14. Using the complex version of the SIMO affine projection CM algorithm to equalize the one-input two-output channel with the transfer function given below. The input signal is a four QAM signal representing a randomly generated bit stream with the signal to noise ratios $\frac{\sigma_{r_i}^2}{\sigma_n^2} = 10$, for $i = 1, 2$, at the receiver end, that is, $r_i(k)$ is the received signal without taking into consideration the additional channel noise. The adaptive filter has 5 coefficients.

$$H_1(z) = (0.27 - 0.34\jmath) + (0.43 + 0.87\jmath)z^{-1} + (0.21 - 0.34\jmath)z^{-2}$$

$$H_2(z) = (0.34 - 0.27\jmath) + (0.87 + 0.43\jmath)z^{-1} + (0.34 - 0.21\jmath)z^{-2}$$

(a) Run the algorithm for $\mu = 0.1$, $\mu = 0.4$, and $\mu = 0.8$. Comment on the convergence behavior in each case.
(b) Plot the real versus imaginary parts of the received signals before equalization and the single output signal after equalization.

15. Repeat problem 14 for the case the adaptive-filter order is one and comment on the results.

A

COMPLEX DIFFERENTIATION

A.1 INTRODUCTION

This appendix briefly describes how to deal with complex signals in adaptive-filtering context in a simple manner, for further details the reader is encouraged to refer to [1]-[2].

A.2 THE COMPLEX WIENER SOLUTION

Environments with complex signals are typical of some communication applications. In order to address these cases, this section describes the complex Wiener solution. In the complex case, the error signal and its complex conjugate are defined as

$$\begin{aligned} e(k) &= d(k) - \mathbf{w}^H(k)\mathbf{x}(k) \\ e^*(k) &= d^*(k) - \mathbf{w}^T(k)\mathbf{x}^*(k) \end{aligned} \tag{A.1}$$

Their product is then described by

$$\begin{aligned} |e(k)|^2 &= e(k)e^*(k) = [d(k) - \mathbf{w}^H(k)\mathbf{x}(k)][d^*(k) - \mathbf{w}^T(k)\mathbf{x}^*(k)] \\ &= |d(k)|^2 - \mathbf{w}^T(k)\mathbf{x}^*(k)d(k) - \mathbf{w}^H(k)\mathbf{x}(k)d^*(k) + \mathbf{w}^H(k)\mathbf{x}(k)\mathbf{x}^H(k)\mathbf{w}(k) \\ &= |d(k)|^2 - 2\mathrm{re}[\mathbf{w}^H(k)\mathbf{x}(k)d^*(k)] + \mathbf{w}^H(k)\mathbf{x}(k)\mathbf{x}^H(k)\mathbf{w}(k) \end{aligned} \tag{A.2}$$

The expression of the error squared of equation (A.2) can be written as a function of the real and imaginary parts of the filter coefficients as

$$\begin{aligned}
|e(k)|^2 &= |d(k)|^2 - \left(\mathrm{re}[\mathbf{w}^T(k)] + \jmath\mathrm{im}[\mathbf{w}^T(k)]\right)\mathbf{x}^*(k)d(k) \\
&\quad - \left(\mathrm{re}[\mathbf{w}^T(k)] - \jmath\mathrm{im}[\mathbf{w}^T(k)]\right)\mathbf{x}(k)d^*(k) \\
&\quad + \left(\mathrm{re}[\mathbf{w}^T(k)] - \jmath\mathrm{im}[\mathbf{w}^T(k)]\right)\mathbf{x}(k)\mathbf{x}^H(k)\left(\mathrm{re}[\mathbf{w}(k)] + \jmath\mathrm{im}[\mathbf{w}(k)]\right) \\
&= |d(k)|^2 - \mathrm{re}[\mathbf{w}^T(k)]\left(\mathbf{x}^*(k)d(k) + \mathbf{x}(k)d^*(k)\right) \\
&\quad -\jmath\mathrm{im}[\mathbf{w}^T(k)]\left(\mathbf{x}^*(k)d(k) - \mathbf{x}(k)d^*(k)\right) \\
&\quad +\mathrm{re}[\mathbf{w}^T(k)]\mathbf{x}(k)\mathbf{x}^H(k)\mathrm{re}[\mathbf{w}(k)] - \jmath\mathrm{im}[\mathbf{w}^T(k)]\mathbf{x}(k)\mathbf{x}^H(k)\mathrm{re}[\mathbf{w}(k)] \\
&\quad +\jmath\mathrm{re}[\mathbf{w}^T(k)]\mathbf{x}(k)\mathbf{x}^H(k)\mathrm{im}[\mathbf{w}(k)] + \mathrm{im}[\mathbf{w}^T(k)]\mathbf{x}(k)\mathbf{x}^H(k)\mathrm{im}[\mathbf{w}(k)]
\end{aligned} \quad \text{(A.3)}$$

where re[·] and im[·] indicate real and imaginary parts of [·], respectively.

For a filter with fixed coefficients, see equation (A.2), the MSE function is given by

$$\begin{aligned}
\xi &= E[|e(k)|^2] \\
&= E[|d(k)|^2] - 2\mathrm{re}\{\mathbf{w}^H E[d^*(k)\mathbf{x}(k)]\} + \mathbf{w}^H E[\mathbf{x}(k)\mathbf{x}^H(k)]\mathbf{w} \\
&= E[|d(k)|^2] - 2\mathrm{re}[\mathbf{w}^H\mathbf{p}] + \mathbf{w}^H\mathbf{R}\mathbf{w}
\end{aligned} \quad \text{(A.4)}$$

where $\mathbf{p} = E[d^*(k)\mathbf{x}(k)]$ is the cross-correlation vector between the desired and input signals, and $\mathbf{R} = E[\mathbf{x}(k)\mathbf{x}^H(k)]$ is the input signal correlation matrix. As before, the objective function ξ is a quadratic function of the tap-weight coefficients which would allow a straightforward solution for $\mathbf{w}$, if vector $\mathbf{p}$ and matrix $\mathbf{R}$ are known.

The gradient with respect to a complex parameter is defined as

$$\mathbf{g}_{\mathbf{w}}\{E[|e(k)|^2]\} = \frac{1}{2}\left\{\frac{\partial E[|e(k)|^2]}{\partial \mathrm{re}[\mathbf{w}(k)]} - \jmath\frac{\partial E[|e(k)|^2]}{\partial \mathrm{im}[\mathbf{w}(k)]}\right\} \quad \text{(A.5)}$$

However, the direction of maximum rate of change of a real-valued scalar function of a complex vector variable, in this case denoted by $\mathbf{w}$, is given by

$$\mathbf{g}_{\mathbf{w}^*}\{E[|e(k)|^2]\} = \frac{1}{2}\left\{\frac{\partial E[|e(k)|^2]}{\partial \mathrm{re}[\mathbf{w}(k)]} + \jmath\frac{\partial E[|e(k)|^2]}{\partial \mathrm{im}[\mathbf{w}(k)]}\right\} \quad \text{(A.6)}$$

Consult the references [1]-[2] for details[1].

---

[1] Any real valued function of a complex complex vector variable $\mathbf{w}$ can be represented by a Taylor series. The first-order term is given by

$$\begin{aligned}
\Delta\xi &= \left[\frac{\partial\xi}{\partial\mathbf{w}(k)}\right]^T \Delta\mathbf{w} + \left[\frac{\partial\xi}{\partial\mathbf{w}^*(k)}\right]^T \Delta\mathbf{w}^*(k) \\
&= 2\mathrm{re}\left\{\left[\frac{\partial\xi}{\partial\mathbf{w}(k)}\right]^T \Delta\mathbf{w}(k)\right\} = 2\mathrm{re}\left\{\left[\frac{\partial\xi}{\partial\mathbf{w}^*(k)}\right]^T \Delta\mathbf{w}^*(k)\right\} = 2\mathrm{re}\left\{\left[\frac{\partial\xi}{\partial\mathbf{w}^*(k)}\right]^H \Delta\mathbf{w}(k)\right\}
\end{aligned}$$

The term within the real part operator is an inner product, as such the maximum change in the objective function occurs when the change in $\Delta\mathbf{w}(k)$ is in the same direction as $\left[\frac{\partial\xi}{\partial\mathbf{w}^*(k)}\right]$. Therefore, the maximum change of the objective function ξ occurs in the direction $\frac{\partial\xi}{\partial\mathbf{w}^*(k)}$.

Note that the partial derivatives are calculated for each element of $\mathbf{w}(k)$. With this definition, the following relations are valid for the complex scalar parameter case

$$\frac{\partial w_i}{\partial w_i} = 1$$
$$\frac{\partial w_i^*}{\partial w_i} = 0$$
$$\frac{\partial E[|e(k)|^2]}{\partial w_i} = 0 \quad \text{if and only if} \quad \frac{\partial E[|e(k)|^2]}{\partial \text{re}[w_i]} = \frac{\partial E[|e(k)|^2]}{\partial \text{im}[w_i]} = 0$$

The gradient of the MSE with respect to the vector $\mathbf{w}^*$ is given by

$$\mathbf{g}_{\mathbf{w}^*} E\{e(k)e^*(k)\} = E\{-e^*(k)\mathbf{x}(k)\} \tag{A.7}$$

Proof:

In order to compute the gradient of the MSE with respect to the coefficients, we need the expressions for the partial derivatives of the error modulus squared with respect to the real and imaginary parts of the coefficients. These equations are

$$\begin{aligned}\frac{\partial E[|e(k)|^2]}{\partial \text{re}[\mathbf{w}(k)]} &= -E\left[\mathbf{x}^*(k)d(k) + \mathbf{x}(k)d^*(k)\right] + E\left[\mathbf{x}(k)\mathbf{x}^H(k) + \mathbf{x}^*(k)\mathbf{x}^T(k)\right]\text{re}[\mathbf{w}(k)] \\ &\quad -E\{\jmath\mathbf{x}^*(k)\mathbf{x}^T(k)\text{im}[\mathbf{w}(k)]\} + E\{\jmath\mathbf{x}(k)\mathbf{x}^H(k)\text{im}[\mathbf{w}(k)]\}\end{aligned} \tag{A.8}$$

and

$$\begin{aligned}\frac{\partial E[|e(k)|^2]}{\partial \text{im}[\mathbf{w}(k)]} &= -E\{\jmath\left[\mathbf{x}^*(k)d(k) - \mathbf{x}(k)d^*(k)\right]\} - E\{\jmath\mathbf{x}(k)\mathbf{x}^H(k)\text{re}[\mathbf{w}(k)]\} \\ &\quad +E\{\jmath\mathbf{x}^*(k)\mathbf{x}^T(k)\text{re}[\mathbf{w}(k)]\} + E\{\left[\mathbf{x}(k)\mathbf{x}^H(k) + \mathbf{x}^*(k)\mathbf{x}^T(k)\right]\text{im}[\mathbf{w}(k)]\}\end{aligned} \tag{A.9}$$

respectively.

The gradient of the error modulus squared with respect to the complex coefficients can then be computed as

$$\begin{aligned}
\mathbf{g}_{\mathbf{w}^*} E[e(k)e^*(k)] &= \frac{\partial E[|e(k)|^2]}{\partial \mathbf{w}^*(k)} \\
&= \frac{1}{2} E\left\{ -\left[\mathbf{x}^*(k)d(k) + \mathbf{x}(k)d^*(k)\right] + \left[\mathbf{x}(k)\mathbf{x}^H(k) + \mathbf{x}^*(k)\mathbf{x}^T(k)\right] \mathrm{re}[\mathbf{w}(k)] \right. \\
&\quad - \jmath \mathbf{x}^*(k)\mathbf{x}^T(k)\mathrm{im}[\mathbf{w}(k)] + \jmath \mathbf{x}(k)\mathbf{x}^H(k)\mathrm{im}[\mathbf{w}(k)] \\
&\quad + \left[\mathbf{x}^*(k)d(k) - \mathbf{x}(k)d^*(k)\right] + \mathbf{x}(k)\mathbf{x}^H(k)\mathrm{re}[\mathbf{w}(k)] - \mathbf{x}^*(k)\mathbf{x}^T(k)\mathrm{re}[\mathbf{w}(k)] \\
&\quad \left. + \jmath \left[\mathbf{x}(k)\mathbf{x}^H(k) + \mathbf{x}^*(k)\mathbf{x}^T(k)\right] \mathrm{im}[\mathbf{w}(k)] \right\} \\
&= \frac{1}{2} E\left\{ -2\mathbf{x}(k)d^*(k) + 2\mathbf{x}(k)\mathbf{x}^H(k)\mathrm{re}[\mathbf{w}(k)] + 2\jmath \mathbf{x}(k)\mathbf{x}^H(k)\mathrm{im}[\mathbf{w}(k)] \right\} \\
&= E\left\{ -\mathbf{x}(k)d^*(k) + \mathbf{x}(k)\mathbf{x}^H(k)\mathbf{w}(k) \right\} \\
&= E\left\{ \mathbf{x}(k)\left[-d^*(k) + \mathbf{x}^H(k)\mathbf{w}(k)\right] \right\} \\
&= -E\left\{ \left[d^*(k) - \mathbf{w}^T(k)\mathbf{x}^*(k)\right] \mathbf{x}(k) \right\} \\
&= -E\left\{ e^*(k)\mathbf{x}(k) \right\}
\end{aligned}$$

□

The gradient vector of the MSE function related to the filter tap-weight coefficients is then given by

$$\begin{aligned}
E\{\mathbf{g}_{\mathbf{w}^*}[e(k)e^*(k)]\} &= \mathbf{g}_{\mathbf{w}^*} \\
&= \frac{\partial \xi}{\partial \mathbf{w}^*} \\
&= E[-e^*(k)\mathbf{x}(k)] \\
&= -\mathbf{p} + \mathbf{R}\mathbf{w} \qquad \text{(A.10)}
\end{aligned}$$

By equating the gradient vector to zero and assuming $\mathbf{R}$ is nonsingular, the optimal values for the tap-weight coefficients that minimize the objective function leads to the Wiener solution for the complex case given by

$$\mathbf{w}_o = \mathbf{R}^{-1}\mathbf{p} \qquad \text{(A.11)}$$

where $\mathbf{R} = E[\mathbf{x}(k)\mathbf{x}^H(k)]$ and $\mathbf{p} = E[d^*(k)\mathbf{x}(k)]$, assuming that $d^*(k)$ and $\mathbf{x}(k)$ are jointly wide-sense stationary.

A.3 DERIVATION OF THE COMPLEX LMS ALGORITHM

The LMS algorithm employs instantaneous estimates of matrix $\mathbf{R}$, denoted by $\hat{\mathbf{R}}(k)$, and of vector $\mathbf{p}$, denoted by $\hat{\mathbf{p}}(k)$, given by

$$\begin{aligned}
\hat{\mathbf{R}}(k) &= \mathbf{x}(k)\mathbf{x}^H(k) \\
\hat{\mathbf{p}}(k) &= d^*(k)\mathbf{x}(k) \qquad \text{(A.12)}
\end{aligned}$$

Using these estimates the objective function actually minimized is the instantaneous square error $|e(k)|^2$ instead of the MSE. As a result, the expression of the gradient estimate is

$$\begin{aligned}
\hat{\mathbf{g}}_{\mathbf{w}^*}\{e(k)e^*(k)\} &= \frac{\partial |e(k)|^2}{\partial \mathbf{w}^*} \\
&= \frac{1}{2}\left\{\frac{\partial |e(k)|^2}{\partial \text{re}[\mathbf{w}(k)]} + \jmath \frac{\partial |e(k)|^2}{\partial \text{im}[\mathbf{w}(k)]}\right\} \\
&= \frac{1}{2}\left[\frac{\partial |e(k)|^2}{\partial \text{re}[w_0(k)]}\ \frac{\partial |e(k)|^2}{\partial \text{re}[w_1(k)]} \cdots \frac{\partial |e(k)|^2}{\partial \text{re}[w_N(k)]}\right]^T \\
&\quad + \jmath\frac{1}{2}\left[\frac{\partial |e(k)|^2}{\partial \text{im}[w_0(k)]}\ \frac{\partial |e(k)|^2}{\partial \text{im}[w_1(k)]} \cdots \frac{\partial |e(k)|^2}{\partial \text{im}[w_N(k)]}\right]^T \\
&= -e^*(k)\mathbf{x}(k)
\end{aligned} \tag{A.13}$$

A.4 USEFUL RESULTS

Table A.1 shows some useful complex differentiation of scalar and vector functions $f(\mathbf{w})$ and $\mathbf{f}(\mathbf{w})$, respectively, with respect to variable vectors $\mathbf{w}$ and $\mathbf{w}^*$.

Table A.1 Complex Differentiation

Type of Function	Function	Variable w	Variable w*
$f(\mathbf{w})$	$\text{re}[\mathbf{w}^H\mathbf{x}]$	$\frac{1}{2}\mathbf{x}^*$	$\frac{1}{2}\mathbf{x}$
$f(\mathbf{w})$	$\mathbf{w}^H\mathbf{x}$	0	$\mathbf{x}$
$f(\mathbf{w})$	$\mathbf{x}^H\mathbf{w}$	$\mathbf{x}^*$	$\mathbf{0}$
$f(\mathbf{w})$	$\mathbf{w}^H\mathbf{R}\mathbf{w}$	$\mathbf{R}^T\mathbf{w}^*$	$\mathbf{R}\mathbf{w}$
$\mathbf{f}(\mathbf{w})$	$\mathbf{H}_1\mathbf{w}+\mathbf{H}_2\mathbf{w}^*$	$\mathbf{H}_1^T$	$\mathbf{H}_2^T$

References

1. D. H. Brandwood, "A complex gradient operator and its application in adaptive array theory," *IEE Proceedings Parts F and G*, vol. 130, pp. 11-16, Feb. 1983.

2. A. Hjørungnes and D. Gesbert, "Complex-valued matrix differentiation: Techniques and key results," *IEEE Trans. on Signal Processing*, vol. 55, pp. 2740-2746, June 2007.

B

QUANTIZATION EFFECTS IN THE LMS ALGORITHM

B.1 INTRODUCTION

In this appendix, several aspects of the finite-wordlength effects in the LMS algorithm are discussed for the cases of implementations in fixed- and floating-point arithmetics [1]-[3].

B.2 ERROR DESCRIPTION

All scalars and vector elements in the LMS algorithm will deviate from their correct values due to quantization effects. The error generated in any individual quantization is considered to be a zero-mean random variable that is independent of any other errors and quantities related to the adaptive-filter algorithm. The variances of these errors depend on the type of quantization and arithmetic that will be employed in the algorithm implementation.

The errors in the quantities related to the LMS algorithm are defined by

$$n_e(k) = e(k) - e(k)_Q \tag{B.1}$$

$$\mathbf{n_w}(k) = \mathbf{w}(k) - \mathbf{w}(k)_Q \tag{B.2}$$

$$n_y(k) = y(k) - y(k)_Q \tag{B.3}$$

where the subscript Q denotes the quantized form of the given value or vector.

It is assumed that the input signal and desired signal suffer no quantization, so that only internal computation quantizations are taken into account. The effects of quantization in the input and desired signals can be easily taken into consideration separately from other quantization error sources. In the case of the desired signal, the quantization error can be added to the measurement noise, while for the input signal the basic effect at the output of the filter is an additional noise as will be discussed later.

Algorithm B.1

LMS Algorithm Including Quantization

Initialization

$\mathbf{x}(0) = \mathbf{w}(0) = [0\,0\ldots 0]^T$

Do for $k \geq 0$

$e(k)_Q = \big(d(k) - \mathbf{x}^T(k)\mathbf{w}(k)_Q\big)_Q$

$\mathbf{w}(k+1)_Q = (\mathbf{w}(k)_Q + 2\mu e(k)_Q \mathbf{x}(k))_Q$

The following relations describe the computational errors introduced in the LMS algorithm implemented with finite wordlength:

$$e(k)_Q = d(k) - \mathbf{x}^T(k)\mathbf{w}(k)_Q - n_e(k) \tag{B.4}$$

$$\mathbf{w}(k+1)_Q = \mathbf{w}(k)_Q + 2\mu e(k)_Q\mathbf{x}(k) - \mathbf{n_w}(k) \tag{B.5}$$

where $n_e(k)$ is the noise sequence due to quantization in the inner product $\mathbf{x}^T(k)\mathbf{w}(k)_Q$, the additional measurement noise $n(k)$ is included in $d(k)$, and $\mathbf{n_w}(k)$ is a noise vector generated by quantization in the product $2\mu e(k)_Q\mathbf{x}(k)$. The generation of quantization noise as described applies for fixed-point arithmetic, whereas for floating-point arithmetic the addition also introduces quantization error that should be included in $n_e(k)$ and $\mathbf{n_w}(k)$.

The objective now is to study the LMS algorithm behavior when internal computations are performed in finite precision. Algorithm B.1 describes the LMS algorithm including quantization and with presence of additional noise.

Define

$$\Delta\mathbf{w}(k)_Q = \mathbf{w}(k)_Q - \mathbf{w}_o \tag{B.6}$$

where $\mathbf{w}_o$ is the optimal coefficient vector, and considering that

$$d(k) = \mathbf{x}^T(k)\mathbf{w}_o + n(k) \tag{B.7}$$

it then follows that

$$e(k)_Q = \big(d(k) - \mathbf{x}^T(k)\mathbf{w}(k)_Q\big)_Q = -\mathbf{x}^T(k)\Delta\mathbf{w}(k)_Q - n_e(k) + n(k) \tag{B.8}$$

and from equation (B.5)

$$\Delta\mathbf{w}(k+1)_Q = \Delta\mathbf{w}(k)_Q + 2\mu\mathbf{x}(k)\left[-\mathbf{x}^T(k)\Delta\mathbf{w}(k)_Q - n_e(k) + n(k)\right] - \mathbf{n_w}(k) \tag{B.9}$$

This equation can be rewritten as

$$\Delta\mathbf{w}(k+1)_Q = \left[\mathbf{I} - 2\mu\mathbf{x}(k)\mathbf{x}^T(k)\right]\Delta\mathbf{w}(k)_Q + \mathbf{n}'_{\mathbf{w}}(k) \tag{B.10}$$

where

$$\mathbf{n}'_{\mathbf{w}}(k) = 2\mu\mathbf{x}(k)(n(k) - n_e(k)) - \mathbf{n}_{\mathbf{w}}(k) \tag{B.11}$$

For the sake of illustration and completeness, the solution of equation (B.10) is

$$\Delta\mathbf{w}(k+1)_Q = \prod_{i=0}^{k}\left[\mathbf{I} - 2\mu\mathbf{x}(i)\mathbf{x}^T(i)\right]\Delta\mathbf{w}(0)_Q + \sum_{i=0}^{k}\left\{\prod_{j=i+1}^{k}\left[\mathbf{I} - 2\mu\mathbf{x}(j)\mathbf{x}^T(j)\right]\mathbf{n}'_{\mathbf{w}}(i)\right\} \tag{B.12}$$

where we define that for $j = k+1$ in the second product, $\prod_{j=k+1}^{k}[\cdot] = 1$.

B.3 ERROR MODELS FOR FIXED-POINT ARITHMETIC

In the case of fixed-point arithmetic, with rounding assumed for quantization, the error after each product can be modeled as a zero-mean stochastic process, with variance given by [4]-[6]

$$\sigma^2 = \frac{2^{-2b}}{12} \tag{B.13}$$

where b is the number of bits after the sign bit. Here it is assumed that the number of bits after the sign bit for quantities representing signals and filter coefficients are different and given by b_d and b_c, respectively. It is also assumed that the internal signals are properly scaled, so that no overflow occurs during the computations and that the signal values lie between -1 and +1 all the time. The error signals consisting of the elements of $n_e(k)$ and $\mathbf{n}_{\mathbf{w}}(k)$ are all uncorrelated and independent of each other. The variance of $n_e(k)$ and the covariance of $\mathbf{n}_{\mathbf{w}}(k)$ are given by

$$E[n_e^2(k)] = \sigma_e^2 \tag{B.14}$$

$$E[\mathbf{n}_{\mathbf{w}}(k)\mathbf{n}_{\mathbf{w}}^T(k)] = \sigma_{\mathbf{w}}^2\mathbf{I} \tag{B.15}$$

respectively. If distinction is made between data and coefficient wordlengths, the above mentioned variances are given by

$$\sigma_e^2 = \sigma_y^2 = \gamma\frac{2^{-2b_d}}{12} \tag{B.16}$$

$$\sigma_{\mathbf{w}}^2 = \gamma'\frac{2^{-2b_c}}{12} \tag{B.17}$$

where $\gamma' = \gamma = 1$ if the quantization is performed after addition, i.e., products are performed in full precision and only after all the additions in the inner product are finished, the quantization is applied. For quantization after each product, $\gamma = N+1$ where $N+1$ is the number of partial products, and $\gamma' = 1$. Those not familiar with the results of the above equations should consult a basic digital signal processing textbook such as [4], [5], or [6].

Note that $\sigma_{\mathbf{w}}^2$ depends on how the product $2\mu e(k)\mathbf{x}(k)$ is performed. In the above equation, it was assumed that the product was available in full precision, and then a quantization to b_c bits in the fractional part was performed, or equivalently, the product $2\mu e(k)$ in full precision was multiplied by $\mathbf{x}(k)$, and only in the last operation quantization was introduced. In case of quantization of partial results, the variance $\sigma_{\mathbf{w}}^2$ is increased due to the products of partial errors with the remaining product components.

B.4 COEFFICIENT-ERROR-VECTOR COVARIANCE MATRIX

Obviously, internal quantization noise generated during the operation of the LMS algorithm affects its convergence behavior. In this section, we discuss the effects of the finite-wordlength computations on the second-order statistics of the errors in the adaptive-filter coefficients. First, we assume that the quantization noise $n_e(k)$ and the vector $\mathbf{n}_{\mathbf{w}}(k)$ are all independent of the data, of the filter coefficients, and of each other. Also, these quantization errors are all considered zero-mean stochastic processes. With these assumptions, the covariance of the error in the coefficient vector, defined by $E[\Delta\mathbf{w}(k)_Q\Delta\mathbf{w}^T(k)_Q]$, can be easily derived from equations (B.10) and (B.11):

$$\begin{aligned}
\text{cov}[\Delta\mathbf{w}(k+1)_Q] &= E[\Delta\mathbf{w}(k+1)_Q\Delta\mathbf{w}^T(k+1)_Q] \\
&= E\left\{\left[\mathbf{I}-2\mu\mathbf{x}(k)\mathbf{x}^T(k)\right]\Delta\mathbf{w}(k)_Q\Delta\mathbf{w}^T(k)_Q\left[\mathbf{I}-2\mu\mathbf{x}(k)\mathbf{x}^T(k)\right]\right\} \\
&\quad +4\mu^2E[\mathbf{x}(k)\mathbf{x}^T(k)]E[n^2(k)]+4\mu^2E[\mathbf{x}(k)\mathbf{x}^T(k)]E[n_e^2(k)] \\
&\quad +E[\mathbf{n}_{\mathbf{w}}(k)\mathbf{n}_{\mathbf{w}}^T(k)]
\end{aligned} \tag{B.18}$$

Each term on the right-hand side of the above equation can be approximated in order to derive the solution for the overall equation. The only assumption made is the independence between $\mathbf{x}(k)$ and $\Delta\mathbf{w}(k)_Q$ that is reasonably accurate in practice.

The first term in equation (B.18) can be expressed as

$$\begin{aligned}
\mathbf{T}_1 &= \text{cov}[\Delta\mathbf{w}(k)_Q]-2\mu\text{cov}[\Delta\mathbf{w}(k)_Q]E[\mathbf{x}(k)\mathbf{x}^T(k)]-2\mu E[\mathbf{x}(k)\mathbf{x}^T(k)]\text{cov}[\Delta\mathbf{w}(k)_Q] \\
&\quad +4\mu^2E\left\{\mathbf{x}(k)\mathbf{x}^T(k)\text{cov}[\Delta\mathbf{w}(k)_Q]\mathbf{x}(k)\mathbf{x}^T(k)\right\}
\end{aligned} \tag{B.19}$$

The element (i,j) of the last term in the above equation is given by

$$\begin{aligned}
&4\mu^2E\left\{\mathbf{x}(k)\mathbf{x}^T(k)\text{cov}[\Delta\mathbf{w}(k)_Q]\mathbf{x}(k)\mathbf{x}^T(k)\right\}_{i,j} \\
&= 4\mu^2\sum_{m=0}^{N}\sum_{l=0}^{N}\text{cov}[\Delta\mathbf{w}(k)_Q]_{m,l}E[x_i(k)x_m(k)x_l(k)x_j(k)]
\end{aligned} \tag{B.20}$$

where $x_i(k)$ represents the ith element of $\mathbf{x}(k)$. If it is assumed that the elements of the input signal vector are jointly Gaussian, the following relation is valid

$$E[x_i(k)x_m(k)x_l(k)x_j(k)] = \mathbf{R}_{i,m}\mathbf{R}_{l,j}+\mathbf{R}_{m,l}\mathbf{R}_{i,j}+\mathbf{R}_{m,j}\mathbf{R}_{i,l} \tag{B.21}$$

where $\mathbf{R}_{i,j}$ is the element (i, j) of $\mathbf{R}$. Replacing this expression in equation (B.20), it can be shown that

$$\sum_{m=0}^{N}\sum_{l=0}^{N} \text{cov}[\Delta\mathbf{w}(k)_Q]_{m,l} E[x_i(k)x_m(k)x_l(k)x_j(k)]$$
$$= 2\left\{\mathbf{R}\text{cov}[\Delta\mathbf{w}(k)_Q]\mathbf{R}\right\}_{i,j} + \mathbf{R}_{i,j}\text{tr}\left\{\mathbf{R}\text{cov}[\Delta\mathbf{w}(k)_Q]\right\} \quad \text{(B.22)}$$

Using this result in the last term of $\mathbf{T}_1$, it follows that

$$\begin{aligned}\mathbf{T}_1 \;=\;& \text{cov}[\Delta\mathbf{w}(k)_Q] - 2\mu\left\{\mathbf{R}\text{cov}[\Delta\mathbf{w}(k)_Q] + \text{cov}[\Delta\mathbf{w}(k)_Q]\mathbf{R}\right\} \\ &+4\mu^2\left(2\mathbf{R}\text{cov}[\Delta\mathbf{w}(k)_Q]\mathbf{R} + \mathbf{R}\text{tr}\left\{\mathbf{R}\text{cov}[\Delta\mathbf{w}(k)_Q]\right\}\right)\end{aligned} \quad \text{(B.23)}$$

Since the remaining terms in equation (B.18) are straightforward to calculate, replacing equation (B.23) in (B.18) yields

$$\begin{aligned}\text{cov}[\Delta\mathbf{w}(k+1)_Q] \;=\;& (\mathbf{I} - 2\mu\mathbf{R})\text{cov}[\Delta\mathbf{w}(k)_Q] - 2\mu\text{cov}[\Delta\mathbf{w}(k)_Q]\mathbf{R} \\ &+4\mu^2\mathbf{R}\text{tr}\left\{\mathbf{R}\text{cov}[\Delta\mathbf{w}(k)_Q]\right\} + 8\mu^2\mathbf{R}\text{cov}[\Delta\mathbf{w}(k)_Q]\mathbf{R} \\ &+4\mu^2(\sigma_n^2 + \sigma_e^2)\mathbf{R} + \sigma_{\mathbf{w}}^2\mathbf{I}\end{aligned} \quad \text{(B.24)}$$

Before reaching the steady state, the covariance of $\Delta\mathbf{w}(k+1)_Q$ presents a transient behavior that can be analyzed in the same form as equation (3.23). It is worth mentioning that the condition for convergence of the coefficients given in equation (3.30) also guarantees the convergence of the above equation. In fact, equation (B.24) is almost the same as equation (3.23) except for the extra excitation terms σ_e^2 and $\sigma_{\mathbf{w}}^2$ that account for the quantization effects, and, therefore, the behavior of the coefficients in the LMS algorithm in finite precision must resemble its behavior in infinite precision, with the convergence curve shifted up in the finite-precision case.

In most cases, the norm of $\mathbf{R}\text{cov}[\Delta\mathbf{w}(k)_Q]\mathbf{R}$ is much smaller than the norm of $\mathbf{R}\text{tr}\left\{\mathbf{R}\text{cov}[\Delta\mathbf{w}(k)_Q]\right\}$ so that the former term can be eliminated from equation (B.24). Now, by considering in equation (B.24) that in the steady state $\text{cov}[\Delta\mathbf{w}(k)_Q] \approx \text{cov}[\Delta\mathbf{w}(k+1)_Q]$ and applying the trace operation in both sides, it is possible to conclude that

$$\text{tr}\left\{\mathbf{R}\text{cov}[\Delta\mathbf{w}(k)_Q]\right\} = \frac{4\mu^2(\sigma_n^2 + \sigma_e^2)\text{tr}[\mathbf{R}] + (N+1)\sigma_{\mathbf{w}}^2}{4\mu - 4\mu^2\text{tr}[\mathbf{R}]} \quad \text{(B.25)}$$

This expression will be useful to calculate the excess MSE in the finite-precision implementation of the LMS algorithm.

If $x(k)$ is considered a Gaussian white noise with variance σ_x^2, it is possible to calculate the expected value of $||\Delta\mathbf{w}(k)_Q||^2$, defined as the trace of $\text{cov}[\Delta\mathbf{w}(k)_Q]$, from equations (B.24) and (B.25). The result is

$$E[||\Delta\mathbf{w}(k)_Q||^2] = \frac{\mu(\sigma_n^2 + \sigma_e^2)(N+1)}{1 - \mu(N+1)\sigma_x^2} + \frac{(N+1)\sigma_{\mathbf{w}}^2}{4\mu\sigma_x^2[1 - \mu(N+1)\sigma_x^2]} \quad \text{(B.26)}$$

As can be noted, when μ is small, the noise in the calculation of the coefficients plays a major role in the overall error in the adaptive-filter coefficients.

B.5 ALGORITHM STOP

The adaptive-filter coefficients may stop updating due to limited wordlength employed in the internal computation. In particular, for the LMS algorithm, it will occur when

$$|2\mu e(k)_Q \mathbf{x}(k)|_i < 2^{-b_c-1} \tag{B.27}$$

where $|(\cdot)|_i$ denotes the modulus of the ith component of $(\cdot)$. The above condition can be stated in an equivalent form given by

$$4\mu^2(\sigma_e^2 + \sigma_n^2)\sigma_x^2 < 4\mu^2 E[e^2(k)_Q]E[x_i^2(k)] < \frac{2^{-2b_c}}{4} \tag{B.28}$$

where in the first inequality it was considered that the variances of all elements of $\mathbf{x}(k)$ are the same, and that $\sigma_e^2 + \sigma_n^2$ is a lower bound for $E[e^2(k)_Q]$ since the effect of misadjustment due to noise in the gradient is not considered. If μ is chosen such that

$$\mu > \frac{2^{-b_c}}{4\sigma_x\sqrt{\sigma_e^2 + \sigma_n^2}} \tag{B.29}$$

the algorithm will not stop before convergence is reached. If μ is small such that the convergence is not reached, the MSE at the output of the adaptive system will be totally determined by the quantization error. In this case, the quantization error is usually larger than the expected MSE in the infinite-precision implementation.

B.6 MEAN-SQUARE ERROR

The mean-square error of the conventional LMS algorithm in the presence of quantization noise is given by

$$\xi(k)_Q = E[e^2(k)_Q] \tag{B.30}$$

By recalling from equation (B.8) that $e(k)_Q$ can be expressed as

$$e(k)_Q = -\mathbf{x}^T(k)\Delta\mathbf{w}(k)_Q - n_e(k) + n(k)$$

it then follows that

$$\begin{aligned}\xi(k)_Q &= E[\mathbf{x}^T(k)\Delta\mathbf{w}(k)_Q\mathbf{x}^T(k)\Delta\mathbf{w}(k)_Q] + \sigma_e^2 + \sigma_n^2 \\ &= E\left\{\text{tr}[\mathbf{x}(k)\mathbf{x}^T(k)\Delta\mathbf{w}(k)_Q\Delta\mathbf{w}^T(k)_Q]\right\} + \sigma_e^2 + \sigma_n^2 \\ &= \text{tr}\left\{\mathbf{R}\text{cov}[\Delta\mathbf{w}(k)_Q]\right\} + \sigma_e^2 + \sigma_n^2\end{aligned} \tag{B.31}$$

If we replace equation (B.25) in (B.31), the MSE of the adaptive system is given by

$$\begin{aligned}\xi(k)_Q &= \frac{\mu(\sigma_n^2 + \sigma_e^2)\text{tr}[\mathbf{R}]}{1 - \mu\text{tr}[\mathbf{R}]} + \frac{(N+1)\sigma_{\mathbf{w}}^2}{4\mu(1 - \mu\text{tr}[\mathbf{R}])} + \sigma_e^2 + \sigma_n^2 \\ &= \frac{\sigma_e^2 + \sigma_n^2}{1 - \mu\text{tr}[\mathbf{R}]} + \frac{(N+1)\sigma_{\mathbf{w}}^2}{4\mu(1 - \mu\text{tr}[\mathbf{R}])}\end{aligned} \tag{B.32}$$

This formula is valid as long as the algorithm does not stop updating the coefficients. However, the MSE tends to increase in a form similar to that determined in equation (B.32) when μ does not satisfy equation (B.29).

In case the input signal is also quantized, a noise with variance σ_i^2 is generated at the input, causing an increase in the MSE. The model for the input signal is then,

$$\mathbf{x}(k)_Q = \mathbf{x}(k) - \mathbf{n}_i(k) \tag{B.33}$$

In this case the quantized error can be expressed as

$$\begin{aligned}
e(k)_Q &= d(k) - \mathbf{w}^T(k)_Q\mathbf{x}(k)_Q - n_e(k) \\
&= \mathbf{w}_o^T\mathbf{x}(k) + n(k) - \mathbf{w}^T(k)_Q[\mathbf{x}(k) - \mathbf{n}_i(k)] - n_e(k) \\
&= \mathbf{w}_o^T\mathbf{x}(k) + n(k) - \mathbf{w}^T(k)_Q[\mathbf{x}(k) - \mathbf{n}_i(k)] - n_e(k) \\
&= -\Delta\mathbf{w}^T(k)_Q\mathbf{x}(k) - [\mathbf{w}_o^T + \Delta\mathbf{w}^T(k)_Q]\mathbf{n}_i(k) - n_e(k) + n(k)
\end{aligned} \tag{B.34}$$

The basic difference between the above expression and equation (B.8) is the inclusion of the term $-[\mathbf{w}_o^T + \Delta\mathbf{w}^T(k)_Q]\mathbf{n}_i(k)$. By assuming this term is uncorrelated to other terms of the error expression, the MSE in equation (B.32) includes an extra term given by

$$E[(\mathbf{w}_o^T + \Delta\mathbf{w}^T(k)_Q)\mathbf{n}_i(k)(\mathbf{w}_o^T + \Delta\mathbf{w}^T(k)_Q)\mathbf{n}_i(k)]$$

that can be simplified as

$$\begin{aligned}
E[(\mathbf{w}_o^T + \Delta\mathbf{w}^T(k)_Q)\mathbf{n}_i(k)\mathbf{n}_i^T(k)(\mathbf{w}_o + \Delta\mathbf{w}(k)_Q)] &= \mathbf{w}_o^T E[\mathbf{n}_i(k)\mathbf{n}_i^T(k)]\mathbf{w}_o \\
&\quad + E[\Delta\mathbf{w}_Q^T(k)E[\mathbf{n}_i(k)\mathbf{n}_i^T(k)]\Delta\mathbf{w}_Q(k)] \\
&= \sigma_i^2\{\mathbf{w}_o^T\mathbf{w}_o + \mathrm{tr}[\mathrm{cov}(\Delta\mathbf{w}_Q(k))]\} \\
&= \sigma_i^2(||\mathbf{w}_o||^2 + \mathrm{tr}\{\mathrm{cov}[\Delta\mathbf{w}(k)_Q]\}) \\
&\approx \sigma_i^2||\mathbf{w}_o||^2
\end{aligned} \tag{B.35}$$

This additional term due to the input signal quantization leads to an increment in the MSE. As a result of this term being fedback in the algorithm through the error signal generates an extra term in the MSE with the same gain as the measurement noise that is approximately given by

$$\frac{\mu\sigma_i^2\mathrm{tr}[\mathbf{R}]}{1 - \mu\mathrm{tr}[\mathbf{R}]}||\mathbf{w}_o||^2$$

Therefore, the total contribution of the input signal quantization is

$$\xi_i \approx \frac{||\mathbf{w}_o||^2\sigma_i^2}{1 - \mu\mathrm{tr}[\mathbf{R}]} \tag{B.36}$$

where in the above analysis it was considered that the terms with $\sigma_i^2 \cdot \sigma_{\mathbf{w}}^2$, $\sigma_i^2 \cdot \sigma_e^2$, and $\sigma_i^2 \cdot \sigma_n^2$ are small enough to be neglected.

B.7 FLOATING-POINT ARITHMETIC IMPLEMENTATION

A succinct analysis of the quantization effects in the LMS algorithm when implemented in floating-point arithmetic is presented in this section. Most of the derivations are given in the section B.8 and follow closely the procedure of the fixed-point analysis.

In floating-point arithmetic, quantization errors occur after addition and multiplication operations. These errors are respectively modeled as follows [7]:

$$\text{fl}[a+b] = a+b-(a+b)n_a \tag{B.37}$$

$$\text{fl}[a\cdot b] = a\cdot b-(a\cdot b)n_p \tag{B.38}$$

where n_a and n_p are zero-mean random variables that are independent of any other errors. Their variances are respectively given by

$$\sigma_{n_p}^2 \approx 0.18\cdot 2^{-2b} \tag{B.39}$$

and

$$\sigma_{n_a}^2 < \sigma_{n_p}^2 \tag{B.40}$$

where b is the number of bits in the mantissa representation.

The quantized error and the quantized filter coefficients vector are given by

$$e(k)_Q = d(k) - \mathbf{x}^T(k)\mathbf{w}(k)_Q - n_e(k) \tag{B.41}$$

$$\mathbf{w}(k+1)_Q = \mathbf{w}(k)_Q + 2\mu\mathbf{x}(k)e(k)_Q - \mathbf{n_w}(k) \tag{B.42}$$

where $n_e(k)$ and $\mathbf{n_w}(k)$ represent computational errors, and their expressions are given in the section B.8. Since $\mathbf{n_w}(k)$ is a zero-mean vector, it is shown in the section B.8 that on the average $\mathbf{w}(k)_Q$ tends to $\mathbf{w}_o$. Also, it can be shown that

$$\Delta\mathbf{w}(k+1)_Q = [\mathbf{I} - 2\mu\mathbf{x}(k)\mathbf{x}^T(k) + \mathbf{N}_{\Delta\mathbf{w}}(k)]\Delta\mathbf{w}(k) + \mathbf{N}_a'(k)\mathbf{w}_o + 2\mu\mathbf{x}(k)[n(k) - n_e(k)] \tag{B.43}$$

where $\mathbf{N}_{\Delta\mathbf{w}}(k)$ combines several quantization-noise effects as discussed in the section B.8, and $\mathbf{N}_a'(k)$ is a diagonal noise matrix that models the noise generated in the vector addition required to update $\mathbf{w}(k+1)_Q$. The error matrix $\mathbf{N}_{\Delta\mathbf{w}}(k)$ can be considered negligible as compared to $[\mathbf{I}-2\mu\mathbf{x}(k)\mathbf{x}^T(k)]$ and therefore is eliminated in the analysis below.

By following a similar analysis used to derive equation (B.24) in the case of fixed-point arithmetic, we obtain

$$\text{tr}\,\{\mathbf{R}\text{cov}[\Delta\mathbf{w}(k)_Q]\} = \frac{4\mu^2(\sigma_n^2+\sigma_e^2)\text{tr}[\mathbf{R}] + ||\mathbf{w}_o||^2\sigma_{n_a}^2 + \text{tr}\{\text{cov}[\Delta\mathbf{w}(k)]\}\sigma_{n_a}^2}{4\mu - 4\mu^2\text{tr}[\mathbf{R}]} \tag{B.44}$$

where it was considered that all noise sources in matrix $\mathbf{N}_a'(k)$ have the same variance given by $\sigma_{n_a}^2$.

If $x(k)$ is considered a Gaussian white noise with variance σ_x^2, it is straightforward to calculate $E[||\mathbf{w}(k)_Q||^2]$. The expression is given by

$$E[||\mathbf{w}(k)_Q||^2] = \frac{\mu(\sigma_n^2+\sigma_e^2)(N+1)}{1-\mu(N+1)\sigma_x^2} + \frac{||\mathbf{w}_o||^2\sigma_{n_a}^2}{4\mu\sigma_x^2[1-\mu(N+1)\sigma_x^2]} + \frac{\sigma_{n_a}^2\sigma_n^2(N+1)}{4\sigma_x^2[1-\mu(N+1)\sigma_x^2]^2} \tag{B.45}$$

where the expression for $\mathrm{tr}\{\mathrm{cov}[\Delta\mathbf{w}(k)]\}$ used in the above equation is given in the section B.8, equation (B.52). For small values of μ, the quantization of addition in the updating of $\mathbf{w}(k)_Q$ may be the dominant source of error in the adaptive-filter coefficients.

The MSE in the LMS algorithm implemented with floating-point arithmetic is then given by

$$\begin{aligned}\xi(k)_Q &= \mathrm{tr}\{\mathbf{R}\mathrm{cov}[\Delta\mathbf{w}(k)_Q]\} + \sigma_e^2 + \sigma_n^2 \\ &= \frac{(\sigma_n^2+\sigma_e^2)}{1-\mu\mathrm{tr}[\mathbf{R}]} + \frac{||\mathbf{w}_o||^2\sigma_{n_a}^2 + \mathrm{tr}\{\mathrm{cov}[\Delta\mathbf{w}(k)]\}\sigma_{n_a}^2}{4\mu(1-\mu\mathrm{tr}[\mathbf{R}])}\end{aligned} \tag{B.46}$$

For $\mu \ll \frac{1}{\mathrm{tr}[\mathbf{R}]}$, using equation (B.52), and again considering $x(k)$ a Gaussian white noise with variance σ_x^2, the above equation can be simplified as follows:

$$\xi(k)_Q = \sigma_n^2 + \sigma_e^2 + \frac{||\mathbf{w}_o||^2\sigma_{n_a}^2}{4\mu} + \frac{(N+1)\sigma_n^2\sigma_{n_a}^2}{4} \tag{B.47}$$

The ith coefficient of the adaptive filter will not be updated in floating-point implementation if

$$|2\mu e(k)_Q\mathbf{x}(k)|_i < 2^{-b_a-1}|\mathbf{w}(k)|_i \tag{B.48}$$

where $|(\cdot)|_i$ denotes the modulus of the ith component of $(\cdot)$, and b_a is the number of bits in the fractional part of the addition in the coefficient updating. In the steady state we can assume that $\sigma_n^2+\sigma_e^2$ is a lower bound for $E[e^2(k)_Q]$ and equation (B.48) can be equivalently rewritten as

$$4\mu^2(\sigma_n^2+\sigma_e^2)\sigma_x^2 < 4\mu^2E[e^2(k)_Q]E[x_i^2(k)] < \frac{2^{-2b_a}}{4}w_{oi}^2 \tag{B.49}$$

The algorithm will not stop updating before the convergence is achieved, if μ is chosen such that

$$\mu > \frac{2^{-b_a}}{4}\sqrt{\frac{w_{oi}^2}{(\sigma_n^2+\sigma_e^2)\sigma_x^2}} \tag{B.50}$$

In case μ does not satisfy the above condition, the MSE is determined by the quantization error.

B.8 FLOATING-POINT QUANTIZATION ERRORS IN LMS ALGORITHM

In this section, we derive the expressions for the quantization errors generated in the implementation of the LMS algorithm using floating-point arithmetic.

The error in the output error computation is given by

$$
\begin{aligned}
n_e(k) &\approx -n_a(k)[d(k) - \mathbf{x}^T(k)\mathbf{w}(k)_Q] \\
&\quad + \mathbf{x}^T(k) \begin{bmatrix} n_{po}(k) & 0 & 0 & \cdots & 0 \\ 0 & n_{p_1}(k) & \cdots & \cdots & 0 \\ \vdots & & \ddots & \vdots & \vdots \\ 0 & 0 & & & n_{p_N}(k) \end{bmatrix} \mathbf{w}(k)_Q \\
&\quad - [n_{a_1}(k)\; n_{a_2}(k)\; \cdots\; n_{a_N}(k)] \begin{bmatrix} \sum_{i=0}^{1} x(k-i)w_i(k)_Q \\ \sum_{i=0}^{2} x(k-i)w_i(k)_Q \\ \vdots \\ \sum_{i=0}^{N} x(k-i)w_i(k)_Q \end{bmatrix} \\
&= -n_a(k)e(k)_Q - \mathbf{x}^T(k)\mathbf{N}_p(k)\mathbf{w}(k)_Q - \mathbf{n}_a(k)\mathbf{s}_i(k)
\end{aligned}
$$

where $n_{p_i}(k)$ accounts for the noise generated in the products $x(k-i)w_i(k)_Q$ and $n_{a_i}(k)$ accounts for the noise generated in the additions of the product $\mathbf{x}^T(k)\mathbf{w}(k)$. Note that the error terms of second- and higher-order have been neglected.

Using similar assumptions one can show that

$$
\mathbf{n_w}(k) = -2\mu n_p'(k)e(k)_Q\mathbf{x}(k) - 2\mu\mathbf{N}_p''(k)e(k)_Q\mathbf{x}(k) - \mathbf{N}_a'(k)[\mathbf{w}(k)_Q + 2\mu e(k)_Q\mathbf{x}(k)] \tag{B.51}
$$

where

$$
N_p''(k) = \begin{bmatrix} n_{po}''(k) & 0 & \cdots & 0 \\ 0 & n_{p_1}''(k) & \cdots & 0 \\ \vdots & \vdots & \ddots & \vdots \\ 0 & \cdots & \cdots & n_{p_N}''(k) \end{bmatrix}
$$

$$
N_a'(k) = \begin{bmatrix} n_{ao}'(k) & 0 & \cdots & 0 \\ 0 & n_{a_1}'(k) & \cdots & 0 \\ \vdots & \vdots & \ddots & \vdots \\ 0 & \cdots & \cdots & n_{a_N}'(k) \end{bmatrix}
$$

and $n_p'(k)$ accounts for the quantization of the product 2μ by $e(k)_Q$, considering that 2μ is already available. Matrix $\mathbf{N}_p''(k)$ models the quantization in the product of $2\mu e(k)_Q$ by $\mathbf{x}(k)$, while $\mathbf{N}_a'(k)$ models the error in the vector addition used to generate $\mathbf{w}(k+1)_Q$.

If we substitute the expression for $e(k)_Q$ of equation (B.8) in $\mathbf{n_w}(k)$ given in equation (B.51), and use the result in equation (B.11), it can be shown that

$$\begin{aligned}\Delta\mathbf{w}(k+1)_Q &= [\mathbf{I}-2\mu\mathbf{x}(k)\mathbf{x}^T(k)]\Delta\mathbf{w}(k)_Q+2\mu\mathbf{x}(k)[n(k)-n_e(k)]-\mathbf{n_w}(k)\\ &\approx [\mathbf{I}-2\mu\mathbf{x}(k)\mathbf{x}^T(k)+2\mu n_p'(k)\mathbf{x}(k)\mathbf{x}^T(k)\\ &\quad +2\mu\mathbf{N}_p''(k)\mathbf{x}(k)\mathbf{x}^T(k)+2\mu\mathbf{N}_a'(k)\mathbf{x}(k)\mathbf{x}^T(k)\\ &\quad +\mathbf{N}_a'(k)]\Delta\mathbf{w}(k)_Q+\mathbf{N}_a'(k)\mathbf{w}_o+2\mu\mathbf{x}(k)[n(k)-n_e(k)]\end{aligned}$$

where the terms corresponding to products of quantization errors were considered small enough to be neglected.

Finally, the variance of the error noise can be derived as follows:

$$\begin{aligned}\sigma_e^2 &= \sigma_{n_a}^2\xi(k)_Q+\sigma_{n_p}^2\sum_{i=0}^{N}\mathbf{R}_{i,i}\text{cov}[\mathbf{w}(k+1)_Q]_{i,i}\\ &\quad +\sigma_{n_a}^2\left\{E[(\sum_{i=0}^{1}x(k-i)w_i(k)_Q)^2]+E[(\sum_{i=0}^{2}x(k-i)w_i(k)_Q)^2]\right.\\ &\quad \left.+\cdots+E[(\sum_{i=0}^{N}x(k-i)w_i(k)_Q)^2]\right\}\end{aligned}$$

where $\sigma_{n_{a_i}}'^2$ was considered equal to $\sigma_{n_a}^2$, and $[\cdot]_{i,i}$ means diagonal elements of $[\cdot]$. The second term can be further simplified as follows:

$$\begin{aligned}\text{tr}\{\mathbf{R}\text{cov}[\mathbf{w}(k+1)_Q]\} &\approx \sum_{i=0}^{N}\mathbf{R}_{i,i}w_{oi}^2+\mathbf{R}_{i,i}\text{cov}[\Delta\mathbf{w}(k+1)]_{i,i}\\ &\quad +\text{first- and higher-order terms}\cdots\end{aligned}$$

Since this term is multiplied by $\sigma_{n_p}^2$, any first- and higher-order terms can be neglected. The first term of σ_e^2 is also small in the steady state. The last term can be rewritten as

$$\begin{aligned}&\sigma_{n_a}^2\left\{E[(\sum_{i=0}^{1}x(k-i)w_{oi})^2]+E[(\sum_{i=0}^{2}x(k-i)w_{oi})^2]+\cdots+E[(\sum_{i=0}^{N}x(k-i)w_{oi})^2]\right\}\\ &=\sigma_{n_a}^2\left\{\sum_{j=1}^{N}\sum_{i=0}^{j}\mathbf{R}_{i,i}\text{cov}[\Delta\mathbf{w}(k+1)]_{i,i}\right\}\end{aligned}$$

where terms of order higher than one were neglected, $x(k)$ was considered uncorrelated to $\Delta\mathbf{w}(k+1)$, and $\text{cov}[\Delta\mathbf{w}(k+1)]$ was considered a diagonal matrix. Actually, if $x(k)$ is considered a zero-mean Gaussian white noise, from equation (3.23) it can be shown that

$$\text{cov}[\Delta\mathbf{w}(k)]\approx\mu\sigma_n^2\mathbf{I}+\frac{\mu^2(N+1)\sigma_x^2\sigma_n^2\mathbf{I}}{1-\mu(N+1)\sigma_x^2}=\frac{\mu\sigma_n^2\mathbf{I}}{1-\mu(N+1)\sigma_x^2}\qquad\text{(B.52)}$$

Since this term will be multiplied by $\sigma_{n_a}^2$ and $\sigma_{n_p}^2$, it can also be disregarded. In conclusion,

$$\sigma_e^2 \approx \sigma_{n_a}^2 \left\{ E[\sum_{j=1}^{N}(\sum_{i=0}^{j} x(k-i)w_{oi})^2] \right\} + \sigma_{n_p}^2 \sum_{i=0}^{N} \mathbf{R}_{i,i}w_{oi}^2$$

This equation can be further simplified when $x(k)$ is as above described and $\sigma_{n_a}^2 = \sigma_{n_p}^2 = \sigma_d^2$, leading to

$$\sigma_e^2 \approx \sigma_d^2 \left[\sum_{i=1}^{N}(N-i+2)\mathbf{R}_{i,i}w_{oi}^2 - \mathbf{R}_{1,1}w_{o1}^2 \right] = \sigma_d^2\sigma_x^2 \left[\sum_{i=1}^{N}(N-i+2)w_{oi}^2 - w_{o1}^2 \right]$$

References

1. M. Andrews and R. Fitch, "Finite wordlength arithmetic computational error effects on the LMS adaptive weights," *Proc. IEEE Intern. Conf. Acoust., Speech, and Signal Processing*, pp. 628-631, May 1977.

2. C. Caraiscos and B. Liu, "A roundoff error analysis of the LMS adaptive algorithm," *IEEE Trans. on Acoust., Speech, and Signal Processing*, vol. ASSP-32, pp. 34-41, Feb. 1984.

3. S. T. Alexander, "Transient weight misadjustment properties for the finite precision LMS algorithm," *IEEE Trans. on Acoust., Speech, and Signal Processing*, vol. ASSP-35, pp. 1250-1258, Sept. 1987.

4. A. V. Oppenheim and R. W. Schaffer, *Discrete-Time Signal Processing*, Prentice Hall, Englewood Cliffs, NJ, 1989.

5. A. Antoniou, *Digital Signal Processing: Signals, Systems, and Filters*, McGraw Hill, New York, NY, 2005.

6. P. S. R. Diniz, E. A. B. da Silva, and S. L. Netto, *Digital Signal Processing: System Analysis and Design*, Cambridge University Press, Cambridge, UK, 2002.

7. A. B. Spirad and D. L. Snyder, "Quantization errors in floating-point arithmetic," *IEEE Trans. on Acoust., Speech, and Signal Processing*, vol. ASSP-26, pp. 456-464, Oct. 1983.

C

QUANTIZATION EFFECTS IN THE RLS ALGORITHM

C.1 INTRODUCTION

In this appendix, several aspects of the finite-wordlength effects in the RLS algorithm are discussed for the cases of implementation with fixed- and floating-point arithmetic [4]-[7], [8]-[11].

C.2 ERROR DESCRIPTION

All the elements of matrices and vectors in the RLS algorithm will deviate from their correct values due to quantization effects. The error generated in any individual quantization is considered to be a zero-mean random variable that is independent of any other error and quantities related to the adaptive-filter algorithm. The variances of these errors depend on the type of quantization and arithmetic that will be applied in the algorithm implementation.

The errors in the quantities related to the conventional RLS algorithm are defined by

$$n_e(k) = e(k) - e(k)_Q \tag{C.1}$$

$$\mathbf{n}_{\psi}(k) = \mathbf{S}_D(k-1)_Q\mathbf{x}(k) - [\mathbf{S}_D(k-1)_Q\mathbf{x}(k)]_Q \tag{C.2}$$

$$\mathbf{N}_{\mathbf{S}_D}(k) = \mathbf{S}_D(k) - \mathbf{S}_D(k)_Q \tag{C.3}$$

$$\mathbf{n}_{\mathbf{w}}(k) = \mathbf{w}(k) - \mathbf{w}(k)_Q \tag{C.4}$$

$$n_y(k) = y(k) - y(k)_Q \tag{C.5}$$

$$n_{\varepsilon}(k) = \varepsilon(k) - \varepsilon(k)_Q \tag{C.6}$$

where the subscript Q denotes the quantized form of the given matrix, vector, or scalar.

It is assumed that the input signal and desired signal suffer no quantization, so only quantizations of internal computations are taken into account. With the above definitions, the following relations describe the computational error in some quantities of interest related to the RLS algorithm:

$$e(k)_Q = d(k) - \mathbf{x}^T(k)\mathbf{w}(k-1)_Q - n_e(k) \tag{C.7}$$

$$\mathbf{w}(k)_Q = \mathbf{w}(k-1)_Q + \mathbf{S}_D(k)_Q\mathbf{x}(k)e(k)_Q - \mathbf{n}_{\mathbf{w}}(k) \tag{C.8}$$

where $n_e(k)$ is the noise sequence due to quantization in the inner product $\mathbf{x}^T(k)\mathbf{w}(k-1)_Q$ and $\mathbf{n_w}(k)$ is a noise vector due to quantization in the product $\mathbf{S}_D(k)_Q\mathbf{x}(k)e(k)_Q$.

The development here is intended to study the algorithm behavior when the internal signals, vectors, and matrices are available in quantized form as happens in a practical implementation. This means that, for example in Algorithm 5.2, all the informations needed from the previous time interval $(k-1)$ to update the adaptive filter at instant k are available in quantized form.

Now we can proceed with the analysis of the deviation in the coefficient vector generated by the quantization error. By defining

$$\Delta\mathbf{w}(k)_Q = \mathbf{w}(k)_Q - \mathbf{w}_o \tag{C.9}$$

and considering that

$$d(k) = \mathbf{x}^T(k)\mathbf{w}_o + n(k)$$

then it follows that

$$e(k)_Q = -\mathbf{x}^T(k)\Delta\mathbf{w}(k-1)_Q - n_e(k) + n(k) \tag{C.10}$$

and

$$\Delta\mathbf{w}(k)_Q = \Delta\mathbf{w}(k-1)_Q + \mathbf{S}_D(k)_Q\mathbf{x}(k)[-\mathbf{x}^T(k)\Delta\mathbf{w}(k-1)_Q - n_e(k) + n(k)] - \mathbf{n_w}(k) \tag{C.11}$$

Equation (C.11) can be rewritten as follows:

$$\Delta\mathbf{w}(k)_Q = [\mathbf{I} - \mathbf{S}_D(k)_Q\mathbf{x}(k)\mathbf{x}^T(k)]\Delta\mathbf{w}(k-1)_Q + \mathbf{n'_w}(k) \tag{C.12}$$

where

$$\mathbf{n'_w}(k) = \mathbf{S}_D(k)_Q\mathbf{x}(k)[n(k) - n_e(k)] - \mathbf{n_w}(k) \tag{C.13}$$

The solution of equation (C.12) can be calculated as

$$\begin{aligned}\Delta\mathbf{w}(k)_Q &= \prod_{i=0}^{k}\left[\mathbf{I} - \mathbf{S}_D(i)_Q\mathbf{x}(i)\mathbf{x}^T(i)\right]\Delta\mathbf{w}(-1)_Q \\ &\quad + \sum_{i=0}^{k}\left\{\prod_{j=i+1}^{k}[\mathbf{I} - \mathbf{S}_D(j)_Q\mathbf{x}(j)\mathbf{x}^T(j)]\right\}\mathbf{n'_w}(i)\end{aligned} \tag{C.14}$$

where in the last term of the above equation for $i = k$, we consider that

$$\prod_{j=k+1}^{k}[\cdot] = 1$$

Now, if we rewrite Algorithm 5.2 taking into account that any calculation in the present updating generates quantization noise, we obtain Algorithm C.1 that describes the RLS algorithm with quantization and additional noise taken into account. Notice that Algorithm C.1 is not a new algorithm.

Algorithm C.1

RLS Algorithm Including Quantization

Initialization
$\mathbf{S}_D(-1) = \delta\mathbf{I}$
where δ can be the inverse of an estimate of the input signal power.
$\mathbf{x}(-1) = \mathbf{w}(-1) = [0\,0 \ldots 0]^T$
Do for $k \geq 0$
$e(k)_Q = d'(k) - \mathbf{x}^T(k)\mathbf{w}(k-1)_Q - n_e(k) + n(k)$
$\boldsymbol{\psi}(k)_Q = \mathbf{S}_D(k-1)_Q\mathbf{x}(k) - \mathbf{n}_{\boldsymbol{\psi}}(k)$
$\mathbf{S}_D(k)_Q = \frac{1}{\lambda}\left[\mathbf{S}_D(k-1)_Q - \frac{\boldsymbol{\psi}(k)_Q\boldsymbol{\psi}^T(k)_Q}{\lambda+\boldsymbol{\psi}^T(k)_Q\mathbf{x}(k)}\right] - \mathbf{N}_{\mathbf{S}_D}(k)$
$\mathbf{w}(k)_Q = \mathbf{w}(k-1)_Q + e(k)_Q\mathbf{S}_D(k)_Q\mathbf{x}(k) - \mathbf{n}_{\mathbf{w}}(k)$
If necessary compute
$y(k)_Q = \mathbf{w}^T(k)_Q\mathbf{x}(k) - n_y(k)$
$\varepsilon(k)_Q = d(k) - y_Q(k)$

C.3 ERROR MODELS FOR FIXED-POINT ARITHMETIC

In the case of fixed-point arithmetic, with rounding assumed for quantization, the error after each product can be modeled as a zero-mean stochastic process, with variance given by [1]-[2]

$$\sigma^2 = \frac{2^{-2b}}{12} \tag{C.15}$$

where b is the number of bits after the sign bit. Here it is assumed that the number of bits after the sign bit for quantities representing signals and filter coefficients are different, and given by b_d and b_c, respectively. It is also assumed that the internal signals are properly scaled, so that no overflow occurs during the computations, and that the signal values are between -1 and $+1$. If in addition independence between errors is assumed, each element in equations (C.1) to (C.6) is on average zero. The respective covariance matrices are given by

$$E[n_e^2(k)] = E[n_\varepsilon^2(k)] = \sigma_e^2 \tag{C.16}$$
$$E[\mathbf{N}_{\mathbf{S}_D}(k)\mathbf{N}^T_{\mathbf{S}_D}(k)] = \sigma^2_{\mathbf{S}_D}\mathbf{I} \tag{C.17}$$
$$E[\mathbf{n}_{\mathbf{w}}(k)\mathbf{n}^T_{\mathbf{w}}(k)] = \sigma^2_{\mathbf{w}}\mathbf{I} \tag{C.18}$$
$$E[\mathbf{n}_{\boldsymbol{\psi}}(k)\mathbf{n}^T_{\boldsymbol{\psi}}(k)] = \sigma^2_{\boldsymbol{\psi}}\mathbf{I} \tag{C.19}$$
$$E[n_y^2(k)] = \sigma_y^2 \tag{C.20}$$

If distinction is made between data and coefficient wordlengths, the noise variances of data and coefficients are respectively given by

$$\sigma_e^2 = \sigma_y^2 = \gamma \frac{2^{-2b_d}}{12} \tag{C.21}$$

$$\sigma_{\mathbf{w}}^2 = \gamma' \frac{2^{-2b_c}}{12} \tag{C.22}$$

where $\gamma' = \gamma = 1$ if the quantization is performed after addition, i.e., the products are performed in full precision and the quantization is applied only after all the additions in the inner product are finished. For quantization after each product, then $\gamma = N+1$ and $\gamma' = N+2$, since each quantization in the partial product generates an independent noise, and the number of products in the error computation is $N+1$ whereas in the coefficient computation it is $N+2$.

As an illustration, it is shown how to calculate the value of the variance $\sigma_{\mathbf{S}_D}^2$ when making some simplifying assumptions. The value of $\sigma_{\mathbf{S}_D}^2$ depends on how the computations to generate $\mathbf{S}_D(k)$ are performed. Assume the multiplications and divisions are performed with the same wordlength and that the needed divisions are performed once, followed by the corresponding scalar matrix product. Also, assuming the inner product quantizations are performed after the addition, each element of the matrix $\mathbf{S}_D(k)_Q$ requires five multiplications[1] considering that $1/\lambda$ is prestored. The diagonal elements of equation (C.17) consist of $N+1$ noise autocorrelations, each with variance $5\sigma_\psi^2$. The desired result is then given by

$$\sigma_{\mathbf{S}_D}^2 = 5(N+1)\sigma_\psi^2 \tag{C.23}$$

where σ_ψ^2 is the variance of each multiplication error.

C.4 COEFFICIENT-ERROR-VECTOR COVARIANCE MATRIX

Assume that the quantization signals $n_e(k), n(k)$, and the vector $\mathbf{n_w}(k)$ are all independent of the data, filter coefficients, and of each other. Also, assuming that these errors are all zero-mean stochastic processes, the covariance matrix of the coefficient-error vector given by $E[\Delta\mathbf{w}(k)_Q \Delta\mathbf{w}^T(k)_Q]$ can be derived from equations (C.12) and (C.13)

$$\begin{aligned}
\text{cov}[\Delta\mathbf{w}(k)_Q] &= E[\Delta\mathbf{w}(k)_Q \Delta\mathbf{w}^T(k)_Q] \\
&= E\{[\mathbf{I} - \mathbf{S}_D(k)_Q \mathbf{x}(k)\mathbf{x}^T(k)]\Delta\mathbf{w}(k-1)_Q \Delta\mathbf{w}^T(k-1)_Q \\
&\quad [\mathbf{I} - \mathbf{x}(k)\mathbf{x}^T(k)\mathbf{S}_D(k)_Q]\} \\
&\quad + E[\mathbf{S}_D(k)_Q \mathbf{x}(k)\mathbf{x}^T(k)\mathbf{S}_D(k)_Q]E[n^2(k)] \\
&\quad + E[\mathbf{S}_D(k)_Q \mathbf{x}(k)\mathbf{x}^T(k)\mathbf{S}_D(k)_Q]E[n_e^2(k)] \\
&\quad + E[\mathbf{n_w}(k)\mathbf{n_w}^T(k)]
\end{aligned} \tag{C.24}$$

[1]One is due to the inner product at the denominator; one is due to the division; one is due to the product of the division result by $1/\lambda$; one is to calculate the elements of the outer product of the numerator; the other is the result of quantization of the product of the last two terms.

The above equation can be approximated in the steady state, where each term on the right-hand side will be considered separately. It should be noted that during the derivations it is implicitly assumed that the algorithm follows closely the behavior of its infinite-precision counterpart. This assumption can always be considered as true if the wordlengths used are sufficiently long. However, under short-wordlength implementation this assumption might not be true as will be discussed later on.

Term 1:

The elements of $\Delta\mathbf{w}(k-1)_Q$ can be considered independent of $\mathbf{S}_D(k)_Q$ and $\mathbf{x}(k)$. In this case, the first term in equation (C.24) can be expressed as

$$\begin{aligned}\mathbf{T}_1 = {} & \text{cov}[\Delta\mathbf{w}(k-1)_Q] - \text{cov}[\Delta\mathbf{w}(k-1)_Q]E[\mathbf{x}(k)\mathbf{x}^T(k)\mathbf{S}_D(k)_Q] \\ & -E[\mathbf{S}_D(k)_Q\mathbf{x}(k)\mathbf{x}^T(k)]\text{cov}[\Delta\mathbf{w}(k-1)_Q] \\ & +E\left\{\mathbf{S}_D(k)_Q\mathbf{x}(k)\mathbf{x}^T(k)\text{cov}[\Delta\mathbf{w}(k-1)_Q]\mathbf{x}(k)\mathbf{x}^T(k)\mathbf{S}_D(k)_Q\right\}\end{aligned} \quad \text{(C.25)}$$

If it is recalled that $\mathbf{S}_D(k)_Q$ is the unquantized $\mathbf{S}_D(k)$ matrix disturbed by a noise matrix that is uncorrelated to the input signal vector, then in order to compute the second and third terms of $\mathbf{T}_1$ it suffices to calculate

$$E[\mathbf{S}_D(k)\mathbf{x}(k)\mathbf{x}^T(k)] \approx E[\mathbf{S}_D(k)]\,E\left[\mathbf{x}(k)\mathbf{x}^T(k)\right] \quad \text{(C.26)}$$

where the approximation is justified by the fact that $\mathbf{S}_D(k)$ is slowly varying as compared to $\mathbf{x}(k)$ when $\lambda \rightarrow 1$. Using equation (5.55) it follows that

$$E\left[\mathbf{S}_D(k)\mathbf{x}(k)\mathbf{x}^T(k)\right] \approx \frac{1-\lambda}{1-\lambda^{k+1}}\mathbf{I} \quad \text{(C.27)}$$

Now we need to use stronger assumptions for $\mathbf{S}_D(k)$ than those considered in the above equation. If the matrix $E[\mathbf{S}_D(k)_Q]$ is assumed to be approximately constant for large k (see the discussions around equation (5.54)), the last term in $\mathbf{T}_1$ can be approximated by

$$\begin{aligned}& E\left\{\mathbf{S}_D(k)_Q\mathbf{x}(k)\mathbf{x}^T(k)\text{cov}[\Delta\mathbf{w}(k-1)_Q]\mathbf{x}(k)\mathbf{x}^T(k)\mathbf{S}_D(k)_Q\right\} \\ & \approx E[\mathbf{S}_D(k)_Q]E\left\{\mathbf{x}(k)\mathbf{x}^T(k)\text{cov}[\Delta\mathbf{w}(k-1)_Q]\mathbf{x}(k)\mathbf{x}^T(k)\right\}E[\mathbf{S}_D(k)_Q]\end{aligned} \quad \text{(C.28)}$$

If it is further assumed that the elements of the input signal vector are jointly Gaussian, then each element of the middle term in the last equation can be given by

$$\begin{aligned}& E\left\{\mathbf{x}(k)\mathbf{x}^T(k)\text{cov}[\Delta\mathbf{w}(k-1)_Q]\mathbf{x}(k)\mathbf{x}^T(k)\right\}_{i,j} \\ & = \sum_{m=0}^{N}\sum_{l=0}^{N}\text{cov}[\Delta\mathbf{w}(k-1)_Q]_{ml}E[x_i(k)x_m(k)x_l(k)x_j(k)] \\ & = 2\{\mathbf{R}\text{cov}[\Delta\mathbf{w}(k-1)_Q]\mathbf{R}\}_{i,j} + [\mathbf{R}]_{i,j}\text{tr}\{\mathbf{R}\text{cov}[\Delta\mathbf{w}(k-1)_Q]\}\end{aligned} \quad \text{(C.29)}$$

where $[\cdot]_{i,j}$ denotes the ith, jth element of the matrix $[\cdot]$. It then follows that

$$\begin{aligned}E\left\{\mathbf{x}(k)\mathbf{x}^T(k)\text{cov}[\Delta\mathbf{w}(k-1)_Q]\mathbf{x}(k)\mathbf{x}^T(k)\right\} = {} & 2\mathbf{R}\text{cov}[\Delta\mathbf{w}(k-1)_Q]\mathbf{R} \\ & +\mathbf{R}\text{tr}\left\{\mathbf{R}\text{cov}[\Delta\mathbf{w}(k-1)_Q]\right\}\end{aligned} \quad \text{(C.30)}$$

The last term of $\mathbf{T}_1$ in equation (C.25), after simplified, yields

$$2\left(\frac{1-\lambda}{1-\lambda^{k+1}}\right)^2 \text{cov}[\Delta\mathbf{w}(k-1)_Q] + \left(\frac{1-\lambda}{1-\lambda^{k+1}}\right)^2 \text{tr}\left\{\mathbf{R}\text{cov}[\Delta\mathbf{w}(k-1)_Q]\right\}\mathbf{R}^{-1}$$
$$+E\left\{\mathbf{N}_{\mathbf{S}_D}(k)\mathbf{x}(k)\mathbf{x}^T(k)\text{cov}[\Delta\mathbf{w}(k-1)_Q]\mathbf{x}(k)\mathbf{x}^T(k)\mathbf{N}_{\mathbf{S}_D}(k)\right\} \quad \text{(C.31)}$$

After a few manipulations, it can be shown that the third term in the above equation is nondiagonal with $\mathbf{N}_{\mathbf{S}_D}(k)$ being symmetric for the RLS algorithm described in Algorithm C.1. On the other hand, if the matrix $\mathbf{R}$ is diagonal dominant, that is in general the case, the third term of (C.31) becomes approximately diagonal and given by[2]

$$\mathbf{T}_S(k) \approx \sigma^2_{\mathbf{S}_D}\sigma_x^4 \text{tr}\{\text{cov}[\Delta\mathbf{w}(k-1)_Q]\}\mathbf{I} \quad \text{(C.32)}$$

where σ_x^2 is the variance of the input signal. This term, which is proportional to a quantization noise variance, can actually be neglected in the analysis, since it has in general much smaller norm then the remaining terms in $\mathbf{T}_1$.

Terms 2 and 3:

Using the same arguments applied before, such as $\mathbf{S}_D(k)$ is almost fixed as $\lambda \to 1$, then the main result required to calculate the terms 2 and 3 of equation (C.24) is approximately given by

$$\begin{aligned} E[\mathbf{S}_D(k)_Q\mathbf{x}(k)\mathbf{x}^T(k)\mathbf{S}_D(k)_Q] &\approx E[\mathbf{S}_D(k)]\mathbf{R}E[\mathbf{S}_D(k)] + E[\mathbf{N}_{\mathbf{S}_D}(k)\mathbf{R}\mathbf{N}_{\mathbf{S}_D}(k)] \\ &\approx \left(\frac{1-\lambda}{1-\lambda^{k+1}}\right)^2 \mathbf{R}^{-1} \end{aligned} \quad \text{(C.33)}$$

where the term $E[\mathbf{N}_{\mathbf{S}_D}(k)\mathbf{R}\mathbf{N}_{\mathbf{S}_D}(k)]$ can be neglected because it is in general much smaller than the remaining term. In addition, it will be multiplied by a small variance when equation (C.33) is replaced back in equation (C.24). From equations (C.24), (C.28), (C.33), (C.16), (C.18), and (C.22) it follows that

$$\begin{aligned} \text{cov}[\Delta\mathbf{w}(k)_Q] &= \left[1 - 2\left(\frac{1-\lambda}{1-\lambda^{k+1}}\right) + 2\left(\frac{1-\lambda}{1-\lambda^{k+1}}\right)^2\right]\text{cov}[\Delta\mathbf{w}(k-1)_Q] \\ &\quad + \left(\frac{1-\lambda}{1-\lambda^{k+1}}\right)^2 \text{tr}\{\mathbf{R}\text{cov}[\Delta\mathbf{w}(k-1)_Q]\}\mathbf{R}^{-1} \\ &\quad + \left(\frac{1-\lambda}{1-\lambda^{k+1}}\right)^2 (\sigma_n^2 + \sigma_e^2)\mathbf{R}^{-1} + \sigma^2_{\mathbf{w}}\mathbf{I} \end{aligned} \quad \text{(C.34)}$$

Now, by considering in equation (C.34) that in the steady state $\text{cov}[\Delta\mathbf{w}(k)_Q] \approx \text{cov}[\Delta\mathbf{w}(k-1)_Q]$, multiplying the resulting expression by $\mathbf{R}$, and calculating the trace of the final equation, it can be shown that

$$\text{tr}\{\mathbf{R}\text{cov}[\Delta\mathbf{w}(k-1)_Q]\} \approx \frac{(1-\lambda)^2(N+1)(\sigma_n^2+\sigma_e^2) + \sigma^2_{\mathbf{w}}\text{tr}(\mathbf{R})}{(1-\lambda)[2\lambda - (1-\lambda)(N+1)]} \quad \text{(C.35)}$$

[2]The proof is not relevant but following the lines of equation (C.30) and considering that its last term is the most relevant, the result follows.

where it was considered that $\lambda^{k+1} \rightarrow 0$. Replacing the equation (C.35) in (C.34), and computing the steady-state solution the following equation results

$$\text{cov}[\Delta\mathbf{w}(k)_Q] \approx \frac{(1-\lambda)(\sigma_n^2+\sigma_e^2)}{2\lambda-(1-\lambda)(N+1)}\mathbf{R}^{-1} + \frac{(1-\lambda)\text{tr}(\mathbf{R})\mathbf{R}^{-1}+[2\lambda-(1-\lambda)(N+1)]\mathbf{I}}{2(1-\lambda)\lambda[2\lambda-(1-\lambda)(N+1)]}\sigma_{\mathbf{w}}^2 \tag{C.36}$$

Finally, if the trace of the above equation is calculated considering that $x(k)$ is a Gaussian white noise with variance σ_x^2, and that $2\lambda \gg (1-\lambda)(N+1)$ for $\lambda \rightarrow 1$, the resulting expected value of $||\Delta\mathbf{w}(k)_Q||^2$ is

$$E[||\Delta\mathbf{w}(k)_Q||^2] \approx \frac{(1-\lambda)(N+1)}{2\lambda}\frac{\sigma_n^2+\sigma_e^2}{\sigma_x^2} + \frac{(N+1)\sigma_{\mathbf{w}}^2}{2\lambda(1-\lambda)} \tag{C.37}$$

As can be noted if the value of λ is very close to one, the square errors in the tap coefficients tend to increase and to become more dependent of the tap coefficient wordlengths. On the other hand, if λ is not close to one, in general for fast tracking purposes, the effects of the additive noise and data wordlength become more disturbing to the coefficient square errors. The optimum value for λ close to 1, as far as quantization effects are concerned, can be derived by calculating the derivative of $E||\Delta\mathbf{w}(k)_Q||^2]$ with respect to λ and setting the result to zero

$$\lambda_{\text{opt}} \approx 1 - \frac{\sigma_{\mathbf{w}}\sigma_x}{\sqrt{\sigma_n^2+\sigma_e^2}} \tag{C.38}$$

where it was assumed that $(2\lambda-1) \approx 1$.

By noting that $\frac{1-\lambda}{1-\lambda^{k+1}}$ should be replaced by $\frac{1}{k+1}$ when $\lambda = 1$, it can be shown from equation (C.34) that the algorithm tends to diverge when $\lambda = 1$, since in this case $||\text{cov}[\Delta\mathbf{w}(k)_Q]||$ is growing with k.

C.5 ALGORITHM STOP

In some cases the adaptive-filter tap coefficients may stop adapting due to quantization effects. In particular, the conventional RLS algorithm will freeze when the coefficient updating term is not representable with the available wordlength. This occurs when its modulus is smaller than half the value of the least significant bit, i.e.,

$$|e(k)_Q\mathbf{S}_D(k)_Q\mathbf{x}(k)|_i < 2^{-b_c-1} \tag{C.39}$$

where $|\;|_i$ denotes the modulus of the ith component. Equivalently it can be concluded that updating will be stopped if

$$E[e(k)_Q^2]E[|\mathbf{S}_D(k)_Q\mathbf{x}(k)\mathbf{x}^T(k)\mathbf{S}_D(k)_Q|_{ii}] \approx \left(\frac{1-\lambda}{1-\lambda^{k+1}}\right)^2\frac{\sigma_e^2+\sigma_n^2}{\sigma_x^2} < 2^{-2b_c-2} \tag{C.40}$$

where $x(k)$ was considered a Gaussian white noise with variance σ_x^2, and the following approximation was made $E[e(k)_Q^2] \approx \sigma_e^2+\sigma_n^2$.

For a given coefficient wordlength b_c, the algorithm can always be kept updating if

$$\lambda < 1 - 2^{-b_c-1}\frac{\sigma_x}{\sqrt{\sigma_e^2 + \sigma_n^2}} \tag{C.41}$$

On the other hand, if the above condition is not satisfied, it can be expected that the algorithm will stop updating in

$$k \approx \frac{\sqrt{\sigma_e^2 + \sigma_n^2}}{\sigma_x}2^{b_c+1} - 1 \tag{C.42}$$

iterations for $\lambda = 1$, and

$$k \approx \frac{\ln[(\lambda - 1)\frac{\sqrt{\sigma_e^2+\sigma_n^2}}{\sigma_x}2^{b_c+1} + 1]}{\ln \lambda} - 1 \tag{C.43}$$

iterations for $\lambda < 1$.

In the case $\lambda = 1$ the algorithm always stops updating. If σ_n^2 and b_c are not large, any steady-state analysis for the RLS algorithm when $\lambda = 1$ does not apply, since the algorithm stops prematurely. Because of that, the norm of the covariance of $\Delta\mathbf{w}(k)_Q$ does not become unbounded.

C.6 MEAN-SQUARE ERROR

The MSE in the conventional RLS algorithm in the presence of quantization noise is given by

$$\xi(k)_Q = E[\varepsilon^2(k)_Q] \tag{C.44}$$

By recalling that $\varepsilon(k)_Q$ can be expressed as

$$\varepsilon(k)_Q = -\mathbf{x}^T(k)\Delta\mathbf{w}(k)_Q - n_e(k) + n(k) \tag{C.45}$$

it then follows that

$$\begin{aligned}\xi(k)_Q &= E[\mathbf{x}^T(k)\Delta\mathbf{w}(k)_Q\mathbf{x}^T(k)\Delta\mathbf{w}(k)_Q] + \sigma_e^2 + \xi_{\min} \\ &= E\left\{\text{tr}[\mathbf{x}(k)\mathbf{x}^T(k)\Delta\mathbf{w}(k)_Q\Delta\mathbf{w}^T(k)_Q]\right\} + \sigma_e^2 + \xi_{\min} \\ &= \text{tr}\left\{\mathbf{R}\text{cov}[\Delta\mathbf{w}(k)_Q]\right\} + \sigma_e^2 + \xi_{\min}\end{aligned} \tag{C.46}$$

By replacing equation (C.35) in (C.46), it can be concluded that

$$\xi(k)_Q = \frac{(1-\lambda)^2(N+1)(\sigma_n^2 + \sigma_e^2) + \sigma_{\mathbf{w}}^2\text{tr}\mathbf{R}}{(1-\lambda)[2\lambda - (1-\lambda)(N+1)]} + \xi_{\min} + \sigma_e^2 \tag{C.47}$$

If it is again assumed that $x(k)$ is a Gaussian white noise with variance σ_x^2 and that $2\lambda \gg (1-\lambda)(N+1)$ for $\lambda \to 1$, the MSE expression can be simplified to

$$\xi(k)_Q \approx \xi_{\min} + \sigma_e^2 + \frac{(N+1)\sigma_{\mathbf{w}}^2\sigma_x^2}{2\lambda(1-\lambda)} \tag{C.48}$$

C.7 FIXED-POINT IMPLEMENTATION ISSUES

The implementation of the conventional RLS algorithm in fixed-point arithmetic must consider the possibility of occurrence of overflow and underflow during the computations. In general, some scaling must be performed in certain quantities of the RLS algorithm to avoid undesired behavior due to overflow and underflow. The scaling procedure must be applied in almost all computations required in the conventional RLS algorithm [7], increasing the computational complexity and/or the implementation control by a large amount. A possible solution is to leave enough room in the integer and fractional parts of the number representation, in order to avoid frequent overflows and underflows and also avoid the use of cumbersome scaling strategies. In other words, a fixed-point implementation does require a reasonable number of bits to represent each quantity.

The error propagation analysis can be performed by studying the behavior of the difference between each quantity of the algorithm calculated in infinite precision and finite precision. This analysis allows the detection of divergence of the algorithm due to quantization error accumulation. The error propagation analysis for the conventional RLS algorithm reveals divergence behavior linked to the fact that $\mathbf{S}_D(k)$ loses the positive definiteness property [7]. The main factors contributing to divergence are:

- Large maximum eigenvalue in the matrix $\mathbf{R}$ that amplifies some terms in propagation error of the $\mathbf{S}_D(k)$ matrix. In this case, $\mathbf{S}_D(k)$ might have a small minimum eigenvalue, being as consequence "almost" singular.
- A small number of bits used in the calculations increases the roundoff noise contributing to divergence.
- The forgetting factor when small, turns the memory of the algorithm short, making the matrix $\mathbf{S}_D(k)$ deviate from its expected steady-state value and more likely to lose the positive definiteness property.

Despite these facts, the conventional RLS algorithm can be implemented without possibility of divergence if some special quantization strategies for the internal computations are used [7]. These quantization strategies, along with adaptive scaling strategies, must be used when implementing the conventional RLS algorithm in fixed-point arithmetic with short wordlength.

C.8 FLOATING-POINT ARITHMETIC IMPLEMENTATION

In this section, a succinct analysis of the quantization effects in the conventional RLS algorithm when implemented in floating-point arithmetic is presented. Most of the derivations are given in section C.9 and follow closely the procedure of the fixed-point analysis.

In floating-point arithmetic, quantization errors are injected after multiplication and addition operations, and are modeled as follows [3]:

$$\text{fl}[a+b] = a + b - (a+b)n_a \tag{C.49}$$
$$\text{fl}[a \cdot b] = a \cdot b - a \cdot b \cdot n_p \tag{C.50}$$

where n_a and n_p are zero-mean random variables that are independent of any other errors. Their variances are given by

$$\sigma_{n_p}^2 \approx 0.18\, 2^{-2b} \tag{C.51}$$

and

$$\sigma_{n_a}^2 < \sigma_{n_p}^2 \tag{C.52}$$

where b is the number of bits in the mantissa representation.

The quantized error and the quantized coefficient vector are given by

$$e(k)_Q = d'(k) - \mathbf{x}^T(k)\mathbf{w}(k-1)_Q - n_e(k) + n(k) \tag{C.53}$$
$$\mathbf{w}(k)_Q = \mathbf{w}(k-1)_Q + \mathbf{S}_D(k)_Q\mathbf{x}(k)e(k)_Q - \mathbf{n_w}(k) \tag{C.54}$$

where $n_e(k)$ and $\mathbf{n_w}(k)$ represent computational errors and their expressions are given in section C.9. Since $\mathbf{n_w}(k)$ is a zero-mean vector, it is shown in section C.9 that on average $\mathbf{w}(k)_Q$ tends to $\mathbf{w}_o$. Also, it can be shown that

$$\begin{aligned}\Delta\mathbf{w}(k)_Q &= [\mathbf{I} - \mathbf{S}_D(k)_Q\mathbf{x}(k)\mathbf{x}^T(k) + \mathbf{N}_{\Delta\mathbf{w}}(k)]\Delta\mathbf{w}(k-1) \\ &\quad + \mathbf{N}'_a(k)\mathbf{w}_o + \mathbf{S}_D(k)_Q\mathbf{x}(k)[n(k) - n_e(k)]\end{aligned} \tag{C.55}$$

where $\mathbf{N}_{\Delta\mathbf{w}}(k)$ combines several quantization noise effects as discussed in section C.9 and $\mathbf{N}'_a(k)$ is a diagonal noise matrix that models the noise generated in the vector addition required to update $\mathbf{w}(k)_Q$.

The covariance matrix of $\Delta\mathbf{w}(k)_Q$ can be calculated through the same procedure previously used in the fixed-point case, resulting in

$$\begin{aligned}\text{cov}[\Delta\mathbf{w}(k)_Q] &\approx \frac{(1-\lambda)(\sigma_n^2 + \sigma_e^2)\mathbf{R}^{-1}}{2\lambda - (1-\lambda)(N+1)} \\ &\quad + \frac{(1-\lambda)\mathbf{R}^{-1}\text{tr}\left\{\mathbf{R}\text{diag}[w_{oi}^2]\right\} + [2\lambda - (1-\lambda)(N+1)]\text{diag}[w_{oi}^2]}{2(1-\lambda)\lambda[2\lambda - (1-\lambda)(N+1)]}\sigma_{n'_a}^2\end{aligned} \tag{C.56}$$

where $\mathbf{N}_{\mathbf{S}_D}(k)$ of equation (C.3) and $\mathbf{N}_{\Delta\mathbf{w}}(k)$ were considered negligible as compared to the remaining matrices multiplying $\Delta\mathbf{w}(k-1)$ in equation (C.55). The expression of $\sigma_{n'_a}^2$ is given by equation (C.52). The term $\text{diag}[w_{oi}^2]$ represents a diagonal matrix formed with the squared elements of $\mathbf{w}_o$.

The expected value of $||\Delta\mathbf{w}(k)_Q||^2$ in the floating-point case is approximately given by

$$E[||\Delta\mathbf{w}(k)_Q||^2] \approx \frac{(1-\lambda)(N+1)}{2\lambda}\frac{\sigma_n^2 + \sigma_e^2}{\sigma_x^2} + \frac{1}{2\lambda(1-\lambda)}||\mathbf{w}_o||^2\sigma_{n'_a}^2 \tag{C.57}$$

where it was considered that $x(k)$ is a Gaussian white noise with variance σ_x^2 and that $2\lambda \gg (1-\lambda)(N+1)$ for $\lambda \rightarrow 1$. If the value of λ is very close to one, the squared errors in the tap coefficients tend to increase. Notice that the second term on the right-hand side of the above equation turns these errors more dependent on the precision of the vector addition of the taps updating. For λ not very close to one, the effects of the additive noise and data wordlength become more pronounced. In floating-point implementation, the optimal value of λ as far as quantization effects are concerned is given by

$$\lambda_{\text{opt}} = 1 - \frac{\sigma_{n'_a}\sigma_x}{\sqrt{\sigma_n^2 + \sigma_e^2}}||\mathbf{w}_o|| \tag{C.58}$$

where this relation was obtained by calculating the derivative of equation (C.57) with respect to λ, and equalizing the result to zero in order to reach the value of λ that minimizes the $E[||\Delta\mathbf{w}(k)_Q||^2]$. For $\lambda = 1$, like in the fixed-point case, $||\text{cov}[\Delta\mathbf{w}(k)_Q]||$ is also a growing function that can make the conventional RLS algorithm diverge.

The algorithm may stop updating if

$$|e(k)_Q\mathbf{S}_D(k)\mathbf{x}(k)|_i < 2^{-b_c-1}w_i(k) \tag{C.59}$$

where $|\;|_i$ is the modulus of the ith component and b_c is the number of bits in the mantissa of the coefficients representation. Following the same procedure to derive equation (C.40), we can infer that the updating will be stopped if

$$\left(\frac{1-\lambda}{1-\lambda^{k+1}}\right)^2 \frac{\sigma_e^2 + \sigma_n^2}{\sigma_x^2} < 2^{-2b_c-2}|w_{oi}|^2 \tag{C.60}$$

where w_{oi} is the ith element of $\mathbf{w}_o$.

The updating can be continued indefinitely if

$$\lambda < 1 - 2^{-b_c-1}\frac{\sigma_x|w_{oi}|}{\sqrt{\sigma_e^2 + \sigma_n^2}} \tag{C.61}$$

In the case λ does not satisfy the above condition, the algorithm will stop updating the ith tap in approximately

$$k = \frac{\sqrt{\sigma_e^2 + \sigma_n^2}}{\sigma_x|w_{oi}|} - 1 \tag{C.62}$$

iterations for $\lambda = 1$, and

$$k \approx \frac{\ln[(\lambda-1)\frac{\sqrt{\sigma_e^2+\sigma_n^2}}{\sigma_x|w_{oi}|}2^{-b_c-1} + 1]}{\ln\lambda} - 1 \tag{C.63}$$

iterations for $\lambda < 1$.

Following the same procedure as in the fixed-point implementation, it can be shown that the MSE in the floating-point case is given by

$$\begin{aligned}\xi(k)_Q &= \text{tr}\{\mathbf{R}\text{cov}[\Delta\mathbf{w}(k)_Q]\} + \sigma_e^2 + \xi_{\min} \\ &\approx \frac{(1-\lambda)^2(N+1)(\sigma_n^2+\sigma_e^2) + \sigma_{n'_a}^2\,\text{tr}\left\{\mathbf{R}\text{diag}[w_{oi}^2]\right\}}{(1-\lambda)[2\lambda-(1-\lambda)(N+1)]} + \sigma_e^2 + \xi_{\min}\end{aligned} \tag{C.64}$$

where σ_{ε}^2 was considered equal to σ_e^2. If $x(k)$ is a Gaussian white noise with variance σ_x^2 and $2\lambda \gg (1-\lambda)(N+1)$ for $\lambda \to 1$, the MSE can be approximated by

$$\xi(k)_Q \approx \xi_{\min} + \sigma_e^2 + \frac{||\mathbf{w}_o||^2 \sigma_{n_a'}^2 \sigma_x^2}{2\lambda(1-\lambda)} \tag{C.65}$$

Note that σ_e^2 has a somewhat complicated expression that is given in section C.9.

Finally, it should be mentioned that in floating-point implementations the matrix $\mathbf{S}_D(k)$ can also lose its positive definite property [10]. In [7], it was mentioned that if no interactions between errors is considered, preserving the symmetry of $\mathbf{S}_D(k)$ is enough to keep it positive definite. However, interactions between errors do exist in practice, so the conventional RLS algorithm can become unstable in floating-point implementations unless some special quantization procedures are employed in the actual implementation. An alternative is to use numerically stable RLS algorithms discussed in Chapters 7, 8, and 9.

C.9 FLOATING-POINT QUANTIZATION ERRORS IN RLS ALGORITHM

The error in the *a priori* output error computation is given by

$$\begin{aligned}
n_e(k) \;\approx\; & -n_a(k)[d(k) - \mathbf{x}^T(k)\mathbf{w}(k-1)_Q] \\
& -\mathbf{x}^T(k) \begin{bmatrix} n_{p_o}(k) & 0 & 0 & \cdots & 0 \\ 0 & n_{p_1}(k) & \cdots & \cdots & 0 \\ \vdots & & \ddots & & \\ 0 & 0 & \cdots & \cdots & n_{p_N}(k) \end{bmatrix} \mathbf{w}(k-1)_Q \\
& -[n_{a_1}(k)\, n_{a_2}(k) \ldots n_{a_N}(k)] \begin{bmatrix} \sum_{i=0}^{1} x(k-i)w_i(k-1)_Q \\ \sum_{i=0}^{2} x(k-i)w_i(k-1)_Q \\ \vdots \\ \sum_{i=0}^{N} x(k-i)w_i(k-1)_Q \end{bmatrix} \\
= \;& -n_a(k)e(k)_Q - \mathbf{x}^T(k)\mathbf{N}_p(k)\mathbf{w}(k-1)_Q - \mathbf{n}_a(k)\mathbf{s}_i(k)
\end{aligned}$$

where $n_{p_i}(k)$ accounts for the noise generated in the products $x(k-i)w_i(k-1)_Q$ and $n_{a_i}(k)$ accounts for the noise generated in the additions of the product $\mathbf{x}^T(k)\mathbf{w}(k-1)$. Please note that the error terms of second- and higher-order have been neglected.

Using similar assumptions one can show that

$$\mathbf{n_w}(k) = -\left\{\mathbf{n}_{Sx}(k)e(k)_Q + \mathbf{S}_D(k)_Q\mathbf{N'}_p(k)\mathbf{x}(k)e(k)_Q \right.$$
$$\left. +\mathbf{N''}_p(k)\mathbf{S}_D(k)_Q\mathbf{x}(k)e(k)_Q + \mathbf{N'}_a(k)[\mathbf{w}(k-1) + \mathbf{S}_D(k)_Q\mathbf{x}(k)e(k)_Q]\right\}$$

where

$$\mathbf{n}_{Sx}(k) = \begin{bmatrix} \sum_{j=1}^{N} n'_{a_{1,j}}(k) \sum_{i=0}^{j} \mathbf{S}_{D_{1,i}}(k)_Q x(k-i) \\ \vdots \\ \sum_{j=1}^{N} n'_{a_{N+1,j}}(k) \sum_{i=0}^{j} \mathbf{S}_{D_{N+1,i}}(k)_Q x(k-i) \end{bmatrix}$$

$$\mathbf{N'}_a(k) = \begin{bmatrix} n'_{a_o}(k) & 0 & \cdots & 0 \\ 0 & n'_{a_1}(k) & & \vdots \\ \vdots & & & \vdots \\ 0 & \cdots & \cdots & n'_{a_N}(k) \end{bmatrix}$$

$$\mathbf{N'}_p(k) = \begin{bmatrix} n'_{p_o}(k) & 0 & \cdots & 0 \\ 0 & n'_{p_1}(k) & & \vdots \\ \vdots & & \ddots & \vdots \\ 0 & \cdots & \cdots & n'_{p_N}(k) \end{bmatrix}$$

$$\mathbf{N''}_p(k) = \begin{bmatrix} n''_{p_{1,1}}(k) & n''_{p_{1,2}}(k) & \cdots & n''_{p_{1,N+1}}(k) \\ n''_{p_{2,1}}(k) & n''_{p_{2,2}}(k) & & \vdots \\ \vdots & & \ddots & \vdots \\ n''_{p_{N+1,1}}(k) & \cdots & \cdots & n''_{p_{N+1,N+1}}(k) \end{bmatrix}$$

The vector $\mathbf{n}_{Sx}(k)$ is due to the quantization of additions in the matrix product $\mathbf{S}_D(k)\mathbf{x}(k)$, while the matrix $\mathbf{N''}_p(k)$ accounts for product quantizations in the same operation. The matrix $\mathbf{N'}_a(k)$ models the error in the vector addition to generate $\mathbf{w}(k)_Q$, while $\mathbf{N'}_p(k)$ models the quantization in the product of $e(k)$ by $\mathbf{S}_D(k)_Q\mathbf{x}(k)$.

By replacing $d'(k)$ by $\mathbf{x}^T(k)\mathbf{w}_o$ in the expression of $e(k)_Q$ given in equation (C.7), it follows that

$$e(k)_Q = -\mathbf{x}^T(k)\Delta\mathbf{w}(k-1)_Q - n'_e(k) + n(k)$$

By using in the above equation the expression of $\mathbf{w}(k)_Q$ of equation (C.8) (after subtracting $\mathbf{w}_o$ in each side of the equation), and neglecting the second- and higher-order errors, after some manipulations the following equality results

$$\Delta\mathbf{w}(k)_Q = [\mathbf{I} - \mathbf{S}_D(k)_Q\mathbf{x}(k)\mathbf{x}^T(k) + \mathbf{n}_{Sx}\mathbf{x}^T(k) + \mathbf{S}_D(k)_Q\mathbf{N'}_p(k)\mathbf{x}(k)\mathbf{x}^T(k)$$
$$+\mathbf{N}''_p(k)\mathbf{S}_D(k)_Q\mathbf{x}(k)\mathbf{x}^T(k) + \mathbf{N'}_a(k)\mathbf{S}_D(k)_Q\mathbf{x}(k)\mathbf{x}^T(k) + \mathbf{N'}_a(k)]\Delta\mathbf{w}(k-1)_Q$$
$$+\mathbf{N'}_a(k)\mathbf{w}_o + \mathbf{S}_D(k)_Q\mathbf{x}(k)[n(k) - n'_e(k)]$$

Since all the noise components in the above equation have zero mean, on average the tap coefficients will converge to their optimal values because the same dynamic equation describes the evolution of $\Delta\mathbf{w}(k)$ and $\Delta\mathbf{w}(k)_Q$.

Finally, the variance of the *a priori* error noise can be derived as follows:

$$\sigma_e^2 = \sigma_\varepsilon^2 = \sigma_{n_a}^2\xi(k)_Q + \sigma_{n_p}^2\sum_{i=0}^{N}\mathbf{R}_{i,i}\text{cov}[\mathbf{w}(k)_Q]_{i,i} + \sigma_{n_a}^2\left\{E\left[\left(\sum_{i=0}^{1}x(k-i)w_i(k-1)_Q\right)^2\right]\right.$$
$$\left.+E\left[\left(\sum_{i=0}^{2}x(k-i)w_i(k-1)_Q\right)^2\right] + \cdots + E\left[\left(\sum_{i=0}^{N}x(k-i)w_i(k-1)_Q\right)^2\right]\right\}$$

where $\sigma_{n'_{ai}}^2 = \sigma_{n_a}^2$ was used and $[\;]_{i,i}$ means diagonal elements of $[\;]$. The second term in the above equation can be further simplified as follows:

$$\text{tr}\{\mathbf{R}\text{cov}[\mathbf{w}(k)_Q]\} \approx \sum_{i=0}^{N}\mathbf{R}_{i,i}w_{oi}^2 + \sum_{i=0}^{N}\mathbf{R}_{i,i}\text{cov}[\Delta\mathbf{w}(k)]_{i,i}$$
$$+\text{first} - \text{ and higher} - \text{order terms } \cdots$$

Since this term is multiplied by $\sigma_{n_p}^2$, any first- and higher-order terms can be neglected. The first term of σ_e^2 is also small in the steady state. The last term can be rewritten as

$$\sigma_{n_a}^2\left\{E\left[\left(\sum_{i=0}^{1}x(k-i)w_{oi}\right)^2\right] + E\left[\left(\sum_{i=0}^{2}x(k-i)w_{oi}\right)^2\right] + \cdots\right.$$
$$\left.+E\left[\left(\sum_{i=0}^{N}x(k-i)w_{oi}\right)^2\right]\right\} = \sigma_{n_a}^2\left\{\sum_{j=1}^{N}\sum_{i=0}^{j}\mathbf{R}_{i,i}[\text{cov}(\Delta\mathbf{w}(k))]_{i,i}\right\}$$

where terms of order higher than one were neglected, $x(k)$ was considered uncorrelated to $\Delta\mathbf{w}(k)$, and $\text{cov}[\Delta\mathbf{w}(k)]$ was considered a diagonal matrix. Actually, if $x(k)$ is considered a zero-mean Gaussian white noise from the proof of equation (5.36) and equation (5.55), it can be shown that

$$\text{cov}[\Delta\mathbf{w}(k)] \approx \frac{\sigma_n^2}{\sigma_x^2}\mathbf{I}$$

Since this term will be multiplied by $\sigma_{n_a}^2$ and $\sigma_{n_p}^2$, it can also be disregarded. In conclusion

$$\sigma_e^2 \approx \sigma_{n_a}^2\left\{E[\sum_{j=1}^{N}(\sum_{i=0}^{j}x(k-i)w_{oi})^2]\right\} + \sigma_{n_p}^2\sum_{i=0}^{N}\mathbf{R}_{i,i}w_{oi}^2$$

This equation can be simplified further when $x(k)$ is as above described and $\sigma_{na}^2 = \sigma_{n_p}^2 = \sigma_d^2$

$$\sigma_e^2 \approx \sigma_d^2 \left[\sum_{i=1}^{N} (N-i+2)\mathbf{R}_{i,i} w_{oi}^2 - \mathbf{R}_{1,1} w_{o1}^2 \right]$$
$$= \sigma_d^2 \sigma_x^2 \left[\sum_{i=1}^{N} (N-i+2) w_{oi}^2 - w_{o1}^2 \right]$$

References

1. A. Antoniou, *Digital Signal Processing: Signals, Systems, and Filters*, McGraw Hill, New York, NY, 2005.

2. P. S. R. Diniz, E. A. B. da Silva, and S. L. Netto, *Digital Signal Processing: System Analysis and Design*, Cambridge University Press, Cambridge, UK, 2002.

3. A. B. Spirad and D. L. Snyder, "Quantization errors in floating-point arithmetic," *IEEE Trans. on Acoust., Speech, and Signal Processing*, vol. ASSP-26, pp. 456-464, Oct. 1983.

4. S. H. Ardalan, "Floating-point analysis of recursive least-squares and least-mean squares adaptive filters," *IEEE Trans. on Circuits and Systems*, vol. CAS-33, pp. 1192-1208, Dec. 1986.

5. J. M. Cioffi, "Limited precision effects in adaptive filtering," *IEEE Trans. on Circuits and Systems*, vol. CAS-34, pp. 821-833, July 1987.

6. S. H. Ardalan and S. T. Alexander, "Fixed-point roundoff error analysis of the exponentially windowed RLS algorithm for time-varying systems," *IEEE Trans. on Acoust., Speech, and Signal Processing*, vol. ASSP-35, pp. 770-783, June 1983.

7. G. E. Bottomley and S. T. Alexander, "A novel approach for stabilizing recursive least squares filters," *IEEE Trans. on Signal Processing*, vol. 39, pp. 1770-1779, Aug. 1991.

8. F. Ling and J. G. Proakis, "Numerical accuracy and stability: two problems of adaptive estimation algorithms caused by round-off error," *Proc. IEEE Intern. Conf. on Acoust., Speech, Signal Processing*, San Diego, CA, pp. 30.3.1-30.3.4, March 1984.

9. G. Kubin, "Stabilization of the RLS algorithm in the absence of persistent excitation," *Proc. IEEE Intern. Conf. on Acoust., Speech, Signal Processing*, NY, pp. 1369-1372, Apr. 1988.

10. M. H. Verhaegen, "Round-off error propagation in four generally applicable, recursive, least squares estimation schemes," *Automatica*, vol. 25, pp. 437-444, Mar. 1989.

11. T. Adali and S. H. Ardalan, "Steady state and convergence characteristics of the fixed-point RLS algorithm," *Proc. IEEE Intern. Symposium on Circuits Systems,* New Orleans, LA, pp. 788-791, May 1990.

D

KALMAN FILTERS

D.1 INTRODUCTION

This section provides a brief description of Kalman filter that can be considered an extension of the Wiener filtering concept [1]. The Kalman filter has as objective the minimization of the estimation square error of a nonstationary signal buried in noise. The estimated signal itself is modeled utilizing the state-space formulation [2] describing its dynamical behavior. In summary, Kalman filtering deals with random processes described using state-space modeling which generate signals that can be measured and processed utilizing time recursive estimation formulas. The presentation here is brief and addresses the case of signals and noises represented in vector form, for more details in this subject the reader can consult many books available presenting Kalman filtering, including [3], [4]. There are many different ways to describe the Kalman filtering problem and to derive its corresponding relations, here we follow the presentations of [5]-[6].

D.2 STATE-SPACE MODEL

A convenient form of representing some dynamic systems is through what is called the state-space representation [2]. In such description, the outputs of the memory elements are considered as the system states. The state signals are collected in a vector denoted as $\mathbf{x}(k)$ which are in turn generated from its previous state $\mathbf{x}(k-1)$ and from an external signal vector denoted as $\mathbf{n}(k)$. The observed or measured signals are collected in another vector denoted as $\mathbf{y}(k)$ whose elements originate from linear combinations of the current state variables and of external signals represented in $\mathbf{n}_1(k)$. If we know the values of the external signals $\mathbf{n}(k)$ and $\mathbf{n}_1(k)$, we can determine the current values of the system states, which will be the delay inputs, and the system observation vector as follows:

$$\begin{cases} \mathbf{x}(k) = \mathbf{A}(k-1)\mathbf{x}(k-1) + \mathbf{B}(k)\mathbf{n}(k) \\ \mathbf{y}(k) = \mathbf{C}^T(k)\mathbf{x}(k) + \mathbf{D}(k)\mathbf{n}_1(k) \end{cases} \tag{D.1}$$

where $\mathbf{x}(k)$ is the $(N+1) \times 1$ vector of the state variables. If M is the number of system inputs and L is the number of system outputs, we then have that $\mathbf{A}(k-1)$ is $(N+1) \times (N+1)$, $\mathbf{B}(k)$ is $(N+1) \times M$, $\mathbf{C}(k)$ is $(N+1) \times L$, and $\mathbf{D}(k)$ is $L \times L$[1].

Fig. D.1 shows the state-space system which generates the observation vector $\mathbf{y}(k)$ having as inputs the noise vectors $\mathbf{n}(k)$ and $\mathbf{n}_1(k)$, where the state variables $\mathbf{x}(k)$ are processes generated with excitation noise $\mathbf{n}(k)$.

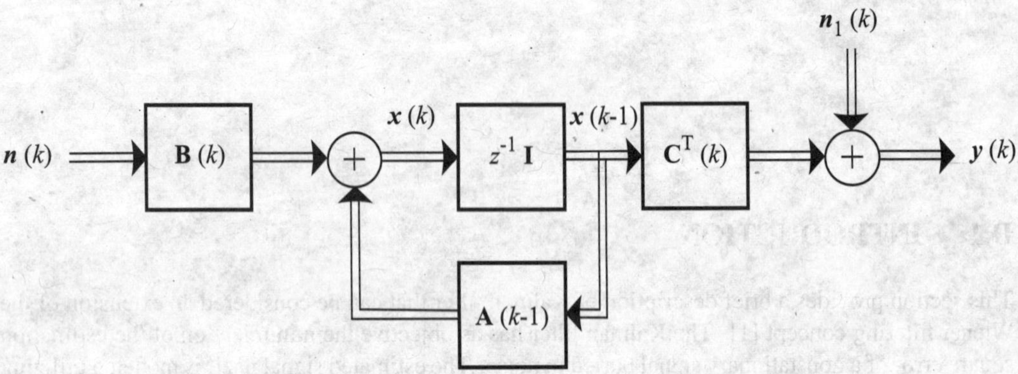

Figure D.1 State-space model for Kalman filtering formulation.

The recursive solution of equation (D.1) can be described as

$$\mathbf{x}(k) = \prod_{l=0}^{k-1} \mathbf{A}(l)\mathbf{x}(0) + \sum_{i=1}^{k} \left[\prod_{l=i}^{k-1} \mathbf{A}(l) \right] \mathbf{B}(i)\mathbf{n}(i) \tag{D.2}$$

where $\prod_{l=k}^{k-1} \mathbf{A}(l) = 1$.

D.2.1 Simple Example

Let's describe a particular example where we assume the signal $x(k)$ is a sample of an autoregressive process generated from the output of a system described by a linear difference equation given by

$$x(k) = \sum_{i=1}^{N} -a_i(k-1)x(k-i) + n(k) \tag{D.3}$$

where $n(k)$ is a white noise. The coefficients $a_i(k-1)$, for $i = 1, 2 \ldots, N$, are the time-varying parameters of the AR process. As part of the Kalman filtering procedure is the estimation of $x(k)$ from noisy measurements denoted as $y_l(k)$ for $l = 1, 2, \ldots, L$.

[1] In standard state-space formulation the matrix $\mathbf{D}(k)$ represents a feedforward connection between the input and the output of the dynamic system, in this discussion this matrix in not a feedforward matrix and is considered to be identity.

We can collect a sequence of signals to be estimated and noise measurements in vector forms as

$$\mathbf{x}(k) = \begin{bmatrix} x(k) \\ x(k-1) \\ \vdots \\ x(k-N) \end{bmatrix}$$

$$\mathbf{y}(k) = \begin{bmatrix} y_1(k) \\ y_2(k) \\ \vdots \\ y_L(k) \end{bmatrix} \tag{D.4}$$

where L represents the number of observations collected in $\mathbf{y}(k)$.

Each entry of the observation vector is considered to be generated through the following model

$$y_l(k) = \mathbf{c}_l^T(k)\mathbf{x}(k) + n_{1,l}(k) \tag{D.5}$$

where $n_{1,l}(k)$ for $l = 1, 2, \ldots, L$ are also white noises uncorrelated with each other and with $n(k)$.

Applying the state-space formulation to the particular set of equations (D.3) and (D.5) leads to a block of state variables originating from an autoregressive process described by

$$\begin{aligned}
\mathbf{x}(k) &= \begin{bmatrix} x(k) \\ x(k-1) \\ \vdots \\ x(k-N) \end{bmatrix} \\
&= \begin{bmatrix} -a_1(k-1) & -a_2(k-1) & \cdots & -a_{N-1}(k-1) & -a_N(k-1) \\ 1 & 0 & \cdots & 0 & 0 \\ 0 & 1 & \cdots & 0 & 0 \\ \vdots & \vdots & \ddots & \vdots & \vdots \\ 0 & 0 & \cdots & 1 & 0 \end{bmatrix} \begin{bmatrix} x(k-1) \\ x(k-2) \\ \vdots \\ x(k-N-1) \end{bmatrix} \\
&\quad + \begin{bmatrix} 1 \\ 0 \\ \vdots \\ 0 \end{bmatrix} n(k) \\
\mathbf{y}(k) &= \begin{bmatrix} \mathbf{c}_1^T(k) \\ \mathbf{c}_2^T(k) \\ \vdots \\ \mathbf{c}_L^T(k) \end{bmatrix} \begin{bmatrix} x(k-1) \\ x(k-2) \\ \vdots \\ x(k-N) \end{bmatrix} + \mathbf{n}_1(k)
\end{aligned} \tag{D.6}$$

where for this case of single-input and multiple-output system $\mathbf{B}(k)$ is $(N+1) \times M$ whose only nonzero element is the entry $(1,1)$ that equals one, $\mathbf{C}(k)$ is $(N+1) \times L$, and $\mathbf{D}(k)$ is just an identity matrix since the measurement noise contributes to the elements of the observation vector in an uncoupled form.

D.3 KALMAN FILTERING

In the following discussion we derive the Kalman filter for the general state-space description of equation (D.1). For that its assumed we know

$$\mathbf{R}_{n_1}(k) = E[\mathbf{n}_1(k)\mathbf{n}_1^T(k)] \tag{D.7}$$
$$\mathbf{R}_{n}(k) = E[\mathbf{n}(k)\mathbf{n}^T(k)] \tag{D.8}$$

$\mathbf{A}(k-1)$ and $\mathbf{C}(k)$, and that $\mathbf{n}(k)$ and $\mathbf{n}_1(k)$ are zero-mean white processes and uncorrelated to each other.

By assuming that we have the measurements $\mathbf{y}(k)$ available and that we employ all the data available up to a given iteration, we seek the optimal estimate of the state vector $\mathbf{x}(k)$, denoted by $\hat{\mathbf{x}}(k|k)$. As justified along the Kalman filtering derivation, the optimal solution has the following general form

$$\hat{\mathbf{x}}(k|k) = \mathbf{A}(k-1)\hat{\mathbf{x}}(k-1|k-1) + \mathbf{K}(k)\left[\mathbf{y}(k) - \mathbf{C}^T(k)\mathbf{A}(k-1)\hat{\mathbf{x}}(k-1|k-1)\right] \tag{D.9}$$

where $\mathbf{K}(k)$ is the $(N+1) \times L$ matrix called Kalman gain. The reader can notice that:

- The term $\mathbf{A}(k-1)\hat{\mathbf{x}}(k-1|k-1)$ tries to bring the contribution of the previous estimation of the state variable to the current one, as suggests the state-space equation (D.1).
- The term $\left[\mathbf{y}(k) - \mathbf{C}^T(k)\mathbf{A}(k-1)\hat{\mathbf{x}}(k-1|k-1)\right]$ is a correction term consisting of the difference between the observation vector and its estimate given by $\mathbf{C}^T(k)\mathbf{A}(k-1)\hat{\mathbf{x}}(k-1|k-1)$, which in turn is a function the previous state-variable estimate.
- The Kalman gain aims at filtering out estimation errors and noise so that the state variable gets the best possible correction term, which minimizes the MSE.

In order to derive the optimal solution for the Kalman gain let's first consider two cases where the estimate of $\mathbf{x}(k)$ is computed using observation data available until iteration k and another until iteration $k-1$, denoted by $\hat{\mathbf{x}}(k|k)$ and $\hat{\mathbf{x}}(k|k-1)$, respectively. The estimation error vectors in these cases are defined by

$$\mathbf{e}(k|k) = \mathbf{x}(k) - \hat{\mathbf{x}}(k|k) \tag{D.10}$$
$$\mathbf{e}(k|k-1) = \mathbf{x}(k) - \hat{\mathbf{x}}(k|k-1) \tag{D.11}$$

These errors have covariance matrices defined as

$$\mathbf{R}_e(k|k) = E[\mathbf{e}(k|k)\mathbf{e}^T(k|k)] \tag{D.12}$$
$$\mathbf{R}_e(k|k-1) = E[\mathbf{e}(k|k-1)\mathbf{e}^T(k|k-1)] \tag{D.13}$$

Given an instant $k-1$ when the informations $\hat{\mathbf{x}}(k-1|k-1)$ and $\mathbf{R}_e(k-1|k-1)$ are available, we first try to estimate $\hat{\mathbf{x}}(k|k-1)$ which does not require the current observation. Whenever a new observation $\mathbf{y}(k)$ is available, $\hat{\mathbf{x}}(k|k)$ is estimated.

According to equation (D.1), at a given iteration the actual state-space vector evolves as

$$\mathbf{x}(k) = \mathbf{A}(k-1)\mathbf{x}(k-1) + \mathbf{B}(k)\mathbf{n}(k) \tag{D.14}$$

Since the elements of $\mathbf{n}(k)$ are zero mean, a possible unbiased MSE estimate for $\mathbf{x}(k)$ is provided by

$$\hat{\mathbf{x}}(k|k-1) = \mathbf{A}(k-1)\hat{\mathbf{x}}(k-1|k-1) \tag{D.15}$$

since the previous estimate $\hat{\mathbf{x}}(k-1|k-1)$ is available and $\mathbf{A}(k-1)$ is assumed known.

As a result, the state-variable estimation error when the last available observation is related to iteration $k-1$ is given by

$$\begin{aligned}\mathbf{e}(k|k-1) &= \mathbf{x}(k) - \hat{\mathbf{x}}(k|k-1)\\ &= \mathbf{A}(k-1)\mathbf{x}(k-1) + \mathbf{B}(k)\mathbf{n}(k) - \mathbf{A}(k-1)\hat{\mathbf{x}}(k-1|k-1)\\ &= \mathbf{A}(k-1)\mathbf{e}(k-1|k-1) + \mathbf{B}(k)\mathbf{n}(k)\end{aligned} \tag{D.16}$$

Assuming that $E[\mathbf{e}(k-1|k-1)] = \mathbf{0}$, meaning that $\hat{\mathbf{x}}(k-1|k-1)$ is an unbiased estimate of $\mathbf{x}(k-1)$, and recalling that the elements of $\mathbf{n}(k)$ are white noise with zero mean, then it is possible to conclude that

$$E[\mathbf{e}(k|k-1)] = \mathbf{0} \tag{D.17}$$

so that $\hat{\mathbf{x}}(k|k-1)$ is also an unbiased estimate of $\mathbf{x}(k)$.

The covariance matrix of $\mathbf{e}(k|k-1)$ can be expressed as follows

$$\begin{aligned}\mathbf{R}_e(k|k-1) &= E[\mathbf{e}(k|k-1)\mathbf{e}^T(k|k-1)]\\ &= \mathbf{A}(k-1)E[\mathbf{e}(k-1|k-1)\mathbf{e}^T(k-1|k-1)]\mathbf{A}^T(k-1)\\ &\quad +\mathbf{B}(k)E[\mathbf{n}(k)\mathbf{n}^T(k)]\mathbf{B}^T(k)\\ &= \mathbf{A}(k-1)\mathbf{R}_e(k-1|k-1)\mathbf{A}^T(k-1) + \mathbf{B}(k)\mathbf{R}_n(k)\mathbf{B}^T(k)\end{aligned} \tag{D.18}$$

The next step is to estimate $\hat{\mathbf{x}}(k|k)$ from $\hat{\mathbf{x}}(k|k-1)$. In this case we use a linear filtering of the most recent estimate of the state variable $\hat{\mathbf{x}}(k|k-1)$ properly combined with another linear filtered contribution of the most recent measurement vector $\mathbf{y}(k)$. The resulting estimation expression for $\hat{\mathbf{x}}(k|k)$ has the following form

$$\hat{\mathbf{x}}(k|k) = \tilde{\mathbf{K}}(k)\hat{\mathbf{x}}(k|k-1) + \mathbf{K}(k)\mathbf{y}(k) \tag{D.19}$$

The challenge now is to compute the optimal expressions for the linear filtering matrices $\tilde{\mathbf{K}}(k)$ and $\mathbf{K}(k)$.

The state-variable estimation error $\mathbf{e}(k|k)$ that includes the last available observation can then be described as

$$\mathbf{e}(k|k) = \mathbf{x}(k) - \tilde{\mathbf{K}}(k)\hat{\mathbf{x}}(k|k-1) - \mathbf{K}(k)\mathbf{y}(k) \tag{D.20}$$

This expression can be rewritten in a more convenient form by replacing $\hat{\mathbf{x}}(k|k-1)$ using the first relation of equation (D.16) and replacing $\mathbf{y}(k)$ by its state-space formulation of equation (D.6). The resulting relation is

$$\begin{aligned}\mathbf{e}(k|k) &= \mathbf{x}(k) + \tilde{\mathbf{K}}(k)\left[\mathbf{e}(k|k-1) - \mathbf{x}(k)\right] - \mathbf{K}(k)\left[\mathbf{C}^T(k)\mathbf{x}(k) + \mathbf{n}_1(k)\right] \\ &= \left[\mathbf{I} - \tilde{\mathbf{K}}(k) - \mathbf{K}(k)\mathbf{C}^T(k)\right]\mathbf{x}(k) + \tilde{\mathbf{K}}(k)\mathbf{e}(k|k-1) - \mathbf{K}(k)\mathbf{n}_1(k) \qquad \text{(D.21)}\end{aligned}$$

We know that $E[\mathbf{n}_1(k)] = \mathbf{0}$ and that $E[\mathbf{e}(k|k-1)] = \mathbf{0}$ since $\hat{\mathbf{x}}(k|k-1)$ is an unbiased estimate of $\mathbf{x}(k)$. However, $\hat{\mathbf{x}}(k|k)$ should also be an unbiased estimate of $\mathbf{x}(k)$, that is, $E[\mathbf{e}(k|k)] = \mathbf{0}$. The latter relation is true if we choose

$$\tilde{\mathbf{K}}(k) = \mathbf{I} - \mathbf{K}(k)\mathbf{C}^T(k) \qquad \text{(D.22)}$$

so that the first term in the last expression of equation (D.21) becomes zero.

By replacing equation (D.22) in (D.19), the estimate of the state variable using the current measurements becomes

$$\begin{aligned}\hat{\mathbf{x}}(k|k) &= \left[\mathbf{I} - \mathbf{K}(k)\mathbf{C}^T(k)\right]\hat{\mathbf{x}}(k|k-1) + \mathbf{K}(k)\mathbf{y}(k) \\ &= \hat{\mathbf{x}}(k|k-1) + \mathbf{K}(k)\left[\mathbf{y}(k) - \mathbf{C}^T(k)\hat{\mathbf{x}}(k|k-1)\right] \qquad \text{(D.23)}\end{aligned}$$

where according to equations (D.21) and (D.22) the corresponding estimation error vector is described by

$$\begin{aligned}\mathbf{e}(k|k) &= \left[\mathbf{I} - \mathbf{K}(k)\mathbf{C}^T(k)\right]\mathbf{e}(k|k-1) - \mathbf{K}(k)\mathbf{n}_1(k) \\ &= \tilde{\mathbf{K}}(k)\mathbf{e}(k|k-1) - \mathbf{K}(k)\mathbf{n}_1(k) \qquad \text{(D.24)}\end{aligned}$$

where the last equality highlights the connection with equation (D.19).

The covariance matrix of $\mathbf{e}(k|k)$ can then be expressed as

$$\begin{aligned}\mathbf{R}_e(k|k) &= E[\mathbf{e}(k|k)\mathbf{e}^T(k|k)] \\ &= \left[\mathbf{I} - \mathbf{K}(k)\mathbf{C}^T(k)\right]\mathbf{R}_e(k|k-1)\left[\mathbf{I} - \mathbf{K}(k)\mathbf{C}^T(k)\right]^T + \mathbf{K}(k)\mathbf{R}_{n_1}(k)\mathbf{K}^T(k) \\ &= \left[\mathbf{I} - \mathbf{K}(k)\mathbf{C}^T(k)\right]\mathbf{R}_e(k|k-1) \\ &\quad - \left\{\left[\mathbf{I} - \mathbf{K}(k)\mathbf{C}^T(k)\right]\mathbf{R}_e(k|k-1)\mathbf{C}(k) - \mathbf{K}(k)\mathbf{R}_{n_1}(k)\right\}\mathbf{K}^T(k) \qquad \text{(D.25)}\end{aligned}$$

The trace of this covariance matrix determines how good is the estimate of the state variables at a given iteration. As a result, the Kalman gain should be designed in order to minimize the trace of $\mathbf{R}_e(k|k)$ as following shown, since it corresponds to the estimation error variance. Defining

$$\xi_K = \mathrm{tr}[\mathbf{R}_e(k|k)] \qquad \text{(D.26)}$$

it then follows that[2]

$$\frac{\partial \xi_K}{\partial \mathbf{K}(k)} = -2\left[\mathbf{I} - \mathbf{K}(k)\mathbf{C}^T(k)\right]\mathbf{R}_e(k|k-1)\mathbf{C}(k) + 2\mathbf{K}(k)\mathbf{R}_{n_1}(k) \qquad \text{(D.27)}$$

[2]It was used the facts that $\frac{\partial \mathrm{tr}[\mathbf{AB}]}{\partial \mathbf{A}} = \mathbf{B}^T$ and $\frac{\partial \mathrm{tr}[\mathbf{ABA}^T]}{\partial \mathbf{A}} = 2\mathbf{AB}$, and that $\mathbf{R}_e(k|k-1)$ and $\mathbf{R}_{n_1}(k)$ are symmetric matrices.

By equating this derivative with zero it is possible to simplify equation (D.25) since its last term becomes zero, allowing the update to the covariance matrix to have a rather simple form given by

$$\mathbf{R}_e(k|k) = \left[\mathbf{I} - \mathbf{K}(k)\mathbf{C}^T(k)\right]\mathbf{R}_e(k|k-1) \tag{D.28}$$

The main purpose of equation (D.27) is of course to calculate the Kalman gain whose expression is given by

$$\mathbf{K}(k) = \mathbf{R}_e(k|k-1)\mathbf{C}(k)\left[\mathbf{C}^T(k)\mathbf{R}_e(k|k-1)\mathbf{C}(k) + \mathbf{R}_{n_1}(k)\right]^{-1} \tag{D.29}$$

We have now available all the expressions required to describe the Kalman filtering algorithm. First we should initialize $\hat{\mathbf{x}}(0|0)$ with $\mathbf{x}(0)$ if available, otherwise generate a zero-mean white Gaussian noise vector. Then initialize the error covariance matrix as $\mathbf{R}_e(0|0) = \mathbf{x}(0)\mathbf{x}^T(0)$. After initialization the algorithm computes $\hat{\mathbf{x}}(k|k-1)$ as per equation (D.15) then the error covariance $\mathbf{R}_e(k|k-1)$ using equation (D.18). Next we calculate the Kalman gain as in equation (D.29) and update the estimate $\hat{\mathbf{x}}(k|k)$ using (D.23) which now takes the form

$$\begin{aligned}\hat{\mathbf{x}}(k|k) &= \hat{\mathbf{x}}(k|k-1) + \mathbf{K}(k)\left[\mathbf{y}(k) - \mathbf{C}^T(k)\hat{\mathbf{x}}(k|k-1)\right] \\ &= \hat{\mathbf{x}}(k|k-1) + \mathbf{K}(k)\left[\mathbf{y}(k) - \hat{\mathbf{y}}(k|k-1)\right]\end{aligned} \tag{D.30}$$

where in the first expression we used equation (D.15), and in the second expression we observe that the term $\mathbf{C}^T(k)\hat{\mathbf{x}}(k|k-1)$ represents an unbiased estimate of $\mathbf{y}(k)$ denoted as $\hat{\mathbf{y}}(k|k-1)$. Finally equation (D.28) updates the error covariance $\mathbf{R}_e(k|k)$ to include the current measurement contribution. Algorithm D.1 describes the Kalman filtering procedure. Fig. D.2 illustrates how the building blocks of the Kalman filtering algorithm interact among themselves. As can be observed, from the measurement signal $\mathbf{y}(k)$ we perform the best possible estimate of the state variable $\hat{\mathbf{x}}(k|k)$. The Kalman filter solution corresponds to the optimal minimum MSE estimator whenever the noise and the state signal are jointly Gaussian, otherwise it is the optimal linear minimum MSE solution, see [3] for details.

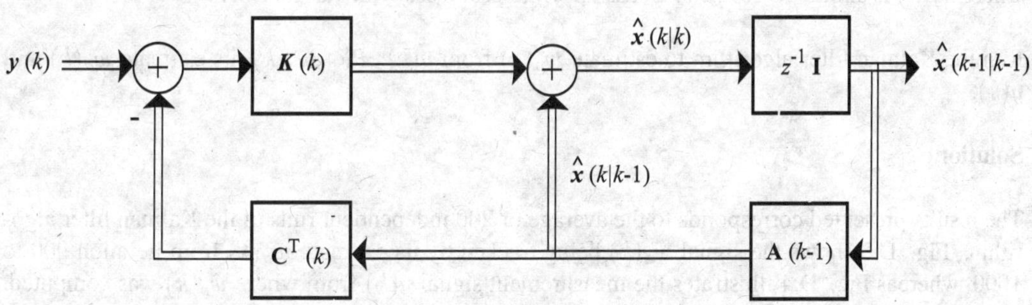

Figure D.2 Kalman filtering structure.

The complex version of the Kalman filter algorithm is almost identical to Algorithm D.1 and can be derived by replacing $\mathbf{x}^T(0)$ by $\mathbf{x}^H(0)$, $\mathbf{C}^T(k)$ by $\mathbf{C}^H(k)$, and $\mathbf{A}^T(k-1)$ by $\mathbf{A}^H(k-1)$.

Algorithm D.1   **Kalman Filter**
Initialization $\hat{\mathbf{x}}(0\|0) = \mathbf{x}(0) \quad \mathbf{R}_e(0\|0) = \mathbf{x}(0)\mathbf{x}^T(0)$ Do for $k \geq 1$ $\hat{\mathbf{x}}(k\|k-1) = \mathbf{A}(k-1)\hat{\mathbf{x}}(k-1\|k-1)$ $\mathbf{R}_e(k\|k-1) = \mathbf{A}(k-1)\mathbf{R}_e(k-1\|k-1)\mathbf{A}^T(k-1) + \mathbf{B}(k)\mathbf{R}_n(k)\mathbf{B}^T(k)$ $\mathbf{K}(k) = \mathbf{R}_e(k\|k-1)\mathbf{C}(k)\left[\mathbf{C}^T(k)\mathbf{R}_e(k\|k-1)\mathbf{C}(k) + \mathbf{R}_{n_1}(k)\right]^{-1}$ $\hat{\mathbf{x}}(k\|k) = \hat{\mathbf{x}}(k\|k-1) + \mathbf{K}(k)\left(\mathbf{y}(k) - \mathbf{C}^T(k)\hat{\mathbf{x}}(k\|k-1)\right)$ $\mathbf{R}_e(k\|k) = \left[\mathbf{I} - \mathbf{K}(k)\mathbf{C}^T(k)\right]\mathbf{R}_e(k\|k-1)$

Example D.1

In a nonstationary environment the optimal coefficient vector is described by

$$w_o(k) = 0.9w_o(k-1) - 0.81w_o(k-2) + n_w(k)$$

for $k \geq 1$, where $n_w(k)$ is a zero-mean Gaussian white processes with variance 0.64. Assume $w_o(0) = w_o(-1) = 0$.

Assume this time-varying coefficient is observed through a noisy measurement described by

$$y(k) = 0.9w_o(k) + n_1(k)$$

where $n_1(k)$ is another zero-mean Gaussian white processes with variance 0.16.

Run the Kalman filter algorithm to estimate $w_o(k)$ from $y(k)$. Plot $w_o(k)$, its estimate $\hat{w}_o(k)$ and $y(k)$.

Solution:

The results presented corresponds to the average of 200 independent runs of the Kalman filter algorithm. Fig. D.3 shows the signal $w_o(k)$ being tracked by its estimate $\hat{w}_o(k)$ from iteration 900 to 1000, whereas Fig. D.4 illustrates the measurement signal $y(k)$ from where $\hat{w}_o(k)$ was computed. As can be observed, the Kalman filter algorithm is able to track quite closely the signal $w_o(k)$ from noisy measurements given by $y(k)$.

□

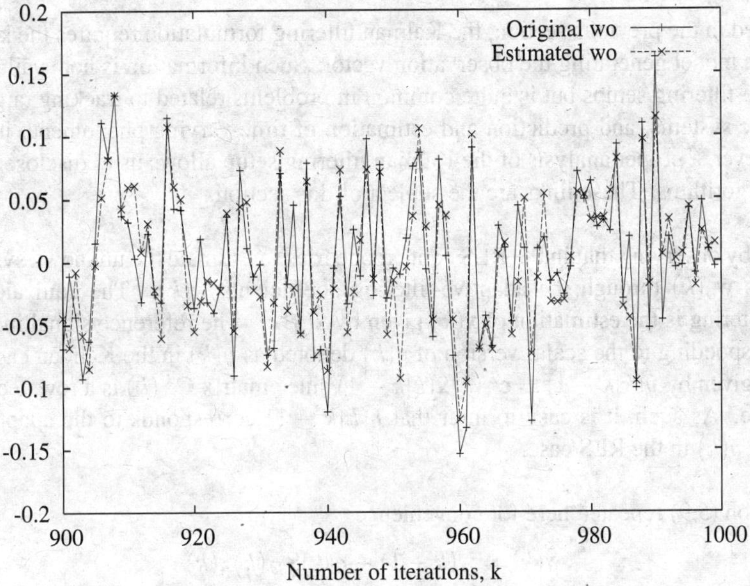

Figure D.3 Tracking performance of the Kalman filter.

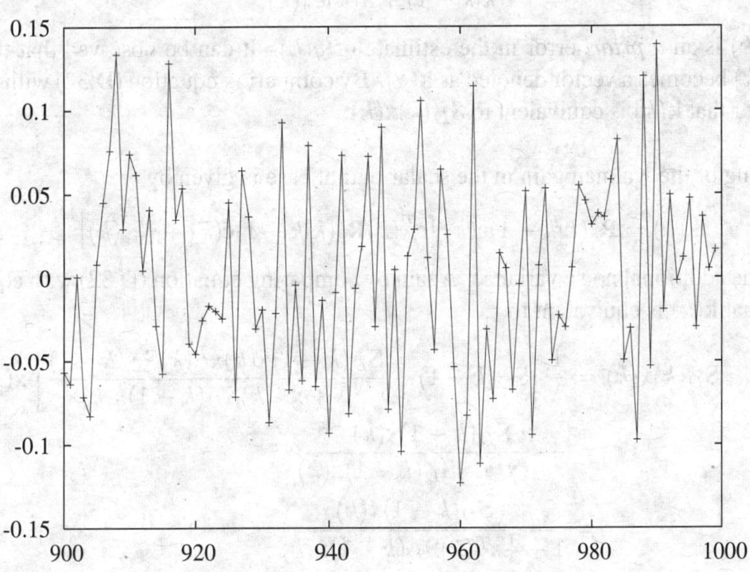

Figure D.4 Noisy measurement signal.

D.4 KALMAN FILTER AND RLS

As observed in the previous section, the Kalman filtering formulation requires the knowledge of the state-space model generating the observation vector. Such information is not available in a number of adaptive-filtering setups but is quite common in problems related to tracking targets, positioning of dynamic systems, and prediction and estimation of time-varying phenomena, just to mention a few. However, a proper analysis of the Kalman filtering setup allows us to disclose some links with the RLS algorithms. These links are the subject of this section.

Let's start by observing that in the RLS context one tries to estimate the unknown system parameters denoted as $\mathbf{w}_o(k)$ through the adaptive-filtering coefficients $\mathbf{w}(k)$. The equivalent operation in Kalman filtering is the estimation of $\mathbf{x}(k)$ given by $\hat{\mathbf{x}}(k|k)$. The reference signal in the RLS case is $d(k)$ corresponding to the scalar version of $\mathbf{y}(k)$ denoted as $y(k)$ in the Kalman case. The estimate of $y(k)$ is given by $\hat{y}(k|k-1) = \mathbf{c}^T(k)\hat{\mathbf{x}}(k|k-1)$ since matrix $\mathbf{C}^T(k)$ is a row vector in the single output case. As such, it is easy to infer that $\hat{y}(k|k-1)$ corresponds to the adaptive-filter output denoted as $y(k)$ in the RLS case.

The equation (5.9) repeated here for convenience

$$\mathbf{w}(k) = \mathbf{w}(k-1) + e(k)\mathbf{S}_D(k)\mathbf{x}(k) \tag{D.31}$$

is meant for coefficient update in the RLS algorithms. This equation is equivalent to

$$\begin{aligned}
\hat{\mathbf{x}}(k|k) &= \hat{\mathbf{x}}(k|k-1) + \mathbf{k}(k)\left(y(k) - \mathbf{c}^T(k)\hat{\mathbf{x}}(k|k-1)\right) \\
&= \hat{\mathbf{x}}(k|k-1) + \mathbf{k}(k)\left(y(k) - \hat{y}(k|k-1)\right) \\
&= \hat{\mathbf{x}}(k|k-1) + \mathbf{k}(k)e_y(k)
\end{aligned} \tag{D.32}$$

where $e_y(k)$ is an *a priori* error in the estimate of $y(k)$. It can be observed that the Kalman gain matrix $\mathbf{K}(k)$ becomes a vector denoted as $\mathbf{k}(k)$. By comparing equation (D.32) with equation (D.31) we can infer that $\mathbf{k}(k)$ is equivalent to $\mathbf{S}_D(k)\mathbf{x}(k)$.

The updating of the Kalman gain in the scalar output case is given by

$$\mathbf{k}(k) = \mathbf{R}_e(k|k-1)\mathbf{c}(k)\left[\mathbf{c}^T(k)\mathbf{R}_e(k|k-1)\mathbf{c}(k) + r_{n_1}(k)\right]^{-1} \tag{D.33}$$

$r_{n_1}(k)$ is the additional noise variance. Again by comparing equation (D.32) with equation (5.5), we can infer that $\mathbf{k}(k)$ is equivalent to

$$\begin{aligned}
\mathbf{S}_D(k)\mathbf{x}(k) &= \frac{1}{\lambda}\left[\mathbf{S}_D(k-1) - \frac{\mathbf{S}_D(k-1)\mathbf{x}(k)\mathbf{x}^T(k)\mathbf{S}_D(k-1)}{\lambda + \mathbf{x}^T(k)\mathbf{S}_D(k-1)\mathbf{x}(k)}\right]\mathbf{x}(k) \\
&= \frac{\mathbf{S}_D(k-1)\mathbf{x}(k)}{\lambda + \mathbf{x}^T(k)\mathbf{S}_D(k-1)\mathbf{x}(k)} \\
&= \frac{\frac{1}{\lambda}\mathbf{S}_D(k-1)\mathbf{x}(k)}{1 + \frac{1}{\lambda}\mathbf{x}^T(k)\mathbf{S}_D(k-1)\mathbf{x}(k)}
\end{aligned} \tag{D.34}$$

Now if we assume that the measurement noise in equation (D.33) has unit variance its straightforward to observe by comparing equations (D.33) and (D.34) that $\mathbf{R}_e(k|k-1)$ plays the role of $\frac{1}{\lambda}\mathbf{S}_D(k-1)$ in the RLS algorithm.

The related quantities in the specialized Kalman filter and the RLS algorithm disclosed so far are

$$\begin{aligned}
\mathbf{x}(k) &\iff \mathbf{w}_o(k)\\
y(k) &\iff d(k)\\
\hat{y}(k|k-1) &\iff y(k)\\
\hat{\mathbf{x}}(k|k) &\iff \mathbf{w}(k)\\
e_y(k) &\iff e(k)\\
\mathbf{k}(k) &\iff \mathbf{S}_D(k)\mathbf{x}(k)\\
\mathbf{R}_e(k|k-1) &\iff \frac{1}{\lambda}\mathbf{S}_D(k-1)
\end{aligned} \tag{D.35}$$

These relations show that given that $\mathbf{x}(k)$ in the Kalman filter algorithm follows the pattern of $\mathbf{w}_o(k)$ and $r_{n_1}(k)$ has unit variance (compare equations (D.33) and (D.34)), the Kalman filter and the RLS algorithms should lead to similar solutions.

As happens with the conventional RLS algorithm, the Kalman filter algorithm faces stability problems when implemented in finite precision mainly related to the ill-conditioning of the estimation error covariance matrix $\mathbf{R}_e(k|k)$. In practical implementations this matrix could be updated in a factorized form such as $\mathbf{U}_e(k|k)\mathbf{D}_e(k|k)\mathbf{U}_e^T(k|k)$ where $\mathbf{U}_e(k|k)$ is upper triangular with ones on the diagonal and $\mathbf{D}_e(k|k)$ is a diagonal matrix.

D.5 REFERENCES

1. R. E. Kalman, "A new approach to linear filtering and prediction problem," *Trans. ASME Journal of Basic Engineering*, vol. 82, pp. 34-45, March 1960.

2. P. S. R. Diniz, E. A. B. da Silva, and S. L. Netto, *Digital Signal Processing: System Analysis and Design*, Cambridge University Press, Cambridge, UK, 2002.

3. S. M. Kay, *Fundamentals of Statistical Signal Processing: Estimation Theory*, Prentice Hall, Englewood Cliffs, NJ, 1993.

4. T. Kailath, A. H. Sayed, and B. Hassibi, *Linear Estimation*, Prentice Hall, Englewood Cliffs, NJ, 2000.

5. M. H. Hayes, *Statistical Digital Signal Processing and Modeling*, John Wiley & Sons, New York, NY, 1996.

6. D. G. Manolakis, V. K. Ingle, and S. M. Kogon, *Statistical and Adaptive Signal Processing*, McGraw Hill, New York, NY, 2000.

INDEX

F

G

H

M

N

O

P